PESTICIDES

Evaluation of Environmental Pollution

PESTICIDES

Evaluation of Environmental Pollution

Edited by
Hamir S. Rathore and Leo M.L. Nollet

CRC Press
Taylor & Francis Group
Boca Raton London New York

CRC Press is an imprint of the
Taylor & Francis Group, an **informa** business

CRC Press
Taylor & Francis Group
6000 Broken Sound Parkway NW, Suite 300
Boca Raton, FL 33487-2742

First issued in paperback 2019

ISBN-13: 978-1-4398-3624-8 (hbk)
ISBN-13: 978-0-367-86519-1 (pbk)

Visit the Taylor & Francis Web site at
http://www.taylorandfrancis.com

and the CRC Press Web site at
http://www.crcpress.com

Contents

Section IV Pesticides Residues in Vegetation

Section V Pesticides Residues in Animals

Section VI Pesticides and Men

Section VII Pesticides and Food

Section VIII Biopesticides

Section IX   Endocrine-Disrupting Pesticides

Preface

One kind word can warm three winter months.

Japanese proverb

The systematic pollution of the environment is one of the biggest hazards that humanity faces today. People are becoming increasingly aware of the threat posed by pollution and governments are enacting legislations aimed at protecting the environment. Recognizing the need for trained technical manpower in this area, many universities and institutions are increasingly offering courses related to environmental pollution and its control.

Pesticide is a chemical used to control, repel, attract, or kill pests, for example, insects, weeds, birds, mammals, fish, or microbes that are considered to be a nuisance. It would be a tough challenge for India and similar developed countries to produce enough food for the growing population (almost 20 million a year), protect plant, animal, and human health and at the same time conserve the environment. Green revolution technologies have doubled the yield of rice and wheat. Green revolution has been made possible only with the help of agrochemicals, particularly pesticides. Still a considerable quantity of food products is destroyed because of pests. The process of chemical crop protection is profit-induced poisoning of the environment. Many surveys have indicated that the pesticide residues in the environment increase day by day due to their repeated and continuous use. Chronic exposure to small amounts of residues through consumption of contaminated foodstuffs may lead to suppression of the immune system, which in turn, may render humans vulnerable to infectious diseases. Laboratory experiments have indicated that some pesticide residues may cause carcinogenicity on long exposure. The pesticide residues also exhibit specific effects on species other than the pests for which they are solely intended. It is also known that bodies of water, air, birds, and aquatic animals are constantly moving and transporting poisons from one region of the globe to another. For example, DDT was used in the fields of East Africa, but in a few months, it was found in the water of the Bay of Bengal, that is, at a distance of 6000 km away. The magnitude of the threat is considerable to humans and the environment through either deliberate or ignorant misuse of pesticides (e.g., using parathion to treat head lice). This threat is considerably greater in developing countries where there is little awareness of the danger of pesticide use and inadequate user protection.

However, even in developed countries, users are at considerable risk as a result of the intensive use of pesticides despite more information of the dangers and widespread user protection. Pesticide residue problems cannot therefore be regarded as specific only to developing countries.

Considerable amount of work is being done to develop nonchemical methods of pest control, but the stage where the use of pesticides can be dispensed with has still not been reached. There is general consensus that the use of chemical pesticides will continue in the foreseeable future. Of course, it is possible to minimize the load of pesticide by selecting or synthesizing suitable compounds that are applied at low dosages. Recently, some new compounds such as bromine-containing pesticides, thiazole insecticides, *Piper nigrum* extract, and phthalamide derivatives, which are biodegradable and possess high mammalian safety and low residual life and are compatible with nontarget organisms and allied traits, have been developed. This is a highly encouraging development.

Therefore, it is worthwhile to publish a book dealing with environmental pollution by pesticides. Keeping in view the importance of this important subject, chapters were invited from eminent scientists working on different research aspects of pesticides. In this book, the 20 chapters are organized systematically in 9 sections; the chapters discuss pesticide residues in detail. Section I entitled *'Biocide'* consists of Chapter 1 on "Uses and Environmental Pollution of Biocides". Section II on *'Transposition and Transport of Pesticides'* has Chapters 2–5, which discuss "Fate and Transport of Pesticide in the Environment," "Degradation of Pesticides," "Pesticide Degradation in Water," and "Microbial Remediation of Pesticides." Section III entitled *'Pesticides Residues in the Environment'* discusses "Pesticides Residues in the Environment," "Pesticide Residues in the Atmosphere," "Response of Soil Micro Flora to Pesticides," and "Pesticide Residues in Surface Water and Groundwater." This section includes Chapters 6–9. Section IV covers *'Pesticides Residues in Vegetation'* in Chapter 10. Section V *'Pesticides Residues in Animals'* comprises 4 chapters; Chapters 11–14 on "Pesticide Residues in Aquatic Invertebrates," "Pesticide Residues in Soil Invertebrates," "Pesticide Residues in fish," and "Pesticide Residues in Birds and Mammals." Section VI, *'Pesticides and Men'*, consists of three chapters, Chapters 15–17 describe "Pesticide Residues in Man," "Pesticides and Skin Diseases in Man," and "Pesticide Residues in Mother's Milk." Section VII has Chapter 18 on *'Pesticides and Food'*, gives details of "Pesticide Residues in Milk and Milk Products" Section VIII is also Chapter 19 which describes "Biopesticides." The last Section IX covers *"Endocrine-Disrupting Pesticides"* in Chapter 20.

We hope that this book will be immensely useful to the large section of the readers in the area of chemical–crop protection and environmental pollution.

The editors wish to thank the contributors for their expeditious response.

H.S. Rathore and Leo M.L. Nollet

Further Readings

The reader is directed to interesting sites with summaries of data about pesticides in groundwater, surface waters, aquatic sediments and/or biota. These data have been assembled by various members of the NAWQA Pesticide National Synthesis (PNS) team over the past 15 years. These publications are cited in the *Publications* section of the PNS team's website: http://water.usgs.gov/nawqa/pnsp/.

The most comprehensive overview of the pesticide occurrence data from the NAWQA program is provided in the 2006 publication "The Quality of Our Nation's Waters – Pesticides in the Nation's Streams and Ground Water, 1992–2001" available from http://pubs.usgs.gov/circ/2005/1291/.

Furthermore, most of the NAWQA data on the occurrence of pesticides and other contaminants in the hydrologic system across the United States may be downloaded from https://infotrek.er.usgs.gov/apex/f?p=NAWQA:HOME:0.

On www.epa.gov/pesticides/index.htm the reader can find massive amount of information about pesticides in the environment and food.

For the European Union, information on pesticides and its residues are available on http://ec.europa.eu/sanco_pesticides/public/index.cfm and http://ec.europa.eu/food/plant/protection/pesticides/index_en.htm.

All cited websites were accessed on July 14, 2011.

Contributors

Ibrahim Abdulwahid Arif
Prince Sultan Research Chair for
    Environment and Wildlife
College of Sciences
King Saud University
Riyadh, Saudi Arabia

Mohammad Abdul Bakir
Prince Sultan Research Chair for
    Environment and Wildlife
College of Sciences
King Saud University
Riyadh, Saudi Arabia

O. P. Bansal
Chemistry Department
Dharam Samaj College
Aligarh
Uttar Pradesh, India

Kristen Bartlett
School of Environmental and Biological
    Science
Rutgers University
New Brunswick, New Jersey

N. C. Basantia
Avon Food Lab
Delhi, India

Marija Borjan
School of Environmental and Biological
    Science
Rutgers University
New Brunswick, New Jersey

S. Chandrasekaran
Agricultural Chemicals
Pesticide Toxicology Laboratory
Department of Agricultural University
Tamil Nadu Agricultural University
Tamil Nadu, India

Hao Chen
Department of Soil and Water Sciences
University of Florida
Gainesville, Florida

Mariusz Cycoń
Department of Microbiology
Faculty of Pharmacy
Medical University of Silesia
Jagiellońska, Sosnowiec, Poland

Nikolina Dujaković
Faculty of Technology and Metallurgy
University of Belgrade
Karnegijeva, Belgrade, Serbia

Prem Dureja
Division of Agricultural Chemicals
Indian Agricultural Research Institute
New Delhi, India

Teresa Vera Espallardo
Euphore Lab
Fundación CEAM
Paterna, Spain

Antonia Garrido Frenich
Department of Analytical Chemsitry
University of Almería
Almería, Spain

Bin Gao
Department of Agricultural and Biological
    Engineering
University of Florida
Gainesville, Florida

Jie Gao
Department of Environmental Engineering
    Sciences
University of Florida
Gainesville, Florida

Svetlana Grujić
Faculty of Technology and Metallurgy
University of Belgrade
Karnegijeva, Belgrade, Serbia

Raimon Guitart
Toxicology Unit
Lab. Toxicology - Fac. Veterinary Sciences
Universitat Autònoma de Barcelona
Bellaterra, Spain

Isabel Mª Campoy Jiménez
Department of Analytical Chemsitry
University of Almería
Almería, Spain

Sarun Keithmaleesatti
Faculty of Science
Khon Kaen University
Khon Kaen, Thailand

Haseeb Ahmad Khan
Prince Sultan Research Chair for
    Environment and Wildlife
College of Sciences
King Saud University
Riyadh, Saudi Arabia

S. Kuttalam
Agricultural Chemicals
Pesticide Toxicology Laboratory
Department of Agricultural University
Tamil Nadu Agricultural University
Tamil Nadu, India

Mila Laušević
Faculty of Technology and Metallurgy
University of Belgrade
Belgrade, Serbia

Naglaa Loutfy
Departmernt of Plant Protection
Faculty of Agriculture
Suez Canal University
Ismailia, Egypt

Mahmoud Farag Mahmoud
Departmernt of Plant Protection
Faculty of Agriculture
Suez Canal University
Ismailia, Egypt

Sameeh A. Mansour
Professor of Pesticides and Environmental
    Toxicology
Environmental Toxicology Research Unit
    (ETRU)
Pesticide Chemistry Department
National Research Centre
Cairo, Egypt

Ma del Mar Bayo Montoya
Department of Analytical Chemsitry
University of Almería
Almería, Spain

Amalia Muñoz
Euphore Lab
Fundación CEAM
Paterna, Spain

Leo M. L. Nollet
Faculty of Applied Engineering Sciences
University College Ghent
Ghent, Belgium

José Luis Palau
Euphore Lab
Fundación CEAM
Paterna, Spain

M. Paramasivam
Agricultural Chemicals
Pesticide Toxicology Laboratory
Department of Agricultural University
Tamil Nadu Agricultural University
Tamil Nadu, India

Zofia Piotrowska-Seget
Department of Microbiology
University of Silesia
Katowice, Poland

Marina Radišić
Faculty of Technology and Metallurgy
University of Belgrade
Belgrade, Serbia

J. Rajeswaran
Agricultural Chemicals
Pesticide Toxicology Laboratory
Department of Agricultural University
Tamil Nadu Agricultural University
Tamil Nadu, India

Hamir Singh Rathore
Applied Chemistry Department
Zakir Husain College of Engineering
Aligarh Muslim University
Aligarh, India

Mark Robson
School of Environmental and Biological
    Science
Rutgers University
New Brunswick, New Jersey

Pilar Sandín-España
Unit of Plant Protection Products
Madrid, Spain

Sushil Kumar Saxena
Export Inspection Council
Department of Commerce
Ministry of Commerce and Industry
    Government of India
New Delhi, India

Beatriz Sevilla-Morán
Unit of Plant Protection Products
Madrid, Spain

Shafiullah
Chemical Research Unit
A. K. Tibbiya College
Aligarh Muslim University
Aligarh, India

Wattasit Siriwong
College of Public Health Sciences
Chulalongkorn University
Bangkok, Thailand

Radoslaw Spiewak
Department of Experimental Dermatology
    and Cosmetology
Faculty of Pharmacy
Jagiellonian University Medical College

and

Scientific Director, Institute
    of Dermatology
Krakow, Poland

Rajendra Singh Tanwar
Division of Agricultural Chemicals
Indian Agricultural Research Institute
New Delhi, India

Tomasz Tuzimski
Department of Physical Chemistry
Chair of Chemistry
Medical University
Lublin, Poland

Tatjana Vasiljević
Faculty of Technology and Metallurgy
University of Belgrade
Belgrade, Serbia

José Luis Martínez Vidal
Department of Analytical Chemsitry
University of Almería
Almería, Spain

Yu Wang
Department of Agricultural and Biological
    Engineering
University of Florida
Gainesville, Florida

Lei Wu
Department of Agricultural and Biological
    Engineering
University of Florida
Gainesville, Florida

Section I

Biocides

1

Uses and Environmental Pollution of Biocides

Mahmoud Farag Mahmoud and Naglaa Loutfy

CONTENTS

1.1 Introduction

Man's endeavors to improve his life have been a long and momentous pursuit throughout. Ecosystems, in which life thrived, have been generous donors of numerous services that man has tamed, manipulated, and extracted copious dividends and values from. Agricultural ecosystems have especially played a pivotal role, providing man with his basic needs for foods, feeds, fibers, and many other raw materials. Each year, forty million people die of hunger and hunger-related diseases. Our world is growing at an enormous rate, and by the year 2050, its population could reach 10 billion. World population growth is projected to reach over 8 billion in 2030 and to level off at 9 billion by 2050. The unprecedented increase in human population that the last century witnessed has triggered a massive increase in man's need for food and feed. Frequent drought and lack of rain in various parts of the world have added one more somber dimension to food shortage. The United Nations Organization has challenged the international community to work together to improve the situation, and one of the main objectives of the Millennium

Development Goal (MDG) is to reduce by half the proportion of people who suffer from hunger by the year 2015. Pesticides have emerged as one of the most powerful tools to secure the provision of food and protect it against the wide mosaic of pests that attack food crops. Owing to their chemical nature, pesticides are biocides, which have the potential for poisoning organisms other than the target insect, microorganism, or plant species that should be controlled as has been mentioned in the Biocidal Products Directive 98/8/EC of the European Parliament and Council. Pesticides used in agricultural production affect environmental quality and human health. These external costs can amplify due to climate change because pest pressure and optimal pesticide application rates vary with weather and climate conditions.

Pesticides have tremendously benefited humanity but at the same time caused considerable negative effects on biodiversity, environment, food quality, and human health. This is because the structure of pesticides and their mode of action and also application, especially the way they are used by farmers, have rendered them serious pollutants of the environment in general. Some pesticides (organochlorine compounds) are classified as persistent organic pollutants (POPs). POPs have the ability to volatilize and travel great distances through the atmosphere to become deposited in remote regions, and also have the ability to bioaccumulate and biomagnify, and can bioconcentrate up to 70,000 times their original concentrations.

The occurrence of persistent organochlorine compounds in the environment is changing relatively slowly over a span of years; similar time trends are characteristic of contents in fish, meat, eggs, and dairy products, which are the foods that make the greatest contribution to the intake of organochlorine compounds (Watterson 1991).

The use of POPs is being phased out, and some are already banned, while others continue to be used. Following the discovery of pesticides' negative effects on all compartments of life, usage patterns have changed dramatically, particularly in recent years with the introduction of some alternative approaches to control pests, such as biopesticides.

It is clear that feeding the world without pesticide use remains elusive, "especially with some 800 million undernourished" people in the world today; however, intensification in agricultural production should be sustainable and should protect human health and the environment as well. Therefore, sustainable agriculture is one of the greatest challenges that the world faces now.

1.2 Pesticides: A Global View

1.2.1 History of Pesticides

The pesticides machinery has been rolling since the early 1940s when dichlorodiphenyltrichloroethane (DDT) was first introduced, bringing about a novel paradigm in man's fight against pests and diseases. The pesticides machinery has continued producing its assorted species of chemicals that secured food supply to meet people's demands; however, flaws and drawbacks were part of that big syndrome that marked pesticide use. Reported episodes of pesticide-caused pollution and other perturbation were rather frequent, and most were illustrated in the touchy somber book, the Silent Spring written in the early 1960s by the renowned environmentalist Rachel Carson. The impacts of pesticide use on man's health and environment welfare constituted a major concern that shed some light on pesticides machinery along with the drivers behind it. In this respect, the unforgettable events of Clear Lake, Dutch elm disease, are landmarks in that undesirable history, let alone the little known sad stories of those innumerable victims of pesticide intoxication in developing countries.

The history of pesticide use dates back many centuries, certainly before 1000 BC, when it was mentioned by Homer, but the real landmark in terms of modern agriculture is the spread of the Colorado beetle (*Leptinotarsa decemlineata*) across the United States in the second half of the nineteenth century. The knowledge and skills for protecting crops against pests and diseases have greatly improved over the centuries. Since an old era, people have used pesticides to prevent damage to their crops. The first known pesticide to be used was sulfur. By the fifteenth century, toxic chemicals such as arsenic, mercury, and lead were being applied to crops to kill pests. In the seventeenth century, nicotine sulfate was extracted from tobacco leaves for use as an insecticide. The nineteenth century saw the introduction of two more natural pesticides, pyrethrum and rotenone. A dramatic breakthrough in insect pest control was achieved in 1939 with the discovery of the insect-killing properties of DDT, which led to the development of chlorinated hydrocarbon and organophosphate pesticides during the Second World War (1940–1945). It quickly became the most widely used pesticide in the world. However, in the 1960s, it was discovered that DDT was preventing many fish-eating birds from reproducing, causing birth defects in animals and humans, which was a huge threat to biodiversity. DDT is now banned in many countries, but is still used in some developing nations to prevent malaria and other tropical diseases. Pesticide use has increased 50-fold since 1950, and 2.5 million tons of industrial pesticides are now used each year. Table 1.1 shows the chronology of pesticide development in the world.

TABLE 1.1

Chronology of Pesticide Development

Period	Example	Source	Characteristics
1800s–1920s	Early organics, nitrophenols, chlorophenols, creosote, naphthalene, petroleum oils	Organic chemistry, by-products of coal gas production, etc.	Often lacked specificity and were toxic to user or nontarget organisms
1945–1955	Chlorinated organics, DDT, hexachlorocyclohexane (HCCH), chlorinated cyclodienes	Organic synthesis	Persistent, good selectivity, good agricultural properties, good public health performance, resistance, harmful ecological effects
1945–1970	Cholinesterase inhibitors, organophosphorus compounds, carbamates	Organic synthesis, good use of structure–activity relationships	Lower persistence, some user toxicity, some environmental problems
1970–1985	Synthetic pyrethroids, avermectins, juvenile hormone mimics, biological pesticides	Refinement of structure activity relationships, new target systems	Some lack of selectivity, resistance, costs, and variable persistence
1985–	Genetically engineered organisms	Transfer of genes from biological pesticides to other organisms and into beneficial plants and animals. Genetic alteration of plants to resist nontarget effects of pesticides	Possible problems with mutations and escapes, disruption of microbiological ecology, monopoly

Source: From Stephenson, G. A. and Solomon, K. R. 1993. *Pesticides and the Environment.* Department of Environmental Biology, University of Guelph, Guelph, ON. www.fao.org/docrep/w2598e/w2598e07.htm.

Due to the tough competition and large demand, many farmers resort to the extensive and irrational use of pesticides to increase yield. A rapid increase in synthetic pesticide use occurred during the second half of the twentieth century, bringing enormous benefits including improved food security and disease vector control. However, a number of drawbacks became apparent. Organochlorine pesticides (aldrin, chlordane, DDT, dieldrin, endrin, heptachlor, mirex, toxaphene, hexachlorobenzene (HCB)) constitute nine of the twelve chemical substances/groups currently defined under the Stockholm Convention on POPs. While use of these chemicals had been either banned or restricted in many developed countries during the 1970s and 1980s, they are still used in developing countries and continue to linger in the environment because of their chemical stability. A global surveillance of DDT levels in human tissues discovered higher levels in Africa, Asia, and Latin America than in Europe and the United States (Jaga and Dharmani 2003).

Many environmental contaminants that are produced and released into the environment at low latitudes tend to accumulate in the polar regions. POPs, for example, are stable, fat-soluble, carbon-based compounds that volatilize at warm temperatures and are transported poleward by wind, water, and wildlife. Atmospheric transport is the most rapid pathway by which POPs, especially volatile or semivolatile compounds, reach the poles. Once in the polar regions, POPs are deposited on particles or exchanged with water, both processes that are enhanced by low temperature. Oceanic transport occurs more slowly but is an equally or more important pathway for compounds such as hexachlorocyclohexane that partition strongly into water (MacDonald et al. 2002). Organochlorine pesticides, which have been classified by the US Environmental Protection Agency (EPA) as POPs, have the ability to bioaccumulate and biomagnify and can bioconcentrate (i.e., become more concentrated) up to 70,000 times their original concentrations. These compounds do not break down in the environment and are classified as probable or possible carcinogens.

For POPs and other kinds of pesticides, there is a need to monitor levels and trends now and into the future in order to assess compliance with the regulations or standards within each country as well as on a regional and global basis and a need for a useful tool for following the impacts on health and environment. In this respect, most developed countries have a regular and routine monitoring program for such contaminants, covering all environmental segments such as air, water, food, soil, and biota (EC 2000, SCOOP, and EC 2001), but this is not the case for most developed countries specially in the Mediterranean region, where the lack of analysis facilities and resources is common. In Egypt, for example, less data exist on POPs in food or environmental samples (e.g., Loutfy et al. 2007, 2008, 2010).

1.2.2 Biocides and Pesticides

Biocidal products are necessary for the control of organisms that are harmful to human or animal health and that cause damage to natural or manufactured products. According to the Directive 98/8/EC of the European Parliament and Council of 16 February 1998, biocide products are defined as active substances and preparations containing one or more active substances, put in the form in which they are supplied to the user, intended to destroy, render harmless, prevent the action of, or otherwise exert a controlling effect on any harmful organism by chemical or biological means. Most biocidal substances are highly reactive. If they are not used under controlled terms, their use can bring risks to human health and the environment in a variety of ways due to their intrinsic properties and associated use patterns. Particularly vulnerable groups like children or pregnant women are threatened by the wide and improper use of hazardous biocides. The Biocidal Products Directive 98/8/EC classified biocides into 4 main groups and 23 product types (i.e., application categories), with several comprising multiple subgroups as follows: disinfectants and general biocidal products, preservatives, pest control (pesticides) and other biocidal products.

Pesticides fall into three main categories—fungicides, herbicides, and insecticides. These three types are used to combat different pests. Most pesticides are toxic to human beings; WHO has classified their toxic effects from class Ia (extremely hazardous) to class III (slightly hazardous) and then to "active ingredients unlikely to present acute hazard" (WHO 2001).

Chemical insecticides are usually divided into four major classes: Chlorinated hydrocarbons or organochlorines—these pesticides break down chemically very slowly and can remain in the environment for long periods of time (dieldrin, chlordane, aldrin, DDT, and heptachlor). Organophosphates—these pesticides are highly toxic to humans but do not remain in the environment for long periods of time (parathion, malathion, thimet, and trichlorphone). Carbamate compounds—these pesticides are considered highly toxic to humans (carbaryl, methomyl, carbofuran, and aldicarb). Pyrethroids—the first synthetic pyrethroids that combined high toxicity to insects with low mammalian toxicity and greatly increased stability were announced in 1973, and since then many new pyrethroids have been synthesized and marketed, such as cypermethrin. In addition to these four classes, there are now a number of new compounds representing other smaller classes, for example, insect growth regulators, imidates, and phenylpyrazoles.

The first synthetic broad-spectrum pesticide, the organochlorine, caused a revolution in the efficacy of pest control, especially of the malaria mosquito and other disease vectors of mankind. Additionally, they offered cheap, sure, and long-lasting control of crop pests. Organochlorine pesticides are substances containing chemically combined chlorine and carbon. They may be grouped into three general classes: dichlorodiphenylethanes (DDT, DDD, dicofol, etc.), chlorinated cyclodienes (aldrin, dieldrin, heptachlor, etc.), and hexachlorocyclohexanes (lindane). These compounds differ substantially between and within groups with respect to toxic doses, skin absorption, fat storage, metabolism, and elimination. The signs and symptoms of toxicity in humans, however, are remarkably similar except for DDT (Hayes 1991).

Organochlorine pesticides persist and tend to bioaccumulate in the environment. These qualities make them the most dangerous group of chemicals to which natural systems can be exposed. It is well known that the widespread use of organochlorine compounds have caused serious problems to man due to their bioaccumulation in several organs. The majority of these compounds are toxic in high levels and few are carcinogenic, as shown by few animal tests (Fytianos et al. 1985).

1.2.3 Pesticide Market in the Worldwide

As many pesticides were determined to be harmful to human health and/or the environment, pesticides have come under more scrutiny and some have been banned by the government. In addition, the number of species resistant to major classes of pesticides has increased dramatically in recent years; some pesticides are becoming obsolete as pests develop genetic resistance to them. These factors have increased the need for alternative approaches to control pests. The 1993–1994 annual report of the German Association of Chemical Industries (GACI) indicates that the value of world chemical pesticides sold was about US $23 billion. The distribution of this value by region is summarized in Table 1.2 and compared by the distribution percentages for the year 2008.

About three quarters of pesticide use occurs in developed countries. In these regions, the pesticide market is dominated by herbicides, which tend to have a lower acute, or

TABLE 1.2

Global Pesticides' Market Share (%) by Region in
1993–1994 and 2008

Region	Market Share (%) (1993–1994)	Market Share (%) (2008)
Africa	3	4[a]
Europe	29	32
Japan	11	—
Latin America	11	21
Asia	16	23
North America	30	20

Source: www.fao.org/docrep/w8419e/W8419e07.htm and
            www.croplife.org.
[a] Middle East, Africa.

immediate, toxicity than insecticides. In most developing nations, the situation is reversed, and insecticide use predominates, with a correspondingly higher level of acute risk. In developed countries, it is nonetheless substantial and is growing.

The pesticide market has changed dramatically since the adoption of the Code of Conduct. Then, around 15 European and US multinational agrochemical companies dominated pesticide sales; following reorganizations and takeovers, just six of them now control 80% of the market. Genetically engineered seeds, based on herbicide- and insect-resistant technology, make up a significant additional element in the profits of these companies. Japanese companies have a lesser share of global sales, while Chinese and Indian companies are important producers, and China is expanding its pesticide exports. The market for agricultural pesticides was US $17 billion at the time the Code was adopted. Global agricultural pesticides sales ranged from US $30 billion in 1999 to US $25.2 billion in 2002, with a reduction percentage of 6%. On the contrary, sales reached US $26.7 million in 2003 and have reached US $40 billion in 2008, showing fluctuation in the last 10 years (PAN AP 2010). Europe has been the largest market for crop protection products and Taiwan has the largest per-hectare pesticide consumption (Koncept Analytics 2010).

Since the 1970s, the trends in pesticide use have been mixed, with herbicide use increasing and insecticide use decreasing. Usage patterns are influenced by changes in pest biology (including the development of resistance to pesticides) and legislative and regulatory action following the discovery of negative effects on water quality and the environment. Expenditures on herbicides accounted for the largest portion of total expenditures (more than 40%), followed by expenditures on insecticides, fungicides, and other pesticides, respectively.

Asian regulators meeting at an FAO workshop in 2005 estimated an annual pesticide use in the region of around 500,000 tons of active ingredients. Some analysts suggest that the Asian market accounted for 43% of agrochemical revenue in 2008 (Agronews 2009), and that China is the world's biggest user, producer, and exporter of pesticides (Yang 2007). India is the second largest pesticide producer in Asia and the twelfth largest producer globally (WHO 2009). In Latin America, pesticide use has shifted dramatically from a 9% share of sales in 1985 to 21% in 2008. Some of the explanation lies in the expansion of soya bean production, which dominates parts of the subcontinent. Soya beans now cover 16.6 million ha or 50% of the cropping area of Argentina. Pesticide application there reached 270 million liters in 2007; in the same year, in the neighboring Brazil, also a major soya bean producer, application reached 650 million liters. Soya beans are mainly exported to

TABLE 1.3

Amount of Pesticide Active Ingredient and World Pesticide Expenditures at User Level by Pesticide Type, 2000 and 2001 Estimate

Pesticide Type	World Market Active Ingredient		World Market Expenditures	
	Million lbs of a.i.	%	Million $	%
2000				
Herbicides and plant growth regulators	1.944	36	14.319	44
Insecticides	1.355	25	9.102	28
Fungicides	516	10	6.384	19
Other[a]	1.536	29	2.964	9
Total	*5351*	*100*	*32.769*	*100*
2001				
Herbicides and plant growth regulators	1.870	37	14.118	44
Insecticides	1.332	24	8.763	28
Fungicides	475	9	6.027	19
Other[a]	1.469	29	2.848	9
Total	*5.046*	*100*	*31.756*	*100*

Source: Modified from EPA estimates based on Croplife America annual surveys and EPA proprietary data.

[a] Other includes nematicides, fumigants, rodenticides, molluscicides, aquatic and fish/bird pesticides, other miscellaneous conventional pesticides, and other chemicals used as pesticides (e.g., sulfur and petroleum oil).

Europe for animal feed and to China for food uses. In Africa, the trends in pesticide use are less clear, but there are few areas where farmers now pass the year without applying pesticides (Williamson 2003).

Each year, millions of tons of pesticides are used around the world. Six major players—*Syngenta, BASF, Bayer, Dow, Du Pont,* and *Monsanto*—control the majority of the toxic chemical market. Although global figures are hard to come by, industry reports suggest that an estimated 1.5–2.5 million tons of pesticides are used annually.

Table 1.3 shows the amount of pesticide active ingredient and world pesticide expenditures at user level by pesticide type.

1.2.4 Challenges of Climate Change and Pesticides

Climate change is already widely considered a reality (IPCC 2007). A two-way interaction between agriculture and climate is known. Agriculture and the world's supply of food and fiber are particularly vulnerable to such climate change. Many sectors will be influenced by changing climate and climate variability, including increasing global temperatures, changing precipitation patterns, and increasing frequency of unusual weather events. Several studies have examined the interaction between pests and climate change (Patterson et al. 1999; Gutierrez et al. 2008), concluding that pest activity, especially of insects, will increase and lead to higher crop losses. Climate change will have important implications for insect conservation and pest status and can substantially influence the development and distribution of insects. Current best estimates of changes in climate indicate an increase in global mean annual temperatures of 1°C by 2025 and 3°C by the end of the next century. Such increases in temperature have a number of implications for temperature-dependent insect pests (Palikhe 2007). Climate change may allow pest migration or population expansions, which may adversely affect agricultural productivity, profitability, and possibly even viability. Crop–pest interactions may shift as the timing of developmental stages in both hosts and pests is altered. The majority of insect pests that currently affect crops are likely to benefit from climate change as a result of increased summer activity and reduced winter mortality. Some insect pests that are currently present at low levels, or that are not considered a threat at this time, may become more prevalent. The natural enemies of pests will have their own climate optima, although not necessarily the same as their hosts. It is pertinent to ask whether climate change will affect natural enemies to the same extent, or in the "same direction," as the pests they control (Thomson et al. 2010).

During the last three decades, agricultural pesticides have been increasingly recognized for their adverse effects on the environment and human health. These external costs can amplify due to climate change because pest pressure and optimal pesticide application rates vary with weather and climate conditions. The possible increases in pest infestations may bring about a greater use of chemical pesticides to control them, a situation that will require the further development and application of integrated pest-management (IPM) techniques. Climate change may affect our ability to control pests. For example, high temperature is reported to reduce the effectiveness of some pesticides. Humidity levels can also modify their efficacy, as can the timing and amount of rain following their application. On a simpler level, rain can affect the growers' ability to apply the pesticide at the time when it is most needed, possibly an increasingly likely scenario. If pests are able to complete more generations in a season, then this may lead to a greater pesticide use, which in turn may lead to the more rapid development of pesticide resistance. Koleva et al (2009) mentioned that weather and climate differences significantly influence the application rates of most pesticides in the United States as farmers may grow different crops, use different rotations, and change the intensity of management related to irrigation, tillage, fertilization, and pesticide use.

The main climate drivers that change pesticide fate and behavior are thought to be changes in rainfall season availability and intensity and increased temperatures, but the effect of climate change on pesticide fate and transport is likely to be very variable. Difficulty in predicting the long-term, indirect impacts such as land-use change driven by changes in climate may have a more significant effect on pesticides in surface waters and groundwater than the direct impacts of climate change on pesticide fate and transport (Bloomfield et al. 2006).

The effects of climate warming on POP transport to polar regions have low certainty. Sources of uncertainty include the dynamics of adsorption to snow and the extent to which POPs currently trapped in sea and glacial ice will be released with warming.

Several studies have examined the interactive effects of climate and chemical contaminants on biota. Some of these studies suggest that climate change will increase the toxicity of contaminants (Monserrat and Bianchini 1995; Capkin et al. 2006; Delorenzo et al. 2009) or that contaminants will reduce tolerances to extreme temperatures (Heath et al. 1994; Gaunt and Barker 2000; Lannig et al. 2006; Patra et al. 2007), perhaps because an organism has finite resources to allocate for competing selection pressures (Rohr et al. 2003) and thus multiple stressors tend to decrease the energy available for detoxification or temperature regulation (Noyes et al. 2009). In contrast, there are several studies revealing that excretion or tolerance of chemicals is positively associated with temperature (Maruya et al. 2005; Paterson et al. 2007; Harwood et al. 2009; Weston et al. 2009; Rohr et al. 2010).

Finally, in responding to the challenges of climate changes toward agriculture, researchers worldwide are developing drought-resistant crops that survive with less water or recover and re-grow after dry conditions. Work is being done to develop biotech crops that can handle the stress of high temperatures and salt-tolerant crops that can grow in saline soils that were traditionally viewed as unproductive. Most importantly, all these areas of research are working to develop biotech crops that can not only survive in extreme climates but also continue to produce high yields (Croplife 2010).

1.2.5 Pesticide Use in Developing Countries

The earth summit of Johannesburg agreed the principle, which states that 2020 should be the year for reducing the hazardous impact of chemicals on health and environment. This target is not achievable in many developing countries.

Pesticides are some of the most stringently regulated chemicals in the world. But some developing countries lack laws and regulations that properly regulate pesticide imports/exports and use. Therefore, they utilize only one fifth of the pesticides applied in the world (Ecobichon 2001). Most countries may still lack strict enforcement; for example, in Nepal, despite having the necessary legislation, farmers continue to succumb to market and peer pressures to buy highly toxic obsolete pesticides. In Egypt, many of the pesticides that have already been classified as probable or possible carcinogens by the US EPA and banned by the Egyptian government in 1996 are still listed among the pesticides registered and recommended for use in Egypt during the 2001/2002 period (Loutfy et al. 2008). Moreover, a number of nonregistered and possibly internationally banned pesticides are smuggled into Egypt and are extensively used, constituting a major concern.

A global surveillance of DDT levels in human tissues discovered higher levels in Africa, Asia, and Latin America than in Europe and the United States (Jaga and Dharmani 2003). Contamination of 2,3,7,8-tetrachlorodibenzo-p-dioxin (TCDD, a contaminant of the phenoxy herbicide 2,4,5-T) has been measured in butter samples from Egypt and found to be several times higher than the reported values in all EU countries (Loutfy et al. 2007). Also, according to WHO, Egypt had the highest PCD D/Fs levels in breast milk among 26 countries (Malish and Van Lewen 2003).

The use of these compounds has been either banned or restricted in many developed countries; however, still many industries from these countries market these products to the developing world. For example, from 1997 to 2000, the US pesticide companies exported over 30,500 metric tons of pesticides forbidden from use in the United States (Raven et al. 2008). Frey (1995) has examined the problem of the flow of pesticides from developed countries to less developed countries in terms of increased human and environmental health risks, and social and economic costs, and argued that political and economic forces characterized the increased flow to the less developed countries.

Additionally, export of chemicals that are banned in Western countries to developing countries without adequate warnings and precautions would cause people to become marginalized. Indeed, developed nations have, in the past, deliberately or otherwise, "dumped" highly toxic and expired chemicals into less developed countries as "aid." For instance, more than 74 metric tons of highly toxic and persistent chemical pesticides were donated by multinational companies to Nepal (Shah and Devkota 2009), which has essentially become an "ecological time bomb" that could go off in the near future. The ingredients of this "ecological time bomb" include DDT, dieldrin, and chlorinated organomercury compounds.

The farmers have inadequate knowledge about pesticides as to their suitability, application techniques, and safety measures. This is one of the reasons for poor pest control, environmental pollution, and health problems in some areas. Programs for guidance of the farmers in this respect are far and few. The pesticide industry does not put sufficient resources on dissemination of knowledge on pests, pesticides, environment, and management techniques. In this area, there is great scope of extension work in the public sector.

Many individuals are unaware that pesticides, freely available and so commonly used, are in fact deadly poisons.

Lack of advanced analytical methods tends to prevail in developing countries. Methodology is available for polychlorinated biphenyls (PCBs) and organochlorine pesticides (OCPs) as a result of a vast amount of environmental analytical chemistry research and development over the past 30–40 years. However, the establishment of an analytical laboratory and the application of this methodology at currently acceptable international standards are relatively expensive undertakings. Furthermore, the current trend to use isotope-labeled analytical standards and high-resolution mass spectrometry for routine POPs analysis is particularly expensive. These costs limit the participation of scientists in developing countries, and this is clear from the relative lack of publications and information on POPs from developing countries. The responsible use of pesticides requires the ability to read and follow label directions. Farmers also often lack the resources to purchase equipment and supplies specified on the label to properly apply a pesticide. Pest identification is lacking, and risks from pests are often not properly assessed. Pesticide and application equipment availability is too often determined by the government's or aid agency's use of "surplus" goods from elsewhere and often not well suited to solve the problems at hand. Improper protective gear for the applicator and improper safeguards for the environment are also the case in developing countries.

1.2.6 Pesticide Movement in the Environment

Pesticides are found at detectable levels in many parts of the environment in the inhabited as well as noninhabited areas of the world. Once pesticides are released into the environment, they tend to build up in the fat tissues of living organisms, causing serious harm to the health and a potential loss of biodiversity.

The introduction of pesticides into these areas can occur in several ways and for a variety of purposes (Westlake and Gunther 1966). It can result from direct application to suppress insects and other pests in agriculture, forestry, home, garden, and greenhouse and pests and vectors affecting man and animals. From the relatively small treated areas, indirect entry into wide areas of the environment can occur through wind, water, and food. A brief synopsis of the main direct and indirect sources of environmental contamination by pesticides is given in Table 1.4. Drift and evaporation during aerial application, volatilization

TABLE 1.4

A Brief of the Direct and Indirect Sources of Environmental Contamination by Pesticides

Direct Sources	Indirect Sources
Application for pest control in agriculture	Drift (air), rain, and snow
Application for pest/disease control in livestock	Animal dips
Soil treatments to control subterranean pests	Soil erosion
Water treatments to control weeds, mosquitoes, and other	Sanitation system carrying pesticides from washing and cleaning of equipment and containers
	Dumping of pesticides
	Industrial wastes from pesticide-manufacturing plants
	Pesticide spills

from crops and agricultural soils, wind erosion of contaminated soils, and emissions from manufacturing and disposal processes constantly add to the level of pesticides in the atmosphere. Both direct and indirect sources and other routes play important roles in pesticidal contamination of the major components of the environment.

The environmental dynamics of pesticides are influenced largely by the various factors operating in the environment and the physicochemical and biological properties of pesticides. In the broad sense, the environment is divided into four major components, namely atmosphere (air), hydrosphere (water), lithosphere (soil), and biosphere (biota), each possessing its own physical and chemical and/or biological properties. The biotic and abiotic elements in each component influence the dynamics of pesticides. The environmental dynamics of pesticides are further influenced by the physicochemical properties of pesticides. Such properties of pesticides as hydrophilicity or lipophilicity, partition coefficients, adsorption, vapor pressure, and volatility determine the ultimate fate of pesticides in the living and nonliving portions of the systems. This part will be explained in more details in Chapter 2.

Trans-boundary transport of chemicals is known since long ago. Once applied, pesticides disperse from the point of application and become redistributed, some on a global scale. Organochlorine insecticides such as DDT have been detected in the Arctic as well as in the mid-Pacific Ocean. They are long lived and spread through the atmosphere mostly as spray drift, which adheres to dust and is transported thousands of kilometers by global wind currents.

Outside the food chain, the atmosphere is probably the most important medium for long-distance dispersal of pesticides. Prevailing air currents may transport these pesticides and their alteration products over a great distance (Wolfe 1979).

Pesticides belong to the category of semivolatile organic compounds and can be transported owing to their aerial/particulate distribution into the atmosphere (Valavanidis 2000). In this manner, pesticides used in one country may reach another in which their use has been banned. Pesticide residues may return from the atmosphere to the surface of Earth by way of rainout, fallout, or direct sorption. In largely agricultural regions, pesticides in the ambient air may constitute a major source of exposure for human beings and wildlife. On a global scale, airborne pesticides may affect the ecological balance in nonagricultural areas.

1.3 Unavoidable Cost

Although the use of synthetic chemical pesticides is unavoidable, the extensive consumption not only constitutes a massive loss of money but also causes a lot of undesirable side effects on human health, environment, food quality, and biodiversity. Pesticide drift occurs when pesticides suspended in the air as particles are carried by wind to other areas, potentially contaminating them. It has been reported that over 98% of sprayed insecticides and 95% of herbicides reach a destination other than their target species, including nontarget species, air, water, and soil (Miller 2004). Pesticides are one of the causes of air and water pollution, and some pesticides are considered POPs and contribute to soil contamination. Some pesticides, even though no longer used, nevertheless, persist in the environment. Residues of these pesticides are sometimes found on food grown on contaminated soil or in the fish that live in contaminated waters. Chlorinated pesticides have been implicated in a variety of human health issues and as causing significant and widespread ecosystem disruption through their toxic effects on organisms.

1.3.1 Human Health

Global use of pesticides creates substantial health impacts in all parts of the world, although the exact toll is difficult to pinpoint, given both the various chemicals and the types of exposure. In short, not all pesticides are equally risky, and not all people are equally at risk. There are two broad categories of health effects caused by pesticides: short-term effects (acute effects), which appear immediately or very soon after exposure, and long-term effects (chronic effects), which may manifest themselves many years later and whose origins are often difficult to trace.

The four routes of exposure are skin (dermal), lungs (inhalation), mouth (oral), and eyes. Suspected chronic effects from exposure to certain pesticides include birth defects, toxicity to a fetus, benign or malignant tumors, genetic changes, blood disorders, nerve disorders, endocrine disruption, and reproduction effects. The symptoms of pesticide poisoning can range from a mild skin irritation to coma or even death. Different classes or families of chemicals cause different types of symptoms. Individuals also vary in their sensitivity to different levels of these chemicals (Lorenz 2009). For instance, mild organophosphate poisoning manifests in the form of malaise, vomiting, nausea, diarrhea, loose stools, sweating, abdominal pain, and salivation. Moderate poisoning includes dyspnea, decreased muscular strength, bronchospasm, miosis, muscle fasciculation, tremor, motor incoordination, bradycardia, and hypotension/hypertension. Severe manifestation could result in coma, respiratory paralysis, extreme hypersecretion, cyanosis, sustained hypotension, extreme muscle weakness, muscular paralysis, and convulsion (Iowa State University 1995). Pesticides can cause a range of adverse effects on human health, including injury to the nervous system, lung damage, reproductive dysfunction, dysfunction of the endocrine and immune systems, and possibly cancer. Breast cancer is major public health concern. A number of studies have pinpointed the causal link between body burden of some organochlorine pesticides and the higher risk of breast cancer in some developing countries (Tawfic Ahmed et al. 2001).

Exposure to pesticides is mainly divided into three categories: occupational exposure (spray operators and farm), accidental exposure, and pesticide residues in food. Workers with occupational exposures to pesticides on average have significantly greater exposure than the rest of the population (Ongley 1996). Although attempts to reduce pesticide use through organic agricultural practices and the use of other technologies to control pests continue, exposures to pesticides occupationally, through home and garden use, through termite control or indirectly through spray drifts, and through residues in household dust, food, and water are common (Fenner-Crisp 2001). Even in developed countries, despite the strict regulations and the use of safer pesticides, occupational exposures may be significant.

The majority of pesticide poisonings and deaths occur in the developing world and account for a staggering 99% of the related deaths, although these countries use only 25% of global pesticide production (FAO Newsroom 2004).

Reasons for this include the following: developing countries have a higher proportion of the population involved in agriculture, their safe-health standards can be inadequate or nonexistent, they have poorer pesticide handling practices, they commonly use unsafe equipment, and their knowledge of health risks and safe use is limited. Also, harmful pesticides are easily accessible (Ongley 1996). The World Health Organization and the United Nations Environment Programme estimate that each year, 3 million workers in agriculture in the developing world experience severe poisoning from pesticides, about 18,000 of whom die (Miller 2004). It is believed that in developing countries, the incidence of pesticide poisoning may even be greater than reported due to under-reporting, lack of data, and misdiagnosis (Wilson and Tisdell 2000).

1.3.2 Environmental Contamination

Pesticides are found at detectable levels in many parts of the environment in the inhabited as well as noninhabited areas of the world. Some pesticides belong to the POPs class and contribute to soil contamination. In addition to ecological impacts in countries of application, pesticides that have been long banned in developed countries (such as DDT, toxaphene, etc.) are consistently found in remote areas such as the high arctic. Chemicals that are applied in tropical and subtropical countries are transported over long distances by global air circulation (Ongley 1996). GESAMP (1986) defines environmental (also known as receiving, absorptive, or assimilative) capacity as "the ability of a receiving system or ecosystem to cope with certain concentrations or levels of waste discharges without suffering any significant deleterious effects" (Cairns 1977, 1989). All activities, including agriculture have some level of impact on air, water, and soil quality; the issue is whether the impact exceeds the thresholds that society deems unacceptable for social, economic, or cultural reasons. For agriculture, there is a need to determine what the environmental capacity is for different types of runoff products in the local context (Ongley 1996).

Agriculture, as the single largest user of freshwater on a global basis and as a major cause of degradation of surface and groundwater resources through erosion and chemical runoff, has a cause to be concerned about the global implications of water quality. The US EPA (1994) identified agriculture as the leading cause of water quality impairment of rivers and lakes in the United States and third in importance for pollution of estuaries. Degradation of water quality by pesticide runoff has two principal human health impacts. The first is the consumption of fish and shellfish that are contaminated by pesticides; the second is the direct consumption of pesticide-contaminated water. WHO (1993) has established drinking water guidelines for 33 pesticides. Many health and environmental protection agencies have established "acceptable daily intake" (ADI) values that indicate the maximum allowed daily ingestion over a person's lifetime without appreciable risk to the individual.

There is no doubt that the environmental pollution has become a real global crisis. The International Code of Conduct on the Distribution and Use of Pesticides (FAO 1985), along with other important regulations such as the Pesticide Risk Reduction Initiative of Organization for Economic Co-operation and Development (OECD) and FAO and Stockholm Convention on POPs, is very relevant to pesticide pollution control and environmental protection in general. Many countries all over the world have taken good steps to control their use of pesticides.

Monitoring of environmental samples is usually done on a regular and routine basis in developed countries to evaluate the risk to environment. Key pesticides are included in the monitoring schedule in these countries. In developing countries, basic data about pesticide releases and environmental contamination levels are missing. Monitoring programs are limited to some organochlorine pesticides. For example, the low levels of pesticide residues reported in Egypt reflect good progress made by the country in reducing pesticide use cit. (Loutfy et al. 2008). The rationale is based on the fact that if aquatic contamination exists, it will be evident in biota. Canada and the United States have agreed on the use of biologically based objectives for the management of water quality of the Great Lakes. These techniques are very useful in developing countries insofar as the measurements require training in biology (usually good in developing countries) and are labor-intensive rather than capital-intensive. Environmental Effects Monitoring (EEM) reduces reliance on high-end analytical chemistry with its large capital costs and lengthy training. EEM can involve many other types of biological measures such as fish health, fecundity, immune suppression, etc. Many of these tests are simple and are of low cost (Ongley 1996).

Based on the limited literature on pesticide use and impacts in Africa, Calamari and Naeve (1994) conclude that:

> The concentrations found in various aquatic compartments, with few exceptions are lower than in other parts of the world, in particular in developed countries which have a longer history of high pesticide consumption and intense use. Generally, the coastal waters, sediments and biota are less contaminated than inland water environmental compartments, with the exception of a few hot spots.

One of the mandates of the Basel Convention is to help countries eliminate their stockpiles of obsolete pesticides. Stockpiles have accumulated largely because some products that were banned for health or environmental reasons were never properly discarded. These pesticides contain some of the most dangerous insecticides produced—members of the POPs group. These dangerous chemicals threaten communities through the potential contamination of food, water, soil, and air. Poor communities are the most vulnerable to environmental degradation of this sort.

1.3.3 Pesticides Residues in Food

Pesticide use may lead to traces of residues in food. When a crop is treated with a pesticide, a very small amount of the pesticide, or indeed what it changes to in the plant (its "metabolites" or "degradation products"), can remain in the crop until after it is harvested (residues). Occasionally, residues may also result from environmental or other "indirect" sources. Many of the broad-spectrum pesticides do not biodegrade readily and remain in the environment for many years, where they contaminate air, water, soil, and other resources. Residues of old pesticides, such as DDT (banned in the developed countries for many years), are an example of such environmental contaminants. Chlorinated pesticides accumulate in animals and humans and biomagnify in the food chain at different levels. Food consumption has been identified as the major pathway of human exposure, accounting for >90% compared to other ways of exposure such as inhalation and dermal contact. The major food sources of POPs have been reported to be fat-containing animal products, fish, and shellfish (Fries 1995). Therefore, dietary intake studies have been carried out extensively in most developed countries.

The monitoring program is the prime means of ensuring that pesticides are used in accordance with *Good Agricultural Practice*. The program is essential to the elimination of abuses in the use of pesticides, such as use of excessive dose rates, failure to respect the minimum periods specified between last application and harvest (i.e., preharvest intervals), and use for purposes for which they are not authorized (i.e., illegal uses). When used in accordance with *Good Agricultural Practice*, unacceptable levels of pesticide residues should not occur in treated produce (IDAFRD 2001). Monitoring programs for EU countries are conducted on a regular basis to evaluate the health risks of pesticides. In 2004, 47% of fruit, vegetables, and cereals consumed in Europe contained pesticide residues according to the annual DG SANCO report (European Commission 2006).

Pesticide residue levels in treated crops are regulated through the establishment of maximum residue levels (MRLs). EPA sets MRLs, or tolerances, for pesticides that can be used on various food and feed commodities. These limits are set to protect humans from harmful levels of pesticides in their food (US EPA 2000). Residue levels found in food are usually below the tolerance levels. Tolerances for a given pesticide may vary for different crops. EPA has set tolerances for about 400 different pesticides. Since a pesticide may be used on many different crops, there are about 9700 tolerances in effect.

Although there is little information on pesticide residues in food for developing countries, particularly food intended for local consumption, export crops are analyzed for residues to protect consumers in developed nations. Vegetables and fruits appear to receive the highest pesticide doses and may contain high levels of residue.

1.3.4 Biodiversity Loss

In 2010, the UN International Year of Biodiversity, the PAN Europe review summarized recent research findings from the scientific literature on the impact of pesticides on biodiversity and concluded that pesticides are a major factor affecting biological diversity globally, along with habitat loss and climate change. They can be directly toxic to organisms or cause changes in their habitat and the food chain (PAN Europe 2010).

Pesticides are toxic products designed to kill a target organism; however, they also kill other nontarget organisms such as the natural predators of the pest and organisms that are beneficial to the health and balance of the ecosystem (WHO 2003).

There is growing concern about the health consequences of biodiversity loss and change. Biodiversity changes affect ecosystem functioning, and significant disruptions of ecosystems can result in loss of life-sustaining ecosystem resources. Biodiversity loss also means that we are losing, before discovery, many of nature's chemicals and genes, of the kind that has already provided humankind with enormous health benefits. Specific pressures and linkages between health and biodiversity include biodiversity that provides numerous ecosystem services that are crucial to human population's well-being at present and in the future.

Over the past 40 years, the use of highly toxic carbamates and organophosphates has increased dramatically and threatens exposed wildlife through their toxicity. Organochlorines such as endosulfan, which are highly persistent, are still used on a large scale. Pesticide poisoning can cause population declines, which may threaten rare species.

Pesticides accumulating in the food chain, particularly those that cause endocrine disruption, pose a long-term risk to mammals, birds, amphibians, and fish. But pesticides can also have indirect effects by reducing the abundance of weeds and insects, which are important food sources for many species. Herbicides can change habitats by altering vegetation structure, ultimately leading to population decline. Fungicide use has also allowed farmers to stop growing "break crops" such as grass or roots. This has led to the decline of some arable weeds (PAN Europe 2010). Miller (2004) reported that pesticide use contributes to pollinator decline, destroys habitat (especially for birds), and threatens endangered species.

1.4 Out of the Impasse

Rapid growth in populations and the demand for food, coupled with degradation in the human and environmental health and, recently, the challenge of climate change, accelerated the search for alternatives to chemical pesticides.

The alternatives to pesticides include plant breeding for resistance, biological control, microbial pesticides, botanical pesticides, and IPM (Mackay 1993). Application of composted yard waste has also been used as a way of controlling pests. These methods are becoming increasingly popular and often are safer than traditional pest control methods.

1.4.1 Pest Control: A Changing Concept

To minimize the risks posed by the indiscriminate and intensive use of the chemical pesticides, most of the developed and developing countries applied the concept of IPM to control pests.

IPM, the internationally acclaimed solution to indiscriminate use of pesticides, has been declared as one of the government policies in sustainable development of agriculture. IPM was adopted as a policy to control pests by various world governments during the 1970s and 1980s. Since then, there have been numerous success stories of successful IPM programs around the world. Table 1.5 shows that IPM strategies are a key component in reducing the amount of pesticides in some developing countries. IPM combines a judicious use of chemicals with various other control strategies. There are, however, many constraints to its adoption (Bottrell 1987).

1.4.2 Pesticide Alternatives

Pesticides cause many problems. Therefore, developing and applying nonpesticide methods as alternatives to chemical control of plant diseases and pests has become the global trend. In recent years, new classes of insecticides have been marketed, none of which

TABLE 1.5

Examples of Pesticide Amounts Reduced Due to IPM in Some Developing Countries

Country	Findings	Source
Indonesia	Application of IPM techniques saved about $1200 a year per farm through reduced pesticide use	ADB (1999)
India	Decreased conventional pesticide use by 50% on average. Incomes increased by Rs. 1000–1250/ha, and rice yields increased by 250 kg/ha	FAO (2002)
Sri Lanka	Reduced insecticide use (from 3 applications to 1 application per season) and increased yields (by 12%–44% for rice) were observed	Administration Report, Department of Agriculture (2002)
Developing countries	A review of 25 impact evaluation studies reported substantial and consistent reductions in pesticide use and convincing increase in yield due to training	Van den Berg (2004)
South America	In Argentina, IPM has reduced the number of insecticide treatments from 2–3 per season to an average 0.3 treatments, corresponding to savings of US $1.2 million per year in pesticides and application costs. Similarly, in Brazil, IPM has been adopted by about 40% of soybean farmers, which has resulted in savings of over US $200 million annually due to reduced use of insecticides, labor, machinery, and fuel	Iles and Sweetmore (1991)

Source: Modified from Atreya, K., Sitaula, B. K., Johnsen, F. H., and Bajracharya R. M. 2011. Continuing issues in the limitations of pesticide use in developing countries. *J. Agric. Environ. Ethics* 24, 49–62.

persist or bioaccumulate. They include juvenile hormone mimics, synthetic versions of insect juvenile hormones that act by preventing immature stages of the insects from molting into an adult, and pesticides based on natural products. Biopesticides include naturally occurring substances that control pests (biochemical pesticides), microorganisms that control pests (microbial pesticides), and pesticidal substances produced by plants containing added genetic material (plant-incorporated protectants). Biopesticides are considered environment friendly, causing less harm to human and animal health and to the ecosystem. They can make important contributions to IPM and help reduce reliance on chemical pesticides. Hence, they have a major role to play in the development of sustainable agriculture.

Biopesticides are considered environmentally safe materials. Biopesticides often have a narrow spectrum of pest activity, which means they have a relatively low direct impact on nontargets, including humans. Their use is often compatible with other control agents, and they produce little or no residue. They are relatively inexpensive to develop. One significant advantage of biopesticides based on natural enemies is that they can reproduce in the pest population. This means that the natural enemy population can respond to changes in the pest population, giving a flexible form of pest management. Biopesticides have been criticized for their higher unit prices and lower efficacy compared to chemical pesticides. However, such comparisons are overly simplistic and may well detract from the beneficial properties of biopesticides (cit. Taha 2000).

Pesticides such as avermectins, spinosad, and azadirachtin are smart, selective, and green molecules and have been introduced for controlling selected groups of insect pests.

1. *Avermectins*: They are families of macrocyclic lactones that have a novel mode of action against a broad spectrum of nematodes and arthropods in doses as low as 10 g/kg (Campell et al. 1983; Shoop et al. 1995). Avermectins were first found in the fermentation broth of a soil-dwelling microorganism, *Streptomyces avermitilis*, at the Kitasato Institute in Japan. After conducting numerous bioassays in Merck laboratories, eight natural avermectin components, namely A1a, A1b, A2a, A2b, B1a, B1b, B2a, and B2b, were discovered.

   Because of its high tolerability, prolonged posttreatment effect, and broad spectrum of antiparasitic activity, avermectin has become a popular drug in the treatment of many animal and human parasite infestations, such as onchocerciasis (Burkhart 2000). Other avermectins, including abamectin, doramectin, and emamectin, were subsequently commercialized and used as agricultural insecticides and miticides in animal health and/or crop protection.

2. *Spinosad (Dow AgroSciences)*: It is a naturally derived insecticide produced by the fermentation of *Saccharopolyspora spinosa*, an actinomycete bacterium originally isolated from a Caribbean soil sample (Sparks et al. 2001). It is highly active by ingestion and to a lesser degree by contact. Spinosad-based products have been registered in more than 30 countries for control of pests. Spinosad has very less mammalian toxicity and is classified by the US EPA as an environmentally and toxicologically reduced-risk material.

3. *Neem (Azadirachtin)*: In recent years, the bioactivity of Neem against insect pests has been particularly investigated in detail. Large numbers of insect pests from different orders have been shown to exhibit different levels of susceptibility to Neem seed extracts, or the most active constituent, azadirachtin (Mahmoud 2007; Mahmoud and Shoeib 2008). Azadirachtin, a mixture of several structurally

related tetranortriterpenoids, has attracted the greatest attention in recent years for modern pest-control strategies (Govindachari et al. 1992).

Neem extract is often described to have minimal toxicity to nontarget organisms such as parasitoids, predators, and pollinators (Raguraman and Singh 1999). Another important feature of Neem is the rapid degradation in the environment (Isman 1999).

   4. *Aphicides*: A number of selective aphicides are being developed that provide selective aphid control, with little or no toxicity or repellency to honey bees.

Aphistar (triazimate, Rohm & Haas) and Pirimor (pirimicarb, Zeneca) have excellent contact and systemic aphid activity on a number of crops. Unfortunately, both products are carbamates, and their future status is not known. Another promising product under consideration is Assail (acetamiprid, Ceragri), which is also safe for pollinators.

1.4.3 Genetic Engineering and Herbicide-Resistant Crops

Genetic engineering involves the insertion of a foreign gene or genes into the genome of another organism. In the process of crop genetic engineering, DNA is removed from one organism (a bacterium, for example) and the specific gene of interest is isolated and manipulated so that it can be inserted into the cells of the crop plant.

Crop genetic engineering is a kind of biotechnology. It is used to develop seeds that will produce crops with specific traits. There are three generations, or phases, of crop biotechnology. With the advent and availability of Roundup Ready soybeans, farmers can plant the Roundup-tolerant crop and spray the entire field including the crop plants with Roundup. The weeds growing in the field will die, but the crop plants will be unaffected. The first generation of crop biotechnology produced crops genetically engineered to tolerate chemical herbicides or to produce pesticides. Roundup Ready (a trade name given to certain varieties of corn, soybean, cotton, or canola, which are genetically engineered to be resistant to the herbicide Roundup) soybeans and Bt corn (corn that has been transformed with the Bt gene and is resistant to European corn borer (ECB)) are the most common examples of these crops. Both are widely used today. Roundup (a herbicide that provides nonselective control of several annual and perennial weeds and also damages crops such as corn and soybeans that are not Roundup resistant) is a very common glyphosate-based herbicide that is effective in killing a wide range of weed species. It is referred to as a broad-spectrum herbicide. Roundup Ready soybeans are genetically modified crop plants that are resistant to Roundup and are an example of herbicide-tolerant crops.

Bacillus thuringiensis (*Bt*) is a naturally occurring soil-borne bacterium that produces a crystalline protein lethal to certain insects. When insects ingest *Bt*, the crystalline protein paralyzes their digestive tracts, causing them to stop eating. The insects die within 12 hours to 5 days. *Bt* was formulated as an insecticide spray in the 1950s, and its use is approved for organic farming because it is a "natural" insecticide in the spray form. In 1996, the first *Bt* corn hybrid was produced using genetic engineering. Scientists isolated the gene coding for one of the *Bt* proteins that was toxic to the ECB and inserted it into a corn plant. *Bt* corn, a genetically modified organism (GMO), produces the crystalline protein that is lethal to ECB larvae. When larvae feed on any tissue of the *Bt* corn plant, they ingest the protein, their digestive tracts are paralyzed, and they die. Because *Bt* corn is a GMO crop, its use is not allowed for organic farming.

1.4.4 Green Chemistry and Green Pesticides

The closing decades of the last century have witnessed the emergence of green chemistry, thanks to the new conceptualized ideas of a number of renowned chemists. New principles in chemistry have emerged, including atom economy, solventless reaction, biocatalysts, design for degradation, and many others. They highlight the sprouting concept of chemistry that amalgamated the basic essence of chemistry with sound environmental and safety codes.

The term Green Chemistry as adopted by the International Union of Pure and Applied Chemistry (IUPAC) is defined as "The invention, design and application of chemical products and processes to reduce or to eliminate the use and generation of hazardous substances." Green chemistry is also defined as "Discovery and application of new chemistry/ technology leading to prevention/reduction of environmental, health and safety impacts at source."

With the trend of green chemistry starting to prevail, a number of novel molecules have been developed and used in pest control. Green pesticides are still in infancy but rigorous efforts are being exerted in a number of international laboratories to develop new chemical species exhibiting the paradigm of green chemistry.

In the highly interrelated, interdependent world of modern technology and trade, the challenge of protecting crops and livestock from insects, diseases, weeds, and other pests without hazards to humans, animals, or the environment requires the combined and sustained efforts of scientists, technicians, and administrators; producers, processors, and distributors; industry and government; and nations working together to establish and administer sound, acceptable standards of food safety and environment quality (FAO 1985).

References

ADB. 1999. The Growth and Sustainability of Agriculture in Asia. Oxford University Press: Oxford, UK.

Administration Report. 2002. Department of Agriculture, Peradeniya.

Agronews. 2009. Global AgroChemicals Market (2009–2014). http://news agropages.com/Report/ReportDetail 53.html.

Atreya, K., Sitaula, B. K., Johnsen, F. H., and Bajracharya, R. M. 2011. Continuing issues in the limitations of pesticide use in developing countries. *J. Agric. Environ. Ethics* 24: 49–62.

Bloomfield, J. P., Williams, R. J., Gooddy, D. C., Cape, J. N., and Guha, P. 2006. Impacts of climate change on the fate and behaviour of pesticides in surface and groundwater – A UK perspective. *Sci. Total Environ.* 369: 163–177.

Bottrell, D. G. 1987. Applications and problems of integrated pest management in the tropics. *J. Plant Prot. Tropics* 4: 1–8.

Burkhart, C. N. 2000. Ivermectin: An assessment of its pharmacology, microbiology and safety. *Vet. Human Toxicol.* 42: 30–35.

Cairns, J. 1977. Aquatic ecosystem assimilative capacity. *Fisheries* 2: 5–7.

Cairns, J. 1989. Applied ecotoxicology and methodology. In: Boudou, A. and Ribeyre, P. (eds), *Aquatic Ecotoxicology: Fundamental Concepts and Methodologies*, Vol. 2, pp. 275–290. CRC Press: Boca Raton, FL.

Calamari, D. and Naeve, H. 1994. Review of Pollution in the African Aquatic Environment. Comité de Pesca Continental para África (CPCA), Documento Técnico No. 25. FAO, Roma.

Campell, W. C., Fisher, M. H., and Stapley, E. O. 1983. Ivermectin: A potent antiparasitic agent. *Science* 221: 823–828.

Capkin, E., Altinok, I., and Karahan, S. 2006. Water quality and fish size affect toxicity of endosulfan, an organochlorine pesticide, to rainbow trout. *Chemosphere* 64: 1793–1800.

Croplife. 2010. Climate change issues: sustainability, climate change. http://www.croplife.org/public/climate_change.

Delorenzo, M. E., Wallace, S. C., Danese, L. E., and Baird, T. D. 2009. Temperature and salinity effects on the toxicity of common pesticides to the grass shrimp, *Palaemonetes pugio*. *J. Environ. Sci. Health B Pestic. Food Contam. Agric. Wastes* 44: 455–460.

EC (European Commission). 2001. Council Regulation (EC) No. 2375/2001 of 29 November 2001 amending Commission Regulation (EC) 2006. Setting maximum levels for certain contaminants in foodstuffs. *Off. J. Eur. Commun.* 321: 1–5.

EC (European Commission) SCOOP. 2000. Reports on tasks for scientific cooperation. Assessment of dietary intake of dioxins and related PCBs by the population of UM Member States. Report of experts participating in Task 3.2.5.

Ecobichon, D. J. 2001. Pesticide use in developing countries. *Toxicology* 160: 27–33.

EPA (US Environmental Protection Agency). 1994. National Water Quality Inventory. 1992 Report to Congress. EPA-841-R-94-001. Office of Water: Washington, DC.

EPA (US Environmental Protection Agency). 2000. Exposure and human health reassessment of 2,3,7,8-tetrachlorodibenzo-p-dioxins and related compounds. EPA/600/P-00/001 Bg.

EPA (US Environmental Protection Agency). 2000–2001. Pesticide market estimates: Sales. http://www.epa.gov/opp00001/pestsales/01pestsales/sales2001.htm.

FAO (Food and Agriculture Organization). 1985. Erosion-induced loss in soil productivity: A research design. M. Stocking. Consultants Working Paper No. 2, Soil Conservation Programme, División de Fomento de Tierras y Aguas, FAO, Roma.

FAO (Food and Agriculture Organization of the United Nations). 2002. Land and Agriculture from UNCED, Rio de Janeiro 1992 to WSSD, Johannesburg 2002: A compendium of recent sustainable development initiatives in the field of agriculture and land management. Rome, Italy.

FAO Workshop (Food and Agriculture Organization). 2005. Proceedings of the Asia Regional Workshop on the Implementation, Monitoring and Observance of the International Code of Conduct on the Distribution and Use of Pesticides, Regional Office for Asia and the Pacific, Bangkok. http://www.fao.org/docrep/008/af340e/af340e00.htm#Contents

FAO Newsroom (Food and Agriculture Organization). 2004. Children face higher risks from pesticide poisoning. http://www.fao.org/newsroom/en/news/2004/51018/index.html

Fenner-Crisp, P. E. 2001. Risk-assessment and risk management: The regulatory process. In: Kreiger, R. (ed.), *Handbook of Pesticide Toxicology*, 2nd edn, pp. 681–690. Academic Press: San Diego, CA.

Frey, R. S. 1995. The international traffic in pesticides. *Technol. Forecast. Social Change* 50: 151–169.

Fries, G. F. 1995. A review of the significance of animal food products as potential pathways of human exposure to dioxins. *J. Animal Sci.* 73: 1639–1650.

Fytianos, K., Vasilikiotis, J., Weil, L., Kavlengis, E., and Laskaridis, N. 1985. Preliminary study of organochlorine compounds in milk products, human milk and vegetables. *Bull. Environ. Contam. Toxicol.* 34: 504–508.

Gaunt, P. and Barker, S. A. 2000. Matrix solid phase dispersion extraction of triazines from catfish tissues; examination of the effects of temperature and dissolved oxygen on the toxicity of atrazine. *Int. J. Environ. Pollut.* 13: 284–312.

Govindachari, T. R., Sandhya, G., and Ganeshraj, S. P. 1992. Azadirachtin H and I: Two new tetranorterpenoids from *Azadirachta indica*. *J. Nat. Prod.* 55: 596–601.

Gutierrez, A. P., Ponti, L., d'Oultremont, T., and Ellis, C. K. 2008. Climate change effects on poikilotherm tritrophic interactions. *Climatic Change* 87: S167–S192.

Harwood, A. D., You, J., and Lydy, M. J. 2009. Temperature as a toxicity identification evaluation tool for pyrethroid insecticides: Toxicokinetic confirmation. *Environ. Toxicol. Chem.* 28: 1051–1058.

Hayes, W. J. 1991. Chlorinated hydrocarbon insecticides. In: Hayes, W. J. and Lawes, E. R. (eds), *Pesticides Studied in Man*, pp. 731–868. Academic Press: San Diego, CA.

Heath, S., Bennett, W. A., Kennedy, J., and Beitinger, T. L. 1994. Heat and cold tolerance of the fathead minnow, *Pimephales promelas*, exposed to the synthetic pyrethroid cyfluthrin. *Can. J. Fish. Aquat. Sci.* 51: 437–440.

IDAFRD (Ireland Department of Agriculture Food & Rural Development). 2001. Pesticide control service: Pesticide residues in food. http://www.pcs.agriculture.gov.ie/Docs/residu00.pdf (August 2001).

Iles, M. J. and Sweetmore, A. 1991. *Constraints on the Adoption of IPM in Developing Countries – A Survey*. National Resources Institute: Kent, UK.

Iowa State University. 1995. Safe farm: Promoting agricultural health and safety.

IPCC (United Nations Intergovernmental Panel on Climate Change). 2007. Impacts, Adaptation and Vulnerability: Scientific-Technical Analyses – Contribution of Working Group II to the Fourth Assessment. *Report of the Intergovernmental Panel on Climate Change*, 976 pp. Cambridge University Press: Cambridge, UK.

Isman, M. B. 1999. Neem and related natural products. In: Hall, F. R. and Menn, J. J. (eds), *Biopesticides: Use and Delivery*, pp. 139–153. Totowa: New Jersey.

Jaga, K. and Dharmani, C. 2003. Global surveillance of DDT and DDE levels in human tissues. *Int. J. Occup. Med. Environ. Health* 16: 7–20.

Koleva, N. G., Schneider, U. A., and Tol Richard, S. J. 2009. The impact of weather variability and climate change on pesticide applications in the US – An empirical investigation Working Paper FNU-171 (March 2009).

Koncept Analytics. 2010. Global Crop Protection (Pesticides) Market Report – 2010 Edition. http://www.pr-inside.com/global-crop-protection-pesticides-market-r2269531.htm.

Lannig, G., Flores, J. F., and Sokolova, I. M. 2006. Temperature-dependent stress response in oysters, *Crassostrea virginica*: Pollution reduces temperature tolerance in oysters. *Aquat. Toxicol.* 79: 278–287.

Lorenz, E. S. 2009. Potential Health Effects of Pesticides. Pesticide Safety Fact Sheet. http//www.pested.psu.edu/.

Loutfy, N., Fuerhacker, M., Tundo, P., Raccanelli, S., and Ahmed, M. T. 2007. Monitoring of polychlorinated dibenzo-p-dioxins and dibenzofurans, dioxin-like PCBs and polycyclic aromatic hydrocarbons in food and feed samples from Ismailia city, Egypt. *Chemosphere* 66: 1962–1970.

Loutfy, N., Fuerhacker, M., Lesueur, C., Gartner, M., Ahmed, M. T., and Mentler, A. 2008. Pesticide and non-dioxin-like polychlorinated biphenyls (NDL-PCBs) residues in foodstuffs from Ismailia city, Egypt. *Food Addit. Contam.* 1–9.

Loutfy, N., Mosleh, Y., and Ahmed, M. T. 2010. Dioxin, Dioxin-Like PCBs and Indicator PCBs in some medicinal plants irrigated with wastewater in Ismailia, Egypt. *Polycyclic Aromat. Compd.* 30: 9–26.

MacDonald, R., Mackay, D., and Hickie, B. 2002. Contaminant amplification in the environment. *Environ. Sci. Technol.* 36: 457A–462A.

MacKay, K. T. 1993. Impact of Pesticide Use on Health in Developing Countries (IDRC). Alternative methods for pest management in developing countries. http://www.nzdl.org/gsdlmod%3Fe%3Dd-00000-00.

Mahmoud, M. F. 2007. Efficacy of combination between botanical insecticides NSK extract, NeemAzal T 5%, Neemix 4.5% and entomopathogenic nematode *Steinernema feltiae* Cross N 33 for control the peach fruit fly *Bactrocera zonata* (Saundres). *Plant Protect. Sci.* 43: 19–25.

Mahmoud, M. F. and Shoeib, M. A. 2008. Sterilant and oviposition deterrent activity of neem formulation on peach fruit fly *Bactrocera zonata* (Saunders) (Diptera: Tephritidae). *J. Biopest.* 1: 177–181.

Malish, R. and Van Lewen, F. X. R. 2003. Results of the WHO coordinated exposure study on the levels of PCBs, PCDDs and PCDFs in human milk. *Organohalog. Compd.* 64: 40–143.

Maruya, K. A., Smalling, K. L. and Vetter, W. 2005. Temperature and congener structure affect the enantioselectivity of toxaphene elimination by fish. *Environ. Sci. Technol.* 39: 3999–4004.

Miller, G. T. 2004. *Sustaining the Earth*, 6th ed, Chapter 9, pp. 211–216. Thompson Learning, Inc.: Pacific Grove, California.

Monserrat, I. and Bianchini, A. 1995. Effects of temperature and salinity on the toxicity of a commercial formulation of methyl parathion to *Chasmagnathus granulata* (Decapoda, Grapsidae). *Braz. J. Med. Biol. Res.* 28: 74–78.

Noyes, P. D., McElwee, M. K., Miller, H. D., et al. 2009. The toxicology of climate change: Environmental contaminants in a warming world. *Environ. Int.* 35: 971–986.

Ongley, E. D. 1996. Control of water pollution from agriculture. FAO Irrigation and Drainage Paper No. 55. Food and Agriculture Organization of the United Nations, Rome. http://www.fao.org/docrep/W2598E/W2598E00.htm.

Palikhe, B. R. 2007. Relationship between pesticide use and climate change for crops. *J. Agric. Environ.* 8: 83–91.

PAN AP (Pesticide Action Network Asia Pacific). 2010. Communities in Peril: Global report on health impacts of pesticide use in agriculture, 182 pp. Edited by Barbara Dinham from regional reports for PAN International.

PAN Europe (Pesticide Action Network Europe). 2010. International year of biodiversity: Biodiversity and pesticides. *Pesticides News* 88: 2010. Available from http://www.pan-europe.info/Campaigns/html.

Paterson, G., Drouillard, K. G. and Haffner, G. D. 2007. PCB elimination by yellow perch (*Perca flavescens*) during an annual temperature cycle. *Environ. Sci. Technol.* 41: 824–829.

Patra, R. W., Chapman, J. C., Lim, R. P., and Gehrke, P. C. 2007. The effects of three organic chemicals on the upper thermal tolerances of four freshwater fishes. *Environ. Toxicol. Chem.* 26: 1454–1459.

Patterson, D. T., Westbrook, J. K., Joyce, R. J. V., Lingren, P. D., and Rogasik, J. 1999. Weeds, insects, and diseases. *Climatic Change* 43: 711–727.

Raguraman, S. and Singh, R. P. 1999. Biological effects of neem (*Azadirachta indica*) seed oil on an egg parasitoid, *Trichogramma chilonis*. *J. Economic Entomol.* 92: 1274–1280.

Raven, P. H., Berg, L. R., and Hassenzahl, D. M. 2008. *Environment.* Wiley: USA.

Rohr, J. R., Madison, D. M., and Sullivan, A. M. 2003. On temporal variation and conflicting selection pressures: A test of theory using newts. *Ecology* 84: 1816–1826.

Rohr, J. R., Timothy, M. S., and Chrisropher, S. 2010. Will climate change reduce the effects of a pesticide on amphibians?: partitioning effects on exposure and susceptibility to contaminants. *Global Change Biol.* 17, 657–666.

Shah, B. P. and Devkota, B. 2009. Obsolete pesticide: Their environmental and human health hazards. *J. Agric. Environ.* 10: 60–66.

Shoop, W. L., Mrozik, H., and Fisher, M. H. 1995. Structure and activity of avermectins and milbemycins in animal health and Veterinary. *Parasitology* 59: 139–156.

Sparks, T. C., Crouse, G. D., and Durst, G. 2001. Natural products as insecticides: the biology, biochemistry and quantitative structure–activity relationship of spinosyns and spinosoids. *Pest Manag. Sci.* 57: 896–905.

Stephenson, G. A. and Solomon, K. R. 1993. Pesticides and the environment. Department of Environmental Biology, University of Guelph, Guelph, ON.

Taha, H. A. A. 2000. Integrated crop broad bean management system by using biotechnology for best control. Annual report for the project No. 16/15. Regional Councils for Agriculture. Research and Extension, Agriculture Research Center, Ministry of Agriculture and land reclamation.

Tawfic Ahmed, M., Loutfy, N., and El Shiekh, E. 2001. Residues level of DDE and PCBs in the blood serum of women in the Port Said region of Egypt. *J. Hazard. Mater.* 27: 1–8.

Thomson, L. J., Macfadyen, S., and Hoffmann, Ary A. 2010. Predicting the effects of climate change on natural enemies of agricultural pests. *Biol. Control* 52: 296–306.

Valavanidis, A. 2000. Fundamental principles of environmental chemistry. *Ecotoxicol. Ecol. Risk Assess.* (in Greek).

Van den Berg, H. 2004. *IPM Farmer Field Schools: A Synthesis of 25 Impact Evaluations.* Wageningen University: The Netherland.

Watterson, A. 1991. *Pesticides and Your Food.* Greenprint: London. New Internationalist 323 May.

Westlake, W. E. and Gunther, F. A. 1966. Occurrence and mode of introduction of pesticides in the environment. In: *Organic Pesticides in the Environment*. Symposium 150th Meeting American Chemistry of Society, pp. 110–121.

Weston, D. P., You, J., Harwood, A. D., and Lydy, M. J. 2009. Whole sediment toxicity identification evaluation tools for pyrethroid insecticides: III temperature manipulation. *Environ. Toxicol. Chem.* 28: 173–180.

Wilson, C. and Tisdell, C. 2000. Why farmers continue to use pesticides despite environmental, health and sustainability costs. Working Paper on *Economic, Ecology and the Environment*. Working Paper No. 53.

WHO (World Health Organization). 1993. Guidelines for Drinking-Water Quality, Volume 1: Recommendations, 2nd edn. WHO: Geneva.

WHO (World Health Organization). 2001. Recommended classification of pesticides by hazard and guidelines to classification, 2000–2001. WHO: Geneva. (Document reference WHO/PCS/01.4.)

WHO (World Health Organization). 2009. Health implications from monocrotophos use: A review of the evidence in India, WHO Regional Office for South-East Asia, New Delhi. http://www.searo.who.int/LinkFiles/Publications_and_Documents_.

WHO (World Health Organization). 2003. Climate change and human health. http://www.who.int/globalchange/ecosystems/biodiversity/en/index.html.

Williamson, S. 2003. The dependency syndrome: Pesticide use by African smallholders, PAN UK, London. http://www.pan-uk.org/.../pn61p3.html.

Wolfe, H. R. 1979. Field exposure to airborne pesticides. In: *Air Pollution from Pesticides and Agriculture Processes*. CRC Press: Boca Raton, FL.

Yang, Y. 2007. *Factsheet: Pesticides and Environmental Health Trends in China*. Woodrow-Wilson International Centre for Scholars: Washington, DC. http://www.wilsoncenter.org/index.cfm?topic_id=1421&fuseaction=topics. item&news_id=225756.

Section II

Transposition and Transport of Pesticides

2

Environmental Fate and Transport of Pesticides

Jie Gao, Yu Wang, Bin Gao, Lei Wu, and Hao Chen

CONTENTS

2.1 Introduction

Pesticides are biologically active substances used for preventing, destroying, or controlling pests by interfering with their metabolic processes (Rice et al. 2007). For decades, pesticides have been used for preventing diseases transmitted by pests such as mosquitoes and fleas in humans and animals, increasing food production by destroying insects and other pests in agricultural areas, and protecting the environment by controlling the growth of molds, weeds, and algae (Whitford 2006). For example, it was found that rice production was reduced by at least 10% without the use of pesticides (Kuniuki 2001).

Clearly, pesticides are playing an important role in improving the quality of our daily life; however, in recent years, there have been continuing concerns about and research on the adverse impacts of pesticides and their degradation products on public health and the environment. Frequent detections of pesticides and their degradation products in surface waters and aquifers in the United States and around the world have resulted in considerable research efforts dedicated to investigating their environmental fate and transport. This information is essential for understanding and assessing the potential exposure and risks of the pesticides to the environment and ecosystem, providing proper guidelines for pesticide registration and management practices, and developing effective remediation strategies. This chapter mainly focuses on current knowledge of the fate and transport of pesticides in the environment.

2.2 Source

A report of Environmental Protection Agency (EPA) estimates that in 2001, more than 5 billion pounds of pesticides were used worldwide, with the United States using over 1.2 billion pounds, which was more than 20% of the total global use (Kiely et al. 2004). In the United States, approximately 888 million pounds of active ingredients were used in the pesticides, and the agricultural sector was the largest user (76% of total active ingredients used), followed by industry/commercial/government sector (12.5%) and home and garden sector (11.5%). Another 2.6 billion pounds of chlorine/hypochlorite-containing pesticides were used as disinfectants.

In addition to the deliberate sources, pesticides can also be present in remote areas by long-distance transport from their origins. This happens for many volatile or semivolatile pesticides that undergo evaporation at higher temperatures and condensation at lower temperatures (Bloomfield et al. 2006). This condensation effect has caused, for example, high levels of persistent bioaccumulative toxicants (i.e., PBTs; many currently used pesticides are considered PBTs) in remote areas far from the source of pollution (Blais et al. 2006). Once having entered the hydrological paths, pesticides can also be transported by flows or via the vehicle of mobile particles (e.g., colloids and nanoparticles) or migratory animals (Blais et al. 2006).

2.3 Fate and Transport

Pesticides undergo many different pathways once they enter the environment (Figure 2.1), including transformation/degradation, sorption–desorption, volatilization, uptake by plants, runoff to surface waters, and transport to groundwater. Among them, transformation/degradation is the most important pathway that eliminates pesticides from the environment, whereas many other pathways merely lead to the migration of pesticides. The following sections provide detailed descriptions of these pathways in different environmental compartments.

2.3.1 Transformation and Degradation

Transformation or degradation is one of the key processes that governs the environmental fate and transport of a pesticide, which also comprises different processes including abiotic degradation (e.g., oxidation, hydrolysis, and photolysis) and biodegradation. During this process, a pesticide is transformed to a degradation product or completely mineralized to carbon dioxide. The faster the degradation takes place, the less time a pesticide stays in the applied

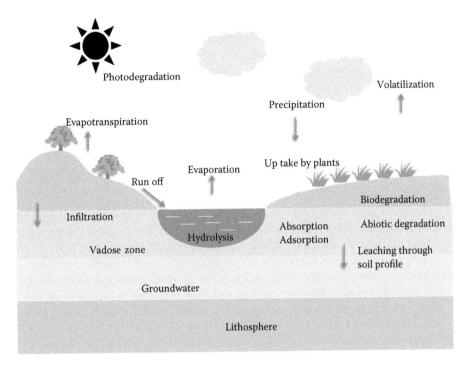

FIGURE 2.1
Overview of the fate and transport pathways of pesticides in the environment.

field, available for pesticidal activity or other movement (Aharonson et al. 1987). Although abiotic degradation plays a role in many cases, biodegradation of pesticides by microorganisms is usually the most important and dominant transformation process (Torstensson 1987, 1990).

2.3.1.1 Abiotic Degradation

In certain conditions, the abiotic pathway for pesticides can be the key mechanism of degradation. For example, under low microbial activity conditions, such as deep soils or deep subsurface conditions, biodegradation is very limited (Lovley and Chapelle 1995), and abiotic transformation may become the dominant pathway. Hydrolysis, oxidation, and photolysis are the major abiotic transformation or degradation processes (Wolfe et al. 1990). Some studies list photolysis and hydrolysis as two top pathways of pesticide abiotic degradation.

Oxidation: Oxidation is a reaction by which pesticides are oxidized. Oxidation of pesticides is affected by various environmental factors, including the amount of oxygen, metal ion concentration, natural organic matter content, and pH of the media. In the upper vadose zone, oxidation takes place primarily due to the abundance of oxygen (Kookana et al. 1998), whereas it becomes negligible once the subsurface depth gets further. Presence of metal ions can also catalyze the oxidation reactions. Nowack and Stone (2000, 2002) described numerous instances where phosphonate pesticide molecules chelate metal ions and are subsequently degraded in aqueous solution. Similarly, studies have shown that mineral phases of manganese (IV) (e.g., manganite, birnessite) may be involved in the degradation of xenobiotics, including chlorophenols and triclosan (Barrett and McBride 2005).

Cordeiro et al. (1986) proposed that the oxidative breakdown of phosphonates involves coordination with a metal oxide via a free-radical pathway.

Hydrolysis: Hydrolysis is a reaction between pesticide and water molecules involving catalysis by proton, hydroxide, or inorganic ions such as phosphate ion in the aquatic environment.

Hydrolysis of pesticides in water is generally believed to follow first-order kinetics, where the rate of pesticide degradation, $-d[P]/dt$, is proportional to the pesticide concentration, $[P]$, and k_{obs} is the hydrolysis rate constant (Equation 2.1).

$$-d[P]/dt = k_{obs}[P] \tag{2.1}$$

Hydrolysis reactivity depends on the chemical structure and functional group(s) of a pesticide molecule, but it also varies within the same class. For example, organophosphorus pesticides are prone to alkaline hydrolysis, whereas acidic hydrolysis is more common for some phosphorodithioates (Davisson et al. 2005). Hydrolysis is also temperature and pH dependent (Auld and Vallee 1971). Ruzicka et al. (1967) studied the hydrolysis rate of organophosphorus pesticides, and they found that, at higher temperatures, the hydrolysis kinetics rate constant is much larger than at lower temperatures. In field application, the residence time of pesticides can significantly vary with temperature (Aharonson et al. 1987). pH also controls the hydrolysis of pesticides. Normally, the pH-rate profile of pesticide hydrolysis is a U- or V-shaped curve (Figure 2.2). As mentioned above, certain pesticides may undergo alkaline or acidic hydrolysis, in which a pH >7 or <7 will prompt the chemical degradation of these pesticides, respectively.

In some cases, metal ions have the ability to catalyze the hydrolysis of pesticides. Mortland and Raman (1967) showed that Cu (II) can catalyze some organic phosphate pesticides that are normally considered relatively stable.

Photolysis: Photolysis or photodegradation becomes a significant degradation pathway when there are high levels of UV radiation. The process begins when the pesticide molecule receives energy and gets excited, after which the molecule either breaks up or forms less stable bonds that can easily break up later. Pesticide molecules can utilize photo energy in two ways: direct, in which the pesticide receives UV light within the spectrum of sunlight (<300 nm), or indirect, in which the energy is transmitted from other compounds that absorb the photo energy. There have been many studies focusing on the different

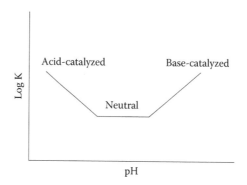

FIGURE 2.2
pH-Rate profile of pesticide hydrolysis.

mechanisms by which pesticides can be photodegraded in soil and water (Zeep and Cline 1977; Dureja and Chattopadhyay 1995; Romero et al. 1995; Cheng and Hwang 1996; Pirisi et al. 1996; Conceiçao et al. 2000; Konstantinou et al. 2001; Frank et al. 2002; Graebing et al. 2003). The occurrence and intensity of the process depend on the depth of pesticide location, presence of catalysts, exposure to radiation, soil pH and aeration, and physicochemical properties of the pesticide molecule. Hebert and Miller (1990) demonstrated that the vertical depth of direct photolysis is restricted to the top 0.2–0.3 mm in the soil, whereas indirect photolysis could go as far as 0.7 mm. According to Konstantinou et al. (2001), the presence of both humic acids and metal oxides could accelerate the pesticide photolysis because humic acids act as photosensitizers by generating reactive oxidative species such as singlet oxygen ($^1O_2$), hydroxyl radicals ($\cdot OH$), hydrogen peroxide (H_2O_2), and peroxy radicals ($ROO\cdot$), and metal oxides such as ZnO, Fe_2O_3, and MnO are able to absorb solar radiation and trigger the photochemical reactions. Readers are referred to Burrows et al. (2002) for a more detailed review on the photoreactivity of pesticides.

2.3.1.2 Biodegradation (Biotic Degradation)

The biodegradation process is usually characterized by half-life ($T_{1/2}$), the time for half of the initial amount of a pesticide to be biodegraded. Theoretically, pesticide degradation follows exponential decline and can be described as (Beulke et al. 2000) (Equation 2.2)

$$C_{(t)} = C_0 \exp(-kt) \tag{2.2}$$

where $C_{(t)}$ is the pesticide concentration at time t (mg/kg soil), C_0 is the concentration at time 0 (mg/kg soil), k is the degradation rate (per day), and t is the time (days). Therefore, in Equation 2.3, $T_{1/2}$ is written as

$$T_{1/2} = 0.693/k \tag{2.3}$$

In real systems, however, the half-life of a pesticide has also been shown to be dependent on temperature, soil and water content, and soil organic carbon content (Walker 1974).

Effects of temperature on the pesticide degradation can be described using the Arrhenius equation (Beulke et al. 2000) (Equation 2.4):

$$H_{(T)} = A_1 \exp\left(E_a/RT\right) \tag{2.4}$$

and (Equation 2.5)

$$H_{(T2)} = H_{(T1)} \exp\left[E_a\left(T_1 - T_2\right)/\left(RT_1T_2\right)\right] \tag{2.5}$$

where $H_{(T)}$ is the half-life at temperature T (days), T is the temperature (K), A_1 is a coefficient (days), E_a is the activation energy (J/mol), R is the gas constant (8.314 J/mol K), $H_{(T1)}$ is the half-life at temperature T_1 (days) and $H_{(T2)}$ is the half-life at temperature T_2 (days).

The effects of soil moisture can be quantified in Equation 2.6:

$$H_{(M)} = A M^{-B} \tag{2.6}$$

where $H_{(M)}$ is the half-life at moisture M (days), A is the half-life at a moisture content of 1 kg H_2O 100/kg soil dry matter (days), M is the moisture content (kg H_2O 100/kg soil), and B is a coefficient (unitless) (Beulke et al. 2000).

2.3.2 Sorption and Desorption

Adsorption is a process by which an aqueous molecule is attracted and retained onto the surface of a solid. Although adsorption can be either a chemical (e.g., electrostatic interaction) or a physical (e.g., van der Waals forces) process, adsorption of pesticide molecules on soil usually takes place as a result of coulombic attraction between the positively charged pesticide molecules and the negatively charged soil particles or organic matter.

Distribution coefficient (K_d) is an important parameter used to quantify the adsorption of pesticide molecules to soils. It is defined as the ratio of the sorbed phase concentration to the solution phase concentration at equilibrium (Equation 2.7).

$$K_d = C_a/C_d \qquad (2.7)$$

where K_d is the distribution coefficient of a pesticide molecule between soil and water (V/M); C_a is the amount of pesticide adsorbed per unit of adsorbent mass (M/M); and C_d is the concentration of pesticide dissolved (M/V). K_d is directly related to the K_{oc} value of the pesticide and the organic matter (OM) and clay content of the soil.

A number of methods have been proposed to determine the distribution coefficient (Karickhoff and Brown 1978; Veith et al. 1979). Karickhoff et al. (1979) corroborated a linear correlation between the distribution coefficient and the soil's organic carbon content (Equation 2.8).

$$K_d = K_{oc} \cdot OC/100 \qquad (2.8)$$

where K_{oc} is soil organic partition coefficient and OC is the organic carbon content (%).

Several other studies have also shown that the values of K_d are directly related to the concentration of organic matter in soil (O'Connor and Connolly 1980; Voice et al. 1983; Gschwend and Wu 1985) and pesticide chemical structure (Lohninger 1994).

There are a variety of factors affecting the adsorption process, including physicochemical properties of soils and the pesticide molecules, such as soil pH, surface charge, surface area and size of the soil particles, chemical functions, solubility, polarity, and octanol-water partition coefficient (K_{ow}) of the pesticide molecules (Senesi 1992; Pignatello and Xing 1996). Soil pH affects not only the solubility and reactivity of the pesticide molecules but also soil ecological functions such as microbial activities. For different pesticides, the effects of pH on sorption are different. For example, Andrade et al. (2010) concluded that soils with low organic matter content and/or high pH showed less ametryn sorption rates. Using a model, Lohninger (1994) summarized that adsorption increases with pH and organic matter content but decreases with ionic strength.

Adsorption plays an important role in determining the fate and transport of a pesticide as it reduces the pesticide's bioavailability and mobility and, consequently, its environmental and health impacts. One of the most widely used experimental techniques for evaluating the adsorption potential of a pesticide by soil is to determine its adsorption isotherm. Adsorption isotherm is a graph between the mass of sorbate per unit dry mass of sorbent (S) and the concentration (C) of the sorbate.

Several mathematical models have been developed to describe the adsorption isotherms of pesticides onto solid surfaces (e.g., soil surfaces). The most commonly used ones are the Langmuir, Freundlich, and Langmuire-Freundlich models.

Langmuir (Equation 2.9)

$$q_e = \frac{K_l Q C_e}{1 + K C_e} \tag{2.9}$$

Freundlich (Equation 2.10)

$$q_e = K_f C_e^n \tag{2.10}$$

Langmuir-Freundlich (Equation 2.11)

$$q_e = \frac{K_l Q C_e^n}{1 + K C_e^n} \tag{2.11}$$

where K_l and K_f represent the Langmuir bonding term related to interaction energies (L/mg) and the Freundlich affinity coefficient ($mg^{(1-n)}L^n/kg$), Q denotes the Langmuir maximum capacity (mg/kg), C_e is the equilibrium solution concentration (mg/L) of the sorbate, and n is the Freundlich linearity constant. The Langmuir model is mechanistic and describes monolayer adsorption onto a homogeneous surface with no interactions between the adsorbed molecules. The Freundlich and Langmuir-Freundlich models, however, are empirical equations, which are often used to describe chemisorptions onto heterogeneous surfaces (Zhou et al. 2003).

2.3.3 Volatilization

Volatilization is the conversion of pesticides into gas from soil, water, and plant surfaces. A pesticide can move into air from all surfaces and thus transport away from the initial application/target sites. This airborne movement is also called "drift" and it can induce damage to nearby crops, livestock, humans, and other living organisms.

Volatilization of pesticides is influenced by environmental conditions, adsorption of pesticides by soils, and physicochemical properties of soils, such as soil moisture and organic matter contents, etc. High temperature, low relative humidity, and air movement tend to favor volatilization, whereas low temperature and high humidity decrease volatilization (Farmer 1971). Soil properties, including texture, organic matter content, and moisture content, can also influence the volatilization of pesticides. Soil texture may affect the binding force of pesticides with soil particles (Gevao et al. 2000). Higher organic matter contents can sorb more pesticides, thus reducing the volatilization (Cox et al. 1998). Overall, a pesticide loosely adsorbed to soil particles is more likely to volatilize (Khan 1982). In addition, formulated pesticides (e.g., granular and wettable powders) are less prone to volatilization than loose powders (Navarro et al. 2007).

Volatilization is also affected by the wind velocity and the application equipment. Pesticides can volatilize more rapidly into the air when they are applied on a windy day. Poorly adjusted equipment (pump pressure, spray height, choice of nozzles, etc.) can lead to an excessive amount of pesticide material being wasted in the vapor and moved away from the target site (Roberts 1984).

The diffusion-based form of volatilization loss of pesticides can be described in Equation 2.12 (Wagenet et al. 1989).

$$Qv = -(KH * DA * Ds * CIL)/(Ds * \delta a + KH * DA * \delta s) \tag{2.12}$$

where Qv is diffusive flux ($mg/mm^2/day$); CIL, the aqueous concentrations of the chemical at the first soil compartment (mg/mL); δs, thickness of the surface soil film of thickness (mm); δa, thickness of the stagnant atmospheric film of thickness (mm); Ds, effective diffusion coefficient in the surface soil segment (mm^2/day); DA, diffusion coefficient in air (mm^2/day); KH, Henry's constant.

2.3.4 Uptake by Plants

Uptake or absorption of pesticides is the movement of pesticides into plants. Although the processes and magnitude of pesticide uptake by plants remain to be fully understood, this process is known to be influenced by environmental conditions and the physicochemical properties of both the pesticides and the soils. Lower pH will increase the uptake of pesticides (Garcinuno et al. 2003). The presence of both biochar and organic matter can decrease the uptake of pesticides by plants (Yu et al. 2009; Saha et al. 1971). The type of mineral in the soils can also affect the uptake of pesticides, depending on its affinity for the pesticide and sorption capacity (Saha et al. 1971). Greater affinity leads to less uptake of the pesticide by plants.

Uptake of pesticides by terrestrial and aquatic plants depends on the hydrophobicity of the pesticide. The uptake can be significant if the pesticide is hydrophobic or lipophilic. Once in the plants, pesticides may be broken down or remain inside the plant until the tissues decay or the crop is harvested. If they persist in the plant, pesticides can be bioaccumulated in the food chain and such process can be quantified by the bioconcentration factor (BCF). The BCF is closely related to the octanol/water partition coefficient (K_{ow}) as it quantifies the lipophilicity or hydrophobicity of a pesticide, and therefore, is a measure of the pesticide's potential to be bioaccumulated in the living organisms. Pesticides with $\log K_{ow} < 3$ are generally considered nonbioaccumulating and those with $\log K_{ow} > 3$ to be high-bioaccumulating chemicals (Vighi and Di Guardo 1995).

2.3.5 Surface Runoff

Runoff is the overflow of water on the soil surface, and it occurs when the precipitation rate exceeds the soil's infiltration rate. As a result, pesticides can be carried into surface water bodies via overland flow. While movement of pesticides can be beneficial as leaching into the plant root zone gives better pest control, pesticide runoff is usually considered harmful. When surface runoff occurs, pesticides are moved away from the target sites and their effectiveness is greatly reduced. Furthermore, runoff with pesticides can contaminate surface water bodies, such as streams, ponds, lakes, and rivers, and harm aquatic plants and animals. Therefore, best management practices (BMPs) are needed to reduce the pesticide runoff.

Pesticide runoff is influenced by the surface slope, the timing, amount and duration of rainfall and irrigation, soil moisture content, and physicochemical properties of the soils and pesticides. Surface slope will increase the runoff significantly. The more the ground slopes, the more likely water will run off the site. The rainfall intensity also contributes to the runoff (Sharpley 1985). Higher intensity will cause more runoff because of the greater total amount of water going through the surface. Furthermore, higher soil moisture content will increase the runoff (Sharpley 1985) because it reduces the infiltration of water.

Some agricultural management practices, such as tillage systems and crop rotation, can also affect the movement of pesticides (Clemente et al. 1993). For example, atrazine and metalochlor were found to persist longer in tillage ridges than in tillage valleys because of lower moisture content of the ridges (Gaynor et al. 1987, 1998).

A number of models and equations have been developed for estimating different parameters in a runoff event. Runoff depth can be calculated using the US Soil Conservation Service (USSCS) Curve Number Method (USSCS 1972) (Equation 2.13):

$$Q = (P - 0.2S)^2 / (P + 0.8S), \quad P > 0.2S \tag{2.13}$$

where Q is storm runoff depth (mm); P is storm rainfall (mm); S is the potential maximum retention after run off begins (mm), in which S = (25400/CN) – 254, where CN is the runoff curve number.

Soil loss can be simulated using the Modified Universal Soil Loss Equation (MUSLE, Wischmeier and Smith 1978) (Equation 2.14):

$$Xt = (11.8/A) * (Vt * qt) 0.56 * Ke * (LS) * Ct * SP \tag{2.14}$$

where A is field area (ha); Vt, runoff volume (m³) given by 100 AQt; qt, peak runoff rate (m³/s); Ke, standard soil erosion factor; LS, topographic factor; Ct, coyer factor; SP, supporting factor.

Pesticide partitioning in runoff can be estimated using the Equation 2.15 (Clemente 1991):

$$Pr = Pt - PXt - PQt \tag{2.15}$$

where Pr is total pesticide remaining in the top 10 mm soil layer after the rainstorm (g/ha); Pt, pesticide level in the surface 10 mm (g/ha); PXt, is solid phase pesticide loss in runoff (g/ha); PQt, loss of dissolved pesticide in runoff (g/ha).

2.3.6 Leaching Through the Soil Profile

Pesticides have the potential to migrate to groundwater under certain conditions and can lead to groundwater contamination. Therefore, understanding the mobility and transport of pesticides is critical to the assessment of environmental distribution and risks associated with the use of pesticides. Leaching of pesticides through soils is undoubtedly the most important route by which these chemicals percolate into the groundwater. Studies examining pesticide leaching have been conducted in field scale or in the laboratory using packed soil columns or undisturbed soil cores (Banks et al. 1979; Van Genuchten and Cleary 1979; White 1985; Flury 1996). The results from these studies show that pesticides leach out below the root zone and the leaching process is controlled by various factors, such as physicochemical and biological properties of pesticides and soils and the timing and amount of rainfall events following pesticide application.

In a study conducted by Kookana (1995) using sandy soils, it was found that the intrinsic mobility of a pesticide through leaching is inversely related to its sorption on soil. Low sorption affinity would lead to more dissolved pesticide molecules in water and thus faster leaching rate. Conversely, pesticides with low aqueous solubility and higher affinity for soils are less likely to leach through the soil profile.

Although still poorly understood, the phenomenon of "preferential flow" has been often noticed in field studies or when real soils are used. In these cases, pesticides move rapidly and preferentially along macropores or decayed root and earthworm channels in soils, and their retention times are comparable with conservative tracer solutes (Andreini and Steenhuis 1990; Flury 1996).

The rainfall events also greatly influence the leaching processes. It was estimated by Flury (1996) that if there is no heavy rainfall shortly after pesticide application, the annual losses of the pesticide are between <0.1% and 1%, whereas up to 5% of applied pesticide can leach out of the soil at the application site in the worst rainfall case.

2.4 Transport Models

Due to the wide application of pesticides in agricultural and industrial practices, their presence has been detected and reported in many aquifers and surface waters. Transport of pesticides encompasses infiltration, runoff, and preferential flow (or macropore flow). Modeling of the transport behavior has been described in a number of reviews and studies. At the end of this chapter, a brief overview of these studied transport models is provided.

2.4.1 HERBSIM and SIMULAT Models

HERBSIM is a computer model that primarily predicts the fate of pesticides and their metabolites in the soil considering processes such as sorption, degradation, and evaporation and their dependencies on environmental variables, whereas SIMULAT Model can also be used for one-dimensional transport (Diekkrüger et al. 1995; Aden and Diekkrüger 2000). Generally, in these two models, water flow, solute transport, potential evapotranspiration, sorption, and degradation can be calculated by Richards' equation, convection–dispersion equation, Penman-Monteith equation, linear/Freundlich/Langmuir adsorption isotherm with up to three different binding sites considered, and Walker equation, respectively (Aden and Diekkrüger 2000). In addition, plant growth, heat flux, and tile drainage can also be simulated in the SIMULAT model.

2.4.2 HYDRUS Model

HYDRUS model applies to water flow and solute transport in vadose zone. It incorporates many physical and chemical nonequilibrium models, such as the mobile–immobile water content model, Dual-Permeability Model, One Kinetic Site Model, and Two-Site Sorption model, to simulate the transport of solutes (Simunek and van Genuchten 2008). Advantageously, HYDRUS has a large capacity and is suitable for one-, two-, and three-dimensional applications. With respect to pesticides, Malone et al. (2004) analyzed the effects of having equilibrium or kinetic sorption models on the simulated degradation in using HYDRUS and summarized the transport of pesticide in soil–water system.

2.4.3 Leaching Estimation and Chemistry Model

Leaching Estimation and Chemistry Model (LEACHM) is a process-based model that simulates water flow, transformation and transport of solute, and plant uptake in the unsaturated zone (Jemison et al. 1994). A complete description of this model was given by Hutson and Wagenet (1992). Specifically, LEACHM-P is designed for simulating dynamics of pesticides and has shown success in many applications (Kohne et al. 2009). However, it does not simulate macropore flow of water, unequal depth increments, runoff, erosion, and management practices and requires a lot of computer run time and data acquisition time (Smith et al. 1991).

2.4.4 MACRO Model

MACRO model simulates macropore flow processes in agricultural systems (Jarvis 1994; Larsbo and Jarvis 2005). Its assumptions include that sorption of chemicals on soil is instantaneous and that adsorption follows the Freundlich isotherm model (Kohne et al. 2009). Brown et al. (1999) applied MACRO 4.0 to simulate isoproturon leaching in a lysimeter experiment and noted that a calibration of fraction of sorption is needed to match the observation curve. In a leaching test of the fungicide iprodione in golf green, MACRO very well predicted its transport in drainflow (Stromqvist and Jarvis 2005).

2.4.5 Root Zone Water Quality Model

Root Zone Water Quality Model (RZWQM) simulates plant growth in agricultural systems, and water and chemical movement through macropores (Ahuja et al. 2000a,b; Kohne et al. 2009). In this model, various physical, chemical, and biological processes, including degradation and adsorption, were integrated and adjusted according to environmental parameters such as soil and water content and temperature. It is capable of predicting water transport by assuming that water radially flows through macropores when rainfall rate exceeds soil infiltration rates. Chemicals move with water and their transport can be simulated by assuming instantaneous partitioning of chemicals between water and soil (Malone et al. 2001). Malone et al. (2004), Wauchope et al. (2004), and Ma et al. (2007) used RZWQM in their soil column studies to simulate pesticide transport.

2.4.6 Studio of Analytical Models

Studio of Analytical Models (STANMOD) is a windows-based computer software package for evaluating solute transport in soils using analytical solutions of the convection-dispersion solute transport equation (Simunek et al. 1999). It contains seven independent models for predicting fate and transport of chemicals (including pesticides) in soils. Five of these models (i.e., CXTFIT, CFITM, CFITIM, CHAIN, and SCREEN) are for one-dimensional transport and the other two (i.e., 3DADE and N3DADE) are for two- and three-dimensional transport.

2.4.7 Saturated–Unsaturated Transport Model

Saturated–Unsaturated Transport (SUTRA) is a computer model for simulating density-dependent saturated or unsaturated fluid flow and transport of either a solute or thermal energy in the subsurface environment (Voss and Provost 2010). Transport of the solute takes into account processes such as equilibrium adsorption and production/degradation. SUTRA flow simulation may be employed for one- or two-dimensional transport that can be either steady state or transient. The latest version of SUTRA also enables three-dimensional modeling of saturated and unsaturated groundwater flow (Voss and Provost 2010).

Other transport models include Pesticide Leaching Model (PLM) developed by Hall (1994) and Nicholls and Hall (1995); Pesticide Root Zone Model (PRZM) developed and described by Carsel et al. (1985); Groundwater Loading Effects of Agricultural Management Systems (GLEAMS), a mathematical model described in Leonard et al. (1987); SPASMO for simulation of field studies (Green et al. 2003); PEARL model studied by Leistra et al. (2001); and Pesticide Leaching Model (PELMO) described by Klein (1994).

2.4.8 Model Application

A number of model applications have been published for predicting the transport and mobility of pesticide molecules and their derivatives. Steady-state flow and preferential pathway were chosen in most of cases. Some laboratory studies used soil column or porous medium column to simulate the transport process, and some studies collected field data in their mathematical prediction.

Porous medium columns and lysimeter with artificial preferential pathway were utilized as well-defined systems to facilitate the pesticide transport. Solute transport models were often picked in predicting the breakthrough curve. During this simulation, processes such as sorption–desorption and degradation are often considered as important factors. A summary of model application examples is given in Table 2.1.

TABLE 2.1

Application of Models Simulating Preferential Pesticide Transport

Model	Application	References
HYDRUS	Atrazine transport and fate in soil column	Cheyns et al. (2010), Pang et al. (2000), Dann et al. (2006), Persicani (1996), and Boivin et al. (2006)
LEACHM	s-Triazine transport in soil lysimeter	Krutz et al. (2010), Petach et al. (1991), and Hancock et al. (2008)
RZWQM	Transformation and transport of cyanazine and metribuzin in the soil profile; pyrethroids transport in soil/sediment	Ahuja et al. (2000b), Luo et al. (2011), and Bayless et al. (2008)
STANMOD	Atrazine and glyphosate transport in soil column	Zhou et al. (2010)
PLM	Bentazone and ethoprophos transport in lysimeter	Nicholls et al. (2000), Armstrong et al. (2000)
GLEAMS	Atrazine-leaching study through lysimeters	Shirmohammadi and Knisel (1994), Sichani et al. (1991), and Knisel and Turtola (2000)
SPASMO	Atrazine, diazinon, hexazinone, procymidone, terbuthylazine transport in soil profile	Sarmah et al. (2005) and Close et al. (2003)
PRZM	Atrazine transport in lower soil depth Alachlor, metribuzin, norflurazon in soil leaching	Sadeghi et al. (1995) and Mueller et al. (1992)
PELMO	^{14}C-Benazolin transport in lysimeter	Fent et al. (1998)

References

Aden, K., and Diekkrüger, B. 2000. Modeling pesticide dynamics of four different sites using the model system SIMULAT. *Agric. Water Manag.* 44(1–3): 337–355.

Aharonson, N., Cohen, S. Z., Drescher, N., and Gish, T. J. 1987. Potential contamination of groundwater by pesticides. *Pure Appl. Chem.* 59(10): 1419–1446.

Ahuja, L. R., Johnsen, K. E., and Rojas, K. W. 2000a. Water and chemical transport in soil matrix and macropores. In: Ahuja, L. R., Rojas, K.W., Hanson, J. D., Shaffer, J. J., and Ma, L. (eds), *The Root Zone Water Quality Model.* pp. 13–50, Water Resources Publications LLC: Highlands Ranch, CO.

Ahuja, L. R., Rojas, K. W., Hanson, J. D., Shaffer, J. J., and Ma, L. 2000b. *The Root Zone Water Quality Model.* Water Resources Publications LLC: Highlands Ranch, CO.

Andrade, S. R. B., Silva, A. A., Queiroz, M. E. L. R., Lima, C. F., and D'Antonino, L. 2010. Ametryn sorption and desorption in Red-Yellow Latosol and Red-Yellow Ultisol with different pH values. *Planta Daninha* 28(1): 177–184.

Andreini, M. S. and Steenhuis, T. S. 1990. Preferential paths of flow under conventional and conservation tillage. *Geoderma* 46(1–3): 85–102.

Armstrong, A., Aden, K., Amraoui, N., Diekkrüger, B., Jarvis, N., Mouvet, C., Nicholls, P., and Wittwer, C. 2000. Comparison of the performance of pesticide-leaching models on a cracking clay soil: Results using the Brimstone Farm dataset. *Agric Water Manage.* 44(1–3): 85–104.

Auld, D. S. and Vallee, B. L. 1971. Kinetics of carboxypeptidase-A – Ph and temperature dependence of tripeptide hydrolysis. *Biochemistry* 10(15): 2892–2897.

Banks, P. A., Ketchersid, M. L., and Merkle, M. G. 1979. Persistence of fluridone in various soils under field and controlled conditions. *Weed Sci.* 27(6): 631–633.

Barrett, K. A. and McBride, M. B. 2005. Oxidative degradation of glyphosate and aminomethylphosphonate by manganese oxide. *Environ. Sci. Technol.* 39(23): 9223–9228.

Bayless, E. R., Capel, P. D., Barbash, J. E., Webb, R. M. T., Hancock, T. L. C., and Lampe, D. C. 2008. Simulated fate and transport of metolachlor in the unsaturated zone, Maryland, USA. *J. Environ. Qual.* 37(3): 1064–1072.

Beulke, S. 2000. Simulation of pesticide persistence in the field on the basis of laboratory data – A review. *J. Environ. Qual.* 29(5): 1371–1379.

Blais, J. M., Charpentie, S., Pick, F., Kimpe, L. E., Amand, A. S., and Regnault-Roger, C. 2006. Mercury, polybrominated diphenyl ether, organochlorine pesticide, and polychlorinated biphenyl concentrations in fish from lakes along an elevation transect in the French Pyrenees. *Ecotoxicol. Environ. Saf.* 63(1): 91–99.

Bloomfield, J. P., Williams, R. J., Gooddy, D. C., Cape, J. N., and Guha, P. 2006. Impacts of climate change on the fate and behaviour of pesticides in surface and groundwater – A UK perspective. *Sci. Total Environ.* 369(1–3): 163–177.

Boivin, A., Simunek, J., Schiavon, M., and van Genuchten, M. T. 2006. Comparison of pesticide transport processes in three tile-drained field soils using HYDRUS-2D. *Vadose Zone J.* 5(3): 838–849.

Brown, C. D., Marshall, V. L., Deas, A., Carter, A. D., Arnold, D., and Jones, R. L. 1999. Investigation into the effect of tillage on solute movement to drains through a heavy clay soil. II. Interpretation using a radio-scanning technique, dye-tracing and modelling. *Soil Use Manag.* 15(2): 94–100.

Burrows, H. D., Canle, M., Santaballa, J. A., and Steenken, S. 2002. Reaction pathways and mechanisms of photodegradation of pesticides. *J. Photochem. Photobiol. B Biol.* 67: 71–108.

Carsel, R. F., Mulkey, L. A., Lorber, M. N., and Baskin, L. B. 1985. The pesticide root zone model (PRZM) – A procedure for evaluating pesticide leaching threats to groundwater. *Ecol. Model.* 30(1–2): 49–69.

Cheng, H. M. and Hwang, D. F. 1996. Photodegradation of benthiocarb. *Chem. Ecol.* 12: 91–101.

Cheyns, K., Mertens, J., Diels, J., Smolders, E., and Springael, D. 2010. Monod kinetics rather than a first-order degradation model explains atrazine fate in soil mini-columns: Implications for pesticide fate modelling. *Environ. Pollut.* 158(5): 1405–1411.

Clemente, R. S. 1991. A mathematical model for simulating pesticide fate and dynamics in the environmental (PESTFADE). Ph.D. Dissertation, McGill University.

Clemente, R. S., Prasher, S. O., and Barrington, S. F. 1993. Pestfade, a new pesticide fate and transport model – Model development and verification. *Trans. ASAE* 36(2): 357–367.

Close, M. E., Pang, L., Magesan, G. N., Lee, R., and Green, S. R. 2003. Field study of pesticide leaching in an allophanic soil in New Zealand. 2: Comparison of simulations from four leaching models. *Aust. J. Soil Res.* 41(5): 825–846.

Cordeiro, M. L., Pompliano, D. L., and Fraost, J. W. 1986. Degradation and detoxification of organophosphonates. Cleavage of the carbon-phosphorus bond. *J. Am. Chem. Soc.* 108(2): 332–334.

Cox, L., Koskinen, W. C., and Yen, P. Y. 1998. Changes in sorption of imidacloprid with incubation time. *Soil Sci. Soc. Am. J.* 62(2): 342–347.

Dann, R. L., Close, M. E., Lee, R., and Pang, L. 2006. Impact of data quality and model complexity on prediction of pesticide leaching. *J. Environ. Qual.* 35(2): 628–640.

Davisson, M. L., Love, A. H., Vance, A., and Reynold, J. G. 2005. *Environmental Fate of Organophosphorus Compounds Related to Chemical Weapons*. Lawrence Livermore National Laboratory: Livermore, CA.

Diekkrüger, B., Nortersheuser, P., and Richter, O. 1995. Modeling pesticide dynamics of a loam site using HERBSIM and SIMULAT. *Ecol. Model.* 81(1–3): 111–119.

Dureja, P. and Chattopadhyay, S. 1995. Photodegradation of pyrethroid insecticide flucythrinate in water and on soil surface. *Toxicol. Environ. Chem.* 52: 97–102.

Farmer, V. C. 1971. Characterization of adsorption bonds in clays by infrared spectroscopy. *Soil Sci.* 112(1): 62–68.

Fent, G., Jene, B., and Kubiak, B. 1998. Performance of the pesticide leaching model PELMO 2.01 to predict the leaching of bromide and 14C-benazolin in a sandy soil in comparison to results of a lysimeter- and field study. IUPAC Congress Book of Abstracts, London.

Flury, M. 1996. Experimental evidence of transport of pesticides through field soils – A review. *J. Environ. Qual.* 25(1): 25–45.

Frank, M. P., Graebing, P., and Chib, J. S. 2002. Effect of soil moisture and sample depth on pesticide photolysis. *J. Agric. Food Chem.* 50: 2607–2614.

Garcinuno, R. M., Fernandez-Hernando, P., and Camara, C. 2003. Evaluation of pesticide uptake by Lupinus seeds. *Water Res.* 37(14): 3481–3489.

Gaynor, J. D., Stone, J. A., and Vyn, T. J. 1987. Tillage systems and atrazine and alachlor residues on a poorly drained soil. *Can. J. Soil Sci.* 67(4): 959–963.

Gaynor, J. D., MacTavish, D. C., and Labaj, A. B. 1998. Atrazine and metolachlor residues in Brookston CL following conventional and conservation tillage culture. *Chemosphere* 36(15): 3199–3210.

Gevao, B., Semple, K. T., and Jones, K. C. 2000. Bound pesticide residues in soils: A review. *Environ. Pollut.* 108(1): 3–14.

Graebing, P., Frank, M. P., and Chib, J. S. 2003. Soil photolysis of herbicides in a moisture- and temperature-controlled environment. *J. Agric. Food Chem.* 51: 4331–4337.

Green, S. R., van den Dijssel, C., Snow, V. O., Clothier, B. E., Webb, T., Russell, J., Ironside, N., and Davidson, P. 2003. SPASMO – A risk assessment model for water, nutrient and chemical fate under agricultural lands. In: Currie, L. D. and Hanly, J. H. (eds), *Tools for Nutrient and Pollutant Management*. pp. 321–335, Fertiliser and Lime Research Centre, Massey University: Palmerston North.

Gschwend, P. M. and Wu, S. C. 1985. On the constancy of sediment water partition-coefficients of hydrophobic organic pollutants. *Environ. Sci. Technol.* 19(1): 90–96.

Hall, D. G. M. 1994. Simulation of dichlorprop leaching in three texturally distinct soils using the pesticide leaching model. *J. Environ. Sci. Health A* 29(6): 1211–1230.

Hancock, T. C., Sandstrom, M. W., Vogel, J. R., Webb, R. M. T., Bayless, E. R., and Barbash, J. E. 2008. Pesticide fate and transport throughout unsaturated zones in five agricultural settings, USA. *J. Environ. Qual.* 37(3): 1086–1100.

Hebert, V. R. and Miller, G. C. 1990. Depth dependence of direct and indirect photolysis on soil surfaces. *J. Agric. Food Chem.* 38(3): 913–918.

Hutson, J. L. and Wagenet, R. J. 1992. LEACHM. Leaching Estimation and Chemistry Model: A process based model of water and solute movement, transformations, plant uptake, and chemical reactions in the unsaturated zone, Version 3. Department of Agronomy, Cornell University, Ithaca, NY.

Jarvis, N. J. 1994. *The MACRO Model (Version 3.1), Technical Description and Sample Simulations.* Reports and Dissert. Department of Soil Sciences, Swedish University of Agricultural Sciences, Uppsala, Sweden.

Jemison, J. M., Jabro, J. D., and Fox, R. H. 1994. Evaluation of LEACHM. 2. Simulation of nitrate leaching from nitrogen-fertilized and manured corn. *Agron. J.* 86(5): 852–859.

Karickhoff, S. W. and Brown, D. S. 1978. Paraquat sorption as a function of particle-size in natural sediments. *J. Environ. Qual.* 7(2): 246–252.

Karickhoff, S. W., Brown, D. S., and Scott, T. A. 1979. Sorption of hydrophobic pollutants on natural sediments. *Water Res.* 13(3): 241–248.

Khan, S. U. 1982. Bound pesticide-residues in soil and plants. *Residue Rev.* 84: 1–25.

Kiely, T., Donaldson, D., and Grube, A. 2004. *Pesticides Industry Sales and Usage 2000 and 2001 Market Estimates.* In: Agency USEP (ed.), USEP: Washington, DC.

Klein, M. 1994. Evaluation and comparison of pesticide leaching models for registration purposes – Results of simulations performed with the pesticide leaching model. *J. Environ. Sci. Health A* 29(6): 1197–1209.

Knisel, W. G. and Turtola, E. 2000. Gleams model application on a heavy clay soil in Finland. *Agric. Water Manag.* 43(3): 285–309.

Kohne, J. M., Kohne, S., and Simunek, J. 2009a. A review of model applications for structured soils: a) Water flow and tracer transport. *J. Contam. Hydrol.* 104(1–4): 4–35.

Kohne, J. M., Kohne, S., and Simunek, J. 2009b. A review of model applications for structured soils: b) Pesticide transport. *J. Contam. Hydrol.* 104(1–4): 36–60.

Konstantinou, I. K., Zarkadis, A. K., and Albanis, T. A. 2001. Photodegradation of selected herbicides in various natural waters and soils under environmental conditions. *J. Environ. Qual.* 30: 121–131.

Kookana, R. S. 1995. A field-study of leaching and degradation of nine pesticides in a sandy soil. *Aust. J. Soil Res.* 33(6): 1019–1030.

Kookana, R. S., Baskaran, S., and Naidu, R. 1998. Pesticide fate and behaviour in Australian soils in relation to contamination and management of soil and water: A review. *Aust. J. Soil Res.* 36(5): 715–764.

Krutz, L. J., Shaner, D. L., Weaver, M. A., Webb, R. M. T., Zablotowicz, R. M., Reddy, K. N., Huang, Y. B., and Thomson, S. J. 2010. Agronomic and environmental implications of enhanced s-triazine degradation. *Pest Manag. Sci.* 66(5): 461–481.

Kuniuki, S. 2001. Effects of organic fertilization and pesticide application on growth and yield of field-grown rice for 10 years. *Jpn. J. Crop Sci.* 70(4): 530–540.

Larsbo, M. and Jarvis, N. 2005. Simulating solute transport in a structured field soil: Uncertainty in parameter identification and predictions. *J. Environ. Qual.* 34(2): 621–634.

Leistra, M., Linden, A., Boesten, A., and Tiktak, A. 2001. PEARL model for pesticide behaviour and emissions in soil-plant systems: Description of the processes in FOCUS PEARL. Alterra, Green World Research.

Leonard, R. A., Knisel, W. G., and Still, D. A. 1987. Gleams – Groundwater loading effects of agricultural management-systems. *Trans. ASAE* 30(5): 1403–1418.

Lohninger, H. 1994. Estimation of soil partition-coefficients of pesticides from their chemical-structure. *Chemosphere* 29(8): 1611–1626.

Lovley, D. R. and Chapelle, F. H. 1995. Deep subsurface microbial processes. *Rev. Geophys.* 33(3): 365–381.

Luo, Y. Z. and Zhang, M. H. 2011. Environmental modeling and exposure assessment of sediment-associated pyrethroids in an agricultural watershed. *PLoS One* 6(1): DOI:10.1371/journal.pone.0015794.

Ma, L., Ahuja, L. R., and Malone, R. W. 2007. Systems modelling for soil and water research and management: Current status and needs for the 21st century. *Trans. ASABE* 50(5): 1705–1713.

Malone, R. W., Shipitalo, M. J., Ma, L., Ahuja, L. R., and Rojas, K. W. 2001. Macropore component assessment of the root zone water quality model (RZWQM) using no-till soil blocks. *Trans. ASAE* 44(4): 843–852.

Malone, R. W., Shipitalo, M. J., and Meek, D. W. 2004. Relationship between herbicide concentration in percolate, percolate breakthrough time and number of active macropores. *Trans. ASAE* 47(5): 1453–1456.

Mortland, M. M. and Raman, K. V. 1967. Catalytic hydrolysis of some organic phosphate pesticides by copper (2). *J. Agric Food Chem.* 15(1): 163–167.

Mueller, T. C., Jones, R. E., Bush, P. B., and Banks, P. A. 1992. Comparison of PRZM and GLEAMS computer-model predictions with field data for alachlor, metribuzin and norflurazon leaching. *Environ. Toxicol. Chem.* 11(3): 427–436.

Navarro, S., Perez, G., Navarro, G., and Vela, N. 2007. Decline of pesticide residues from barley to malt. *Food Addit. Contam.* 24(8): 851–859.

Nicholls, P. H. and Hall, D. G. M. 1995. Use of the pesticide leaching model (PLM) to simulate pesticide movement through macroporous soils, pesticide movement to water. *BCPC Monograph* 62: 187–192.

Nicholls, P. H., Harris, G. L., and Brockie, D. 2000. Simulation of pesticide leaching at Vredepeel and Brimstone farm using the macropore model PLM. *Agric. Water Manag.* 44(1–3): 307–315.

Nowack, B. and Stone, A. T. 2000. Degradation of nitrilotris(methylenephosphonic acid) and related (amino)phosphonate chelating agents in the presence of manganese and molecular oxygen. *Environ. Sci. Technol.* 34(22): 4759–4765.

Nowack, B. and Stone, A. T. 2002. Homogeneous and heterogeneous oxidation of nitrilotrismethylenephosphonic acid (NTMP) in the presence of manganese (II, III) and molecular oxygen. *J. Phys. Chem. B* 106(24): 6227–6233.

O'connor, D. J. and Connolly, J. P. 1980. The effect of concentration of adsorbing solids on the partition-coefficient. *Water Res.* 14(10): 1517–1523.

Pang, L. P., Close, M. E., Watt, J. P. C., and Vincent, K. W. 2000. Simulation of picloram, atrazine, and simazine leaching through two New Zealand soils and into groundwater using HYDRUS-2D. *J. Contam. Hydrol.* 44(1): 19–46.

Persicani, D. 1996. Pesticide leaching into field soils: Sensitivity analysis of four mathematical models. *Ecol. Model.* 84(1–3): 265–280.

Petach, M. C., Wagenet, R. J., and Degloria, S. D. 1991. Regional water-flow and pesticide leaching using simulations with spatially distributed data. *Geoderma* 48(3–4): 245–269.

Pignatello, J. J. and Xing, B. S. 1996. Mechanisms of slow sorption of organic chemicals to natural particles. *Environ. Sci. Technol.* 30(1): 1–11.

Pirisi, F. M., Cabras, P., Garau, V. L., Melis, M., and Secchi, E. 1996. Photodegradation of pesticides. Photolysis rates and half-life of pirimicarb and its metabolites in reactions in water and in solid phase. *J. Agric. Food Chem.* 44(8): 2417–2422.

Rice, P. J., Rice, P. J., Arthur, E. L., and Barefoot, A. C. 2007. Advances in pesticide environmental fate and exposure assessments. *J. Agric. Food Chem.* 55(14): 5367–5376.

Roberts, T. R. 1984. Iupac reports on pesticides. 17. Non-extractable pesticide-residues in soils and plants. *Pure Appl. Chem.* 56(7): 945–956.

Romero, E., Schmitt, P. H., and Mansour, M. 1995. Photodegradation of pirimicarb in natural water and in different aqueous solutions under simulated sunlight conditions. *Fresenius Environ. Bull.* 4: 649–654.

Ruzicka, J. H. A., Simmons, J. H., and Tatton, J. O. 1967. Pesticide residues in foodstuffs in Great Britain. 4. Organochlorine pesticide residues in welfare foods. *J. Sci. Food Agric.* 18(12): 579–585.

Sadeghi, A. M., Isensee, A. R., and Shirmohammadi, A. 1995. Atrazine movement in soil: Comparison of field observations and PRZM simulations. *J. Soil Contam.* 4(2): 151–161.

Saha, J. G., Karapall, J. C., and Janzen, W. K. 1971. Influence of the type of mineral soil on the uptake of dieldrin by wheat seedlings. *J. Agric. Food Chem.* 19(5): 842–848.

Sarmah, A. K., Close, M. E., Pang, L. P., Lee, R., and Green, S. R. 2005. Field study of pesticide leaching in a Himatangi sand (Manawatu) and a Kiripaka bouldery clay loam (Northland). 2. Simulation using LEACHM, HYDRUS-1D, GLEAMS, and SPASMO models. *Aust. J. Soil Res.* 43(4): 471–489.

Senesi, N. 1992. Binding mechanisms of pesticides to soil humic sub-stances. *Sci. Total Environ.* 123–124: 63–76.

Sharpley, A. N. 1985. Depth of surface soil-runoff interaction as affected by rainfall, soil slope, and management. *Soil Sci. Soc. Am. J.* 49(4): 1010–1015.

Shirmohammadi, A. and Knisel, W. G. 1994. Evaluation of the GLEAMS model for pesticide leaching in Sweden. *J. Environ. Sci. Health A* 29(6): 1167–1182.

Sichani, S. A., Engel, B. A., Monke, E. J., Eigel, J. D., and Kladivko, E. J. 1991. Validating GLEAMS with pesticide field data on a Clermont silt loam soil. *Trans. ASAE* 34(4): 1732–1737.

Simunek, J. and van Genuchten, M. T. 2008. Modeling nonequilibrium flow and transport processes using HYDRUS. *Vadose Zone J.* 7(2): 782–797.

Simunek, J., van Genuchten, M. T., Sejna, M., Toride, N., and Leij, F. J. 1999. The STANMOD computer software for evaluating solute transport in porous media using analytical solutions of the convection–dispersion equation. U.S. Salinity Laboratory, USDA/ARS, Riverside, CA.

Smith, W. N., Prasher, S. O., and Barrington, S. F. 1991. Evaluation of PRZM and LEACHMP on intact soil columns. *Trans. ASAE* 34(6): 2413–2420.

Stromqvist, J. and Jarvis, N. 2005. Sorption, degradation and leaching of the fungicide iprodione in a golf green under Scandinavian conditions: Measurements, modelling and risk assessment. *Pest Manag. Sci.* 61(12): 1168–1178.

Torstensson, N. T. L. 1987. Microbial decomposition of herbicides in soil. In: Hutson, H. D. and Roberts, R. T. (eds), *Herbicides*, pp. 249–270. John Wiley & Sons: Chichester.

Torstensson, N. T. L. 1990. Role of microorganisms in decomposition. In: Hance, R. J. (ed.), *Interactions Between Herbicides and the Soil*, pp. 159–178. Academic Press: London.

USSCS. 1972. *National Engineering Handbook, Section 4 Hydrology*. USSCS: Washington, DC.

Van Genuchten, M. T. and Cleary, R. W. 1979. Movement of solutes in soil: Computer-simulated and laboratory results. In: Singh, B. and Grafe, M. (eds), *Synchrotron-Based Techniques in Soils and Sediments*. pp. 23–67, Elsevier, Amsterdam, The Netherlands.

Veith, G. D., Kuehl, D. W., Leonard, E. N., Puglisi, F. A., and Lemke, A. E. 1979. Fish, wildlife, and estuaries – Polychlorinated biphenyls and other organic-chemical residues in fish from major watersheds of the United States, 1976. *Pestic. Monit. J.* 13(1): 1–11.

Vighi, M. and Di Guardo, A. 1995. Predictive approaches for the evaluation of pesticide exposure. In: Vighi, M. and Funari, E. (eds), *Pesticide Risk in Groundwater*. pp. 73–85, CRC Press, Tampa, USA.

Voice, T. C., Rice, C. P., and Weber, W. J. 1983. Effect of solids concentration on the sorptive partitioning of hydrophobic pollutants in aquatic systems. *Environ. Sci. Technol.* 17(9): 513–518.

Voss, C. I. and Provost, A. M. 2010. SUTRA, a model for saturated-unsaturated, variable-density ground-water flow with solute or energy transport. Water-Resources Investigations Report 02-4231, U.S. Geological Survey, Reston, VA.

Wagenet, R. J., Hutson, J. L., and Biggar, J. W. 1989. Simulating the fate of a volatile pesticide in unsaturated soil – A case-study with DBCP. *J. Environ. Qual.* 18(1): 78–84.

Walker, A. 1974. A simulation model for prediction of herbicide persistence. *J. Environ. Qual.* 3: 396–401.

Wauchope, R. D., Rojas, K. W., Ahuja, L. R., Ma, Q., Malone, R. W., and Ma, L. W. 2004. Documenting the pesticide processes module of the ARS RZWQM agroecosystem model. *Pest Manag. Sci.* 60(3): 222–239.

White, R. E. 1985. The influence of macropores on the transport of dissolved and suspended matter through soil. *Adv. Soil Sci.* 3: 95–120.

Whitford, F. 2006. *The Benefits of Pesticide*. Purdue University, West Lafayette, IN.

Wischmeier, W. H. and Smith, D. D. 1978. Predicting rainfall erosion losses from cropland – A guide to conservation planning. *Agriculture Handbook*. Department of Agriculture, Washington, DC.

Wolfe, N. L., Mingelgri, U., and Miller, G.C. 1990. Abiotic transformations in water, sediments, and soil. In: *Pesticides in the Soil Environment: Proccesses, Impacts and Modeling*, pp. 103–168. SSSA: Madison, WI.

Yu, X. Y., Ying, G. G., and Kookana, R. S. 2009. Reduced plant uptake of pesticides with biochar additions to soil. *Chemosphere* 76(5): 665–671.

Zeep, R. G. and Cline, D. M. 1977. Rates of direct photolysis in aquatic environments. *Environ. Sci. Technol.* 11: 359–366.

Zhou, D., Kaczmarski, K., Cavazzini, A., Liu, X., and Guiochon, G. 2003. Modeling of the separation of two enantiomers using a microbore column. *J. Chromatogr. A* 1020(2): 199–217.

Zhou, Y., Wang, Y., Hunkeler, D., Zwahlen, F., and Boillat, J. 2010. Differential transport of atrazine and glyphosate in undisturbed sandy soil column. *Soil Sediment Contam.* 19(3): 365–377.

3

Degradation of Pesticides

O.P. Bansal

CONTENTS

3.1 Introduction

The ever-growing human population, which is expected to be 9.1 billion by 2050 (UN 2005), will require more agricultural production, especially in tropical regions. Increasing agricultural production in these regions relies on the use of agrochemicals, as shown by the growth of agrochemical market by approximately 4.5% annually. On the one hand, use of agrochemicals helps in meeting with the dietary needs of the ever-increasing world population, but on the other hand, many agrochemicals used in agricultural activities con-taminate soil, air, and water resources. Therefore, it becomes imperative to identify and quantify the chemical and biological processes that control the behavior of organic chemicals in the environment to improve pesticidal management for minimizing contamination of our natural resources and remediating contaminated environment. To an estimate, we have an annual production of approximately 10^6 tons of active ingredients worldwide. Assuming a total arable land area of 13.8×10^6 km^2 on earth, though unrealistic, the even distribution of pesticides would amount to about 0.7 kg/ha year. So, much use of organic chemicals by all sectors of the economy is affecting human health and environment as the result of soil, air, and water contamination. These effects may range from growth retardation to physiological or behavioral deficiencies, which substantially affect the most sensitive individuals or species (Atterby et al. 2002; Bollag and Liu 1990). Subsequently, it may lead to population shift resulting in ecological changes. Pesticides, when used in agriculture, are likely to enter various compartments of the environment. They may undergo leaching by water, volatilization into atmosphere, or sorption to various surfaces. When introduced into the environment, the fate of pesticides is influenced by three major processes: transfer (sorption), transport (runoff, leaching, and volatilization), and transformation (chemical degradation (hydrolysis and photolysis) and microbial degradation).

Pesticide degradation: Pesticide degradation or the breakdown of pesticides is usually considered beneficial. Pesticide-destroying reactions change most pesticide residues in the environment to nontoxic or harmless compounds. However, degradation is detrimental when a pesticide is destroyed before the target pest has been controlled. There are three types of pesticide degradation: (a) chemical degradation, (b) photodegradation, and (c) microbial degradation.

During pesticide transformation brought about by physical, chemical, and biological forces, following types of reactions generally occur.

1. *Oxidation*: Physical and chemical oxidation reactions involve molecular oxygen or peroxide or singlet oxygen. Biological oxidation occurs by mixed-function oxidase enzyme. Specific oxidative reactions in pesticide degradation are

    a. Ring hydroxylation

    b. Side-chain oxidation

    c. Epoxidation

    d. O-dealkylation

e. N-dealkylation

$$>NCH_3 \longrightarrow >NCH_2OH \longrightarrow >NH$$

f. S-dealkylation

$$- SR \longrightarrow SH$$

g. Oxidative desulfuration

$$>P = S \longrightarrow >P = O$$

h. Amine oxidation

$$- NH_2 \longrightarrow - NHOH \longrightarrow NO_2$$

i. Sulfoxidation

$$- SR \longrightarrow - \overset{\overset{\displaystyle O}{\|}}{\underset{\underset{\displaystyle O}{\|}}{S}} - R$$

2. *Hydrolysis*: Ester group in pesticides undergoes hydrolytic cleavage. Hydrolysis occurs chemically or enzymatically as:

a. Hydrolysis of ester

$$- \overset{}{\underset{\underset{\displaystyle O}{\|}}{C}} - OR \longrightarrow - COOH$$

b. Hydrolysis of epoxides

$$>C \overset{}{\underset{\displaystyle O}{-}} C< \longrightarrow \underset{\underset{\displaystyle OH \quad OH}{|\quad|}}{>C - C<}$$

c. Hydrolysis of amide

$$>\overset{}{\underset{\underset{\displaystyle O}{\|}}{C}} - NH_2 \longrightarrow - \overset{}{\underset{\underset{\displaystyle O}{\|}}{C}} - OH$$

d. Hydrolysis of halogens

$$\equiv C\text{–}Cl \rightarrow \equiv C\text{–}OH$$

3. *Reductions*

a. Nitro group

$$-NO_2 \rightarrow -NH_2$$

b. Ketonic group

$$>C=O \rightarrow >CHOH$$

c. Reductive dehalogenation

$$\equiv C\text{–}Cl \rightarrow \equiv C\text{–}H$$

4. *Conjugation*
   a. Methylation

$$\equiv C\text{-}OH \rightarrow \equiv C\text{-}OCH_3$$

   b. Acetylation

$$\equiv C\text{-}NH_2 \rightarrow \equiv C\text{-}NH\text{-}\overset{\overset{\displaystyle O}{\|}}{C}\text{-}CH_3$$

5. *Rearrangement*: Transformation of one isomer to another, occurring by physical or chemical agents.
   a. Cyclization

$$\text{Open} \rightarrow \text{Closed chain}$$

   b. Dimerization

$$2A \rightarrow [A\text{-}A]$$

3.2 Chemical Degradation

Chemical degradation is the breakdown of pesticides by processes in which living organisms are not involved. Hydrolysis, oxidation–reduction, substitution, elimination, dehalogenation, and reduction, without the influence of microbial activities, are processes involved in chemical degradation. Hydrolysis is an important reaction that takes place in water and soil for pesticide degradation.

3.2.1 Factors Influencing Chemical Degradation

1. *Organic matter content and clay content*: Organic matter and clay content are the two major components that influence chemical degradation of pesticides, as organic matter content and clay content provide larger surface area for enhancing hydrolytic degradation.
2. *pH*: Soil pH or pH of the medium affects the hydrolytic process of pesticide dissipation, which depends on the nature of pesticides, for example, some pesticides are acid hydrolyzed and others are base hydrolyzed.
3. *Temperature*: With the increasing temperature, the molecules in solution have more energy, causing them to move and react faster, which causes hydrolysis reaction to occur at a faster rate.
4. *Nature of substituents*: The reactivity of pesticides depends on the substituents present in the pesticides used. Some substituents are easily replaced from the substrate by hydrolysis reaction since the products formed are very stable in water. The substituents that pull electron density away from the substrate facilitate hydrolysis.
5. *Effluent irrigation*: Effluent irrigation enhances the chemical degradation of pesticides by altering the pH of the soil solution and increasing dissolved organic matter content (Muller et al. 2007).

6. *Crop residue and pesticide concentration* are two other factors that influence chemical degradation.

3.2.2 Organochlorine Pesticides

Organochlorine compounds are resistant to chemical degradation in soil. Dichlobenil and chlorothalonil undergo substitution of Cl with OH or hydration of CN group. Dichlorodiphenyltrichloroethane (DDT) and methoxychlor undergo reductive dehalogenation at the trichloromethyl group and hydroxylation at the tertiary carbon to form alcohol (Pirnie et al. 2006). Cyclodiene-chlorinated compound endosulfan hydrolyzes to endosulfan alcohol and oxidizes to endosulfan sulfate in soil and water.

Dichlorvos undergoes hydrolytic decomposition as (Musa et al. 2010).

3.2.3 Organophosphorus Pesticides

Organophosphorus pesticides undergo oxidation by chlorine present in soil or water to convert P=S to P=O group. Organophosphorus esters such as parathion or methidathion undergo hydrolysis to form alcohol and phosphoric acid (Washburn 2003; Fan and Lio 2009).

Methiodothion

Chlorpyrifos undergoes hydrolysis easily in alkaline medium (Gilani et al. 2010).

3.2.4 Carbamate Pesticides

Oxamyl and methomyl, widely used oxime pesticides, undergo rapid degradation in anoxic and suboxic soil suspensions by a redox pathway involving Fe(II). During the mechanism, Fe(II) forms a precursor complex that forms nitrile and thiol (Strathman and Stone 2000).

N-Methylcarbamates undergo successive oxidation in soil and are converted to CO_2 as the final product. Aldicarb or thiobencarb undergoes oxidative degradation easily in chlorinated water.

During the studies on carbamate pesticides, the author himself found that the degradation of carbamate pesticides increased with soil organic matter, soil surface area, percentage clay content, temperature, and soil-moisture content (Bansal 2005), amending the soil with farm yield manure (FYM). It was also found that degradation increased with pH and was faster in alkaline medium (Bansal 2004).

3.2.5 Pyrethroid Pesticides

Pyrethroid pesticides undergo hydrolysis on the ester linkage and produce a number of metabolites. Hydroxylation of alcohol moiety at 2′, 4′, 5′, or 6′ position also occurs. Fenvalerate underwent hydration of the cyano group on soil with a trace formation of decarboxylated derivative. Flucythrinate and fluvalinate similarly underwent hydration of the cyano group and cleavage of the ester. Permethrin undergoes ester cleavage and forms 3-phenoxybenzyl alcohol and the dichlorovinyl chrysanthemic acid.

3.2.6 Miscellaneous

Imidacloprid undergoes a more rapid hydrolysis in alkaline medium than in acidic medium (Liu et al. 2006).

It was also reported that chemical hydrolysis is slower than photodegradation.

Phenyl propionate on hydrolysis forms 2-phenyl ethanol and propionic acid (Hu and Coats 2008).

Sulfonyl urea on chemical degradation forms sulfonamides and heterocyclic amines (Sarmah and Sabadie 2002).

3.3 Photodegradation

Photodegradation is the breakdown of pesticides by sunlight. Photodegradation can destroy pesticides on foliage, on the surface of the soil, in water, and even in the air. The photodegradation of pesticides is influenced by the intensity of sunlight, properties of the application site, exposure time, properties of the pesticides, pH of medium, water depth, and presence of other commonly occurring ions.

Photochemical reactions play a key role in the environmental degradation of pesticides that contain organic chromophore or metal–organic complexes capable of absorbing light energy directly. Pesticides having no chromophore group undergo photosensitive reactions. Photodegradation of the pesticides in soils is increased in the presence of soil organic matter such as fulvic acid and humic acid, dyes such as rose Bengal, pigments such as chlorophyll and xanthone, secondary plant metabolites such as riboflavin, tyrosine, etc. Surfactants used in various pesticide formulations act as photosensitizers.

During a photochemical reaction, homolytic bond cleavage with the formation of a free radical occurs. During a photodegradation reaction of pesticides, free radicals $^{\cdot}CH_3$, $^{\cdot}R$, $RO^{\cdot}$, $ROO^{\cdot}$, $NO^{\cdot}$, $^{\cdot}OH$, $^{\cdot}NO_3$, etc. are formed, which require energy of approximately 400 kJ/mole (of ~300 nm wavelength). Photo-oxidation is one of the most prominent means of photodegradation initiated by various reactive species such as singlet oxygen, free radicals, organic hydroxyperoxidase, hydrogen peroxidase, etc.

Direct photolysis reaction (Equations 3.1 and 3.2) occurs as

$$\text{Pesticide} \rightarrow \text{Pesticide}^{\cdot} \tag{3.1}$$

$$\text{Pesticide}^{\cdot} \rightarrow \text{Photoproducts} \tag{3.2}$$

(Pesticide = Pesticide in ground state and pesticide$^{\cdot}$ = Pesticide in excited state) (Equation 3.3)

$$\text{Pesticide}^{\cdot} + X \rightarrow \text{Photoproducts} \tag{3.3}$$

where X is solvent or other molecules in solution.

Indirect photochemical reaction (Equations 3.4 and 3.5) occurs as

$$H_2O_2 + hv \rightarrow 2^{\cdot}OH \tag{3.4}$$

$$^{\cdot}OH + \text{Pesticide} \rightarrow \text{Photoproducts} \tag{3.5}$$

Another method based on photo-Fenton reaction (Equations 3.6 through 3.8) is

$$Fe^{3+} + hv \rightarrow Fe^{2+} + H^+ + {}^{\cdot}OH$$

(3.6)

$$Fe^{2+} + H_2O_2 \rightarrow Fe^{3+} + {}^-OH + {}^{\cdot}OH$$

(3.7)

$${}^{\cdot}OH + Pesticide \rightarrow Photoproduct$$

(3.8)

During photodegradation of chemical compounds including pesticides, following types of photoreactions are possible:

1. Norrish type-I reaction

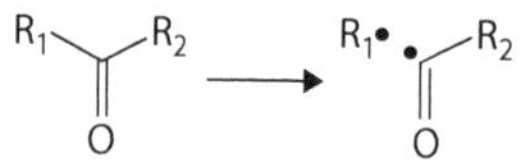

2. Norrish type-II reaction

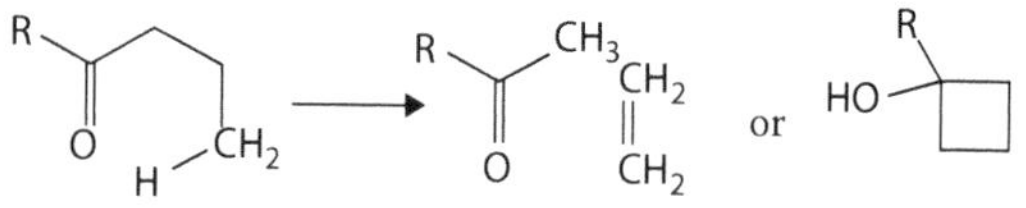

3. Isomerization (geometrical or optical)

(a)

(b)

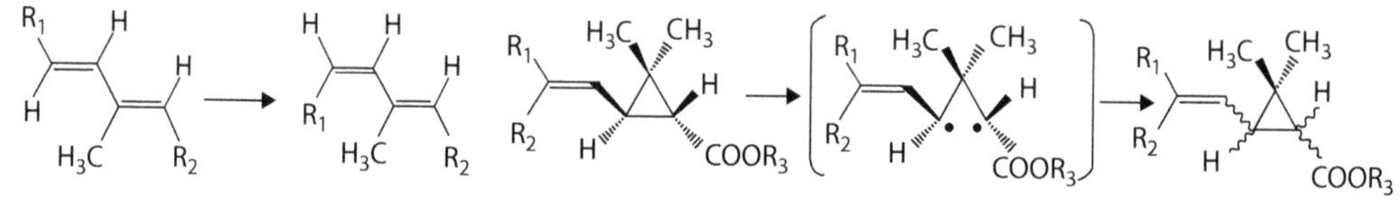

4. Ester cleavage, decarboxylation, decarbonylation

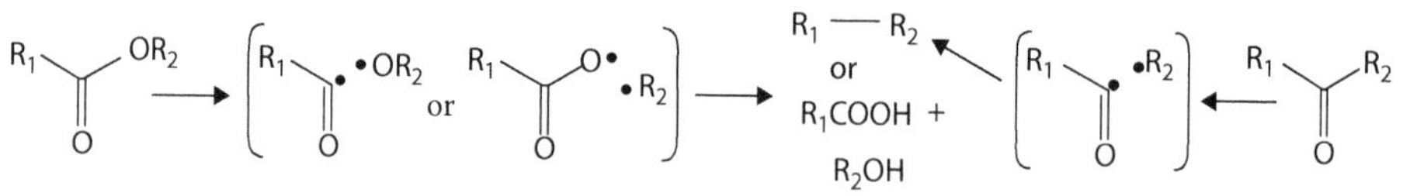

5. Dehalogenation

6. Dealkylation

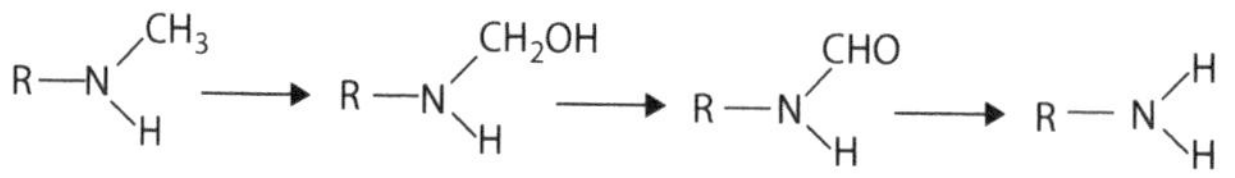

7. C-Oxidation

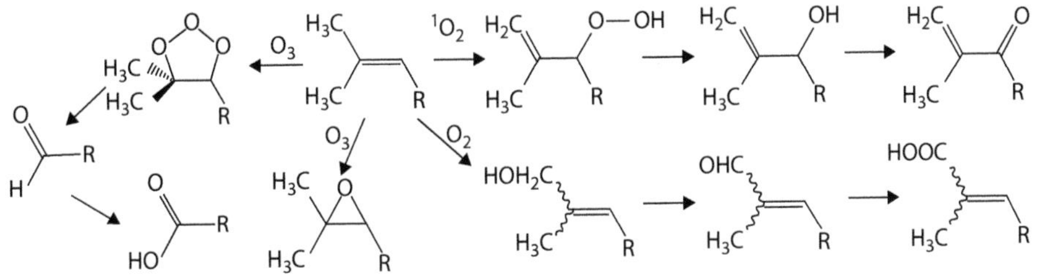

8. S-Oxidation

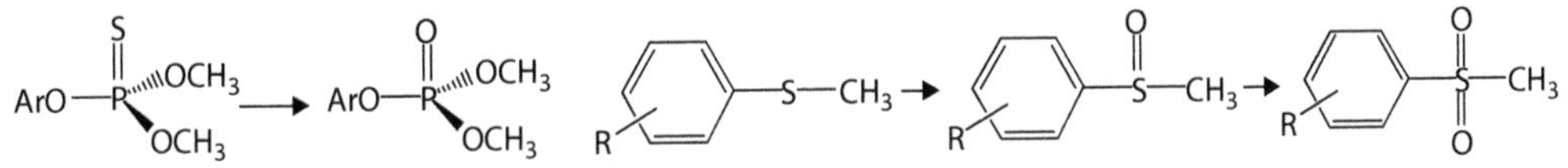

9. Rearrangement

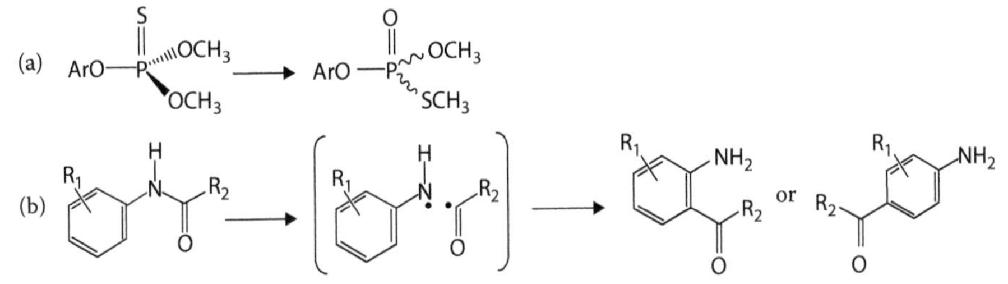

(a)

(b)

10. Cyclization

(a)

(b)

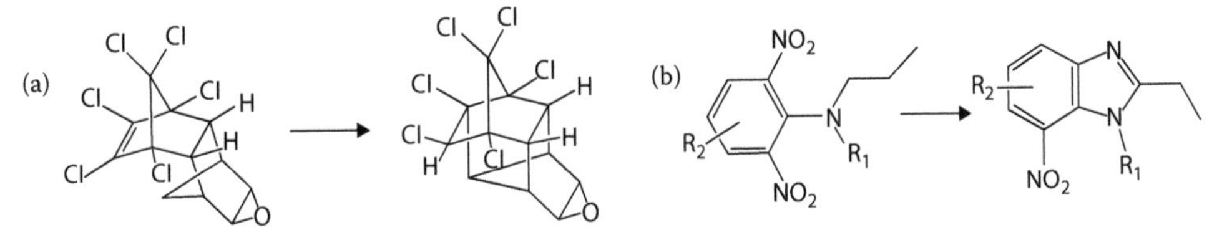

3.3.1 Photolysis of Organochlorine Pesticides

This class includes some of the well-known and popular compounds such as DDT, lindane, chlordane, aldrin, endrin, endosulfan, chlorocamphene, etc. As these pesticides have high persistence in environment, they pose a threat to the ecosystem.

Photochemically induced isomerization, dimerization, and dechlorination are common reactions that occur in organochlorine compounds.

p,p'-DDT: This popular insecticide, on photodegradation, produces the following compounds by dechlorination, dimerization, and oxidation (Zayed et al. 1994).

p,p'-DDT by photolysis in the presence of methanol and amines, which act as sensitizers, undergoes dehydrohalogenation (Miller and Narang 1970).

Aldrin: Aldrin, an insecticide, by photolysis in sunlight, produces photoaldrin, pentachloroaldrin, dieldrin, and photodieldrin (Dureja and Mukherjee 1986). Further studies of literature also revealed that under the influence of ultraviolet light, dimers of photoaldrin and tetrachloroaldrin are also formed.

Dieldrin: When dieldrin (an insecticide) is irradiated under sunlight, it forms pentachlorodieldrin by the replacement of a chlorine atom of the vinylic group by hydrogen and photodieldrin.

Endosulfan: Endosulfan, one of the most environmentally safe organochlorine insecticides, on photodegradation undergoes cleavage of the sulfite ester group to produce diol, ether, and lactone, and there is a dechlorination from the double bond to form pentachloroendosulfan and tetrachloroendosulfan. The photodegradation of endosulfan is rapid under ultraviolet light.

Fenarimol: (±)2,4'-Dichloro-α-(pyrimidine-5yl) benzhydryl alcohol, a broad-spectrum fungicide, by photodegradation in aqueous alcoholic solution, produces the following compounds.

Chlorothalonil: 2,4,5,6-tetrachloroisophthalonitrile, a broad-spectrum fungicide, on photo-transformation in ethanolic solution under UV irradiation (λ ~ 200 nm) produces three products, which might be formed due to cleavage of 4–Cl–C bond followed by alkylation of α-radicals (Kwon and Armbrust 2006).

Imidacloprid: Imidacloprid, 1-(6-chloro-3-pyridylmethyl) N-Nitroimidazolidin-2-ylideneamine, an insecticide, on photodegradation ($\lambda \sim 254$ nm) in water, produces the following compounds which might be formed via the bond cleavage of N–NO$_2$ to imine (=NH), followed by oxidation.

2,4-D: 2,4-Dichlorophenoxy acetic acid, a weedicide, on photodegradation by UV irradiation, undergoes degradation as

3.3.2 Photodegradation of Organophosphate Pesticides

Most organophosphorus pesticides have the following general formula.

where X is aliphatic, homocyclic, or heterocyclic groups; R is alkyl group.

Organophosphorus pesticides undergo photochemical degradation mainly by ester cleavage, reduction, oxidation of thio-ether group, isomerization, dehalogenation, dehydrohalogenation, dealkylation, cyclization, dimerization, etc.

Malathion, Parathion, and *dimethoate,* on photodecomposition in air, undergo isomerization as:

Parathion → Isoparathion

which, on further decomposition, are converted into phosphoric acid and p-nitrophenol.

Parathion undergoes photoreduction of the nitro group via nitroso and hydroxyamino groups and forms azoxy dimer as

Quinalphos is widely used as an insecticide. Photodegradation is one of the main modes of degradation in the environment. Quinalphos, in alcoholic solution, undergoes photochemical degradation as

Parathion

Azoxy

Azo

2-Hydroxyazo

$$R = -O-P(=S)(OC_2H_5)(OC_2H_5)$$

Alcohol hv

C_2H_5OH | uv

hv

Photodegradation is faster in soil (t½ 2–5 days)

Fenitrothion, an organophosphorus insecticide, on photodegradation in oxygenated solution by UV irradiation undergoes isomerization, oxidation, and hydrolysis to form products as (Derbalah et al. 2004)

Chlorpyrifos, a broad-spectrum insecticide, on photolysis in the presence of UV light undergoes oxidative desulfuration, dehalogenation, and hydrolytic reactions as

Phosalone, 5-6-chloro-2,3-dihdyro-2-oxobenzoxazol-3-yl methyl O,O-diethyl phosphoro-dithioate, a broad-spectrum nonionic insecticide, on photochemical reaction undergoes oxidation readily to form disulfide as a major product and a number of minor products due to N–CH$_2$ or S–CH$_2$ cleavage in addition to dehalogenation.

Coumaphos, in the presence of oxygen and UV light, undergoes rearrangement to form degraded products as:

Methidathion, S-2,3-dihydro-5-methoxy-2-oxo-1,3,4-thiadiazol-3yl methyl O,O dimethyl phosphorodithioate, an insecticide, on photochemical degradation in water and soil undergoes hydrolysis as

Degradation is faster in soil ($t_{1/2} = 1.54$ day) than in water ($t_{1/2} = 8.2$ days) (Washburn 2003).

Monocrotophos, Dimethyl (E)-1-methyl-2 (methylcarbamoyl) vinyl (phosphate), a foliar insecticide, on photochemical degradation forms dimethylphosphate, and its degradation in soil depends on soil moisture.

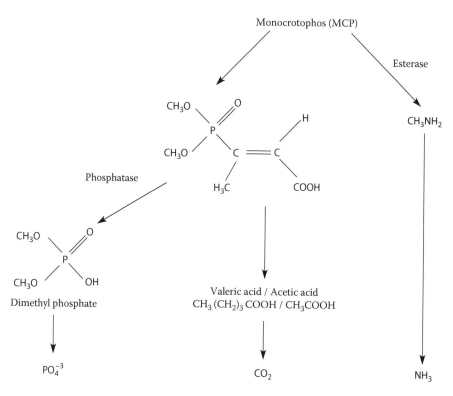

3.3.3 Carbamate Pesticides

Carbamate pesticides, which include insecticides, molluscicides, and nematicides, show a high biological activity and have medium to high polarity.

Carbamate pesticides possess the following general structure:

The three subgroups of carbamate pesticides include the following: (i) N-methylcarbamates and esters of phenols, for example, carbaryl, methiocarb, and propoxur; (ii) N-methyl and N-dimethyl carbamate esters of heterocyclic phenols, for example, carbofuran or pirimicarb; and (iii) oxime derivatives of aldehydes, for example, aldicarb, methomyl, oxamyl, etc.

Carbamate pesticides undergo hydrolysis, aliphatic side-chain oxidation, thioether oxidation, methylation, N-dealkylation, and rearrangement reactions when exposed to light (Tamini et al. 2006). In the presence of UV light, carbamate pesticides in water undergo cleavage of the ester bond, resulting in the production of the phenol or heterocyclic enol of carbamate ester. The hydrolysis products produced further undergo photodecomposition to form a number of products.

Methomyl, oxamyl, and oxime derivatives used as insecticides, on photolysis, form a number of products.

Carbofuran: The photodegradation of broad-spectrum insecticide of N-methyl carbamate ester series does not occur predominantly, but when it occurs, it undergoes hydrolysis, oxidation, and methyl and photo-Fries rearrangements to form a number of products as

N-methylcarbamates (Carbaryl or propoxur) by UV irradiation undergo cleavage of the ester bond, resulting in the production of the phenol or heterocyclic enol of the carbamate ester. The rate of photolysis depends on the pH of the medium. Similar results were reported for photodecomposition of other carbamate esters, for example, bendiocarb, pirimiphos-methyl in aqueous solution, etc.

3.3.4 Pyrethroid Pesticides

Pyrethroid pesticides are the derivatives of chrysanthemic acid, containing an isobutenyl group, and are very much susceptible to photochemical oxidation. Pyrethroid pesticides undergo photolysis by hydrolysis of the ester bond, cis–trans isomerization, carboxylation, and reductive dehalogenation.

Cypermethrin, deltamethrin, permethrin, and fenvalerate undergo photodegradation by cleavage of the ester or diphenyl ether linkage, oxidation of the –C=O group to –COOH group, hydration of the CN group, hydrolysis of $CONH_2$ group to –COOH group, oxidative cleaving of the halogenated side chain, and dehalogenation and intramolecular cyclization to form γ or δ lactone. In distilled water, the photolysis was in the order deltamethrin > cypermethrin > permethrin > fenvalerate and $t_{1/2}$ ranged from 1.4 to 10 days. The photodegradation rate on soil surface was correlated with the content of organic matter in the soil.

Esfenvalerate, a pyrethroid insecticide, undergoes photo-induced decarboxylation and ester cleavage in water under natural sunlight as

3.3.5 Miscellaneous

Pyriproxyfen, 2-[1-methyl-2-4-(phenoxyphenoxy) ethoxy] pyridine, effective against public health insect pests such as houseflies, cockroaches, and mosquitoes, on photo-degradation in water and soil undergoes oxidative cleavage to form various phenolic products as:

Isoproturon, an herbicide, on irradiation forms a number of products by the successive loss of N CH$_3$ group, followed by slow oxidative demethylation.

Pendimethalin, a selective herbicide of the dinitroaniline group, on photodecomposition in soil and water undergoes oxidative dealkylation, nitroreduction, and cyclization.

Propiconazole, a broad-spectrum fungicide, on photodegradation in soil and water, forms a stable aromatic moiety, 1,2,4 triazole.

Thiamethoxam, an efficient synthetic agrochemical, neonicotinoid, on photodegradation forms 3-(2-chloro-1,3-thiazol-5-yl methyl) 1,3,5-oxaziazinan-4-ylidene (nitrosamine).

The photochemical degradation of most of the pesticides follows first-order chemical kinetics. The first-order rate equation is generally expressed as:

$$-\ln(CA/CA0) = k1t,$$

where k_1 is the degradation constant. Plot of $\ln(C_A/C_{A0})$ vs t would therefore be a straight line passing through origin, and the slope will give the rate constant, k_1.

3.4 Microbial Degradation

Microorganisms are known to play major roles in metabolizing pesticides in the environment, mainly in water and soil. The process can take several steps to degrade pesticides into CO_2, H_2O, and mineral salts. There are four types of microbes: bacteria, fungi, protozoa, and algae. Bacteria and fungi are the most abundant in nature, so they are the most important microorganisms for biological degradation. There will always be one or more microorganisms that are able to degrade any organic compound. Due to an enormous number of species, mutations of a microorganism may create an organism that can degrade the compound. So, the rate of degradation increases with time.

In the microbial degradation process, the pesticide is absorbed into the cell membrane of the microbe. Enzymes present in the microbe breakdown the pesticide into smaller fragments with minerals as the final end-product. The degradation process follows different pathways depending on the microbes present.

The distribution of microbes in soil is not homogeneous. The population of microbes in the proximity of the surface is high, which may be due to the fact that microbes naturally live on the material excreted from a plant's roots. Pesticides are generally degraded by microbes along with material excreted by the roots of plants.

Bacterial degradation dominates in soils and water at pH > 5.5, while in acidic medium, soil fungal degradation dominates.

Microbial degradation is of two types, aerobic and anaerobic, and depends on the surrounding conditions. Aerobic metabolism is a process in which oxygen is utilized to oxidize the pesticide and occurs in soil, not in water, as less oxygen is available. While in anaerobic degradation oxygen is not present, degradation occurs via some other pathway. In soils, the degradation of pesticides in the first layer is aerobic, while below this layer, it is anaerobic.

During microbial degradation reaction of pesticides, a large number of reactions have been observed viz. reduction, oxidation, hydrolysis, dealkylation, hydroxylation, ring cleavage, and conjugation.

3.4.1 Factors Influencing Microbial Degradation

1. *Water depth*: At different water depths, different microbes exist, and a chemical may or may not be metabolized depending on the microbes present at that depth. At lower depths, the rate of metabolism will be slower because of the lower temperature.

2. *Mobility*: If a pesticide is strongly bound to the soil, the biodegradation process cannot occur. The organic content in soil strongly influences the binding of nonpolar pesticides in soil and their biodegradation. As microbes move more freely in water than in soil, pesticides in water will be degraded faster in water than in soil.

3. *Primary/secondary metabolism*: Primary metabolism occurs when microbes are able to derive energy from the metabolism of a pesticide. The microbial population is directly proportional to the pesticide concentration.

   If the microbial population does not use pesticide for energy to live, secondary metabolism occurs. The pesticide is degraded by the microbes as it degrades other material in the soil.

4. *Temperature*: High temperature increases the degradation rate, as with the rise in temperature sorption decreases and metabolic activity of microbes' increases.

5. *pH*: The microbial degradation increases with pH.

6. *Soil moisture*: Waterlogged conditions in combination with a high number of nutrients promote growth of anaerobic microbial species. Bacteria, in general, require a high level of moisture for degradation.

7. *Organic matter*: In soils rich with organic matter, usually degradation of pesticides increases due to cometabolism.

8. *Pesticide*: The microbial degradation of pesticides depends on the structure of the pesticide as:

   a. The microbial degradation of polar, water-soluble pesticides is more than nonpolar, water-insoluble pesticides. As cations are more easily adsorbed on soil particles, microbial degradation of anionic pesticides is faster than cationic ones.

   b. Degradation of aliphatic moiety is faster than aromatic ones.

c. Substances that are highly toxic to microorganisms are not easily degraded.

d. Compounds with a high oxidation state are resistant to further oxidation and hence degrade slowly.

The most important degraders in soil are within the genuses *Arthrobacter, Aspergillus, Alcaligenes, Bacillus, Corynebacterium, Flavobacterium, Fusarium, Nocardia, Penicillium, Pseudomonas,* and *Trichoderma. Alcaligenes* and *Pseudomonas* are very good to degrade chloroaromatic pesticides (Hoostal et al. 2002). Fungus *Phanerochaete chrysosporium* is used efficiently to degrade persistent pesticides such as DDT or lindane.

3.4.2 Organophosphorus Pesticides

Organophosphorus pesticides form a major and most widely used group accounting for more than 36% of the total world market. The organophosphorus pesticides are degraded by bacteria such as *Pseudomonas* sp., *Agrobacter, Arthrobacter* sp., *Flavobacterium* sp., fungi, *Fusarium, Aspergillus niger, Geobacillus caldoxylosilyticus, Actinomycetes, Streptomycetes,* etc.

Organophosphorus pesticides undergo microbial degradation via hydrolysis and/or microbial cleavage. Enzymes such as phosphatase, esterase, hydrolyase, and oxygenase participate actively in the degradation of organophosphorus pesticides.

Parathion: On microbial hydrolysis by enzyme oxidases or hydrolyases, parathion is decomposed to p-nitrophenol and ionic diethylthiophosphate. Parathion, in the presence of microorganisms, also undergoes transformation to form amino parathion.

Methyl parathion: It gets hydrolyzed on soil and also in flooded soil by *Pseudomonas* sp. and *Flavobacterium* sp. to p-nitrophenol. The optimum enzymatic activity of hydrolyases occurs at pH range 7.5–9.5.

Malathion: Malathion on degradation in the presence of *Pseudomonas* strain forms diethyl phosphorothioate, which is subsequently converted to dimethyl malate or salt of succinic acid.

Chlorpyrifos and quinalphos: As these organophosphorus pesticides are less soluble, they are difficult to degrade. In the presence of lipid-producing microorganisms, chlorpyrifos is degraded to 3,5,6-trichloro-2-pyridinol (Gilani et al. 2010). The half-life period of chlorpyrifos ranges from 34 to 46 days (Swati and Singh 2002). Microbial degradation contributes significantly to the dissipation of chlorpyrifos in fresh water but is inhibited in sea water (Bondarenko et al. 2004). The degradation of chlorpyrifos occurs by nonspecific and noninducible enzyme systems produced in high pH soils. Quinalphos is rapidly degraded in flooded soils, besides, other kind of degradation of quinalphos also occurs by green algae, *Chlorella vulgaris* and *Scenedesmus*, and blue algae, *Synechococcus*.

Monocrotophos (MCP): MCP is most susceptible to bacterial degradation. Monocrotophos on microbial degradation forms valeric acid and acetic acid in addition to dimethyl phosphate, which is not used by the bacteria as a sole source of phosphorus. Degradation of MCP occurs by *Bacillus, Azospirillum lipoferum, Arthrobacter,* and *Pseudomonas*.

Dimethoate: The microbial degradation by fungi and bacteria is the means of disappearance of dimethoate from soil, as it is used as a source of carbon and energy or source of phosphorus by *Aspergillus sydowii, A. niger, A. flavus,* and *Fusarium oxysporum*. Dimethoate on microbial hydrolysis forms methylamine, phosphoric acid, and other products.

Fenitrothion: By microbial degradation, like methyl parathion, fenitrothion undergoes hydrolytic cleavage to form 4-nitro-m-cresol in addition to phosphoric acid.

Coumaphos: By microbial degradation, coumaphos undergoes hydrolysis as:

Dichlorvos (DDVP): By microbial degradation, DDVP forms desmethyldichlorvos, which on hydrolysis forms phosphate and carbon dioxide.

Methidathion: Microbial degradation is the dominating route for methidathion degradation in soil and water. The degradation occurs via isomerization, hydrolysis, and oxidation.

3.4.3 Organochlorine Pesticides

Microbial degradation of organochlorine pesticides occurs by *Pseudomonas*, *Moraxella* sp., *Actinobacter*, etc. Hydrolytic cleavage and dechlorination are the two major mechanisms of microbial degradation.

Generally, methylated and halogenated aromatics are dissimilated via oxygenases. However, nucleophile attack catalyzed by dehalogenases has also been reported. Bacteria transform aromatic compounds into dihydroxy derivatives, which serve as substrates for oxygenolytic cleavage of the aromatic ring. Chlorophenoxy acids (2,4-D, 2,4,5-T, 2-methyl-4-chlorophenoxy acetic acid (MCPA)) are degraded by bacteria such as *Pseudomonas* sp., *Arthrobacter* sp., *Mycoplana* sp., and *Flavobacterium*. The degradation of 2,4-D occurs via cleavage of the ether linkage between oxygen and the aliphatic side chain to form 2,4-dichlorophenol and glyoxalate. 2,4-Dichlorophenol is degraded to succinic acid via ketoadipate. MCPA is also metabolized by *Pseudomonas* sp. and *Arthrobacter* sp. through an oxidative cleavage. 2,4,5-T undergoes mineralization by *Pseudomonas cepacia* to form 2,4,5-trichlorophenol, which is degraded to succinate derivative.

Herbicides such as monuron, diuron, linuron, and propanil are metabolized through catechol intermediates. Endosulfan undergoes degradation with the help of mycobacterium and forms endosulfan diol and endosulfan lactone.

3.4.4 Carbamate Pesticides

Carbamate pesticides are transformed metabolically by a variety of chemical reactions into more water-soluble molecules with increased polar properties. The principal route of metabolism of carbamate esters is

1. *via oxidation*: Generally associated with the mixed-function oxidase enzymes. Depending on the functional groups in the molecules, oxidative reactions include (a) hydroxylation of aromatic ring, or epoxidation, (b) O-dealkylation, (c) N-methyl hydroxylation, (d) N-dealkylation, (e) hydroxylation and subsequent oxidation of aliphatic side chain, and (f) thio ether oxidation to sulfoxides and sulfones.
2. *via hydrolysis*: Carbamates are hydrolyzed by esterases to form amine, CO_2, alcohol, or phenol.
3. via conversion of conjugated compounds to hydroxy products.

Aldicarb, methomyl, and oxamyl on hydrolytic metabolism form pesticide oxime as a major product, which is then degraded to carbon dioxide in soil.

Propham and chlorpropham undergo hydroxylation at position 2 and 3. Oxidation and hydrolysis form 1, 3 hydroxy and 1-carboxy derivates, which are converted into hydroxy acetanilide.

Phenmedipham and desmedipham are hydrolyzed mainly into methyl-N or ethyl-N-(3-hydroxy phenyl carbamate), which are hydrolyzed to 3-amino phenol.

Carbaryl undergoes hydroxylation and hydrolysis in the presence of microbes to form 1-Naphthyl N-hydroxy methyl carbamate (due to the hydroxylation), 4-Hydroxy-1-naphthyl-N-methyl carbamate, 5-Hydroxy-1-naphthyl-N-methyl carbamate, and 1-Naphthol via hydrolysis. Microbial degradation of benomyl, carbendazim, and carbofuran involves hydroxylation at positions 4 and 5.

3.4.5 Pyrethroid Pesticides

The microbial degradation of pyrethroid pesticides is very slow. Degradation of pyrethroids is brought about by a system of esterase and oxidase enzymes. Ester hydrolyzes the ester linkage between the acid and the alcohol moieties. Hydroxylation occurs at 2' or 5' or 6' position of the alcohol moiety. The aerobic degradation is faster than the anaerobic one.

3.4.6 Miscellaneous

Simazine: Simazine is not degraded easily by microbes. Strong et al. (2002) found that *Arthrobacter aurescens* strain is capable of degrading s-triazines, including simazine. The simazine during degradation undergoes dehalogenation, dealkylation, and hydroxylation without cleavage of triazine ring. Kodama et al. (2001) found that *Penicillium steckii*, a fungal strain, can also degrade simazine.

Pyriproxyfen: Pyriproxyfen undergoes degradation via biological catalysis and serves as a carbon source for soil microorganisms. Hydroxylation followed by degradation to phenol occurs during microbial decay.

Phenylurea: These pesticides undergo stepwise microbial degradation by demethylation, demethoxylation, deamination, and decarboxylation and form phenol in the presence of *Pseudomonas striata*, *Fusarium solani*, and *Penicillium* sp.

Butachlor: This chloroacetanilide on microbial degradation by *Fusarium solani* and *F. oxysporum* undergoes dechlorination, hydroxylation, dehydrogenation, C-dealkylation, N-dealkylation, and debutoxymethylation to form a number of products.

Phenylethyl propionate: Phenylethyl propionate undergoes microbial biotransformation to form 2-phenyl ethanol (Hu and Coats 2008).

3.5 Pesticide Degradation and Environmental Pollution

Pesticides have been extensively applied to the agricultural activities as well as environmental sanitation implementation due to their chemical characteristics of toxicity, bioaccumulation, and persistence. The residual pesticides have become the contamination sources and pose a serious threat to the soil and groundwater. This contamination becomes more serious when persistent pesticides are involved. The abundant use of less-persistent pesticides such as organophosphate and carbamate pesticides and their residuals in the environment have also posed a great threat to the environment. Some studies also have reported a relationship between the amounts of pesticides used and their concentration in groundwater and surface water.

A number of researchers have reported that surface water and groundwater all over the world are contaminated by pesticides and their metabolites. Though the use of DDT and benzene hexachloride (BHC) has been terminated globally in the last decade, still a number of groundwater and surface water samples contain DDT, dichlorodiphenyldichloroethylene (DDE), and dichlorodiphenyldichloroethane (DDD) in addition to benzene hexachloride benzene hexachloride (BHC) above permissible limit, particularly in developing countries. The pesticides and their metabolites from domestic, industrial, and agricultural effluents enter the food chain through groundwater or surface water. A number of studies have reported that the concentration of pesticides and their metabolites in vegetables is above permissible limits.

TABLE 3.1

Hazardous Effects of Pesticides and Their Degraded Products

Types of Pesticides	Hazardous Effects
Dieldrin, BHC, DDT	Headache, dizziness, gastrointestinal disturbances, numbness and weakness of the extremities, apprehension, and hyperirritability DDT produces serious functional and morphological changes in every organ of the body; most affected is the nervous system. Metabolites of DDT are carcinogenic
2,4-D and 2,4,5-T	Irritation to skin and mucus membranes, vomiting, headache, diarrhea, confusion, bizarre or aggressive behavior, and muscle weakness
Organophosphate	Headache, excessive salivation and tearing, respiratory depression, blurred vision, cramps, nausea, diarrhea, rapid or slow heart rate, weakness, and giddiness
Carbamate	Malaise, muscle weakness, dizziness, sweating, headache, salivation, nausea, vomiting, abdominal pain, diarrhea, nervous system depression, blurred vision, and weight loss
Pyrethroids	Abnormal facial sensation, dizziness, salivation, headache, fatigue, vomiting, diarrhea, numbness, irritation to skin and upper respiratory tract, and asthma

Stable airborne pesticide residues and their degradation products may move from the application site and be deposited in dew, rainfall, or dust. This may result in pesticide redistribution within the application site or movement of some compounds off site. Strongly sorbed and high-persistent pesticides, that is, large $t_{1/2}$ values remain near the ground surface for a long period, increasing the chances of being carried to a stream or lake or into groundwater via runoff or leaching. Organochlorine compounds, which are highly stable, pose a threat to nontarget animals inhabiting the water bodies, which can get polluted through groundwater. Though most of the organochlorine compounds, especially DDT and BHC, have been banned in the last decade, most of the groundwater samples still contain DDT or BHC or both.

In Table 3.1, some hazardous effects of pesticides and degraded products are depicted.

3.5.1 Adverse Effects

- Some pesticide degradation products show much more toxicity to animals and humans than the parent compounds (Table 3.1).
- Pesticide degradation products may enhance or decrease microbial population, inhibiting the growth of one or more specific microbes.
- Sulfonylureas and their degraded products show acute toxicity to the cladoceran *Daphnia magna*, a primary consumer in freshwater ecosystems.
- Pesticide degradation products in water may block the action of hormones in fish and amphibians, causing reproductive dysfunction and abnormal development, which cause diminishing of productivity and fecundity.

3.6 Conclusions

- Biotic degradation is the major pathway for pesticide degradation in soil and water, followed by photolysis.
- Oxidation, hydrolysis, reduction, conjugation, and rearrangement are the major reactions that occur during pesticide degradation.

- Chemical degradation is influenced by organic matter content, pH, temperature, nature of pesticide, crop residue, and effluent irrigation.
- Generally, hydrolysis of pesticides in alkaline medium is faster.
- Photolysis of pesticides is influenced by the intensity of sunlight, exposure time, properties of the application site and pesticides, pH of the medium, water depth, and presence of other commonly occurring ions.
- Pesticides having organic chromophore or metal–organic complexes can absorb light easily and undergo photodegradation.
- Photodegradation of pesticides in soils increases in the presence of soil organic matter such as fulvic acid and humic acid, dyes such as rose Bengal, pigments such as chlorophyll and xanthone, secondary plant metabolites such as riboflavin, tyrosine, etc.
- During photodegradation of pesticides, $\cdot CH_3$, $\cdot R$, $RO\cdot$, $ROO\cdot$, $NO\cdot$, $\cdot OH$, $\cdot NO_3$, etc. are formed. The energy required for this process is approximately 400 kJ/mole.
- Due to the heterogeneity of soil, together with soil properties varying with meteorological conditions, mechanisms of photolysis of pesticides in soil are difficult to understand.
- Photo-induced generation of reactive species under limited mobility modifies degradation mechanisms.
- Microbial degradation is of two types, aerobic and anaerobic, and carried out by bacteria and fungi.
- Bacterial microbial degradation dominates in soils and water at pH > 5.5, while in acidic medium, soil fungal degradation dominates.
- Water depth, mobility, temperature, pH, soil moisture, organic matter, primary and secondary metabolism, and structure of pesticides are the major factors that influences biotic degradation.
- The most important degraders in soil are within the genuses, *Arthrobacter, Aspergillus, Alcaligenes, Bacillus, Corynebacterium, Flavobacterium, Fusarium, Nocardia, Penicillium, Pseudomonas,* and *Trichoderma*.
- Pesticide degradation generally follows first-order kinetics, and $t_{1/2}$ ranges from 0.02 to 4 years (Tables 3.2 and 3.3).

TABLE 3.2

Half-Life Period for Different Groups of Pesticides in Soil

Types of Pesticides	Approximate Half-Life (years)
Dieldrin, BHC, DDT	2–4
2,4-D and 2,4,5-T	0.1–0.4
Organophosphate	0.02–0.2
Carbamate	0.02–0.1
Pyrethroids	0.005–0.01

TABLE 3.3

Grouping of Pesticides by Persistence in Soils

Low Persistence (Half-Life < 30 Days)	Moderate Persistence (Half-Life 30–100 Days)	High Persistence (Half-Life > 100 Days)
Aldicarb	Aldrin	Bromacil
Dalapon	Atrazine	Chlordane
Malathion	Carbaryl	Paraquat
Methyl parathion	Diazinon	Trichloroacetic acid (TCA)
2,4-D	Endosulfan	Trifluralin
2,4,5-T	Fenvalerate	Lindane
	Glyphosate	Propazine

References

Atterby, H., Smith, N., Chaudhry, Q., and Stead, D. 2002. The ernvironmental impact of livestock production – Review of research and literature. *Pestic. Outlook* 13: 9–13.

Bansal, O. P. 2004. Effects of various soil factors and amendments on the persistence and degradation of carbamate pesticides. *Pollut. Res.* 23: 507–513.

Bansal, O. P. 2005. Degradation studies of three carbamate pesticides in soils of Aligarh district as influenced by temperature, water content, concentration of pesticide, FYM and Nitrogen. *Proc. Natl. Acad. Sci. India* 75B: 19–27.

Bollag, J. M. and Liu, S. Y. 1990. Biological transformation processes of pesticides. In: Cheng, H. H. (ed.), *Pesticides in the Soil Environment: Processes, Impacts and Modeling, SSSA Book Ser. 2*, pp. 169–211. Science Society of America, Inc.: Madison, WI.

Bondarenko, S., Gan, J., Haver, D. L., and Kabashima, J. N. 2004. Persistence of selected organophosphate and carbamate insecticides in waters from a coastal watershed. *Environ. Toxicol. Chem.* 23: 2649–2654.

Derbalah, A. S. H., Wakatusuki, H., Yamazaki, T., and Sakugawa, H. 2004. Photodegradation kinetics of Fenitrothion in various aqueous medium and its effect on steroid hormones biosynthesis. *Geochem. J.* 38: 201–213.

Dureja, P. and Mukerjee, S. K. 1986. Amine induced photodehalogenation of cycldiene insecticides. *Tetrahed. Lett.* 26: 5211–5212.

Fan, C. and Liao, M.-C. 2009. The mechanistic and oxidative study of methomyl and parathion degradation by fenton process. *World Acad. Sci. Eng. Technol.* 59: 87–91.

Gilani, S. T. S., Ageen, M., Shah, H., and Raza, S. 2010. Chlorpyrifos degradation in soil and its effect on soil microorganisms. *J. Animal Plant Sci.* 20: 99–102.

Hoostal, M. J., Bullerjahn, G. S., and Mckay, R. M. L. 2002. Molecular assessment of the potential for in situ bioremediation of PCBs from aquatic sediments. *Hydrobiologia* 469: 59–65.

Hu, D. and Coats, J. 2008. Evaluation of the environmental fate of thymol and phenyl propionate in the laboratory. *Pest. Manag. Sci.* 64: 775–779.

Kodama, T., Ding, L., Yoshida, M., and Yajima, M. 2001. Biodegradation of an s-triazine herbicide, simazine. *J. Mol. Catal. B: Enzymatic* 11(4–6): 1073–1078.

Kwon, J. W. and Armbrust, K. L. 2006. Degradation of chlorothalonil in irradiated water/sediment systems. *J. Agricul. Food Chem.* 54: 3651–3657.

Liu, W., Zheng, W., Ma, Y., and Liu, K. K. 2006. Sorption and degradation of imidacloprid in soil and water. *J. Environ. Sci. Health B* 41: 623–634.

Miller, L. L. and Narang, R. S. 1970. Induced photolysis of DDT. *Science* 169: 368–370.

Muller, K., Magesan, G. N., and Bolan, N. S. 2007. A critical review of the influence of effluent irrigation on the fate of pesticides in soil. *Agricult. Ecosyst. Environ.* 120: 93–116.

Musa, U., Hati, S. S., Mustapha, A., and Magaji, G. 2010. Dichlorvos concentrations in locally formulated pesticide (ota-piapia) utilized in northeastern Nigeria. *Sci. Res. Essay* 5: 49–54.

Pirnie, E. F., Talley, J. W., and Hundal, L. S. 2006. Transformation of DDT and its metabolites by various abiotic methods. *J. Environ. Eng.* 132: 560–564.

Sarmah, A. K. and Sabadie, J. 2002. Hydrolysis of sulfonylurea herbicides in soils and aqueous solutions. *J. Agricult. Food Chem.* 50: 6253–6265.

Strathmann, T. J. and Stone, A. T. 2000. Abiotic reduction of the pesticides oxamyl and methomyl by Fe(II): Reaction kinetics and mechanism, pp 684–687. Symposia papers presented before the division of Environmental Chemistry, American Chemical Society, Washington, DC.

Strong, L. C., Rosendahl, C., Johnson, G., Sadwosky, M. J., and Wackett, L. P. 2002. Arthrobacter aurescens TC1 metabolizes diverse s-triazine ring compounds. *Appl. Environ. Microbiol.* 68: 5973–5980.

Swati, A. and Singh, D. K. 2002. Utilization of chlorpyrifos by *Aspergillus niger* and *A. flavus* as carbon and phosphorus source. *17th World Congress of Soil Science,* 14–21 August 2001, Bangkok, Thailand.

Tamini, M., Qourzal, S., Assabbane, A., Chovelon, J. M., Ferronato, C., and Ait-Ichou, Y. 2006. Photocatalytic degradation of pesticide methomyl: Determination of the reaction pathway and identification of intermediate products. *Phytochem. Photobiol. Sci.* 5: 477–482.

UN. 2005. *World Population Prospects. The 2004 Revision, Highlights.* Population division, Department of Economic and Social Affairs, United Nations: New York.

Washburn, A. D. 2003. *The Environmental Fate of Methdathion.* Sacramento Department of Pesticide Regulation, Environmental Monitoring Branch.

Zayed, S., Mostafa, I. Y., and El-Arab, A. E. 1994. Degradation and fate of ^{14}C–DDT and ^{14}C–DDE in Egyptian soil. *J. Environ. Sci. Health B* 29: 47–56.

4

Pesticide Degradation in Water

Pilar Sandín-España and Beatriz Sevilla-Morán

CONTENTS

4.1 Routes for Pesticides and Transformation Products to Reach Aquatic Environment

Since many years, intensive agriculture has led to a large-scale use of herbicides, fungicides, and insecticides in developed countries. As a result, some of these pesticides, which are widely used in rural areas for agricultural purposes, constitute important pollutants of natural water. In this sense, environmental contamination of natural water by pesticide residues has been a great concern since 1940 (Coats 1993). Freshwater is considered the most precious of all natural resources. Furthermore, groundwater is the main source of drinking water in many European countries and in the United States (D'Archivio et al. 2007). In rural areas, groundwater is often the only source of water with acceptable quality for human consumption without any treatment.

During the last 30–40 years, there has been a general trend toward the introduction of agrochemicals with lower application rates, decreased environmental persistence, and reduced nontarget organism toxicity. For example, whereas the average rate of use for chlorinated hydrocarbon insecticides was estimated at 3 Kg/ha, more recently introduced pyrethroid or neonicotinoid insecticides may be applied at 0.01–0.1 Kg/ha (Casida and Quistad 1998). Similar trends exist for herbicides also, the use of which, for many crops, has shifted from Kg/ha in the 1950s to g/ha by the 1990s, with herbicides belonging to families such as sulfonylurea and cyclohexanedione oxime.

Complete mineralization of pesticides into inorganic constituents such as carbon dioxide, ammonia, water, mineral salts, and humic substances often occurs slowly in the environment. Different compounds, however, are formed before the pesticides can be completely degraded. The organic compounds formed by the different transformation processes are referred to by several names such as degradates and by-products. The most widely used names are "transformation product" and "degradation product." "Metabolites" is a term usually inappropriately employed, as it should be used only when the transformation product is a result of biological transformation. The products formed by sunlight-induced transformations are known as "photoproducts" or "photolysis products". In general, the term "residue" is used for the parent compound and the by-products formed (Somasundaram and Coats 1991a).

Extensive research has been conducted regarding the occurrence of pesticides in the aquatic compartment and their fate and their effects on human health (Bintein and Devillers 1996; Carbonell et al. 1995; Walls et al. 1996). However, the understanding of all the consequences of pesticide use is limited to the fact that most studies have focused on the parent compound and generally did not consider their transformation products.

Nowadays, some pesticide transformation products are considered as "emerging pollutants" (Richardson 2009; Rodríguez-Mozaz et al. 2007), as most of them have been present in the environment for a long time, but their significance and presence have only now been elucidated and, therefore, they are generally not included in the legislation. In addition, there is still a lack of knowledge about the long-term risks that the presence of these emerging pollutants may pose for organisms as well as for human health. Consequently, transformation products have become a new environmental problem and have raised great concern among scientists in the last few years (Richardson 2006, 2007).

In some cases, these degradation products are more prevalent for water due to their recalcitrant character, polarity, toxicity, and/or continuous input into the environment that favors their spread along water (Coats 1993; Kolpin et al. 2004; Reemtsma and Jekel 2006). Research has shown that pesticide degradates are persistent in groundwater (Green and Young 2006; Kolpin et al. 2000a, 2004; Postle et al. 2004; Spalding et al. 2003) and, sometimes, are more frequently detected than their parent compounds. Therefore, the statement that relatively few detections of parent compounds are observed in natural water provides a false impression that little chemical transport to water occurs from pesticide application. In addition, pesticide degradates can either be less toxic or have similar or greater toxicity than their parent compounds (Belfroid et al. 1998; Chiron et al. 1995; Kawahigashi et al. 2003; Tuxhorn et al. 1986). Thus, obtaining data on parent compounds and their primary degradation products is critical for understanding the fate of the pesticides in the aquatic compartment.

Once the pesticide is applied, the highest initial concentrations are generally present in plant leaves, soil, and water to which direct applications are made. Relatively few pesticide applications are made directly and exclusively to the target pests, and most application methods rely on application of enough pesticide to the environment so that exposure to the pest species reaches efficacious levels. It is known that less than 0.1% of the applied pesticide actually reaches the targeted pests, while the rest, 99.9%, has the potential to move into other environmental compartment, including groundwater and surface water (Racke 2003; Younos and Weigmann 1988).

Once a pesticide is introduced into the environment, physical, chemical, and biological processes make the organic compounds to be transported to groundwater or surface water and/or, in many cases, transformed into reaction products. Figure 4.1 shows the main

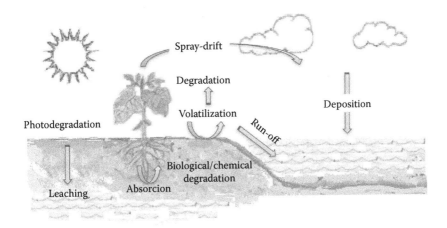

FIGURE 4.1
Main pathways of pesticides to degrade and to enter into aquatic compartment.

pathways for pesticides to degrade and to enter into aquatic compartments once they are applied. Once the transformation products are formed in the environment, they are subjected to the same processes as their precursor.

Pesticides can degrade by three common processes that take place in the surface environment and include the following:

1. Biological degradation, which can lead, to either transformation products or complete mineralization to form inorganic constituents.
2. Chemical degradation (hydrolysis, thermolysis).
3. Photochemical reactions, which require solar energy.

Surface-water concentrations of pesticides are highest for those associated with direct applications to water bodies such as for aquatic weed control. For example, initial herbicide concentrations of 0.2–3 mg/L are generally targeted for aquatic weed. Indirect or unintentional entries of pesticides into water generally result in much lower concentrations, in the range of 0.001–0.01 mg/L or lower (Racke 2003). Such entries can occur due to the following physical processes:

1. Runoff movement in either a dissolved or sorbed state
2. Atmospheric deposition after volatilization
3. Leaching movement as soluble constituents of the aqueous phase
4. Spray drift.

Understanding these dissipation routes is the key to ensuring the accurate assessment of environmental fate and the behavior of pesticides.

Surface *runoff* occurs whenever the rate of water application to the ground surface exceeds the rate of infiltration and the surface storage capacity is exceeded. Losses of chemicals in runoff depends primarily on the duration and intensity of rainfall and watershed characteristics such as soil properties, land use, vegetation cover, soil moisture, and topography (Wolfe 2001). Surface runoff may become more significant under very heavy storms (Bronstert et al. 2002). During runoff, pesticide residues can be both transported in solution and sorbed to soil particles. The magnitude of the latter loss path is generally small compared to that transported in the water phase because of the relative amounts of water moved compared to eroded soil. Losses of molecules with high solubility via surface runoff are more important than losses of molecules with K_{oc} greater than ca. 1000 L/Kg (e.g., synthetic pyrethroids), where erosion is considered to be the main loss pathway (Dabrowski et al. 2002; Holvoet et al. 2007; Wu et al. 2004).

The amount of organic chemicals that move off soil into surface waters is normally less than 1% or 2% of that applied (Bloomfield et al. 2006; Wauchope 1996). Studies of large agricultural watersheds have shown that the atrazine flux in rivers varied between 0.25% and 1.5% of the amount spread on the land (Richardson 2007; Schottler and Eisenreich 1994). This suggests that, under most circumstances, the runoff process plays a minor role in the transport and fate of organic compounds. However, exceptional fields with steep slopes and heavy precipitation can make the runoff a dominant transport process and, in some cases, up to 5% has been measured (Burgoa and Wauchope 1995).

The fate of a pesticide from its point of application through surface soil and into subsoil is governed by the interactive processes of adsorption, transformation, and transport (Weber and Miller 1989). *Leaching* is the vertical movement of the dissolved pesticide residues and transformation products down soil profiles as a result of a downward potential gradient caused by the infiltration of rain at the surface. Xenobiotics dissolved in water tend to move more slowly than the water, as they are subject to sorption to the soil as well as degradation processes. Moreover, slow groundwater movement and pesticide residue attenuation in the groundwater due to sorption and degradation further diminish the residue concentrations (Röpke et al. 2004). The rate of movement of water through the soil profile is controlled by the hydraulic conductivity of the soil, which depends on the type and water content of the soil. Wetter soils have higher hydraulic conductivities and therefore, the organic compounds can move more rapidly from the surface of the fields through the soil. The chemical properties of the soil particles, their distribution and size, and the amount of organic matter will influence the capacity of the soil to retain more hydrophobic compounds. The mobility of the organic compounds through the soil is largely dominated by sorption processes. These processes are controlled by the lipophilicity of the molecules, the chemical properties, distribution, and size of the soil particles, organic matter, and soil humidity. In general, it has been stated that an increase in water content reduces adsorption and increases the mobility, whereas an increase in clay and organic matter content increases adsorption and reduces mobility. Mobilities of some compounds are directly related to pH, showing higher mobility in soils with higher pH (Somasundaram et al. 1991).

Likewise, irrigation practices and rainfall frequency and intensity also influence the leachability of pesticides and transformation products; a higher water input favors the transport of compounds to groundwater. Furthermore, physicochemical properties of pesticides and transformation products play an important role in their transport to the subsoil. Hydrophobic compounds with a high organic carbon partition coefficient (K_{oc}) have a high affinity to be retained in the soil and therefore, their lixiviating only takes place under conditions when residues have a very high half-life.

According to the literature, an organic compound is able to contaminate the groundwater if its solubility in water is higher than 30 mg/L, its adsorptivity K_{oc} (organic carbon partition coefficient) is lower than 300–500, its K_d (distribution adsorption constant) is lower than 5 mL/g, and its soil half-life is longer than 3 weeks (Aharonson 1987; Demoliner et al. 2010).

One relevant parameter linked to chemical properties is the Gustafson Ubiquity Score (GUS) factor (Gustafson 1989), which makes a leaching classification, taking into account soil mobility and soil persistence (Demoliner et al. 2010).

$$GUS = \log\left(t_{1/2\,soil}\right) \times \left(4 - \log K_{oc}\right)$$

Molecules with a GUS above 2.8 are very likely to reach groundwater and those with a GUS below 1.8 are considered nonleacher to groundwater. Molecules with a GUS higher than 1.8 and lower than 2.8 are considered transition leachers.

Volatilization of pesticides and transformation products from soil surface and plant leaves to the atmosphere and further deposition in aquatic environments represent another diffuse source of entry of pesticide residues in surface waters. An increase in temperature increases volatilization as the vapor pressure of most compounds increases with temperature. However, many of the pesticides currently in use have low vapor pressures, and volatilization will not be a significant loss route. Those pesticides known as volatile compounds have special usage instructions to minimize loss by this route, for example, soil incorporation of trifluralin, chloropicrine, and 1,3-dichloropropene.

During the application of pesticides to the field crops, a part of the spray liquid may be carried out of the treated area by wind and reach water bodies (Bach et al. 2001). The extent of contamination of aquatic media is highly dependent on the method of application and on factors such as wind direction and velocity (Gil and Sinfort 2005). *Spray drift* may be an important means of organic compound losses in orchard regions or areas with a network of ditches (Bach et al. 2001; Holvoet et al. 2007). A study revealed an offsite deposition of aerially applied pesticide at 30, 60, and 120 m of 5%, 2%, and 0.8% of the material, respectively (Bird 1995).

Likewise, entry of pesticides and/or transformation products into water bodies may occur via accidental spills during filling of the sprayer, cleaning of the spraying equipment, washing of the measuring utilities, leaks from the sprayer, or improper waste handling and disposal.

Finally, nonagricultural use of pesticides may also represent a significant source of entry of pesticide residues into the environment. In the United States, for example, it has been estimated to account for approximately 25% of pesticide use by volume (Aspelin and Grube 1999). In the urban environment, significant pesticide uses include applications for ornamental pests and for control of aquatic weed management in recreational water, which may involve direct application of herbicides to surface waters (Racke 2003).

4.2 Pesticide Degradation in the Water Compartment

When an active substance and/or its transformation products enter into the aquatic compartment by the different processes detailed in the previous section, it is subject to different physical, chemical, and biological processes. A combination of factors influences the transformation of xenobiotics over a time period of hours, days, or even years and through different processes. Although some pesticides are biodegradable, by the act of microorganisms ubiquitous in environmental waters, this is not the main route of pesticide transformation in the aquatic environment (Boxall et al. 2004). Physical processes in the water environment include dilution by the dispersion or diffusion of the pesticide residue in the water body, its movement to the atmosphere through volatilization from the water surface, and sorption and desorption to the particles present in the water.

Abiotic transformation may include chemical (mainly hydrolysis and thermolysis) and photochemical reactions. Both types of reactions are degradative pathways for many pesticides in the aquatic environment. Information about these degradation routes is necessary to estimate the persistence of these compounds and to identity the factors that influence their behavior in the aquatic media.

Regarding hydrolysis reactions, pH of water is responsible for the transformation of some of the pesticides in solution, especially in conjunction with pH extremes (García-Repetto et al. 1994). Furthermore, organic compounds sensitive to pH give rise to their rapid degradation, even with a slight variation of pH. For example, many organophosphorus pesticides (Dannenberg and Pehkonen 1998; García-Repetto et al. 1994; Santos et al. 1998), carbamate pesticides, (Sanz-Asencio et al. 1997), and pyrethroid pesticides (Al-Mughrabi et al. 1992) present a chemical structure that makes them susceptible to hydrolysis in an aqueous media, and therefore, they have a low environmental persistence. The cleavage of the ester group, present in these classes of pesticides, leads to the formation of two new degradation products.

Photochemical reactions are one of the major transformation processes affecting the fate of the pesticides in natural water (Dimou et al. 2004a; Marcheterre et al. 1988; Neilson and Allard 2008; Saha and Kulshrestha 2002). Two ways of photodegradation reactions occur in sunlit natural water. In direct photolysis, organic compounds absorb light and, as a consequence, undergo transformation. For this to occur in water, the Sun's emission (290–800 nm) needs to fit the adsorption spectrum of the pesticide. In indirect photochemical reactions, organic chemicals are transformed by energy transfer from another excited species (e.g., components of natural organic matter) or by reaction with very reactive, short-lived species formed in the presence of light (e.g., hydroxyl radicals, single oxygen, ozone, and peroxy radicals). Absorption of actinic radiation by nitrate and dissolved organic matter (DOM) leads to the formation of most of these species. Therefore, the composition of the aquatic media plays an important role in the phototransformation of pesticides in this compartment. Dissolved substances present in natural waters are responsible for the different photolysis rates of pesticides observed between natural and distilled water (Dimou et al. 2005; Durand et al. 1991; Sevilla-Morán et al. 2010b).

The hydroxyl radical, $OH^{\cdot}$, is the most reactive of the aforementioned reactive intermediates due to its nonselective and highly electrophilic nature. Its concentration in surface water is reported to be at 10^{-14}–10^{-18} M (Brezonik and Fulkerson-Brekken 1998). The types of reactions where $OH^{\cdot}$ is involved include H-abstraction and addition to double bonds.

Different parameters affect the photolysis of pesticides in the aquatic environment. Thus, the photodegradation behavior of many pesticides in aquatic media has been investigated under various experimental conditions such as different types of aqueous matrix (ground, surface, and mineral water) (Konstantinou et al. 2001; Lin et al. 1999; Sakkas et al. 2002a), light sources (natural sunlight, xenon arc lamps, etc.) (Elazzouzi et al. 1999; Sakellarides et al. 2003; Zamy et al. 2004), and light intensities, as well as under the influence of substances dissolved in natural water (humic acid (HA), nitrate, and iron ions) (Bachman and Patterson 1999; Dimou et al. 2004b; Wilson and Mabury 2000). Ideally, solar radiation should be used in the studies of environmental photochemistry; however, meteorological conditions in most countries and a slow degradation rate do not permit reproducible experimentation. In a first approach to the study of photochemical behavior of organic compounds in different matrices, it is common to conduct the degradation under controlled conditions. Generally, the use of xenon arc lamp, with light above 290 nm (provided by a filter), is preferred as its spectral emission distribution is very close to the solar radiation spectrum (Marcheterre et al. 1988). It has been demonstrated that the use of different light sources under identical aqueous conditions can produce similar degradation products, with the only difference being in their kinetics of formation (Barceló et al. 1996; Marcheterre et al. 1988). As for the experimental equipment, quartz glass is preferred instead of other glass material since it permits a greater transmission of radiation (Peñuela et al. 2000).

The photodegradation of organophosphorus, phenylurea, and cyclohexanedione oxime pesticides has been studied in a variety of different aquatic media. In many cases, significant differences in the degradation rates were obtained, suggesting a strong dependence on the type of the aquatic media (Durand and Barceló 1990; Durand et al. 1990).

Sevilla-Morán and coworkers (2010b) investigated the photodegradation of the herbicide sethoxydim-lithium in ultrapure and natural waters (mineral, well, and river). Half-lives of the herbicides were higher in natural waters than in ultrapure waters, with the photolysis rate decreasing in the following order: river < well ~ mineral < ultrapure water. Sethoxydim-lithium almost totally degraded after 5 h in ultrapure water versus 10 h in river water. The composition of aquatic media also plays an important role in the phototransformation of pesticides. Various authors point out that particulate matter, such as sediment particles, and dissolved substances present in natural waters could be responsible for the different photolysis rates observed between natural and distilled water (Dimou et al. 2004a; Schwarzenbach Rene et al. 2002; Tchaikovskaya et al. 2007). The most important light-absorbing species that may induce indirect photolytic transformation of organic pollutants in natural waters are the chromophores present in DOM, where HAs (Figure 4.2) and, to a lesser extent, fulvic acids (FAs) are the important absorbing constituents. The complex structure of HA is basically composed of phenolic, carbonylic, and carboxylic acids, and they absorb radiation in the range of 300–600 nm (Sevilla Morán 2010). Their concentration in natural waters ranges from less than 5 ml/L in groundwater and sea water to 25 mg/L in some rivers (Barceló and Hennion 1997; Dimou et al. 2005).

Diverse studies are available from the literature, in which HA act by enhancing (Sakkas et al. 2002a,b; Santoro et al. 2000; Vialaton and Richard 2002) or inhibiting (Bachman and Patterson 1999; Dimou et al. 2004a, 2005; Elazzouzi et al. 1999; Sevilla-Morán et al. 2008, 2010a) the degradation of pesticides. In the first case, HA behave as a "sensitizer," in which the excited states of HA can participate in a charge-transfer interaction with pesticides or generate reactive intermediates, such as hydroxyl radicals, singlet oxygen, solvated electrons, or hydrogen peroxide. In the second case, HAs act as photon trap (optical filter effect), decreasing the photodegradation rate of pesticides.

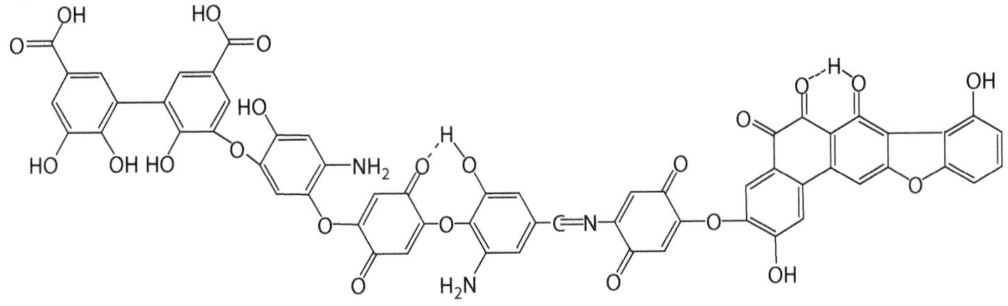

FIGURE 4.2
Chemical structure of humic acids.

The influence of different concentrations of HA in the photodegradation of alloxydim has been studied by Sevilla-Morán and coworkers (2010a). Photolysis of this cyclohexane-dione herbicide, in the presence of different concentrations of HA (1–20 mg/L), showed that increasing concentrations of HA decrease the photolysis rate of this herbicide, indicating that it absorbed most of the emitted photons, thereby slowing down the direct photochemical reaction of alloxydim. On the contrary, Sakkas and coworkers (2002a) studied the photodegradation of the fungicide chlorothalonil in the presence of HA and FA under simulated solar irradiation. Results showed that, for both substances, an increased concentration increased the photodegradation rate of chlorothalonil.

In addition to DOM, there are other water constituents that influence the photolysis of pesticides. The most prominent examples are nitrate (NO_3^-), nitrite (NO_2^-), and iron ions (Fe(II) and Fe(III)). NO_3^- concentrations in the environment range between 1 and 30 mg/L, whereas NO_2^- concentrations are lower (less than 1 mg/L) (Somasundaram and Coats 1991b; Wilson and Mabury 2000). In many freshwaters, photolysis of NO_3^- and NO_2^- appears to be the major source of OH•. In this sense, pesticides such as metolachlor (Dimou et al. 2005), carboxin (DellaGreca et al. 2004), and diuron (Shankar et al. 2007) showed an increase in the degradation rates. However, not always does the presence of nitrate accelerate the degradation of pesticides; and no effect was observed in the photodegradation of other pesticides (Andreozzi et al. 2003; Chaabane et al. 2007; Sevilla-Morán et al. 2008, 2010a). The irradiating power available in solution can be reduced when the pesticide itself strongly absorbs in the same UV range, diminishing the amount of radiation absorbed by nitrate ions.

The chemical transformation of pesticides is mediated by different types of reactions such as oxidation, hydrolysis, reduction, elimination, cyclation, and rearrangements.

Bavcon and coworkers (2007) studied the aqueous degradation of different organophosphorus compounds. Oxidation of chlorpyrifos to chlorpyrifos-oxon was the main degradation route of the pesticide (Figure 4.3). The oxidative desulfuration is very common in compounds that present a P=S group. Oxidation pathways are extremely important transformation reactions.

The sulfoxidation of sulfides to sulfoxide and sulfone (sulfoxidation) is often rapid and one of the most important degradation pathways in pesticides containing a sulfur atom (Somasundaram and Coats 1991b). Pesticides such as aldicarb, carboxin, disulfoton, fenamiphos (Miles 1991), and clethodim (Sevilla-Morán et al. 2010a) are known to degrade to their corresponding sulfoxide and/or sulfone in environmental waters. Photocyclization after HCl elimination was one of the main reaction pathways of the photolysis of propiconazole in pure water at 254 nm (Vialaton et al. 2001). Detomaso and coworkers (2005a,b) identified up to seven photoproducts during high-pressure UV degradation of carbofuran in aqueous solution, mainly resulting from photo-Fries rearrangement, hydroxylation

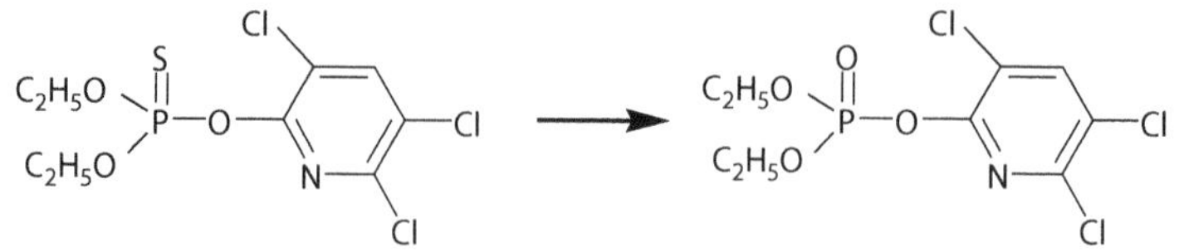

Chlorpyrifos　　　　　　　　　　　　　Chlorpyrifos-oxon

FIGURE 4.3
Main photodegradation products of chlorpyrifos.

of the benzene ring, oxidation, cleavage of a carbamate group, and hydrolysis of the ether group.

The aquatic photochemistry of trifluralin in natural water (Dimou et al. 2004b) was investigated under simulated solar irradiation (Suntest apparatus). Photodecomposition of trifluralin generally involves three major routes: oxidative dealkylation of propylamines, cyclization, and reduction of nitro groups.

Consequently, information about photodegradation of pesticides is necessary to estimate their persistence and to identify the factors that influence their behavior in the environment. Furthermore, it is important to investigate what these compounds degrade into, the persistence of the by-products relative to the parent compounds, and whether the degradation products retain the activity of the active substance to cause a toxicological effect on nontarget organisms in aqueous systems.

4.3 Environmental Relevance of Transformation Products

4.3.1 Properties Affecting the Fate of Transformation Products

Many pesticides applied to the field are completely degraded or mineralized into innocuous compounds. However, different transformation products are formed before the pesticide is complete degraded. Sometimes, one or two transformations in its molecular structure are enough to modify its properties. In fact, for some pesticides, biological activity and/or environmental contamination attributed to the parent compound can be due to the degradation products.

Historically, some of the most serious concerns about the safety of pesticides have been raised from its transformation products which can cause detrimental side effects. The majority of the environmental concerns about 1,1,1-trichloro-2,2-bis(*p*-chlorophenyl)ethane (DDT) were attributable to one of its breakdown products, 1,1-dichloro-2,2-bis(*p*-chlorophenyl) ethylene (DDE). This compound lacked insecticidal activity, but it was extremely persistent, mobile in the environment, and bioaccumulative.

Different studies confirm that many degradation products are more mobile and some others are more persistent than their respective parent compound (Boxall et al. 2004; Coats 1993; Khan and Saidak 1981; Richards and Baker 1993).

Some physicochemical properties of transformation products are characteristic of their own molecule, such as water solubility, vapor pressure, volatility, and water–octanol partition coefficient. However, other parameters such as half-life in soil and water compartments depend not only on the chemical structure of the molecule but also on the environmental conditions. However, calculation of these properties is sometimes difficult due to the absence of analytical standards.

As mentioned before, the different abiotic/biotic reactions that take place in the environment modify the physicochemical properties of the parent molecule. Biotic transformation processes generally produce transformation products that are more polar and water-soluble than the parent compound. Organisms transform oxidized xenobiotics to more soluble and polar molecules to facilitate their elimination from their organism. Hence, the resulting transport behavior of metabolites may be different.

Most of the oxidative reactions (hydroxylation, sulfoxidation, dealkylation, etc.) and hydrolysis contribute, also to some degree, to the increase in polarity and hence, water solubility of the molecule. Therefore, the new xenobiotics are more mobile in the soil.

Reduction reactions are characterized in environments with low oxygen concentrations, low pH, and anaerobic microorganisms. These reactions are less commonly observed and generally give rise to products with lower polarity.

Pesticides with reactive moieties in their structures are known to give rise to degradation products of environmental concern. For example, sulfide pesticides undergo oxidation of the sulfur atom and parent compound is converted into products that are more water-soluble and thus more mobile in soil. Thus, sulfoxidation is one of the most important pesticide degradation pathways since the sulfoxide product has a much higher groundwater contamination potential (Bavcon Kralj and Trebše 2007; Lacorte et al. 1995; Miles 1991). In soil, sulfide pesticides are often rapidly oxidized to sulfoxides, while sulfones are formed more slowly (Miles 1991) and for this reason, sulfoxides are the major compounds found in soil. This oxidation is so rapid and complete that sulfoxides are often the dominant species found in soil, shortly after application of the parent compound. In most cases, sulfoxide and sulfone also have pesticidal activity. Some sulfur pesticides known to suffer this reaction are aldicarb, malathion, parathion, methomyl, fenamiphos, and methiocarb (Barceló et al. 1996; Dabestani et al. 2000; García de Llasera and Bernal-González 2001; Priddle et al. 1992; Wang and Hoffman 1991). Recent studies indicate that about one third of the degradation products derived from a range of pesticide types have an organic carbon absorption coefficient (K_{oc}) of at least one order of magnitude lower than that of the corresponding parent compound. Thus, these transformation products may be more likely to be transported to surface waters and groundwater.

Nowadays, great efforts are being made to predict the movement of pesticides and transformation products through the soil to the groundwater. However, physicochemical data of the degradation products are not always available, and different methods are used to estimate the unknown parameters. Quantitative structure–activity relationships (QSAR) permit to calculate the physicochemical properties of a molecule from its chemical structure, such as K_{oc}, Henry´s law constant, and persistence. This data would be used in models for estimating the exposure or for assessing the ecotoxicity (Cronin and Livingstone 2004).

There are several basic physicochemical properties of the transformation products that allow partly assessing their behavior and fate in the environment. Water solubility indicates the tendency of a by-product to be removed from soil, for example, by runoff, and to reach the surface water. This parameter cannot by itself be used to predict its mobility in groundwater. The processes of degradation of pesticides generally increase the water solubility and the polarity of the compounds and, as a consequence, favor their mobility in groundwater. The increase in solubility is caused by the loss of carbon, the incorporation of oxygen, and the addition of carboxylic acid functional groups. For every carbon atom that is removed, the water solubility increases two or three times. For example, atrazine solubility is 33 mg/L; deethylatrazine solubility, which implies the loss of two carbon atoms, is 670 mg/L; and deisopropylatrazine solubility, which has lost three carbon atoms, is 3200 mg/L (Mills and Thurman 1994). Addition of oxygen in the molecule of aldicarb, by sulfoxidation, gives rise to sulfoxide and sulfone metabolites that are 55 and 1.4 times more soluble, respectively, than the parent aldicarb (Somasundaram and Coats 1991b). Increases in water solubility is dramatic when the degradation product is a carboxylic acid, which can occur when there is an alkyl chain, a ketone group, or an aldehyde group that can be degraded (Hornsby et al. 1996).

Water-soluble products are frequently the primary metabolites formed by biological and chemical oxidations and hydrolysis in the soil. Those polar molecules demonstrate a greater tendency to leach into and to be dissolved in the runoff water, as do compounds with low binding constants.

Water–octanol partition coefficient (K_{ow}) is characteristic of the lipophilicity of the molecule and indicates that it may accumulate in living organisms. In terms of polarity, log K_{ow} above 4–5 is specific for nonpolar compounds, whereas log K_{ow} below 1–1.5 corresponds to polar compounds. Between these two values, compounds are classified as moderately polar.

Other important factors are the *vapor pressure* of the transformation product and the *Henry law constant* (H), which allow prediction of the by-product volatilization. It is generally considered that compounds with values below 10^{-5} Pa m^3/mol have little tendency to volatilize. The vapor pressure of pentachlorocyclohexene (PCCH), a metabolite of lindane, is 14 times greater than that of lindane (Cliath and Spencer 1972). Therefore, a majority of the lindane volatized from soil would be in the form of PCCH.

The soil sorption coefficient (K_{oc}) represents the compound distribution between the soil and liquid phases of the soil. This parameter has a great environmental relevance for assessing the potential leaching of pesticides and their transformation products in soil. This coefficient is normalized as a function of the organic carbon content, which plays a very important role for the nonionized compounds at natural pH in soils. The higher is the K_{oc}, the more sorbed is the molecule. It is considered that compounds with K_{oc} values below 50 are considered to be highly mobile; values of 150–500 signify moderately mobile, and above 2000, slightly mobile compounds. Half-life is another factor to predict the fate of metabolites in soil and water. As mentioned before, there is not a single half-life for a transformation product, and the measurements depend on the environmental conditions. However, laboratory studies permit to carry out experiments under controlled conditions.

Recent studies have shown that some of the degradation products of the pesticides from different chemical classes (carbamates, triazines, organophosphorus compounds, and sulfonylureas) exhibit half-lives in soil higher than those of their degradates, thus being more persistent than their corresponding parent compounds (Boxall et al. 2004). Half-life and K_{oc} are related to the GUS factor (Gustafson 1989), which is used to assess the leaching potential of a compound.

4.3.2 Biological Activity of Transformation Products

Except for propesticides designed to be converted after application to a new compound with pesticidal activity, the activity of transformation products on target pests is generally significantly lower than that of the parent compound. However, in some cases, the degradation products can be more toxic (Sinclair and Boxall 2003). Furthermore, evidence shows that for some pesticides, the pesticidal activity attributed to the parent compound is partly due to the products formed (Bresnahan et al. 2004; Tuxhorn et al. 1986). In some cases, pesticides are formed as degradation products of other pesticides. For instance, chlorthiamid, a benzonitrile herbicide, is the parent compound and the precursor of dichlobenil, which is a degradation product formed in soil and which is also a herbicide. Boxall and coworkers (2004) showed that among different classes of pesticides and their transformation products, 41% of the transformation products were less toxic than the parent compounds, and 39% had a toxicity level similar to that of their parents, but 20% were over 3 times more toxic, and 9% were over 10 times more toxic than their parent compounds. In general, the biggest increases in toxicity were observed for parent compounds that had a low toxicity. Four hypotheses tried to explain these results: (a) the active moiety of the parent compound is still present in the degradation product; (b) the transformation product is the active component of a propesticide; (c) the bioconcentration factor of the degradation product is greater than that of the parent; and (d) the transformation results in a compound with a more potent mode of action than that of the parent.

Oxidation reactions occur frequently in the soil and are extremely important in the transformation pathway. Sulfur-containing herbicides are often rapidly oxidized to sulfoxide and afterward more slowly to sulfones. Sulfoxidation can occur in soil and water mediated chemically or biologically (Ankumah et al. 1995; Bronstert et al. 2002; Turgut 2007).

This oxidation is so rapid and complete that sulfoxides are often the compounds found in soil, shortly after application of the parent sulfide compound. Furthermore, in some cases, sulfoxides and sulfones are suspected to have the pesticidal activity (Campbell and Penner 1985).

Not only degradative reactions but also rearrangements in the molecule can alter the toxicity of active substances. For example, isomers of malathion and fenitrothion are several orders of magnitude more active inhibitors of acetylcholinesterase than the parent compound (Ryu et al. 1991).

Organophosphate and organosulfur insecticides commonly have degradation products with insecticidal activity, often of greater potency than the parent compounds. Sulfoxide and sulfone of sulfide insecticides are usually quite active on a spectrum of pests similar to the parent compounds. Formation of aldicarb sulfoxide and sulfone is an example of this. This sulfoxidation increases the acetylcholinesterase-inhibiting capacity of the organophosphate compounds (Miles 1991). Even though few insecticides have previously been found to extensively impact the groundwater, the potential for persistent and mobile degradation products to impact the groundwater may be significant.

Organophosphorus insecticides have traditionally been designed to undergo transformation reactions that activate them to more potent forms in organisms or environment. The thione moiety (P=S) is oxidized to the oxon form (e.g., malathion to maloxon). This is often three orders of magnitude more potent as an inhibitor of acetylcholinesterase. In most cases, this toxic form is considerably less stable in the environment. Ethylene bisdithiocarbamate (EBDC) and daminozide are two fungicides whose degradation products, ethylene thiourea (ETU) and dimethylhydrazine (UDMH), are known to be carcinogenic (Kobayashi 1992).

Some degradation products of pesticides have shown toxicity to aquatic organisms. Herbicides such as alachlor, metolachlor, and diuron showed that the toxicity to the bacteria, *Vibrio fischeri*, was enhanced upon degradation (Osano et al. 2002; Tixier et al. 2000, 2001). Degradation of acifluorfen shows toxicity to *Daphnia magna* after 36 h of irradiation, where 98% of acifluorfen is degraded (Scrano et al. 2002). On the contrary, other studies showed that herbicides and their degradation products cannot be considered a risk to the environment. This is the case of some sulfonylurea herbicides, in which neither the active substance nor the metabolites are toxic to *D. magna* and *V. fischeri* (Martins et al. 2001; Vulliet et al. 2004).

Major degradation products of some herbicides also have herbicidal activity against target and/or nontarget pests. Some herbicide degradation products are of significance in crop protection by being effective against the target pest. It has been demonstrated that the formation of the sulfoxide by-product of thiocarbamate herbicides, such as butylate, increased the herbicidal activity (Tuxhorn et al. 1986). On the contrary, some can be responsible for inadequate weed control by inducing rapid degradation of their parent compounds.

The herbicidal activity of carbamothiate herbicide sulfoxides has been previously reported (Warner and Morrow 2007). In soils treated with butylate, herbicide residues of the parent compound were not detected in significant amounts within a few weeks after application. However, good control of weeds was observed in these fields. The good performance of this herbicide, despite its lack of persistence, was probably due to the by-products formed. Another degradation product that is effective in controlling target

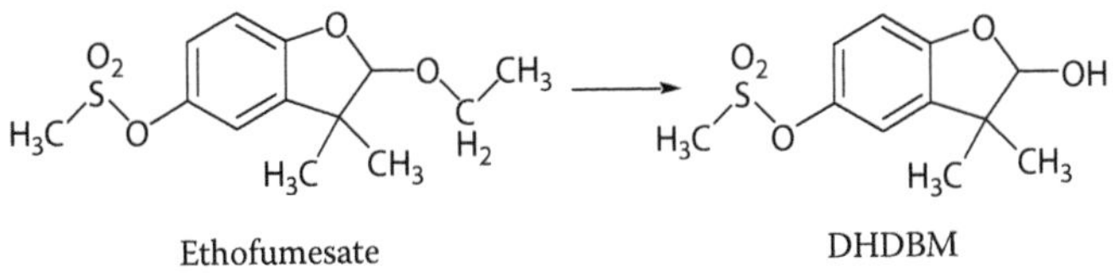

FIGURE 4.4
Chemical structures of ethofumesate and its degradation product DHDBM.

weeds is S-ethyl-N,N-dipropylthiocarbamate (EPTC) sulfoxide, which is the oxidation degradation product of the herbicide thiocarbamate S-ethyl-N,N-dipropylthiocarbamate (EPTC) (Somasundaram and Coats 1991a).

Just as herbicides can be selective among plant species, metabolites can differ in their phytotoxicity pattern. Metabolites can have different mechanisms of action and selectivities than the parent compounds. For instance, bromoxynil is biologically degraded in soil into 2,6-dibromophenol, which is a potent growth regulator (Frear 1976). Kawahigashi and coworkers (2002) showed that the phytotoxicity of the deethylated metabolite of ethofumesate 2,3-dihydro-2-hydroxy-3,3-dimethyl-5-benzofuranyl methanesulfonate (DHDBM) (Figure 4.4) to rice plants was at least four times greater than that of the parent compound.

Reddy and coworkers (2004) suggested that injury to glyphosate-resistant soybean from glyphosate is due to its degradation product formed in plants, aminomethylphosphonic acid (AMPA). The degradation product of irgarol 1051, M1 in the root-elongation inhibition bioassay, showed phytotoxicity at least 10 times greater than that of irgarol and six other triazine herbicides (Okamura et al. 2000).

For herbicides, bioassays are important tools to screen herbicide residues and can be useful to exclude the occurrence of low levels of phytotoxic residues in soil (Hsiao and Smith 1983; Sandín-España et al. 2003). In this sense, Sandín-España and coworkers studied the phytotoxicity of alloxydim, and its main metabolite was studied with hydroponic bioassays on wheat (Sandín-España et al. 2005). The effect of the degradation product of alloxydim (Figure 4.5) on root growth occurred at 10 mg/L, causing a 32% reduction in root growth. Root system control presented normal growth (main tap root plus secondary roots), while those from injured plants were increasingly deformed (main tap root twisted and lack of secondary roots).

It is also important to highlight that a part of the degradation products formed in the soil may remain as bound residues (Albers et al. 2008; Bresnahan et al. 2004; Rice et al. 2002). This nonextractable residue retained by organic matter in soil is bioavailable. Therefore, this portion of residue as degradation products and/or metabolites is

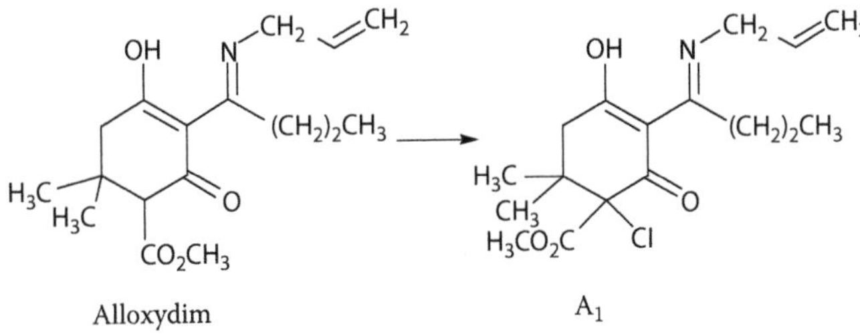

FIGURE 4.5
Chemical structures of alloxydim and its chlorinated degradation product A1.

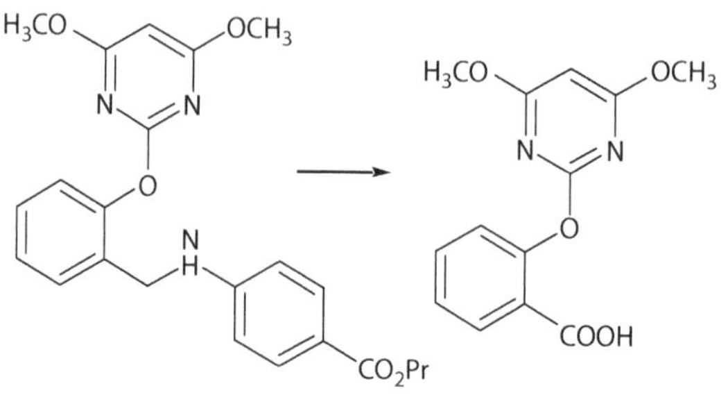

ZJ0273　　　　　　　　　　　　　　M2

FIGURE 4.6
Herbicide ZJ0273 and its main metabolite M2.

underestimated if bound residues may be released from soil. For example, a study on the phytotoxicity of soil-bound residues of the herbicide ZJ0273, a novel acetolactate synthase potential inhibitor, to rice and corn, revealed that one of its main metabolite (M2) (Figure 4.6) played a dominant role in the inhibition effect on the growth of rice seedlings. In the extractable residues released from bound residues, the most biologically active M2 accounted for the largest fraction in all soils. Therefore, it was concluded that the main cause of phytotoxicity from exposure to soil-bound residues of ZJ0273 is related to the release of ZJ0273 and its degradation products and the subsequent inhibition of acetolactate synthase (ALS) by M2 (Han et al. 2009).

4.4 Legislative Framework of Pesticide Residues in Water

In the scope of legislation about pesticide residues in water, it is necessary to differentiate between groundwater and surface water. Generally, there is less information on the presence of pesticides in groundwater than in surface water. In addition, in contrast to food analysis, where there is more precise and extensive legislation about the definition of residues and maximum acceptable residue levels, the regulation of residues in water is still poor with respect to the definition of residues and only gives general reference to pesticide-related compounds. Although it is worth mentioning that the legislation has evolved during the last few years to increasingly consider metabolites and degradation products as part of pesticide residues to be considered for risk assessment and monitoring.

4.4.1 European Union Water Legislation

In 2002, in the European Union (EU), the Water Framework Directive (WFD) (European Commission 2000) established a new legislative approach to manage and protect water by the coordination of different EU policies, based on geographical and hydrological formations (river basins), and made it the responsibility of Member States to draw up river basin management plans. The WFD takes account of all aspects of water use and consumption. The objective is that all European waters have to achieve "good ecological and chemical status" to protect human health, water supply, natural ecosystems, and biodiversity. The directive sets out a precise timetable for action, with 2015 as the target date. That means that not only low levels of chemical pollution but also sustaining healthy aquatic ecosystems have to be achieved. In this sense, the WFD is complemented by other legislations regulating specific aspects of water.

Particularly in terms of chemical pollution, there are differences between surface water and groundwater. Regarding surface water, to define good chemical status, environmental quality standards have to be established for chemical pollutants of high concern across the EU, that is, the concentration of these compounds should not exceed in order to protect human health and the environment. The rules for groundwater are slightly different; groundwater should not be polluted at all and any pollution must be detected and stopped.

Groundwater is the main source of public drinking water. The EU Groundwater Directive (GWD) 2006/118/EC (European Commission 2006) on the protection of groundwater against chemical pollution and deterioration was developed under Article 17 of the WFD and sets out criteria for the assessment of the chemical status of groundwater. The GWD establishes groundwater quality standards for pesticides. For individual pesticides, including active substances and their relevant metabolites and degradation and reaction products, the quality standard is 0.1 µg/L and for the total pesticides, it is 0.5 µg/L. Furthermore, the directive specifies that "total" means the sum of all individual pesticides detected and quantified in the monitoring procedure, including their relevant metabolites, degradation, and reaction products. They are the same criteria of the previous Drinking Water Directive (DWD) (European Commission 1998), where not only the active ingredient but also its relevant metabolites and degradation products had already been considered "pesticide." These general quality standard values are not based on any scientific findings but mean that pesticides should not be present in water or be below the detection limit. It is not possible to determine how many pesticides have to be considered in total as this will vary among catchment areas and could add up to hundreds of pesticides.

But there is an additional input in the GWD; if the groundwater quality standards for pesticides are not sufficient to meet the objectives of the WFD, for example, if there is an identified risk for associated terrestrial ecosystems or for other legitimate uses or functions of groundwater, Member States need to establish more stringent values, known as threshold values. Two criteria can be considered when deriving threshold values: environmental criteria when the protection of associated aquatic ecosystems and groundwater-dependent terrestrial ecosystems is the concern; and usage criteria, where the threshold values aim at the protection of drinking water in Drinking Water Protected Areas. The list of threshold values established by Member States should be regularly reviewed within the river basin management planning framework. As a consequence, additional substances can be considered in case of new identified risks, or some substances can be deleted in case formerly identified risks no longer exist.

The GWD also states that there should be consistency with other EU legislation and, in particular, for pesticides, consistency with Council Directive 91/414/EEC (European Commission 1991) concerning the placing of plant protection products on the market has to be guaranteed. This directive was substituted by the regulation 1107/2009 (European Commission 1991) in June 2011 by which the directive was revised and improved.

The strategy against pollution of *surface waters* is set out in Article 16 of the WFD (European Commission 2000), which requires the establishment of a list of priority substances and a procedure for the identification of priority substances/priority hazardous substances as well as the adoption of specific measures against pollution with these substances. The objective is to protect the aquatic environment from effects such as acute and chronic toxicity and to achieve good ecological status and healthy ecosystems. The Commission will propose a list of priority substances that shall be prioritized for action on the basis of risk to or via the aquatic environment and will set environmental quality standards for these substances. The prioritization will be based on a risk assessment carried out, following the relevant Community legislation regarding hazardous substances or relevant international agreements, particularly under Council Directive 91/414/EEC for assessment of pesticides, and Member States may need to ensure that additional pollutants of national relevance are controlled. The list has to be reviewed from time to time, and it was last amended by Directive 2008/105/EC (European Commission 2008). In this case, different environmental quality standards can be set for every pesticide in surface water. The situation is quite different for pesticides in groundwater, in which a general quality standard is set.

The WFD also sets a common approach for monitoring water quality across all Member States but does not specify the methods to be used. It is up to the Member States to decide the best method based on local conditions and existing national approaches. Where pesticide controls include a review of the relevant authorizations issued under Directive 91/414/EEC, such reviews shall be carried out in accordance with the provisions of this Directive.

The risk assessment conducted in the framework of directive 91/414/EEC (European Commission 1991) at EU level for the authorization of plant protection products is aimed to guarantee that the residues, generated when good agricultural practices (GAPs) are used, have no hazard for human or animal health and for groundwater and the environment. Since its publication, this directive has been repeatedly amended. In particular, Directive 95/36/EC (European Commission 1995) develops data requirements in relation to the fate and behavior in the environment; Directive 96/12/EC (European Commission 1996) develops data requirements in relation to the ecotoxicological behavior; and Directive 97/57/EC (European Commission 1997) lays down the Uniform Principles for the evaluation. In the frame of these directives, different guidance documents have been developed in order to harmonize the evaluation process. The identification of relevant metabolites is one of the tasks to be considered in the risk assessment. It is important to point out that, within the evaluation, a residue definition is set for the risk assessment, but for monitoring purpose, the residue definition can be simpler for a practical enforcement.

4.4.2 Other International Legislation

All over the world, national governments have established guidelines and residue limits for environmental waters (groundwater and surface waters), although most attention has been given to the set of limits for drinking water. The World Health Organization (WHO) published International Standards for Drinking Water from 1958 to 1982, when it started the publication of Guidelines for Drinking-Water Quality (WHO 2010). The guidelines are based on a system called "Water Safety Plans" that integrate risk assessment and risk management and allow the establishment of particular national standards and regulations in each country. WHO uses a different approach than the EU legislation and it set guideline values for a large number of individual pesticides in drinking water.

The International Union of Pure and Applied Chemistry (IUPAC) has also published "recommendations" for the establishment of regulatory limits for pesticide residues in water (Hamilton et al. 2003). This IUPAC technical report also gives an overview of the water legislation in different countries. In some cases, the residue limit in water is expressed as the sum of the parent compound and toxicologically relevant transformation products, when it is derived from a toxicological risk assessment, but for monitoring purpose, it is more convenient to set a limit for the parent compound or a marker residue in order to take a simple residue definition for enforcement purpose. But it is also suggested that in certain cases, when the degradation products are separated from the parent compound in the environment because of different mobilities, separate limits should be set for degradation products.

In the United States, the threshold values for pesticide residues in water (standards) are legally enforceable by agencies of the US government, for example, Environmental Protection Agency Maximum concentration levels (EPA MCLs) for drinking water. Federal guidelines for these values are provided as advice or recommendation for the protection of human health and/or aquatic organisms. Nowell and Resek (1994) provided a detailed review of US standards for pesticide residues in water.

The overall legislation and guidelines for pesticide residues in water are derived from various criteria in different countries, and the harmonization of criteria is a difficult task. Although valuable progress has been achieved in the last few years to guarantee a good quality of the environmental water and drinking water, it is still necessary to work harder to improve the enforcement of legislation in view of the scientific progress.

4.5 Occurrence of Transformation Products in Environmental Waters

Both, surface waters and groundwater are the main sources for drinking water production (Ibáñez et al. 2004). In many countries, drinking water is obtained from groundwater, which undergoes only simple treatment, such as aeration and filtration. Consequently, the numerous detections of pesticides in groundwater constitute a severe threat to the drinking water supplies (Barceló and Hennion 1997; Tuxen et al. 2000). The growing concern of preserving the quality of groundwater and surface water promoted the research on the presence of pesticides in natural waters in the 1980s. However, until recently, little research has been available on the residues of degradation products in waters, and monitoring studies have only focused on the parent compounds. This limited the understanding of the contamination of waters by pesticides. The detection of some of the degradation products of some pesticides in groundwater stimulated research on these xenobiotics as a contaminant of water set aside for human consumption.

Agrochemical industries focus the synthesis of new pesticides toward degradable compounds since the ban of recalcitrant pesticides such as organochlorine DDT and lindane. Many of the replacement pesticides are organophosphorus, pyrethroid, or urea compounds that, in general, degrade in the environment in a relatively short period of time.

However, this tendency presents some disadvantages. Transformation products generated in the environment have, in general, little biological activity on target pests, but these products are often more mobile and can be more persistent than their parent compounds, leading to concerns about water contamination. It is known that the most frequent reactions in the environment increase the polarity of the molecule and, therefore, the tendency to leach into groundwater or to be dissolved in the runoff water. This is particularly applicable to some of the soil-applied herbicides.

Overall, monitoring data for pesticide degradation products are still relatively sparse. Table 4.1 summarizes monitoring data, where pesticide degradation products were detected and quantified in groundwater and surface water. Most studies include only a few known degradation products of some pesticides. In some cases, these degradation products are obvious to be included due to their toxicity on target or nontarget pests, such as aldicarb, sulfoxide, and sulfone (see Table 4.1).

In some monitoring studies, degradation products occurred in groundwater in which the parent compound was no longer detected (Spalding et al. 2003). These findings reveal that degradation products of some pesticides are important contaminants of groundwater.

With the exception of some insecticides such as aldicarb, endosulfan, and methiocarb, the large majority of detections of currently used pesticides have been of the soil-applied herbicides (Sancho et al. 2004), as it can be observed in Table 4.1.

Triazine herbicides, particularly atrazine and its degradation products, are among the most monitored and the most commonly detected pesticides in groundwater. They have moderate soil mobility and relatively high soil persistence. The degradation of atrazine through hydrolysis, UV radiations, and microbial activity leads mainly to the formation of dealkylated and hydroxylated species. As a result of the widespread use of atrazine and the persistence of its degradation products, these have been frequently detected in surface waters and groundwater. The determination of atrazine and its degradation products in water is of huge interest as these compounds are suspected to be carcinogenic. Di Corcia and coworkers (1997) carried out a survey of atrazine and its degradation products by analyzing three river (Tiber) water samples collected between June and August 1995. Except for deethyl-deisopropyl-atrazine (DEDIA) (Figure 4.7 and Table 4.1), all the other atrazine metabolites were present in the three samples analyzed. Therefore, these data confirm that even deisopropyl-hydroxy-atrazine (DIHA) and deethyl-hydroxy-atrazine (DEHA) (Figure 4.7) can contaminate surface waters. Concentration levels of atrazine measured in the Tiber river were only three to four times lower than that measured 8 years before, and the metabolite concentrations were comparable with that of the parent compound. These results give further evidence of the persistence of atrazine and its metabolites in environmental compartments.

Data show that some degradation products of atrazine have a higher leaching potential than the active substance (Barrett 1996; Wauchope 1996). Deisopropyl-atrazine (DIA) and deethyl-atrazine (DEA) are the degradation products of atrazine more frequently detected, with concentrations that can reach a maximum of 20.9 mg/L for DEA and 7.4 mg/L for DIA in groundwater (Spalding et al. 2003). All monitoring studies (Table 4.1) confirm that atrazine degradation products contribute to the total triazine residues in groundwater and surface water. In the same study, the total residue of these degradation products and atrazine constitute two or three times the residue of atrazine alone (Di Corcia et al. 1997).

Metribuzin is the most frequently applied herbicide in potato crops in various countries and its dissipation is usually fast with DT50 values ranging from 30 to 60 days in both laboratory and field studies (Tomlin 2006). The primary degradation products of metribuzin are desaminometribuzin (DA), desaminodiketometribuzin (DADK), and diketometribuzin (DK). Low sorption capacity of both metribuzin and its metabolites has been reported in lysimeters studies (Bowman 1991), concluding that metribuzin and its metabolites are considered more mobile than atrazine. Kjaer and coworkers (2005) evaluated the leaching of metribuzin and its primary metabolites DA, DADK, and DK at a sandy test site in Denmark. Soil water and groundwater were sampled monthly over a 4-year period. Results showed that while metribuzin and DA were not detected in any of the analyzed groundwater samples, monitoring did reveal marked groundwater contamination

TABLE 4.1

Surface Waters and Groundwater Monitoring Results of the Detection of Pesticide Transformation Products

Parent Compound	Chemical Group[a]	Degradation Products[b]	Concentration (μg/L)[c] Groundwater	Concentration (μg/L)[c] Surface Water	Analysis Method[d]	Country	References
Acetochlor	Chloroacetamide (H)	—	n.d.	—	GC/MS	United States	Postle et al. (2004)
		Acetochlor ESA	0.104–0.809	—			
		Acetochlor OA	0.155	—			
Alachlor	Chloroacetamide (H)	—	0.63 (max)	—	SPE-GC/MS	United States	Kolpin et al. (1997)
Alachlor		Alachor ESA	14.8 (max)	—	SPE-ELISA	United States	Spalding et al. (2003)
		—	n.d.	—	SPE-GC/MS		
Alachlor		Alachor ESA	4.3 (max)	—	SPE-LC/MS/MS	United States	Postle et al. (2004)
		Alachor OA	0.3 (max)	—			
Alachlor		—	0.69	—	GC/MS	United States	Postle et al. (2004)
		Alachor ESA	0.101–14.8	—			
		Alachor OA	0.145–13.5	—			
Alachlor		—		0.555	LLE-GC/MS	United States	Coats (1993)
		2-Chloro-2′,6′-diethylacetanilide		0.008			
		2-Hydroxy-2′,6′-diethylacetanilide		0.085			
Aldicarb	Carbamate (I)	—		0.069	LLE-LC/FLD	Greece	Fytianos et al. (2006)
		Aldicarb sulfoxide		0.107			
		Aldicarb sulfone		n.d.			

(continued)

TABLE 4.1 (Continued)

Surface Waters and Groundwater Monitoring Results of the Detection of Pesticide Transformation Products

Parent Compound	Chemical Group[a]	Degradation Products[b]	Concentration (µg/L)[c] Groundwater	Surface Water	Analysis Method[d]	Country	References
Aldicarb		—	—	0.01	SPE-GC/MS	United States	Kolpin et al. (2000b)
		Aldicarb sulfoxide	—	1.80			
		Aldicarb sulfone	—	0.32			
Aldicarb			36–249	—	SPE-LC/MS	United States	Miles and Delfino (1984)
		Aldicarb nitrile	n.d.	—			
		Aldicarb oxime	3.2–32	—			
		Aldicarb sulfone	25–50	—			
		Aldicarb sulfone nitrile	n.d.	—			
		Aldicarb sulfone oxime	7.8–78	—			
		Aldicarb sulfoxide	20–200	—			
		Aldicarb sulfoxide nitrile	22–220	—			
		Aldicarb sulfoxide oxime	3.8–38	—			
Atrazine	1,3,5-Triazine (H)	—	$(6.55–13.92) \times 10^{-3}$	—	MIPS-LC/MS/MS	Spain	García-Galán et al. (2010)
		DIA	n.d.	—			
		DEA	$(0.53–2.56) \times 10^{-3}$				
Atrazine		—		0.054–4.735	LLE-GC/MS	United States	Chiron et al. (1993)
		DEA		0.005–0.855			
		DIA		0.011–0.355			
Atrazine		—		n.d.	SPE-LC/MS/MS	France	Bintein and Devillers (1996)
		DIA		n.d.			
		DEA		0.022–0.051			

Pesticide	Degradation product			Detection method	Location	Reference
Atrazine	—	0.01–2.46	0.01–0.63	SPE-GC/MS	Spain	Hildebrandt et al. (2008)
Atrazine	—	0.01–1.98	0.01–0.12	SPE-GC/MS	United States	Spalding et al. (2003)
	DEA	119 (max)	—			
Atrazine	—	20.9 (max)	7.5 (max)	SPE-GC/MS	United States	Thurman et al. (1994)
	DEA	7.4 (max)	0.85 (max)			
	DIA	—	0.4 (max)			
Atrazine	—	—	0.09	SPE-GC/MS	China	Gfrerer et al. (2002)
Atrazine	—	0.098 (max)	0.042	SPE-GC/MS	Greece	Albanis et al. (1998)
	DEA	—	0.310 (max)			
Atrazine	—	0.205 (max)	0.526 (max)	N/A	United States	Whitall et al. (2010)
	DEA	—	1.89 (max)			
Atrazine	—	—	0.64 (max)	GC/NPD	Canada	Maguire and Tkacz (1993)
	DEA	—	0.80 (max)			
	DIA	—	0.043–6.78			
Atrazine	—	—	0.0055–0.0312	SPE-LC/MS	Italy	Di Corcia et al. (1997)
	DEA	—	$(3.4-6.7) \times 10^{-3}$			
	DIHA	—	n.d.			
	DEDIA	—	$(3.9-16) \times 10^{-3}$			
	DEHA	—	$(2.4-6.6) \times 10^{-3}$			
	DIA	—	$(2-5) \times 10^{-3}$			
	DEA	—	$(4.3-6.5) \times 10^{-3}$			
	HA	—	$(3.7-9.8) \times 10^{-3}$			

(continued)

TABLE 4.1 (Continued)

Surface Waters and Groundwater Monitoring Results of the Detection of Pesticide Transformation Products

Parent Compound	Chemical Group[a]	Degradation Products[b]	Concentration (μg/L)[c]		Analysis Method[d]	Country	References
			Groundwater	Surface Water			
Atrazine		—	0.056	0.073	SPE-LC/MS/MS	Portugal	Carvalho et al. (2008)
Atrazine		DEA	n.d.	n.d.	GC/MS	Spain	Quintana et al. (2001)
			0.025–0.059	—			
Atrazine		DIA	0.025–0.063	—	SPE-GC/MS	United States	Kolpin et al. (1997)
		DEA	0.025–0.036	—			
Atrazine			2.13 (max)	—			
		DIA	0.44 (max)	—			
		DEA	0.59 (max)	—			
Atrazine			2.5	—	SPE-GC/MS	United States	Cai et al. (1995)
		DDA	0.14–1.1	—			
Atrazine			—	0.147–0.232	SPE-GC/NPD	Italy	Galassy (2006)
		DEA	—	0.030–0.045			
		DIA	—	0.018–0.058			
Atrazine			5.3	—	SPE-LC/DAD	France	Baran et al. (2008)
		DEA	1.86	—	SPE-LC/MS/MS		
Atrazine			—	4.735 (max)	LLE-GC/MS	United States	Coats (1993)
		DEA	—	0.855 (max)			
		DIA	n.d.	0.335 (max)			
Atrazine			—	0.43 (max)	LLE-LC/MS	France	Baran et al. (2007)
		DEA	—	1.16 (max)			

Pesticide (class)	TP / metabolite	Conc. 1	Conc. 2	Method	Country	Reference
Atrazine	DIA	—	0.025 (max)	SPE-LC/MS/MS	Spain/Italy	Benvenuto et al. (2010)
	—	—	0.004–0.009			
	DIA	—	0.008–0.018	SPE-LC/DAD, SPE-LC/MS	Spain	Santos et al. (2000)
	DEA	—	0.009–0.008			
	HA	—	0.038–0.151			
Bentazone (Benzothiadiazinone (H))	—	—	0.02–8.8			
	6-HBtzn	—	n.d.			
	8-HBtzn	—	0.05–0.8			
Carbofuran (Carbamate (I))	—	—	—	SPE-LC/FLD	Spain	Chiron et al. (1995)
	3-Hydroxycarbofuran	0.32	—			
Carbofuran	—	—	0.206	LLE-LC/FLD	Greece	Fytianos et al. (2006)
	3-Hydroxycarbofuran	0.25	—			
Carbofuran	—	—	0.117	SPE-GC/MS	United States	Kolpin et al. (2000b)
	3-Hydroxycarbofuran	1.3 (max)	—			
Cyanazine (1,3,5-Triazine (H))	—	0.014 (max)	—	SPE-GC/MS	United States	Kolpin et al. (1997)
	Cyanazine amide	0.30 (max)	—			
Cyanazine	—	0.58 (max)	0.849 (max)	LLE-GC/MS	United States	Coats (1993)
	Cyanazineamide	—	—			
Dichlobenil (Benzonitrile (H))	—	n.d.	0.222 (max)	SPE-GC/MS	Italy	Porazzi et al. (2005)
	2,6-DCBA	0.049	—			
	BAM	3.11 (max)	—			
Dichlofluanid (Sulfamide (H))	—	—	n.d.	SPE-GC/MS	Greece	Hamwijk et al. (2005)
	DMSA	—	<0.003–0.036			

(continued)

TABLE 4.1 (Continued)

Surface Waters and Groundwater Monitoring Results of the Detection of Pesticide Transformation Products

Parent Compound	Chemical Group[a]	Degradation Products[b]	Concentration (μg/L)[c]		Analysis Method[d]	Country	References
			Groundwater	Surface Water			
Dimethenamid	Chloroacetanilide (H)	—	—	0.23 (max)	SPE-GC/MS (p.c.)	United States	Zimmerman et al. (2002)
		Dimethenamid ESA	—	0.08 (max)	SPE-LC/MS (d.p.)		
		Dimethenamid OA	—	0.11 (max)			
Diuron	Urea (H)	—	0.14 (max)	—	SPE-LC/DAD	England	Lapworth and Gooddy (2006)
		DCPMU	0.162	—			
		DCPU	0.356	—			
		3,4-DCA	0.902	—			
Diuron		—	—	2	SPE-LC/DAD	United States	Green and Young (2006)
		DCPMU	—	0.02–0.15			
		DCPU	—	< 0.1			
α-Endosulfan	Cyclodiene (I)	—	—	0.0028–0.22	N/A	United States	Pfeuffer and Rand (2004)
β-Endosulfan		—	—	0.0045–0.078			
		Endosulfan sulfate	—	0.003–0.45			
Endosulfan		—	—	—	SPE-LC/MS/MS	Thailand	Chusaksri et al. (2006)
		Endosulfan sulfate	—	0.31–0.50			
Endosulfan		—	—	0.813	LLE-GC/ECD	Turkey	Willis and McDowell (1982)
		Endosulfan sulfate	—	0.847			
Endosulfan		—	—	0.023–0.025	LLE-GC/ECD	Turkey	Boxall et al. (2004)
		Endosulfan sulfate	—	0.120–0.123			
Endrin	Organochlorine (I)	—	—	n.d.	LLE-GC/ECD	Turkey	Willis and McDowell (1982)
		Endrin aldehyde	—	n.d.			

Endrin		Endrin ketone	0.617	—	LLE-GC/ECD	Turkey	Boxall et al. (2004)
		—	n.d.	—			
Endrin		Endrin aldehyde	n.d.–0.070	—	SPE-GC/ECD	Greece	Vagi et al. (2003)
		Endrin ketone	n.d.–0.069	—			
Endrin		Endrin aldehyde	0.013	—	GC/ECD	Canada	Maguire and Tkacz (1993)
Heptachlor	Organochlorine (I)	Heptachlor epoxide	0.027	—	LLE-GC/MS	Argentina	Weber and Miller (1989)
		Heptachlor epoxide	0.0016–0.0067	—			
Heptachlor		Heptachlor epoxide	0.0005–0.0076	—	LLE-GC/ECD	Turkey	Boxall et al. (2004)
		Heptachlor epoxide	1.064	—			
Heptachlor		Heptachlor epoxide	0.421	—	LLE-GC/ECD	Turkey	Willis and McDowell (1982)
		Heptachlor epoxide	0.121–0.139	—			
Heptachlor		Heptachlor epoxide	0.246–0.316	—	SPE-GC/ECD	Greece	Vagi et al. (2003)
		Heptachlor epoxide	0.2	—			
Hexazinone	1,2,4-Triazinone (H)	Metabolite A	0.233	7.6×10^{-3} (max)	HPLC-DAD	Canada	Keizer et al. (2001)
		Metabolite B	0.006	2×10^{-3} (max)			
		—	0.007	3×10^{-3} (max)			
Methiocarb	Carbamate (I)	Methiocarb sulfone	—	0.39	SPE-LC/FLD	Spain	Chiron et al. (1995)
		—	—	0.40			
Methiocarb		—	—	0.28	LLE-LC/UV	Spain	Chiron et al. (1993)
		Methiocarb sulfone	—	0.38			

(continued)

TABLE 4.1 (Continued)

Surface Waters and Groundwater Monitoring Results of the Detection of Pesticide Transformation Products

Parent Compound	Chemical Group[a]	Degradation Products[b]	Concentration (µg/L)[c]		Analysis Method[d]	Country	References
			Groundwater	Surface Water			
Metolachlor	Chloroacetamide (H)	—	—	0.8 (max)	SPE-GC/MS (p.c.)	United States	Zimmerman et al. (2002)
		Metolachlor ESA	—	0.9 (max)	SPE-LC/MS (d.p.)		
		Metolachlor OA	—	0.5 (max)			
Metolachlor		—	6.7 (max)	—	SPE-GC/MS	United States	Spalding et al. (2003)
		Metolachlor ESA	24 (max)	—			
		Metolachlor OA	3.8 (max)	—			
Metolachlor		—	n.d.	—	GC/MS	United States	Postle et al. (2004)
		Metolachlor ESA	0.103–10.2	—			
		Metolachlor OA	0.103–5.89	—			
Metolachlor		—	—	1.19 (max)	N/A	United States	Whitall et al. (2010)
		Metolachlor ESA	—	5.29 (max)			
		Metolachlor OA	—	0.89 (max)			
Metribuzin	1,2,4-Triazinone (H)	—	n.d.	—	LC/MS (p.c.) GC/MS (d.p.)	Denmark	Kjaer et al. (2005)
		DA	n.d.	—			
		DADK	2.1 (max)	—			
		DK	0.6 (max)	—			

Pesticide	Class	d.p.	Conc.	Conc.	Method	Country	Reference
Propanil	Anilide (H)	3,4-DCA	—	16.5–18.2	SPE-LC/MS	Spain	Santos et al. (2000)
			—	28.9–43.0	SPE-LC/DAD		
Simazine	1,3,5-Triazine (H)		0.01–0.54	0.03–0.35	SPE-GC/MS	Spain	Hildebrandt et al. (2008)
Terbumeton	1,3,5-Triazine (H)	DES	0.01–0.64	0.01–0.11	SPE-LC/MS/MS	Spain/Italy	Benvenuto et al. (2010)
			—	0.002–0.009			
Terbuthylazine	1,3,5-Triazine (H)	DETer	0.027	0.003–0.042	SPE-LC/MS/MS	Portugal	Carvalho et al. (2008)
				0.132			
Terbuthylazine	1,3,5-Triazine (H)	DETbzne	0.094	n.d.	SPE-LC/MS/MS	Spain/Italy	Benvenuto et al. (2010)
			—	0.002–0.053			
		DETbzne	—	0.008–0.008			
		HTbzne	—	0.068–0.787			
Terbuthylazine	1,3,5-Triazine (H)	DET	0.01–1.27	0.01–0.24	SPE-GC/MS	Spain	Hildebrandt et al. (2008)
			0.01–0.27	0.01–0.11			

[a] H, herbicide; I, insecticide; F, fungicide.
[b] Chemical structures of degradation products are available in Figure 4.7.
[c] n.d., not detected.
[d] p.c., parent compound; d.p., degradation products.

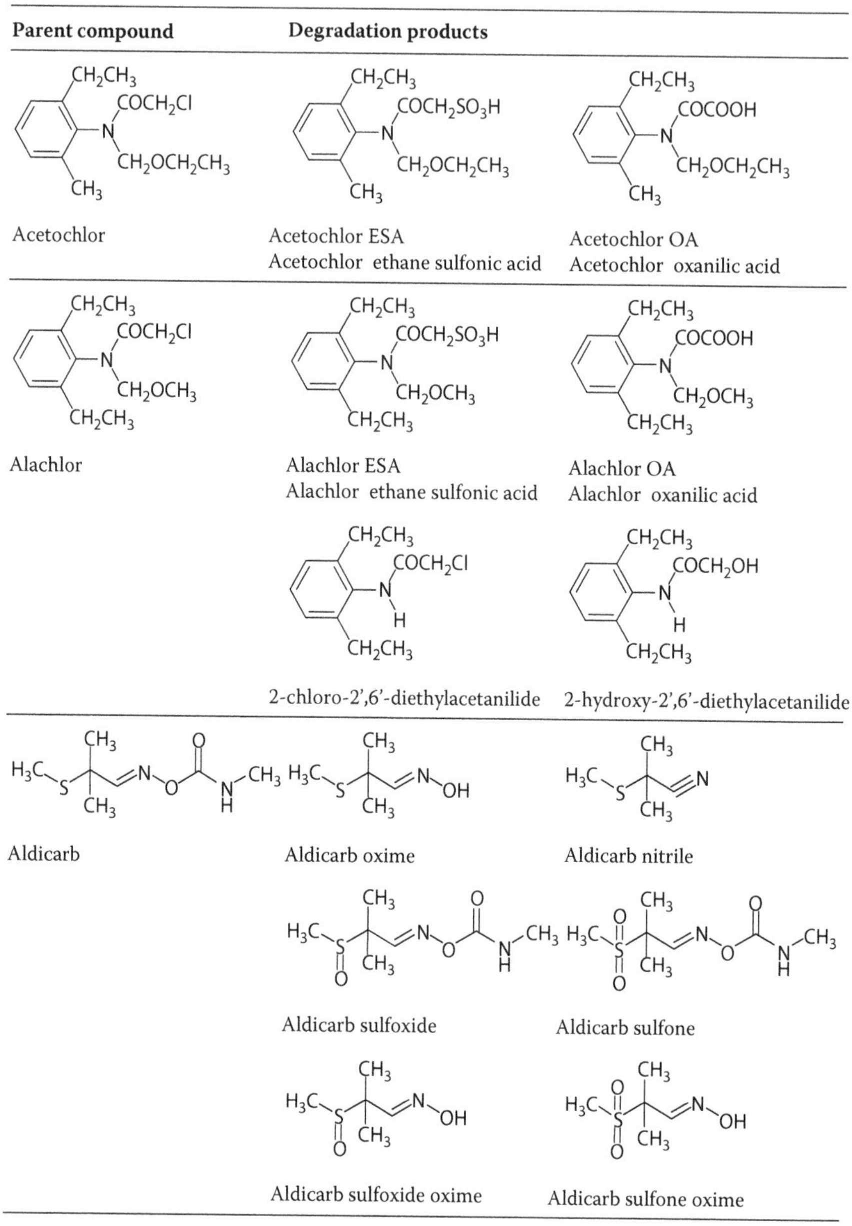

Parent compound	Degradation products	

Acetochlor

Acetochlor ESA
Acetochlor ethane sulfonic acid

Acetochlor OA
Acetochlor oxanilic acid

Alachlor

Alachlor ESA
Alachlor ethane sulfonic acid

Alachlor OA
Alachlor oxanilic acid

2-chloro-2′,6′-diethylacetanilide

2-hydroxy-2′,6′-diethylacetanilide

Aldicarb

Aldicarb oxime

Aldicarb nitrile

Aldicarb sulfoxide

Aldicarb sulfone

Aldicarb sulfoxide oxime

Aldicarb sulfone oxime

FIGURE 4.7
Chemical structures of pesticides and degradation products cited in Table 4.1.

with other degradation products of metribuzin; DK and DADK leached from the root zone and were detected in 99% and 48%, respectively, of the groundwater samples analyzed in average concentrations, considerably exceeding the EU limit value for drinking water (0.1 μg/L). Both metabolites appear to be relatively stable and persisted in soil water and groundwater several years after application. Groundwater contamination with metribuzin metabolites has also been reported by Kjaer and coworkers (2002), based on the groundwater samples from a sandy aquifer in Southern Denmark. While metribuzin and DA

Parent Compound	Degradation products	
	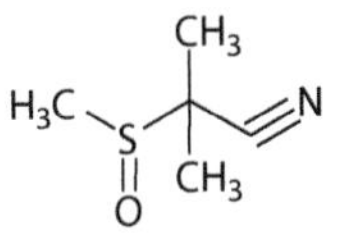 Aldicarb sulfoxide nitrile	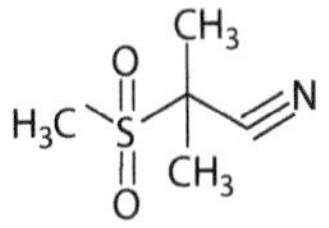 Aldicarb sulfone nitrile
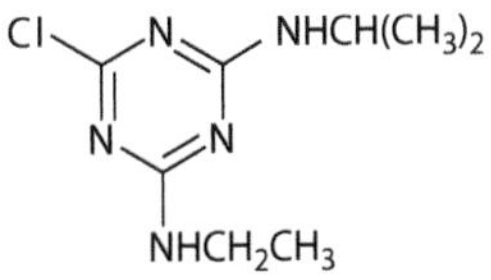 Atrazine	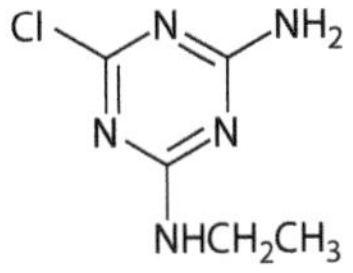 DIA Deisopropyl-atrazine	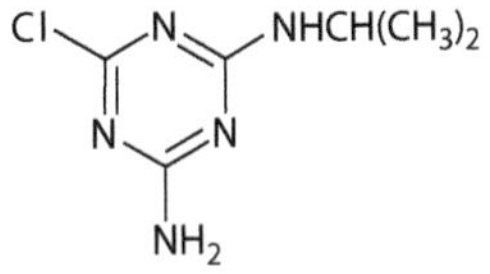 DEA Deethyl-atrazine
	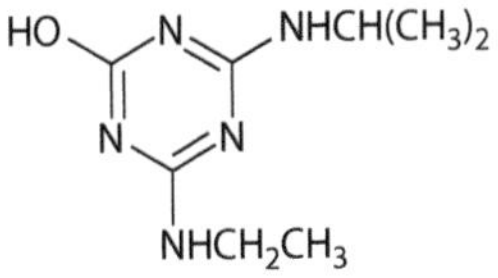 HA Hydroxy-atrazine	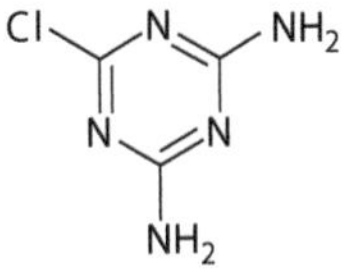 DDA Didealkyl-atrazine
	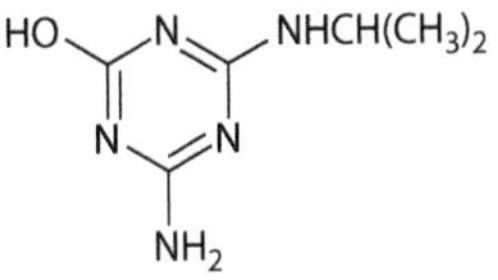 DEHA Deethyl-hydroxy-atrazine	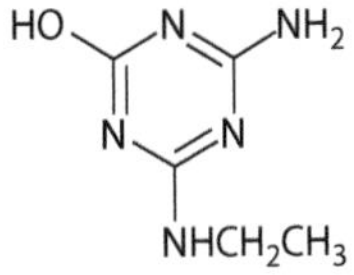 DIHA Deisopropyl-hydroxy-atrazine
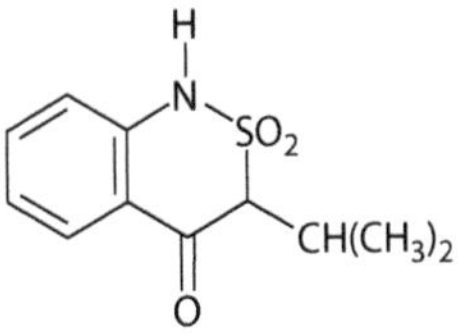 Bentazone	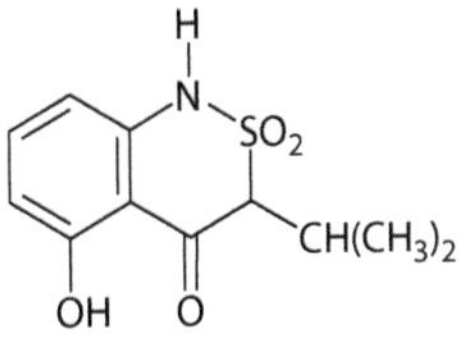 6-HBtzn 6-hydroxy-bentazone	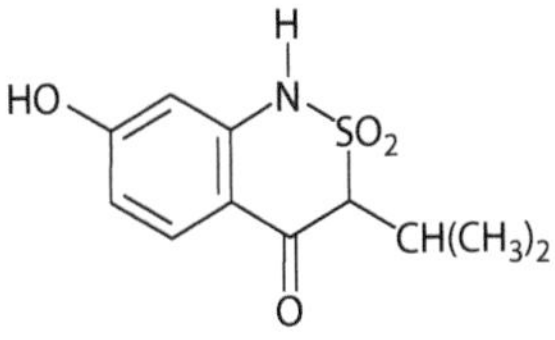 8-HBtzn 8-hydroxy-bentazone
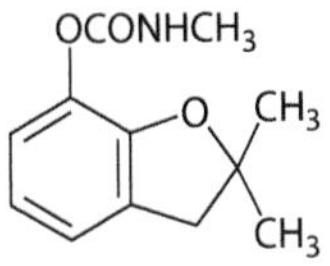 Carbofuran	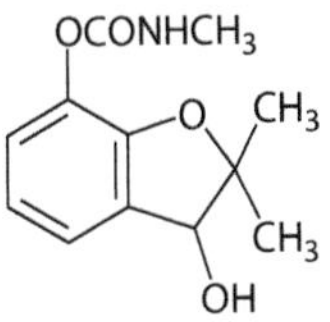 3-hydroxycarbofuran	

FIGURE 4.7 (Continued)

Parent Compound	Degradation products	

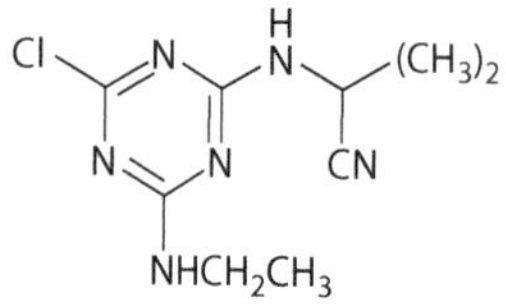

Cyanazine

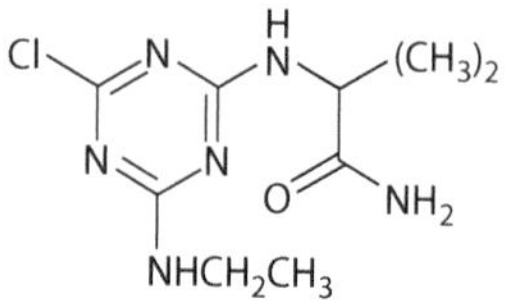

Cyanazine amide

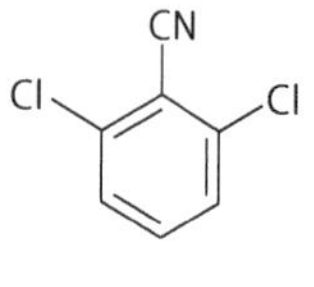

Dichlobenil

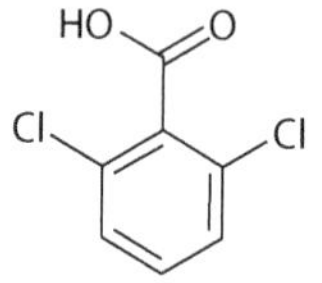

2,6-DCBA
2,6-dichlorobenzoic acid

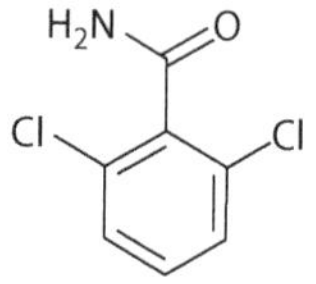

BAM
2,6-dichlorobenzamide

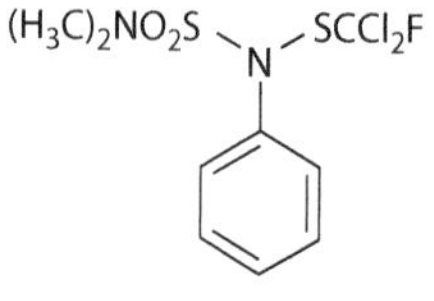

Dichlofluanid

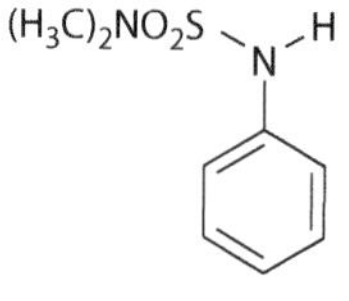

DMSA
N,N-dimethyl-N'-phenylsulphamide

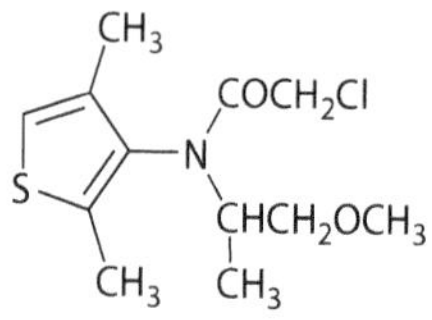

Dimethenamid

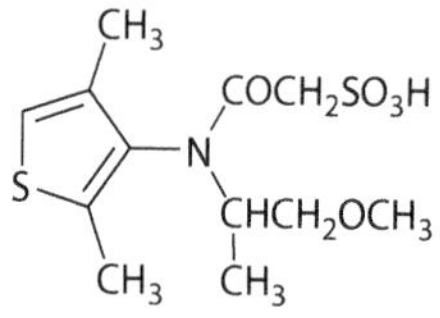

Dimethenamid ESA
Dimethenamid ethane sulfonic acid

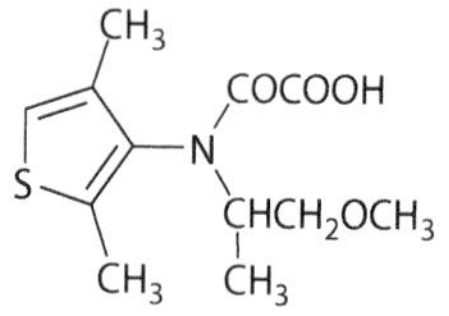

Dimethenamid OA
Dimethenamid oxanilic acid

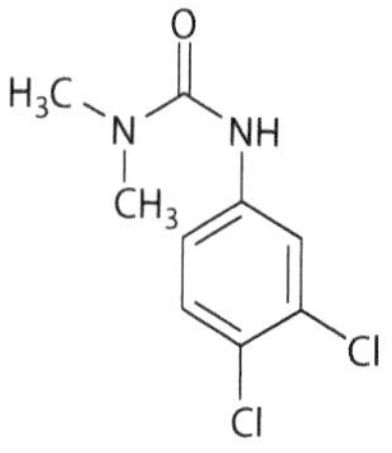

Diuron

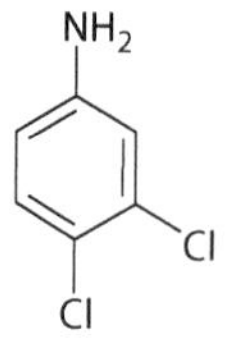

3,4-DCA
3,4-Dichloroaniline

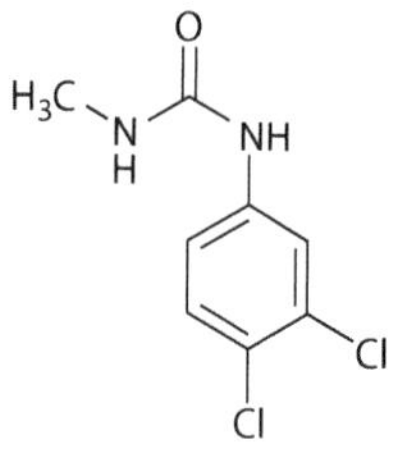

DCPMU
Dichlorophenyl monomethyl urea

FIGURE 4.7 (Continued)

Parent Compound	Degradation products

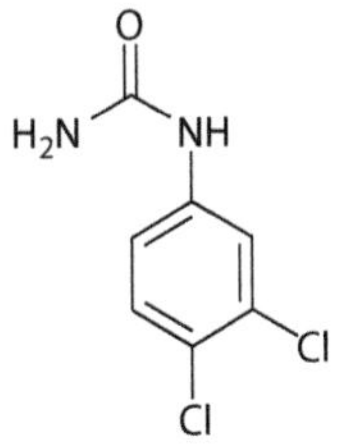

DCPU
Dichlorophenyl urea

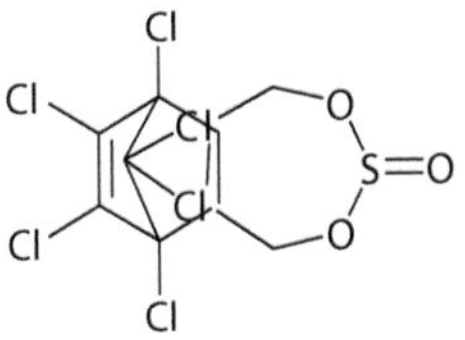

Endosulfan

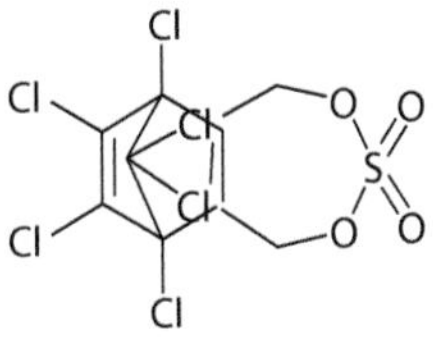

Endosulfan sulfate

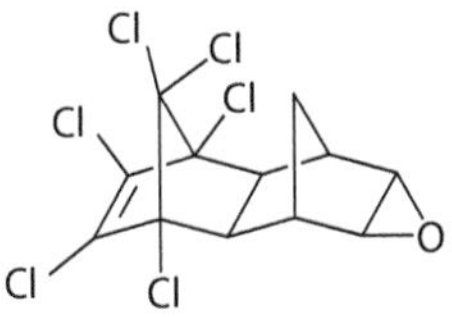

Endrin

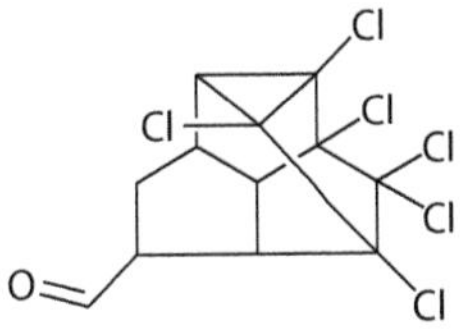

Endrin aldehyde

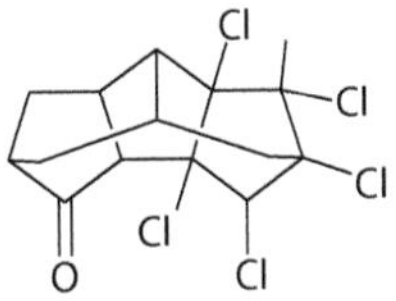

Endrin ketone

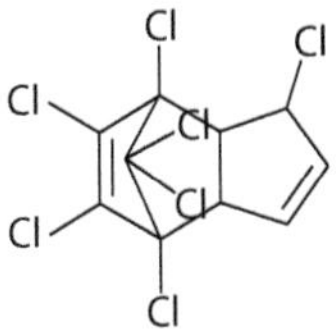

Heptachlor

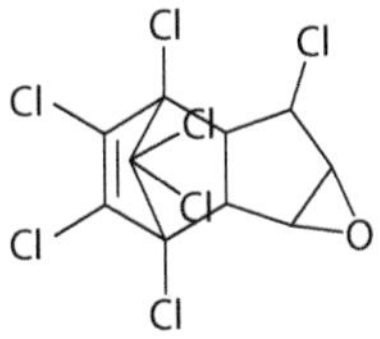

Heptachlor epoxide

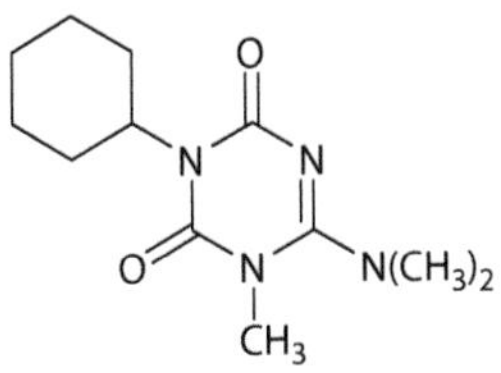

Hexazinone

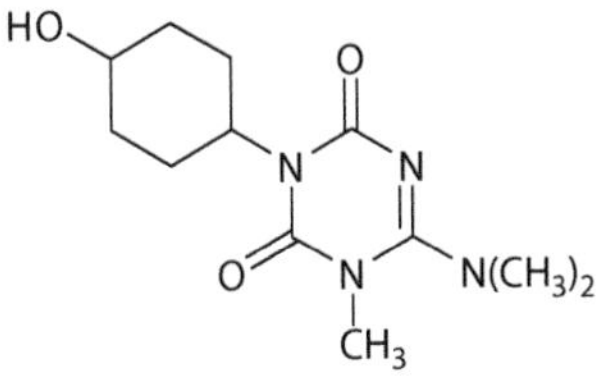

Metabolite A
Hydroxyhexazinone

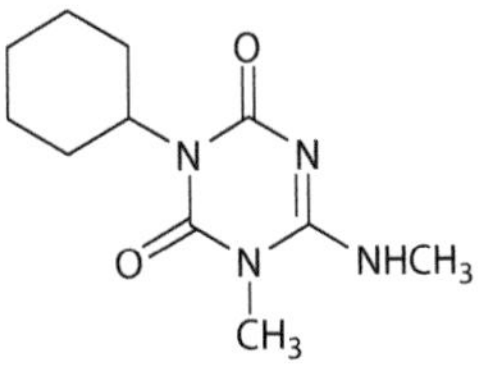

Metabolite B
Demethylhexazinone

FIGURE 4.7 (Continued)

Parent Compound	Degradation products	

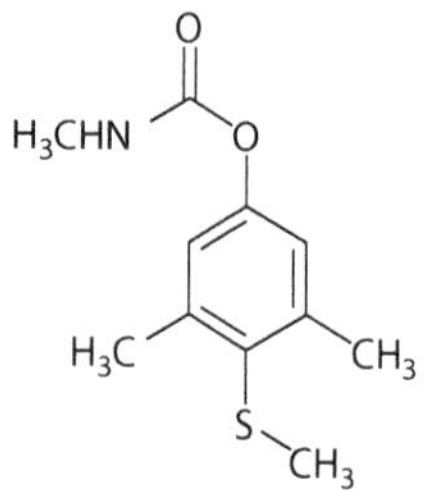

Methiocarb

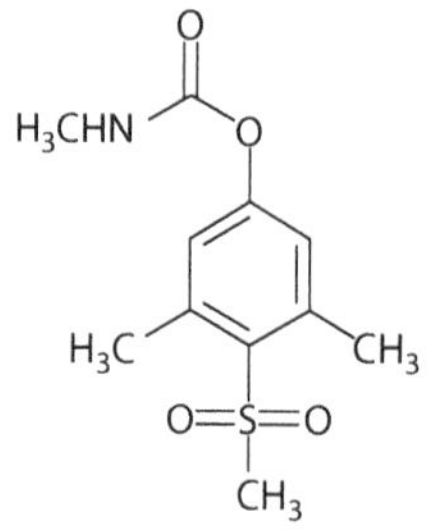

Methiocarb sulfone

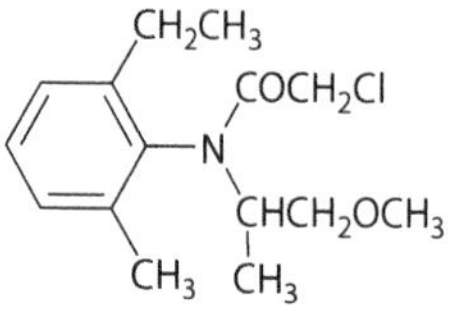

Metolachlor

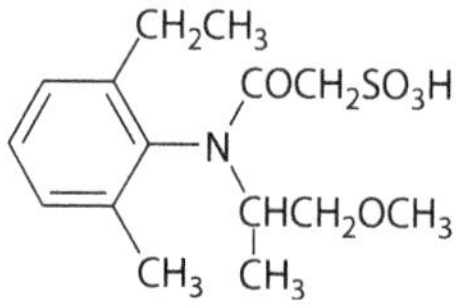

MESA
Metolachlor ethane sulfonic acid

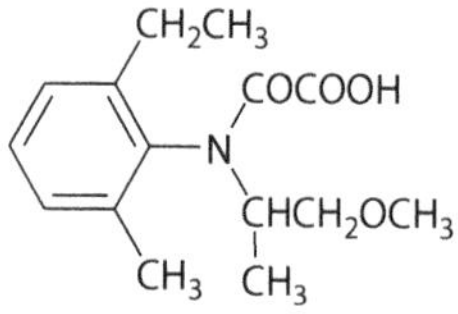

MOA
Metolachlor oxanilic acid

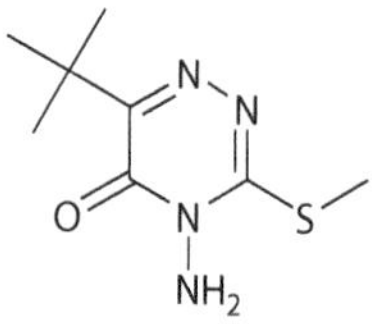

Metribuzin

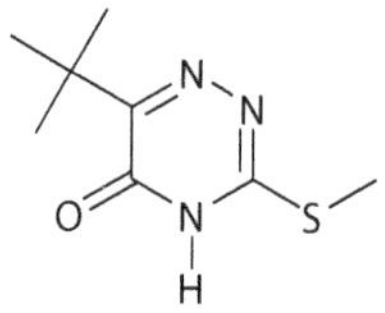

DA
Desaminometribuzin

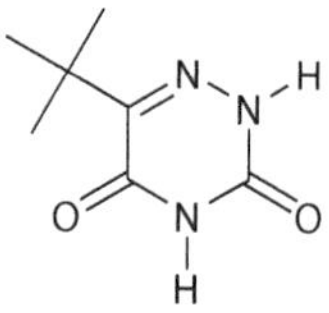

DADK
Desaminodiketometribuzin

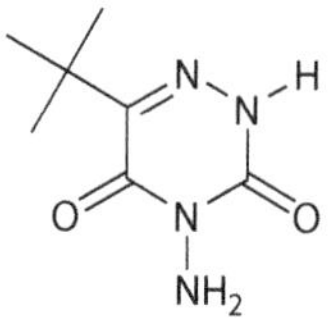

DK
Diketometribuzin

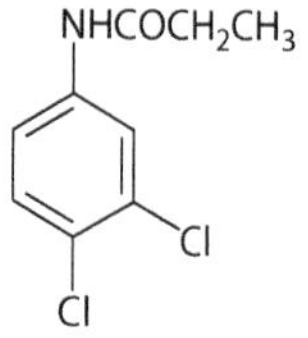

Propanil

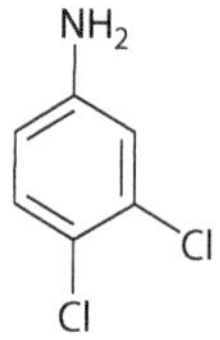

3,4-DCA
3,4-dichloroaniline

FIGURE 4.7 (Continued)

Parent Compound	Degradation products	
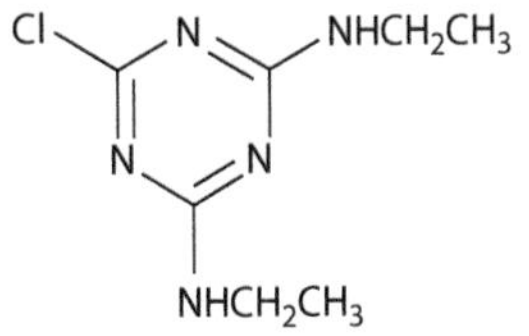 Simazine	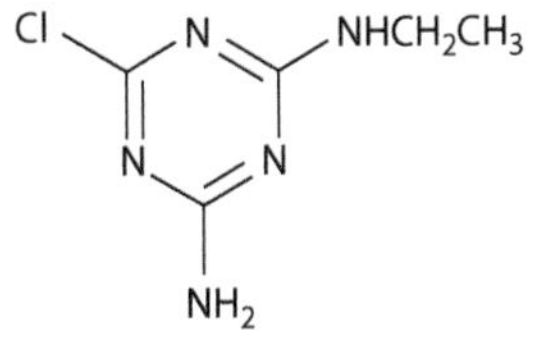 DES Deethylsimazine	
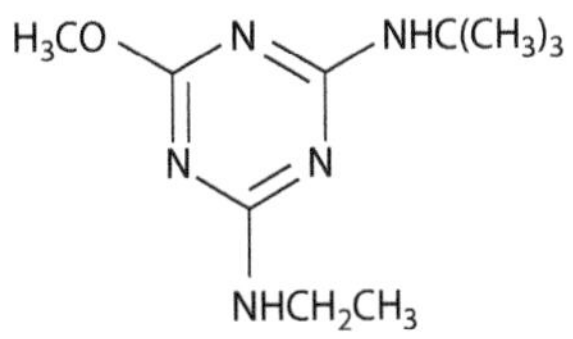 Terbumeton	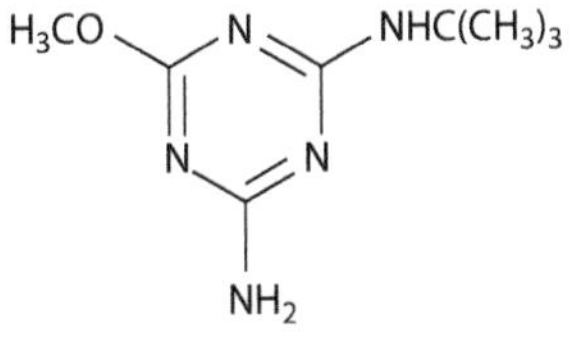 Deter Deethylterbumeton	
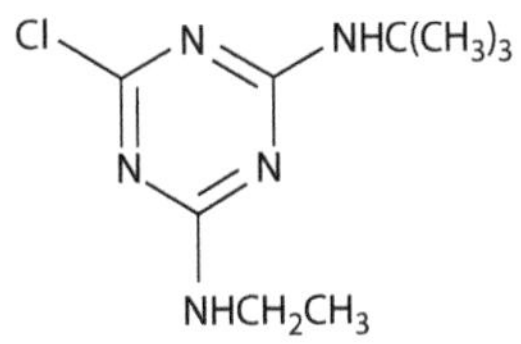 Terbuthylazine	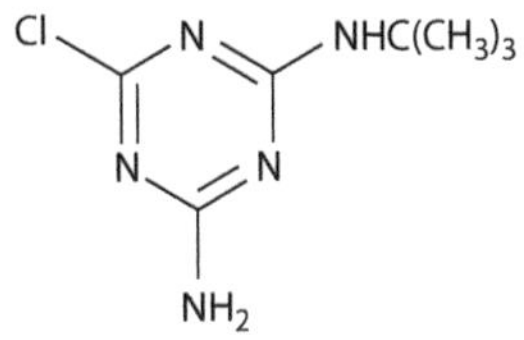 DETbzne Deethylterbuthylazine	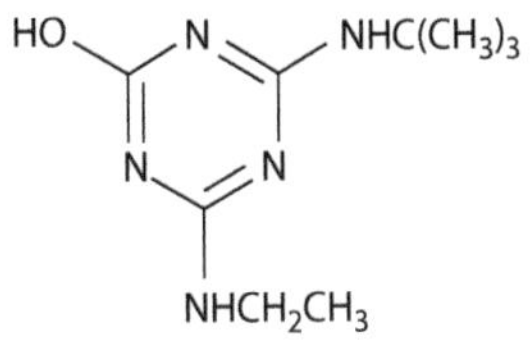 HTbzne 2-hydroxyterbuthylazine

FIGURE 4.7 (Continued)

were not detected, DK was found in concentrations as high as 1.37 µg/L and DADK was detected in concentrations as high as 1.83 µg/L.

Alachlor, metolachlor, and acetochlor acetamide are three similar chloroacetanilide herbicides and are among the most commonly used preemergence herbicides of corn and soybean crops in the United States (Graham et al. 1999). Each of these three parent compounds breaks down into their ethane sulfonic acid (ESA) and oxanilic acid (OA) metabolites (Figure 4.7). The chloroacetanilide herbicides are transformed to the ESA and OA metabolites primarily by microbial activity in soil. The degradation of the parent compounds to the ESA metabolites, for example, results in the removal of a chlorine atom and the addition of a sulfonic acid functional group to the molecule. This greatly increases the water solubility relative to the parent compound. In addition to the solubility, the apparent higher stability than that of the parent compound contributes to the increase in their leaching potential in groundwater. Furthermore, degradation products of chloroacetanilide herbicides are considered not only more mobile but also potentially persistent than the respective parent compounds (Barrett 1996). Therefore, their extensive use and moderate persistence make these herbicides and their degradation products accumulate in many surface waters and groundwater (Kolpin et al. 1996, 1997; Spalding et al. 2003; Thurman et al. 1996). Some monitoring studies in which the degradation products have been detected, the parent compounds were not found (Postle et al. 2004). Kolpin and coworkers (1997) studied the occurrence of the degradation product of alachlor, ESA, and it was detected frequently and very often at concentrations up to 10 times the concentration of its parent compound, alachlor. In this same study, cyanazine amide was detected more than three times as frequently as cyanazine and its concentration was double that of cyanazine. This is in agreement with the high stability of this by-product.

Diuron is one of most widely used herbicides in the United States. This herbicide is known for its mobility in water, with an aqueous solubility of 42 mg/L, relative persistence in soil with half-life ranging from 30 to 180 days, and a modest sorption coefficient ($K_{oc} = 485$ L/Kg) (El Imache et al. 2009; Green and Young 2006). Consistent with these transport predictions, diuron is among the leading groundwater pesticide problems. As for toxicity, diuron is highly toxic at the cellular level though its metabolites may be more toxic than the parent compound (Giacomazzi and Cochet 2004). The main degradation products of diuron are dichlorophenyl monomethyl urea (DCPMU, from the loss of one methyl group), dichlorophenyl urea (DCPU, from the loss of both methyl groups), and 3,4-dichloroaniline (3,4-DCA) (Figure 4.7). Beyond any possible direct effects of diuron exposure, a serious concern to human health is the possible formation of N-nitrosodimethylamine (NDMA, a potent carcinogen) during water treatment disinfection (Mitch et al. 2003). NDMA can be formed from dimethylamine (DMA). This by-product (DMA) can also be formed during water chlorination or during diuron degradation in the environment by biotic and abiotic processes.

DCPMU, DCPU, and 3,4-DCA have been recently detected in groundwater (Lapworth and Gooddy 2006) at concentrations higher than the permitted levels for pesticides 0.1 µg/L according to the European DWD (European Commission 1998).

Another family of pesticides, carbamates, possess a high polarity and solubility in water and thermal instability (Chiron et al. 1995). In addition, these pesticides have high acute toxicity; they are of great environmental concern because of their high acute toxicity (Gupta 1994). Their presence in a wide variety of waters is mainly due to their leachability, confirmed by their groundwater ubiquity score (GUS) that identifies carbamates as potential leachers (Gustafson 1989). The importance of the presence of their transformation products in water must be considered. Their by-products are formed as a consequence of several processes such as hydrolysis, biodegradation, oxidation, photolysis, and biotransformation (Chiron et al. 1996; Climent and Miranda 1996, 1997; Hidalgo et al. 1998). These processes lead to compounds generally more toxic than the parent pesticides (de Bertrand and Barceló 1991) and are more persistent in the environment, for example, 1-naphthol is more toxic to aquatic organisms than carbaryl, its parent compound (Barrett 1996).

Sulfonylurea herbicides are widely used as they are active at very low application rates (10–90 g/ha) and their environmental and toxicological properties are relatively favorable (Beyer et al. 1988). In general, these herbicides are moderately persistent in soil and water (Berger et al. 2002; Saha and Kulshrestha 2002; Sarmah and Sabadie 2002; Tomlin 2006). Rimsulfuron is one of the most recent sulfonylurea herbicides and is among the four most used herbicides for potato crops in many countries. Studies on the fate of rimsulfuron and its primary degradation products IN70941 and IN70942 (Figure 4.8) generally show that rimsulfuron is moderately persistent to nonpersistent in aqueous solutions/soil suspensions under anaerobic/aerobic conditions, with half-lives of 6–40 days in soil (Martins et al. 2001; Schneiders et al. 1993).

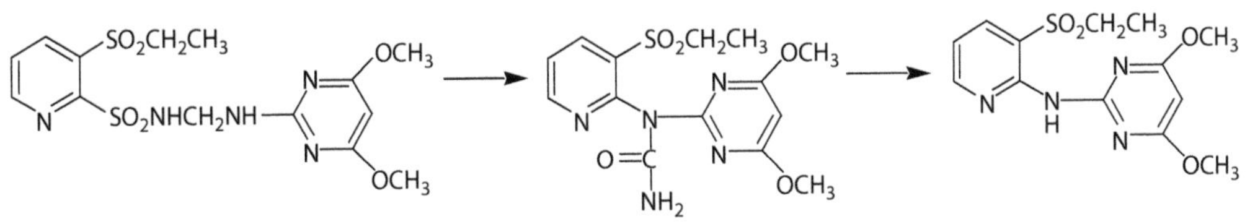

FIGURE 4.8
Major pathway of degradation of rimsulfuron in soil.

Rosenbom and coworkers (2010) studied the leaching of rimsulfuron and its by-products IN70941 and IN70942 (Figure 4.8) at two sandy research fields in Denmark. Water was sampled every month during a 4–6-year period following the application of the herbicide. No rimsulfuron was detected in the water beneath the two sandy soil research fields. By contrast, the degradation products were detected in the water or the vadose and groundwater zones for several years following the application of rimsulfuron, thus indicating that they are relatively stable and persist in the soil, from where they leach into groundwater. Both degradation products exhibited continuous leaching during the 6-year monitoring period and were still detected toward the end of 2008, that is, 6 years after the application of rimsulfuron. The observed leaching patterns seem to be consistent with fate studies suggesting that rimsulfuron is nonpersistent and that the two degradation products IN70491 and IN70942 are more persistent, with half-lives of 30–1100 and 101–214 days, respectively.

These authors suggest that the long-term leaching of low-dose pesticide degradation products such as IN70941 and IN70942 can pose a potential risk to the aquatic environment, thus indicating a need to further study/monitor the long-term leaching and ecotoxicological effects of low-dose pesticides and their degradation products.

The identification of new degradation products formed from the primary by-products is gaining importance, as is shown by the identification of new degradation products of aldicarb, such as aldicarb sulfoxide oxime and aldicarb sulfoxide nitrile, by-products derived from the previous degradate aldicarb sulfoxide (Table 4.1). These degradation products were detected at a concentration in the range of 22–220 µg/L for aldicarb sulfoxide nitrile and 3.8–38 µg/L for aldicarb sulfoxide oxime (Miles and Delfino 1984).

4.6 Chemical Analysis of Pesticide Degradation Products

Analysis of polar pesticides and even more polar transformation products is quite challenging because they are extremely water-soluble and therefore difficult to be enriched and detected by common methods such as gas chromatography (GC) coupled with mass spectrometry (MS). Furthermore, the difficulty in determining the organic by-products in raw and finished water is enhanced by taking into account the following issues: (a) By-products and metabolites of some pesticides, formed in the different environmental conditions, are still unknown; (b) A scarce number of analytical standards for the degradates are available, hence there are no analytical methods for the analysis of transformation products in the aquatic environment; (c) The concentrations expected in the aquatic environment are not known and no data exist on their physicochemical properties, environmental fate, and toxicity; and (d) Among the numerous metabolites and degradation products detected in laboratory experiments, which are the most likely to be monitored? There is no information about which of them will be significant in the environment, that is, to occur in surface waters and groundwater and/or have biological activity.

Although GC has been used to determine pesticides and their transformation products, liquid chromatography (LC) is found to be more adequate as the majority of the transformation products are more polar than their parent pesticides, less volatile, and often also thermolabile. Undoubtedly, MS is one of the best techniques for the elucidation of transformation products due to its sensitivity, universality, and the sample information provided. LC-MS has permitted to achieve the restricted levels required by the European Directive (Alder et al. 2006; European Commission 1998).

One of the main problems in developing multiresidue methods is due to not only the wide range of polarities that present the great variety of pesticides but also the analysis of transformation products that are generally more polar than the parent pesticides.

Quadrupole mass analyzers have been frequently employed in the analysis of pesticide residues in water (Sandín-España et al. 2005). However, the main drawback is the lack of accuracy for the determination of pesticides and their degradation products in a complex matrix with interferences that coelute. Triple quadrupole mass analyzer is, nowadays, one of the most widely used techniques for the routine analysis of pesticides in the environment (El Atrache et al. 2005; Hernández et al. 2004; Lehotay et al. 2005; Marín et al. 2009; Rodrigues et al. 2007). This technique is highly selective and sensitive but it does not permit structural elucidation of nontarget compounds (Bobeldijk et al. 2001).

The development of hybrid quadrupole time-of-flight (QTOF) instruments represents an attractive new tool for the determination of transformation products in the environment (Ferrer et al. 2004; Meng et al. 2010). The high resolution and accurate mass measurements provided by the TOF analyzer allow for the assignment of a highly probable empirical formula for each compound, the differentiation between nominal isobaric compounds, and the possibility of performing MS/MS spectra with accurate mass measurement of the identification of the transformation products (Grimalt et al. 2010; Keizer et al. 2001). Thus, QTOF analyzer is a great advantage in the elucidation of unknown compounds. In addition, it permits the possibility of obtaining the elemental compositions of all of the product ions obtained, which is very helpful to elucidate the transformation products in the environment (Barceló and Petrovic 2007; Bobeldijk et al. 2001; Brix et al. 2009; Trovó et al. 2009).

The increase in polarity makes the isolation and analysis of transformation products more difficult than the analysis of the parent compounds. Thus, the analysis of pesticide degradation products almost begins with the problem of the isolation of metabolites from the water sample (Bagheri et al. 1993; Guenu and Hennion 1996; Sandín-España et al. 2002).

Following isolation of the degradation products, the compounds are identified usually by a chromatographic method. This identification is sometimes hampered due to the lack of the analytical standard available. Therefore, the synthesis may be required. The effort and cost are sometimes great, and this fact has discouraged studies of transformation products in groundwater and surface waters. For these reasons, the analysis of pesticide transformation products has not proceeded as quickly as the detection of the parent compounds. In addition, knowing which of the degradation products of a pesticide may be important in surface waters and groundwater is a difficult question and requires different types of understanding, such as transport, persistence, and toxicity.

Detomaso and coworkers (2005a,b) reported the structural elucidation of polar degradation products of carbofuran photodegradation employing a QTOF. This technique provided the exact mass measurement of the [M+H]+ ions of the by-products and of the products ions, obtaining the elemental formula and related structures of seven photodegradation by-products. In addition, fragmentation patterns were proposed. Ibáñez and coworkers (2006) identified, by means of LC-QTOF-MS, the transformation product of the insecticide diazinon. Hydrolysis, hydroxylation, and oxidation were the most important processes found to occur in the degradation of diazinon, and the degradation pathway could be proposed. In the same way, LC-QTOF-MS was employed to study the transformation products of the pesticides terbuthylazine, simazine, terbutryn, and terbumeton in the environment (Ibáñez et al. 2004). The high sensitivity in full scan mode allowed the identification of minor metabolites below 2% of the total peak area. The mass errors were below 2 mDa, allowing the assignment of a highly probable empirical formula for each degradation product. Identification of the photodegradation products of cyclohexanedione oxime herbicides by employing this technique has been reported recently (Sevilla-Morán et al. 2008, 2010a). Clethodim photodegradation gave rise to the formation of nine by-products, some of them described for the first time. Main reactions observed were photoisomerization, sulfur-oxidation of isomers leading to the formation of sulfoxide diastereoisomers, and reduction of the oxime moiety. Table 4.2 shows the chemical structure of the identified transformation products and the accurate mass measurements and elemental composition of clethodim and its photodegradation products by LC-ESI-QTOF.

However, it is important to highlight that when there is no available analytical standard, the employment of high resolution analyzers for the complete and unequivocal structural characterization of metabolites and transformation products is not enough. The use of complementary and more powerful techniques able to unequivocally confirm the identity of organic compounds is required. The analytical techniques applied for the elucidation of organic compounds in addition to MS are UV, IR elemental analysis and the monodimensional ^{1}H and ^{13}C-NMR and the bidimensional techniques such as COSY (correlation spectroscopy [homonuclear chemical shift]), HMQC (heteronuclear multiple bond coherence), and HMBC (heteronuclear multiple quantum coherence) spectroscopy (Boschin et al. 2007; Kodama et al. 1999; Vialaton et al. 1998, 2001; Werres et al. 1996).

Pesticides and their transformation products are generally present in environmental waters in trace amounts at and often below the ppb levels. As a result, pesticide residues cannot be analyzed without some previous sample preparation. The sample preparation is, nowadays, a key factor in the analysis of organic compounds and therefore, there is a considerable interest in the development of new selective methods for extracting pesticide residues from environmental matrices (Barceló and Hennion 1997; Pichon et al. 1996). Sample preparation, which includes extraction, concentration, and isolation of analytes, influences the accuracy of the methods. Different solid-phase extraction (SPE) techniques can be employed by sample preparation as an alternative to the classical liquid–liquid extraction that presented multiple disadvantages such as low recovery of polar pesticides and transformation products and use of large volumes of solvents. These include SPE, solid-phase microextraction (SPME) (Aulakh et al. 2005; Beltran et al. 2000), matrix solid-phase dispersion (MSPD) (Kristenson et al. 2006), and stir-bar sortive extraction (SBSE) (Sandra et al. 2001, 2003). SPE is the technique most employed for the monitoring of pesticides and their transformation products in water (see Table 4.1; Carabias-Martínez et al. 2004; De la Pena et al. 2003; Sabik et al. 2000).

Nowadays, a large number of SPE materials are available for the analysis of pesticide residues (Rodríguez-Mozaz et al. 2007; Soriano et al. 2001). The most common SPE sorbents

TABLE 4.2

Accurate Mass Measurements and Elemental Compositions of Clethodim, Its Photodegradation Products, and Their Major Ions Obtained by LC-ESI-QTOF

Compound	Experimental Mass ([M + H]+)	Major Ions	Calculated Mass	Elemental Composition	DBE	Error (ppm)
E-clethodim	360.1384	382.1215	382.1214	$C_{17}H_{26}NO_3SClNa^+$	4.5	0.22
		360.1384	360.1394	$C_{17}H_{27}NO_3SCl^+$	4.5	−2.97
		268.1378	268.1365	$C_{14}H_{22}NO_2S^+$	4.5	4.56
		240.1049	240.1052	$C_{12}H_{18}NO_2S^+$	4.5	−1.57
		206.1152	206.1175	$C_{12}H_{16}NO_2^+$	5.5	−11.43
		164.0711	164.0706	$C_9H_{10}NO_2^+$	5.5	3.02
Ketone of clethodim imine (C1)	224.1264	246.1121	246.1100	$C_{12}H_{17}NO_3Na^+$	4.5	8.27
		224.1264	224.1281	$C_{12}H_{18}NO_3^+$	4.5	−7.67
		196.1329	196.1332	$C_{11}H_{18}NO_2^+$	3.5	−1.56
		166.0854	166.0862	$C_9H_{12}NO_2^+$	4.5	−5.15
		111.0434	111.0440	$C_6H_7O_2^+$	3.5	−5.91
Clethodim imine sulfoxide (C2&C3)	286.1452	308.1267	308.1290	$C_{14}H_{23}NO_3SNa^+$	3.5	−7.75
		286.1452	286.1471	$C_{14}H_{24}NO_3S^+$	3.5	−6.79
		208.1319	208.1332	$C_{12}H_{18}NO_2^+$	4.5	−1.30
		166.0853	166.0862	$C_9H_{12}NO_2^+$	4.5	−6.27
Clethodim sulfoxide (C4&C5, C7&C8)	376.1327	398.1161	398.1163	$C_{17}H_{26}NO_4SClNa^+$	4.5	0.57
		376.1327	376.1343	$C_{17}H_{27}NO_4SCl^+$	4.5	−4.48
		298.1202	298.1204	$C_{15}H_{21}NO_3Cl^+$	5.5	−0.83
		284.1301	284.1314	$C_{14}H_{22}NO_3S^+$	4.5	−4.90
		206.1168	206.1175	$C_{12}H_1NO_2^+$	5.5	−3.66
		164.0702	164.0706	$C_9H_{10}NO_2^+$	5.5	−2.47
Clethodim imine (C6)	270.1520	292.1331	292.1341	$C_{14}H_{23}NO_2SNa^+$	3.5	−1.27
		270.1520	270.1522	$C_{14}H_{24}NO_2S^+$	3.5	−0.84
		208.1316	208.1332	$C_{12}H_{18}NO_2^+$	4.5	−7.71
		180.1373	180.1382	$C_{11}H_{18}NO^+$	3.5	−5.50
		166.0851	166.0862	$C_9H_{12}NO_2^+$	4.5	−6.95
Z-Clethodim (C9)	360.1382	382.1215	382.1214	$C_{17}H_{26}NO_3SClNa^+$	4.5	0.22
		360.1382	360.1394	$C_{17}H_{27}NO_3SCl^+$	4.5	−3.53
		268.1369	268.1365	$C_{14}H_{22}NO_2S^+$	4.5	1.20
		240.1050	240.1052	$C_1H_{18}NO_2S^+$	4.5	−1.15
		206.1166	206.1175	$C_{12}H_{16}NO_2^+$	5.5	−4.63
		164.0700	164.0706	$C_9H_{10}NO_2^+$	5.5	−3.69

Source: Data from Sevilla-Morán, B., Alonso-Prados, J. L., García-Baudín, J. M., and Sandín-España, P. 2010a. Indirect photodegradation of clethodim in aqueous media. By-product identification by quadrupole time-of-flight mass spectrometry. *J. Agric. Food Chem.* 58: 3068–3076.
DBE, double bond equivalent.

reported in the literature for the trace enrichment of transformation products of pesticides from water are alkyl-bonded silicas, especially C-18 (Hernández et al. 2001; Huang et al. 2003; Steinheimer 1993), and copolymer sorbents such as cross-linked polystyrene divinylbenzene (D'Archivio et al. 2007; Gervais et al. 2008; Picó et al. 2007; Sandín-España et al. 2002). However, other sorbents are also applied for polar pesticides, obtaining good recoveries. Cai and coworkers (1995) determine trace levels of didealkylatrazine (DDA), one of the dealkylated products of atrazine degradation with graphitized carbon black cartridges, with recoveries greater than 80%. These recoveries were one order of magnitude greater than those achieved with C-18 bonded silica cartridges due to the higher polarity of DDA than atrazine. Santos and coworkers (2000) developed a method for the trace determination of acidic and neutral herbicides (bentazone, 2,4-D, propanil, (4-chloro-2-methylphenoxy)acetic acid (MCPA), 4-(4-chloro-2-methylphenoxy)butanoic acid (MCPB), molinate) and some of their transformation products (8-hydroxybentazone, 6-hydroxybentazone, 2,4-dichlorophenol, 3,4-dichlorophenol, and 4-chloro-2-methylphenol). Due to the wide range of polarities of the studied compounds (log K_{ow} ranged from 1.26 to 5.84), the stationary phase selected for the extraction of the analytes was the styrene-divinylbenzene copolymer, PLRP-S. Recoveries for the analytes were higher than 76%.

The drawback of these sorbents is the limited selectivity. In addition to the analytes, many water constituents such as HAs and FAs are enriched in the sorbent and interfere in the chromatographic separation. This occurs especially in the case of transformation products (Chiron et al. 1993; Hennion 1991; Hogendoorn et al. 1999) where the matrix peak coeluted with these compounds which are generally more polar than the active substance.

The increase in selectivity has been obtained by the development of molecularly imprinted polymer materials. These sorbents are extensively cross-linked polymers containing specific recognition with a predetermined selectivity for analytes of interest (Bjarni Bjarnason et al. 1999; Hogendoorn and Zoonen 2000; Koeber et al. 2001). The preconcentration of the target compound and the simultaneous elimination of sample matrix interferences result both in an excellent chromatographic resolution and in an increase in sensitivity of the analysis (Rodríguez-Mozaz et al. 2007). However, few MIPS sorbents have already been developed for the analysis of the transformation products of the pesticides. Carabias-Martínez and coworkers (2005) obtained a molecularly imprinted polymer obtained by precipitation using propazine as template for the determination of triazine herbicides and some of their hydroxylated and dealkylated metabolites from river water. The LODs obtained were lower than 0.1 mg/L, the limit established by EU legislation as the maximum concentration allowed for individual pesticide residue in drinking water (European Commission 1998).

4.7 Future Trends

Nowadays, the number of existing active substances currently employed as pesticides is quite large, belonging to different families such as organophosphate, pyrethroid, and neonicotinoid insecticides; phenoxy, dinitroaniline, and sulfonylurea herbicides; and dithiocarbamate and strobulirin fungicides.

Over the last few years, the production and application of pesticides have changed. Polar and more easily degradable pesticides have replaced nonpolar and persistent ones in order to decrease environmental persistence and reduce nontarget organism toxicity.

Such polar pesticides tend to decompose into smaller molecules that have the potential to be similarly or even more mobile, persistent, or toxic than their parent compounds. Hence, their transport and behavior as well as biological activity may be different. In general, the degradation products thus formed do not accumulate in soil. They are often not effectively removed by water treatment processes (e.g., activated carbon) and even new toxic degradation products are formed when chlorination or chloramination is employed.

Such products of biological and chemical transformation have been detected in all environmental compartments, including groundwater and surface waters. Several cases have been reported where single transformation products are present in higher concentrations or detected more frequently than their parent compounds. Important examples include pesticides such as organochlorine compounds, triazines, or chloroacetanilides.

The discovery of the occurrence of degradation products in drinking water provides important insight into how our vision of surface water and groundwater contamination by pesticides is limited when the only samples of parent compounds are analyzed. Therefore, there is a need to understand better the impact of degradation products on the environment and to have knowledge of the occurrence and significance of pesticide residues in water. Detection of transformation products will spur investigation of pesticide degradation pathways in an aquatic environment to identify other possible toxic transformation products.

In the past, such polar metabolites could hardly be detected with available analytical methods. The lack of suitable instrumentation was a handicap for the detection of these degradation products in the environment when the market launched these polar pesticides. Nowadays, development of analytical strategies is essential for monitoring these very polar transformation products in natural waters and, consequently, for understanding the fate of the parent compounds in the environment. As monitoring efforts have increased and analytical methods have become more sensitive, there have been many more detections of pesticides in water and more public concern about the possible health effects of these residues. In this sense, development of new analytical methods is needed to determine trace concentrations of very polar pesticide analytes.

Concerns will increase further if it turns out that some of the environmental degradation products of the pesticides have toxicological effects. Yet, for many of the currently used pesticides, the fate and significance of their degradation products in the aquatic environment are not clearly understood and therefore, continuous research is still needed. Studies should be aimed at the identification of major degradation pathways and mechanisms of degradation of pesticides and identification of new degradation products and their behavior in the environment. The determination of their physico-chemical properties would also be desirable in order to know their persistence and mobility in the environment. In addition, the study of their toxicity, the development of long-term studies to determine the effects of persistent substances, and studies about interactions with their parent compounds and with other degradates and their effects on soil and microorganisms will be helpful to know their significance and impact in the environment. To achieve this task, predictive approaches such as QSARs or QSPRs may be employed.

These studies could lead researchers to consider including selected degradation products in environmental monitoring programs and to a better enforcement of legislation. All these issues are formidable challenges that environmental chemists will have to face in the forthcoming years.

Acknowledgments

The authors wish to thank the Spanish Ministry of Science and Innovation for the financial support (CICYT Project RTA 2008-00027-00-00).

References

Aharonson, N. 1987. Potential contamination of groundwater by pesticides. *Pure Appl. Chem.* 59: 1419–1446.

Albanis, T. A., Hela, D. G., Sakellarides, T. M., and Konstantinou, I. K. 1998. Monitoring of pesticide residues and their metabolites in surface and underground waters of Imathia (N. Greece) by means of solid-phase extraction disks and gas chromatography. *J. Chromatogr. A* 823: 59–71.

Albers, C. N., Banta, G. T., Hansen, P. E., and Jacobsen, O. S. 2008. Effect of different humic substances on the fate of diuron and its main metabolite 3,4-dichloroaniline in soil. *Environ. Sci. Technol.* 42: 8687–8691.

Alder, L., Greulich, K., Kempe, G., and Vieth, B. 2006. Residue analysis of 500 high priority pesticides: Better by GC–MS or LC–MS/MS? *Mass Spectrom. Rev.* 25: 838–865.

Al-Mughrabi, K. I., Nazera, I. K., and Al-Shuraiqi, Y. T. 1992. Effect of pH of water from the King Abdallah Canal in Jordan on the stability of Cypermethrin. *Crop Protect.* 11: 341–344.

Andreozzi, R., Raffaele, M., and Nicklas, P. 2003. Pharmaceuticals in STP effluents and their solar photodegradation in aquatic environment. *Chemosphere* 50: 1319–1330.

Ankumah, R. O., Dick, W. A., and McClung, G. 1995. Metabolism of carbamothioate herbicide, EPTC, by Rhodococcus strain JE1 isolated from soil. *Soil Sci. Soc. Am. J.* 59: 1071–1077.

Aspelin, A. L. and Grube, A. H. 1999. *Pesticides Industry Sales and Usage 1996 and 1997 Market Estimates*, U.S. Environmental Protection Agency: Washington, DC.

Aulakh, J. S., Malik, A. K., Kaur, V., and Schmitt-Kopplin, P. 2005. A review on solid phase micro extraction-high performance liquid chromatography (SPME-HPLC) analysis of pesticides. *Crit. Rev. Anal. Chem.* 35: 71–85.

Bach, M., Huber, A., and Frede, H.-G. 2001. Input pathways and river load of pesticides in Germany: A national scale modeling assessment. *Water Sci. Technol.* 43: 261–268.

Bachman, J. and Patterson, H. H. 1999. Photodecomposition of the carbamate pesticide carbofuran: Kinetics and the influence of dissolved organic matter. *Environ. Sci. Technol.* 33: 874–881.

Bagheri, H., Brouwer, E. R., Ghijsen, R. T., and Brinkman, U. A. T. 1993. On-line low-level screening of polar pesticides in drinking and surface waters by liquid chromatography thermospray mass spectrometry. *J. Chromatogr.* 647: 121–129.

Baran, N., Mouvet, C., and Negrel, P. 2007. Hydrodynamic and geochemical constraints on pesticide concentrations in the groundwater of an agricultural catchment (Brevilles, France). *Environ. Pollut.* 148: 729–738.

Baran, N., Lepiller, M., and Mouvet, C. 2008. Agricultural diffuse pollution in a chalk aquifer (Trois Fontaines, France): Influence of pesticide properties and hydrodynamic constraints. *J. Hydrol.* 358: 56–69.

Barceló, D. and Hennion, M.-C. 1997. *Trace Determination of Pesticides and their Degradation Products in Water*. Elsevier Science: Amsterdam.

Barceló, D. and Petrovic, M. 2007. *Challenges and Achievements of LC-MS in Environmental Analysis: 25 Years on.*

Barceló, D., Chiron, S., Fernandez, A., Valverde, A., and Alpendurada, M. F. 1996. Monitoring pesticides and metabolites in surface water and groundwater in Spain. In: Meyer, M. T. and Thurman, E. M. (eds), *Herbicide Metabolites in Surface Water and Groundwater*, 237–253. ACS Symposium Series: Washington, DC.

Barrett, M. R. 1996. The environmental impact of pesticide degradates in groundwater. In: Meyer, M. T. and Thurman, E. M. (eds), *Herbicide Metabolites in Surface Water and Groundwater*, ACS Symposium Series, pp. 200–225. ACS: Washington, DC.

Bavcon Kralj, M. and Trebše, P. 2007. Photodegradation of organophosphorus insecticides – Investigations of products and their toxicity using gas chromatography–mass spectrometry and AChE-thermal lens spectrometric bioassay. *Chemosphere* 67: 99–107.

Belfroid, A. C., van Drunen, M., Beek, M. A., Schrap, S. M., van Gestel, C. A. M., and van Hattum, B. 1998. Relative risks of transformation products of pesticides for aquatic ecosystems. *Sci. Total Environ.* 222: 167–183.

Beltran, J., Lopez, F., and Hernandez, F. 2000. Solid-phase micro-extraction in pesticide residue analysis. *J. Cromatogr. A* 885: 389–404.

Benvenuto, F., Marin, J. M., Sancho, J. V., Canobbio, S., Mezzanotte, V., and Hernández, F. 2010. Simultaneous determination of triazines and their main transformation products in surface and urban wastewater by ultra-high-pressure liquid chromatography-tandem mass spectrometry. *Anal. Bioanal. Chem.* 397: 2791–2805.

Berger, B. M., Müller, M., and Eing, A. 2002. Quantitative structure–transformation relationships of sulfonylurea herbicides. *Pest Manag. Sci.* 58: 724–735.

Beyer, E. M. J., Duffy, M. J., Hay, J. V., and Schlueter, D. D. 1988. Sulfonylureas. In: Kearney, P. and Kaufman, D. (eds), *Herbicides: Chemistry, Degradation, and Mode of Action*, 117–89. New York: Marcel Dekker.

Bintein, S. and Devillers, J. 1996. Evaluating the environmental fate of atrazine in France. *Chemosphere* 32: 2441–2456.

Bird, S. L. 1995. A compilation of aerial spray drift field study data for low-flight agricultural application of pesticides. In: Leng, M. L., Leovey, E. M., and Zubkoff, P. L. (eds), *Agrochemical Environmental Fate: State of the Art*, pp. 195–207. Lewis Publishers: Boca Raton, FL.

Bjarnason, B., Chimuka, L., and Ramström, O. 1999. On-line solid-phase extraction of triazine herbicides using a molecularly imprinted polymer for selective sample enrichment. *Anal. Chem.* 71: 2152–2156.

Bloomfield, J. P., Williams, R. J., Gooddy, D. C., Cape, J. N., and Guha, P. 2006. Impacts of climate change on the fate and behaviour of pesticides in surface and groundwater – A UK perspective. *Sci. Total Environ.* 369: 163–177.

Bobeldijk, I., Vissers, J. P. C., Kearney, G., Major, H., and van Leerdam, J. A. 2001. Screening and identification of unknown contaminants in water with liquid chromatography and quadrupole-orthogonal acceleration-time-of-flight tandem mass spectrometry. *J. Chromatogr. A* 929: 63–74.

Boschin, G., D'Agostina, A., Antonioni, C., Locati, D., and Arnoldi, A. 2007. Hydrolytic degradation of azimsulfuron, a sulfonylurea herbicide. *Chemosphere* 68: 1312–1317.

Bowman, B. T. 1991. Mobility and dissipation studies of metribuzin, atrazine and their metabolites in plainfield sand using field lysimeters. *Environ. Toxicol. Chem.* 10: 573–579.

Boxall, A. B. A., Sinclair, C. I., Fenner, K., Kolpin, D. W., and Maund, S. J. 2004. When synthetic chemicals degrade in the environment. *Environ. Sci. Technol.* 1: 369A–75A.

Bresnahan, G. A., Koskinen, W. C., Dexter, A. G., and Cox, L. 2004. Sorption–desorption of 'aged' isoxaflutole and diketonitrile degradate in soil. *Weed Res.* 44: 397–403.

Brezonik, P. L. and Fulkerson-Brekken, J. 1998. Nitrate-induced photolysis in natural waters: Controls on concentrations of hydroxyl radical photo-intermediates by natural scavenging agents. *Environ. Sci. Technol.* 32: 3004–3010.

Brix, R., Bahi, N., Alda, M. J. L. D., Farré, M., Fernandez, J.-M., and Barceló, D. 2009. Identification of disinfection by-products of selected triazines in drinking water by LC-Q-ToF-MS/MS and evaluation of their toxicity. *J. Mass Spectrom.* 44: 330–337.

Bronstert, A., Niehoff, D., and Bürger, G. 2002. Effects of climate and land-use change on storm runoff generation: Present knowledge and modelling capabilities. *Hydrol. Process.* 16: 509–529.

Burgoa, B. and Wauchope, R. D. 1995. Pesticides in run-off and surface waters. In: Roberts, T. R. and Kearney, P. C. (eds), *Environmental Behaviour of Agrochemicals*, pp. 221–55. John Wiley & Sons: Chichester.

Cai, Z. W., Gross, M. L., and Spalding, R. F. 1995. Determination of didealkylatrazine in water by graphitized carbon-black extraction followed by gas-chromatography high-resolution mass-spectrometry. *Anal. Chim. Acta* 304: 67–73.

Campbell, J. R. and Penner, D. 1985. Abiotic transformations of sethoxydim. *Weed Sci.* 33: 435–439.

Carabias-Martínez, R., Rodríguez-Gonzalo, E., and Herrero-Hernández, E. 2005. Determination of triazines and dealkylated and hydroxylated metabolites in river water using a propazine-imprinted polymer. *J Cromatogr. A* 1085: 199–206.

Carabias-Martínez, R., Rodríguez-Gonzalo, E., Herrero-Hernández, E., and Hernández-Méndez, J. 2004. Simultaneous determination of phenyl- and sulfonylurea herbicides in water by solid-phase extraction and liquid chromatography with UV diode array or mass spectrometric detection. *Anal. Chim. Acta* 517: 71–79.

Carbonell, E., Xamena, N., Creus, A., and Marcos, R. 1995. Chromosomal aberrations and sister-chromatid exchanges as biomarkers of exposure to pesticides. *Clin. Chem.* 41: 1917–1919.

Carvalho, J. J., Jeronimo, P. C. A., Goncalves, C., and Alpendurada, M. F. 2008. Evaluation of a multiresidue method for measuring fourteen chemical groups of pesticides in water by use of LC-MS-MS. *Anal. Bioanal. Chem.* 392: 955–968.

Casida, J. E. and Quistad, G. B. 1998. Golden age of insecticide research: Past, present, or future? *Ann Rev. Entomol.* 43: 1–16.

Chaabane, H., Vulliet, E., Joux, F., et al. 2007. Photodegradation of sulcotrione in various aquatic environments and toxicity of its photoproducts for some marine micro-organisms. *Water Res.* 41: 1781–1789.

Chiron, S., Fernández-Alba, A., and Barceló, D. 1993. Comparison of on-line solid-phase disk extraction to liquid-liquid extraction for monitoring selected pesticides in environmental waters. *Environ. Sci. Technol.* 27: 2352–2359.

Chiron, S., Valverde, A., Fernández-Alba, A., and Barceló, D. 1995. Automated sample preparation for monitoring groundwater pollution by carbamate insecticides and their transformation products. *J. AOAC Int.* 78: 1346–1352.

Chiron, S., Torres, J. A., FernandezAlba, A., Alpendurada, M. F., and Barceló, D. 1996. Identification of carbofuran and methiocarb and their transformation products in estuarine waters by on-line solid phase extraction liquid chromatography – Mass spectrometry. *Int. J. Environ. Anal. Chem.* 65: 37–52.

Chusaksri, S., Sutthivaiyakit, S., and Sutthivaiyakit, P. 2006. Confirmatory determination of organo-chlorine pesticides in surface waters using LC/APCI/tandem mass spectrometry. *Anal. Bioanal. Chem.* 384: 1236–1245.

Cliath, M. M. and Spencer, W. 1972. Dissipation of pesticides from soil by volatilization of degradation products. I. Lindane and DDT. *Environ. Sci. Technol.* 6: 910–914.

Climent, M. J. and Miranda, M. A. 1996. Gas chromatographic-mass spectrometric study of photo-degradation of carbamate pesticides. *J. Chromatogr. A* 738: 225–231.

Climent, M. J. and Miranda, M. A. 1997. Erattum to "Gas chromatographic-mass spectrometric study of photodegradation of carbamate pesticides". [*J. Chromatogr. A*, 738 (1996) 225–231] *J. Chromatogr. A* 761: 341.

Coats, J. R. 1993. What happens to degradable pesticides? *Chemtech* 23: 25–29.

Cronin, M. T. D. and Livingstone, D. J. 2004. *Predicting Chemical Toxicity and Fate.* CRC Press: Boca Raton, FL.

Dabestani, R., Higgin, J., Stephenson, D., Ivanov, I. N., and Sigman, M. E. 2000. Photophysical and photochemical processes of 2-methyl, 2-ethyl, and 2-tert-butylanthracenes on silica gel. A substituent effect study. *J. Phys. Chem. B* 104: 10235–10241.

Dabrowski, J. M., Peall, S. K. C., Van Niekerk, Z., Reinecke, A. J., Day, J. A., and Schulz, R. 2002. Predicting runoff-induced pesticide input in agricultural sub-catchment surface waters: Linking catchment variables and contamination. *Water Res.* 36: 4975–4984.

Dannenberg, A. and Pehkonen, S. O. 1998. Investigation of the heterogeneously catalyzed hydrolysis of organophosphorus pesticides. *J. Agric. Food Chem.* 46: 325–334.

D'Archivio, A. A., Fanelli, M., Mazzeo, P., and Fabrizio, R. 2007. Comparison of different sorbents for multiresidue solid-phase extraction of 16 pesticides from groundwater coupled with high-performance liquid chromatography. *Talanta* 71: 25–30.

de Bertrand, N., and Barceló, D. 1991. Photodegradation of the carbamate pesticides aldicarb, carbaryl and carbofuran in water. *Anal. Chim. Acta* 254: 235–244.

De la Pena, A. M., Mahedero, M. C., and Baustista-Sanchez, A. 2003. Monitoring of phenylurea and propanil herbicides in river water by solid-phase-extraction high performance liquid chromatography with photoinduced-fluorimetric detection. *Talanta* 60: 279–285.

DellaGreca, M., Iesce, M. R., Cermola, F., Rubino, M., and Isidori, M. 2004. Phototransformation of carboxin in water. Toxicity of the pesticide and its sulfoxide to aquatic organisms. *J. Agric. Food Chem.* 52: 6228–6232.

Demoliner, A., Caldas, S. S., Costa, F. P., et al. 2010. Development and validation of a method using SPE and LC-ESI-MS-MS for the determination of multiple classes of pesticides and metabolites in water samples. *J. Braz. Chem. Soc.* 21: 1424–1433.

Detomaso, A., Mascolo, G., and Lopez, A. 2005a. Characterization of carbofuran photodegradation by-products by liquid chromatography/hybrid quadrupole time-of-flight mass spectrometry. [Erratum to document cited in CA143:253330]. *Rapid Commun. Mass Spectrom.* 19: 2480.

Detomaso, A., Mascolo, G., and Lopez, A. 2005b. Characterization of carbofuran photodegradation by-products by liquid chromatography/hybrid quadrupole time-of-flight mass spectrometry. *Rapid Commun. Mass Spectrom.* 19: 2193–2202.

Di Corcia, A., Crescenzi, C., Guerriero, E., and Samperi, R. 1997. Ultratrace determination of atrazine and its six major degradation products in water by solid-phase extraction and liquid chromatography-electrospray/mass spectrometry. *Environ. Sci. Technol.* 31: 1658–1663.

Dimou, A. D., Sakkas, V. A., and Albanis, T. A. 2004a. Photodegradation of trifluralin in natural waters and soils: Degradation kinetics and influence of organic matter. *Int. J. Environ. Anal. Chem.* 84: 173–182.

Dimou, A. D., Sakkas, V. A., and Albanis, T. A. 2004b. Trifluralin photolysis in natural waters and under the presence of isolated organic matter and nitrate ions: Kinetics and photoproduct analysis. *J. Photochem. Photobiol. A* 163: 473–480.

Dimou, A. D., Sakkas, V. A., and Albanis, T. A. 2005. Metolachlor photodegradation study in aqueous media under natural and simulated solar irradiation. *J. Agric. Food Chem.* 53: 694–701.

Durand, G. and Barceló, D. 1990. Determination of chlorotriazines and their photolysis products by liquid chromatography with photodiode-array and thermospray mass spectrometric detection. *J. Chromatogr.* 502: 275–286.

Durand, G., Barceló, D., Albaigés, J., and Mansour, M. 1990. Utilisation of liquid chromatography in aquatic photodegradation studies of pesticides: A comparison between distilled water and seawater. *Chromatographia* 29: 120–124.

Durand, G., Barceló, D., Albaigés, J., and Mansour, M. 1991. On the photolysis of selected pesticides in the aquatic environment. *Toxicol. Environ. Chem.* 31–32: 55–62.

El Atrache, L. L., Sabbah, S., and Morizur, J. P. 2005. Identification of phenyl-N-methylcarbamates and their transformation products in Tunisian surface water by solid-phase extraction liquid chromatography-tandem mass spectrometry. *Talanta* 65: 603–612.

El Imache, A., Dahchour, A., Elamrani, B., Dousset, S., Pozzonni, F., and Guzzella, L. 2009. Leaching of diuron, linuron and their main metabolites in undisturbed field lysimeters. *J. Environ. Sci. Health B* 44: 31–37.

Elazzouzi, M., Bensaoud, A., Bouhaouss, A., et al. 1999. Photodegradation of imazapyr in the presence of humic substances. *Fresenius Environ. Bull.* 8: 478–485.

European Commission. 1991. Council Directive 91/414/EC concerning the placing of plant protection products on the market. *Off. J. Eur. Commun.* L230: 1–32.

European Commission. 1995. Commission Directive 95/36/EC of 14 July 1995 amending Council Directive 91/414/EC concerning the placing of plant protection products on the market. *Off. J. Eur. Commun.* L172: 8–20.

European Commission. 1996. Commission Directive 96/12/EC of 8 March 1996 amending Council Directive 91/414/EC concerning the placing of plant protection products on the market. *Off. J. Eur. Commun.* L65: 20–37.

European Commission. 1997. Council Directive 97/57/EC of 22 September 1997 establishing Annex VI to Directive 91/414/EEC concerning the placing of plant protection products on the market. *Off. J. Eur. Commun.* L265: 87–109.

European Commission. 1998. Council Directive 98/83/EC of 3 November 1998 on the quality of water intended for human consumption. *Off. J. Eur. Commun.* L330: 32–54.

European Commission. 2000. Directive 2000/60/EC of the European Parliament and of the Council of 23 October 2000 establishing a framework for Community action in the field of water policy. *Off. J. Eur. Commun.* L327: 1–72.

European Commission. 2006. Directive 2006/118/EC of the European Parliament and of the Council of 12 December 2006 on the protection of groundwater against pollution and deterioration. *Off. J. Eur. Commun.* L372: 19–31.

European Commission. 2008. Directive 2008/105/EC of the European Parliament and of the Council of 16 December 2008 on environmental quality standards in the field of water policy, amending and subsequently repealing Council Directives 82/176/EEC, 83/513/EEC, 84/156/EEC, 84/491/EEC, 86/280/EEC and amending Directive 2000/60/EC of the European Parliament and of the Council. *Off. J. Eur. Commun.* L348: 84–97.

European Commission. 2009. Regulation (EC) N° 1107/2009 of the European Parliament and of the Council of 21 October 2009 concerning the placing of plant protection products on the market and repealing Council Directives 79/117/EEC and 91/414/EEC. *Off. J. Eur. Commun.* L309: 1–50.

Ferrer, I., Mezcua, M., Gómez, M. J., et al. 2004. Liquid chromatography/time-of-flight mass spectrometric analyses for the elucidation of the photodegradation products of triclosan in wastewater samples. *Rapid Commun. Mass Spectrom.* 18: 443–450.

Frear, D. S. 1976. The benzoic acid herbicides. In: Kearney, P. C. and Kaufman, D. D. (eds), *Herbicides: Chemistry, Degradation, and Mode of Action*, pp. 541–607. Marcel Dekker: New York.

Fytianos, K., Pitarakis, K., and Bobola, E. 2006. Monitoring of N-methylcarbamate pesticides in the Pinios river (central Greece) by HPLC. *Int. J. Environ. Anal. Chem.* 86: 131–145.

Galassi, S., Provini, A., and Halfon, E. 2006. Risk assessment for pesticides and their metabolites in water. *Inter. J. Environ. Anal. Chem.* 65: 331–344.

García de Llasera, M. P. and Bernal-González, M. 2001. Presence of carbamate pesticides in environmental waters from the northwest of Mexico: Determination by liquid chromatography. *Water Res.* 35: 1933–1940.

García-Galán, M. J., Díaz-Cruz, M. S., and Barceló, D. 2010. Determination of triazines and their metabolites in environmental samples using molecularly imprinted polymer extraction, pressurized liquid extraction and LC-tandem mass spectrometry. *J. Hydrol.* 383: 30–38.

García-Repetto, R., Martínez, D., and Repetto, M. 1994. The influence of pH on the degradation kinetics of some organophosphorus pesticides in aqueous solutions. *Vet. Hum. Toxicol.* 36: 202–204.

Gervais, G., Brosillon, S., Laplanche, A., and Helen, C. 2008. Ultra-pressure liquid chromatography-electrospray tandem mass spectrometry for multiresidue determination of pesticides in water. *J. Chromatogr. A* 1202: 163–172.

Gfrerer, M., Wenzl, T., Quan, X., Platzer, B., and Lankmayr, E. 2002. Occurrence of triazines in surface and drinking water of Liaoning province in Eastern China. *J. Biochem. Biophys. Methods* 53: 217–228.

Giacomazzi, S. and Cochet, N. 2004. Environmental impact of diuron transformation: A review. *Chemosphere* 56: 1021–1032.

Gil, Y. and Sinfort, C. 2005. Emission of pesticides to the air during sprayer application: A bibliographic review. *Atmos. Environ.* 39: 5183–5193.

Graham, W. H., Graham, D. W., Denoyelles, F., Smith, V. H., Larive, C. K., and Thurman, E. M. 1999. Metolachlor and alachlor breakdown product formation patterns in aquatic field mesocosms. *Environ. Sci. Technol.* 33: 4471–4476.

Green, P. G. and Young, T. M. 2006. Loading of the herbicide diuron into the California water system. *Environ. Eng. Sci.* 23: 545–551.

Grimalt, S., Sancho, J. V., Pozo, O. J., and Hernández, F. 2010. Quantification, confirmation and screening capability of UHPLC coupled to triple quadrupole and hybrid quadrupole time-of-flight mass spectrometry in pesticide residue analysis. *J. Mass Spectrom.* 45: 421–436.

Guenu, S. and Hennion, M. C. 1996. Evaluation of new polymeric sorbents with high specific surface areas using an on-line solid-phase extraction liquid chromatographic system for the trace-level determination of polar pesticides. *J. Chromatogr. A* 737: 15–24.

Gupta, R. C. 1994. Carbofuran toxicity. *J. Toxicol. Environ. Health* 43: 383–418.

Gustafson, D. I. 1989. Groundwater ubiquity score: A simple method for assessing pesticide leachability. *Environ. Toxicol. Chem.* 8: 339–357.

Hamilton, D. J., Amabrus, Á., Dieterle, R. M., et al. 2003. Regulatory limits for pesticide residues in water. *Pure Appl. Chem.* 75: 1123–1155.

Hamwijk, C., Schouten, A., Foekema, E. M., et al. 2005. Monitoring of the booster biocide dichlofluanid in water and marine sediment of Greek marinas. *Chemosphere* 60: 1316–1324.

Han, A., Li, Z., Wang, H., et al. 2009. Plant availability and phytotoxicity of soil bound residues of herbicide ZJ0273, a novel acetolactate synthase potential inhibitor. *Chemosphere* 77: 955–961.

Hennion, M. C. 1991. Sample handling strategies for the analysis of organic compounds in environmental water samples. *Trend. Anal. Chem.* 10: 317–323.

Hernández, F., Ibáñez, M., Sancho, J. V., and Pozo, Ó. J. 2004. Comparison of different mass spectrometric techniques combined with liquid chromatography for confirmation of pesticides in environmental water based on the use of identification points. *Anal. Chem.* 76: 4349–4357.

Hernández, F., Sancho, J. V., Pozo, O., Lara, A., and Pitarch, E. 2001. Rapid direct determination of pesticides and metabolites in environmental water samples at sub-mu g/l level by on-line solid-phase extraction-liquid chromatography-electrospray tandem mass spectrometry. *J. Chromatogr. A* 939: 1–11.

Hidalgo, C., Sancho, J. V., Lopez, F. J., and Hernandez, F. 1998. Automated determination of phenylcarbamate herbicides in environmental waters by on-line trace enrichment and reversed-phase liquid chromatography diode array detection. *J. Chromatogr. A* 823: 121–128.

Hildebrandt, A., Guillamon, M., Lacorte, S., Tauler, R., and Barceló, D. 2008. Impact of pesticides used in agriculture and vineyards to surface and groundwater quality (North Spain). *Water Res.* 42: 3315–3326.

Hogendoorn, E. and Zoonen, P. V. 2000. Recent and future developments of liquid chromatography in pesticide trace analysis. *J. Chromatogr. A* 892: 435–453.

Hogendoorn, E. A., Dijkman, E., Baumann, B., Hidalgo, C., Sancho, J.-V., and Hernandez, F. 1999. Strategies in using analytical restricted access media columns for the removal of humic acid interferences in the trace analysis of acidic herbicides in water samples by coupled column liquid chromatography with UV detection. *Anal. Chem.* 71: 1111–1118.

Holvoet, K. M. A., Seuntjens, P., and Vanrolleghem, P. A. 2007. Monitoring and modeling pesticide fate in surface waters at the catchment scale. *Ecol. Model.* 209: 53–64.

Hornsby, A. G., Wauchope, R. D., and Herner, A. E. 1996. *Pesticide Properties in the Environment.* Springer-Verlag: New York.

Hsiao, A. I. and Smith, A. E. 1983. A root bioassay procedure for the determination of chlorsulfuron, diclofop acid and sethoxydim residues in soils. *Weed Res.* 23: 231–236.

Huang, S. B., Stanton, J. S., Lin, Y., and Yokley, R. A. 2003. Analytical method for the determination of atrazine and its dealkylated chlorotriazine metabolites in water using SPE sample preparation and GC-MSD analysis. *J. Agric. Food Chem.* 51: 7252–7258.

Ibáñez, M., Sancho, J. V., Pozo, O. J., and Hernández, F. 2004. Use of quadrupole time-of-flight mass spectrometry in environmental analysis: Elucidation of transformation products of triazine herbicides in water after UV exposure. *Anal. Chem.* 76: 1328–1335.

Ibáñez, M., Sancho, J. V., Pozo, O. J., and Hernández, F. 2006. Use of liquid chromatography quadrupole time-of-flight mass spectrometry in the elucidation of transformation products and metabolites of pesticides. Diazinon as a case study. *Anal. Bioanal. Chem.* 384: 448–457.

Kawahigashi, H., Hirose, S., Hayashi, E., Ohkawa, H., and Ohkawa, Y. 2002. Phytotoxicity and metabolism of ethofumesate in transgenic rice plants expressing the human CYP2B6 gene. *Pestic. Biochem. Physiol.* 74: 139–147.

Keizer, J. P., MacQuarrie, K. T. B., Milburn, P. H., McCully, K. V., King, R. R., and Embleton, E. J. 2001. Long-term groundwater quality impacts from the use of hexazinone for the commercial production of lowbush blueberries. *Ground Water Monit. Remediat.* 21: 128–135.

Khan, S. U. and Saidak, W. J. 1981. Residues of atrazine and its metabolites after prolonged usage. *Weed Res.* 21: 9–12.

Kjaer, J., Ullum, M., Olsen, P., et al. 2002. The Danish pesticide leaching assessment programme: Monitoring results, May 1999–July 2001. http://www.pesticidvarsling.dk.

Kjaer, J., Olsen, P., Henriksen, T., and Ullum, M. 2005. Leaching of metribuzin metabolites and the associated contamination of a sandy Danish aquifer. *Environ. Sci. Technol.* 39: 8374–8381.

Kobayashi, H. 1992. A study on analytical methods for determination of pesticide-residue in animals, plants, and the environment. *J. Pestic. Sci.* 17: S125–S36.

Kodama, S., Yamamoto, A., Ohto, M., and Matsunaga, A. 1999. Major degradation pathway of Thiuram in tap water processed by oxidation with sodium hypochlorite. *J. Agric. Food Chem.* 47: 2914–2919.

Koeber, R., Fleischer, C., Lanza, F., Boos, K. S., Sellergren, B., and Barcelo, D. 2001. Evaluation of a multidimensional solid-phase extraction platform for highly selective on-line cleanup and high-throughput LC-MS analysis of triazines in river water samples using molecularly imprinted polymers. *Anal. Chem.* 73: 2437–2444.

Kolpin, D. W., Thurman, E. M., and Goolsby, D. A. 1996. Occurrence of selected pesticides and their metabolites in near-surface aquifers of the Midwestern United States. *Environ. Sci. Technol.* 30: 335–340.

Kolpin, D. W., Thurman, E. M., and Linhart, M. S. 2000a. Finding minimal herbicide concentrations in groundwater? Try looking for their degradates. *Sci. Total Environ.* 248: 115–22.

Kolpin, D. W., Schnoebelen, D. J., and Thurman, E. M. 2004. Degradates provide insight to spatial and temporal trends of herbicides in groundwater. *Ground Water* 42: 601–608.

Kolpin, D. W., Kalkhoff, S. J., Goolsby, D. A., and Sneck-Fahrer, D. A. 1997. Occurrence of selected herbicides and herbicide degradation products in Iowa's groundwater, 1995. *Ground Water* 35: 679–688.

Kolpin, W., Barbash, J. E., and Gilliom, R. J. 2000b. Pesticides in groundwater of the United States, 1992–1996. *Ground Water* 38: 858–863.

Konstantinou, I. K., Zarkadis, A. K., and Albanis, T. A. 2001. Photodegradation of selected herbicides in various natural waters and soils under environmental conditions. *J. Environ. Qual.* 30: 121–130.

Kristenson, E., Ramos, L., and Brinkman, U. 2006. Recent advances in matrix solid-phase dispersion. *Trends Anal. Chem.* 25: 96–111.

Lacorte, S., Lartiges, S. B., Garrigues, P., and Barceló, D. 1995. Degradation of organophosphorus pesticides and their transformation products in estuarine waters. *Environ. Sci. Technol.* 29: 431–438.

Lapworth, D. J. and Gooddy, D. C. 2006. Source and persistence of pesticides in a semi-confined chalk aquifer of Southeast England. *Environ. Pollut.* 144: 1031–1044.

Lehotay, S. J., Kok, A. D., Hiemstra, M., and Bodegraven, P. V. 2005. Validation of a fast and easy method for the determination of residues from 229 pesticides in fruits and vegetables using gas and liquid chromatography and mass spectrometric detection. *J. AOAC Int.* 88: 595–614.

Lin, Y. J., Karuppiah, M., Shaw, A., and Gupta, G. 1999. Effect of simulated sunlight on atrazine and metolachlor toxicity of surface waters. *Ecotoxicol. Environ. Safe.* 43: 35–37.

Maguire, R. J. and Tkacz, R. J. 1993. Occurrence of pesticides in the Yamaska river, Québec. *Arch. Environ. Contam. Toxicol.* 25: 220–226.

Marcheterre, L., Choudhry, G. G., and Webster, G. R. B. 1988. Environmental photochemistry of herbicides. *Rev. Environ. Contam. Toxicol.* 103: 61–126.

Marín, J. M., Gracia-Lor, E., Sancho, J. V., López, F. J., and Hernández, F. 2009. Application of ultra-high-pressure liquid chromatography–tandem mass spectrometry to the determination of multi-class pesticides in environmental and wastewater samples Study of matrix effects. *J. Chromatogr. A* 1216: 1410–1420.

Martins, M. F. J., Chevre, N., Spack, L., Tarradellas, J., and Mermoud, A. 2001. Degradation in soil and water and ecotoxicity of rimsulfuron and its metabolites. *Chemosphere* 45: 515–522.

Meng, C. K., Zweigenbaum, J., Furst, P., and Blanke, E. 2010. Finding and confirming nontargeted pesticides using GC/MS, LC/quadrupole-time-of-flight MS, and databases. *J. AOAC Int.* 93: 703–711.

Miles, C. J. 1991. Degradation products of sulfur-containing pesticides in soil and water. In: Somasundaram, L. and Coats, J. R. (eds), *Pesticide Transformation Products. Fate and Significance in the Environment*, 61–74. ACS Symposium Series: Washington, DC.

Miles, C. J. and Delfino, J. J. 1984. Determination of aldicarb and its derivatives in groundwater by high performance liquid chromatography with UV detection. *J. Chromatogr. A* 299: 275–280.

Mills, M. S. and Thurman, E. M. 1994. Reduction of nonpoint source contamination of surface water and groundwater by starch encapsulation of herbicides. *Environ. Sci. Technol.* 28: 73–79.

Mitch, W. A., Sharp, J. O., Trussell, R. R., Valentine, R. L., Alvarez-Cohen, L., and Sedlak, D. L. 2003. N-Nitrosodimethylamine (NDMA) as a drinking water contaminant: A review. *Environ. Eng. Sci.* 20: 389–404.

Neilson, A. H. and Allard, A.-S. 2008. *Environmental Degradation and Transformation of Organic Chemicals*. CRC Press: Boca Raton, FL.

Nowell, L. H. and Resek, E. A. 1994. National standards and guidelines for pesticides in water, sediment, and aquatic organisms: Application to water-quality assessments. *Rev. Environ. Contam. Toxicol.* 140: 1–164.

Okamura, H., Aoyama, I., Liu, D., Maguire, R. J., Pacepavicius, G. J., and Lau, Y. L. 2000. Fate and ecotoxicity of the new antifouling compound irgarol 1051 in the aquatic environment. *Water Res.* 34: 3523–3530.

Osano, O., Admiraal, W., Klamer, H. J. C., Pastor, D., and Bleeker, E. A. J. 2002. Comparative toxic and genotoxic effects of chloroacetanilides, formamidines and their degradation products on Vibrio fischeri and Chironomus riparius. *Environ. Pollut.* 119: 195–202.

Peñuela, G. A., Ferrer, I., and Barceló, D. 2000. Identification of new photodegradation by-products of the antifouling agent Irgarol in seawater samples. *Int. J. Environ. Anal. Chem.* 78: 25–40.

Pfeuffer, R. J. and Rand, G. M. 2004. South Florida ambient pesticide monitoring program. *Ecotoxicology* 13: 195–205.

Pichon, V., Cau dit Coumes, C. C. D., Chen, L., and Hennion, M.-C. 1996. Solid-phase extraction, clean-up and liquid chromatography for routine multiresidue analysis of neutral and acidic pesticides in natural waters in one run. *Int. J. Environ. Anal. Chem.* 65: 11–25.

Picó, Y., Fernández, M., Ruiz, M. J., and Font, G. 2007. Current trends in solid-phase-based extraction techniques for the determination of pesticides in food and environment. *J. Biochem. Biophys. Methods* 70: 117–131.

Porazzi, E., Martinez, M. P., Fanelli, R., and Benfenati, E. 2005. GC-MS analysis of dichlobenil and its metabolites in groundwater. *Talanta* 68: 146–154.

Postle, J. K., Rheineck, B. D., Allen, P. E., et al. 2004. Chloroacetanilide herbicide metabolites in Wisconsin groundwater: 2001 survey results. *Environ. Sci. Technol.* 38: 5339–5343.

Priddle, M. W., Mutch, J. P., and Jackson, R. E. 1992. Long-term monitoring of aldicarb residues in groundwater beneath a Canadian potato field. *Arch. Environ. Contam. Toxicol.* 22: 183–189.

Quintana, J., Martí, I., and Ventura, F. 2001. Monitoring of pesticides in drinking and related waters in NE Spain with a multiresidue SPE-GC-MS method including an estimation of the uncertainty of the analytical results. *J. Chromatogr. A* 938: 3–13.

Racke, K. D. 2003. Release of pesticides into the environment and initial concentrations in soil, water, and plants. *Pure Appl. Chem.* 75: 1905–1916.

Reddy, K., Rimando, A., and Duke, S. 2004. Aminomethylphosphonic acid, a metabolite of glyphosate, cause injury in glyphosate-treated, glyphosate-resistant soybean. *J. Agric. Food Chem.* 52: 5139–5143.

Reemtsma, T. and Jekel, M. 2006. *Organic Pollutants in the Water Cycle*. Wiley-VCH: Weinheim.

Rice, P. J., Anderson, T. A., and Coats, J. R. 2002. Degradation and persistence of metolachlor in soil: Effects of concentration, soil moisture, soil depth, and sterilization. *Environ. Toxicol. Chem.* 21: 2640–2648.

Richards, R. P. and Baker, D. B. 1993. Pesticide concentration patterns in agricultural drainage networks in the lake Erie basin. *Environ. Toxicol. Chem.* 12: 13–26.

Richardson, S. D. 2006. Environmental mass spectrometry: Emerging contaminants and current issues. *Anal. Chem.* 78: 4021–4045.

Richardson, S. D. 2007. Water analysis: Emerging contaminants and current issues. *Anal. Chem.* 79: 4295–4324.

Richardson, S. D. 2009. Water analysis: Emerging contaminants and current issues. *Anal. Chem.* 81: 4645–4677.

Rodrigues, A. M., Ferreira, V., Cardoso, V. V., Ferreira, E., and Benoliel, M. J. 2007. Determination of several pesticides in water by solid-phase extraction, liquid chromatography and electrospray tandem mass spectrometry. *J. Chromatogr. A* 1150: 267–278.

Rodríguez-Mozaz, S., López de Alda, M. J., and B. D. 2007. Advantages and limitations of on-line solid phase extraction coupled to liquid chromatography-mass spectrometry technologies versus biosensors for monitoring of emerging contaminants in water. *J. Chromatogr. A* 1152: 97–115.

Röpke, B., Bach, M., and Frede, H.-G. 2004. DRIPS-aDSS for estimating the input quantity of pesticides for German river basins. *Environ. Model. Softw.* 19: 1021–1028.

Rosenbom, A. E., Kjaer, J., and Olsen, P. 2010. Long-term leaching of rimsulfuron degradation products through sandy agricultural soils. *Chemosphere* 79: 830–838.

Ryu, S. M., Lin, J., and Thompson, C. M. 1991. Comparative anticholinesterase potency of chiral isoparathion methyl. *Chem. Res. Toxicol.* 4: 517–520.

Sabik, H., Jeannot, R., and Rondeau, B. 2000. Multiresidue methods using solid-phase extraction techniques for monitoring priority pesticides, including triazines and degradation products, in ground and surface waters. *J. Chromatogr. A* 885: 217–236.

Saha, S. and Kulshrestha, G. 2002. Degradation of sulfosulfuron, a sulfonylurea herbicide, as influenced by abiotic factors. *J. Agric. Food Chem.* 50: 4572–4575.

Sakellarides, T. M., Siskos, M. G., and Albanis, T. A. 2003. Photodegradation of selected organophosphorus insecticides under sunlight in different natural waters and soils. *Int. J. Environ. Anal. Chem.* 83: 33–50.

Sakkas, V. A., Lambropoulou, D. A., and Albanis, T. A. 2002a. Study of chlorothalonil photodegradation in natural waters and in the presence of humic substances. *Chemosphere* 48: 939–945.

Sakkas, V. A., Lambropoulou, D. A., and Albanis, T. A. 2002b. Photochemical degradation study of irgarol 1051 in natural waters: Influence of humic and fulvic substances on the reaction. *J. Photochem. Photobiol. A* 147: 135–141.

Sancho, J. V., Pozo, O. J., and Hernández, F. 2004. Liquid chromatography and tandem mass spectrometry: A powerful approach for the sensitive and rapid multiclass determination of pesticides and transformation products in water. *Analyst* 129: 38–44.

Sandín-España, P., Magrans, J. O., and García-Baudín, J. M. 2005. Study of clethodim degradation and by-product formation in chlorinated water by HPLC. *Chromatographia* 62: 133–137.

Sandín-España, P., González-Blázquez, J. J., Magrans, J. O., and García-Baudín, J. M. 2002. Determination of herbicide tepraloxydim and main metabolites in drinking water by solid-phase extraction and liquid chromatography with UV detection. *Chromatographia* 55: 681–686.

Sandín-España, P., Llanos, S., Magrans, J. O., Alonso-Prados, J. L., and García-Baudín, J. M. 2003. Optimization of hydroponic bioassay for herbicide tepraloxydim by using water free from chlorine. *Weed Res.* 43: 451–457.

Sandra, P., Tienpont, B., and David, F. 2003. Stir bar sorptive extraction (Twister) RTL-CGC-MS. A versatile method to monitor more than 400 pesticides in different matrices (water, beverages, fruits, vegetables, baby food). Paper presented at New Horizons and Challenges in Environmental Analysis and Monitoring, Gdansk.

Sandra, P., Tienpont, B., Vercammen, J., Tredoux, A., Sandra, T., and David, F. 2001. Stir bar sorptive extraction applied to the determination of dicarboximide fungicides in wine. *J. Chromatogr. A* 928: 117–126.

Santoro, A., Scopa, A., Bufo, S. A., Mansour, M., and Mountacer, H. 2000. Photodegradation of the triazole fungicide hexaconazole. *Bull. Environ. Contam. Toxicol.* 64: 475–480.

Santos, T. C. R., Rocha, J. C., and Barceló, D. 2000. Determination of rice herbicides, their transformation products and clofibric acid using on-line solid-phase extraction followed by liquid chromatography with diode array and atmospheric pressure chemical ionization mass spectrometric detection. *J. Chromatogr. A* 879: 3–12.

Santos, T. C. R., Rocha, J. C., Alonso, R. M., Martínez, E., Ibáñez, C., and Barceló, D. 1998. Rapid degradation of propanil in rice crop fields. *Environ. Sci. Technol.* 32: 3479–3484.

Sanz-Asencio, J., Plaza-Medina, M., and Martínez-Soria, M. T. 1997. Kinetic study of the degradation of ethiofencarb in aqueous solutions. *Pestic. Sci.* 50: 187–194.

Sarmah, A. K. and Sabadie, J. 2002. Hydrolysis of sulfonylurea herbicides in soils and aqueous solutions: A review. *J. Agric. Food Chem.* 50: 6253–6265.

Schneiders, G. E., Koeppe, M. K., Naidu, M. V., Horne, P., Brown, A. M., and Mucha, C. F. 1993. Fate of rimsulfuron in the environment. *J. Agric. Food Chem.* 41: 2404–2410.

Schottler, S. P. and Eisenreich, S. J. 1994. Herbicides in the Great Lakes. *Environ. Sci. Technol.* 28: 2228–2232.

Schwarzenbach Rene, P., Gschwend, P. M., and Imboden, M. D. 2002. *Environmental Organic Chemistry*. John Wiley & Sons: Hoboken, NJ.

Scrano, L., Bufo, S. A., D'Auria, M., Meallier, P., Behechti, A., and Shramm, K. W. 2002. Photochemistry and photoinduced toxicity of acifluorfen, a diphenyl-ether herbicide. *J. Environ. Qual.* 31: 268–274.

Sevilla Morán, B. 2010. Photodegradation study of cyclohexanedione herbicides in aquatic media. Ph.D. Dissertation, Universidad Autónoma de Madrid.

Sevilla-Morán, B., Alonso-Prados, J. L., García-Baudín, J. M., and Sandín-España, P. 2010a. Indirect photodegradation of clethodim in aqueous media. By-product identification by quadrupole time-of-flight mass spectrometry. *J. Agric. Food Chem.* 58: 3068–3076.

Sevilla-Morán, B., Sandín-España, P., Vicente-Arana, M. J., Alonso-Prados, J. L., and García-Baudín, J. M. 2008. Study of alloxydim photodegradation in the presence of natural substances: Elucidation of transformation products. *J. Photochem. Photobiol. A* 198: 162–168.

Sevilla-Morán, B., Mateo-Miranda, M. M., Alonso-Prados, J. L., García-Baudín, J. M., and Sandín-España, P. 2010b. Sunlight transformation of sethoxydim-lithium in natural waters and effect of humic acids. *Int. J. Environ. Anal. Chem.* 90: 487–496.

Shankar, M. V., Nélieu, S., Kerhoas, L., and Einhorn, J. 2007. Photo-induced degradation of diuron in aqueous solution by nitrites and nitrates: Kinetics and pathways. *Chemosphere* 66: 767–774.

Sinclair, C. J. and Boxall, A. B. A. 2003. Assessing the ecotoxicity of pesticide transformation products. *Environ. Sci. Technol.* 37: 4617–4625.

Somasundaram, L. and Coats, J. R. 1991a. Pesticide transformation products in the environment. In: Somasundaram, L. and Coats, J. R. (eds), *Pesticide Transformation Products. Fate and Significance in the Environment*, 2–9. ACS Symposium Series: Washington, DC.

Somasundaram, L. and Coats, J. R. (eds.) 1991b. *Pesticide Transformation Products. Fate and Significance in the Environment*. ACS Symposium Series: Washington, DC.

Somasundaram, L., Coast, J. R., Shanbhag, V. M., and Racke, K. D. 1991. Mobility of pesticides and their hydrolysis metabolites in soil. *Environ. Toxicol. Chem.* 10: 185–194.

Soriano, J. M., Jimenez, B., Font, G., and Molto, J. C. 2001. Analysis of carbamate pesticides and their metabolites in water by solid phase extraction and liquid chromatography: A review. *Crit. Rev. Anal. Chem.* 31: 19–52.

Spalding, R. F., Exner, M. E., Snow, D. D., Cassada, D. A., Burbach, M. E., and Monson, S. J. 2003. Herbicides in groundwater beneath Nebraska's management systems evaluation area. *J. Environ. Qual.* 32: 92–99.

Steinheimer, T. R. 1993. HPLC determination of atrazine and principal degradates in agricultural soils and associated surface and groundwater. *J. Agric. Food Chem.* 41: 588–595.

Tchaikovskaya, O., Sokolova, I., Svetlichnyi, V., Karetnikova, E., Fedorova, E., and Kudryasheva, N. 2007. Fluorescence and bioluminescence analysis of sequential UV-biological degradation of p-cresol in water. *Luminescence* 22: 29–34.

Thurman, E. M., Gooddy, D. A., Aga, D. S., Pomes, M. L., and Meyer, M. T. 1996. Occurrence of ala-chlor and Its sulfonated metabolite in rivers and reservoirs of the Midwestern United States: The importance of sulfonation in the transport of chloroacetanilide herbicides. *Environ. Sci. Technol.* 30: 569–574.

Thurman, E. M., Meyer, M. T., Mllls, M. S., Zimmerman, L. R., Perry, C. A., and Goolsby, D. A. 1994. Formation and transport of deethylatrazine and deisopropylatrazine in surface water. *Environ. Sci. Technol.* 28: 2267–2277.

Tixier, C., Sancelme, M., Bonnemoy, F., Cuer, A., and Veschambre, H. 2001. Degradation products of a phenylurea herbicide, diuron: Synthesis, ecotoxicity, and biotransformation. *Environ. Toxicol. Chem.* 20: 1381–1389.

Tixier, C., Bogaerts, P., Sancelme, M., et al. 2000. Fungal biodegradation of a phenylurea herbicide, diuron: Structure and toxicity of metabolites. *Pest Manag. Sci.* 56: 455–462.

Tomlin, C. D. S. 2006. *The Pesticide Manual: A World Compendium.* BCPC Publications: Hampshire.

Trovó, A. G., Nogueira, R. F. P., Agüera, A., Sirtori, C., and Fernández-Alba, A. R. 2009. Photodegradation of sulfamethoxazole in various aqueous media: Persistence, toxicity and photoproducts assessment. *Chemosphere* 77: 1292–1298.

Turgut, C. 2007. Organochlorine insecticide residues in Turkish mineral waters. *Fresenius Environ. Bull.* 16: 252–255.

Tuxen, N., Tüchsen, P. L., Rügge, K., Albrechtsen, H.-J., and Bjerg, P. L. 2000. Fate of seven pesticides in an aerobic aquifer studied in column experiments. *Chemosphere* 41: 1485–1494.

Tuxhorn, G. L., Roeth, F. W., Martin, A. R., and Wilson, R. G. 1986. Butylate persistence and activity in soils previously treated with thiocarbamates. *Weed Sci.* 34: 961–965.

Vagi, M. C., Petsas, A. S., Kostopoulou, M. N., and Lekkas, T. D. 2003. Monitoring of pesticide residues in the surface waters of Greece. Paper presented at XII Symposium Pesticide Chemistry, Piacenza.

Vialaton, D. and Richard, C. 2002. Phototransformation of aromatic pollutants in solar light: Photolysis versus photosensitized reactions under natural water conditions. *Aquat. Sci.* 64: 207–215.

Vialaton, D., Richard, C., Baglio, D., and Paya-Perez, A. B. 1998. Phototransformation of 4-chloro-2-methylphenol in water: Influence of humic substances on the reaction. *J. Photochem. Photobiol. A* 119: 39–45.

Vialaton, D., Pilichowski, J. F., Baglio, D., Paya-Perez, A., Larsen, B., and Richard, C. 2001. Phototransformation of propiconazole in aqueous media. *J. Agric. Food Chem.* 49: 5377–5382.

Vulliet, E., Emmelin, C., Chovelon, J. M., Chouteau, C., and Clement, B. 2004. Assessment of the toxicity of trisulfuron and its photoproducts using aquatic organisms. *Environ. Toxicol. Chem.* 23: 2837–2843.

Walls, D., Smith, P. G., and Mansell, M. G. 1996. Pesticides in grounwater in Britain. *Int. J. Environ. Health Res.* 6: 55–62.

Wang, T. C. and Hoffman, M. E. 1991. Degradation of organophosphorus pesticides in coastal water. *J. Assoc. Off. Anal. Chem.* 74: 883–886.

Warner, K. L. and Morrow, W. S. 2007. Pesticide and transformation product detections and age-dating relations from till and sand deposits. *J. Am. Water Resour. Assoc.* 43: 911–922.

Wauchope, R. D. 1996. Pesticides in runoff: Measurement, modeling, and mitigation. *J. Environ. Sci. Health B* 31: 337–344.

Weber, J. B. and Miller, C. T. 1989. Organic chemical movement over and through soil. In: Sawheney, B. L. and Brown, K. (eds), *Reactions and Movement of Organic Chemicals in Soil*, 305–34. SSSA Spec. Publ: Madison.

Werres, F., Fastabend, A., Balsaa, P., and Overath, H. 1996. Studies on the degradation of pesticides in water by the exposure to chlorine. *Vom Wasser* 87: 39–49.

Whitall, D., Hively, W. D., Leight, A. K., et al. 2010. Pollutant fate and spatio-temporal variability in the choptank river estuary: Factors influencing water quality. *Sci. Total Environ.* 408: 2096–2108.

WHO. 2010. WHO guidelines for drinking-water quality. http://www.who.int/water_sanitation_health/dwq/gdwq3rev/en/.

Willis, G. H. and McDowell, L. L. 1982. Pesticides in agricultural runoff and their effects on downstream water quality. *Environ. Toxicol. Chem.* 1: 267–279.

Wilson, R. I. and Mabury, S. A. 2000. Photodegradation of metolachlor: Isolation, identification, and quantification of monochloroacetic acid. *J. Agric. Food Chem.* 48: 944–950.

Wolfe, M. L. 2001. Hydrology. In: Ritter, F. and Shirmohammadi, A. (eds), *Agricultural Nonpoint Source Pollution: Watershed Management and Hydrology*, 1–27. CRC press LLC: Florida.

Wu, Q., Riise, G., Lundekvam, H., Mulder, J., and Haugen, L. E. 2004. Influences of suspended particles on the runoff of pesticides from an agricultural field at Askim, SE-Norway. *Environ. Geochem. Health* 26: 295–302.

Younos, T. M. and Weigmann, D. L. 1988. Pesticides: A continuing dilema. *J. WPCF* 60: 1199–1205.

Zamy, C., Mazellier, P., and Legube, B. 2004. Phototransformation of selected organophosphorus pesticides in dilute aqueous solutions. *Water Res.* 38: 2305–2314.

Zimmerman, L. R., Schneider, R. J., and Thurman, E. M. 2002. Analysis and detection of the herbicides dimethenamid and flufenacet and their sulfonic and oxanilic acid degradates in natural water. *J. Agric. Food Chem.* 50: 1045–1052.

5

Microbial Remediation of Pesticides

Ibrahim Abdulwahid Arif, Mohammad Abdul Bakir, and Haseeb Ahmad Khan

CONTENTS

5.1 Introduction

As a generalized definition, a pesticide is any substance or mixture of substances intended for preventing, destroying, repelling, or mitigating any pest. Though often misunderstood to refer only to insecticides, the term pesticide also applies to herbicides, fungicides, and various other substances used to control pests (US EPA 2010c). Pesticides are often referred to according to the type of pest they control. Major types of pesticides are listed in Table 5.1 (modified from US EPA 2010b). Pesticides have been used to control, eliminate, or destroy pests in order to meet the continuously increasing food demand of the world. Therefore, contamination of soils with persistent pesticides is a widespread problem arising from the extensive use of pesticides in agriculture or from industrial chemical wastes. They have harmful effects directly or indirectly on soil, environment, surface water and groundwater, natural flora and fauna, and aquatic life, which will ultimately adversely influence human beings and livestock. So, the impact of pesticides on atmosphere and community health is of great significance regardless of their noticeable benefits (Rashid et al. 2010). Use of pesticides worldwide exceeded 5.0 billion pounds in 2000 and 2001. Herbicides accounted for the largest portion of total use, followed by other pesticides, insecticides, and fungicides (Table 5.2; modified from US EPA 2010a). Pesticides have been used for agriculture, gardening, and controlling of home-based pests. Despite environmental disturbances, harmful effects of pesticides come from direct use. Ironically, pesticide poisoning causes more deaths than infectious diseases in

TABLE 5.1

Major Groups of Pesticides

Chemical Pesticides	Biopesticides
1. Organophosphate pesticides (e.g., parathion, malathion) 2. Carbamate pesticides (e.g., aldicarb, carbofuran (Furadan), fenoxycarb, carbaryl (Sevin), ethienocarb, and fenobucarb) 3. Organochlorine insecticides (e.g., DDT, dicofol, heptachlor, endosulfan, chlordane, aldrin, dieldrin, endrin, mirex, and pentachlorophenol) 4. Pyrethroid pesticides (e.g., permethrin)	1. Microbial pesticides (e.g., subspecies and strains of *Bacillus thuringiensis* or Bt. 2. Plant-incorporated-protectants (PIPs) (e.g., introduction of the gene for the Bt. pesticidal protein into the plant's own genetic material. The plant, instead of the Bt. bacterium, manufactures the substance that destroys the pest) 3. Biochemical pesticides (include substances, such as insect sex pheromones, that interfere with mating, as well as various scented plant extracts that attract insect pests to traps)

certain parts of the developing world. Use of pesticides in certain parts of the developing world is poorly regulated and often dangerous; their easy availability also makes them a popular method of self-harm (Eddleston et al. 2002).

Bioremediation is defined as an accelerated process using microorganisms (indigenous or introduced) and other manipulations to degrade and detoxify organic substances to harmless compounds (such as carbon dioxide and water) in a confined and controlled environment (EPA 2005). Bioremediation is suitable for the treatment of a variety of organic chemicals, including volatile organic compounds, benzene, toluene, ethyl benzene, xylene (together known as BTEX), phenolic compounds, polycyclic aromatic hydrocarbons (PAHs), other petroleum hydrocarbons, and nitroaromatic compounds. Numerous studies have been conducted for bioremediation purposes. However, despite extensive studies, microbial remediation system of pesticides is less understood. Microbes that have been very recently used for the remediation of pesticides are discussed briefly in this chapter. Readers of this chapter interested in microbial remediation of pesticides will have a glimpse of very recent ideas on the microbes that have been used for bioremediation.

TABLE 5.2

Worldwide Estimated Use of Pesticides in 2000 (Amount of Active Ingredient at User Level)

Pesticide Type	Millions of Pounds	Percentage
Herbicides (including plant growth regulators)	1944	36
Insecticides	1355	25
Fungicides	516	10
Others (including nematicides, fumigants, rodenticides, molluscicides, aquatic and fish/bird pesticides, other miscellaneous conventional pesticides, and other chemicals used as pesticides, e.g., sulfur and petroleum oil)	1536	29
Total	*5351*	*100*

5.2 General Considerations for Microbial Remediation of Pesticides

Bioremediation of a contaminated site depends on several factors: (i) types and amounts of harmful chemicals present, (ii) size and depth of the polluted area, (iii) type of soil and the conditions present, and (iv) whether the remediation process occurs above ground or underground. These factors vary from site to site. It can take a few months or even several years for microbes for the partial or complete degradation of harmful chemicals to clean up the site (US EPA 2001). A flowchart outlining the strategy involved in bioremediation activities is illustrated in Figure 5.1 (modified from EPA 2005). Bioremediation strategy employs modification of the existing environment to encourage the growth and reproduction of natural or exogenous microorganisms. To grow and reproduce, these microorganisms require a source of energy (i.e., electron donor) and a means of extracting this energy from the electron donor via an appropriate electron acceptor as the following (Equation for heterotrophic organisms) (US EPA 1998):

*Microbes + Electron Donor (Energy and Carbon Source) + Nutrients + Electron Acceptor →
More Microbes + Oxidized (Aerobic) or Reduced (Anaerobic) End Products*

Biological systems generally treat contaminated media by using waste contaminants of concern as the electron donor and supply microorganisms with the required electron acceptors and nutrients. The anaerobic chlorination of chlorinated solvents is the exception. The chlorinated solvents serve as the electron acceptor under highly reducing conditions. Generally, the limiting factor in full-scale, engineered biological treatment systems is the rate of transfer of the electron acceptor to the reaction site. Electron acceptors are those chemicals that can be used by biological systems to extract energy from electron donors for microbial cell growth and replication. The main electron acceptors of interest include oxygen, nitrate, sulfate, iron, manganese, carbon dioxide, and organic carbon. If oxygen is

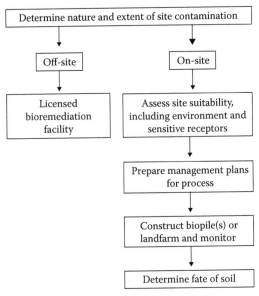

FIGURE 5.1
Strategic flowchart for bioremediation activities.

present, it will be utilized preferentially over the other electron acceptors because it provides maximum energy yield to the microorganism, resulting in the maximum possible amount of cell production and organism growth per unit amount of electron donor used. Once oxygen is depleted at a site, facultative and anaerobic microorganisms use other electron acceptors. Microorganisms use acceptors in sequence based on the relative energy yield of each—nitrate, manganese, iron, sulfate, carbon dioxide, and, finally, organic carbon (US EPA 1998).

The effectiveness of biodegradation is highly dependent on the toxicity and initial concentrations of the contaminants, their biodegradability, the properties of the contaminated soil, and the type of microorganism used for remediation purpose. In general, there are mainly two types of microorganisms that are used for bioremediation: (i) indigenous and (ii) exogenous. The indigenous ones are those microorganisms that are already present at a contaminated site. Exogenous ones are the effective microorganisms that are not present in the pesticide-contaminated site but are added to the contaminated soil. To stimulate the growth of indigenous microorganisms, the proper soil temperature and oxygen and nutrient content may need to be provided. Bioremediation can take place under aerobic and anaerobic conditions. With sufficient oxygen, microorganisms can convert many organic contaminants to carbon dioxide and water. Anaerobic conditions support biological activity in which no oxygen is present, so the microorganisms break down chemical compounds in the soil to release the energy they need. A key difference between aerobic (oxidative) and anaerobic breakdown is that the former is predominantly used for lower-chlorinated congeners and the latter for high-chlorinated congeners (hydrodechlorination). Bioremediation is natural. It is happening on earth without human help. However, sometimes it is too slow to meet the demand. Generally, bioremediation technologies can be classified as in situ or ex situ. In situ bioremediation involves treating the contaminated material at the site, while ex situ bioremediation involves the removal of the contaminated material to be treated elsewhere. Some of the bioremediation technologies include bioventing, land farming, bioreactor, composting, bioaugmentation, and biostimulation. Bioremediation is best accomplished with bioaugmentation. In addition to bioaugmentation, several physical and chemical parameters must be controlled in order to obtain optimal growth and maximum degradation of contaminants. Microbial population, nutrient concentration, oxygen supply, temperature, moisture content, pH, toxicity, heavy metal, molecular structure, and cometabolism are important for optimum performance of the microorganisms (Table 5.3, adapted from Vidali 2001). Microbial activity can be also affected by competition from undesirable organisms (Mulligan 2002).

TABLE 5.3

Optimal Environmental Factors for the Degradation of Contaminants by Soil Microbes

Factors	Optimal Condition for Microbial Activity
pH	5.5–8.8
Temperature	15°C–45°C
Moisture	25%–28% of the soil water-holding capacity
Soil type	Low clay or silt content
Oxygen	For aerobic, >10% of air-filled space of the soil
Nutrient	N&P for growth
Contaminants	<toxicity level
Heavy metal	Total 2000 ppm

5.3 Bacteria

Extensive work has been conducted on the remediation of pesticides by bacterial strains and microbial consortia. Various bacterial strains have been recently proved to be useful for the removal of pesticides. A list of recently used taxa under bacteria and archaea, which are capable of degrading various persistent pesticides, is provided in Table 5.4 and briefly discussed here. Five bacterial strains (*Pseudomonas* sp., *Pseudomonas putida, Micrococcus lylae, Pseudomonas aureofaciens,* and *Acetobacter liquefaciens*) are known to show carboxylesterase activity and degrade malathion to malathion monocarboxylic and dicarboxylic acids (Goda et al. 2010). A Gram-positive *Micrococcus* sp. strain PS-1 is found capable of utilizing phenylurea herbicide diuron and a range of its analogs (monuron, linuron, monolinuron, chlortoluron, and fenuron) as the sole carbon source at a high concentration (up to 250 ppm). The relative degradation profile by the isolate was in the order of fenuron > monuron > diuron > linuron > monolinuron > chlortoluron. The degradation proceeds via formation of dealkylated metabolites to form 3,4-dichloroaniline (3,4-DCA)

TABLE 5.4

List of Bacterial and Archaeal Taxa Capable of Degrading Pesticides

Organism Used	Pesticides	References
Bacteria		
Pseudomonas sp., *Pseudomonas putida, Micrococcus lylae, Pseudomonas aureofaciens,* and *Acetobacter liquefaciens*	Malathion	Goda et al. (2010)
Micrococcus sp.	Phenylurea (herbicide diuron) and a range of its analogs (monuron, linuron, monolinuron, chlortoluron, and fenuron)	Sharma et al. (2010)
Sphingobacterium sp.	Dichlorodiphenyltrichloroethane (DDT), dichlorodiphenyldichloroethane (DDD), and dichlorodiphenyldichloroethylene (DDE)	Fang et al. (2010)
Pseudomonas sp.	Propargite (miticide)	Sarkar et al. (2010)
Klebsiella jilinsis	Chlorimuron-ethyl, sulfonylurea herbicides ethametsulfuron, metsulfuron-methyl, nicosulfuron, rimsulfuron, and tribenuron-methyl	Zhang et al. (2010)
Burkholderia cepacia	Carbofuran	Plangklang and Reungsang (2010)
Azotobacter chroococcum	Lindane	Anupama and Paul (2010)
Paenibacillus polymyxa	Beta-cypermethrin, chlorpyrifos, and imidacloprid	An and Zhao (2009)
Providencia stuartii	Chlorpyrifos	Rani et al. (2008)
Klebsiella sp.	Opera (epoxiconazole and pyraclostrobin)	Lopes et al. (2010)
Sphingomonas, Acidovorax, and *Chryseobacterium* spp.	Dichlorvos (an organophosphorus pesticide)	Ning et al. (2010)
Genetically engineered *Stenotrophomonas* sp. + *Pseudomonas syringae*	A mixture of six organophosphate pesticides	Yang et al. (2010)
Arthrobacter protophormiae	p-Nitrophenol	Paul et al. (2006)
Archaea		
Haloarcula sp. and *Haloferax* sp.	Heptadecane	Tapilatu et al. (2010)

(Sharma et al. 2010). Dichlorodiphenyltrichloroethane (DDT) is still persisting in some parts of the developing countries. An isolate of *Sphingobacterium* sp. is observed to utilize DDT, dichlorodiphenyldichloroethane (DDD), and dichlorodiphenyldichloroethylene (DDE) in soil. Based on the metabolite's detection, a pathway is proposed for DDT degradation by the isolated strain in which it undergoes dechlorination, hydrogenation, dioxygenation, decarboxylation, hydroxylation, and phenyl-ring cleavage reactions to complete the mineralization process (Fang et al. 2010). A *Pseudomonas* strain isolated from tea rhizosphere demonstrated capacity for the degradation of miticide propargite (Sarkar et al. 2010). *Klebsiella jilinsis* strain 2N3, which is isolated from industrial wastewater treatment pond, degraded chlorimuron-ethyl, sulfonylurea herbicides ethametsulfuron, metsulfuron-methyl, nicosulfuron, rimsulfuron, and tribenuron-methyl (Zhang et al. 2010). Immobilization of *Burkholderia cepacia* strain PCL3 on corncob increased the percentage of carbofuran removal (96.97%) (Plangklang and Reungsang 2010). *Azotobacter chroococcum* strain JL 102 exhibited lindane degradation at 10 ppm concentration on the 8th week of incubation (Anupama and Paul 2010). A bacillus strain B3 (*Paenibacillus polymyxa*) isolated from peanut rhizosphere is observed to be capable of utilizing beta-cypermethrin, chlorpyrifos, and imidacloprid as the sole source of carbon for growth. The isolate showed bioremediation capabilities and a broad inhibition spectrum against various soil-borne phytopathogenic fungi as well (An and Zhao 2009). A soil bacterium *Providencia stuartii* strain MS09 utilized chlorpyrifos as the sole carbon source at 50–700 mg/L (Rani et al. 2008). *Klebsiella* sp. strain 1805, isolated from the soil of soybean field, is found capable of degrading Opera (a fungicide containing epoxiconazole and pyraclostrobin) (Lopes et al. 2010). Evidence shows that bacteria on leaves also can degrade organophosphate pesticides and demonstrated that phyllosphere bacteria have great potential for the bioremediation of pesticides in situ, where the environment is hostile to nonepiphytic bacteria. Bacterial species from the genera *Sphingomonas*, *Acidovorax*, and *Chryseobacterium* were found to degrade dichlorvos, an organophosphorus pesticide sprayed on rape phyllosphere (Ning et al. 2010).

5.3.1 Genetically Engineered Bacteria

With the advent of latest molecular techniques, genetically engineered strains have been showing promise for in situ remediation of organophosphate-contaminated sites with broader substrate specificity in combination with the rapid degradation rate. A native soil bacterium, *Stenotrophomonas* sp. (strain YC-1), that produces methyl parathion hydrolase (MPH) was genetically engineered with the ice nucleation protein from *Pseudomonas syringae*. The genetically engineered bacterial strain was able to degrade a mixture of six organophosphate pesticides (0.2 mM each) completely within 5 h (Yang et al. 2010).

5.3.2 Culture-Independent Study

A complete picture of the microbial community structure and their subsequent persistence for the remediation of a pesticide-contaminated site can only be measured through culture-independent study of the metagenome. However, experimental reports based on culture-independent techniques are scarce. Paul et al. (2006) conducted a culture-independent study to remediate p-nitrophenol (PNP)-contaminated soil site using the efficient PNP-degrading organism *Arthrobacter protophormiae* RKJ100. PNP is a metabolite arising from the conversion of the organophosphorus pesticides, parathion and methyl parathion. The bacterial community structure was profiled using terminal restriction fragment length polymorphism (T-RFLP) so that any changes during implementation of the bioremediation could be assessed. Microbial community composition was also determined by 16S rRNA gene clone library, followed by sequencing of the individual clones. Sequencing of representative clones of each phylotype showed that the community structure of the pesticide-contaminated soil was mainly constituted by the members of the Proteobacteria and Actinomycetes. The 16S rRNA gene pool amplified from the soil metagenome and T-RFLP studies revealed 46 different phylotypes on the basis of similar banding patterns. The study demonstrated that bioaugmentation of contaminated soil with *A. protophormiae* stimulated complete degradation of PNP under field conditions. The study also demonstrated that none of the treatments (strain inoculation and/or addition of corncob powder) significantly influenced the parameters, species richness and evenness of the community composition, in the soils. In another culture-independent study, the dominant benzene degraders (responsible for 13C uptake) were determined by comparing relative abundance of T-RFLP phylotypes in heavy fractions of labeled benzene (13C) amended samples to the controls (from unlabeled benzene amended samples). Two phylotypes (*Polaromonas* sp. and *Acidobacterium* sp.) were the major benzene degraders in the microcosms constructed from the soil of the contaminated site; whereas one phylotype ("candidate" phylum TM7; unclassified Sphingomonadaceae) incorporated the majority of the benzene-derived 13C in each of the agricultural soils (Xie et al. 2010).

5.3.3 Microbial Consortia

Microbial remediation is a complex biological system including synergism and antagonism. For example, pyrene degradation in the rhizosphere commonly involves the activity of bacterial consortia in which various species of bacteria interact to achieve PAH degradation (Balcom and Crowley 2010). Methanogenic granular sludge and wastewater fermented sludge were used as inocula for batch tests of anaerobic bioremediation of chlorinated-pesticide–contaminated soil. Results showed that 80%–90% of γ-hexachlorocyclohexane (γ-HCH), 1,1,1-trichloro-2,2-bis-(4-methoxyphenyl)ethane (methoxychlor), and DDT were removed in 4–6 weeks. Residual fractions of these pesticides persisted till the end of the 16-week experiment. DDT was degraded through DDD (Baczynskia and Pleissner 2010). Microbial consortium present in the cow-dung slurry was used for the remediation of fenvalerate (a synthetic pyrethroid) amended soil (Geetha and Fulekar 2010).

5.4 Archaea

Little information exists about the ability of archaea to degrade pesticides. Recently, alkane-degrading halophilic archaeal strains were isolated. One (strain MSNC 2) was closely related to *Haloarcula* sp. and three (strains MSNC 4, MSNC 14, and MSNC 16) were closely related to *Haloferax* sp. Biodegradation assays showed that depending on the strain, 32%–95% (0.5 g/l) of heptadecane was degraded after 30 days of incubation at 40°C in 225 g/l NaCl artificial medium. One of the strains of *Haloarcula* (MSNC 14) was also able to degrade phenanthrene (Tapilatu et al. 2010).

5.5 Fungi

Soil bioremediation using fungi is a promising technique to clean up soil contaminated especially with organic compounds. Fungal remediation techniques rely not only on suitable organisms obtained through screening but also on their inoculation and interaction with natural soil microflora. Therefore, several feasible application techniques have been tested (Table 5.5), but they are still not largely in use. Modern techniques utilize fungi to degrade larger and more recalcitrant molecules and combine them with the natural ability of the soil microflora to handle part of the cleaning process (Steffen and Tuomela 2010).

Persistence of pesticides depends on their binding to other components and subsequent degradation. Environmental conditions at the time of application, such as pH, temperature, soil, and especially water content, have a great impact on the ability of fungi to degrade pesticides (Amauri et al. 2010). For example, use of laccase from white-rot fungi *Panus conchatus* showed that soil oxygen conditions and soil pH have significant effects on the remediation of DDT-contaminated soil. This study found a positive correlation between the concentration of oxygen in soil and the degradation of DDT by laccase. The residues of DDTs in soils treated with laccase were lower at pH range 2.5–4.5 (Zhao and Yi 2010). A recent study showed that *Trametes versicolor* and *Pleurotus ostreatus* were very efficient degraders and degraded the pesticides imazalil, thiophanate methyl, ortho-phenylphenol, diphenylamine, and chlorpyrifos (Karas et al. 2010). The endophytic fungus *Ceratobasidium stevensii* (strain B6) isolated from *Bischofia polycarpa* showed a high degradation efficiency of phenanthrene (Dai et al. 2010). Not only the fungal strain but even the spent mushroom waste from *Pleurotus ostreatus* degraded DDT by 40% and 80% during a 28-day incubation in sterilized and unsterilized soil, respectively (Purnomo et al. 2010). *Trichoderma* sp. Gc1 was observed to be capable of degrading 58% of DDD after 14 days (Ortega et al. 2010).

TABLE 5.5

List of Fungal and Cyanobacterial Taxa Capable of Degrading Pesticides

Organism Used	Pesticides	References
Fungi		
Panus conchatus	1,1,1-Trichloro-2,2-bis (4-chlorophenyl) ethane (DDT)	Zhao and Yi (2010)
Trametes versicolor and Pleurotus ostreatus	imazalil, thiophanate methyl, ortho-phenylphenol, diphenylamine, and chlorpyrifos	Karas et al. (2010)
Ceratobasidium stevensii	Phenanthrene	Dai et al. (2010)
Pleurotus ostreatus	1,1,1-trichloro-2,2-bis (4-chlorophenyl) ethane (DDT)	Purnomo et al. (2010)
Trichoderma sp.	Dichlorodiphenyldichloroethane (DDD)	Ortega et al. (2010)
Ganoderma australe	Lindane	Dritsa et al. (2009)
Aspergillus oryzae	Organophosphorus pesticides such as monocrotophos	Bhalerao and Puranik (2009)
Phlebia brevispora	Dieldrin	Kamei et al. (2010)
Cyanobacteria		
Spirulina platensis	Chlorpyrifos	Thengodkar and Sivakami (2010)
Microcystis novacekii	Methyl parathion (an organophosphorus pesticide)	Fioravante et al. (2010)

A polypore fungus, *Ganoderma australe*, was found to have the potential to degrade lindane. The optimum lindane biodegradation (3.11 mg biodegradation of lindane per gram of biomass) was observed in liquid-agitated sterile cultures with a nitrogen content of 1.28 g/L, lindane concentration of 7.0 ppm, temperature of 18.0°C, and cultivation time of 5 days (Dritsa et al. 2009).

Organophosphorus pesticides are widely used for protection of agricultural yields. However, these pesticides pose various threats to organisms, including humans, and hamper soil microbial activity. *Aspergillus oryzae* strain ARIFCC 1054 was found to possess phosphatase activity and capable of degrading organophosphorus pesticides such as monocrotophos (Bhalerao and Puranik 2009). Wood-rotting fungi were also used for the degradation of a wide spectrum of recalcitrant organopollutants. Transformation of dieldrin to 9-hydroxydieldrin was found by a wood-rotting fungus, *Phlebia brevispora* strain YK543. The *P. brevispora* strain YK543 degraded 39.1 ± 8.8% of dieldrin during 30 days of incubation (Kamei et al. 2010).

5.5.1 Hybridization

Studies have demonstrated the ability to genetically augment the metabolic capabilities of naturally occurring fungi. Genetically engineered strains were capable of ex situ environmental remediation of a variety of widely used neurotoxic organophosphate pesticides. A hybrid genetic system formed by the combination of a bacterial structural gene in a fungal expression system has been developed and productively introduced into the common soil fungus, *Gliocladium virens*. The hybrid gene was formed by combining the constitutive promoter and translational signals from a fungal gpd gene (encoding glyceraldehyde-3-phosphate-dehydrogenase) with the structural region of the bacterial opd gene (organophosphate-degrading gene). Organophosphate hydrolase (OPH, encoded by the opd gene) activity was detected in several of the genetic transformants, and the levels of expression were greatly enhanced (>20 fold) as compared to transformants with nonhomologous fungal genetic sequences (Xu et al. 1996). Because of its structure, atrazine is not easily degraded. Its partial degradation is possible by fungus, but its total mineralization is not possible by a single microorganism (Levanon 1993). In fact, it is necessary to have two or more different kinds of microorganisms capable of the degradation of atrazine (Radosevich et al. 1995) to achieve total mineralization, although some reports point out that, in some cases, even the presence of various microorganisms is not enough to attain this (Korpraditskul et al. 1993).

5.6 Cyanobacteria

Cyanobacteria are photoautotrophic, and some can fix atmospheric nitrogen. Therefore, their use for bioremediation of surface waters would circumvent the need to supply biodegradative heterotrophs with organic nutrients (Kuritz and Wolk 1995). Cyanobacteria can derive energy from sunlight and carbon from the air. Some cyanobacteria are also able to fix atmospheric nitrogen (van-Baalen 1987) and are therefore especially inexpensive to maintain. Filamentous cyanobacteria, including nitrogen-fixing strains that combine aerobic metabolism in their vegetative cells with anaerobic metabolism in their differentiated cells called heterocysts (Wolk et al. 2004), are widespread in many ecosystems, including polluted ones (Fogg 1987; Gibson and Smith 1982; Sorkhoh et al. 1992). Unlike those of heterotrophic microorganisms, the viability and metabolic activity of these cyanobacteria are not subject to reduction by the decrease in the concentration of the pollutants that they may break down (Kuritz and Wolk 1995). A recent report showed that a blue-green micro-alga *Spirulina platensis* produces the enzyme, alkaline phosphatase. The purified enzyme from the isolate, alkaline phosphatase, degraded 100 ppm chlorpyrifos to 20 ppm in 1 h, transforming it into its primary metabolite, 3,5,6-trichloro-2-pyridinol (Thengodkar and Sivakami 2010). The cyanobacterium *Microcystis novacekii* was found capable of removing methyl parathion, an organophosphorus pesticide, with an extraction rate higher than 90%, from the culture medium. This study showed that spontaneous degradation was not significant, which indicates a high efficiency level of biological removal. No metabolites of methyl parathion were detected in the culture medium when the concentration level was evaluated (0.10 to 2.00 mg/dm^3) (Fioravante et al. 2010).

The synergistic coupling of microalgae propagation with carbon sequestration and wastewater treatment potential for mitigation of environmental impacts associate with energy conversion and utilization (Brennan and Owende 2010). Therefore, based on current knowledge and technology projections, bioremediation using microalgae is considered to be a technically viable alternative that is devoid of the major drawbacks (gradual depletion of nutrients) associated with other microorganisms such as bacteria and fungi.

5.7 Concluding Remarks

There are numerous biotic and abiotic factors acting as synergistic and antagonistic effects in the microenvironment of the remediation site. Effects are interrelated while the presence or absence of specific microbes may alter the procedure. For example, endosulfan had a short-lived inhibitory effect on soil fungi, but bacteria increased in number in response to endosulfan application (Joseph et al. 2010). Screening of functional genes in coral-associated microbial communities revealed 6700 genes, providing evidence that the coral microbiome contains a diverse community of archaea, bacteria, and fungi capable of fulfilling numerous functional niches. These included carbon, nitrogen, and sulfur cycling; metal homeostasis and resistance; and xenobiotic contaminant degradation (Kimes et al. 2010).

The literature on bioremediation is huge, and only selected examples of microbial strains' potential for use in bioremediation processes of pesticides are provided in this chapter. Some of the microbial strains described could be combined or improved through genetic manipulation. In other cases, appropriate detailed knowledge of microbes, pesticides, and environment of the specific site may allow further optimization of the desired process by altering the physicochemical conditions of the contaminated area. A combination of genetic engineering with appropriate ecoengineering of polluted sites may be relevant to some future bioremediation strategies (Valls and de-Lorenzo 2002). All kinds of microbes, including prokaryotes and eukaryotes and their symbiotic associations with each other and "higher organisms" including humans, play remarkably wide and diverse geoactive roles in the biosphere. Increasing our understanding of microbiology and exploiting it in bioremediation and other areas of biotechnology clearly require a multidisciplinary approach (Gadd 2010). A vast diversity of uncultivable or yet to culture microorganisms points towards the necessity for the identification of microbial communities of polluted sites. This has led to community analyses using total community DNA extracted from the contaminated sites. Comparative genomic analysis and microarray technology may be used to determine patterns of gene expression and to detect novel metabolic pathways for pesticide remediation. This will provide functional information about genes of unknown function related with specific pesticide cleaning. Microarray technology is very useful in functional diversity studies to track highly expressed genes and genes critical in biogeochemical pathways (Torsvik and Øvreås 2002). It is also important to understand how microbial cells are regulated under varying conditions such as carbon supply, energy source, and electron acceptor availability and how microbial community responds to environmental changes at different stages of pesticide remediation. Studies on microbial systematics, microbial sequences, comparative genomics, and microarray technology will certainly improve our understanding of the structure–function relationships and the effects of abiotic and biotic factors on microbial communities of polluted sites. It is conceivable that the research field dealing with the interaction of microbial genomes with the environment will play an important role to develop sustainable microbial remediation of pesticides in the future.

References

Amauri, G. S., Leiliane, C. A. A., and Zenilda, L. C. 2010. Studies of the analysis of pesticides degradation in environmental samples. *Curr. Anal. Chem.* 6(3): 237–248.

An, X. and Zhao, L. 2009. Isolation and identification of a pesticide-degrading and biocontrol bacterium. *Microbiology/Wei Sheng Wu Xue Tong Bao* 36(12): 1838–1841.

Anupama, K. S. and Paul, S. 2010. *Ex situ* and *in situ* biodegradation of lindane by *Azotobacter chroococcum*. *J. Environ. Sci. Health* 45(1): 58–66.

Baczynskia, T. P. and Pleissner, D. 2010. Bioremediation of chlorinated pesticide-contaminated soil using anaerobic sludges and surfactant addition. *J. Environ. Sci. Health B* 45(1): 82–88.

Balcom, I. N. and Crowley, D. E. 2010. Isolation and characterization of pyrene metabolizing microbial consortia from the plant rhizoplane. *Int. J. Phytoremediation* 12(6): 599–615.

Bhalerao, T. S. and Puranik, P. R. 2009. Microbial degradation of monocrotophos by *Aspergillus oryzae*. *Int. Biodeterior. Biodegradation* 63(4): 503–508.

Brennan, L. and Owende, P. 2010. Biofuels from microalgae – A review of technologies for production, processing, and extractions of biofuels and co-products. *Renew. Sustain. Energ. Rev.* 14(2): 557–577.

Dai, C. C., Tian, L. H., Zhao, Y. T., Chen, Y., and Ie, H. 2010. Degradation of phenanthrene by the endophytic fungus *Ceratobasidum stevensii* found in *Bischofia polycarpa*. *Biodegradation* 21(2): 245–255.

Dritsa, V., Rigas, F., Doulia, D., Avramides, E. J., and Hatzianestis, I. 2009. Optimization of culture conditions for the biodegradation of lindane by the polypore fungus *Ganoderma australe*. *Water Air Soil Pollut.* 204(1–4): 19–27.

Eddleston, M., Karalliedde, L., Buckley, N., Fernando, R., Hutchinson, G., Isbister, G., Konradsen, F., Murray, D., Piola, J. C., Senanayake, N., Sheriff, R., Singh, S., Siwach, S. B., and Smit, L. 2002. Pesticide poisoning in the developing world—A minimum pesticides list. *Lancet* 360: 1163–1167.

EPA. 2005. *EPA Guidelines, Soil Bioremediation*. Environment Protection Authority, Publication # EPA 589/05, Adelaide, SA. http://www.epa.sa.gov.au/xstd_files/Site%20contamination/Guideline/guide_soil.pdf, accessed on October 15, 2010.

Fang, H., Dong, B., Yan, H., Tang, F., and Yu, Y. 2010. Characterization of a bacterial strain capable of degrading DDT congeners and its use in bioremediation of contaminated soil. *J. Hazard. Mater.*, DOI:10.1016/j.jhazmat.2010.08.034.

Fioravante, I. A., Barbosa, F. A. R., Augusti, R., and Magalhaes, S. M. S. 2010. Removal of methyl parathion by cyanobacteria *Microcystis novacekii* under culture conditions. *J. Environ. Monit.* 12: 1302–1306.

Fogg, G. E. 1987. Marine planktonic cyanobacteria. In: Fay, P. and Van Baalen, C. (eds), *The Cyanobacteria*, pp. 393–413. Elsevier Biomedical Press: Amsterdam.

Gadd, G. M. 2010. Metals, minerals and microbes: Geomicrobiology and bioremediation. *Microbiology* 156: 609–643.

Geetha, M. and Fulekar, M. H. 2010. A remediation technique for removal of fenvalerate from contaminated soil. *Asian J. Water Environ. Pollut.* 7(3): 85–91.

Gibson, C. E. and Smith, R. V. 1982. Freshwater plankton. In: Carr, N. G. and Whitton, B. A. (eds), *The Biology of Cyanobacteria*, pp. 463–489. Blackwell Scientific Publications Ltd: Oxford.

Goda, S. K., Elsayed, I. M., Khodair, T. A., El-Sayed, W., and Mohamed, M. E. 2010. Screening for and isolation and identification of malathion-degrading bacteria: Cloning and sequencing a gene that potentially encodes the malathion-degrading enzyme, carboxylestrase in soil bacteria. *Biodegradation*, in press, DOI:10.1007/s10532-010-9350-3.

Joseph, R., Reed, S., Jayachandran, K., Cuadrado, C. C., and Dunn, C. 2010. Endosulfan has no adverse effect on soil respiration. *Agric. Ecosyst. Environ.* 138(3–4): 181–188.

Kamei, I., Takagi, K., and Kondo, R. 2010. Bioconversion of dieldrin by wood-rotting fungi and metabolite detection. *Pest Manag. Sci.* 66(8): 888–891.

Karas, P. A., Perruchon, C., Exarhou, K., Ehaliotis, C., and Karpouza, D. G. 2010. Potential for bioremediation of agro-industrial effluents with high loads of pesticides by selected fungi. *Biodegradation*, in press, DOI:10.1007/s10532-010-9389-1.

Kimes, N. E., van-Nostrand, J. D., Weil, E., Zhou, J., and Morris, P. J. 2010. Microbial functional structure of *Montastraea faveolata*, an important Caribbean reef-building coral, differs between healthy and yellow-band diseased colonies. *Environ. Microbiol.* 12(2): 541–556.

Korpraditskul, R., Katayama, A., and Kuwatsuka, S. 1993. Chemical and microbiological degradation of atrazine in Japanese and Thai soils. *J. Pest Sci.* 18: 77–83.

Kuritz, T. and Wolk, C. P. 1995. Use of filamentous cyanobacteria for biodegradation of organic pollutants. *Appl. Environ. Microbiol.* 61(1): 234–238.

Levanon, D. 1993. Roles of fungi and bacteria in the mineralization of the pesticides atrazine, alachlor, malathion and carbofuran in soil. *Soil Biol. Biochem.* 25: 1097–1105.

Lopes, F. M., Batista, K. A., Batista, G. L. A., Mitidieri, S., Bataus, L. A. M., and Fernandes, K. F. 2010. Biodegradation of epoxyconazole and piraclostrobin fungicides by *Klebsiella* sp. from soil. *World J. Microbiol. Biotechnol.* 26(7): 1155–1161.

Mulligan, C. N. 2002. Environmental Biotreatment: Technologies for Air, Water, Soil and Wastes, Government Institutes: Rockville, MD.

Ning, J., Bai, Z., Gang, G., Jiang, D., Hu, Q., He, J., Zhang, H., and Zhuang, G. 2010. Functional assembly of bacterial communities with activity for the biodegradation of an organophosphorus pesticide in the rape phyllosphere. *FEMS Microbiol. Lett.* 306: 135–143.

Ortega, S. N., Nitschke, M., Mouad, A. M., Landgraf, M. D., Rezende, M. O. O., Seleghim, M. H. R., Sette, L. D., and Porto, A. L. M. 2010. Isolation of Brazilian marine fungi capable of growing DDD pesticide. *Biodegradation*, in press, DOI:10.1007/s10532-010-9374-8.

Paul, D., Pandey, G., Meier, C., van-der-Meer, J. R., and Jain, R. K. 2006. Bacterial community structure of a pesticide-contaminated site and assessment of changes induced in community structure during bioremediation. *FEMS Microbiol. Ecol.* 57: 116–127.

Plangklang, P. and Reungsang, A. 2010. Bioaugmentation of carbofuran by *Burkholderia cepacia* pcl3 in a bioslurry phase sequencing batch reactor. *Process Biochem.* 45(2): 230–238.

Purnomo, A. S., Mori, T., Kamei, I., Nishii, T., and Kondo, R. 2010. Application of mushroom waste medium from *Pleurotus ostreatus* for bioremediation of DDT-contaminated soil. *Int. Biodeterior. Biodegradation* 64(5): 397–402.

Radosevich, M., Hao, Y. L., Traina, S. J., and Tuovinen, O. H. 1995. Degradation and mineralization of atrazine by a soil bacterial isolate. *Appl. Environ. Microbiol.* 61: 297–302.

Rani, M. S., Lakshmi, K. V., Devi, P. S., Madhuri, R. J., Aruna, S., Jyothi, K., Narasimha, G., and Venkateswarlu, K. 2008. Isolation and characterization of a chlorpyrifos degrading bacterium from agricultural soil and its growth response. *Afr. J. Microbiol. Res.* 2: 026–031.

Rashid, B., Husnain, T., and Riazuddin, S. 2010. Herbicides and pesticides as potential pollutants: A global problem. In: Ashraf, M., Ozturk, M., and Ahmad, M. S. A. (eds), *Plant Adaptation and Phytoremediation*, Part 2, pp. 427–447. Springer: Netherlands.

Sarkar, S., Seenivasan, S., and Asir, R. P. 2010. Biodegradation of propargite by *Pseudomonas putida*, isolated from tea rhizosphere. *J. Hazard. Mater.* 174(1–3): 295–298.

Sharma, P., Chopra, A., Cameotra, S. S., and Suri, C. R. 2010. Efficient biotransformation of herbicide diuron by bacterial strain *Micrococcus* sp. PS-1. *Biodegradation*, in press, DOI:10.1007/s10532-010-9357-9.

Sorkhoh, N., Al-Hasan, R., Radwan, S., and Höpner, T. 1992. Self-cleaning of the Gulf. *Nature (London)* 359: 109.

Steffen, K. and Tuomela, M. 2010. Fungal soil bioremediation: Developments towards large-scale applications. *Mycota* 10(4): 451–467.

Tapilatu, Y. H., Grossi, V., Acquaviva, M., Militon, C., Bertrand, J. C., and Cuny, P. 2010. Isolation of hydrocarbon-degrading extremely halophilic archaea from an uncontaminated hypersaline pond (Camargue, France). *Extremophiles* 14(2): 225–231.

Thengodkar, R. R. M. and Sivakami, S. 2010. Degradation of chlorpyrifos by an alkaline phosphatase from the cyanobacterium *Spirulina platensis*. *Biodegradation* 21(4): 637–644.

Torsvik, V. and Øvreås, L. 2002. Microbial diversity and function in soil: From genes to ecosystems. *Curr. Opin. Microbiol.* 5: 240–245.

US EPA. 1998. *Innovative Site Remediation Technology: Design & Application, Volume 1: Bioremediation, Volume 1 of Innovative Site Remediation Technology*, Publication # EPA 542-B-97-004. US Environmental Protection Agency: Washington, DC.

US EPA. 2001. *A Citizen's Guide to Bioremediation*, Publication # EPA 542-F-01-001. US Environmental Protection Agency: Washington, DC. http://www.epa.gov/tio/download/citizens/bioremediation.pdf, visited on October 15, 2010.

US EPA. (2010a). *2000–2001 Pesticide Market Estimates: Usage*. US Environmental Protection Agency: Washington, DC. http://www.epa.gov/opp00001/pestsales/01pestsales/usage2001.htm#3_1, visited on October 15, 2010.

US EPA. (2010b). *Types of Pesticides*. US Environmental Protection Agency: Washington, DC. http://www.epa.gov/pesticides/about/types.htm, visited on October 15, 2010.

US EPA. (2010c). *What Is a Pesticide?* US Environmental Protection Agency: Washington, DC. http://www.epa.gov/pesticides/about/index.htm, visited on October 15, 2010.

Valls, M. and de-Lorenzo, V. 2002. Exploiting the genetic and biochemical capacities of bacteria for the remediation of heavy metal pollution. *FEMS Microbiol. Rev.* 26: 327–338.

van-Baalen, C. 1987. Nitrogen fixation. In: Fay, P. and Van Baalen, C. (eds), *The Cyanobacteria*, pp. 187–198. Elsevier: Amsterdam.

Vidali, M. 2001. Bioremediation. An overview. *Pure Appl. Chem.* 73(7): 1163–1172.

Wolk, C. P., Ernst, A., and Elhai, J. 2004. Heterocyst metabolism and development. In: Bryant, D. (ed.), *Molecular Biology of Cyanobacteria: Advances in Photosynthesis and Respiration*, vol. 1, pp. 770–795. Kluwer Academic Publishers: Dordrecht, The Netherlands.

Xie, S., Sun, W., Luo, C., and Cupples, A. M. 2010. Novel aerobic benzene degrading microorganisms identified in three soils by stable isotope probing. *Biodegradation*, in press, DOI:10.1007/S10532-010-9377-5.

Xu, B., Wild, J. R., and Kenerley, C. M. 1996. Enhanced expression of a bacterial gene for pesticide degradation in a common soil fungus. *J. Ferment. Bioeng.* 81(6): 473–481.

Yang, C., Song, C., Mulchandani, A., and Qiao, C. 2010. Genetic engineering of *Stenotrophomonas* strain yc-1 to possess a broader substrate range for organophosphates. *J. Agric. Food Chem.* 58(11): 6762–6766.

Zhang, H., Zhang, X., Mu, W., Wang, J., Pan, H., and Li, Y. 2010. Biodegradation of chlorimuron-ethyl by the bacterium *Klebsiella jilinsis* 2N3. *J. Environ. Sci. Health* 45(6): 501–507.

Zhao, Y. and Yi, X. 2010. Effects of soil oxygen conditions and soil pH on remediation of DDT-contaminated soil by laccase from white rot fungi. *Int. J. Environ. Res. Public Health* 7(4): 1612–1621.

Section III

Pesticides Residues in the Environment

6

Pesticide Residues in the Environment

Tomasz Tuzimski

CONTENTS

6.1 Introduction

The majority of pesticides exhibit limited stability in the environment and in plants and animals; it is an extremely important property of these compounds, which, on the one hand, determines the efficiency of their action, and, on the other, permits the safe use of agricultural products by humans. The fate of pesticides in the soil depends on the chemical transformations in which the living organisms participate (biotic, biochemical transformations) and on the physical, chemical, and photochemical processes. Biotic transformations catalyzed by the enzymes of soil microorganisms predominate. The highest ability of efficient degradation of pesticides is exhibited by bacteria, *Actinomycetes,* and mushrooms. The contribution of mushrooms is the greatest, reaching even up to 80%. There are at least two causes of this high activity of soil mushrooms: greater resistance to unfavorable vegetation conditions in comparison to bacteria (mushrooms exhibit high liveliness in acidic environment, even at pH < 5) and greater ability of degradation of pesticides by mushroom enzymes. Soil microorganisms exhibiting highest activity in degradation of pesticides comprise bacteria from the species *Arthrobacter, Bacillus, Corynebacterium, Flavobacterium,* and *Pseudomonas; Actinomycetes* of the *Nocardia* and *Streptomyces* species; and mushrooms of the *Penicillium, Aspergillus, Fusarium,* and *Trichoderma* species.

Photochemical transformation of pesticides is restricted to soil and plant surfaces and only to compounds sensitive to solar radiation (in the range 290–450 nm). It results in the formation of radicals, which are highly reactive, owing to the presence of a single electron, which leads, for example, in aqueous solutions, to numerous reactions of breaking of bonds as well as to recombination reactions.

In this chapter, the readers will find essential information concerning the causes and transformations of pesticides from various chemical groups and their degradation and fragmentation in the environment.

6.2 The Concept of Degradation and Fragmentation of Pesticides

Many environmental fate processes, including sorption, hydrolysis, volatilization, transport, and accumulation of bound residues, are coupled with degradation; each of these processes may respond differently to environmental conditions, thus making the comprehension of factors controlling degradation challenging. The importance of these processes has been recognized recently, leading to studies in which several key fate processes were investigated in the same experiment system.

6.3 Factors Influencing Residual Characteristics

The continually growing amount of information on the behavior of pesticides in soil, and in the environment in general, has advanced our understanding of these phenomena. For this reason, the parameterization and testing of increasingly sophisticated mathematical models and the corresponding computer simulation programs were described by Larson et al. (1997), Azevedo (1998), Reichman et al. (2000), Chen et al. (2001), and Li et al. (2001).

The degradation and fragmentation of pesticides can be described by different transport models.

- CREAMS (Knisel 1980)
- AGNPS (Young et al. 1986)
- RZWQM (USDA-ARS 1995).

The models of pesticide transport in soil are

- SWACRO (Belmans et al. 1983)
- MACRO (Jarvis 1991)
- LEACHP (Hutson and Wagenet 1992)
- PRZM-2 (Mullins et al. 1993).

The advantage of these models is the possibility of application to many pesticides, though generally within limited spatial and temporal windows. Mathematical models that simulate the fate of pesticides in the environment are used for developing the environmental estimated concentrations (EECs) or predicted environmental concentrations (PECs) (Wania 1998). The generic estimated environmental concentration (GENEEC) model, developed by the U.S. Environmental Protection Agency (EPA), determines the generic EEC for aquatic environments under worst-case conditions (Wania 1998). Such models are the pesticide root zone model (PRZM), edge of field runoff/leaching the exposure analysis modeling system (EXAMS), fate in surface waters, and AgDrift (spray drift) (Wania 1998; Whitford 2002) that applies additional parameters, more descriptive of the site studies (Wania 1998). PRZM simulates the leaching, runoff, and erosion from an agricultural field and EXAMS simulates the fate in a receiving water body (Wania 1998).

The development of geographic information system (GIS) technology and remote sensing offers perspectives for the imminent evolution of comprehensive pesticide transport models. As one of the first, the screening model, attenuation factor (AF) (Equations 6.1 and 6.2) developed in 1985 by Rao et al. (1985), serves as an index for pesticide mass emission from the vadose zone and is defined as

$$AF = \frac{M_1}{M_0} = \exp\left[\frac{-0.693 \times L \times RF \times FC}{q \times t_{1/2}}\right] \qquad (6.1)$$

$$RF = \left[1 + \frac{BD \times OC \times K_{oc}}{FC} + \frac{AC \times K_H}{FC}\right] \qquad (6.2)$$

where M_0 is the amount of pesticide applied at the soil surface; M_1, the amount of pesticide leaching to groundwater; RF the retardation factor, which represents the retardation of pesticide leaching through soil due to sorption and partitioning of pesticide between the vapor and liquid phases; FC, the field capacity of soil (fraction); BD, the bulk density of the soil (kg/m^3); AC, the air-filled porosity (fraction) (P-FC); P, the porosity (1-BD/PD); PD, the particle density (kg/m^3); OC, the organic carbon (fraction); K_{oc}, the sorption coefficient of pesticide (m^3/kg); $t_{1/2}$, the degradation half-life (year); K_H, the dimensionless Henry's constant; q, the groundwater recharge (m/year); L, the depth to the groundwater or the depth (m) at which AF is to be calculated.

AF serves as an index for mass emission from the vadose zone and its value ranges between 0 and 1. A value of 1 indicates that all of the surface-applied pesticide is likely to leach to the groundwater, whereas a value of 0 suggests that none of the applied pesticide will leach to the groundwater (Rao et al. 1985). For each soil, depth-weighted values of soil properties are calculated based on the values available for different layers (e.g., 0–20, 20–40 cm, etc.).

For evaluating the impact of management practices on pesticide leaching, the groundwater loading effect of agricultural management systems (GLEAMS) is a widely used field-scale model, which assumes that a field has homogeneous land use, soils, and precipitation (Konstantinou et al. 2008). Effective prediction of pesticide fate using mathematical models requires good process descriptions in the models and good choice of parameter values by the user. Garratt et al. (2002) examined the ability of seven pesticide leaching models (LEACHP, MACRO, PELMO, PESTLA, PLM, PRZM, and VARLEACH) to describe an arable field environment. The models were evaluated in terms of their ability to reproduce field data of soil water content and pesticide residues (aclonifen and ethoprophos) in the soil water and groundwater. The models varied in their ability to predict soil water content in summer and their ability to simulate the persistence of the pesticides in the soil (Garratt et al. 2002).

To recapitulate, the factors influencing the residual characteristics are as follows

- A knowledge of the physicochemical properties of the analytes, that is, structure, vapor pressure (v.p.), acid–base character (pK$_a$), octanol–water partition coefficient (K$_{ow}$, expressed in the logarithmic form log P), and soil–water partition coefficient (K$_{oc}$); solubility in water allows the fate and behavior of pesticides in the environment to be predicted;

- Persistence, which is usually expressed in terms of half-life, that is, the time required for one half of the pesticide to decompose to form products other than the parent compound;

- A knowledge of the physical and chemical characteristics of the soil system, such as moisture content, organic matter, clay contents, pH, etc.;

- Chemical transformations in which living organisms participate: on the sorption, desorption, and degradation of pesticides (biotic and biochemical transformations, physical, chemical, and photochemical processes) and their access to groundwater and surface waters (all pesticides in groundwater and most residues present in surface waters enter via the soil);

- Photochemical transformation of pesticides, which are sensitive to solar radiation (in the range 290–450 nm) and are in soil and on plant surfaces;

- The formulation (e.g., granules or suspended powder or liquid) and technique of application of pesticides in agriculture (e.g., precision band spraying can reduce the dose, which can be a very effective way to minimize transport and emission and also to avoid a buildup of resistance in target organisms).

TABLE 6.1

Classifications of Pesticides

By Target		By Mode or Time of Action		By Chemical Structure
Type	Target	Type	Action	
Bactericide (sanitizers or disinfected)	Bacteria	Contact	Kills by contact with pest	Pesticides can be either organic or inorganic chemicals. Most of today's pesticides are organic
Defoliant	Crop foliage	Eradicant	Effective after infection by pathogen	
Desiccant	Crop foliage	Fumigants	Enters pest as a gas	Commonly used inorganic pesticides include copper-based fungicides. Lime-sulfur used to control fungi and mites, boric acid used for cockroach control, and ammonium sulfamate herbicides
Fungicide	Fungi	Nonselective	Toxic to both crop and weed	
Herbicide	Weeds	Postemergence	Effective when applied after crop or weed emergence	
Insecticide	Insects	Preemergence	Effective when applied after planting and before crop or weed emergence	
Miticide (acaricide)	Mites and ticks	Preplant	Effective when applied prior to planting	Organic insecticides can either be natural (usually extracted from plants or bacteria) or synthetic. Most pesticides used today are synthetic organic chemicals. They can be grouped into chemical families based on their structure
Molluscicide	Slugs and snails	Protectants	Effective when applied before pathogen infects the plant	
Nematicide	Nematodes	Selective	Toxic only to weed	
Plant growth regulator	Crop growth processes	Soil sterilant	Toxic to all vegetation	
Rodenticide	Rodents	Stomach poison	Kills animal pests after ingestion	
Wood preservative	Wood-destroying organisms	Systemic	Transported through crop or pest following absorption	

Source: Reprinted from *Agric. Ecosyst. Environ.*, 123, Arias-Estévez, M., López-Periago, E., Martínez-Carballo, E., Simal-Gándara, J., Mejuto, J.-C., and García-Río, L., The mobility and degradation of pesticides in soils and the pollution of groundwater resources, 247–260, (2008), with permission from Elsevier.

6.4 Degradation of Pesticides by Microorganisms and Soil

Pesticides can be classified according to their targets, modes or periods of action as shown in Table 6.1 (Arias-Estévez et al. 2008), and their chemistry as shown in Table 6.2.

The behavior of pesticides in soils is governed by a variety of complex dynamic physical, chemical, and biological processes, including sorption–desorption, volatilization, chemical and biological degradation, uptake by plants, runoff, and leaching as shown in Table 6.3 (Arias-Estévez et al. 2008). How long the pesticide remains in the soil depends on how strongly it is bound by soil components and how readily it is degraded. Degradation is fundamental for attenuating pesticide residue levels in soil (Arias-Estévez et al. 2008; Guo et al. 2000). It is governed by both abiotic and biotic factors (the latter including enzymatic catalysis by microorganisms). These factors can follow complex pathways involving a

TABLE 6.2

Chemical Classification of Pesticides

Chemical Type	Example	Structure	Typical Action
Organochlorines	p,p'-DDT		Insecticide
Organophosphates	Malathion		Insecticide
Carbamates	Carbaryl		Insecticide
Dithiocarbamates	Thiram	$(CH_3)_2N-CS-S-CNSN(CH_3)_2$	Fungicide
Carboxylic acid derivatives	2,4-D		Herbicide
Substituted ureas	Diuron		Herbicide
Triazines	Simazine		Herbicide
Pyrethroids	Cypermethrin		Insecticide
Neem products	Nimbidin (Azadirachtin)	$C_{35}H_{44}O_{16}$	Insecticide
Others			
Organometalics	Phenylmercury acetate		Fungicide

TABLE 6.2 (Continued)

Chemical Classification of Pesticides

Chemical Type	Example	Structure	Typical Action
Thiocyanates	Lethane 60	$CH_3(CH_2)_{10}\overset{O}{\overset{\|}{C}}-OCH_2CH_2SCN$	Insecticide
Phenols	Dinitrocresol		Insecticide
Formamides	Chlordimeform Insecticide	$Cl-\text{(ring)}-N=CH-N(CH_3)_2$	Insecticide

TABLE 6.3

Factors Influencing the Persistence of Pesticides in Soil

Pesticide	Soil/Site	Climate	Experimental Variables
Chemical nature Volatility Solubility Formulation Concentration Application Method Time (of year and day) Frequency Amount	Site • Elevation, slope, aspect, geographical location • Plant cover (species, density, distribution, history at site) • Fauna (species, density, distribution, history at site) • Microbial populations (species density, distribution, history at site) • Use of "fertilizers," lime, mulches, and green manures • Use of other pesticides and chemicals • Tillage, cultivation, drainage, irrigation (type, depth, amount, timing, frequency) • Fire (e.g., burning of crop residues) • Adjacent environments (hedges, field borders, woodlots, water bodies) • Presence of pollutants Soil type • Texture, especially clay content • Structure, compaction • Organic matter and humus contents • Soil moisture, leaching • pH • Mineral ion content	Wind, air movements Temperature, solar radiation Rainfall, relative humidity, evaporation	Plot size, arrangement Number of replicates Frequency of sampling Sample size, shape Techniques for measuring variables

Source: From Arias-Estévez, M., López-Periago, E., Martínez-Carballo, E., Simal-Gándara, J., Mejuto, J.-C., and García-Río, L. 2008. The mobility and degradation of pesticides in soils and the pollution of groundwater resources. *Agric. Ecosyst. Environ.* 123: 247–260.

variety of interactions among microorganisms, soil constituents, and the pesticides (Arias-Estévez et al. 2008; Topp et al. 1997). Thus, degradation rates depend on many microbiological, physical, and chemical properties of the soil and physicochemical properties of the pesticides (Arias-Estévez et al. 2008; Rao et al. 1983). Degradation of pesticides can be through several mechanisms, such as oxidation, reduction, or hydrolysis.

Sorption plays a fundamental role in the advective–dispersive transport dynamics, persistence, transformation, and bioaccumulation of pesticides (Arias-Estévez et al. 2008; De Jonge et al. 1996). The sorption processes depend on the properties of the analytes, for example, base–acid properties, pK_a. The sorption process is different for nonionic and ionic analytes or weakly acidic, weakly basic, and neutral pesticides because the sorption of these pesticides depends on the soil pH, electric charge, and ionic strength. Other soil constituents than organic matter, including clays and iron oxides, are important sorbents for the sorption of ionic compounds. The molecular nature of soil organic matter has been proved to be important in determining sorption on nonionic pesticides. Numerous studies have determined sorption isotherms in order to investigate the influence of soil parameters (organic matter content, clay content, pH, etc.) on the sorption of weakly acidic, basic, or neutral pesticides by a wide array of soils (Arias-Estévez et al. 2008). Figure 6.1 shows a schematic diagram of the sorption reactions (small k values) and instantaneous sorption equilibria (capital K), which may occur with a pesticide in soil water, as neutral, basic, or acidic molecules. Slow sorption reactions may also occur with cationic or anionic species.

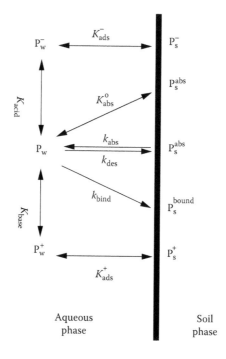

FIGURE 6.1

Schematic diagram of the sorption reactions (small k) and instantaneous sorption equilibria (capital K), which may occur with a pesticide in soil water, as neutral, basic, or acidic molecules. Slow sorption reactions may also occur with cationic or anionic species. Ps and Pw are the potentially available adsorbed-phase and dissolved-phase pesticide levels. (Reprinted from *Agric. Ecosyst. Environ.*, 123, Arias-Estévez M., López-Periago, E., Martínez-Carballo, E., Simal-Gándara, J., Mejuto, J.-C., and García-Río, L., The mobility and degradation of pesticides in soils and the pollution of groundwater resources, 247–260, (2008), with permission from Elsevier.)

P_s and P_w are the potentially available adsorbed-phase and dissolved-phase pesticide levels as shown in Figure 6.1 (Arias-Estévez et al. 2008).

Kinetic studies have revealed several interactions between sorption and degradation (Arias-Estévez et al. 2008). It is commonly accepted that sorbed chemicals are less accessible to microorganisms and that sorption accordingly limits their degradation as well as their transport (Arias-Estévez et al. 2008). Kinetically, the sorption of most organic pesticides is a two-step process: an initial fast step that accounts for the greater part of total sorption is followed by a much slower step tending toward final equilibrium (Arias-Estévez et al. 2008; Pignatello 1998). This means that batch equilibrium partitioning coefficients based on freshly treated samples under slurry conditions can seriously overestimate the availability of aged pesticide: biological availability and biodegradation rate of pesticides in soil will often decrease markedly with increasing time since application (Arias-Estévez et al. 2008; Barriuso et al. 1997; Kristensen et al. 2001; Park et al. 2003, 2004). Furthermore, with longer contact times between the soil and the chemicals, the fraction of strongly bound residues increases at the expense of extractable residues (Arias-Estévez et al. 2008; Boivin et al. 2004).

Soil is a matrix that is not generally monitored on a regular basis, and there is a gap in knowledge at the national and global levels regarding the pesticide residue levels. Regarding Europe, recent discussions have taken place to consider regulation of persistence of soil residues beyond the guidelines given in the Directive 91/414/EEC (European Economic Community 1991). In this regard, stronger emphasis should be given to soil monitoring programs such as Monitoring the State of European Soils (MOSES; http://projects-2004.jrc.cec.eu.int/) and Environmental Indicators for Sustainable Agriculture (ELISA) (ELISA; http://www.ecnc.nl/CompletedProjects/Elisa_119.html).

Rapidity of degradation of pesticides in soils depends in essential degree on the chemical and biological properties of the soils. Even though soils are very diverse, there exist general regularities and characteristics of degradation and fragmentation of pesticides in soils.

The general regularities are as follows.

- More polar pesticides degrade in soils faster than nonpolar pesticides;
- Anionic Pesticides degrade in soils easily than cationic pesticides;
- Aromatic pesticides are more durable than aliphatic pesticides;
- An increase in temperature usually accelerates the degradation and fragmentation of pesticides;
- A decrease in the moisture content of the soil usually decreases the degradation and fragmentation rate of pesticides (but excessive increase in the moisture content of the soil causes the formation of oxygen-free soil);
- A change in the pH of soil toward the alkaline direction (increase of pH values) causes acceleration of chemical processes without enzymes and vice versa; acidification of the soil (decrease of pH values) usually increases the durability of pesticides in soil;
- Fe^{3+}, Cu^{2+}, and other cations of metals and aluminum oxide are also inorganic catalysts of transformation processes.

6.4.1 Organochlorine Insecticides

Organochlorine pesticides (OCPs) constitute the most important group of pesticides. The best-known OCP in the world is 1,1,1-trichloro-2,2-bis(p-chlorophenyl)ethane (DDT). It is an organochlorine pesticide that has been used extensively since World War II to control hundreds of insect pests associated with agricultural practices: as a mosquito larvicide, as a residual spray for the eradication of malaria, and as a delousing dust for typhus control. The persistence of DDT in the environment has been attributed to its low volatility, its low water solubility, and the presence of chlorine substituents in the molecule. The major transformation produces compounds such as 1,1-dichloro-2,2-bis(p-chlorophenyl) ethane (DDD) and 2,2-bis(p-chlorophenyl)-1,1-dichlorethylene (DDE) that are more toxic and recalcitrant than the parent compound. In 1972, the use of DDT was banned in the United States; while in developed countries, it has been progressively restricted or phased out. However, in a number of tropical, less rich countries, DDT is still being used today and some of the OCPs are still detected in the ecosystems of various countries. From the monitoring results obtained so far, the presence of persistent OCPs is still widespread, that is, in Central and North East Asia, Korea, China, Romania, Greece, and the United States (United Nations Environment Programme 2002; Lee et al. 2001; Fu et al. 2003; Zhu et al. 2005b; Covaci et al. 2001; Galanopoulou et al. 2005; Ouyang et al. 2003).

The fate and behavior of OCPs in the environment are governed by their physicochemical properties. Most OCPs are semivolatile, have low water solubility, and have high values of log K_{ow} and log K_{oc} (see Table 6.4).

Especially for the hydrophobic pesticides, their mobility and therefore the risk of their leaching into groundwater have been correlated with weak sorption on the soil matrix, as quantified by K_{oc}, the ratio of adsorbed to solution-phase pesticide normalized with respect to organic matter content (Arias-Estévez et al. 2008). However, pesticides with $K_{oc} \geq 1000$ have also been observed in groundwater and drainage water (Elliott et al. 2000), presumably as the result of leaching: transport to groundwater may be caused by heavy rainfall shortly after application of the pesticide to wet soils with preferential flow paths (Arias-Estévez et al. 2008).

After withdrawal of DDT from use in 1960, it was replaced with 1,1,1-trichloro-2,2-bis(4-metoxyphenylo)-ethane, that is, methoxychlor. Its transformations are presented in Figure 6.2.

The most important reactions of them are O-dealkylation, dehydrochlorination, dechlorination, hydroxylation in the aromatic ring, and coupling reactions of metabolites containing hydroxyl group. Transformations occurring in the ethyl fragment of the molecule are analogous to transformations of DDT. In abiotic systems, no O-dealkylation reactions, for example, during photochemical reactions, were reported.

Faster dissipation and degradation of 1,1,1-trichloro-2,2-bis(4-chlorophenyl)-ethane (p,p'-DDT) in tropical regions has been reported previously (United Nations Environment Programme 2002; Larsson et al. 1995; Carvalho et al. 1998; Allsopp and Johnston 2000). Since p,p'-DDT and its degradation products are known to have high K_{ow} values (see Table 6.4), they are more easily associated with the organic matter of the soil than with the mineral content, as observed by the linear correlation between log K_{ow} and log K_{oc} (Noegrohati and Hammers 1992). A good correlation between the level of OCPs (chlordanes, DDT and its related compounds) and the total organic carbon of the soil was reported by Lee et al. (2001). However, further research is required to elucidate the exact mechanisms that govern their fate and distribution in the tropical environment. The fate and behavior of organochlorine pesticides in the Indonesian tropical climate, which is characterized by heavy rainfall in the rainy season and low rainfall in the dry season, were described (Noegrohati et al. 2008). Since OCPs have a high affinity for soil, a field study on the dissipation and degradation pattern of soil-applied 1,1,1-trichloro-2,2-bis(4-chlorophenyl)-ethane (p,p'-DDT) and 1,1-dichloro-2,2-bis(4-chlorophenyl)-ethylene (p,p'-DDE) as model

TABLE 6.4

Properties of Different Classes of Pesticides

Common Name	Vapor Pressure (mPa) (25°C)	K_{ow} log P (25°C)	Water Solubility (mg/L) (25°C)	Half-Life in Soil (days)
Amide herbicides				
Acetochlor	0.005	4.14	223	8–18
Alachlor	2.0	3.09	170[a]	1–30
Butachlor	0.24	—	23[a]	12
Metolachlor	4.2	2.9	488	20
Propachlor	10	1.4–2.3	580	4
Propanil	0.05	3.3	130[a]	2–3
Benzoic acid herbicides				
Chloramben	—	—	0.7	14–21
Chlorthal-dimethyl	0.21	4.28	0.5×10^{-3}	33
Dicamba	1.67	−1.88	6.1	<14
Carbamate herbicides				
Chlorpropham	1.3	3.76	89	30–65
Desmedipham	4×10^{-5}	3.39	7[a]	34
EPTC	0.01	3.2	375	6–30
Molinate	746	2.88	970	8–25
Phenmedipham	1.3×10^{-6}	3.59	4.7	25
Propham	Sublimes slowly	—	250[a]	5–15
Thiobencarb	2.93	3.42	30[a]	14–21
Triallate	16	4.6	4	56–77
Nitrile herbicides				
Bromoxynil	6.3×10^{-3}	2.8	130	10
Ioxynil	<1	3.43	50	10
Nitroaniline herbicides				
Butralin	0.77	4.93	1	14
Ethalfluralin	11.7	5.11	0.3	25–46
Pendimethalin	4	5.18	0.3[a]	90–120
Trifluralin	6.1	4.83[a]	0.22	57–126
Organophosphorus herbicides				
Glyphosate	1.3×10^{-2}	<−3.2	11.6	3–174
Glufosinate-ammonium	<0.1[a]	<0.1	1370	7–20
Phenoxy acid herbicides				
2,4-D	1.86×10^{-2}	0.04	23,180[a]	<7
Diclofop	9.7×10^{-6}	2.81	122,700[a]	30
Fenoxaprop-p	1.8×10^{-1} [a]	1.83	61,000[a]	1–10
Fluazifop-p	7.9×10^{-4} [a]	−0.8	780[a]	<32
MCPA	2.3×10^{-2} [a]	−0.71	274	<7
Mecoprop-p	0.4[a]	0.02	860[a]	3–13
Quizalofop-p-ethyl	1.1×10^{-4} [a]	4.66	0.61[a]	≤1
Triclopyr	0.2	−0.45	8.10[a]	46

(continued)

TABLE 6.4 (Continued)

Properties of Different Classes of Pesticides

Common Name	Vapor Pressure (mPa) (25°C)	K_{ow} log P (25°C)	Water Solubility (mg/L) (25°C)	Half-Life in Soil (days)
Pyridine herbicides				
Diquat dibromide	<0.013[a]	−4.6	700[a]	<7
Paraquat dichloride	<0.01	−4.5	620[a]	<7
Chlormequat chloride	<0.01[a]	−1.59	1000[a]	1–28
Mepiquat chloride	<0.01[a]	−2.82	500	10–97
Pyridazinone herbicides				
Chloridazon	<0.01[a]	1.19	340[a]	21–76
Norflurazon	3.8×10^{-3}	2.45	34	45–180
Pyridate	4.8×10^{-4} [a]	4.01	Ca. 1.5[a]	<3
Triazine herbicides				
Atrazine	3.8×10^{-2}	2.5	33[a]	35–50
Cyanazine	2.0×10^{-4} [a]	2.1	171	Ca. 14
Metribuzin	0.058[a]	1.6[a]	1050[a]	40
Prometryn	0.165	3.1	33	50
Simazine	2.9×10^{-3}	2.1	6.2[a]	27–102
Terbutryn	0.225	3.65	22	14–50
Phenyl urea herbicides				
Chlorotoluron	0.005	2.5	74	30–40
Diuron	1.1×10^{-3}	2.85	36	90–180
Fenuron	21[a]	—	3850	60
Isoproturon	8.1×10^{-3}	2.5[b]	65	6–28
Linuron	0.051[b]	3.0	63.8[b]	38–67
Sulfonylurea herbicides				
Azimsulfuron	4.0×10^{-6}	−1.37	1050[a]	—
Chlorsulfuron	3×10^{-6}	−0.99	7000	28–42
Flazasulfuron	<0.013	−0.06	2100	<7
Imazosulfuron	4.5×10^{-5}	0.049	308	—
Metsulfuron-methyl	3.3×10^{-7}	−1.74	2790	7–35
Rimsulfuron	1.5×10^{-3}	−1.47	7300	10–20
Thifensulfuron-methyl	1.7×10^{-5}	0.02	6270	6–12
Triasulfuron	$<2 \times 10^{-3}$	−0.59	815	19
Tribenuron-methyl	5.2×10^{-5}	−0.44	2040[a]	1–7
Benzoylurea insecticides				
Diflubenzuron	1.2×10^{-4}	3.89	0.08	<7
Hexaflumuron	5.9×10^{-2}	5.68	0.027[c]	50–64
Teflubenzuron	0.8×10^{-6} [a]	4.3	0.019[c]	14–84
Triflumuron	4×10^{-5} [a]	4.91	0.025[c]	112
Carbamate insecticides				
Aldicarb	13[a]	—	4930[a]	30
Carbaryl	4.1×10^{-2}	1.59	120[a]	7–28

TABLE 6.4 (Continued)

Properties of Different Classes of Pesticides

Common Name	Vapor Pressure (mPa) (25°C)	K_{ow} log P (25°C)	Water Solubility (mg/L) (25°C)	Half-Life in Soil (days)
Carbofuran	0.031[a]	1.52	320[a]	30–60
Carbosulfan	0.041		0.35	2–5
Fenoxycarb	8.67×10^{-4}	4.07	7.9	31
Methomyl	0.72	0.093	57,900	5–45
Oxamyl	0.051	−0.44	280,000	7
Pirimicarb	0.4[a]	1.7	3000[a]	7–234
Organochlorine insecticides				
Aldrin	3.08[a]	5.0–7.4	<0.05	365
p,p'-DDT	0.025[a]	6.19–6.91	0.0055	2000
Dieldrin	0.4[a]	4.32–5.40	0.186	1000
Dicofol	0.053[a]	4.28	0.8	45
Endosulfan	0.83[a]	4.74	0.32[d]	30–70
γ-HCH	4.4[e]	3.5	8.5	400
Methoxychlor	<1[a]	—	0.1	120
Tetradifon	3.2×10^{-5} [a]	4.61	0.078[a]	—
Organophosphorus insecticides				
Azinphos-methyl	5×10^{-4} [a]	2.96	28[a]	10–40
Chlorfenvinphos	1	3.85	145	—
Chlorpyrifos	2.7	4.7	1.4	35–56
Chlorpyrifos-methyl	3	4.24	2.6[a]	1.5–33
Coumaphos	0.013[a]	4.13	1.5[a]	—
Diazinon	12	3.30	60[a]	11–21
Dichlorvos	2.1×10^3	1.9	18,000[a]	0.5
Dimethoate	0.25	0.704	23,800[a]	2–4
Fenitrothion	18[a]	3.43	14	12–28
Fenthion	1.4	4.84	4.2[a]	34
Malation	1.1[a]	2.75	145	1
Methamidophos	2.3[a]	−0.8	$>2 \times 10^5$ [a]	6
Methidathion	0.25[a]	2.2	200	3–18
Oxydemeton-methyl	3.8[a]	−0.74	1×10^6 [a]	2–20
Phosmet	0.065	2.95	25	4–20
Pirimiphos-methyl	2[a]	4.2	10[a]	3.5–25
Profenofos	0.12	4.44	28	7
Trichlorfon	0.5	0.43	1.2×10^5 [a]	5–30
Pyrethroid insecticides				
Acrinathrin	4.4×10^{-5} [a]	5.6	≤0.02	5–100
Cyfluthrin	Diastereoisomer			56–63
	I: 9.6×10^{-4} [a]	I: 6	I: 2.2×10^{-3} [a]	
	II: 1.4×10^{-5} [a]	II: 5.9	II: 1.9×10^{-3}	
	III: 2.1×10^{-5} [a]	III: 6	III: 2.2×10^{-3}	
	IV: 8.5×10^{-5} [a]	IV: 5.9	IV: 2.9×10^{-3}	
Cypermethrin	2×10^{-4} [a]	6.6	0.004	60

(continued)

TABLE 6.4 (Continued)

Properties of Different Classes of Pesticides

Common Name	Vapor Pressure (mPa) (25°C)	K_{ow} log P (25°C)	Water Solubility (mg/L) (25°C)	Half-Life in Soil (days)
Deltamethrin	1.24×10^{-5}	4.6	$<0.2 \times 10^{-3}$	23
Esfenvalerate	2×10^{-4}	6.22	0.002	35–88
Tau-fluvalinate	9×10^{-8}	4.26	0.001[a]	12–92
Permethrin	cis: 0.0025 trans: 0.0015	6.1[a]	$6 \times 10^{-3\,a}$	<38
Azole fungicides				
Cyproconazole	0.04[a]	2.91	140	90
Flusilazole	0.04	3.74	54[a]	95
Hexaconazole	0.018[a]	3.9	17	—
Imazalil	0.158[a]	3.82	180[a]	150
Prochloraz	0.09[a]	4.12	34.4	120
Propiconazole	0.03[a]	3.72	100[a]	110
Tebuconazole	0.002[a]	3.7	36[a]	—
Triadimefon	0.02[a]	3.11	64[a]	6–60
Benzimidazole fungicides				
Benomyl	<0.005	1.37	0.003	67
Carbendazim	0.15	1.51	8	120
Thiabendazole	4.6×10^{-4}	2.39	30[a]	33–120
Dithiocarbamate fungicides				
Mancozeb	<1[a]	1.8	6.2	6–15
Maneb	<0.01[a]	−0.45	257	25
Metiram	<0.01[a]	0.3	0.1	20
Nabam	Negligible	—	2×10^{5}	—
Zineb	<0.01[a]	≤1.3	10	23
Ziram	<0.001[a]	1.23	1.58–18.3[a]	2

Sources: Data from Tomlin, C. (ed.) 2000. In: *The Pesticide Manual*, British Crop Protection Council: Surrey; Hornsby, A. G., Wauchope, R. D., and Herner, A. E. 1996. *Pesticide Properties in the Environment*. Springer-Verlag: New York; De Liñan, C. 1997. In: *Farmacología Vegetal*, Ediciones Agrotecnicas S.L.; http://ec.europa.eu/food/plant/protection/evaluation/exist_subs_rep_en.htm; http://www.epa.gov/opprd001/factsheets/.
[a]20°C, [b]60°C, [c]18°C–23°C, [d]22°C, [e]24°C.

was carried out (Noegrohati et al. 2008). They occurred at a faster rate in the biphasic mode in wet conditions and at a slower rate in dry conditions. In wet conditions, the conversion from p,p'-DDT to p,p'-DDE and p,p'-DDD (1,1-dichloro-2,2-bis(4-chlorophenyl)-ethane) was governed by parallel reactions. In dry conditions, only p,p'-DDE was formed. The fate and behavior of OCPs in sediment estuary are similar to those in soil under wet conditions, except that their sorption–desorption constants are influenced by estuarine surface water salinity (Noegrohati et al. 2008).

The data obtained from the p,p'-DDT column show that p,p'-DDT and its degradation products, p,p'-DDE and p,p'-DDD, were detected only up to a 20-cm depth (the fourth layer of 5 cm thickness). A similar result was observed for the p,p'-DDE column (Noegrohati et al. 2008).

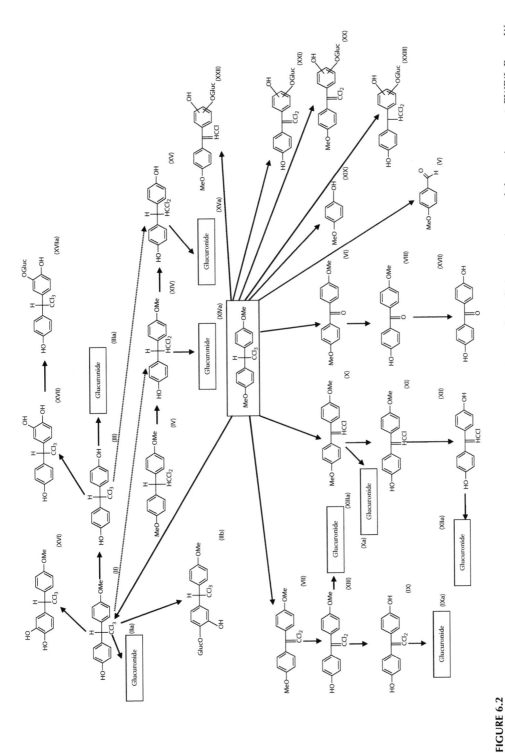

FIGURE 6.2

Transformation of methoxychlor. (Adapted from Różański, L., *Transformations of pesticides in living organisms and the environment.* PWRiL Press, Warsaw, 275, (in Polish). 1992.)

The analytical results of p,p′-DDT residues in the soil column, obtained by Noegrohati et al. (2008), on the one hand, showed the presence of p,p′-DDE and p,p′-DDD, while on the other hand, showed no degradation products in the soil column treated with p,p′-DDE. These data indicate that p,p′-DDD detected in the soil column treated with p,p′-DDT was not the conversion product of p,p′-DDE. Therefore, it is reasonable to assume that in this type of soil environment, a parallel mode of reaction governs the conversion of p,p′-DDT into p,p′-DDE and p,p′-DDD. The following model (Equation 6.3) was applied for the reactions (Noegrohati et al. 2008).

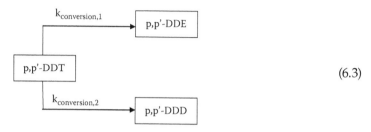
(6.3)

The change in the concentration of p,p′-DDT is given by Noegrohati et al. (2008) (Equation 6.4).

$$\frac{d[p,p'-DDT]}{dt} = k_{conv,1}[p,p'-DDT]_0 + k_{conv,2}[p,p'-DDT]_0 = k[p,p'-DDT]_0 \quad (6.4)$$

where k is the degradation rate constant of p,p′-DDT and $k = k_{conv,1} + k_{conv,2}$.

The change in concentration for p,p′-DDE or p,p′-DDD can be described in Equations 6.5 and 6.6 (Noegrohati et al. 2008):

$$\frac{d[p,o'-DDE]}{dt} = k_{conv,1}[p,p'-DDT]_0 e^{-kt} \quad (6.5)$$

$$\frac{d[p,p'-DDD]}{dt} = k_{conv,2}[p,p'-DDT]_0 e^{-kt} \quad (6.6)$$

The results obtained by Noegrohati et al. (2008) showed that in wet conditions, the conversion from p,p′-DDT to p,p′-DDE follows a first-order kinetics model, with an r^2 value of 0.7039, giving a rate constant, $k_{conv,1}$, of 0.0140/week. Similarly, the conversion from p,p′-DDT to p,p′-DDD also follows a first-order kinetics model with an r^2 value of 0.5024, giving a rate constant, $k_{conv,2}$, of 0.0128/week. The total of the two constants, $k_{conv,1} + k_{conv,2} = 0.0268$, that is, the degradation rate constant for p,p′-DDT during wet conditions demonstrates that, in wet conditions, the conversion from p,p′-DDT into p,p′-DDE and p,p′-DDD is governed by a parallel reaction (Noegrohati et al. 2008).

In dry conditions, the conversion from p,p′-DDT to p,p′-DDE follows a first-order kinetics model, with an r^2 value of 0.7170, giving a rate constant, $k_{conv,1}$, of 0.0016/week. The calculated rate constant for the conversion from p,p′-DDT to p,p′-DDD was negative, indicating that no conversion from p,p′-DDT to p,p′-DDD occurred in dry conditions (Noegrohati et al. 2008). The data obtained by Noegrohati et al. (2008) support the observations of Tolosa et al. (1995) and Lee et al. (2001). The main metabolite of DDT in dry conditions, which are more aerobic than wet conditions, is DDE, whereas the main metabolite under anaerobic conditions is DDD. The closeness of $k_{conv,1}$ to the degradation rate constant for the conversion from p,p'-DDT to p,p'-DDE and p,p'-DDD is governed by a parallel reaction (Noegrohati et al. 2008).

In another study, Juhasz and Naidu (2000) screened the contaminated soils and uncontaminated plant and fecal material for microorganisms with the ability to degrade DDT. Five soils were used for the isolation of xenobiotic-degrading microorganisms. Microbial communities enriched on the basal salt medium—BSM (contained per liter: 0.4 g K_2HPO_4, 0.4 g KH_2PO_4, 0.4 g $(NH4)_2SO_4$, 0.3 g NaCl, 5 mL trace elements solution, 5 mL vitamin solution and 5 mL magnesium/calcium solution (0.4 g $MgSO_4 \times 7 H_2O$ and 0.4 g $CaCl_2 \times 2 H_2O$))—containing peptone (1 g/l) and DDT (100 mg/l) were tested for their ability to degrade DDT, DDD, and DDE after three successive transfers in the enrichment medium (Juhasz and Naidu 2000). A decrease in the concentration of DDT over the incubation period was observed for all microbial communities; a 5%–14.6% decrease was observed after 28 days of exposure to DDT. Small decreases in the concentration of DDD were also observed for microbial communities in the range of 7.3%–9.8%. DDE was generally recalcitrant to microbial attack (Juhasz and Naidu 2000). The degradation of DDT (100 mg/l) by isolate AJR[39],504 resulted in a 35% decrease in DDT concentration after 28 days with a concomitant increase in DDD concentration (Juhasz and Naidu 2000). Guenzi and Beard (1968) reported the accumulation of DDD in anaerobic soil systems seeded with alfalfa-amended Pawnee silt loam; only traces of six other degradation products were detected. Although strain AJR[39],504 could not be positively identified, other Gram-negative organisms and "Pseudomonas-like" organisms have been shown to transform DDT to DDD. In pure culture systems, dechlorination of DDT by *Escherichia coli*, *Enterobacter aerogenes*, *Enterobacter cloacae*, *Klebsiella pneumoniae*, *Pseudomonas aeruginosa*, *Ps. Putida*, *Hydrogenomonas*, and *Aerobacter aerogenes* has been reported (Wedemeyer 1967; Plimmer et al. 1968; Aislable et al. 1997). DDD formation by AJR[39],504 probably resulted from the reduction dechlorination of the aliphatic part of the DDT molecule. In another paper, Plimmer et al. (1968) proposed that in the case of *Aerobacter aerogenes*, conversion of DDT to DDD occurred with the replacement of chlorine by hydrogen or hydride ion.

Other authors (Strompl and Thiele 1997) demonstrated that DDE remained refractory for 105 days when added to aerobic or anaerobic batch reactors inoculated with bacterial consortia from, for example, anaerobic granular sludge and DDE-contaminated soil and suggest that DDE is extremely recalcitrant to microbial degradation.

One of the OCPs with strong sorption characteristics is hexachlorocyclohexane (HCH), a mixture of isomers: α-, β-, γ-, and δ-HCH. The use of surfactants may increase the pollutant's desorption from the soil. Quintero et al. (2005) described the use of three surfactants, Triton X100, Tween 80, and sodium dodecyl sulfate (SDS), on the HCH desorption from a sandy loam soil. Triton X100 exerted the best desorption of HCH isomers, followed by Tween 80, whereas SDS caused no significant desorption of the isomers (Quintero et al. 2005). The desorption assays show that β- and δ-HCH have stronger hydrophobic characteristics than α- and γ-HCH, as they present a lower desorption percentage. This could be influenced by the fact that β-HCH is less soluble than the other isomers. The surfactant Triton X100 showed the highest extraction efficiency among the three surfactants considered and the concentrations corresponding to the two phases well correlated with a linear model (Quintero et al. 2005). Other papers (Park et al. 2003; Sun et al. 1995) report the capability of Triton X100 to desorb other OCPs present in the soil: DDT and pentachlorophenol (PCP) correlated with linear models. The inhibitory effect of a surfactant on the biodegradation capability may have its origin in the following (Quintero et al. 2005; Volkering et al. 1998)

- Toxicity
- Preferential use as substrate, as carbon source alternative to the pollutant
- Low pollutant concentration in the aqueous phase by a micellation effect
- The possibility that the surfactant acts as an interference in the microbial metabolic processes.

The models proposed by Quintero et al. (2005) for the HCH degradation, considering the different surfactant concentrations, suggest that different microbial consortia are involved in the degradation of α- and γ-HCH while the degradation of β- and δ-HCH is affected by diffusional limitations.

Biodegradation and detoxification of pentachlorophenol (PCP) by the white-rot fungus *Phanerochaete chrysosporium* was described (Mendoza-Cantú et al. 2000). Results obtained by them (Mendoza-Cantú et al. 2000) show that *Phanerochaete chrysosporium* application to bioremediate PCP-contaminated soil is a feasible option, owing to its efficiency and the production of innocuous intermediates during the process.

Another very important compound from OCPs is lindane (γ-HCH). An adaptive mechanism and efflux-transport system of *Pseudomonas* sp. LE2 counteracting lindane were investigated to understand the survival mechanism of this microorganism in lindane-contaminated environment (Kim et al. 2002). The results obtained by Kim et al. suggest that *Pseudomonas* sp. LE2 capable of growing in the presence of lindane is able to adapt to the fluidizing action of lindane by incorporating the saturated fatty acid into membrane lipids and eliminate lindane from the cell by an energy-dependent efflux mechanism.

Dieldrin (1,2,3,4,10,10-hexachloro-6,7-epoxy-1,4,4a,5,6,7,8,8a-octahydro-1,4,-endo-exo-5,8-dimethanonaphthalene) is one of the most persistent organochlorine pesticides, having an average half-life of between 5 and 25 years in soil sediment (Vilanova et al. 2001). Therefore, it is listed as one of the 12 persistent organic pollutants in the Stockholm Convention. Microbial degradation is an effective way to remediate environmental pollutants, including persistent organic pollutants. Thirty-four isolates of wood-rotting fungi were investigated to test their ability to degrade dieldrin. Strain YK543 was selected as a fungus capable of degrading dieldrin. The metabolic pathway includes 9-hydroxylation reported in rat's metabolism catalyzed by liver microsomal monooxygenase (Kamei et al. 2010). Kamei et al. (2010) described the transformation of dieldrin to 9-hydroxydieldrin by a microorganism first.

6.4.2 Organophosphate Insecticides

Organophosphate insecticides play an important role in the success of modern farming and food production. Organophosphate insecticides have been widely used throughout the world since the use of organochlorine compounds was discontinued in the 1970s and 1980s in most regions. The objectives of the study described by Bondarenko and Gan (2004) were to evaluate the degradation kinetics of diazinon, chlorpyrifos, malathion, and carbaryl in urban stream sediments under aerobic and anaerobic conditions and to understand the time dependence of pesticide sorption in the sediments. The decline of pesticide concentration over time was fitted to a first-order decay model to estimate the rate constant (d^{-1}) and $t_{1/2}$ (d). The fit obtained by the authors in the experiment was generally good, as evident from the correlation coefficient, except for degradation of carbaryl and chlorpyrifos under anaerobic conditions. The overall persistence for aerobic degradation followed the order: chlorpyrifos > diazinon > carbaryl > malathion. The sorption coefficient consistently increased with time for all pesticides, and chlorpyrifos displayed a greater sorption potential than the other pesticides (Bondarenko and Gan 2004). Chai et al. (2009) described the dissipation of acephate, chlorpyrifos, cypermethrin, and their metabolites studied in green mustard (*Brassica juncea* (L.) Coss.) and in soils. Dissipation of acephate, chlorpyrifos, and cypermethrin in green mustards and topsoils followed first-order kinetics, with half-lives between 1.1 and 3.1 days in green mustards and between 1.4 and 9.4 days in topsoils (26°C) (Chai et al. 2009). The metabolites methamidophos and 3,5,6-trichloropyridinol derived from acephate and chlorpyrifos amounted to less than 10% and 25% by mass of the parent compounds in soils. The metabolites methamidophos and 3,5,6-trichloropyridinol were observed at low levels of below 3.22 mg/kg in topsoils and dissipated completely in 28 and 63 days, respectively. The degradation rates of pesticides in the soil were equal to or slightly faster than the degradation rates in temperate soils (Chai et al. 2009). The degradation of chlorpyrifos in poultry-, swine-, and cow-derived effluents and effluent-soil matrices was studied using batch- and column-incubation studies (Huang et al. 2000). Chlorpyrifos was degraded by aerobic microbial processes in animal-derived lagoon effluents. Microbial community analysis by denaturing gradient gel electrophoresis of polymerase chain reaction-amplified 16S ribosomal ribonucleic acid genes showed that the single band became dominant in effluent during chlorpyrifos degradation. In soils, both biotic and abiotic degradation contributed significantly to the overall dissipation of chlorpyrifos. Large differences in the degradation rates were observed between soils, with the fastest rate observed in soil with higher pH values and cation-exchange capacity. The results indicate that effluent-induced increases in soil-solution pH for low-pH soils may enhance degradation through hydrolysis by a small percent (Huang et al. 2000). Kravvariti et al. (2010) described the degradation and adsorption of chlorpyrifos and terbuthylazine in three different biomixtures composed of composted cotton crop residues, soils, and straw in various proportions and also in sterilized and nonsterilized soils. The results obtained by the authors show that compost biomixtures degraded the less hydrophobic terbuthylazine at a faster rate than soil, while the opposite was evident for more hydrophobic chlorpyrifos. The results were attributed to the rapid abiotic hydrolysis of chlorpyrifos in the alkaline soil (pH 8.5), compared with the lower pH of the compost (6.6), but also to the increasing adsorption (K_d = 746 mL/g) and reduced bioavailability of chlorpyrifos in the biomixtures compared with soil (K_d = 17 mL/g), as verified by the adsorption studies (Kravvariti et al. 2010). A pesticide mixture containing chlorpyrifos, isoproturon, pendimethalin, chlorothalonil, epoxiconazole, and dimethoate was incubated in biomix and topsoil at concentrations to simulate pesticide disposal (Fogg et al. 2003). Degradation was significantly quicker in biomix than in topsoil. While degradation of pesticides applied in mixture to biomix was slower than when applied alone, DT_{90} values indicate that, even in a mixture, pesticides will be degraded within 12 months (Fogg et al. 2003). An effective chlorpyrifos-degrading bacterium (named strain YC-1) was isolated by Yang et al. (2006) from the sludge of the wastewater treating system of an organophosphorus pesticide manufacturer. Based on the phenotypic features, phylogenetic similarity of 16S rRNA gene sequences, and BIOLOG test, the strain YC-1 was identified as the genus *Stenotrophomonas* (Yang et al. 2006). The isolate utilized chlorpyrifos as the sole source of carbon and phosphorus for its growth and hydrolyzed chlorpyrifos to 3,5,6-trichloro-2-pyridinol. Parathion, methyl parathion, and fenitrothion also could be degraded by the strain YC-1 when provided as the sole source of carbon and phosphorus (Yang et al. 2006). The inoculation of the strain YC-1 (10^6 cells/g) to the soil treated with 100 mg/kg chlorpyrifos resulted in a higher degradation rate than in noninoculated soils. The results confirmed that the newly isolated chlorpyrifos-degrading bacterium can be successfully used for bioremediation of contaminated soils (Yang et al. 2006).

Another paper published by Paul et al. (2006) demonstrated for the first time that bio-augmentation of contaminated soil with *A. protophormiae* stimulated complete degradation of p-nitrophenol under field conditions. Ortiz-Hernández et al. (2003) described biotrans-formation by *Flavobacterium* sp. of the following organophosphate pesticides: phorate, tet-rachlorvinphos, methyl-parathion, terbufos, trichloronate, ethoprophos, phosphamidon, fenitrothion, dimethoate, and DEF (S,S,S-tributyl phosphorotrithioate). The *Flavobacterium* sp. ATCC 27551 strain bearing the organophosphate-degradation gene was used. Bacteria were incubated in the presence of each pesticide for a duration of 7 days. Parent pesticides were identified and quantified by means of a gas-chromatography mass spectrometer sys-tem (GC-MS). The activity was defined as the amount (μmol) of each pesticide degraded by *Flavobacterium* sp. Pesticides were hydrolyzed at the bond between phosphorus and hetero-atom, producing phosphoric acid and three metabolites. Enzymatic activity (Equation 6.7) was expressed by the following multiple linear relationship (Ortiz-Hernández et al. 2003).

$$\text{Enzymatic activity} = 162.2 - 9.5\big(\text{dihedral angle energy}\big)$$

$$- 25.0\big(\text{total energy}\big) - 0.51\big(\text{molecular weight}\big) \quad (6.7)$$

Posphotriesterase (PTE) isolated from bacteria can catalyze the cleavage of the P—O bond in a variety of organophosphate triesters (Tsugawa et al. 2000) and is therefore highly efficient in hydrolyzing this kind of pesticides (Masson et al. 1998). Assuming that the enzymatic process was driven by the *Flavobacterium* sp. PTE, the following mechanism to hydrolyze pesticides, divided into three steps for illustrative purpose only, was proposed by Ortiz-Hernández et al. (2003). Considering that the two atoms of zinc, naturally pres-ent in PTE (Benning et al. 2001), are participating in pesticide hydrolysis, the first step starts with the coordination of the electronegative atom X (oxygen or sulfur) with one zinc, resulting in a polarization of the double bond P = X (Figure 6.3a; Ortiz-Hernández et al. 2003). Thereafter, phosphorus undergoes a nucleophilic attack by the bridging hydroxyl, producing a penta-coordinated geometry undergoing a regeneration of the double bond P = X by eliminating a substituent RX- in order to stabilize its charges (Figure 6.3b; Ortiz-Hernández et al. 2003). Thereafter, an OH- obtained from the ambient is coordinated with the metallic atom while the pesticide remains bonded to the PTE (Figure 6.3c; Ortiz-Hernández et al. 2003). At the second step (Figure 6.5; Ortiz-Hernández et al. 2003), the nucleophilic attack is repeated until all RX- groups are eliminated from the phosphorus (Figure 6.4a through 4c; Ortiz-Hernández et al. 2003). Finally, the last step of the mecha-nism of action of PTE is dependent on the heteroatom X (Ortiz-Hernández et al. 2003). If the pesticide contains

- Oxygen, the hydrolyzed pesticide is delivered from the active center to the PTE (Figure 6.5a)
- Sulfur, a last nucleophilic attack occurs to the phosphorus (Figure 6.5b) and a penta-coordinated intermediate compound is formed again (Figure 6.5c).

The aerobic degradation of ethion by mesophilic bacteria isolated from contaminated soils surrounding the disused cattle dip sites was investigated by Foster et al. (2004). Two isolates, identified as *Pseudomonas* and *Azospirillum* species, were capable of biodegrada-tion of ethion when cultivated in minimal salt medium. The abiotic hydrolytic degrada-tion products of ethion such as ethion dioxon, ethion monoxon, O,O-diethylthiophosphate,

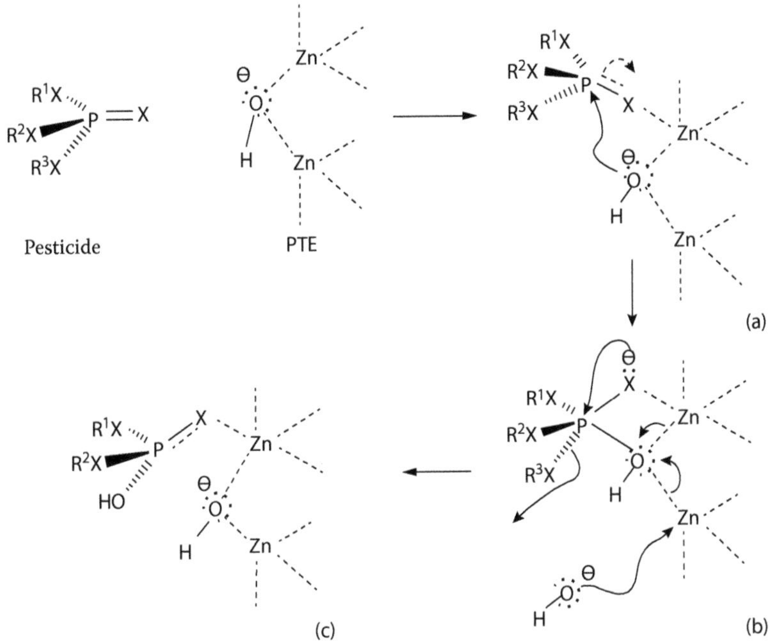

FIGURE 6.3
(a–c) Initial steps of the proposed mechanism of action of *Flavobacterium* sp. PTE, naturally containing two atoms of zinc at the active site, for hydrolyzing organophosphate pesticides. (From Ortiz-Hernández, M. L., Quintero-Ramírez, R., Nava-Ocampo, A. A., and Bello-Ramírez, A. M., Study of the mechanism of *Flavobacterium* sp. for hydrolyzing organophosphate pesticides. *Fundam. Clin. Pharmacol.*, 2003, 17: 717–723. Copyright Wiley-VCH Verlag GmbH & Co. KGaA. Reproduced with permission.)

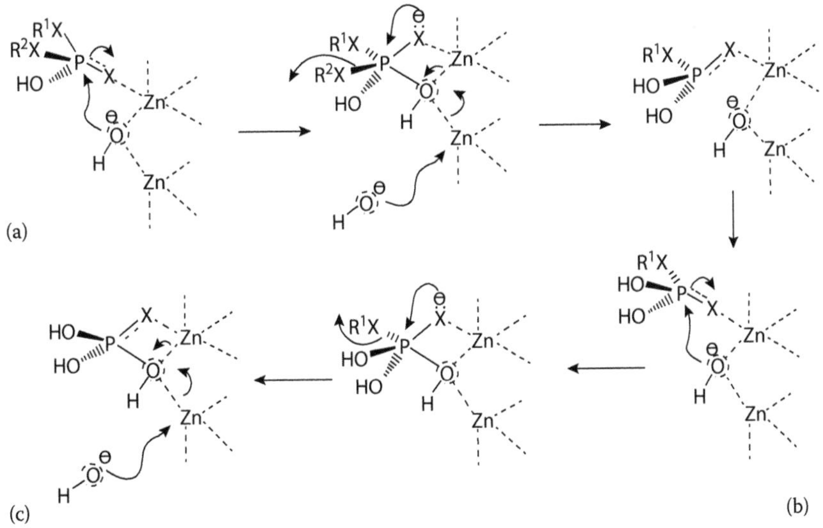

FIGURE 6.4
(a–c) Intermediate steps of the hydrolysis of organophosphate pesticides by *Flavobacterium* sp. PTE. The substituents at the phosphoric acid are eliminated by hydrolysis. (From Ortiz-Hernández, M. L., Quintero-Ramírez, R., Nava-Ocampo, A. A., and Bello-Ramírez, A. M., Study of the mechanism of *Flavobacterium* sp. for hydrolyzing organophosphate pesticides. *Fundam. Clin. Pharmacol.*, 2003, 17: 717–723. Copyright Wiley-VCH Verlag GmbH & Co. KGaA. Reproduced with permission.)

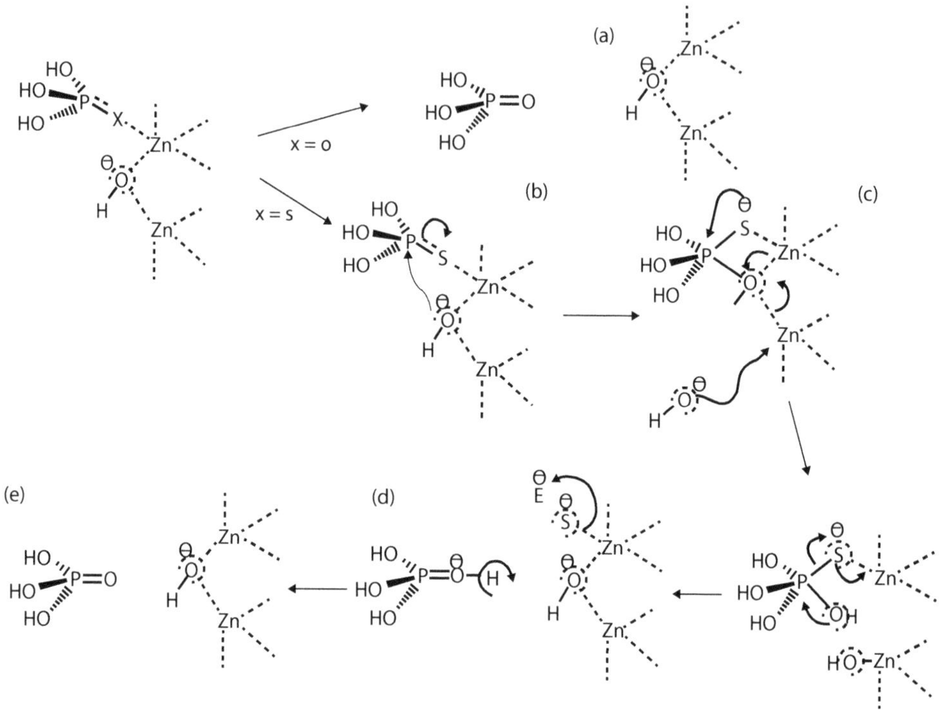

FIGURE 6.5

(a–e) Last step of the mechanism of action of *Flavobacterium* sp. PTE. Phosphoric acid is obtained as the final product of pesticide hydrolysis, and the active site of *Flavobacterium* sp. PTE is regenerated. The $E^{\oplus}$ represents any electrophilic species naturally occurring in the microambient. (From Ortiz-Hernández, M. L., Quintero-Ramírez, R., Nava-Ocampo, A. A., and Bello-Ramírez, A. M., Study of the mechanism of *Flavobacterium* sp. for hydrolyzing organophosphate pesticides. *Fundam. Clin. Pharmacol.*, 2003, 17: 717–723. Copyright Wiley-VCH Verlag GmbH & Co. KGaA. Reproduced with permission.)

and thioformaldehyde were not identified. Their absence may be due to a number of factors; however, it seems most likely that ethion and its degradation products were rapidly utilized for microbial growth (Foster et al. 2004). Factors influencing the ability of *Pseudomonas putida* strains (epI and epII) to degrade the organophosphate ethoprophos were described (Karpouzas and Walker 2000). Degradation of ethoprophos was most rapid when bacterial cultures were incubated at 25°C and 37°C. *Pseudomonas putida* epI was capable of completely degrading ethoprophos at a slow rate at 5°C compared with *Pseudomonas putida* epII, which could not completely degrade ethoprophos at the same rate. Degradation by *Pseudomonas putida* epI in a mineral-salt medium plus nitrogen was accompanied by concurrent bacterial growth, suggesting that *Pseudomonas putida* epI was growing at the expense of ethoprophos (Karpouzas and Walker 2000). Both isolates were able to rapidly degrade concentrations of ethoprophos as high as 50 mg/l, which is 10 times the recommended field rate, in less than 3 days. Several studies have previously reported the isolation of microorganisms capable of rapid degradation of other pesticides at similar concentrations (Parekh et al. 1994; Struthers et al. 1998). Another factor that, according to Zaidi et al. (1988), can influence the degradation activity of microorganisms capable of degrading pesticides is the pH value. Ethoprophos-degrading isolates were capable of rapidly degrading ethoprophos in the range of 5.5–7.6. Similar data were obtained by Mandelbaum et al. (1993), who reported that degradation of atrazine by a *Pseudomonas* strain was not affected by pH in the range 5.5–8.5. Degradation of ethoprophos by both cultures was slow at pH 5.0. Low pH levels are not favorable for the survival and growth of bacteria, in general (Karpouzas and Walker 2000).

Ning et al. (2010) report that three epiphytic bacterial species, from the genera *Sphingomonas*, *Acidovorax*, and *Chryseobacterium*, can degrade organophosphorus compounds (e.g., dichlorvos). The results obtained by Ning et al. (2010) demonstrate that phyllosphere bacteria have great potential for bioremediation of pesticides in situ, where the environment is hostile to nonepiphytic bacteria. Kambiranda et al. (2009) reported that the esterase enzyme (Est5S) was expressed in yeast to demonstrate the organophosphorus hydrolytic activity from a metagenomic library of cow rumen bacteria. The esterase enzyme was tested for degradation of chlorpyrifos. Chlorpyrifos degradation was tested with the purified enzyme using thin-layer chromatography (TLC) and high-performance liquid chromatography (HPLC). Chlorpyrifos could not be detected in the samples treated with the Est5S enzyme. 3,5,6-Trichloro-2-pyridinol (TCP), the primary degradation product of chlorpyrifos (CP), could be detected in the chlorpyrifos-treated samples (Figure 6.6; Kambiranda et al. 2009). The other eight pesticides were also degraded in the range of 45%–100%, when used as a substrate with Est5S protein (Kambiranda et al. 2009).

In another paper (De Roffignac et al. 2008), it has been shown that the use of bagasse compost in a biological bed system is efficient in degrading malathion, glyphosate, and λ-cyhalothrin. The biological bed degraded more than 99% of malathion and glyphosate after 6 months. For malathion, which has the lowest K_{oc} value, the volatilization process should artificially increase the calculated degradation rate (De Roffignac et al. 2008). In the

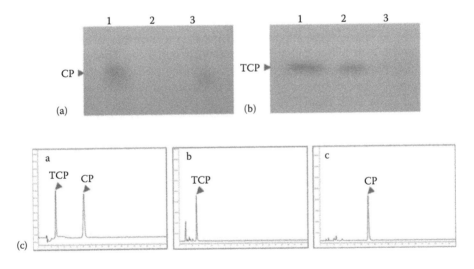

FIGURE 6.6
TLC (a and b) and HPLC (c) profiles of chlorpyrifos (CP) and 3,5,6-trichloro-2-pyridinol (TCP) by the purified Est5S enzyme of *P. pastoris* incubated at temperature 40°C at pH 7.0 for 12 h. Lane A1, B1, and C1, CP, and TCP standard, respectively; lane A2, B2, and C2, Est5S enzyme; lane A3, B3, and C3, inactivation Est5 enzyme (negative control). The TLC plates were developed with chloroform and hexane (4:1, v/v) solvent system for the detection of CP and ethyl acetate, isopropanol, and NH_4OH (5:3:2, v/v) system for the detection of TCP, respectively. The HPLC was eluted with 0.5% acetic acid and methanol (1:4 v/v). The arrowheads indicate the TCP (R_f value: approximately 0.66, retention time: approximately 5.5 min) and CP (R_f value: approximately 0.57, retention time: approximately 14 min) band and peak, respectively. (Reprinted from *Pestic. Biochem. Physiol.*, 94, Kambiranda, D. M., Asraful-Islam, S. Md., Cho, K. M., et al., Expression of esterase gene in yeast for organophosphates biodegradation, 15–20, (2009), with permission from Elsevier.)

biological bed, the DT_{50} value for malathion was 17 days, that for glyphosate was 33 days and that for λ-cyhalothrin was 43 days (De Roffignac et al. 2008). Degradation of the pirimiphos-methyl and benalaxyl in the topsoil was monitored by Patakioutas et al. (2002). The slope of the soil surface and the different sorption capacities of the compounds are the main parameters that influenced the transportation of the studied pesticides, pirimiphos-methyl and benalaxyl residues via surface water in soil–water systems (Patakioutas et al. 2002). Sorption isotherms of pirimiphos-methyl and benalaxyl on clay soil from experimental plots obtained from batch experiments were linear over the range of concentrations studied (0.1–20 mg/l); the slopes of the log-transformed Freundlich isotherms were 1.26 for pirimiphos-methyl and 1.0 for benalaxyl with correlation coefficients 0.97 and 0.99, respectively (Patakioutas 2000; Patakioutas and Albanis 2002). A higher sorption capacity was obtained for pirimiphos-methyl; the K_d values were 274.4 for pirimiphos-methyl and 41.5 for benalaxyl (Patakioutas 2000; Patakioutas and Albanis 2002). The estimated K_{oc} values from the log-transformed Freundlich equation (Equation 6.8)

$$\log S = \log K_d + 1/n \log C_e \tag{6.8}$$

where S is concentration of compound in supernatant (mg/l[1]) were 13,800 for pirimiphos-methyl and 2090 for benalaxyl (Patakioutas 2000; Patakioutas and Albanis 2002).

The disappearance rates indicate that potato planting decreased the half-lives of the two pesticides compared with the control fields: for pirimiphos-methyl from 16.7 to 9.2 days and for benalaxyl from 26.7 to 12.6 days (Patakioutas et al. 2002).

6.4.3 Carbamate Pesticides

During the last 30 years, there has been an increasing number of reports correlating the reduced efficacy of soil-applied pesticides with enhanced biodegradation induced by application to the same soil. However, for carbamate pesticides, there are only a few papers, for example, for carbofuran (2,3-dihydro-2,2-dimethyl-benzofuran-7-yl methylcarbamate) (Felsot et al. 1981; Read 1986; Suett et al. 1993). The effect of the initial concentration on the development and stability of its enhanced biodegradation in topsoil and subsoil was described by Karpouzas et al. (2001). The results obtained by these authors suggest that normal carbofuran applications to topsoil are unlikely to activate the subsoil microbiota for its accelerated degradation, but show that some subsoils already possess an inherent ability for relatively rapid biodegradation of the insecticide. Application of the VARLEACH model to simulate carbofuran movement through the soil profile indicated that approximately 0.01 mg/kg of carbofuran may reach a depth of 70 cm 400 days after a standard field application. The results therefore imply that adaptation of the subsoil microflora (ca. 1 m depth) by normal field rate applications of carbofuran is unlikely to occur. The results further confirmed that some subsoil samples have an inherent capacity for rapid biodegradation of carbofuran (Karpouzas et al. 2001). The high levels of variability observed between the replicates in some of the subsoil samples were attributed to the uneven distribution of a low population of carbofuran-degrading microorganisms in the subsurface soil (Karpouzas et al. 2001). Ahmad et al. (2004) investigated the bioavailability and biodegradation of carbaryl (1-naphthyl methylcarbamate) in soils. Carbaryl is weakly sorbed and generally considered to be easily degradable in soils. Extraction studies revealed that 49% of the total carbaryl in soils (88 mg/kg) was not water-extractable and also not bioavailable, as demonstrated by inoculation of the contaminated soil with a carbaryl-degrading mixed bacterial culture. Inoculation of the contaminated soil with the carbaryl-degrading culture showed that the bacteria were capable of degrading only the available (i.e., water-extractable) fraction of the pesticide. When the soil was pulverized in a ball mill to enhance the release of the residue, an additional 19% of the carbaryl became bioavailable. A significant proportion of the residue (~33%) remained unavailable. The long (>12 years) contact time between the pesticide and the soil (i.e., aging), allowing possible sequestration into the soil nanopores and the organic matter matrices, is suggested to have rendered the pesticide unavailable for microbial degradation (Ahmad et al. 2004). Results obtained by the authors (Ahmad et al. 2004) demonstrated that even a weakly sorbed and easily degradable pesticide, carbaryl, is effectively sequestrated in soil with time, rendering it partly inaccessible to microorganisms and affecting the bioavailability of the compound. Dissipation of ^{14}C carbaryl and quinalphos in the soil under a groundnut crop (*Arachis hypogaea* L.) was also described (Menon and Gopal 2003). The movement of carbaryl was limited to a 15 cm depth in the loamy sand of Jaipur (India) and was not detected till 120 days (DT_{50} of 14.93 days) after application (Menon and Gopal 2003).

6.4.4 Herbicides

6.4.4.1 Chlorotriazine Herbicides

Triazines are the oldest and the most commonly used herbicides, representing around 30% of the pesticide market in the world. These compounds have an appreciable persistence in soil (Table 6.4). The half-life in soil in days is the shortest for cyanazine (ca. 14 days) and the longest for simazine (27–102 days). Atrazine, a chlorotriazine herbicide, is widely used to control broadleaf and grassy weeds in corn, sorghum, sugarcane, and rangeland; it has been detected in groundwater and soil. The reported levels of atrazine in soils at pesticide mix-load sites can vary between 7.9×10^{-5} mM and 1.9 mM (mmole/kg soil) (Krapac et al. 1993). Grigg et al. (1997) reported on a mixed microbial culture, capable of degrading concentrations of atrazine in excess of 1.9 mM. At initial concentrations of 0.046 and 0.23 M, the mixed population degraded 78% and 21% of atrazine in soil (100 days), respectively. At the same initial concentrations in liquid cultures, 90% and 56% of atrazine was degraded (80 days), respectively. Decreased degradation in soil samples may have resulted from atrazine sorption to soil surfaces or decreased contact between the population and the herbicide (Grigg et al. 1997). In the 0.23 M system, Grigg et al. (1997) attribute incomplete degradation to phosphorus depletion. Data obtained by these authors for carbon dioxide evolution was fitted to a three-half-order regression model, but they feel that there are limitations of application of this model to atrazine degradation. The population uses the herbicide as a nitrogen source and little carbon is incorporated into the biomass, as the energy status of carbons in the ring leads to their direct evolution as [^{14}C]carbon dioxide. Grigg et al. (1997) suggest that the mixed culture could be used for bioremediation of atrazine at concentrations up to and exceeding those currently reported for agrochemical mixing-loading facilities. Microorganisms, including *Pseudomonas* and *Actinomycetes* species, are known to degrade atrazine and other pesticides in soil. The objective of the research by Gupta and Baummer III (1996) was to study the degradation of atrazine (2 or 3 ppm) in soil using poultry litter, which has a large number of microorganisms along with many nutrients. Atrazine in the soil was extracted with water and methanol and analyzed by pesticide immunoassay (ELISA) 1, 5, 10, 30, and 60 days after poultry litter treatment. Atrazine was significantly (86%) degraded in soil with untreated poultry litter within 30 days. Degradation was virtually complete within 60 days. The rate of atrazine biodegradation with poultry litter was almost two times faster than that without the litter (Gupta and Baummer III 1996). The toxicity of the soil + atrazine mixture treated with poultry litter (both the untreated and the γ-irradiated) was the same as that of the soil + litter mixture; no significant concentrations of toxic by-products were produced from the biodegradation of atrazine (Gupta and Baummer III 1996).

Simazine is a chlorotriazine herbicide widely used in agriculture. Simazine may be hydrolyzed to 2-hydroxy-simazine via both biotic and abiotic processes, which explains its detection in both sterilized and nonsterilized soils. Urea promoted the formation of desethyl-simazine. Degradation experiments were performed to assess whether the bio-degradation of simazine in soil may be influenced by the presence of urea (Caracciolo et al. 2005). Simazine degradation rates under different experimental conditions (presence or absence of urea, microbiologically active/sterilized soil) were assessed together with formation, degradation, and transformation of its main metabolites in soil. Simazine degradation was affected by the presence of urea, in terms of both a smaller half-life ($t_{1/2}$) and a higher amount of the desethyl-simazine formed. The soil bacterial community was also studied by the authors (Caracciolo et al. 2005), who applied fluorescent in situ hybridization (FISH) and detected all isolated strains as β-proteobacteria, confirming the active role of this group in simazine degradation. Moreover, specific molecular probes have recently been designed (Martínez-Íñigo et al. 2003), which take into account the sequence data on the genes involved in s-triazine catabolism: hydrolytic dechlorination, hydrolytic deamination, and cyanuric acid degradation. The application of these probes by the authors (Caracciolo et al. 2005) to their soil samples and to isolated pure cultures is in progress and will monitor directly the presence of bacteria that have specific metabolic capacities in some steps of triazine degradation. Caracciolo et al. (2005) suggest that in situ hybridization technique is a promising method to determine the structures and functions of the bacterial communities and may be useful to assess, together with the chemical analysis, the effects of pollutants in the environment. Albarrán et al. (2003) described the effects of adding an intermediary by-product of olive oil extraction on the sorption, degradation, and leaching of simazine in a sandy loam soil. Results obtained by these authors (Albarrán et al. 2003) showed that simazine sorption isotherms showed a great increase in herbicide sorption after the addition of solid olive-mill waste (SOMW) to soil; sorption increased with the amount of SOMW added. Incubation studies (Albarrán et al. 2003) showed extended persistence by reduced biodegradation of simazine in soil amended with SOMW compared with the unamended soil. Although the addition of SOMW to soil increased the total porosity, breakthrough curves of simazine in hand-packed soil columns showed that SOMW addition retarded the vertical movement of the herbicide through the soil and reduced the total amount of herbicide leached (Albarrán et al. 2003). It appeared that the longer residence time of simazine in the amended soil columns (>20 days) compared with that in the unamended soil column (<20 days) allowed enhanced degradation and/or irreversible sorption under column-leaching conditions (Albarrán et al. 2003). Results described by these authors indicate that SOMW application to agricultural land is likely to increase the residence time of simazine in the topsoil by enhancing sorption and reducing leaching and degradation losses. Fragoeiro and Magan (2005) examined the extracellular enzymatic activity of two white-rot fungi (*Phanerochaete chrysosporium* and *Trametes versicolor*) in a soil extract broth in relation to different concentrations (0–30 ppm) of simazine and pesticides from other groups (dieldrin–cyclodiene insecticide and trifluralin–dinitroaniline herbicide) under different osmotic stress (−0.7 and −2.8 MPa) and quantified enzyme production, relevant to P and N release (phosphomonoesterase, protease), carbon cycling (β-glucosidase, cellulase), and laccase activity involved in lignin degradation. Results obtained by Fragoeiro and Magan (2005) suggest that *Trametes versicolor* and *Phanerochaete chrysosporium* have the ability to degrade different groups of pesticides, supported by the capacity for expression of a range of extracellular enzymes at both −0.7 and −2.8 MPa water potentials. *Phanerochaete chrysosporium* was able to degrade this mixture of pesticides independent of the laccase activity. In soil extract, *Trametes versicolor* was able to produce the same range of enzymes as *Phanerochaete chrysosporium* plus laccase, even when present as 30 ppm of the pesticide mixture (Fragoeiro and Magan 2005). Complete degradation of dieldrin and trifluralin was observed by Fragoeiro and Magan (2005), while about 80% of the simazine was degraded regardless of osmotic stress treatment in a nutritionally poor soil extract broth. The capacity of tolerance and degradation of high concentrations of mixtures of pesticides and production of a range of enzymes, even under osmotic stress, is advantageous and the authors (Fragoeiro and Magan 2005) suggest potential bioremediation applications.

6.4.4.2 Triazinone Herbicides

Metribuzin (4-amino-6-tert-butyl-4,5-dihydro-3-methylthio-1,2,4-triazin-5-one) is weakly sorbed in soils and therefore leaches easily to lower soil levels, which results in the loss of activity.

Metribuzin degradation and transport described by Malone et al. (2004) were simulated using three pesticide sorption models available in the Root Zone Water Quality Model (RZWQM):

- Instantaneous equilibrium-only (EO)
- Equilibrium-kinetic (EK, includes sites with slow desorption and no degradation)
- Equilibrium-bound (EB, includes irreversibly bound sites with relatively slow degradation).

The result obtained by the authors (Malone et al. 2004) indicate that

- Simulated metribuzin persistence was more accurate using the EK (root mean square error, RMSE = 0.03 kg/ha) and EB (RMSE = 0.03 kg/ha) sorption models compared to the EO (RMSE = 0.08 kg/ha) model because of slowing metribuzin degradation rate with time.
- Simulating macropore flow resulted in prediction of metribuzin transport in percolate over the simulation period within a factor of 2 of that observed using all three pesticide sorption models.

Moreover, little difference in simulated daily transport was observed between the three pesticide sorption models, except that the EB model substantially underpredicted metribuzin transport in runoff and that when macropore flow and hydrology are accurately simulated, metribuzin transport in the field may be adequately simulated using a relatively simple, EO pesticide model (Malone et al. 2004).

Soil amendments play an important role in the management of runoff and leaching losses of pesticides from agricultural fields. Therefore, the effect of biocompost from the sugarcane distillery effluent on metribuzin degradation and mobility was studied by Singh (2008) in a sandy loam soil. Metribuzin was more persistent in biocompost-unamended (T-0) flooded soil ($t_{1/2}$ – 41.2 days) than in nonflooded ($t_{1/2}$ – 33.4 days) soil. Freundlich adsorption constants (K_f) for treatments T-0, T-1, and T-2 were 0.43, 0.64, and 1.13, respectively, suggesting that biocompost application caused increased metribuzin sorption

(Singh 2008). Leaching losses of metribuzin were drastically reduced from 93% in control soil (T-0) to 65% (T-1) and 31% (T-2) in biocompost-amended soils (Singh 2008). Biocompost from sugarcane distillery effluent can be used effectively to reduce downward mobility of metribuzin in low-organic-matter sandy loam soil (Singh 2008). Bending et al. (2002) described degradation of contrasting pesticides by white-rot fungi and its relationship with ligninolytic potential. There was considerable variation among the white-rot fungi in their ability to degrade the pesticides. Moreover, there were no relationships between presumptive ligninolytic activity and the degradation of any of the pesticides (e.g., chlorotriazine herbicides: atrazine and terbuthylazine; phenylurea herbicides: diuron). The three members of the Polyporaceae used in Bending et al. (2002) study showed differences in the extent to which they were able to degrade the pesticides, with *Coriolus versicolor* showing strong degradative abilities and *Dichomitus squalens* and *Pleurotus ostreatus* having limited degradative potential. When *Coriolus versicolor*, *Hypholoma fasciculare*, and *Stereum hirsutum* were grown in biobed matrix, they were all able to degrade the pesticides, although there were differences in the relative degradative capacities of the fungi in liquid and biobed media (Bending et al. 2002).

6.4.4.3 Phenylurea Herbicides

Phenylureas are important herbicides and are used worldwide to control weeds in various crops. Rates of degradation of phenylurea herbicides (isoproturon and diuron) and triazinylsulfonylurea herbicides (metsulfuron-methyl) were measured in two soils incubated at two temperatures (5°C and 25°C) with soil moisture at a metric potential −5 kPa (Walker and Jurando-Exposito 1998). With isoproturon and diuron, the changes in time were similar at the two incubation temperatures. This indicates that the apparent changes in the adsorption with these two compounds were not caused by preferential degradation in the soil solution, but by a slow equilibration with the adsorption sites (Walker and Jurando-Exposito 1998). The results obtained by Walker and Jurando-Exposito (1998) with the weakly adsorbed herbicides, metsulfuron-methyl, however, suggested the possibility of preferential degradation in the solution phase because, when degradation was slow, the absolute amounts adsorbed remained constant or increased slightly, even although solution concentrations declined. Saison et al. (2010) applied 3-(3,4-dichlorophenyl)-1,1-dimethylurea (^{14}C-diuron) to the soil surface to trace its residues in water-extractable (W), methanol-only-extractable (MO), and nonextractable (N) fractions and in runoff waters. Mass-balance calculations were fitted to the observed data to determine the quantities degraded or transferred within and between each of the compartments studied by Saison et al. (2010). Schematic model used for the calculation of degradation and transfers in and between soil fractions is presented in Figure 6.7 (Saison et al. 2010). The authors (Saison et al. 2010) considered the following mass-balance approach, which differentiates several compartments (Figure 6.7): W, water-extractable fraction in the 0–2 cm soil layer; MO, methanol-only-extractable fraction in the 0–2 cm soil layer; N, nonextractable fraction in the 0–2 cm soil layer; M, methanol-extractable fraction in the 2–6 cm soil layer; and R, fraction lost in runoff (Saison et al. 2010). Pesce et al. (2009) described a study that provides a strong suggestive evidence for high diuron biodegradation potential throughout its course, from the pollution source to the final receiving hydrosystem, and suggests that, after microbial adaptation, grass strips may represent an effective environmental tool for mineralization and attenuation of the intercepted pesticides (Pesce et al. 2009). Degradation of phenylurea herbicide isoproturon and herbicide bentazon in peat- and compost-based biomixtures was described by Coppola et al. (2011). Compost-based

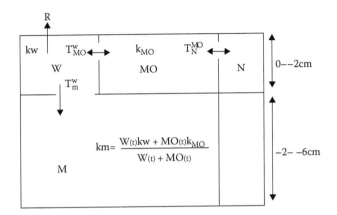

FIGURE 6.7
Schematic model used for the calculation of degradation and transfers in and between soil fractions: W (water-extractable fraction), MO (methanol-only-extractable fraction), N (nonextractable fraction), M (methanol-extractable fraction), R (runoff). (From Saison, C., Louchart, X., Schiavon, M., and Voltz, M., Evidence of the role of climate control and reversible aging processes in the fate of diuron in a Mediterranean topsoil. *Eur. J. Soil Sci.*, 2003, 61, 576–587. Copyright Wiley-VCH Verlag GmbH & Co. KGaA. Reproduced with permission.)

substrates showed faster pesticide dissipation in the presence of lignocellulosic materials, as in garden compost and vine branch straw (Coppola et al. 2011). The adsorption and the rate of degradation of ^{14}C-labeled isoproturon and mecoprop (MCPP) at concentrations from 0.0005 to 25,000 mg/kg were determined in biobed soil (Henriksen et al. 2003). The biobed material showed enhanced ability to adsorb the two herbicides. K_d was 5.2 L/kg for isoproturon and 1.6 L/kg for MCPP in biobed material, which is higher than in natural soil (Henriksen et al. 2003). The influence of lipophilicity and formulation on the distribution of eight pesticides (isoproturon, chlorotoluron, triasulfuron, phenmedipham, pendimethalin, chlorpyrifos, difenconazole, and permethrin) in laboratory-scale sediment–water systems was described by Bromilow et al. (2003). Sorption equilibrium was reached between 15 and 30 days, the proportion of pesticide in the sediment ranging from 20% for acidic and therefore polar triasulfuron to 97% for the lipophilic permethrin (Bromilow et al. 2003). Some degradation was observed for all compounds over 90 days; for some compounds and formulations, enhanced degradation occurred after 20–60 days. Degradation rates ranging from relatively high values for phenmedipham (DT_{50} 2.5–4 days) to low values for triasulfuron and difenconazole ($DT_{50} \gg 90$ days) were observed. It is concluded that lipophilicity is the chief determinant of pesticide distribution in sediment–water systems (Bromilow et al. 2003). Phenmedipham was degraded, presumably by alkaline hydrolysis, much more quickly in the pond water than in the sediment, illustrating the role of sorption and also, possibly, the lower surface pH of clay particles in slowing breakdown (Bromilow et al. 2003). The ability of microorganisms to degrade isoproturon, atrazine, and mecoprop within aerobic aquifer systems was described by Johnson et al. (2003). Biodegradation was confirmed by the formulation of monodesmethyl- and didesmethyl-isoproturon. Mineralization of plant-incorporated residues of ^{14}C-isoproturon in arable soils originating from different farming systems was described by von Wirén-Lehr et al. (2002). ^{14}C-isoproturon residues were incorporated in wheat plants by growing seedlings for 18 days in quartz sand with nutrient solution that was treated with ring-labeled ^{14}C-isoproturon, resulting in ^{14}C-concentration equivalent to 15.4 nmol isoproturon per g of dry shoot mass. The residues were characterized by extraction and HPLC analysis and were shown to consist of unchanged isoproturon, soluble metabolites (monodemethyl-isoproturon, didemethyl-isoproturon, 1-OH-isoproturon, 2-OH-isoproturon, 2-OH-monodemethyl-isoproturon, 2-OH-didemethyl-isoproturon, ispropenyl-isoproturon and unidentified metabolites), and nonextractable residues (von Wirén-Lehr et al. 2002). von Wirén-Lehr and Dörfler found that biomineralization of plant-incorporated ^{14}C-isoproturon residues and soil microbial biomass and activity parameters were related in three soils, whereas soil samples from a former hops garden did not show these relationships. This indicates that the overall soil microbial parameters in many cases are insufficient to describe the influence of biotic factors on the fate of pesticides in soils (von Wirén-Lehr et al. 2002). The biotransformation of phenylurea herbicides and some of their most prominent degradations by different soil microorganisms were also investigated (Berger 1998). In some cases described by the authors (Berger 1998), the different biotransformation rates correlated with high octanol–water partition coefficients and were transformed faster than compounds with a high water solubility. Degradation of phenylurea herbicides (diuron and isoproturon) and other herbicides (diclofop-methyl, bentazon, and pendimethalin) in soils was also described (Piutti et al. 2002). The results of the principal component analysis (PCA) based on kinetic parameters obtained by fitting the modified Gompertz model to mineralization kinetics after two cropping cycles in the greenhouse are presented in Figure 6.8a (Piutti et al. 2002). This figure showed the positioning of the different soils (planted or not, treated or not) in the first principal plane

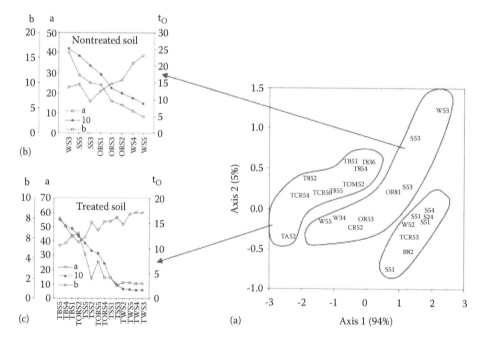

FIGURE 6.8
(a–c) PCA on diclofop-methyl mineralization kinetics parameters obtained in soil samples after two cropping cycles in the greenhouse: BS, bulk soil; WS, wheat-planted soil; ORS, oilseed rape–planted soil; SS, soybean-planted soil. TBS, TWS, TORS, and TSS, the same but treated with diclofop-methyl in the greenhouse. 1–5, treatment replicate in the greenhouse experiment. (From Piutti, S., Marchand, A.-L., Lagacherie, B., Martin-Laurent, F., and Soulas, G. Effect of cropping cycles and repeated herbicide applications on the degradation of diclofop-methyl, bentazon, diuron, isoproturon and pendimethalin in soil. *Pest Manag. Sci.*, 2002, 58, 303–312. Copyright Wiley-VCH Verlag GmbH & Co. KGaA. Reproduced with permission.)

as deduced from PCA. Ninety-two percent of the total variance was explained by the first principal component. In the first principal plane, it is possible to discriminate between three groups of soil samples (Piutti et al. 2002). The first group can easily be separated in the lower right side of the plane. It contains a few treated and a majority of untreated samples showing no or low accelerated degradation. Among other soil samples, Piutti et al. (2002) have distinguished those that have received diclofop-methyl in the greenhouse from those that have not. For those two groups, the authors have reported the values of kinetic parameters as functions of their positions along the first component axis in Figure 6.8b and 8c (Piutti et al. 2002).

6.4.4.4 Phenoxyacetic Herbicides

The herbicide 2,4-dichlorophenoxyacetic acid (2,4-D) is known to inhibit methanotrophic bacteria. Methane oxidation was therefore used as a parameter to evaluate the residual 2,4-D after bioaugmentation of an agricultural soil by Top et al. (1999). Several strains harboring catabolic plasmids that code for the degradation of 2,4-D were compared for their potentials to alleviate the negative impact of 2,4-D on methane oxidation by soil microorganisms. All data obtained by Top et al. (1999) show that bioaugmentation of a 2,4-D-treated soil with strains with the necessary genetic catabolic information shortens the time needed for the complete removal of 2,4-D and its toxicity towards methanotrophs. The transfer of the plasmid pEMT1k, resulting in increased 2,4-D removal, confirms the results that dissemination of catabolic genes can be an alternative to the classical bioaugmentation approach, which relies on the activity of the inoculated strain (Top et al. 1999). Top et al. (1999) have shown that measuring the methane oxidation in soil can be a very good bioassay to evaluate the success of bioaugmentation if the pollutant or a possible metabolite inhibits this microbial process. Gonod et al. (2006) described 2,4-D impact on bacterial communities and the activity and genetic potential of 2,4-D degrading communities in soil. The study of Kumar et al. (2010) carried out to investigate the chronic response of the cyanobacteria *Anabaena fertilissima* to 2,4-D ethyl ester at different concentrations: 15, 30, and 60 ppm. Based on the inhibitory effect and growth arrest, the release of certain products, such as carbohydrates, proteins, amino acids, and phenols was also affected at 15, 30, and 60 ppm of 2,4-D. The results of the experiments on enzyme activity obtained by Kumar et al. (2010) revealed that 2,4-D inhibits the synthesis of nitrate reductase and glutamine synthetase. On the basis of the results described by Kumar et al. (2010), it can be stated that the application of 2,4-D to rice fields should be considered because of its toxicity to the heterocystous filamentous cyanobacteria. Adsorption and degradation of four acidic herbicides (2,4-D, dicamba, metsulfuron-methyl, and flupyrsulfuron-methyl sodium) in soils was described by Villaverde et al. (2008). Adsorption and degradation of four pesticides were measured in four soils characterized by small organic matter (OM) contents (0.3%–1.0%) and varying clay contents (3%–66%). In general, sorption increased in the order: dicamba < metsulfuron-methyl < 2,4-D < flupyrsulfuron-methyl sodium (Villaverde et al. 2008). Both OM and clay content were found to be important in determining adsorption, but relative differences in clay content between soils were much larger than those in OM content, and therefore clay content was the main property determining the extent of herbicide adsorption for soils of this type. pH was negatively correlated with adsorption for all pesticides apart from metsulfuron-methyl. The contrasting behavior shown by these four acidic herbicides indicates that chemical degradation in soil is more difficult to predict than adsorption (Villaverde et al. 2008).

6.4.4.5 Organophosphorus Herbicides

Glyphosate, N-(phosphonomethyl) glycine, is a herbicide frequently used for controlling perennial weeds through application after harvest. The effect of freeze-thaw activity on the availability of glyphosate in soil, and consequently its mineralization by soil microorganisms, was studied by Stenrød et al. (2005). Results obtained by these authors showing low concentrations of glyphosate in soil water are consistent with earlier findings of high K_d values for glyphosate in these soils but do not give any conclusive evidence as to how the bioavailability of glyphosate in the soils affects the biodegradation patterns observed. The different glyphosate degradation rates between soils observed by Stenrød et al. (2005) suggest that bioavailability is involved and the length of time of freezing was an important factor governing total glyphosate mineralization. Amorós et al. (2007) described the assessment of toxicity of a glyphosate-based formulation using bacterial systems in lake water and showed that when the pH of the sample increases, the EC_{50} (mg/l) significantly increases, showing the direct effect of the pH on the toxicity values obtained. Desorption and time-dependent sorption of five herbicides (a broad-spectrum herbicide, glyphosate, and four commonly used selective herbicides, trifluralin, metazachlor, metamitron, and sulcotrione) in three soils were described by Mamy and Barriuso (2007). Desorption of glyphosate and its metabolite aminomethylphosphonic acid (AMPA) was greater in some soils (high pH and phosphate contents, little copper and amorphous iron contents), comparable with that of trifluralin and sulcotrione (Mamy and Barriuso 2007).

6.4.4.6 Chloroacetanilide Herbicides

Si et al. (2009) described adsorption, desorption, and dissipation of metolachlor in surface and subsurface soils. Adsorption of metolachlor was greater in the high-organic-matter surface soil than in subsoils. Lower adsorption distribution coefficient (K_{ads}) values with increasing depth indicated less adsorption at lower depths and greater leaching potential of metolachlor after passage through the surface horizon. Desorption of metolachlor showed hysteresis, indicated by the higher adsorption slope ($1/n_{ads}$) compared with the desorption slope ($1/n_{des}$). Soils that adsorbed more metolachlor also desorbed less metolachlor. The first-order dissipation rate was highest at the 0–50 cm depth (0.140/week) and lowest at the 350–425 cm depth (0.005/week) (Si et al. 2009). The results obtained by Si et al. (2009) indicate that degradation of the herbicide was significantly correlated with microbial activity in soils. Mills et al. (2001) described the quantification of the chloroacetanilide herbicide acetochlor degradation in the unsaturated zone using two novel in situ field techniques. Two different in situ field techniques and traditional laboratory-incubation studies have indicated that acetochlor is rapidly degraded and is mineralized in subsoils to a depth of approximately 5 meters (Mills et al. 2001).

6.4.4.7 Nitrile Herbicides

Biodegradation of the nitrile herbicide bromoxynil was described by Rosenbrock et al. (2004). The mineralization and formation of metabolites and nonextractable residues of the herbicide [^{14}C]bromoxynil octanoate ([^{14}C]3,5-dibromo-4-octanoylbenzonitrile) and the corresponding agent substance [^{14}C]bromoxynil ([^{14}C]3,5-dibromo-4-hydroxybenzonitrile) were investigated in soil. The mineralization of [^{14}C]bromoxynil and [^{14}C]bromoxynil octanoate in soil within 60 days amounted up to 42% and 49%, respectively (Rosenbrock et al. 2004). After the experiment, 52% of the originally applied [^{14}C]bromoxynil and 44% of the [^{14}C]bromoxynil octanoate formed nonextractable residues in the soil. Plant cell-wall–bound [^{14}C]bromoxynil residues were also mineralized to an extent of about 21% within 70 days; the main portion of 76% persisted as nonextractable residues in soil. In bacterial enrichment cultures, two polar metabolites were observed by these authors (Rosenbrock et al. 2004): one of them could be identified as 3,5-dibromo-4-hydroxybenzoate and the other could be described tentatively as 3,5-dibromo-4-hydroxybenzamide.

6.4.4.8 Quaternary Ammonium Herbicides

The effects of paraquat dichloride on the laccase activity and some biochemical parameters of *Trametes versicolor* and *Abortiporus biennis* strains belonging to white-rot *Basidiomycetes* fungi were examined by Jaszek et al. (2006). Exposure of *Trametes versicolor* and *Abortiporus biennis* to paraquat dichloride enhanced their natural ligninolytic metabolism as a consequence of adaptation to oxidative stress conditions of this group of fungi. Results obtained by Jaszek et al. (2006) are interesting because although oxidative stress defenses have been very often described in the recent literature, very few studies estimate the possible adaptive oxidative stress response in the white-rot fungi. The increased activity of laccase in the presence of paraquat dichloride may also suggest that it has an important role in fungal oxidative stress defense mechanism (Jaszek et al. 2006). The results obtained by Jóri et al. (2007) indicate that paraquat is able to enter the chloroplasts in the resistant biotype, causing transitional inhibition that can be clearly detected on the basis of fluorescence quenching. Investigations proved that paraquat entered the cells of both resistant and susceptible biotypes, approached the maximum within the first hour in chloroplasts, and then declined in all organelle fractions. In the resistant biotype, paraquat was located in the vacuoles a day after treatment. Selective transporter inhibitors blocked the sequestration of paraquat, suggesting the participation of transporters not directly energized (Jóri et al. 2007).

6.4.5 Others

Among modern pesticides, several different classes of fungicides are very persistent, though no deleterious effects on soil microbial processes have been reported. The behaviors of five such fungicides (flutriafol, epoxiconazole, propiconazole, triadimefon, and triadimenol) have been examined by Bromilow et al. (1999) in two field trials utilizing different agronomic treatments. Microbial degradation of chlorothalonil in agricultural soil was described by van Eden et al. (2000). Interaction between the following factors could have influenced the rate of biodegradation (van Eden et al. 2000).

- Soil conditions that determined the degree to which soil bacteria were conditioned by preexposure for in vitro aerobic utilization of chlorothalonil.
- The number of viable bacteria in inoculums, as well as their in vitro culturability.
- The composition of indigenous soil populations.
- Soil characteristics that influenced the bioavailability of chlorothalonil.

Structural and functional approach to studying pesticide (quinine fungicide dithianon) side effects on specific soil functions was proposed by Liebich et al. (2003). Aerobic biodegradation of [^{14}C]3-chloro-p-toluidine hydrochloride in loam soil was described (Spanggord et al. 1996). The results obtained by Arshad et al. (2008) demonstrated that bioaugmentation of the contaminated soils with α- and β-endosulfan degrading bacterium under optimized conditions provides an effective bioremediation strategy. Zhang et al. (2007) determined the response of antioxidative enzymes of cucumber (*Cucumis sativus* L.) when carbendazim was applied as soil drench at 0, 5, 50, and 100 mg/kg. On the basis of the results, Zhang et al. (2007) concluded that increased superoxide dismutase, catalase and glutathione peroxidase activity provides plants with increased carbendazim stress tolerance.

6.5 Degradation of Pesticides by Sunlight

Sunlight at the Earth's surface consists of radiation in the so-called UV-B and UV-A regions (295–400 nm), in addition to visible light (~400–800 nm) and infrared (IR) radiation. Approximately, 4% of the total energy in sunlight occurs in the UV band, but the intensity varies greatly with latitude, season, time of day, and thickness of the atmosphere and the ozone layer. The sun produces 0.2–0.3 mol photons/m^2 h in the range of 300–400 nm with a typical UV flux of 20–30 W/m^2. Natural radiation from the sun has also been found to fade the color and reduce the concentration of dissolved organic matter (DOM).

Heterogeneous photocatalytic oxidation process employing catalysts such as TiO_2, ZnO, etc., and UV light has demonstrated promising results for the degradation of pesticides and producing more biologically degradable and less toxic substances (Garcia et al. 2006, 2008; Vora et al. 2009). This process mainly relies on the in situ generation of hydroxyl radicals under ambient conditions, which are capable of converting a wide spectrum of toxic organic compounds including the nonbiodegradable ones into relatively innocuous end-products such as CO_2 and H_2O (Ahmed et al. 2011).

Photocatalytic degradation of pesticides depends on the type and composition of the photocatalyst and light intensity, initial substrate concentration, amount of catalyst, pH of the reaction medium, ionic compounds of wastewater, solvent types, oxidizing agents/ electron catalyst, catalyst application mode, and calcination temperature (Shaktivel et al. 2003).

In the photocatalytic oxidation process, organic pollutants are destroyed in the presence of semiconductor photocatalysts (e.g., TiO_2, ZnO), an energetic light source, and an oxidizing agent such as oxygen or air (Ahmed et al. 2011). Figure 6.9 illustrates that only photons with energies greater than the band-gap energy (ΔE) can result in the excitation of valence band (VB) electrons, which then promote possible reactions with organic pollutants (Ahmed et al. 2011). The absorption of photons with energy lower than ΔE or longer wavelengths usually causes energy dissipation in the form of heat. The illumination of the photocatalytic surface with sufficient energy leads to the formation of a positive hole (h^+) in the valence band and an electron (e^-) in the conduction band (CB). The positive hole oxidizes either the pollutant directly or water to produce hydroxyl radical •OH, whereas the electron in the conduction band reduces the oxygen adsorbed on the photocatalyst (TiO_2) (Ahmed et al. 2011). Ahmed et al. (2011) proposed the following steps (Equations 6.9 through 6.13) for the activation of TiO_2 by UV light

$$TiO_2 + h\nu \ (\lambda < 387 \text{ nm}) \rightarrow e^- + h^+ \qquad (6.9)$$

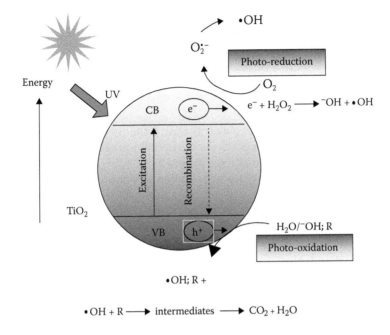

FIGURE 6.9
Schematic diagram illustrating the principle of TiO_2 photocatalysis. (Reprinted from *J. Environ. Manag.*, 92, Ahmed, S., Rasul, M. G., Brown, R., and Hashib, M. A., Influence of parameters on the heterogeneous photocatalytic degradation of pesticides and phenolic contaminants in wastewater: A short review, 311–330, (2011), with permission from Elsevier.)

$$e^- + O_2 \rightarrow O_2 \bullet^- \tag{6.10}$$

In this reaction, h^+ and e^- are powerful oxidizing and reducing agents, respectively. The oxidative and reductive reaction steps are expressed as (Ahmed et al. 2011)
Oxidative reaction

$$h^+ + Organic(R) \rightarrow Intermediates \rightarrow CO_2 + H_2O \tag{6.11}$$

$$h^+ + H_2O \rightarrow \bullet OH + H^+ \tag{6.12}$$

Reductive reaction

$$\bullet OH + Organic(R) \rightarrow Intermediates \rightarrow CO_2 + H_2O \tag{6.13}$$

Hydroxyl radical generation by the photocatalytic oxidation process is shown in the above steps. The resulting intermediates further react with $\bullet OH$ to produce final degradation products such as CO_2 and H_2O. The details are described in the review paper by Ahmed et al. (2011). The photocatalytic activity of TiO_2 depends on the surface and structural semiconductor properties such as crystal composition, surface area, particle size distribution, porosity, band gap, and surface hydroxyl density (Ahmed et al. 2011). At low substrate concentrations, for example, in the case of pesticides, the photocatalytic degradation rate of organic compounds can be explained by a pseudo first-order pattern, with Equation 6.14 demonstrating the relationship between C and t

$$\ln C/C_0 = k_{obs}t \qquad (6.14)$$

where, k_{obs} is the apparent reaction rate constant; t, the reaction time
C_0, the initial concentration of target pesticide in aqueous solution
C, the residual concentration of pesticide at time *t*
Successful application of photocatalytic oxidation system requires the degradation rate on the substrate concentration (C_0).

Table 6.5 summarizes a variety of pesticides studied under various initial concentrations, and Table 6.6 shows the influence of catalyst loading on the photocatalytic degradation of various groups of pesticides.

Electrostatic interaction among semiconductor surface, solvent molecules, substrate, and charged radicals formed during photocatalytic oxidation strongly depends on the solution pH. The pH of the solution can play a key role in adsorption and photocatalytic oxidation of pesticides. The influence of pH on the photocatalytic degradation of pesticides is summarized in Table 6.7. The ionization state of the surface of the photocatalyst can also be protonated and deprotonated under acidic and alkaline conditions, respectively, as shown in the following reactions (Ahmed et al. 2011) (Equations 6.15 and 6.16)

$$pH < P_{zc} \qquad TiOH + H^+ \rightarrow TiOH_2^+ \qquad (6.15)$$

$$pH > P_{zc} \qquad TiOH + OH^- \rightarrow TiO^- + H_2O \qquad (6.16)$$

where, P_{zc} is the point of zero charge of the TiO_2 (Degussa 25); TiO_2 (Degussa 25) is widely investigated and reported at pH ~ 6.25 (Zhu et al. 2005a).

While under acidic conditions, the positive charge of the TiO_2 surface increases as the pH decreases (Equation (6.15)), above pH 6.25, the negative charge at the surface of TiO_2 increases with increasing pH. Moreover, the pH of the solution affects the formation of hydroxyl radicals by the reaction between the hydroxide ions and the photo-induced holes on the TiO_2 surface (Ahmed et al. 2011). The positive holes are considered the major oxidation steps at low pH, whereas the hydroxyl radicals are considered the predominant species at neutral or high pH levels (Shifu and Gengyu 2005; Mathews 1988). The generation of •OH is higher due to the presence of more available hydroxyl ions on the TiO_2 surface. The degradation efficiency of the process is enhanced at higher pH.

TiO_2 has been suggested to be an efficient and viable photocatalytic degradation of compounds for various pesticides (Ahmed et al. 2011).

6.5.1 Organochlorine Insecticides

Organochlorine pesticides have been widely used in growth fields in order to eliminate pests and in this way increase the crops' outputs, improve the quality of the products, and decrease the incidence of illness propagated by insects. Because of their toxic potential,

TABLE 6.5

Influence of the Initial Pesticide Concentration on the Photocatalytic Degradation of Various Pesticides

Pesticide	Light Source	Photocatalyst	Co Range (mM)	Optimum Co (mM)	Group of Pesticide	References
Phosphamidon	UV	TiO_2	0.1–0.6	0.45	Organophosphate insecticide	Rahman and Muneer (2005a)
Acephate	UV	TiO_2	0.7–1.0	1	Phosphoramidothioate insecticide	Rahman et al. (2006)
Diphenamid	UV	TiO_2	0.1–0.6	0.6	Amide herbicide	Rahman et al. (2003)
Carbofuran	UV	TiO_2	0.023–0.113	0.09	Benzofuranyl methylcarbamate insecticide	Mahalakshmi et al. (2007)
Thiram	Solar	TiO_2	4.2×10^{-4}–16.6×10^{-4}	4.2×10^{-4}	Dithiocarbamate fungicide	Kaneco et al. (2009)
Lindane	Visible	N–TiO_2	3.45×10^{-5}–2.07×10^{-4}	3.45×10^{-5}	Organochlorine insecticide	Senthilnathan and Philip (2010)
Dimethoate	UV	TiO_2	0.0195–0.49	0.0195	Aliphatic amide organothiophosphate insecticide	Chen et al. (2007)
Isoproturon	Solar	TiO_2	0.25–0.75	0.75	Phenylurea herbicide	Haque and Muneer (2003)
Triclopyr	UV	TiO_2	0.25–1.0	0.75	Pyridine herbicide	Qamar et al. (2006)
Daminozide	UV	TiO_2	0.50–1.5	0.75	Growth retardant	Qamar et al. (2006)
Diazinon	UV	ZnO	0.003–0.005	0.003	Pyrimidine organothiophosphate insecticide	Daneshvar et al. (2007)
Methamidophos	UV	Re–TiO_2	0.1–0.5	0.1	Phosphoramidothioate insecticide	Zhang et al. (2009)
Bentazon	UV	TiO_2	0.02–0.062	0.02	Unclassified herbicide	Pourata et al. (2009)
Propham	UV	TiO_2	0.25–1.3	0.75	Carbanilate herbicide	Muneer et al. (2005)
Propachlor	UV	TiO_2	0.2–1.35	1.35	Chloroacetanilide herbicide	Muneer et al. (2005)
Tebuthiuron	UV	TiO_2	0.25–1.5	1.0	Thiadiazolylurea herbicide	Muneer et al. (2005)

Source: Reprinted from *J. Environ. Manag.*, 92, Ahmed, S., Rasul, M. G., Brown, R., and Hashib, M. A., Influence of parameters on the heterogeneous photocatalytic degradation of pesticides and phenolic contaminants in wastewater: A short review, 311–330, (2011), with permission from Elsevier.

TABLE 6.6

Influence of Catalyst Loading on the Photocatalytic Degradation of Various Pesticides

Pesticide	Light Source	Photocatalyst	Catalyst dose range, g/L	Optimum dose, g/L	Group of Pesticide	References
Diazinon	UV	ZnO	0.5–4.0	3.0	Pyrimidine organothiophosphate insecticide	Daneshvar et al. (2007)
Glyphosate	UV	TiO_2	1.0–9.0	6.0	Organophosphorus herbicide	Shifu and Yunzhang (2007)
Thiram	Solar	TiO_2	0–0.3	0.20	Dithiocarbamate fungicide	Kaneco et al. (2009)
Tebuthiuron	UV	TiO_2	0.5–7.5	5.0	Thiadiazolylurea herbicide	Muneer et al. (2005)
2,4,5-TCPAA	UV	TiO_2	0.5–5.0	3.0	Phenoxyacetic herbicide	Singh et al. (2007a)
Propham	UV	TiO_2	0.5–5.0	5.0	Carbanilate herbicide	Muneer et al. (2005)
β-Cypermethrin	UV	RuO_2-TiO_2	2.0–6.0	5.0	Pyrethroid ester insecticide	Yao et al. (2007)
Methamidophos	UV	Re-TiO_2	2–12	1.0	Phosphoramidothioate insecticide	Zhang et al. (2009)
Triclopyr	UV	TiO_2	0.5–5.0	2.0	Pyridine herbicide	Qamar et al. (2006)
Picloram	UV	TiO_2	0.2–2.0	2.0	Pyridine herbicide	Rahman and Muneer (2005b)
Carbofuran	UV	TiO_2	0.025–0.125	0.1	Benzofuranyl methylcarbamate insecticide	Mahalakshmi et al. (2007)
Carbendazim	UV	TiO_2	0–0.09	0.07	Benzimidazole fungicide	Saien and Khezrianjoo (2008)
Imazapyr	UV	TiO_2	0.315–2.5	2.5	Imidazolinone herbicide	Pizarro et al. (2005)

Source: Reprinted from *J. Environ. Manag.*, 92, Ahmed, S., Rasul, M. G., Brown, R., and Hashib, M. A., Influence of parameters on the heterogeneous photocatalytic degradation of pesticides and phenolic contaminants in wastewater: A short review, 311–330, (2011), with permission from Elsevier.

TABLE 6.7

Influence of pH on the Photocatalytic Degradation of Various Pesticides

Pesticide	Light Source	Photocatalyst	Tested pH Range	Optimum pH	Group of Pesticide	References
Atrazine	Solar	TiO_2	2.0–9.99	4–5	Chlorotriazine herbicide	Parra et al. (2004)
Diazinon	UV	ZnO	3.5–11.3	5.2	Pyrimidine organothiophosphate insecticide	Daneshvar et al. (2007)
Dimethoate	UV	TiO_2	2.04–11.05	11.05	Aliphatic amide organothiophosphate insecticide	Chen et al. (2007)
Propham	UV	TiO_2	3.0–11.0	5.0	Carbanilate herbicide	Muneer et al. (2005)
β-Cypermethrin	UV	RuO_2–TiO_2	2.4–10.8	6.42	Pyrethroid ester insecticide	Yao et al. (2007)
Propachlor	UV	TiO_2	3.0–11.0	3.0	Chloroacetanilide herbicide	Muneer et al. (2005)
Methamidophos	UV	Re–TiO_2	4.0–10.0	6.0	Phosphoramidothioate insecticide	Zhang et al. (2009)
Thiram	Solar	TiO_2	2.5–11.4	8.0	Dithiocarbamate fungicide	Kaneco et al. (2009)
Isoproturon	Solar	TiO_2	3.0–10.0	7.0	Phenylurea herbicide	Sharma et al. (2009)
4-CPAA	UV	TiO_2	3.0–9.0	3.0	Phenoxyacetic herbicide	Singh et al. (2007b)
Carbofuran	UV	TiO_2	4.0–9.0	7.0	Benzofuranyl methylcarbamate insecticide	Mahalakshmi et al. (2007)
Carbendazim	UV	TiO_2	3.0–9.0	9.0	Benzimidazole fungicide	Saien and Khezrianjoo (2008)
Monocrotophos	UV	La–ZnO	8.0–10.0	10.0	Organophosphate insecticide	Anandan et al. (2007)

Source: Reprinted from *J. Environ. Manag.*, 92, Ahmed, S., Rasul, M. G., Brown, R., and Hashib, M. A., Influence of parameters on the heterogeneous photocatalytic degradation of pesticides and phenolic contaminants in wastewater: A short review, 311–330, (2011), with permission from Elsevier.

their persistence, and their tendency to bioconcentrate, the widespread use of pesticides in the environment generates increasing risks to human health (e.g., cancer and disruption of hormonal functions). Therefore, control over these compounds in water, plants, foodstuffs, and soils is necessary.

The results obtained by Kulovaara et al. (1995) show that the sunlight-induced photochemical degradation rate of DDT is higher in water with high DOM contents, which means that the effects on the bioavailability will be further accentuated. Natural DOM is a complex mixture and the light-initiated reactions are also complex in humic surface waters. DDT was degraded both by the UV_{254} radiation and the amplified artificial solar radiation (Kulovaara et al. 1995). DDT was degraded faster during UV irradiations (minutes) than during the simulated sunlight irradiations (hours). Different reaction pathways are involved in the two experiments (Kulovaara et al. 1995).

6.5.2 Organophosphate Insecticides

Organophosphorus (OP) pesticides have been widely used as an alternative to organochlorine compounds for the control of insects in a wide range of fruits, vegetables, and grains all over the world. Organophosphorus pesticides in water are easily degraded by UV light and immobilized by the TiO_2 system. The photocatalytic degradation of the three pesticides, acephate (phosphoramidothioate insecticide), dimethoate (aliphatic amide organothiophosphate insecticide), and glyphosate (organophosphorus herbicide), in water has been investigated using UV light and TiO_2 immobilized on silica gel as photocatalyst and described by Echavia et al. (2009). The results obtained by Echavia et al. (2009) show that pesticides can be efficiently degraded by the UV–TiO_2 system. Complete (100%) decomposition of dimethoate and glyphosate was attained within 60 min of irradiation, while total degradation of acephate occurred after 105 min of photocatalytic treatment (Echavia et al. 2009). Decomposition of acephate and dimethoate was found to be photocatalytic in nature, while glyphosate degradation appeared to be governed by both adsorption and photocatalytic reactions. As the results of the study by Echavia et al. (2009) demonstrate, adsorption of organic substrates on the surface of TiO_2 is not a precondition for effective photocatalysis to proceed since both weakly and strongly adsorbed pesticides were efficiently degraded by the system. Results obtained by the authors (Echavia et al. 2009) indicate that the 120 min treatment time employed was not sufficient to completely mineralize the parent pesticide molecules and toxic intermediates such as methamidophos and methoate were not detected, revealing good advantage of the UV and immobilized TiO_2 system over other treatment processes. The photocatalytic degradation pathway for each pesticide was tentatively proposed by Echavia et al. (2009), based on the reduction of TOC and the formation of major by-products during the course of UV–TiO_2 photocatalysis, as shown in Figures 6.10 through 6.12.

The photocatalytic degradation of a small concentration of an organophosphorus insecticide, phosphamidon, in water on ZnO and TiO_2 was described by Rabindranathan et al. (2003). In a paper published by Shifu and Gengyu (2005), the floating TiO_2-SiO_2 photocatalyst beads by the dip-coating method was applied; hollow glass microbeads were used as the carrier and titanium tetraisopropoxide [$Ti(iso-OC_3H_7)_4$] and ethyl silicate were used as the raw materials. The results obtained by the authors (Shifu and Gengyu 2005) indicate that the best heat treatment condition for TiO_2-SiO_2 beads is at 650°C for 5 h. After 420 min illumination by sunlight, 0.65×10^{-4} mol/dm^3 of four organophosphorus pesticides of three structures can be completely photocatalytically degraded into PO_4^{3-}. The addition of a small amount of Cu^{2+} (up to 0.05 mmol/dm^3) increases the photodegradation of

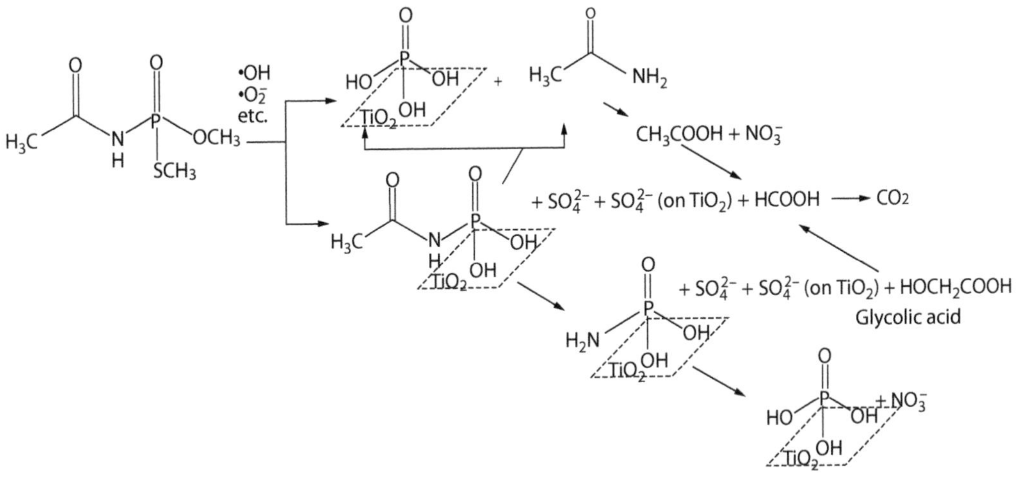

FIGURE 6.10
Proposed tentative pathway for the photocatalytic degradation of acephate. (Reprinted from *Chemosphere*, 76, Echavia, G. R. M., Matzusawa, F., and Negishi, N., Photocatalytic degradation of organophosphate and phosphonoglycine pesticides using TiO_2 immobilized on silica gel., 595–600, (2009), with permission from Elsevier.)

phosphate ester pesticides rapidly; acidic and alkaline solutions are favorable for the photocatalytic degradation (Shifu and Gengyu 2005).

Photocatalytic degradation of phosphate ester pesticides takes place first in the ester group that has strong acidity. The pesticide is degraded into trimethyl phosphate ester, formic acid, acetic acid, and some small molecule organics, and then the trimethyl phosphate ester is photocatalytically degraded directly into PO_4^{3-} and formic acid (Shifu and Gengyu 2005). Photochemical effects of humic substances on the degradation of organophosphorus pesticides were described by Kamiya and Kameyama (1998). The effects of selected ions on photodegradation of organophosphorus pesticides sensitized by humic acid were described by Kamiya and Kameyama (2001). Selected metal ions having paramagnetic property were found to exert inhibition effects on aquatic photodegradation of organophosphorus pesticides sensitized by humic acids, according to the increasing ord er:Cr(III) < Co(II) < Mn(II) < Cu(II). Basic factors dominating the metal-ion effects were classified on the basis of the fluorescence quenching as well as radical scavenging abilities of the metal ions complexed with humic acids (Kamiya and Kameyama 2001). In another

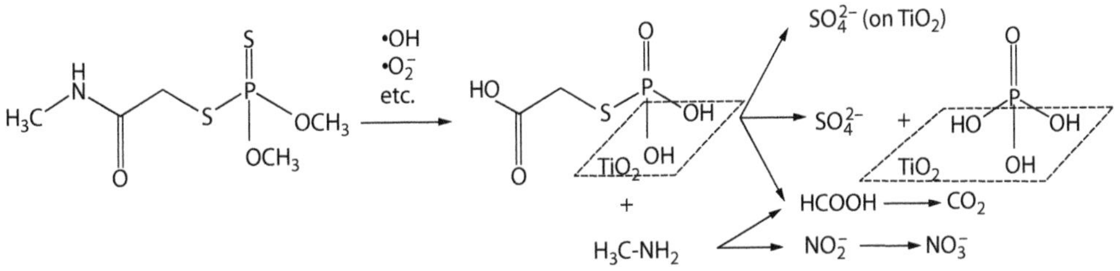

FIGURE 6.11
Proposed tentative pathway for the photocatalytic degradation of dimethoate. (Reprinted from *Chemosphere*, 76, Echavia, G. R. M., Matzusawa, F., and Negishi, N., Photocatalytic degradation of organophosphate and phosphonoglycine pesticides using TiO_2 immobilized on silica gel., 595–600, (2009), with permission from Elsevier.)

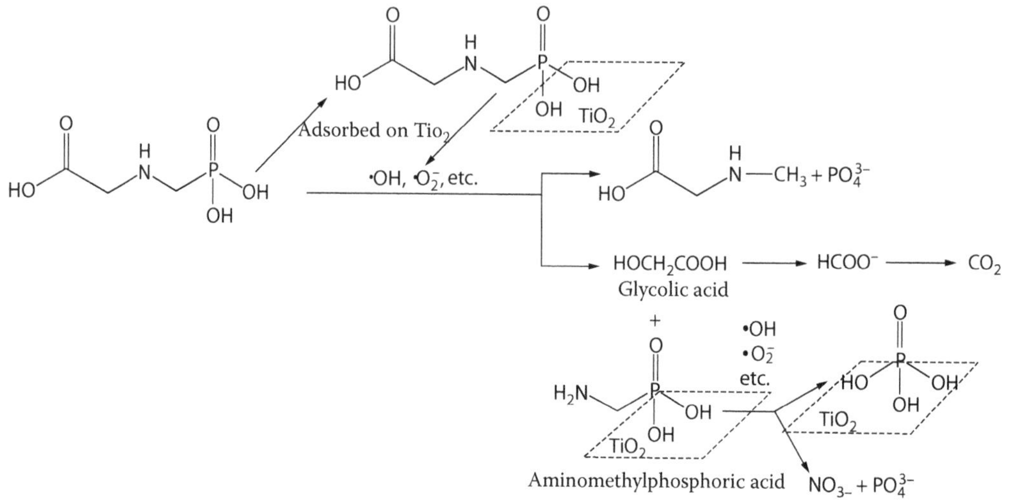

FIGURE 6.12

Proposed tentative pathway for the photocatalytic degradation of glyphosate. (Reprinted from *Chemosphere*, 76, Echavia, G. R. M., Matzusawa, F., and Negishi, N., Photocatalytic degradation of organophosphate and phosphonoglycine pesticides using TiO_2 immobilized on silica gel., 595–600, (2009), with permission from Elsevier.)

paper, Kamiya et al. (2001) also described the effects of cyclodextrins on photodegradation of organophosphorus pesticides in humic water. The decomposition of ^{14}C-fenitrothion on silica gel chromatoplates as well as in polar and nonpolar solvents under sunlight and ultraviolet light was described by Zayed and Mahdy (2008). Its stability in sunlight on the leaf surfaces of bean plants and on different surfaces was also determined by Zayed and Mahdy (2008). The main photoproducts were identified as carboxyfenitrothion, fenitrooxon, carboxyfenitrooxon, and 3-methyl-4-nitrophenol and small amounts of 3-carboxy-4-nitrophenol and methyl parathion. The addition of carbaryl and deltamethrin insecticides slightly accelerated the photodecomposition of fenitrothion on silica gel plates and in solution (Table 6.8; Zayed and Mahdy 2008). Synthesis of ^{14}C-fenitrothion and its main degradation products by photodecomposition is shown in Figure 6.13.

TABLE 6.8

Effect of Carbaryl and Deltamethrin on the Photodecomposition of ^{14}C-Fenitrothion Exposed to Sunlight in Solution and on Silica Gel

Pesticide Chemical	Percentage of Recovered Radiocarbon as Indicated Products													
	I		II		III		IV		V		Origin		Total	
	A	B	A	B	A	B	A	B	A	B	A	B	A	B
Fenitrothion	54	70	10	4	Detected		5	3	2	3	10	7	81	87
Fenitrothion + carbaryl	46	60	5	3	by color		3	1.5	3	2	15	12	72	78.5
Fenitrothion + deltamethrin	40	45	5	5			5	4	2	2	20	18	72	74

Source: Reprinted from *Chemosphere*, 70, Zayed, S. M. A. D. and Mahdy, F., Decomposition of ^{14}C-fenitrothion under the influence of UV and sunlight under tropical and subtropical conditions, 1653–1659, (2008), with permission from Elsevier.

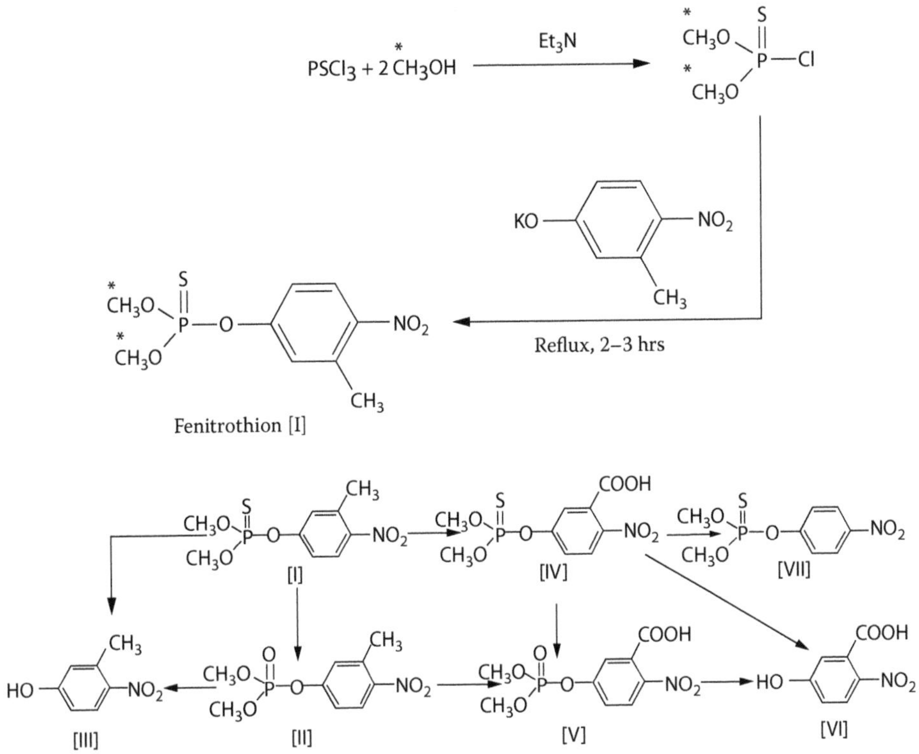

FIGURE 6.13
Synthesis of ^{14}C-fenitrothion and its main degradation products by photodecomposition. (Reprinted from *Chemosphere*, 70, Zayed, S. M. A. D. and Mahdy, F., Decomposition of ^{14}C-fenitrothion under the influence of UV and sunlight under tropical and subtropical conditions, 1653–1659, (2008), with permission from Elsevier.)

A detailed UV-A photolysis mechanism of azinphos-methyl was proposed by Yeasmin et al. (2009), involving two pathways, the major one leading to benzazimide as a stable photoproduct and the other to N-methylanthranilic acid as an intermediate and aniline as a final stable photoproduct. This photolysis has implications for fluorescence-based trace analysis of this pesticide, as controlled UV exposure results in significant fluorescence enhancement of azinphos-methyl in solution via the formation of the highly fluorescent intermediate N-methylanthranilic acid (Yeasmin et al. 2009).

Kralj et al. (2007) described applications of bioanalytical techniques in evaluating advanced oxidation processes in pesticide degradation. The widely accepted Microtox test appears to be the most versatile and sufficiently rapid and sensitive method. However, the bioluminescence of *Vibrio fischeri* used in the Microtox test is not always an adequate indicator of neurotoxic compounds that can be produced during advanced oxidation degradation processes of some organophosphorus pesticides. Kralj et al. (2007) demonstrated that the acetyl-cholinesterase bioassay, based on thermal lens spectrometric determination of enzymatic activity in a flow-injection analysis system is a useful tool. The phototransformations of two organophosphorus pesticides, parathion and chlorpyrifos, by hydroxyl radicals and carbonate radicals in aqueous solution were studied by Wu and Linden (2010). Addition of hydrogen peroxide increased the UV degradation rates of both pesticides and the data were simulated through kinetic modeling. The second-order rate constants of parathion and chlorpyrifos with hydroxyl radical were determined to be $9.7 \pm 0.5 \times 10^9$

and $4.9 \pm 0.1 \times 10^9$/M s, respectively (Wu and Linden 2010). The presence of bicarbonate ions reduced the pesticide degradation rates via scavenging hydroxyl radical, but the formation of carbonate radical also contributed to the degradation of the pesticides with second-order reaction rate constants of $2.8 \pm 0.2 \times 10^6$ and $8.8 \pm 0.4 \times 10^6$/M s, for parathion and chlorpyrifos, respectively (Wu and Linden 2010). Wu and Linden (2010) described the dual roles of bicarbonate ion in UV/H_2O_2 treatment system, that is, scavenging hydroxyl radicals and formation of carbonate radicals by using a simulated kinetic model. Results obtained by Bai et al. (2010) indicates that oxygen plasma treatment has noticeable effects on organophosphorus pesticide with satisfactory degradation efficiency, which mainly depends on related operating parameters including plasma treatment time, discharge power, distance from the center of the induction coil, and concentration of organophosphorus pesticide.

6.5.3 Carbamate Pesticides

Thiram is an alkyldithiocarbamate compound, which has been widely used as a fungicide in agriculture. The photocatalyzed degradation conditions of thiram in aqueous titanium dioxide–suspended solution were optimized under sunlight illumination (Kaneco et al. 2009). Kaneco et al. (2009) investigated the effect of various factors, such as photocatalyst loading, initial substrate concentration, temperature, pH, sunlight intensity, and illumination time on the photocatalytic degradation of thiram. The photocatalytic degradation of thiram is, apparently, a very complex process whose mechanism is not yet fully clarified. The photocatalytic degradation pathway of thiram in water with titanium dioxide on the basis of experimental data obtained by Kaneco et al. (2009) and that reported in the literature is illustrated in Figure 6.14. Evaluation of operational parameters involved in solar photo-Fenton degradation of a pesticide mixture including two carbamate insecticides (oxamyl and methomyl) was described by Zapata et al. (2009a,b).

6.5.4 Herbicides

6.5.4.1 Chlorotriazine Herbicides

Persistence of three chlorotriazine herbicides, terbuthylazine, simazine, and atrazine, and methylthiotriazine herbicide prometryn was studied in the river water, seawater, and groundwater samples exposed to sunlight and darkness under laboratory conditions (127 days) by Navarro et al. (2004). The results showed that light had little effect on the removal of the four herbicides in river water but had a marked effect on their removal from seawater and groundwater. Surprisingly, this removal appeared to be inversely proportional to the concentration of dissolved organic materials (Navarro et al. 2004). Prosen and Zupančič-Kralj (2005) described the evaluation of photolysis and hydrolysis of atrazine and its first degradation products in the presence of humic acids. Under exposure to sunlight, atrazine, desethylatrazine, and desisopropylatrazine were converted to 2-hydroxy analogs only at pH 2 because of acid hydrolysis and possible contribution of photolysis (Prosen and Zupančič-Kralj 2005). Kiss et al. (2007) described the photolytic degradation of atrazine, simazine, prometryn, and terbutryn. Hernández et al. (2008) report the photodegradation of chlorotriazine herbicides (terbuthylazine and simazine), methylthiotriazine herbicides (terbutryn and terbumeton), uracil herbicides (bromacil and terbacil), phenylurea herbicide (diuron), and other pesticides.

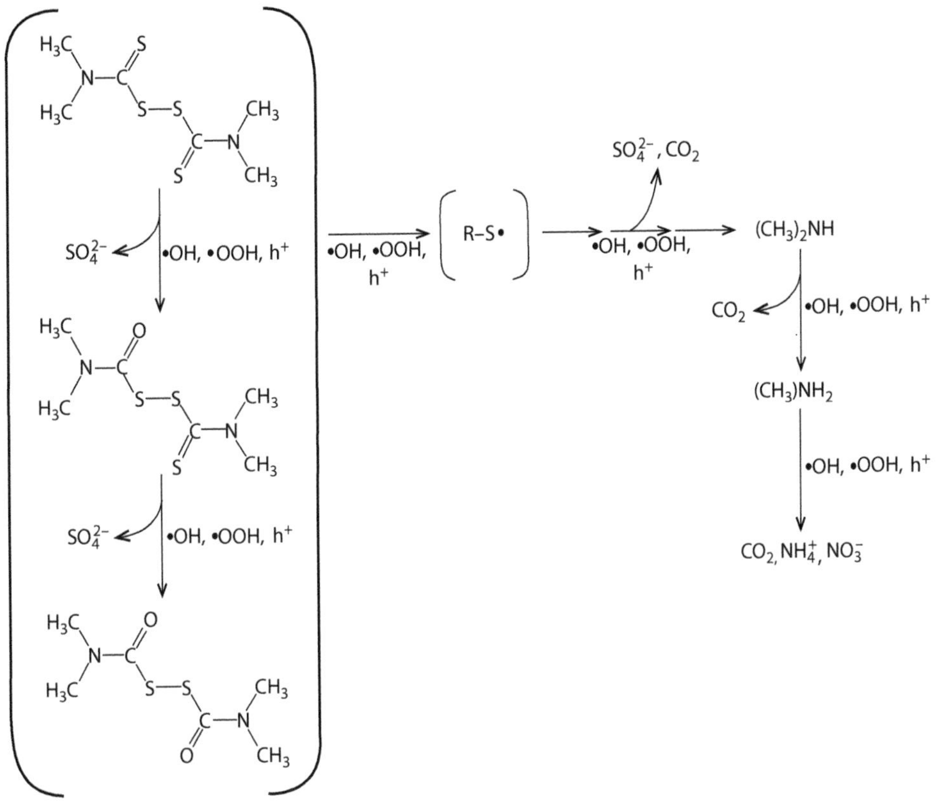

FIGURE 6.14
Proposed photocatalytic degradation pathway of thiram. (Reprinted from *Chem. Eng. J.*, 148, Kaneco, S., Li, N., Itoh, K.-K., Katsumata, H., Suzuki, T., and Ohta, K., Titanium dioxide–mediated solar photocatalytic degradation of thiram in aqueous solution: Kinetics and mineralization, 50–56, (2009), with permission from Elsevier.)

6.5.4.2 Phenylurea Herbicides

The nitrate-induced photodegradation of phenylurea herbicides in water was described by Shankar et al. (2008) to occur efficiently using natural sunlight irradiation. Sharma et al. (2008) described photocatalytic degradation and mineralization of pesticides including the phenylurea herbicide isoproturon over TiO_2-supported mesoporous SBA-15 composite system using sunlight. Silva et al. (2010) described photodegradation of the phenylurea herbicide diuron and phosphoramide insecticide fenamiphos, and the photoproducts were identified by LC-MS.

6.5.4.3 Phenoxypropionic Herbicides

Photodegradation of mecoprop and dichlorprop on dry, moist, and amended soil surfaces exposed to sunlight was described by Romero et al. (1998). Conclusions proposed by the authors (Romero et al. 1998) were the following:

- A slow rate of disappearance of mecoprop and dichlorprop occurs in the three soils exposed to sunlight in the absence of water. Nevertheless, in dry conditions, photolysis may dominate over other transformation pathways

- The soil texture and its adsorption properties may interfere with the kinetics of these herbicides
- On moist soil surfaces exposed to sunlight, photolysis plays an important role during the first two days of exposure with regard to other fate processes observed in dark controls.

6.5.4.4 Dinitroaniline and Nitrile Herbicides

Photolysis of trifluralin has been studied under sunlight conditions and its reaction rate constant with HO• radicals was measured using the relative rate method (Person et al. 2007).

The kinetics and the mechanism of the degradation of dinitroaniline herbicide trifluralin and nitrile herbicide bromoxynil in aqueous solutions by ozonation were described by Chelme-Ayala et al. (2010).

6.5.4.5 Pyridazinone and Nitrophenyl Ether Herbicides

Thomas et al. (2009) described the photodegradation in soils of pyridazinone herbicide norflurazon and nitrophenyl ether herbicide oxyfluorfen. Thomas et al. (2008) also described the photodegradation of norflurazon deposited or adsorbed on solid material.

6.5.4.6 Pyridine Herbicides

Heterogenous photocatalyzed degradation of two pesticide derivatives, triclopyr (pyridine herbicide) and daminozide (growth retardant), in aqueous suspensions of titanium dioxide was described by Qamar et al. (2006).

6.5.5 Others

Photodegradation of pesticides from different classes and chemical groups was described (Bandala et al. 2002; Guillard et al. 2003; Malato et al. 1999, 2000; Navarro et al. 2009; Samsonov and Pokrovskii 2000; Devipriya and Yesodharan 2005).

6.6 Conclusions

The fate of the pesticides in soils depends on chemical transformations in which living organisms participate (biotic, biochemical transformations) and on physical, chemical, and photochemical processes. Photochemical transformations of pesticides are restricted to soil and plant surfaces and only to compounds sensitive to solar radiation. In this chapter, the readers may find essential information concerning the causes and transformations of pesticides from various chemical groups, their degradation and fragmentation in the environment. It is hoped that this chapter will serve as an integrating factor both in focusing attention upon those many residue matters requiring further attention and in collating, for variously trained readers, the present knowledge in specific important areas of analysis of pesticides and related endeavors.

References

Ahmad, R., Kookana, R. S., Megharaj, M., and Alston, A. M. 2004. Aging reduces the bioavailability of even a weakly sorbed pesticide (carbaryl) in soil. *Environ. Toxicol. Chem.* 23(9): 2084–2089.

Ahmed, S., Rasul, M. G., Brown, R., and Hashib, M. A. 2011. Influence of parameters on the heterogeneous photocatalytic degradation of pesticides and phenolic contaminants in wastewater: A short review. *J. Environ. Manag.* 92: 311–330.

Aislable, J. M., Richards, N. K., and Boul, H. L. 1997. Microbial degradation of DDT and its residues: A review. *N. Z. J. Agric. Res.* 40: 269–282.

Albarrán, A., Celis, R., Hermosín, M. C., López-Piñeiro, A., Ortega-Calvo, J. J., and Cornejo, J. 2003. Effects of solid olive-mill waste addition to soil on sorption, degradation and leaching of the herbicide simazine. *Soil Use Manag.* 19: 150–156.

Allsopp, M. and Johnston, P. 2000. *Unseen Poisons in Asia: A Review of Persistent Organic Pollutant Levels in South and Southeast Asia and Oceania.* ISBN: 90-73361-64-8.

Amorós, I., Alonso, J. L., Romaguera, S., and Carrasco, J. M. 2007. Assessment of toxicity of a glyphosate-based formulation using bacterial systems in lake water. *Chemosphere* 67: 2221–2228.

Anandan, S., Vinu, A., Lovely, K. L. P. S., et al. 2007. Photocatalytic activity of Ladoped ZnO for the degradation of monocrotophos in aqueous suspension. *J. Mol. Catal. A Chem.* 266: 148–157.

Arias-Estévez, M., López-Periago, E., Martínez-Carballo, E., Simal-Gándara, J., Mejuto, J.-C., and García-Río, L. 2008. The mobility and degradation of pesticides in soils and the pollution of groundwater resources. *Agric. Ecosyst. Environ.* 123: 247–260.

Arshad, M., Hussain, S., and Saleem, M. 2008. Optimization of environmental parameters for biodegradation of alpha and beta endosulfan in soil slurry by *Pseudomonas aeruginosa*. *J. Appl. Microbiol.* 104: 364–370.

Azevedo, A. S. O. N. 1998. Assessment and simulation of atrazine as influenced by drainage and irrigation. An interference between RZWQM and ArcView GIS. Doctor Thesis, Iowa State University, Ames, IA.

Bai, Y., Chen, J., Yang, Y., Guo, L., and Zhang, C. 2010. Degradation of organophosphorus pesticide induced by oxygen plasma: Effects of operating parameters and reaction mechanisms. *Chemosphere* 81: 408–414.

Bandala, E. R., Gelover, S., Leal, M. T., Arancibia-Bulnes, C., Jimenez, A., and Estrada, C. A. 2002. Solar photocatalytic degradation of aldrin. *Catal. Today* 76: 189–199.

Barriuso, E., Houot, S., and SerraWittling, C. 1997. Influence of compost addition to soil on the behaviour of herbicides. *Pestic. Sci.* 49: 65–75.

Belmans, C., Wesseling, J. G., and Feddes, R. A. 1983. Simulation model of the water balance of a cropped soil: SWATRE. *J. Hydrol.* 63: 271–286.

Bending, G. D., Friloux, M., and Walker, A. 2002. Degradation of contrasting pesticides by white rot fungi and its relationship with ligninolytic potential. *FEMS Microbiol. Lett.* 212: 59–63.

Benning, M. M., Shim, H., Raushel, F. M., and Holden, H. M. 2001. High resolution X-ray structures of different metal-substitued forms of phosphotriesterase from *Pseudomonas diminuta*. *Biochemistry* 40: 2712–2722.

Berger, B. M. 1998. Parameters influencing biotransformation rates of phenylurea herbicides by soil microorganisms. *Pestic. Biochem. Physiol.* 60: 71–82.

Boivin, A., Cherrier, R., Perrin-Ganier, C., and Schiavon, M. 2004. Time effect on bentazone sorption and degradation in soil. *Pest. Manag. Sci.* 60: 809–814.

Bondarenko, S. and Gan, J. 2004. Degradation and sorption of selected organophosphate and carbamate insecticides in urban stream sediments. *Environ. Toxicol. Chem.* 23(8): 1809–1814.

Bromilow, R. H., Evans, A. A., and Nicholls, P. H. 1999. Factors affecting degradation rates of five triazole fungicides in two soils types: 2. Field studies. *Pest Manag. Sci.* 55: 1135–1142.

Bromilow, R. H., Evans, A. A., and Nicholls, P. H. 2003. The influence of lipophilicity and formulation on the distribution of pesticides in laboratory-scale sediment/water systems. *Pest Manag. Sci.* 59: 238–244.

Caracciolo, A. B., Grenni, P., Cicolli, R., Di Landa, G., and Cremisini, C. 2005. Simazine biodegrada-tion in soil: Analysis of bacterial community structure by *in situ* hybridization. *Pest Manag. Sci.* 61: 863–869.

Carvalho, F. P., et al. 1998. Tracking pesticides in the tropics. *IAEA Bull.* 40(3): 24–30.

Chai, L.-K., Mohd-Tahir, N., and Hansen, H. C. B. 2009. Dissipation of acephate, chlorpyrifos, cyper-methrin and their metabolites in a humid-tropical vegetable production system. *Pest Manag. Sci.* 65: 189–196.

Chelme-Ayala, P., El-Din, M. G., and Smith, D. W. 2010. Kinetic and mechanism of the degradation of two pesticides in aqueous solutions by ozonation. *Chemosphere* 78: 557–562.

Chen, B., Huang, G., Li, J., Li, Y. R., and Li, Y. F. 2001. Integration of GIS with pesticides losses runoff model. In: *Proceedings of DMGIS & Geoinformatics Conference*, ISPRS, vol. XXXIV, Part 2W2, pp. 37–44, Bangkok, Thailand, 25–28 May.

Chen, J., Wang, D., Zhu, M., and Gao, C. 2007. Photocatalytic degradation of dimethoate using nano-sized TiO_2 powder. *Desalination* 207: 87–94.

Coppola, L., del Pilar Castillo, M., and Vischetti, C. 2011. Degradation of isoproturon and bentazone in peat- and compost-based biomixtures. *Pest Manag. Sci.* 67: 107–113.

Covaci, A., Hura, C., and Schepens, P. 2001. Selected persistent organochlorine pollutants in Romania. *Sci. Total Environ.* 280(1–3): 143–152.

Daneshvar, N., Aber, S., Dorraji, M. S. S., Khataee, A. R., and Rasoulifard, M. H. 2007. Photocatalytic degradation of the insectide diazinon in the presence of prepared nanocrystalline ZnO pow-ders under irradiation of UV-C light. *Separ. Purif. Technol.* 58: 91–98.

De Jonge, R. J., Breure, A. M., and van Andel, J. G. 1996. Reversibility of adsorption of aromatic com-pounds onto powdered activated carbon (PAC). *Water Res.* 30(4): 883–892.

De Liñan Y. V. C., 1997. In: *Farmacología Vegetal*, Ediciones Agrotécnicas S.L, Madrid.

De Roffignac, L., Cattan, P., Mailloux, J., Herzog, D., and Le Bellec, F. 2008. Efficiency of a bagasse substrate in a biological bed system for the degradation of glyphosate, malathion and lambda-cyhalothrin under tropical climate conditions. *Pest Manag. Sci.* 64: 1303–1313.

Devipriya, S. and Yesodharan, S. 2005. Photocatalytic degradation of pesticide contaminants in water. *Sol. Energ. Mater. Sol. Cell.* 86: 309–348.

Echavia, G. R. M., Matzusawa, F., and Negishi, N. 2009. Photocatalytic degradation of organophosphate and phosphonoglycine pesticides using TiO_2 immobilized on silica gel. *Chemosphere* 76: 595–600.

ELISA. http://www.ecnc.nl/CompletedProjects/Elisa_119.html.

Elliott, J. A., Cesna, A. J., Best, K. B., Nicholaichuk, W., and Tollefson, L. C. 2000. Leaching rates and preferential flow of selected herbicides through tilled and untilled soil. *J. Environ. Qual.* 29: 1650–1656.

European Economic Community. 1991. Council directive concerning the placing of plant protection products on the market, 91/414/EEC, OJ L 230, Brussels, Belgium.

Felsot, A. S., Maddox, J. V., and Bruce, W. 1981. Enhanced degradation of carbofuran in soils with histories of furadan use. *Bull. Environ. Contam. Toxicol.* 26: 781–788.

Fogg, P., Boxall, A. B. A., Walker, A., and Jukes, A. A. 2003. Pesticide degradation in a 'biobed' composting substrate. *Pest Manag. Sci.* 59: 527–537.

Foster, L. J. R., Kwan, B. H., and Vancov, T. 2004. Microbial degradation of the organophosphate pes-ticide, Ethion. *FEMS Microbiol. Lett.* 240: 49–53.

Fragoeiro, S. and Magan, N. 2005. Enzymatic activity, osmotic stress and degradation of pesticide mixtures in soil extract liquid broth inoculated with *Phanerochaete chrysosporium* and *Trametes versicolor*. *Environ. Microbiol.* 7(3): 348–355.

Fu, J., et al. 2003. Persistent organic pollutants in environmental of the Pearl River Delta, China: An overview. *Chemosphere* 52: 1411–1422.

Galanopoulou, S., Vgenopoulos, A., and Conispoliatis, N. 2005. DDTs and other chlorinated organic pesticides and polychlorinated biphenyls pollution in the surface sediments of Keratsini har-bour, Saronikos gulf, Greece. *Mar. Pollut. Bull.* 50: 520–525.

Garcia, A., Amat, A. M., and Arques, A., et al. 2006. Detoxification of aqueous solution of herbicide "Sevnol" by solar photocatalysis. *Environ. Chem. Lett.* 3: 169–172.

Garcia, A., Arques, A., Vicente, R., Domenech, A., Amat, A. M., et al. 2008. Treatment of aqueous solutions containing four commercial pesticides by means of TiO₂ solar photocatalysis. *J. Sol. Energ. Eng.* 130: 041011–041015.

Garratt, J. A., Capri, E., Trevisan, M., Errera, G., and Wilkins, R. M. 2002. Parameterisation, evaluation and comparison of pesticide leaching models to data from a Bologna field site, Italy. *Pest Manag. Sci.* 58: 3–20.

Gonod, L. V., Martin-Laurent, F., and Chenu, C. 2006. 2,4-D impact on bacterial communities, and the activity and genetic potential of 2,4-D degrading communities in soil. *FEMS Microbiol. Ecol.* 58: 529–537.

Grigg, B. C., Assaf, N. A., and Turco, R. F. 1997. Removal of atrazine contamination in soil and liquid systems using bioaugmentation. *Pestic. Sci.* 50: 211–220.

Guenzi, W. D. and Beard, W. E. 1968. Anaerobic conversion of DDT to DDD and aerobic stability of DDT in soil. *Proc. Soil Sci. Soc. Am.* 32: 522–524.

Guillard, C., Disdier, J., Monnet, C., et al. 2003. Solar efficiency of a new deposited titania photocatalyst: Chlorphenol, pesticide and dye removal applications. *Appl. Catal. B Environ.* 46: 319–332.

Guo, L., Jurry, W. A., Wagenet, R. J., and Flury, M. 2000. Dependence of pesticide degradation on sorption: Nonequilibrium model and application to soil reactors. *J. Contam. Hydrol.* 43(1): 45–62.

Gupta, G. and Baummer, J. III. 1996. Biodegradation of atrazine soil using poultry litter. *J. Hazard. Mater.* 45: 185–192.

Haque, M. M. and Muneer, M. 2003. Heterogeneous photocatalysed degradation of a herbicide derivative, isoproturon in aqueous suspension of titanium dioxide. *J. Environ. Manag.* 69: 169–176.

Henriksen, V. V., Helweg, A., Spliid, N. H., Felding, G., and Stenvang, L. 2003. Capacity of model biobeds to retain and degrade mecoprop and isoproturon. *Pest Manag. Sci.* 59: 1076–1082.

Hernández, F., Ibáñez, M., Pozo, Ó. J., and Sancho, J. V. 2008. Investigating the presence of the pesticide transformation products in water by using liquid chromatography-mass spectrometry with different mass analyzers. *J. Mass Spectrom.* 43: 173–184.

Hornsby, A. G., Wauchope, R. D., and Herner, A. E. 1996. *Pesticide Properties in the Environment.* Springer-Verlag: New York.

Huang, X., Lee, L. S., and Nakatsu, C. 2000. Impact of animal waste lagoon effluents on chlorpyrifos degradation in soils. *Environ. Toxicol. Chem.* 19(12): 2864–2870.

Hutson, J. L. and Wagenet, R. J. 1992. Leaching estimation and chemistry model. Version 3. Department of Soil, Crop and Atmospheric Sciences, Research Series No. 92-3. Cornell University, New York.

http://ec.europa.eu/food/plant/protection/evaluation/exist_subs_rep_en.htm.

http://www.epa.gov/opprd001/factsheets/.

Jarvis, N. J. 1991. *MACRO – A model of water movement and solute transport in macroporous soils.* Reports and Dissertations No. 9. Department of Soil Sciences, Swedish University of Agricultural Sciences, Uppsala.

Jaszek, M., Grzywnowicz, K., Malarczyk, E., and Leonowicz, A. 2006. Enhanced extracellular laccase activity as a part of the response system of white rot fungi: *Trametes versicolor* and *Abortiporus biennis* to paraquat-caused oxidative stress conditions. *Pestic. Biochem. Physiol.* 85: 147–154.

Johnson, A. C., White, C., Bhardwaj, C. L., and Dixon, A. 2003. The ability of indigenous microorganisms to degrade isoproturon, atrazine and mecoprop within aerobic UK aquifer systems. *Pest Manag. Sci.* 59: 1291–1302.

Jóri, B., Soós, V., Szegö, D., et al. 2007. Role of transporters in paraquat resistance of horseweed *Conyza Canadensis* (L.) Cronq. *Pestic. Biochem. Physiol.* 88: 57–65.

Juhasz, A. L. and Naidu, R. 2000. Enrichment and isolation of non-specific aromatic degraders from unique uncontaminated (plant and fecal material) sources and contaminated soils. *J. Appl. Microbiol.* 89: 642–650.

Kambiranda, D. M., Asraful-Islam, S. Md., Cho, K. M., et al. 2009. Expression of esterase gene in yeast for organophosphates biodegradation. *Pestic. Biochem. Physiol.* 94: 15–20.

Kamei, I., Takagi, K., and Kondo, R. 2010. Bioconversion of dieldrin by wood-rotting fungi and metabolite detection. *Pest Manag. Sci.* 66: 888–891.

Kamiya, M. and Kameyama, K. 1998. Photochemical effects of humic substances on the degradation of organophosphorus pesticides. *Chemosphere* 36(10): 2337–2344.

Kamiya, M. and Kameyama, K. 2001. Effect of selected metal ions on photodegradation of organophosphorus pesticides sensitized by humic acids. *Chemosphere* 45: 231–235.

Kamiya, M., Kameyama, K., and Ishiwata, S. 2001. Effect of cyclodextrins on photodegradation of organophosphorus pesticides in humic water. *Chemosphere* 42: 251–255.

Kaneco, S., Li, N., Itoh, K.-K., Katsumata, H., Suzuki, T., and Ohta, K. 2009. Titanium dioxide mediated solar photocatalytic degradation of thiram in aqueous solution: Kinetics and mineralization. *Chem. Eng. J.* 148: 50–56.

Karpouzas, D. G. and Walker, A. 2000. Factors influencing the ability of *Pseudomonas putida* strains epl and II to degrade the organophosphate ethoprophos. *J. Appl. Microbiol.* 89: 40–48.

Karpouzas, D. G., Walker, A., Drennan, D. S. H., and Froud-Williams, R. J. 2001. The effect of initial concentration of carbofuran on the development and stability of its enhanced biodegradation in top-soil and sub-soil. *Pest Manag. Sci.* 57: 72–81.

Kim, I. S., Park, R. D., and Suh, Y. T. 2002. Involvement of adaptive mechanism and an energy-dependent efflux system of *Pseudomonas* sp. LE2 counteracting insecticide lindane. *Pestic. Biochem. Physiol.* 73: 140–148.

Kiss, A., Rapi, S., and Csutorás, Cs. 2007. GC/MS studies on revealing products and reaction mechanism of photodegradation of pesticides. *Microchem. J.* 85: 13–20.

Knisel, W. G. (ed.) 1980. *CREAMS: A field scale model for chemical, runoff, and erosion from agricultural management system.* Conservation Research Report No. 26. U.S. Department of Agriculture, Washington, DC.

Konstantinou, I., Hela, D., Lambropoulou, D., and Albanis, T. 2008. Monitoring of pesticides in the environment. In: Tadeo, J. L. (ed.), *Analysis of Pesticides in Food and Environmental Samples*, pp. 319–357. CRC Press Taylor & Francis Group: Boca Raton, FL, ISBN-13: 978-0-8493-7552-1.

Kralj, M. B., Trebše, P., and Franko, M. 2007. Applications of bioanalytical techniques in evaluating advanced oxidation processes in pesticide degradation. *Trends Anal. Chem.* 26(11): 1020–1031.

Krapac, I. G., Roy, W. R., Smyth, C. A., and Barnhardt, M. L. 1993. Occurrence and distribution of pesticides in soil at agrichemical facilities in Illinois. In: *Agrichemical Facility Site Contamination Study*, Illinois Department of Agriculture.

Kravvariti, K., Tsiropoulos, N. G., and Karpouzas, D. G. 2010. Degradation and adsorption of terbuthylazine and chlorpyrifos in biobed biomixtures from composted cotton crop residues. *Pest Manag. Sci.* 66: 1122–1128.

Kristensen, G. B., Johannesen, H., and Aamand, J. 2001. Mineralization of aged atrazine and mecoprop in soil and aquifer chalk. *Chemosphere* 45: 927–934.

Kulovaara, M., Backlund, P., and Corin, N. 1995. Light-induced degradation of DDT in humic water. *Sci. Total Environ.* 170: 185–191.

Kumar, N. J. I., Amb, M. K., and Bora, A. 2010. Chronic response of *Anabaena fertilissima* Rao, C.B. on growth, metabolites and anzymatic activities by chlorophenoxy herbicide. *Pestic. Biochem. Physiol.* 98: 168–174.

Larson, S. J., Capel, P. D., and Majewski, M. S. 1997. Pesticides in surface waters-distribution, trends, and governing factors. In: Gilliom, R. J. (ed.), *Series of Pesticides in Hydrologic System*, vol. 3: p. 373. Ann Arbor Press: Chelsea, MI.

Larsson, P., et al. 1995. Fate in tropical and temperate regions. *Naturwissenschaften* 82: 559–561.

Lee, K. T., Tanabe, S., and Koh, C. H. 2001. Distribution of organochlorine pesticides in sediments from Kyeonggi Bay and nearby areas, Korea. *Environ. Pollut.* 114(2): 207–213.

Li, Y. R., Struger, J., Fisher, J. D., Li, Y. F., and Huang, G. H. 2001. Predicting runoff losses of atrazine from agricultural lands in the Kintore Creek watershed using two statistical models. In: *The 36th Central Canadian Symposium on Water Pollution, Research*, Canada Centre for Inland Waters, Burlington, ON.

Liebich, J., Schäffer, A., and Burauel, P. 2003. Structural and functional approach to studying pesticide side-effects on specific soil functions. *Environ. Toxicol. Chem.* 22(4): 784–790.

Mahalakshmi, M., Arabindoo, B., Palanichamy, M., and Murugesan, V. 2007. Photocatalytic degradation of carbofuran using semiconductor oxides. *J. Hazard. Mater.* 143: 240–245.

Malato, S., Blanco, J., Richter, C., Milow, B., and Maldonado, M. I. 1999. Solar photocatalytic mineralization of commercial pesticides: Methamidophos. *Chemosphere* 38(5): 1145–1156.

Malato, S., Blanco, J., Fernández-Alba, A. R., and Agüera, A. 2000. Solar photocatalytic mineralization of commercial pesticides: Acrinathrin. *Chemosphere* 40: 403–409.

Malone, R. W., Ma, L., Don Wauchope, R., et al. 2004. Modeling hydrology, metribuzin degradation and metribuzin transport in macroporous tilled and no-till silt loam soil using RZWQM. *Pest Manag. Sci.* 60: 253–266.

Mamy, L. and Barriuso, E. 2007. Desorption and time-dependent sorption of herbicides in soils. *Eur. J. Soil Sci.* 58: 174–187.

Mandelbaum, R. T., Wackett, L. R., and Allan, D. L. 1993. Mineralization of the s-triazine ring of atrazine by stable bacterial mixed cultures. *Appl. Environ. Microbiol.* 59: 1695–1701.

Martínez-Íñigo, M. J., Lobo M. C., Garbi, C., and Martin, M. 2003. Applicability of fluorescence *in situ* hybridization to monitor target bacteria in soil samples. In: Del Re, A. A. M., Capri, E., Padovani, L., La Goliardica Pavese, T. M. (eds), *Proceeding of XII Symposium Pesticide Chemistry*, pp. 609–615, Piacenza, Italy, 4–6 June. Italy.

Masson, P., Josse, D., Lockridge, O., Vigule, N., Taupin, C., and Buhler, C. 1998. Enzymes hydrolyzing organophosphates as potential catalytic scavengers against organophosphates poisoning. *J. Physiol. Paris* 92: 357–362.

Mathews, R. W. 1988. Kinetics of photocatalytic oxidation of organic solutes over titanium dioxide catalysis. *J. Catal.* 111: 264–272.

Mendoza-Cantú, A., Albores, A., Fernández-Linares, L., and Rodriguez-Vázquez, R. 2000. Pentachlorophenol biodegradation and detoxification by the white-rot fungus *Phanerochaete chrysosporum*. *Environ. Toxicol.* 15: 107–113.

Menon, P. and Gopal, M. 2003. Dissipation of ^{14}C carbaryl and quinalphos in soil under a groundnut crop (*Arachis hypogaea* L.) in semi-arid India. *Chemosphere* 53: 1023–1031.

Mills, M. S., Hill, I. R., Newcombe, A. C., Simmons, N. D., Vaughan, P. C., and Verity, A. A. 2001. Quantification of acetochlor degradation in the unsaturated zone using two novel *in situ* field techniques: Comparisons with laboratory-generated data and implications for groundwater risk assessments. *Pest Manag. Sci.* 57: 351–359.

MOSES. http://projects-2004.jrc.cec.eu.int/.

Mullins, J. A., Carsel, R. F., Scarbrough, J. E., and Ivery, A. M. 1993. PRZM-2, a model for predicting pesticide fate in the crop root and unsaturated soil zones: Users manual for release 2.0. US EPA, Athens, GA.

Muneer, M., Qamar, M., Saquib, M., and Bahnemann, D. 2005. Heterogenous photocatalysed reaction of three selected pesticide derivatives, propham, propachlor and tebuthiuron in aqueous suspensions of titanium dioxide. *Chemosphere* 61: 457–468.

Navarro, S., Vela, N., Giménez, M. J., and Navarro, G. 2004. Persistence of four s-triazine herbicides in river, sea and groundwater samples exposed to sunlight and darkness under laboratory conditions. *Sci. Total Environ.* 329: 87–97.

Navarro, S., Fenoll, J., Vela, N., Ruiz, E., and Navarro, G. 2009. Photocatalytic degradation of eight pesticides in leaching water by use of ZnO under natural sunlight. *J. Hazard. Mater.* 172: 1303–1310.

Ning, J., Bai, Z., Gang, G., et al. 2010. Functional assembly of bacterial communities with activity for the biodegradation of an organophosphorus pesticide in the rape phyllosphere. *FEMS Microbiol. Lett.* 306: 135–143.

Noegrohati, S. and Hammers, W. E. 1992. Regression models for some solute distribution equilibria in the terrestrial environment. *Toxicol. Environ. Chem.* 34: 175–185.

Noegrohati, S., Hadi, N., and Sanjayadi, S. 2008. Fate and behavior of organochlorine pesticides in the Indonesian tropical climate: A study in the Segara Anakan Estuarine ecosystem. *Clean* 36(9): 767–774.

Ortiz-Hernández, M. L., Quintero-Ramírez, R., Nava-Ocampo, A. A., and Bello-Ramírez, A. M. 2003. Study of the mechanism of *Flavobacterium* sp. for hydrolyzing organophosphate pesticides. *Fundam. Clin. Pharmacol.* 17: 717–723.

Ouyang, Y., Nkedi-Kizza, P., Mansell, R. S., and Ren, J. Y. 2003. Spatial distribution of DDT in sediments from Estuarine Rivers of Central Florida. *J. Environ. Qual.* 32: 1710–1716.

Parekh, N. R., Suett, D. L., Roberts, S. J., McKeown, T., Shaw, E. D., and Jukes, A. A. 1994. Carbofuran-degrading bacteria from previously treated field soils. *J. Appl. Bacteriol.* 76: 559–567.

Park, S. K. and Bielefeldt, A. R. 2003. Aqueous chemistry and interactive effects on non-ionic surfactant and pentachlorophenol sorption to soil. *Water Res.* 37: 4663–4672.

Park, J. H., Feng, Y. C., Ji, P. S., Voice, T. C., and Boyd, S. A. 2003. Assessment of bioavailability of soil-sorbed atrazine. *Appl. Environ. Microbiol.* 69: 3288–3298.

Park, J. H., Feng, Y. C., Cho, S. Y., Voice, T. C., and Boyd, S. A. 2004. Sorbed atrazine shifts into non-desorable sites of soil organic matter during aging. *Water Res.* 38: 3881–3892.

Parra, S., Stanca, S. E., Guasaquillo, I., and Thampi, K. R. 2004. Photocatalytic degradation of aytrazine using suspended and suported TiO$_2$. *Appl. Catal. B Environ.* 51: 107–116.

Patakioutas, G. I. 2000. Study of novel pesticide degradation, distribution and transportation in water and soil systems. Ph.D. Thesis, vol. 58, University of Ioannina.

Patakioutas, G. I. and Albanis, T. A. 2002. Adsorption-desorption of alachlor, metolachlor, EPTC, chlorothalonil and pirimiphos-methyl in contrasting soils. *Pest Manag. Sci.* 58: 352–362.

Patakioutas, G. I., Karras, G., Hela, D., and Albanis, T. A. 2002. Pirimiphos-methyl and benalaxyl losses in surface runoff from plots cultivated with potatoes. *Pest Manag. Sci.* 58: 1194–1204.

Paul, D., Pandey, G., Meier, C., Meer, J. R., and Jain, R. K. 2006. Bacterial community structure of a pesticide-contaminated site and assessment of changes induced in community structure during bioremediation. *FEMS Microbiol. Ecol.* 57: 116–127.

Person, A. L., Mellouki, A., Muñoz, A., Borras, E., Martin-Reviejo, M., and Wirtz, K. 2007. Trifluralin: Photolysis under sunlight conditions and reaction with HO• radicals. *Chemosphere* 67: 376–383.

Pesce, S., Martin-Laurent, F., Rouard, N., and Montuelle, B. 2009. Potential for microbial diuron mineralisation in a small wine-growing watershed: From treated plots to lotic receiver hydrosystem. *Pest Manag. Sci.* 65: 651–657.

Pignatello, J. J. 1998. Soil organic matter as a nonoporous sorbent of organic pollutants. *Adv. Colloid Interface Sci.* 76–77: 445–467.

Piutti, S., Marchand, A.-L., Lagacherie, B., Martin-Laurent, F., and Soulas, G. 2002. Effect of cropping cycles and repeated herbicide applications on the degradation of diclofop-methyl, bentazone, diuron, isoproturon and pendimethalin in soil. *Pest Manag. Sci.* 58: 303–312.

Pizarro, P., Guillard, C., Perol, N., and Hermann, J. M. 2005. Photocatalytic degradation of imazapyr in water: Comparison of activities of different supported and unsupported TiO$_2$-based catalyst. *Catal. Today* 101: 211–218.

Plimmer, J. R., Kearney, P. C., and von Endt, D. W. 1968. Mechanism of conversion of DDT to DDD by *Aerobacter aerogenes*. *J. Agric. Food Chem.* 21: 397–399.

Pourata, R., Khataee, A. R., Aber, S., and Daneshar, N. 2009. Removal of the herbicide bentazon from contaminated water in the presence of synthesized nanocrystalline TiO$_2$ powders under irradiation of UV-C light. *Desalination* 249: 301–307.

Prosen, H. and Zupančič-Kralj, L. 2005. Evaluation of photolysis and hydrolysis of atrazine and its first degradation products in the presence of humic acids. *Environ. Pollut.* 133: 517–529.

Qamar, M., Muneer, M., and Bahnemann, D. 2006. Heterogeneous photocatalysed degradation of two selected pesticide derivatives, triclopyr and daminozid in aqueous suspensions of titanium dioxide. *J. Environ. Manag.* 80: 99–106.

Quintero, J. C., Moreira, M. T., Feijoo, G., and Lema, J. M. 2005. Effect of surfactants on the soil desorption of hexachlorocyclohexane (HCH) isomers and their anaerobic biodegradation. *J. Chem. Technol. Biotechnol.* 80: 1005–1015.

Rabindranathan, S., Devipriya, S., and Yesodharan, S. 2003. Photocatalytic degradation of phosphamidon on semiconductor oxides. *J. Hazard. Mater.* B102: 217–229.

Rahman, M. A. and Muneer, M. 2005a. Photocatalysed degradation of two selected pesticide derivatives, dichlorvos and phosphamidon, in aqueous suspensions of titanium dioxide. *Desalination* 181: 161–172.

Rahman, M. A. and Muneer, M. 2005b. Heterogeneous photocatalytic degradation of picloram, dicamba and fluometuron in aqueous suspensions of titanium dioxide. *J. Environ. Sci. Health* 40: 247–267.

Rahman, M. A., Muneer, M., and Bahnemann, D. 2003. Photocatalysed degradation of a herbicide derivative, a diphenamid in aqueous suspensions of titanium dioxide. *J. Adv. Oxid. Technol.* 6(1): 100–107.

Rahman, M. A., Qamar, M., Muneer, M., and Bahnemann, D. 2006. Semiconductor mediated photocatalysed degradation of a pesticide derivative, acephate in aqueous suspensions of titanium dioxide. *J. Adv. Oxid. Technol.* 9(1): 103–109.

Rao, P. S. C., Mansell, R. S., Baldwin, L. B., and Laurent, M. F. 1983. Pesticides and their behaviour in soil and water. *Soils Science Fact Sheer*. Florida Cooperative Extension Service, Institute of Food and Agricultural Sciences, University of Florida.

Rao, P. S. C., Hornsby, A. G., and Jessup, R. E. 1985. Indices for ranking the potential for pesticide contamination of groundwater. *Soil Crop Sci. Soc. Florida Proc.* 44: 1–8.

Read, D. C. 1986. Accelerated microbial breakdown of carbofuran in soil from previously treated fields. *Agric. Ecosyst. Environ.* 15: 51–61.

Reichman, R., Wallach, R., and Mahrer, Y. 2000. A combined soil-atmosphere model for evaluating the fate of surface-applied pesticides. 1. Model development and verification. *Environ. Sci. Technol.* 34: 1313–1320.

Romero, E., Dios, G., Mingorance, M. D., Matallo, M. B., Peña, A., and Sánchez-Rasero, F. 1998. Photodegradation of mecoprop and dichlorprop on dry, moist and amended soil surfaces exposed to sunlight. *Chemosphere* 37(3): 577–589.

Rosenbrock, P., Munch, J. C., Scheunert, I., and Dörfler, U. 2004. Biodegradation of the herbicide bromoxynil and its plant cell wall bound residues in an agricultural soil. *Pestic. Biochem. Physiol.* 78: 49–57.

Różański, L. 1992. Transformations of pesticides in living organisms and the environment, p. 275. PWRiL Press, Warsaw, (in Polish).

Saien, J. and Khezrianjoo, S. 2008. Degradation of the fungicide carbendazim in aqueous solution with UV/TiO$_2$ process: Optimization, kinetics and toxicity studies. *J. Hazard. Mater.* 157: 269–276.

Saison, C., Louchart, X., Schiavon, M., and Voltz, M. 2010. Evidence of the role of climate control and reversible aging processes in the fate of diuron in a Mediterranean topsoil. *Eur. J. Soil Sci.* 61: 576–587.

Samsonov, Y. N. and Pokrovskii, L. M. 2000. Sensitized photodecomposition of high disperse pesticide chemicals exposed to sunlight and irradiation from halogen or mercury lamp. *Atmos. Environ.* 35: 2133–2141.

Senthilnathan, J. and Philip, L. 2010. Photocatalytic degradation of lindane under UV and visible light using N-doped TiO$_2$. *Chem. Eng. J.* 161: 83–92.

Shaktivel, S., Neppolian, B., Shankar, M. V., Arabindo, B., Palanichamy, M., and Murugesan, V. 2003. Solar photocatalytic degradation of azo dye: Comparison of photocatalytic efficiency of ZnO and TiO$_2$. *Sol. Energ. Mater. Sol. Cell* 77: 65–82.

Shankar, M. V., Nélieu, S., Kerhoas, L., and Einhorn, J. 2008. Natural sunlight NO$_3^-$/NO$_2^-$ – Induced photo-degradation of phenylurea herbicides in water. *Chemosphere* 71: 1461–1468.

Sharma, M. V. P., Kumari, D., and Subrahmanyam, M. 2008. TiO$_2$ supported over SBA-15: An efficient photocatalyst for the pesticide degradation using solar light. *Chemosphere* 73: 1562–1569.

Sharma, M. V. P., Sadanandam, G., Ratnamala, A., Kumari, D., and Subrahmanyam, M. 2009. An efficient and novel porous nanosilica supported TiO$_2$ photocatalyst for the pesticide degradation using solar light. *J. Hazard. Mater.* 171: 626–633.

Shifu, C. and Gengyu, C. 2005. Photocatalytic degradation of organophosphorus pesticides using floating photocatalyst TiO$_2$. SiO$_2$/beads by sunlight. *Sol. Energ.* 79: 1–9.

Shifu, C. and Yunzhang, L. 2007. Study on the photocatalytic degradation of glyphosate pesticides by TiO$_2$ photocatalyst. *Chemosphere* 67: 1010–1017.

Si, Y., Takagi, K., Iwasaki, A., and Zhou, D. 2009. Adsorption, desorption and dissipation of metolachlor in surface and subsurface soils. *Pest Manag. Sci.* 65: 956–962.

Silva, M., Azenha, M. E., Pereira, M. M., et al. 2010. Immobilized of halogenated porphyrins and their copper complexes in MCM-41: Environmental friendly photocatalysts for the degradation of the pesticides. *Appl. Catal. B Environ.* 100: 1–9.

Singh, N. 2008. Biocompost from sugar distillery effluent: Effect on metribuzin degradation, sorption and mobility. *Pest Manag. Sci.* 64: 1057–1062.

Singh, H. K., Saquib, M., Haque, M., Muneer, M., and Bahnemann, D. 2007a. Titanium dioxide mediated photocatalysed degradation of phenoxyacetic acid and 2,4,5-trichlorophenoxyacetic acid in aqueous suspensions. *J. Mol. Catal. A Chem.* 264: 66–72.

Singh, H. K., Saquib, M., Haque, M., and Muneer, M. 2007b. Heterogeneous photocatalysed degradation of 4-chlorophenoxyacetic acid in aqueous suspensions. *J. Hazard. Mater.* 142: 374–380.

Spanggord, R., Gordon, R. G., Starr, R. I., and Elias D. J. 1996. Aerobic biodegradation of [^{14}C]3-chloro-*p*-toluidine hydrochloride in a loam soil. *Environ. Toxicol. Chem.* 15(10): 1664–1670.

Stenrød, M., Eklo, O. M., Charnay, M.-P., and Benoit, P. 2005. Effect of freezing and thawing on microbial activity and glyphosate degradation in two Norwegian soils. *Pest Manag. Sci.* 61: 887–898.

Strompl, C. and Thiele, J. H. 1997. Comparative fate of 1,1-diphenylethylene (DPE), 1,1-dichloro-2,2-*bis* (4-chlorophenyl) ethylene (DDE) and pentachlorophenol (PCP) under alternating aerobic and anaerobic conditions. *Arch. Environ. Contam. Toxicol.* 33: 350–356.

Struthers, J. K., Jayachandran, K., and Moorman, T. B. 1998. Biodegradation of atrazine by *Agrobacterium radiobacter* J14a and use of this strain in bioremediation of contaminated soil. *Appl. Environ. Microbiol.* 64: 3368–3375.

Suett, D. L., Jukes, A. A., and Phelps, K. 1993. Stability of accelerated degradation of soil-applied insecticides: Laboratory behaviour of aldicarb and carbofuran in relation to their efficacy against cabbage root fly (*Delia radicum*) in previously treated field soils. *Crop Protect.* 12: 431–442.

Sun, S. B., Inskeep, W. P., and Boyd, S. A. 1995. Sorption of nonionic organic-compounds in soil-water systems containing a micelle-forming surfactant. *Environ. Sci. Technol.* 29: 903–913.

Thomas, J. P., Bejjani, A., Nsouli, B., Gardon, A., and Chovelon, J. M. 2008. Investigation of norflurazon pesticide photodegradation using plasma desorption time-of-flight mass spectrometry analysis. *Rapid Commun. Mass Spectrom.* 22: 2429–2435.

Thomas, J. P., Bejjani, A., Nsouli, B., Gardon, A., and Chovelon, J. M. 2009. In situ studies of pesticides photodegradation on soils using PD-TOFMS technique: Application to norflurazon and oxyfluorfen. *Int. J. Mass Spectrom.* 279: 59–68.

Tolosa, I., Bayona, J. M., and Albaige's, J. 1995. Spatial and temporal distribution, fluxes, and budgets of organochlorinated compounds in northwest Mediterranean sediments. *Environ. Sci. Technol.* 29: 2519–2527.

Tomlin, C. (ed.) 2000. In: *The Pesticide Manual*. British Crop Protection Council: Surrey.

Top, E. M., Maila, M. P., Clerinx, M., Goris, J., De Vos, P., and Verstraete, W. 1999. Methane oxidation as a method to evaluate the removal of 2,4-dichlorophenoxyactic acid (2,4-D) from soil by plasmid-mediated bioaugmentation. *FEMS Microbiol. Ecol.* 28: 203–213.

Topp, E., Vallaeys, T., and Soulas, G. 1997. Pesticides: Microbial degradation and effects on microorganisms. In: van Elsas, J. D., Trevors, J. T., and Wellington, E. M. H. (eds), Microbial degradation and effects on microorganisms. *Modern Soil Microbiology*, pp. 547–575. Marcel Dekker: New York.

Tsugawa, W., Nakamura, H., Sode, K., and Ohuchi, S. 2000. Improvement of enatioselectivity of chiral organophosphate insecticide hydrolysis by bacterial phosphotriesterase. *Appl. Biochem. Biotechnol. A* 84–86: 311–317.

United Nations Environment Programme. *Chemicals, regionally based assessment of persistent toxic substances*. Regional Report, Central and North East Asia, December 2002.

USDA-ARS. 1995. *Root Zone Water Quality Model (RZWQM) V.3.0*. User's Manual. GPSR Technical Report No. 5. USDA-ARS Great Plains Systems Research Unit, Ft. Collins, CO.

van Eden, M., Potgieter, H. C., and van der Walt, A. M. 2000. Microbial degradation of chlorothalonil in agricultural soil: A laboratory investigation. *Environ. Toxicol.* 15: 533–539.

Vilanova, R., Fernandez, P., Martinez, C., and Grimalt, J. O. 2001. Organochlorine pollutants in remote mountain lake waters. *J. Environ. Qual.* 30: 1286.

Villaverde, J., Kah, M., and Brown, C. D. 2008. Adsorption and degradation of four acidic herbicides in soils from southern Spain. *Pest Manag. Sci.* 64: 703–710.

Volkering, F., Breure, A. M., and Rulkens, W. H. 1998. Microbiological aspects of surfacant use for biological soil remediation. *Biodegradation* 8: 401–417.

von Wirén-Lehr, S., Scheunert, I., and Dörfler, U. 2002. Mineralization of plant-incorporated residues of ^{14}C-isoproturon in arable soils originating from different farming systems. *Geoderma* 105: 351–366.

Vora, J. J., Chauchan, S. K., Parmar, K. C., Vasava, S. B., Sharma, S., and Bhutadiya, L. S. 2009. Kinetic study of application of ZnO as a photocatalyst in heterogeneous medium. *E-Journal of Chemistry* 6(2): 531–536.

Walker, A. and Jurando-Exposito, M. 1998. Adsorption of isoproturon, diuron and metsulfuron-methyl in two soils at high soil: Solution ratios. *Water Res.* 38: 229–238.

Wania, F. 1998. Multi-compartmental models of contaminants fate in the environmental. *Biotherapy* 11: 65.

Wedemeyer, G. 1967. Dechlorination of 1,1,1-trichloro-2,2-*bis*(*p*-chlorophenyl)ethane by *Aerobacter aerogenes*. *Appl. Microbiol.* 15: 569–574.

Whitford, F. 2002. *The Complete Book of Pesticide Management. Science, Regulation, Stewardship and Communication.* John Wiley & Sons: New York.

Wu, C. and Linden, K. G. 2010. Phototransformation of selected organophosphorus pesticides: Roles of hydroxyl and carbonate radicals. *Water Res.* 44: 3585–3594.

Yang, C., Liu, N., Guo, X., and Qiao, C. 2006. Cloning of *mpd* gene from a chlorpyrifos-degrading bacterium and use of this strain in bioremediation of contaminated soil. *FEMS Microbiol. Lett.* 265: 118–125.

Yao, B.-H., Wang, C., Wang, Y.-X., and Zhao, G.-Y. 2007. Preparation of performances of RuO$_2$/TiO$_2$ films photocatalyst supported on float pearls. *Chin. J. Chem. Phys.* 20(6): 789–795.

Yeasmin, L., MacDougall, S. A., and Wagner, B. D. 2009. UV-A photochemistry of the pesticide azinphos-methyl: Generation of the highly fluorescent intermediate N-methylanthranilic acid. *J. Photochem. Photobiol. A Chem.* 204: 217–223.

Young, R. A., Onstad, C. A., Bosh, D. D., and Anderson, W. P. 1986. Agricultural nonpoint source pollution models: A watershed analysis tool, model documentation. Agricultural Research Service, U.S. Department of Agriculture, Morris, NN.

Zaidi, B. R., Stucki, G., and Alexander, M. 1988. Low chemical concentration and pH as factors limiting the success of inoculation to enhance biodegradation. *Environ. Toxicol. Chem.* 7: 143–151.

Zapata, A., Velegraki, T., Sánchez-Pérez, J. A., Mantzavinos, D., Maldonado, M. I., and Malato, S. 2009a. Solar photo-Fenton treatment of pesticides in water: Effect of iron concentration on degradation and assessment of ecotoxicity and biodegradability. *Appl. Catal. B Environ.* 88: 448–454.

Zapata, A., Oller, I., Bizani, E., Sánchez-Pérez, J. A., Maldonado, M. I., and Malato, S. 2009b. Evaluation of operational parameters involved in solar photo-Fenton degradation of a comercial pesticide mixture. *Catal. Today* 144: 94–99.

Zayed, S. M. A. D. and Mahdy, F. 2008. Decomposition of ^{14}C-fenitrothion under the influence of UV and sunlight under tropical and subtropical conditions. *Chemosphere* 70: 1653–1659.

Zhang, L. Z., Wei, N., Wu, Q. X., and Ping, M. L. 2007. Anti-oxidant response of *Cucumis sativus* L. to fungicide carbendazim. *Pestic. Biochem. Physiol.* 89: 54–59.

Zhang, L., Yan, F., Su, M., Han, G., and Kang, P. 2009. A study on the degradation of methamidophos in the presence of nano-TiO$_2$ catalyst doped with Re. *Russ. J. Inorg. Chem.* 54(8): 1210–1216.

Zhu, X., Yuan, C., Bao, Y., Yang, J., and Wu, Y. 2005a. Photocatalytic degradation of pesticide pyridaben on TiO2 particles. *J. Mol. Catal. A Chem.* 229: 95–105.

Zhu, Y., et al. 2005b. Organochlorine pesticides (DDTs and HCHs) in soils from the outskirts of Beijing, China. *Chemosphere* 60(6): 770–778.

7

Pesticide Residues in the Atmosphere

Teresa Vera Espallardo, Amalia Muñoz, and José Luis Palau

CONTENTS

7.1 Introduction

Pesticides are one of the most important chemical groups in use worldwide. The term pesticide includes herbicides—to prevent the growth of unwanted plants, fungicides—to eliminate or inhibit fungi, and insecticides—to control pests that infest crops and plants.

Although pesticides have been used for centuries, their use has increased drastically in the last 50 years as a consequence of intensive agricultural techniques. Because the control of plagues and diseases in crops and farms is currently viewed as very important for the economic development of a country, the use of pesticides in crops and cultivated plants could be regarded as one of the main factors contributing to the increased agricultural productivity in the last century. Nevertheless, this intensive use can imply significant contamination not only of the application areas but also of other remote areas. The contamination can affect soil and water resources and even the atmosphere.

The intensive use of pesticides has led to a great number of studies about their risks with respect to human health and toxicity to other plants, animals, and the environment (Wilson et al. 2007; Lopez-Espinosa et al. 2008; Burns et al. 2008).

When a pesticide is applied, a fraction of the dosage can enter the different environmental compartments: the soil, the surface waters and groundwater, the target plant, and also the atmosphere (Bedos et al. 2002). The atmosphere represents the largest environmental compartment. At the same time, due to the characteristics and properties of the atmosphere, gas-phase compounds and particles can either be released directly or undergo different kinds of atmospheric removal processes. These atmospheric removal processes include enhanced transport (at the local or large scale), photochemical reactions with other pollutants in the atmosphere, for example, nitrogen oxides (NO_x) or volatile organic compounds (VOCs), and physical deposition mechanisms. These processes are affected by the meteorological conditions as well as the physicochemical properties of the pesticides (Foreman et al. 2000).

Pesticides can enter the air by means of spray drift during their application, wind erosion, or volatilization from plants, soils, or surface waters. Some of the volatilization can take place immediately after the application (Van den Berg et al. 1999). Another part of the volatilization from the treated surface can occur sometime after the application. Postapplication emissions represent an important means of input into the troposphere for days or even weeks after the application. In general terms, pesticides are considered low-volatile or semivolatile compounds (only fumigants are considered volatile compounds), and they can therefore be distributed into gas, aqueous, and solid phases. The distribution between the different phases is related to their physicochemical properties as well as to other factors, such as the manner in which the pesticide is applied or the meteorological conditions during the application. The main factor that governs the distribution of semivolatile compounds between the gas phase and the particle phase is vapor pressure (Bidleman 1999; Bedos et al. 2002). Vapor pressures higher than 10^{-2} Pa imply that the compounds are mainly observed in the vapor phase, whereas vapor pressures lower than 10^{-5} Pa are almost always present in the particle phase. Bearing in mind that most pesticides have vapor pressures in between these values, their presence in the atmosphere is also partitioned between gas and particle phases.

Once a pesticide enters into the atmosphere, it tends to become well-mixed and dispersed throughout the surface boundary layer. Depending on its time of residence in the atmosphere, a pesticide will be able to travel short or long distances. At the same time, meteorological factors also play an important role in this issue.

Although the main route of pesticide removal from air is considered to be wet or dry deposition, depending on the compound (e.g., Asman et al. 2003), other removal mechanisms are also important, for example, the chemical reactions that can occur in the atmosphere as well as other processes that take place because the chemical changes do not always result in the detoxification of the initial compound. Regarding the gas phase, it is important to take into consideration the presence of oxidants such as ozone and hydroxyl radicals and the action of reactive intermediates such as NO_x. All these reactions enhance the photochemical conversion of the pesticides. At the same time, photolysis is also important in some cases. Among all these chemical reactions, the main loss process for a large number of pesticides is the reaction with hydroxyl radicals (OH) (e.g., Atkinson et al. 1999).

7.2 The Atmosphere

The atmosphere is a *dynamic complex system*, which is in seasonal equilibrium with the vacuum of outer space, the gravitational and centripetal forces of the Earth, and the energy coming from the Sun. In general terms, of the total energy received by the Earth every year, only 69% is retained by the atmosphere–Earth system, while the remaining 31% is reflected back into space (27% reflected by the atmosphere and 4% reflected by the Earth's surface). The 69% retained is distributed inhomogeneously within the system, with the atmosphere retaining 24% and the Earth's surface retaining the remaining 45%.

Despite there being a global balance between the total energy received and the total energy absorbed, this balance does not occur in every place. The amount of energy absorbed by the Earth's surface depends, among other factors, on the nature of the surface (e.g., soil compared with sea), the altitude of the site considered, and the degree of cloudiness. The heterogeneous distribution of energy, as a result of the different albedo values across the surface of the Earth, is the cause of the atmospheric movements of air masses. This is the reason for considering air movements as energy redistribution mechanisms that tend toward homogenization.

The atmosphere is also a *physical filter* that protects living beings from high-energy radiation (x-rays and UV light), a *mechanical filter* that traps and burns (to some extent) the meteorites that impact the Earth, and a *greenhouse* that protects the Earth's surface from the cold of the outer space (maintaining a mean temperature gradient of about 300 degrees).

Finally, the atmosphere is a *physicochemical reactor*, where a huge amount of circulating water produces constant physical changes, and a lot of chemical compounds combine, dissociate, and recombine with sunlight, constituting a mixture of substances, which, depending on their elevation above ground, play a different role in some of the phenomena that attract our attention today, for example, pollution (excess) by tropospheric ozone, the hole (deficiency) in the stratospheric ozone layer, acid rain, etc.

The chemical composition of the atmosphere has gradually changed during the last few million years. Nowadays, 99.95% of the total atmospheric volume is composed of three species: oxygen, nitrogen, and argon. The remaining atmospheric components (excluding water vapor) are present in concentrations of the order of parts per million (in volume); these include carbon dioxide, neon, helium, methane, hydrogen, carbon monoxide, ozone, etc. Most of these gases remain in fairly constant proportions up to altitudes of around 80 km. However, there are gases whose concentrations vary substantially not only with height but also over time (for some cases, even on a scale of hours). In this last group of gases (the so-called "variable gases") are water vapor, carbon dioxide, stratospheric and tropospheric ozone, aerosols,* and other VOCs (which include pesticides).

These variable gases play an essential role in meteorological processes because they drive the energy balance of the Earth system. As a matter of fact, the origin of "Climate Change" lies mainly in the abrupt change in the mean concentrations of these gases in the atmosphere by the direct or indirect action of man.

From a thermodynamic point of view, the atmosphere is essentially stable[†] (the rate of global change can be measured in centuries), but locally it is unstable because of its stratified nature and because of the combination of cyclic isolation, the Coriolis force, the content of water and its reiterated phase transition, volcanic eruptions, etc. This combination of global organization and local chaos is what makes a physicochemical description of the atmosphere so difficult.

---

* Although aerosols are not gases, they have been included here because their variability plays a crucial role in the physicochemical properties of the atmosphere and the global radiative balance.
† The terms stable atmosphere and unstable atmosphere relate to whether vertical exchanges of matter, momentum, and energy are favored or inhibited.

7.2.1 Structure of the Atmosphere

The surface of the Earth plays a critical role in the thermodynamic processes that take place in the lower layers of the atmosphere; as a matter of fact, it is responsible for the temperature decrease with increasing heights within the lowest layers of the atmosphere.

From a thermodynamic point of view, a perfectly stable atmosphere would be stratified into layers, with the warmer layers closer to the surface and the colder ones aloft as a result of conductive and radiative warming from the surface. Considering the temperature variation with height, the atmosphere can be divided into several regions. The troposphere, or the region of the atmosphere closest to the Earth's surface, is where the temperature decreases with increasing height. This evolution of temperature with height changes around 11 km above the surface, in the so-called tropopause. From this height, and through the next region, called the stratosphere, temperature increases with height. The stratosphere extends up to, roughly, 50 km above sea level. Over the stratosphere, in the mesosphere, the temperature variation changes again, decreasing with height up to 80 km. Above this and over the mesopause, the thermosphere begins. In this outer region of the atmosphere, the temperature again increases with height.

The part of the troposphere that is directly influenced by the Earth's surface, on a time scale of an hour or less, is called the boundary layer. The region of the troposphere over the boundary layer is the free atmosphere (and over it, the tropopause). The boundary layer responds to the changes in the physical and thermodynamic characteristics of the Earth's surface (roughness, heating and cooling, etc.); thus, the diurnal variation in the temperature of the air near the ground is one of the differential characteristics of the lower troposphere or boundary layer with respect to the upper regions of the atmosphere (Stull 1989). Boundary layer depth varies frequently and ranges from a few meters under highly stratified (stable) atmospheric conditions to more than 3 km under strong convective conditions (Holton 1990).

All the exchanges of momentum between the atmosphere and the surface are made through the boundary layer by turbulent motions with spatial scales of the order of the depth of the boundary layer or less. The same applies to all the exchanges of water vapor and many heat exchanges. This is why the concept of boundary layer is crucial in the study of both the dynamics and the thermodynamics of the atmosphere.

On the other hand, most anthropogenic and biogenic atmospheric pollutants are released into the boundary layer. Therefore, boundary layer meteorology will determine whether the pollutants will be dispersed and diluted, exported to the free atmosphere, or settled on the ground (Stewart 1979).

Under strong convective conditions, generally produced by the heating of the Earth's surface, heat and momentum exchanges are dominated by convective turbulence, and the boundary layer is frequently referred to as the mixed layer (Garratt 1999). An unstable boundary layer usually increases its depth due to the entrainment of air from the free atmosphere aloft toward the mixed layer. This drag process is responsible for the so-called fumigation of the atmospheric pollutants aloft to the ground.

7.2.2 Meteorological Scales

Meteorological processes, driving the meteorology and the atmospheric pollutant dispersion, are traditionally classified according to their characteristic spatial (or temporal) scale. In 1975, Orlanski introduced a classification describing the following three meteorological scales.

- *Macroscale* (spatial scale larger than 1000 km or temporal scales ranging from a few days to months). Atmospheric movements are associated with inhomogeneities of the surface energy balance at the Earth (global or synoptic) scale. In these scales, the hydrostatic approach is assumable (neglecting vertical air movements and favoring horizontal advections). Studies on the long-range transport (LRT) of atmospheric pollutants are framed in this scale.

- *Mesoscale* (spatial scales between 1 and 1000 km or temporal scales ranging from one hour to a few days). On this meteorological scale, the atmospheric flows are driven by the hydrodynamic processes (as, e.g., channeling through valleys, surface roughness effects, compensatory subsidences, etc.) and also by the inhomogeneities of the surface energy balance (mainly due to spatial variations in features such as the physiographic properties of the ground, land use, water availability, slope, and orientation of terrain). From the atmospheric pollutant point of view, thermodynamic effects are the most relevant ones, because they drive the boundary layer dynamics under low synoptic forcing (i.e., under poor ventilating conditions). The hydrostatic approach is no longer valid in these scales. Mesoscale meteorological models (nonhydrostatic models) must be able to reproduce the local circulations, such as breeze or/and upslope circulations. Studies on medium- and short-range transport of atmospheric pollutants are framed in this scale.

- *Microscale* (spatial scales under 1 km or temporal scales ranging from a few hours to seconds). On this meteorological scale, air flow is strongly determined by surface features such as geometry of buildings, their orientation with respect to the average wind direction, etc. Wind flows are generated as the result of both thermal effects and hydrodynamic effects (channeling, roughness effects, etc.). Studies on the short-range transport of atmospheric pollutants are framed in this scale.

7.3 Entry of Pesticides into the Atmosphere

The highest pesticide concentrations are usually found near the application area and near the moment of the application. Nevertheless, pesticides can be found far away from the application site (e.g., Unsworth et al. 1999; Coscollà et al. 2010) and at other times of the year after the application (e.g., Scheyer et al. 2007a,b).

There are several ways in which pesticides can enter the atmosphere:

- Due to the drift after application
- Due to volatilization from the soil
- Due to volatilization from the leaves
- From suspended particulate matter from wind erosion
- Due to losses during the application
- From evaporation in the surface water

Figure 7.1 shows a general scheme of the different ways in which pesticides enter, disperse, and interact in the environment. Atmospheric compartment, on which this chapter is focused, is presented in a more detailed way.

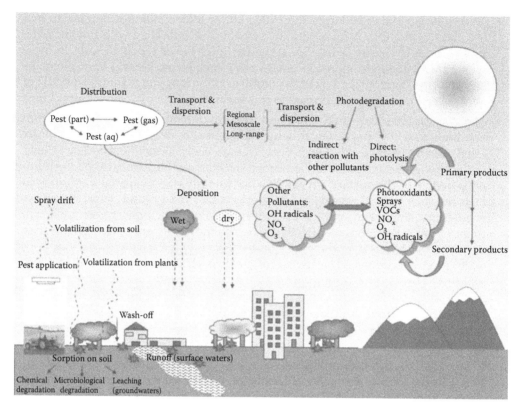

FIGURE 7.1
Scheme of the processes that can affect pesticide entry, dispersion, and interaction with other pollutants in the environment.

In the following sections, some of the ways in which pesticides enter the atmosphere are briefly described.

7.3.1 Entry During Application

In conventional agriculture, one of the most common ways to apply pesticides is by using sprayers; in some cases, spraying is done from aircrafts; in other cases, spraying is done manually by workers who use different kinds of equipment filled with the pesticide. With these sprayers, the formulated pesticides, together with some carrier liquid, usually water, are converted into droplets. Droplets can have different sizes, depending on the type of equipment used. Larger droplets are less susceptible to spray drift than smaller ones, but they require more carrier liquid per unit of land covered. In order to minimize losses due to spray drift, sprayer application systems are designed to dispense a coarse droplet spectrum, whereby droplets can settle and make rapid contact with the surfaces they find (land surface or leaves). Nevertheless, although the equipment nozzles are carefully designed to optimize the droplet size, a fraction of the pesticide applied will always be lost and passed into the air. This fraction of the applied pesticide may have the potential to cause nontarget crop damage as well as pose risks to human health and more sensitive ecosystems (Ramaprasad et al. 2004).

According to Van den Berg et al. (1999 and references therein), data from field measurements indicate that losses or emissions during the application of pesticides can range from a few percent to 30% or more than 50%.

For instance, malathion can be transported in air by drift from the application site to nontarget locations. The California Food and Agriculture Department (Segawa et al. 1991) monitored malathion in 1990 and found concentrations from 1 $\mu g/m^2$ to 101 $\mu g/m^2$ on locations where the aircrafts specifically avoided spraying. The study pointed out that 92% of the malathion applied remained on the surface of the target areas, while 8% was either degraded before arriving at the ground, volatilized before sample collection, or drifted outside the target area. Another study (Sanders 1997; Bradley et al. 1997) related to malathion involved the collection of outdoor air samples near the center of the spray area during the application of pesticides on five sites for four consecutive periods (24 h immediately before the spraying, during the spray application, and two consecutive 24 h periods after the application). The highest average malathion level, 0.067 $\mu g/m^3$ (5.0 ppt), was found during the spray application. During the 24 h following the application, the malathion level stayed high, as it also did in the last two 24 h periods after the application, registering 0.049 and 0.042 $\mu g/m^3$, respectively.

The main factors that affect pesticide loss during application are the manner in which the pesticide is applied, that is, the spray characteristics (size of nozzles, etc.) together with the equipment used and the viscosity and volatility of the formulation used, the environmental conditions (wind speed, temperature, and humidity), and, finally, the operators skills. Thus, losses can be reduced by improving the operators' skills in using the best application technique together with the best formulation for any given problem (Gil and Sinfort 2005; Gil et al. 2007).

Generally, spray drift is understood as the fraction of pesticide losses, in the form of droplets (they can be dry particles or vapors), which move off-target through the air.

In the 1990s, many studies assessing spray drift were performed, for example, in the United States (Bird et al. 2002; Bradley et al. 1997) or in Germany (Van den Berg et al. 1999). In the last few decades, different kinds of models offer the possibility of estimating losses during the application of a pesticide (Klein 1995; Jarvis 1994; Vanclooster et al. 2000). Although the models can be very useful, their critical parameters must be kept in mind if one is to obtain values close to real ones. In this sense, it is essential to have both good meteorological data and a droplet distribution assessment (Gil et al. 2007). On the other hand, models must be fed experimental data to check their predictability (Ferrari et al. 2003). And, at the same time, the development of new technical methodologies or equipment may lead to reductions in spray drift losses.

7.3.2 Volatilization from Soils

Once the pesticide is applied on the target area or soil surface or injected into the soil, it can volatilize. Depending on the climatic conditions and the compound used, the volatilization can reach as high as 90% of the application dose. For example, for atrazine, the cumulative losses after 24 h are around 2% of the application dose (Bedos et al. 2002). Other pesticides, such as lindane (nowadays banned in most of Europe, the United States, and Canada) and trifluralin, present higher volatilization losses, probably because of the differences in their Henry's law constants (see Table 7.1).

In a study conducted in the Italian province of Bologna, under agricultural conditions, the volatilization rates of procymidone, malathion, and ethoprophos were determined after soil application (Ferrari et al. 2003). As the field experiments were carried out in different seasons, the authors could confirm that temperature and soil moisture affected

TABLE 7.1

Physicochemical Properties of Selected Pesticides

Pesticide	CAS	log K_{ow}	VPr (mmHg)	Henry LC (atm m³/mol)	log K_{oa}	K_{OH} (cm³/molec s)	J (per s)	K_{o_3} (cm³/molec s)
Chlorpyrifos	2921-88-2	4.96 (n)	2.03E – 05 (j)	2.93E – 06 (c)	8.88	7.00E – 11 (x)	2.20E – 05 (v)	—
Chlorpyrifos-methyl	5598-13-0	4.31 (l)	4.20E – 05 (i)	1.43E – 06 est	8.12	4.10E – 11 (u)		—
Chlorfenvinphos	470-90-6	3.81 (n)	7.50E – 06 (j)	5.17E – 08 est	9.74	5.78E – 11 est		3.00E – 18
Propachlor	1918-16-7	2.18 (l)	2.30E – 04 (o)	5.17E – 08 est	9.74	5.78E – 11 est		3.00E – 18
Pyriproxyfen	95737-68-1	5.55 est	2.18E – 06 (j)	6.34E – 10 est	13.14	5.22E – 11 est		—
Azoxystrobin	131860-33-8	2.50 (j)	8.25E – 13 (j)	8.01E – 14 est	14.03	4.72E – 11 est		1.05E – 16
Malathion	121-75-5	2.36 (l)	3.38E – 06 (o)	4.89E – 09 (d)	9.06	7.74E – 11 est		—
Imidacloprid	138261-41-3	0.57 (k)	1.68E – 06 est	1.04E – 13 est	13.74	1.52E – 10 est		—
Thiabendazole	148-79-8	2.47 (m)	4.00E – 09 (p)	2.00E – 11 est	11.53	6.50E – 11 est		—
Tebuconazole	107534-96-3	3.70 (j)	1.28E – 08 (j)	5.18E – 10 est	11.93	1.15E – 11 est		—
Hexythiazox	78589-05-0	5.57 est	2.55E – 08 (k)	2.29E – 09 est	11.58	3.54E – 11 est		—
Imazalil	35554-44-0	3.82 (k)	1.19E – 06 (k)	7.25E – 08 est	10.80	8.49E – 11 est		1.20E – 17
Metalaxyl	57837-19-1	1.65 (l)	5.62E – 06 (p)	8.05E – 10 est	8.57	2.69E – 11 est		—
Carbendazim	10605-21-7	1.52 (l)	7.50E – 10 (q)	1.49E – 12 est	10.58	2.01E – 10 est		—
Propanil	709-98-8	3.07 (l)	9.08E – 07 (o)	4.50E – 09 est	10.23	3.78E – 12 est		—
Buprofezin	69327-76-0	4.30 (j)	9.40E – 06 (j)	8.95E – 09 est	8.07	5.38E – 11 est		—

Compound	CAS							
Omethoate	1113-02-6	-0.74 (i)	2.48E + 05 (j)	4.56E – 14 est	10.99	2.60E – 11 est		—
Fenthion	55-38-9	4.09 (l)	1.05E – 05 (j)	1.37E – 06 est	8.31	7.11E – 11 est		—
Methidathion	950-37-8	2.20 (j)	3.37E – 06 (p)	7.10E – 09 est	8.73	1.50E – 06 est		—
Bitertanol	55179-31-2	4.16 (l)	1.89E – 10 est	1.63E – 12 est	15,12	4.43E – 11 est		—
Lindane	58-89-9	3.72 (l)	4.20E – 05 (j)	5.14E – 06 (f)	7,40	1.90E – 13 (a)		—
Trifluralin	1582-09-8	5.34 (l)	4.58E – 05 (j)	1.03E – 04 (c)	7,72	1.70E – 11 (w)	1.20E – 03 (w)	—
Atrazine	1912-24-9	2.61 (l)	2.89E – 07 (j)	4.47E – 09 est	9,63	2.73E – 11 est		—
Parathion-methyl	298-00-0	2.86 (l)	3.50E – 06 (j)	1.00E – 07 (g)	8,25	5.90E – 11 est		—
Quinoxyfen	124495-18-7	4.66 (k)	1.50E – 07 (k)	9.65E – 09 est	11.06	5.35E – 12 est		—
Fenpropimorph	67306-03-0	5.50 est	2.50E – 05 (r)	2.15E – 07 est	8.93	1.38E – 10 est		—
Diazinon	333-41-5	3.81 (l)	9.01E – 05 (i)	1.13E – 07 (d)	9.15	9.67E – 11 est		—
Chloropicrin	076-06-2	2.09 (l)	2.38E + 01 (s)	2.05E – 03 (h)	3.17	1.30E – 13 est	4.50E – 05 (t)	—
Dichlorvos	62-73-7	1.43 (l)	1.58E – 02 est	8.58E – 07 (i)	6.06	2.60E – 11 (b)		3.58E – 20
Methamidophos	1065-92-6	-0.80 (j)	3.53E – 05 (j)	8.68E – 10 (j)	6.65	3.31E – 11 est		—

est: estimated value obtained using EPIweb 4.0. (a): Brubaker and Hites (1998); (b): Feigenbrugel et al. (2006); (c): Rice and Chernyak (1995); (d): Fendinger and Glotfelty (1990); (f): Altschuh et al. (1999); (g): Metcalfe et al. (1980); (h): Kawamoto and Urano (1989); (i): Tomlin (1997); (k): Tomlin (2003); (l): Hansch et al. (1995); (m): Nielsen et al. (1992); (n): Sangster (1994); (o): USDA Pesticide properties database; (p): Wauchope et al. (1991); (q): Augustijn-Beckers et al. (1994); (r): Merck Index (1996); (s): Yalkowsky and Dannenfelser (1990); (t): Vera et al. (2010); (u): Muñoz et al. (2011); (v): Hebert et al. (2000a); (w): Le Person et al. (2007); (x): Hebert et al. (2000b). K_{ow} octanol–water partitioning coefficient; VPr, vapor pressure; Henry LC, Henry's law constant; K_{oa}, octanol–air partitioning coefficient; K_{OH}, reaction coefficient with OH radicals; J, reaction coefficient of light photolysis constant. Ref. (v) and (x) at 60°C.

the volatilization of the pesticides used. They concluded that, with all other factors being constant, when the temperature is higher, more volatilization occurs. On the other hand, an increase in the soil moisture implies an increase in the gas–liquid distribution of the pesticide and its diffusive transport. Finally, these authors pointed out that for pesticides with vapor pressures in the range from 5×10^{-3} to 5×10^{-2} Pa, volatilization can represent up to 22.6% of the total fate in the environment. This shows its important influence in the air quality of the surrounding agricultural areas.

Despite the fact that, initially, vapor pressure could be considered the best parameter to assess the volatilization of a compound, this factor is a measurement of the volatilization of the pure compound in its condensed state. For this reason, Henry's law constant may be a better indicator as it is a measure of the volatilization tendency of a pesticide from dilute aqueous solutions (Unsworth et al. 1999). Taking into account that water is always present in soils and on plant surfaces, even a pesticide with a low vapor pressure can have an appreciable Henry's law constant and thus be subject to volatilization. Unsworth et al. (1999 and references therein) pointed out that cultivation practices and formulations can affect the extent of volatilization. In the same sense, soil characteristics also influence volatilization. For example, dry soils reduce volatilization due to their higher adsorption of the compounds when compared with moist surfaces. Van den Berg et al. (1999) also point out that temperature and soil moisture conditions have an important influence on the volatilization rate of pesticides at the soil surfaces. In a chamber study simulating field conditions, Wolters et al. (2004) observed that the cumulative volatilization of parathion-methyl ranged from 2.4% under dry conditions to nearly 33% under moist conditions.

7.3.3 Volatilization from Plants

Volatilization from plants usually exceeds volatilization from soils. Most losses occur within the first few days after the application and follow a diurnal pattern, with the largest losses at midday, when temperature and solar irradiation are higher. The main factors influencing pesticide volatilization from crops and plants are physicochemical properties, pesticide persistence on plant surfaces, and meteorological conditions during and some days after the application (Van den Berg et al. 1999). Several studies showed the relationship between the vapor pressure and the volatilization rate, mostly during the first 24 h after the application (Woodrow and Seiber 1997; Smit et al. 1998). The information and data on the postapplication volatilization of pesticides are mainly focused on the initial periods after the application, and little is written about emissions further from the time of application. Ramaprasad et al. (2004) pointed out that compounds with vapor pressures higher than 9.75×10^{-5} mmHg could sustain a total loss of 58% in the first 30 days after the application, although this data is only an estimation as other degradation processes can also participate in the loss of the pesticides (such as runoff, plant uptake, and soil adsorption).

Pesticide emission into the air depends on other factors as well, for example, molecular interaction forces in the deposit and adsorption by plant foliage. The adsorption on the leaf surface can be described by the octanol–water partitioning coefficient (K_{ow}). On the other hand, pesticide persistence on leaves can suffer different kinds of dissipation processes such as wash-off due to rain or photodegradation on the leaf surface.

Wolters et al. (2004) studied the volatilization of parathion-methyl, quinoxyfen, and fenpropimorph from plants in a wind tunnel during a 10-day period. The highest volatilization, 29.2%, was observed for parathion-methyl. Quinoxyfen showed a volatilization of 15%, and fenpropimorph had the lowest volatilization, with only 6%. In another experiment under simulated field conditions in a wind tunnel, volatilization rates up to 50% were measured for fenpropimorph and parathion-methyl (Stork et al. 1998). These authors observed that volatilization started immediately after the application of the pesticides at relatively high rates and decreased considerably after the first few days, following a diurnal profile. They also observed that the volatilization rate was strongly dependent on the soil moisture. In addition, unknown metabolites in the air samples led them to conclude that other processes, such as photolysis reactions on plant or soil surfaces, could play an important role in the dissipation of the pesticides in the environment. Another field study (Leistra et al. 2006), which included a characterization of the meteorological conditions, estimated the cumulative volatilization of chlorpyrifos from a potato crop during daylight hours to be 65% of the dosage. This was reduced to 7% for fenpropimorph because other competing processes on the plant surface reduced its dissipation for volatilization during daylight hours.

In a study carried out in Washington by Ramaprasad et al. (2004), the air concentrations of methamidophos were higher in the afternoon when the temperatures were higher and the vapor pressure increased. These authors concluded that volatilization from fields is not a linear process, even though the most intense volatilization occurs immediately after the spraying. Nevertheless, postspray volatilization could release important emissions into the air during the 30 days after the spraying. This fact should be taken into account due to its associated health risks, mainly because people do not usually think that they are at a risk of pesticide exposure after the spraying has already been carried out.

The FOCUS working group concluded by saying that plant volatilization is up to three times higher than soil volatilization under similar conditions. On the other hand, the many complex interactions that take place on plant surfaces, involving variations in vapor pressure, climatic conditions, Henry's law constants, and formulation, make volatilization from plants highly variable. Finally, as these interactions are not fully understood, there are no mathematical models available to describe the phenomenon (FOCUS 2008).

7.3.4 Volatilization from Water

Vapor pressure, water solubility, and Henry's law constant dominate the volatilization of pesticides from water. The physical conditions of the water carrying the pesticide, such as temperature or turbulence, are also important (Bidleman 1999). However, the water–air exchange of pesticides has not been extensively studied. A few studies on flooded rice fields were performed in the late 1980s by the California Department of Pesticide Regulation (e.g., Siebers and McChesney 1988).

7.4 Assessment of the Concentrations of Pesticides in the Atmosphere

Once the pesticide is emitted to the atmosphere, it can be moved by wind, fog, and rain. Depending on the meteorological conditions as well as the physicochemical properties of the pesticide, it can be transported to either long distances (if it is persistent) or short distances and then react with other pollutants present in the atmosphere, that is, OH radicals, ozone, VOCs, etc.

Nowadays, increasing amount of information is available on the monitoring of pesticides in the atmosphere. Up to a few years ago, the largest amount of data had been collected about pesticides in rainwater; however, in the last years, several studies dealing with pesticides in air—in the gas phase and/or particle phase—have also been published (see Table 7.2; e.g.,

TABLE 7.2

Examples of Concentrations of Pesticides and Some Degradation Products Measured in Air

Compound	Range (pg/m³)	Sampling Phase	Site	References
Acetochlor	300–8870	PUF + filters	Urban, rural	Coscollà et al. (2010)
Aclonifen	230–4150	PUF + filters	Urban, rural	Coscollà et al. (2010)
Alachlor	120–6030	PUF + filters	Urban, rural	Coscollà et al. (2010)
Atrazine	ND–6100	PUF (air)	Rural	Yao et al. (2007)
Atrazine	ND–2000	Filters	Rural	Yao et al. (2007)
Atrazine	645–1905	PUF/XAD-2	Rural	Aulagnier et al. (2008)
Azinphos-methyl	<LD–2444	PUF/XAD-2	Rural	Aulagnier et al. (2008)
Azoxystrobin	660–1790	PUF + filters	Urban, rural	Coscollà et al. (2010)
Bitertanol	6.5–118	Filters (air)	Urban, rural	Coscollà et al. (2008)
Bromoxynil	<LD–211	PUF/XAD-2	Rural	Aulagnier et al. (2008)
Captan	1190–67620	PUF + filters	Urban, rural	Coscollà et al. (2010)
Captan	111–5074	PUF/XAD-2	Rural	Aulagnier et al. (2008)
Carbaryl	<LD–1289	PUF/XAD-2	Rural	Aulagnier et al. (2008)
Carbendazim	41–572	Filters (air)	Urban, rural	Coscollà et al. (2008)
Carbofuran	<LD–787	PUF/XAD-2	Rural	Aulagnier et al. (2008)
Clopyralid	<LD–235	PUF/XAD-2	Rural	Aulagnier et al. (2008)
trans-Chlordane	5–838	PUF	Rural	Bidleman et al. (2006)
trans-Chlordane	3–18	PUF/XAD-2	Rural	Aulagnier et al. (2008)
cis-Chlordane	5–479	PUF	Rural	Bidleman et al. (2006)
cis-Chlordane	3–20	PUF/XAD-2	Rural	Aulagnier et al. (2008)
Chlorothalonil	ND–1930	XAD-4 (air)	Rural	LeNoir et al. (1999)
Chlorothalonil	110–107930	PUF + filters	Urban, rural	Coscollà et al. (2010)
Chlorothalonil	35.3–17318	XAD-2	Rural	Daly et al. (2007)
Chlorpyrifos	ND–17500	XAD-4 (air)	Rural	LeNoir et al. (1999)
Chlorpyrifos	<LD–868	PUF/XAD-2	Rural	Aulagnier et al. (2008)
Chlorpyrifos	<LD–6	PUF/XAD-2/PUF	Rural	Primbs et al. (2008)
Chlorpyrifos oxon	100–30370	XAD-4 (air)	Rural	LeNoir et al. (1999)
Chlorpyrifos oxon	<LD–58	PUF/XAD-2/PUF	Rural	Primbs et al. (2008)
Chlorpyriphos ethyl	110–97770	PUF + filters	Urban, rural	Coscollà et al. (2010)
Chlorthal	<LD–7	PUF/XAD-2	Rural	Aulagnier et al. (2008)
Cyhalothrin	<LD–32	PUF/XAD-2	Rural	Aulagnier et al. (2008)
Cyprodinil	120–3290	PUF + filters	Urban, rural	Coscollà et al. (2010)
2,4-D	<LD–1306	PUF/XAD-2	Rural	Aulagnier et al. (2008)
Dacthal	<LD–352	PUF/XAD-2/PUF	Rural	Primbs et al. (2008)
Dacthal	ND–73,8	XAD-2	Rural	Daly et al. (2007)
2,4-DB	<LD–477	PUF/XAD-2	Rural	Aulagnier et al. (2008)
4,4'-DDD	370–550	PUF + filters	Urban, rural	Coscollà et al. (2010)
2,4-DDD	<LD–5	PUF/XAD-2	Rural	Aulagnier et al. (2008)
4,4'-DDE	270–360	PUF + filters	Urban, rural	Coscollà et al. (2010)
4,4'-DDE	8.1–4540	PUF	Rural	Bidleman et al. (2006)
4,4'-DDE	6–57	PUF/XAD-2	Rural	Aulagnier et al. (2008)
4,4'-DDT	10–534	PUF	Rural	Bidleman et al. (2006)

TABLE 7.2 (Continued)

Examples of Concentrations of Pesticides and Some Degradation Products Measured in Air

Compound	Range (pg/m³)	Sampling Phase	Site	References
2,4′-DDT	10–164	PUF	Rural	Bidleman et al. (2006)
Deethylatrazine	<LD–132	PUF/XAD-2	Rural	Aulagnier et al. (2008)
Diazinon	ND–240	XAD-4 (air)	Rural	LeNoir et al. (1999)
Diazinon	280–1490	PUF + filters	Urban, rural	Coscollà et al. (2010)
Diazinon	<LD–1171	PUF/XAD-2	Rural	Aulagnier et al. (2008)
Dicamba	<LD–477	PUF/XAD-2	Rural	Aulagnier et al. (2008)
Dichlobenil	100–2390	PUF + filters	Urban, rural	Coscollà et al. (2010)
Dichlobenil	<LD–328	PUF/XAD-2	Rural	Aulagnier et al. (2008)
Dieldrin	5–1.190	PUF	Rural	Bidleman et al. (2006)
Dieldrin	<LD–14	PUF/XAD-2	Rural	Aulagnier et al. (2008)
Diflufenican	110–560	PUF + filters	Urban, rural	Coscollà et al. (2010)
Dimethenamide	160–740	PUF + filters	Urban, rural	Coscollà et al. (2010)
Dimethenamide	<LD–2588	PUF/XAD-2	Rural	Aulagnier et al. (2008)
Dimethomorph II	130–650	PUF + filters	Urban, rural	Coscollà et al. (2010)
Endosulfan I	300–3670	XAD-4 (air)	Rural	LeNoir et al. (1999)
Endosulfan I	370–81310	PUF + filters	Urban, rural	Coscollà et al. (2010)
Endosulfan I	22–940	PUF	Rural	Bidleman et al. (2006)
Endosulfan I	<LD–1177	PUF/XAD-2	Rural	Aulagnier et al. (2008)
Endosulfan I	<LD–255	PUF/XAD-2/PUF	Rural	Primbs et al. (2008)
Endosulfan I	1.64–1876	XAD-2	Rural	Daly et al. (2007)
Endosulfan II	70–640	XAD-4 (air)	Rural	LeNoir et al. (1999)
Endosulfan II	10–251	PUF	Rural	Bidleman et al. (2006)
Endosulfan II	<LD–27	PUF/XAD-2	Rural	Aulagnier et al. (2008)
Endosulfan II	<<LD–33	PUF/XAD-2/PUF	Rural	Primbs et al. (2008)
Endosulfan II	ND–380	XAD-2	Rural	Daly et al. (2007)
Endosulfan sulfate	10–70	XAD-4 (air)	Rural	LeNoir et al. (1999)
Endosulfan sulfate	5–26	PUF	Rural	Bidleman et al. (2006)
Endosulfan sulfate	<LD–4	PUF/XAD-2/PUF	Rural	Primbs et al. (2008)
Endosulfan sulfate	0.48–97.4	XAD-2	Rural	Daly et al. (2007)
EPTC	<LD–9545.5	PUF/XAD-2	Rural	Aulagnier et al. (2008)
Epoxiconazole	120–3990	PUF + filters	Urban, rural	Coscollà et al. (2010)
Ethofumesate	540–1160	PUF + filters	Urban, rural	Coscollà et al. (2010)
Ethoprophos	210–480	PUF + filters	Urban, rural	Coscollà et al. (2010)
Fenoprop	<LD–26	PUF/XAD-2	Rural	Aulagnier et al. (2008)
Fenpropidin	110–3540	PUF + filters	Urban, rural	Coscollà et al. (2010)
Fenpropimorph	120–13200	PUF + filters	Urban, rural	Coscollà et al. (2010)
Fluazinam	120–2200	PUF + filters	Urban, rural	Coscollà et al. (2010)
Fludioxonil	530–2080	PUF + filters	Urban, rural	Coscollà et al. (2010)
Folpet	140–82220	PUF + filters	Urban, rural	Coscollà et al. (2010)
Hexachlorobenzene	<LD–55	PUF/XAD-2	Rural	Aulagnier et al. (2008)
α-HCH	14–181	PUF	Rural	Bidleman et al. (2006)

(continued)

TABLE 7.2 (Continued)

Examples of Concentrations of Pesticides and Some Degradation Products Measured in Air

Compound	Range (pg/m³)	Sampling Phase	Site	References
β-HCH	5–225	PUF	Rural	Bidleman et al. (2006)
γ-HCH (Lindane)	5–368	PUF	Rural	Bidleman et al. (2006)
γ-HCH (Lindane)	120–1210	PUF + filters	Urban, rural	Coscollà et al. (2010)
γ-HCH (Lindane)	<LD–2623	PUF/XAD-2	Rural	Aulagnier et al. (2008)
γ-HCH (Lindane)	<LD–29	PUF/XAD-2/PUF	Rural	Primbs et al. (2008)
Heptachlor exo-epoxide	5–65	PUF	Rural	Bidleman et al. (2006)
Hexythiazox	6.5–434.9	Filters (air)	Urban, rural	Coscollà et al. (2008)
Imazalil	0–637	Filters (air)	Urban, rural	Coscollà et al. (2008)
Malathion	ND–400	XAD-4 (air)	Rural	LeNoir et al. (1999)
(2-Methyl-4-chlorophenoxy) acetic acid	<LD–53	PUF/XAD-2	Rural	Aulagnier et al. (2008)
MCPB	<LD–114	PUF/XAD-2	Rural	Aulagnier et al. (2008)
Mecoprop	<LD–38	PUF/XAD-2	Rural	Aulagnier et al. (2008)
Metazachlor	170–3130	PUF + filters	Urban, rural	Coscollà et al. (2010)
Metolachlor	120–2490	PUF + filters	Urban, rural	Coscollà et al. (2010)
Metolachlor	3031.1–19061.7	PUF/XAD-2	Rural	Aulagnier et al. (2008)
Metribuzin	<LD–7	PUF/XAD-2/PUF	Rural	Primbs et al. (2008)
Myclobutanil	<LD–358	PUF/XAD-2	Rural	Aulagnier et al. (2008)
1-Naphthol	<LD–44	PUF/XAD-2	Rural	Aulagnier et al. (2008)
trans-Nonachlor	5–135	PUF	Rural	Bidleman et al. (2006)
Oxyfluorfen	650–3010	PUF + filters	Urban, rural	Coscollà et al. (2010)
Parathion-methyl	670–750	PUF + filters	Urban, rural	Coscollà et al. (2010)
Pendimethalin	70–117330	PUF + filters	Urban, rural	Coscollà et al. (2010)
Phosalone	<LD–419	PUF/XAD-2	Rural	Aulagnier et al. (2008)
Phosmet	150–1330	PUF + filters	Urban, rural	Coscollà et al. (2010)
Phosmet	<LD–1913	PUF/XAD-2	Rural	Aulagnier et al. (2008)
Propachlor	140–3480	PUF + filters	Urban, rural	Coscollà et al. (2010)
Propargite	2800–45600	PUF + filters	Urban, rural	Coscollà et al. (2010)
Pyriproxyfen	6.5–333.3	Filters (air)	Urban, rural	Coscollà et al. (2008)
Simazine	<LD–55	PUF/XAD-2	Rural	Aulagnier et al. (2008)
Spiroxamine	150–9010	PUF + filters	Urban, Rural	Coscollà et al. (2010)
4,4-TDE	<LD–28	PUF/XAD-2	Rural	Aulagnier et al. (2008)
Thiabendazole	6.5–1371	Filters (air)	Urban, rural	Coscollà et al. (2008)
Tolylfluanid	100–5870	PUF + filters	Urban, rural	Coscollà et al. (2010)
Triallate	<LD–36	PUF/XAD-2/PUF	Rural	Primbs et al. (2008)
Trifluralin	30–640	XAD-4 (air)	Rural	LeNoir et al. (1999)
Trifluralin	507–1935	PUF/XAD-2	Rural	Aulagnier et al. (2008)
Trifluralin	<LD–2	PUF/XAD-2/PUF	Rural	Primbs et al. (2008)
Vinclozolin	140–190	PUF + filters	Urban, rural	Coscollà et al. (2010)

Range covers minimum to maximum concentration detected.

LD, limit of detection (for more details, see references).

Unsworth et al. 1999; Bidleman et al. 2006; Hageman et al. 2006). Fewer studies have focused on pesticides in fog and snow (e.g., Hageman et al. 2006).

Except for persistent pesticides, the highest pesticide concentration, both in air and in rainwater, will generally be located in the immediate surroundings of the application area. In this respect, however, one must always keep in mind the importance not only of the meteorology (wind direction, wind speed, etc.), which can help distribute or dissipate the pesticide, but also of the physicochemical properties of the pesticide itself.

Moreover, studies on pesticides in the atmosphere indicate a seasonal behavior related to application periods and use patterns in areas within tens of kilometers of the sampling sites. This implies limited local transport (Scheyer et al. 2007b; Coupe et al. 2000). LRT is observed mainly in the case of persistent compounds, such as a large number of organochlorines.

The agricultural use of pesticides is well documented. By contrast, the urban use of pesticides, for example, their applications in houses, gardens, cemeteries, roadways, or industrial settings, are poorly documented. As a result, comparisons between urban and rural pesticide monitoring are usually complicated. In many cases, the distribution between air and precipitation in urban and rural areas is quite different. Most studies point to higher pesticide concentrations on agricultural or rural sites than in urban areas, as expected. Apart from this, there are also differences in the pesticides detected, except on urban sites close to the agricultural areas where the same pesticides are usually detected for both rural and urban sites (e.g., Peck and Hornbuckle 2005; Chevreuil et al. 1996; Shummer et al. 2010; see Tables 7.2 and 7.3).

The fate, behavior, and transport of pesticides in the atmosphere are strongly affected by the distribution of the compounds in the different phases: air, water, and particles. As commented in the Introduction section, a great number of pesticides can undergo photochemical transformations in relation to the presence and concentration of other pollutants, such as ozone or VOCs, which are typically present in higher concentrations on urban sites. The partitioning between the different phases of the atmosphere has an important influence on the removal rate and the LRT or medium-range transport of the pesticides (Schummer et al. 2010; Shen et al. 2005).

The presence of pesticides in urban air and rainwater is strongly related to atmospheric transport from rural areas of significant agricultural importance. The persistent pesticides, for example, the organochlorine pesticides, dacthal, and triazines (atrazine, cyanizine, etc.), represent a special case because of their well-known potential to undergo LRT (Wania et al. 2005; Yao et al. 2008; Daly et al. 2007; Chevreuil et al. 1996).

7.4.1 Pesticides in the Air

Air samples are collected with high-volume samplers or passive samplers. The particle phase and the gas phase can or cannot be sampled together, depending on the study. The most used system for the gas phase is Amberlite XAD™ adsorbent resins (XAD-2). Some studies used polyurethane foam cartridges (PUF) with XAD-2 or XAD-4 resin. The particle phase is usually sampled on glass fiber filters. For active sampling, the sampling time is usually 24 h. The treatment and analysis methods change from one study to another. Depending on the compounds, Gas chromatography coupled with Mass spectrometry (GC-MS) or Liquid chromatography coupled with mass spectrometry (LC-MS) can be chosen for the analysis. Sampling sites and the period of the year are very important factors to be considered because they can determine significant differences in the concentration levels detected (e.g., Scheyer et al. 2007a; Yao et al. 2008). A more detailed review of the air sampling methods used for pesticides can be found in Yusà et al. (2009).

Table 7.2 shows pesticide measurements from different sites around the world. It is indicative of the current worldwide interest in ambient air monitoring of pesticides, but it cannot be considered an exhaustive review of the measurements taken in recent years.

7.4.2 Pesticides in Rainwater

Precipitation samples are usually collected with wet-only precipitation collectors. In some cases, the samples are adsorbed on-line on XAD-2 resin columns during each precipitation event (Aulagnier et al. 2008). In other studies, solid phase extraction (SPE) cartridges are used to prepare the sample for analysis (Bossil et al. 2002; Vogel et al. 2008). In the last few years, solid phase microextraction (SPME) techniques have been developed for the sampling of pesticides in rainwater (i.e., Sauret-Szczepanski et al. 2006; Scheyer et al. 2007b). After the sample is treated, the extracts are analyzed by GC-MS or LC-MS.

In addition to the meteorological conditions and the geographical situation of the analyzed sampling sites, the physicochemical properties of the pesticides are also very important for rainwater sample analyses. For example, Scheyer et al. (2007b) observed that parathion-methyl was detected episodically in Strasbourg, and its low Henry's law constant (9.6×10^{-4} Pa m^3/mol) together with its good solubility in water (60 mg/L) permitted a good washout by precipitation. By contrast, alachlor, atrazine, and metolachlor were detected seasonally, and their low Henry's law constants and high water solubility enhanced their presence in rainwater, because pesticides with low Henry's law constants and high water solubility are mainly present in the particle phase and can be easily deposited by precipitation.

Table 7.3 shows some examples of the measurements carried out during the last few years. It is not an exhaustive data review, but, rather, it reflects the current interest and the importance of this kind of measurement in relation to air quality issues.

7.4.3 Measurements of Persistent Organic Pollutants

During the last 40 years, a great number of pesticides in the group of persistent organic pollutants (POPs) have been banned in North America and Europe. For example, in 2001 during the Stockholm Convention, more than 160 governments agreed to add lindane to the list of POPs to be banned for agricultural usage during the following years, although its use for pharmaceutical purposes was permitted until the end of 2015. Nevertheless, compounds of this type have half-lives of more than 2 days in air (Pacheco Ferreira 2008) and can move in the environment during long periods of time—depending on the half-life time of each compound—due to winds, wet and/or dry deposition, etc.

Within the US Integrated Atmospheric Deposition Network, 24 organochlorine pesticides classified as POPs have been measured since 1990 and evaluated to analyze their time trends in the Great Lakes region (Venier and Hites 2010). Gas and particle phases together with precipitation were sampled every year and analyzed using GC-MS. The authors found that endosulfans showed the slowest decreasing rate, between 11 and 14 years. For chlordane, they observed a halving time of around 6 years for the particle phase, 11 years for the gas phase, and nearly 4 years for the precipitation. Although they could not find any explanation for this great difference in the halving time for precipitation, they concluded that their results were consistent with the fact that chlordane, although banned in the United States in 1988, still had several reservoirs resulting from past uses, which continue to contaminate the atmosphere. These authors also reported that lindane showed the fastest decline rates in all phases, between 3 and 5 years, in comparison with the other organochlorine pesticides.

TABLE 7.3

Selected Pesticide and Degradation Product Concentrations in Rainwater

Pesticide in Rainwater	Range (µg/L)	Site	References
Acetochlor	0.006–4.39	Rural	Vogel et al. (2008)
Acetochlor ESA	0–0.282	Rural	Vogel et al. (2008)
Alachlor	0.005–0.038	Rural	Vogel et al. (2008)
Alachlor	<LQ–3.169	Urban (near crops)	Sauret-Szczepanski et al. (2006)
Alachlor ESA	0–0.07	Rural	Vogel et al. (2008)
Atrazine	0.014–6.58	Rural	Vogel et al. (2008)
Atrazine	<LQ–0.794	Urban (near crops)	Sauret-Szczepanski et al. (2006)
Atrazine	0.002–0.06	Rural	Bossil et al. (2002)
Atrazine	<LD–237	Rural	Aulagnier et al. (2008)
Atrazine	0.15–1.99	Rural	Trevisan et al. (1993)
Atrazine	<LQ–0.498	Rural, urban	Chevreuil et al. (1996)
Atrazine	0.006–0.096	Rural	Coupe et al. (2000)
Atrazine	0.036–1.031	Urban	Scheyer et al. (2007b)
Atrazine	<LQ–6.248	Rural	Scheyer et al. (2007b)
Azinphos-methyl	0.005–0.082	Rural	Vogel et al. (2008)
Azinphos-methyl	<LD–0.045	Rural	Aulagnier et al. (2008)
Benfluralin	0–0.006	Rural	Vogel et al. (2008)
Bromoxynil	<LD–0.091	Rural	Aulagnier et al. (2008)
Captan	<LD–0.225	Rural	Aulagnier et al. (2008)
Carbaryl	0.02–0.093	Rural	Vogel et al. (2008)
Carbaryl	<LD–0.0484	Rural	Aulagnier et al. (2008)
Carbofuran	<LD–0.0252	Rural	Aulagnier et al. (2008)
2-Chloro-2,6-diethylacetanilide	0–0.005	Rural	Vogel et al. (2008)
4-Chloro-2-methylphenol	0.006–0.059	Rural	Vogel et al. (2008)
Chlorpyrifos	0.005–1.84	Rural	Vogel et al. (2008)
Chlorpyrifos OA	0–0.014	Rural	Vogel et al. (2008)
Chlorothalonil	<LD–0.083	Rural	Aulagnier et al. (2008)
Chlorthal	0.0004–0.0011	Rural	Aulagnier et al. (2008)
Cymoxanil	<LQ–14.78	Urban (near crops)	Sauret-Szczepanski et al. (2006)
2,4-D	<LD–7	Rural	Aulagnier et al. (2008)
Dacthal	0.002–0.022	Rural	Vogel et al. (2008)
Deethyl atrazine	0–0.339	Rural	Vogel et al. (2008)
Deethyl atrazine	0–0.273	Rural	Vogel et al. (2008)
Deethyl atrazine	<LQ–0.498	Urban (near crops)	Sauret-Szczepanski et al. (2006)
Deethylterbuthylazine	0.038–0.525	Urban (near crops)	Sauret-Szczepanski et al. (2006)
Desulfinyl	0–3	Rural	Vogel et al. (2008)
Diazinon	0.005–1.2	Rural	Vogel et al. (2008)
Diazinon OA	0–0.118	Rural	Vogel et al. (2008)
Dicamba	<LD–0.003	Rural	Aulagnier et al. (2008)
Dichlorvos	0.012-0.517	Rural	Vogel et al. (2008)
Diflufenicanil	<LQ–2.823	Urban (near crops)	Sauret-Szczepanski et al. (2006)

(continued)

TABLE 7.3 (Continued)

Selected Pesticide and Degradation Product Concentrations in Rainwater

Pesticide in Rainwater	Range (mg/L)	Site	References
Diflufenicanil	<LQ–0.057	Urban	Scheyer et al. (2007b)
Diflufenicanil	<LQ–0.762	Rural	Scheyer et al. (2007b)
Dimethenamid	0.005–0.04	Rural	Vogel et al. (2008)
Dimethenamid	<LD–0.080	Rural	Aulagnier et al. (2008)
3,4-Dichloroaniline	0–0.033	Rural	Vogel et al. (2008)
Diuron	<LQ–1.025	Urban	Scheyer et al. (2007b)
Diuron	<LQ–1.317	Rural	Scheyer et al. (2007b)
EPTC	<LD–0.085	Rural	Aulagnier et al. (2008)
Fenoxaprop-p-ethyl	<LQ–0.3	Urban (near crops)	Sauret-Szczepanski et al. (2006)
Fipronil	0–0.017	Rural	Vogel et al. (2008)
Flufenacet	0–0.05	Rural	Vogel et al. (2008)
γ-HCH (Lindane)	0.014–0.35	Urban, rural	Chevreuil et al. (1996)
γ-HCH (Lindane)	<LQ–0.132	Urban	Scheyer et al. (2007b)
γ-HCH (Lindane)	<LQ–0.174	Rural	Scheyer et al. (2007b)
Iprodione	0.07–3.44	Rural	Vogel et al. (2008)
Iprodione	<LQ–0.124	Urban (near crops)	Sauret-Szczepanski et al. (2006)
Iprodione	0.111–0.56	Rural	Trevisan et al. (1993)
Iprodione	<LQ–5.59	Urban	Scheyer et al. (2007b)
Iprodione	<LQ–3.33	Rural	Scheyer et al. (2007b)
Isoproturon	0.768–1.273	Urban (near crops)	Sauret-Szczepanski et al. (2006)
Isoproturon	0.002–0.073	Rural	Bossil et al. (2002)
Malathion	0.014–0.087	Rural	Vogel et al. (2008)
Malathion OA	0–0.308	Rural	Vogel et al. (2008)
Mechlorprop	0.002–0.015	Rural	Bossil et al. (2002)
Methidathion	0–0.691	Rural	Vogel et al. (2008)
(2-Methyl-4-chlorophenoxy)acetic acid	0.002–0.032	Rural	Bossil et al. (2002)
(2-Methyl-4-chlorophenoxy)acetic acid	0.006–0.035	Rural	Aulagnier et al. (2008)
Methyl parathion	0–0.056	Rural	Vogel et al. (2008)
Methyl-paraoxon	0–0.037	Rural	Vogel et al. (2008)
Metolachlor	0.007–1.76	Rural	Vogel et al. (2008)
Metolachlor	<LQ–1.443	Urban (near crops)	Sauret-Szczepanski et al. (2006)
Metolachlor	<LD–0.644	Rural	Aulagnier et al. (2008)
Metolachlor	<LQ–0.122	Urban	Scheyer et al. (2007b)
Metolachlor	<LQ–0.799	Rural	Scheyer et al. (2007b)
Metribuzin	0–0.103	Rural	Vogel et al. (2008)
Myclobutanil	0–0.113	Rural	Vogel et al. (2008)
Myclobutanil	<LD–0.009	Rural	Aulagnier et al. (2008)
1-Naphthol	0–0.088	Rural	Vogel et al. (2008)
Parathion-methyl	0.024–0.3	Rural	Coupe et al. (2000)
Pendimethalin	0–0.484	Rural	Vogel et al. (2008)

TABLE 7.3 (Continued)

Selected Pesticide and Degradation Product Concentrations in Rainwater

Pesticide in Rainwater	Range (mg/L)	Site	References
Phosalone	<LD–0.010	Rural	Aulagnier et al. (2008)
Phosmet	0–0.047	Rural	Vogel et al. (2008)
Phosmet	<LD–0.062	Rural	Aulagnier et al. (2008)
Prometon	0–0.0122	Rural	Vogel et al. (2008)
Prometryn	0–0.031	Rural	Vogel et al. (2008)
Pronamide	0–0.017	Rural	Vogel et al. (2008)
Propachlor	0–0.163	Rural	Vogel et al. (2008)
Simazine	0–15.6	Rural	Vogel et al. (2008)
Terbuthylazine	0–0.014	Rural	Vogel et al. (2008)
Terbuthylazine	<LQ–1.06	Urban (near crops)	Sauret-Szczepanski et al. (2006)
Terbuthylazine	0.005–0.025	Rural	Bossil et al. (2002)
Trifluralin	0–0.148	Rural	Vogel et al. (2008)
Trifluralin	0.05–3.44	Rural	Trevisan et al. (1993)
Trifluralin	<0.002–0.01	Rural	Coupe et al. (2000)

Range covers minimum to maximum concentration detected.
LD, limit of detection (for more details, see references); LQ, limit of quantification (for more details, see references).

The question of whether some kinds of soils are significant long-term sources of persistent pesticides to the atmosphere lacks a consensus answer. Shunthirasingham et al. (2010) concluded that the arid soils of the subtropical regions cannot be considered either major reservoirs of pesticides or important long-term sources of pesticides. They investigated the concentration levels, spatial trends, and seasonal variability of pesticides in air and soil in Botswana. Even in areas that had been sprayed with DDT and endosulfan repeatedly in the past, these authors found low levels of pesticides in the gas phase, with α-endosulfan and lindane being the most abundant. In fact, they concluded that the spatial and seasonal patterns observed were more likely related to pesticide usage than to environmental factors or historical use.

In India, between December 2006 and March 2007, some authors (Chakraborty et al. 2010) studied samples from seven different cities and found the following mean concentration levels of organochlorine pesticides: 5400 pg/m^3 of Hexachlorocyclohexane (HCHs), 1470 pg/m^3 of DDTs, 1530 pg/m^3 of chlordanes, 1040 pg/m^3 of endosulfans, and 790 pg/m3 of hexachlorobenzene. These authors concluded that the data showed a decrease in the concentrations of HCHs and DDTs for most regions. Lindane dominated the HCHs due to its widespread use in India. In fact, lindane is one of the most studied and measured persistent compounds.

7.5 Fate in the Atmosphere

Pesticides behave in the atmosphere in the same way as other volatile and semivolatile organic compounds. As explained in the previous sections, pesticides can be distributed between the gas, particle, and aqueous phases, and this partitioning depends on both physicochemical properties such as equilibrium vapor pressures and Henry's law constants (Seiber 2002) and environmental conditions (e.g., temperature, wind direction, height of cloud base, etc.; Tsal and Cohen 1991; Wania et al. 1998). The processes of pesticide removal from the atmosphere will be different for the different phases. As most pesticides are considered semivolatile, they could exist in both the gas phase and the particle phase (Majewski and Capel 1995; Bidleman 1998).

The main route for the removal of pesticides in air is wet and/or dry deposition; however, the chemical reactions that can occur in the atmosphere must also be considered because, contrary to what most people think, chemical changes do not always result in the detoxification of the pesticide compounds.

Moreover, the fact that pesticides can be transported or dispersed at regional or LRT scale depending on their properties needs to be taken into account as well.

Recent pesticide measurements in Southern Africa and the subtropics (Shunthirasingham et al. 2010) show behaviors that seem to be influenced mainly by the patterns of use and not by meteorological factors or historical uses. By contrast, in northern countries (United States, Canada, Northern Europe), the measured pesticide concentrations are more influenced by meteorological factors than by other factors (i.e., Venier and Hites 2010). This is due to dispersion processes and the long- and medium-range transport of the pesticides from the south to the north. In these cases, more seasonality is observed.

7.5.1 Degradation in Air

The gas-phase degradation of pesticides in the atmosphere may be controlled by photolysis and/or reaction with ozone, OH, and NO_3 radicals (Atkinson 1994; Atkinson et al. 1999; Woodrow et al. 1983; Finlayson-Pitts and Pitts 1986; Seinfeld 1986).

Photolysis is a chemical process by which molecules are broken down into smaller units through the absorption of sunlight.

Ozone arrives in the troposphere by diffusion from the stratosphere, and it is also formed in the troposphere from the interaction of VOCs and NO_x in the presence of sunlight (Logan 1985; Roelofs and Lelieveld 1997). The 24 h average atmospheric concentration of ozone is 7×10^{11} molecule cm^3 (Logan 1985).

The OH radical is the most important reactive species in the troposphere; it reacts with all organic compounds, including pesticides. It is formed during daylight hours mainly by the photolysis of O_3 in the presence of water, although other sources of OH are also important, such as the photolysis of nitrous acid (HONO), the photolysis of formaldehyde and other carbonyls in the presence of NO, and the reactions of alkenes with O_3 (Atkinson 2000). The OH radical can also be formed during nighttime from the last-mentioned source. The average 12 h daytime concentration of OH radicals is 2×10^6 molecule cm^3 (Prinn et al. 2001).

NO_3 is formed in the troposphere by the reaction of NO_2 with O_3. Nevertheless, as it photolyses rapidly, its possible reactions are only important during nighttime. The average nighttime concentration of NO_3 radicals is 5×10^8 molecule cm^3 (Atkinson 1991).

A parameter widely used by atmospheric chemists to parameterize the persistence of a pollutant in air is atmospheric lifetime (τ). Atmospheric lifetime is the ratio between the time needed for a pesticide to disappear and the time $1/e$ of its initial value (e is the base of natural logarithms 2.718). The lifetime for each pesticide present in the gas phase due to gaseous removal can be derived from the expression (Equation 7.1)

$$\tau = 1/\left(J(\text{pest}) + 1/k_{O_3}(\text{pest})[O_3] + 1/k_{OH}(\text{pest})[OH] + 1/k_{NO_3}(\text{pest})[NO_3]\right) \quad (7.1)$$

where J(pest), k_{O_3}(pest), k_{OH}(pest), *and* k_{NO_3}(VOC) are the rate constants for the photolysis, reaction with ozone, hydroxyl radical, and nitrate radical for each pesticide, respectively, and $[O_3]$, $[OH]$, and $[NO]$ are the concentration values of these compounds previously specified.

Despite its importance, there are few studies on pesticide photodegradation in air (Atkinson et al. 1999). More studies on the photodegradation or photolysis of pesticides in liquid phase are available, as for methanol, hexane, or water, or on soil or leaf surfaces, as shown in other chapters of this book (i.e., Barceló et al. 1993; Bavcon Kralja et al. 2007; Burkhard and Guth 1979; Chukwudebe et al. 1989; Floesser-Mueller and Schwack 2001). Some examples of the degradation of pesticides in air are the studies on the degradation of chloropicrin (Vera et al. 2010), chlorpyrifos (Hebert et al. 2000a,b), chlorpyrifos-methyl (Muñoz et al. 2011), dichlorvos (Feigenbrugel et al. 2006), trifluraline (Le Person et al. 2007; Hebert et al. 2000a), tertbuthylazine (Palm et al. 1997), chlorotoluron, and isoproturon (Millet and Zetzsch 1998). In general terms, modern pesticides have lifetimes in air somewhere between 15 min and 2 days, whereas organoclorine pesticides have lifetimes of more than 2 days. Table 7.4 shows atmospheric lifetimes of selected pesticides in gas phase.

In most of these cases, the values of the constants are estimated using the structure–activity relationship (SAR) developed by Kwok and Atkinson (1995; EPIweb).

Because degradation products are expected to have considerably longer atmospheric lifetimes than pesticides, they will undergo LRT and their oxidation products will thus merit investigation. To date, degradation products are not well determined for a large number of pesticides; however, there are batches of pesticides for which degradation products are well known. This is the case of 1,1-Dichloro-2,2-bis(4-chlorophenyl)ethene and 2-(2-Chlorophenyl)-2-(4-chlorophenyl)-1,1-dichloroethene (DDEs), as products of the degradation of DDTs (e.g., Daly et al. 2007); phosgene, from the photolysis of chloropicrin (Vera et al. 2010) or from the photoxidation of dichlorvos (Feigenbrugel et al. 2006); and oxones, from the organothiophosphate pesticides (Muñoz et al. 2011), among others. Nevertheless, more studies focused on the degradation products of pesticides in air and their implications for human health and the environment are advisable.

TABLE 7.4

Lifetime of Selected Pesticides, Considering the Average 12 h Daytime Concentration of OH Radicals 2×10^6 Molecule cm^3

Pesticide	Lifetime	Main Degradation Process in Gas Phase	References
Chloropicrin	5.4 h	Photolysis	Vera et al. (2010)
Chlorpyrifos	2 h	Photolysis and OH reaction	Hebert et al. (2000a), Hebert et al. (2000b)
Chlorpyrifos-methyl	3.5 h	OH reaction	Muñoz et al. (2011)
Dichlorvos	6 h	OH reaction	Feigenbrugel et al. (2006)
α-hexachlorocyclohexane	41 days	OH reaction	Brubaker and Hites (1998)
Lindane	304 days	OH reaction	Brubaker and Hites (1998)
Malathion	1.8 h	OH reaction	Estimated from SAR method (Kwok and Atkinson 1995, EPIweb)
Propachlor	2.4 h	OH reaction	Estimated from SAR method (Kwok and Atkinson 1995, EPIweb)
Trifluralin	15 min	Photolysis	Le Person et al. (2007)

Source: From Prinn, R. G., Huang, J., Weiss, R. F., Cunnold, D. M., Fraser, P. J., Simmonds, P. G., McCulloch, A., Harth, C., Salameh, P., O'Doherty, S., Wang, R. H. J., Porter, L., Miller, B. R. 2001. Evidence for substantial variations of atmospheric hydroxyl radicals in the past two decades. *Science* 292: 1882–1888.

In the future, this lack of experimental studies on the atmospheric degradation of pesticides and their degradation products can be partially remedied by the use of large chambers, such as the EUPHORE (EUropean PHOtoREactor) facility (Vera et al. 2007). One advantage of large simulation chambers, such as those at EUPHORE with a volume of 200 m³, is that compounds with low vapor pressures can be introduced into the chamber in the gas phase. Furthermore, as these chambers use real sunlight to produce the reactions, it is possible to simulate almost real conditions without the meteorological influences of the real environment (Vera et al. 2010; Muñoz et al. 2011; Feigenbrugel et al. 2006).

7.5.2 Dry and Wet Deposition

Wet deposition is the uptaking of the pesticide at the Earth's surface by precipitation, mainly in rainwater, but also in fog and snow. Wet deposition is determined by the precipitation rate, the cloud and rain drop distribution, the Henry's law constant, the air–water partition coefficient, and the washout ratio for particles (Van Pul et al. 1999; FOCUS 2008 and references therein). Several studies have been published on the detection of pesticides in rainwater, as shown in the section Pesticides in Rainwater.

Dry deposition is the uptaking of the pesticide at the Earth's surface by water, soil, and vegetation. This means that dry deposition is affected by the nature of the receiving body. There are several studies on the dry deposition of selected pesticides, especially in relation to the terrestrial environment, that is, bentazon, dichlorprop-p, chlorothalonil, fenpropimorph, parathion-ethyl, and chlorothalonil in barley fields (Klöpel and Kördel 1997); chlorpyrifos in pine needles (Aston and Seiber 1997); and lindane, parathion, and pirimicarb in cereals and also in water (Siebers et al. 2003). Duyzer and Van Oss (1997) studied the uptake of a number of pesticides by water, soil, and vegetation. They found that the most important parameters for describing dry deposition and further re-emission are the Henry's law constant, the partitioning coefficients of the pesticide in octanol–water (K_{ow}), soil organic carbon–water (K_{oc}), and octanol–air (K_{oa}), the leaf area index (LAI), and the organic carbon content of the soil.

Pesticides adsorbed to aerosols are found mainly in wet deposition (Unsworth et al. 1999), while pesticides in the vapor phase are probably more evenly divided between wet and dry depositions.

Deposition is not always the last step in the elimination of persistent pesticides from the atmosphere, because under certain conditions they could be re-emitted into the atmosphere (Bidleman 1999).

7.5.3 Transport

Once in the atmosphere, the pesticide is dispersed and transported like any other air pollutant, for example, ozone or SO_2 (Van Pul et al. 1999). The actual distance traveled by an air pollutant strongly depends on the amount of time it resides in the atmosphere. There are several models used worldwide for describing the transport and deposition of pesticides through the air (FOCUS 2008 and references there in). According to the FOCUS air group, the EVA model is one of the most appropriate for estimating transport and deposition at short-range distances, whereas the CalTox, CEMC Level III, Chemrange, ELPOS, and Simple Box are the best models for medium and long ranges. Nevertheless, other models are also valid.

With respect to pesticides transported in the particulate phase, on the lower troposphere, they will have LRT capability if the sizes of their aerosol particles are below 1 μm. Because larger-sized particles sediment, they have only a short-range transport capability.

In the case of pesticides in the gas phase, the potential for their long- or short-range transport will depend on their lifetimes, as described in the section Degradation in Air and calculable from Table 7.1.

An illustrative example of the transport potential of the pesticides in the gas phase is shown (see Figure 7.2).

For this example, we used the mesoscale meteorological model, Regional Atmospheric Modeling System (RAMS version 6.0; Pielke et al. 1992), coupled to the RAMS HYbrid Particle and Concentration Transport model (HYPACT version 1.5.0; Tremback et al. 1993). The RAMS model is quite flexible and includes different options for parameterizing the physical processes. In this study, a previously checked and validated configuration for the Mediterranean region was employed (Palau et al. 2005; Pérez-Landa et al. 2007).

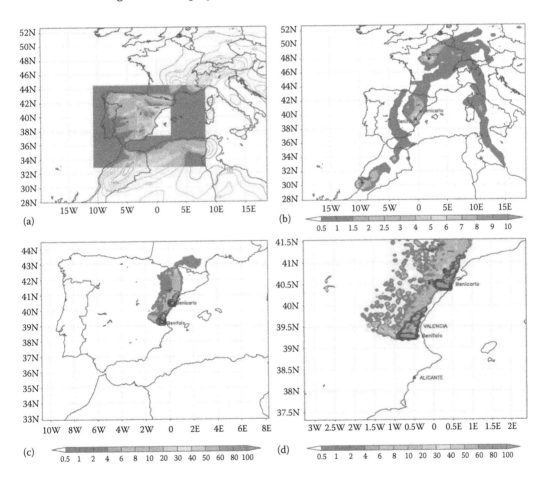

FIGURE 7.2
Illustrative example of potential transport of pesticides in Western European and North African areas, considering only 4 points of emission over the period 19–22 May 2009. (a) Three two-way nested domains centered over the Valencia region. (b) Long-range transport of pesticides with lifetime >2 days. (c) Short- and medium-range transport of pesticides with lifetimes <8 and <2 h through the Iberian Peninsula. (d) Short- and medium-range transport of pesticides with lifetimes <8 and <2 h through the eastern coast of the Iberian Peninsula.

Simulations of the RAMS coupled to the HYPACT model were performed on 19, 20, 21, and 22 May 2009, to check the effect of meteorological processes on the simulation of pollutant fate and dispersion in Western European and North African areas. For this, we used three two-way nested domains centered over the Valencia region, as shown in Figure 7.2a, with horizontal grid sizes of: G1, (grid one) 54 km; G2, 13.5 km; and G3 4.5 km. The HYPACT model was run using the Lagrangian dispersion scheme with 3-D wind and turbulence fields provided by the three RAMS outputs. This model was employed to both simulate pesticide atmospheric dispersion emitted "only" from four different areas around the Western Mediterranean and analyze the long-, medium-, and short-range transport of pesticides according to their lifetimes (more than 10 days, less than 8 h, and less than 2 h).

Emission areas were defined as four identical boxes measuring 1×1 km^2 and 200 m high (with their center at 100 m above ground level). Each box was located at a different area: Orleans (France), Benicarló and Benifaio (Spain), and Agadir (Morocco). The atmospheric dispersion was simulated as emitting a mass of 1 ton per day at a variable rate for 96 h. The emission rates were identical for the four sources and were parameterized for each source as follows: 10% of the total daily emission in the period from 00:00 to 7:00 UTC; 80% of the total daily emission between 7:00 and 18:00 UTC; and the remaining 10% between 18:00 and 23:59 UTC.

Rough estimates were used for emission rates and temporal evolution because the only aim of this simulation was to describe the broadness and differences in the fate and spatial distribution of the pesticides (depending on their lifetimes) in areas around the Western Mediterranean under typical late-spring meteorological conditions.

To show the fate of persistent pesticides (lifetime longer than 2 days), Figure 7.2b shows their LRT for the selected period. Shaded colors indicate the spatial distribution of pesticides emitted simultaneously from the four sources 58 h after the start of emissions. Concentrations correspond to the total mass of pesticide per unit volume within the first 8000 m above ground level.

To show the medium- and short-range transport during the selected time period, Figure 7.2c and 7.2d show the pesticide distribution throughout the Iberian Peninsula and on its Mediterranean side, respectively. The shaded color indicates the spatial distribution for pesticides with a lifetime of 8 h, while the contour lines indicate spatial distributions for pesticides with a lifetime of 2 h. Concentrations correspond to total pesticide mass per unit volume within the first 2000 m above ground level.

This modeling exercise confirms the importance of pesticides as a global problem. The amount of pesticides applied in one region changes continuously, as do the meteorological conditions and the mixtures applied. Furthermore, a pesticide applied in one region can affect the air of another region near or far from the application site. In Figure 7.2b, the application of pesticides with lifetimes of more than 2 days (persistent pesticides) in Morocco affects the air in Spain, and the application of pesticides in Spain affects the air of Western countries. (not only France, but also Italy). Therefore, forbidding the use of certain types of pesticides, especially the persistent ones, should be globally accepted in both developed and underdeveloped countries because of the potential of these pesticides to travel far from their emission points, as is shown in Figure 7.2 and also explained in Section 7.5.3.

On the other hand, the persistent pesticides are not the only ones capable of disturbing other regions. Figure 7.2c and 7.2d show the effect of applying pesticides with short and medium lifetimes in areas not so close to the application sites. It has also been pointed out that the detection of pesticides in mountain areas—without agricultural sites—could be due to the transport of the pesticides across regions and even across countries.

References

Altschuh, J., Bruggemann, R., Santl, H., Eichinger, G., and Piringer, O. G. 1999. Henry's law constants for a diverse set of organic chemicals: Experimental determination and comparison of estimation methods. *Chemosphere* 39: 1871–1887.

Asman, W., Jorgensen, A., and Jensen, P. K. 2003. Dry deposition and spray drift of pesticides nearby water bodies. Pesticide Research Nr 66 2003. Danish Environmental Protection Agency.

Aston, L. S. and Seiber, J. N. 1997. Fate of summertime airborne organophosphate pesticide residues in the Sierra Nevada Mountains. *J. Environ. Qual.* 26: 1483–1492.

Atkinson, R. 1991. Kinetics and mechanisms of the gas-phase reactions of the NO_3 radical with organic compounds. *J. Phys. Chem. Ref. Data* 20: 459–507.

Atkinson, R. 1994. Monograph 2. *J. Phys. Chem. Ref. Data* 1–216.

Atkinson, R. 2000. Atmospheric chemistry of VOCs and NO_x. *Atmos. Environ.* 34: 2063–2101.

Atkinson, R., Guicherit, R., Hites, R. A., Palm, W. U., Seiber, J. N., and de Voogt, P. 1999. Transformations of pesticides in the atmosphere: A state of the art. *Water Air Soil Pollut.* 115: 219–243.

Augustijn-Beckers, P. W. M., Hornsby, A. G., and Wauchope, R. D. 1994. The SCS/ARS/CES pesticide properties database for environmental decision making II. Additional compounds. *Rev. Environ. Contam. Toxicol.* 137: 1–82.

Aulagnier, F., Poissant, L., Brunet, D., Beauvais, C., Pilote, M., Deblois, C., and Dassylva, N. 2008. Pesticides measured in air and precipitation in the Yamaska Basin (Québec): Occurrence and concentrations in 2004. *Sci. Total Environ.* 394: 338–348.

Barceló, D., Durand, G., and De Bertrand, N. 1993. Photodegradation of the organophosphorus pesticides chlorpyrifos, fenamiphos and vamidothion in water. *Toxicol. Environ. Chem.* 38: 183–199.

Bavcon Kralja, M., Franko, M., and Trebse, P. 2007. Photodegradation of organophosphorus insecticides – Investigations of products and their toxicity using gas chromatography-mass spectrometry and AChE-thermal lens spectrometric bioassay. *Chemosphere* 67: 99–107.

Bedos, C., Cellier, P., Calver, R., Barriuso, E., and Gabrielle, B. 2002. Mass transfer of pesticides into the atmosphere by volatilization from soils and plants: Overview. *Agronomie* 22: 21–33.

Bidleman, T. F. 1998. Atmospheric processes: Wet and dry deposition of organic compounds are controlled by their vapor-particle partitioning. *Environ. Sci. Technol.* 22: 361–367.

Bidleman, T. F. 1999. Atmospheric transport and air–surface exchange of pesticides. *Water Soil Air Pollut.* 115: 115–166.

Bidleman, T. F., Leone, A. D., Wong, F., Van Vliet, L., Szeto, S., and Ripley, B. D. 2006. Emission of legacy chlorinated pesticides from agricultural and orchard soils in the British Columbia, Canada. *Environ. Toxicol. Chem.* 25: 1448–1457.

Bird, S. L., Perry, S. G., Ray, S. L., and Teske, M. E. 2002. Evaluation of the AgDISP aerial spray algorithms in the AgDRIFT model. *Environ. Toxicol. Chem.* 21(3): 672–681.

Bossil, R., Vejrup, K. V., Mogensen, B. B., and Asman, W. A. H. 2002. Analysis of polar pesticides in rainwater in Denmark by liquid chromatography-tandem mass spectrometry. *J. Chromatogr. A* 957: 27–36.

Bradley, A., Wofford, P., Gallava, R., Lee, P., and Troiano, J. 1997. Environmental Monitoring Results of the Mediterranean Fruit Fly Eradication Program, Ventura County, 1994–1995. EH97-05.

Brubaker, W. W. and Hites, R. A. 1998. OH reaction kinetic of gas-phase α- and γ-hexachlorocyclohexane and hexachlorobenzene. *Environ. Sci. Technol.* 32: 766–769.

Burkhard, N. and Guth, J. A. 1979. Photolysis of organophosphorus insecticides on soil surfaces. *Pestic. Sci.* 10: 313–319.

Burns, C. J., Collins, J. J., Budinsky, R. A., Bodner, K., Wilken, M., Rowlands, J. C., Martin, G. D., and Carson, M. L. 2008. Factors related to dioxin and furan body levels among Michigan workers. *Environ. Res.* 106: 250–265.

Chakraborty, P., Zhang, G., Li, J., Xu, Y., Liu, X., Tanabe, S., and Jones, K. C. 2010. Selected organo-chlorine pesticides in the atmosphere of major Indian cities: Levels, regional versus local variations, and sources. *Environ. Sci. Technol.* 44: 8038–8043.

Chevreuil, M., Garmouma, M., Teil, M. J., and Chesterikoff, A. 1996. Occurrence of organochlorines (PCBs, pesticides) and herbicides (triazines, phenylureas) in the atmosphere and in the fallout from urban and rural station of the Paris area. *Sci. Total Environ.* 182: 25–37.

Chukwudebe, A., March, R. B., Othman, M., and Fukuto, T. R. 1989. Formation of trialkyl phosphoro-thioate esters from organophosphorus insecticides after exposure to either ultraviolet light or sunlight. *J. Agric. Food Chem.* 37: 539–545.

Coscollà, C., Yusà, V., Martí, P., and Pastor, A. 2008. Analysis of currently used pesticides in fine air-borne particulate matter (PM 2.5) by pressurized liquid extraction and liquid chromatography-tandem mass spectrometry. *J. Chromatogr. A* 1200: 100–107.

Coscollà, C., Colin, P., Yahyaoui, A., Petrique, O., Yusà, V., Mellouki, A., and Pastor, A. 2010. Ocurrence of currently used pesticides in ambient air of Centre Region (France). *Atmos. Environ.* 44: 3915–3925.

Coupe, R. H., Manning, M. A., Foreman, W. T., Goolsby, D. A., and Majewski, M. S. 2000. Occurrence of pesticides in rain and air in urban and agricultural areas of Mississippi, April–September 1995. *Sci. Total Environ.* 248: 227–240.

Daly, G. L., Lei, Y. D., Teixeira, C., Muir, D. C. G., Castillo, L. E., and Wania, F. 2007. Accumulation of current-use pesticides in neotropical montane forest. *Environ. Sci. Technol.* 41: 1118–1123.

Duyzer, J. H. and Van Oss, R. F. 1997. Determination of deposition parameters of a number of organic pollutants by laboratory experiments. TNO Report R97/150, Apeldoorn, The Netherlands.

EPIweb 4.0. http://www.epa.gov/opptintr/exposure/pubs/episuitedl.htm.

Feigenbrugel, V., Le Person, A., Le Calvé, S., Mellouki, A., Muñoz, A., and Wirtz, K. 2006. Atmospheric fate of dichlorvos. *Environ. Sci. Technol.* 40: 850–857.

Fendinger, N. J. and Glotfelty, D. E. 1990. Henry's law constants for selected pesticides, PAHs and PCBs. *Environ. Toxicol. Chem.* 9: 731–735.

Ferrari, F., Trevisan, M., and Capri, E. 2003. Predicting and measuring environmental concentration of pesticides in air after soil application. *J. Environ. Qual.* 32: 1623–1633.

Finlayson-Pitts, B. J. and Pitts, J. N. Jr. 1986. *Atmospheric Chemistry. Fundamentals and Experimental Techniques.* Wiley: New York.

Floesser-Mueller, H. and Schwack, W. 2001. Photochemistry of organophosphorus insecticides. *Rev. Environ. Contam. Toxicol.* 172: 129–228.

FOCUS. 2008. Pesticides in air: Considerations for Exposures Assessment. *Report on the FOCUS Working Group on Pesticides in Air*, 327 pp. EC Document Reference SANCO/10553/2006 Rev, 2 June 2008.

Foreman, W. T., Majewski, M. S., Goolsby, D. A., Wiebe, F. W., and Coupe, R. H. 2000. Pesticides in the atmosphere of the Mississippi river valley, part II – Air. *Sci. Total Environ.* 248: 213–226.

Garratt, J. R. 1999. *The Atmospheric Boundary Layer.* Cambridge University Press: Australia.

Gil, Y. and Sinfort, C. 2005. Emission of pesticides to the air during sprayer application: A bibliographic review. *Atmos. Environ.* 39: 5183–5193.

Gil, Y., Sinfort, C., Brunet, Y., Polveche, V., and Benicelli, B. 2007. Atmospheric loss of pesticides above an artificial vineyard during spray-assisted spraying. *Atmos. Environ.* 41: 2945–2957.

Hageman, K. J., Simonich, S. L., Campbell, D. H., Wilson, G. R., and Landers, D. H. 2006. Atmospheric deposition of current-use and historic-use pesticides in snow at National Parks in the Western United States. *Environ. Sci. Technol.* 40: 3174–3180.

Hansch, C., Leo, A., and Hoekman, D. 1995. *Exploring QSAR. Hydrophobic, Electronic, and Steric Constants. ACS Professional Reference Book.* American Chemical Society: Washington, DC.

Hebert, V. R., Hoonhout, C., and Miller, G. C. 2000a. Use of stable tracer studies to evaluate pesticide photolysis at elevated temperatures. *J. Agric. Food Chem.* 48: 1916–1921.

Hebert, V. R., Hoonhout, C., and Miller, G. C. 2000b. Reactivity of certain organophosphorus insecti-cides toward hydroxyl radicals at elevated air temperatures. *J. Agric. Food Chem.* 48: 1922–1928.

Holton, J. R. 1990. *An Introduction to Dynamic Meteorology.* Elsevier Academic Press: New York.

Jarvis, N. J. 1994. The MACRO model (version 3.1) technical description and sample simulations. *Rep. and Diss. 19*. Department of Soil Science, Swedish University of Agricultural Sciences, Uppsala.

Kawamoto, K. and Urano, K. 1989. Parameters for predicting fate of organochlorine pesticides in the environment. (I) Octanol–water and air–water partition coefficients. *Chemosphere* 18: 1987–1996.

Klein, M. 1995. *PELMO: Pesticide Leaching Model Version 2.01*. Fraunhofer Institut für Umweltchemie und Ökotoxikologie: Schmallenberg, Germany.

Klöpel, H. and Kördel, W. 1997. Pesticide volatilization and exposure of terrestrial ecosystems. *Chemosphere* 35: 1271–1289.

Kwok, E. S. C. and Atkinson, R. 1995. Estimation of hydroxyl radical rate constants for gas-phase organic compounds using a structure–reactivity relationship: An update. *Atmos. Environ.* 29: 1685–1695.

Le Person, A., Mellouki, A., Muñoz, A., Borras, E., Martin-Reviejo, M., and Wirtz Trifluralin, K. 2007. Photolysis under sunlight conditions and reaction with HO radicals. *Chemosphere* 67: 376–383.

Leistra, M., Smelt, J. H., Weststrate, J. H., van der Berg, F., and Aalderink, R. 2006. Volatilization of the pesticides chlorpyrifos and fenpropimorph from a potato crop. *Environ. Sci. Technol.* 40: 96–102.

LeNoir, J. S., McConnell, L. L., Fellers, G. M., Cahill, T. M., and Seiber, J. N. 1999. Summertime transport of current-use pesticides from California's central valley to the Sierra Nevada mountain range, USA. *Environ. Toxicol. Chem.* 12: 2715–2722.

Logan, J. A. 1985. Tropospheric ozone: Seasonal behavior, trends, and anthropogenic influence. *J. Geophys. Res.* 90: 10463–10482.

Lopez-Espinosa, M. J., Lopez-Navarrete, E., Rivas, A., Fernandez, M. F., Nogueras, M., Campoy, C., Olea-Serrano, F., Lardelli, P., and Olea, N. 2008. Organochlorine pesticide exposure in children living in southern Spain. *Environ. Res.* 106: 1–6.

Majewski, M. S. and Capel, P. D. 1995. *Pesticides in the Atmosphere, Distribution, Trends, and Governing Factors*, 214 pp. Ann Arbor Press: Chelsea, MI.

Merck Index. 1996. *The Merck Index*, 12th edn. Merck: New York. ISBN 0911910-12-3.

Metcalfe, C. D., McLeese, D. W., and Zitko, V. 1980. Rate of volatilization of fenitrothion from fresh water. *Chemosphere* 9: 151–155.

Millet, M. and Zetzsch, C. 1998. Investigation of the photochemistry of urea herbicides (chlorotoluron and isoproturon) and quantum yields using polychromatic irradiation. *Environ. Toxicol. Chem.* 17: 258–264.

Muñoz, A., Vera, T., Sidebottom, H., Mellouki, A., Borrás, E., Ródenas, M., Clemente, E., and Vázquez, M. 2011. Studies on the atmospheric degradation of chlorpyrifos-methyl. *Environ. Sci. Technol.* 45: 1880–1886.

Nielsen, L. S., Bundgaard, H., and Falch, E. 1992. Prodrugs of thiabendazole with increased water-solubility. *Acta Pharm. Nord.* 4: 43–49.

Orlanski, J. 1975. A rational subdivision of scales for atmospheric processes. *Bull. Am. Meteorol. Soc.* 56: 527–530.

Pacheco Ferreira, A. 2008. Environmental fate of bioaccumulative and persistent substances – A synopsis of existing and future actions. *Rev. Gerenc. Polit. Salud.* 7(15): 14–23.

Palau, J. L., Pérez-Landa, G., Dieguez, J. J., Monter, C., and Millan, M. M. 2005. The importance of meteorological scales to forecast air pollution scenarios on coastal complex terrain. *Atmos. Chem. Phys.* 5: 2771–2785.

Palm, W. U., Elend, M., Krueger, H. U., and Zetzsch, C. 1997. OH-radical reactivity of airborne terbuthylazine adsorbed on inert aerosol. *Environ. Sci. Technol.* 31: 3389–3396.

Peck, A. M. and Hornbuckle, K. C. 2005. Gas-phase concentrations of current-use pesticides in Iowa. *Environ. Sci. Technol.* 39: 2952–2959.

Pérez-Landa, G., Ciais, P., Sanz, M. J., Gioli, B., Miglietta, F., Palau, J. L., Gangoiti, G., and Millán, M. M. 2007. Mesoscale circulations over complex terrain in the Valencia coastal region, Spain – Part 1: Simulation of diurnal circulation regimes. *Atmos. Chem. Phys.* 7: 1835–1849.

Pielke, R. A., Cotton, W. R., Walko, R. L., Tremback, C. L., Lyons, W. A., Grasso, D., Nichols, M. E., Moran, M. D., Wesley, D. A., Lee, T. L., and Copelands, J. H. 1992. A comprehensive meteorological modeling system-RAMS. *Meteorol. Atmos. Phys.* 49: 69–91.

Primbs, T., Wilson, G., Schmedding, D., Higginbotham, C., and Simonich, S. M. 2008. Influence of Asian and Western United States agricultural areas and fires on the atmosphric transport of pesticides in the Western United States. *Environ. Sci. Technol.* 42: 6519–6525.

Prinn, R. G., Huang, J., Weiss, R. F., Cunnold, D. M., Fraser, P. J., Simmonds, P. G., McCulloch, A., Harth, C., Salameh, P., O'Doherty, S., Wang, R. H. J., Porter, L., and Miller, B. R. 2001. Evidence for substantial variations of atmospheric hydroxyl radicals in the past two decades. *Science* 292: 1882–1888.

Ramaprasad, J., Tsai, M. Y., Elgethun, K., Hebert, V. R., Felsot, A., Yost, M. G., and Fenske, R. A. 2004. The Washington aerial spray drift study: Assessment of off-target organophosphorus insecticide atmospheric movement by plant surface volatilization. *Atmos. Environ.* 38: 5703–5713.

Rice, C. P. and Chernyak, S. M. 1995. Determination of Henry's law constants of halogenated current-used pesticides as a function of environmental conditions. In: *Organohalogen Compounds*, vol. 24, pp. 439–444. Dioxin 95, 15th International Symposium.

Roelofs, G. J. and Lelieveld, J. 1997. Model Study of the influence of cross-tropopause IO3 transports on tropospheric O3 levels. *Tellus B* 49: 38–55.

Sanders, J. S. 1997. Environmental Monitoring Results of the Mediterranean Fruit Fly Eradication Program, Ventura County, 1994–95. California Environmental Protection Agency, Department of Pesticide Regulation, Sacramento, California.

Sangster, J. 1994. LOGKOW Databank. A databank of evaluated octanol–water partition coefficients (Log P). *Sangster Research Laboratories*. http://logkow.cisti.nrc.ca/logkow/index.jsp.

Sauret-Szczepanski, N., Mirabel, P., and Wortham, H. 2006. Development of SPME – GC-MS/MS method for the determination of pesticides in rainwater: Laboratory and field experiments. *Environ. Pollut.* 139: 133–142.

Scheyer, A., Graeff, C., Morville, S., Mirabel, P., and Millet, M. 2005. Analysis of some organochlorine pesticides in an urban atmosphere (Strasbourg, east of France). *Chemosphere* 58: 1517–1524.

Scheyer, A., Morville, S., Mirabel, P., and Millet, M. 2007a. Variability of atmospheric pesticide concentrations between urban and rural areas during intensive pesticide application. *Atmos. Environ.* 41: 3604–3618.

Scheyer, A., Morville, S., Mirabel, P., and Millet, M. 2007b. Pesticides analysed in rainwater in Alsace region (Eastern France): Comparison between urban and rural sites. *Atmos. Environ.* 41: 7241–7252.

Segawa, R. T., Sitts, J. A., White, J. H., Marade, S. J., and Powell, S. J. 1991. Environmental monitoring of malathion aerial applications used to eradicate Mediterranean fruit flies in Southern California, 1990. California Environmental Protection Agency, Department of Pesticide Regulation, Sacramento, California.

Seiber, J. N. 2002. Environmental fate of pesticides. In: Wheeler, W. B. (ed.), *Pesticides in Agriculture and the Environment*, pp. 127–161. Marcel Dekker Inc. New York.

Seinfeld, J. H. 1986. *Atmospheric Chemistry and Physics of Air Pollution*. Wiley: New York.

Shen, L., Wania, F., Lai, Y. D., Teixeira, C., Muir, D. C. G., and Bidleman, T. F. 2005. Atmospheric distribution and long-range transport behavior of organochlorine pesticides in North America. *Environ. Sci. Technol.* 39: 409–420.

Shummer, C., Mothiron, E., Appenzeller, B. M. R., Rizet, A.-L., Wennig, R., and Millet, M. 2010. Temporal variations of concentration of currently used pesticides in the atmosphere of Strasbourg, France. *Envion. Pollut.* 158: 576–584.

Shunthirasingham, C., Mmereki, B. T., Masamba, W., Oyiliagu, C. E., Lei, Y. D., and Wania, F. 2010. Fate of pesticides in the arid subtropics, Botswana, Southern Africa. *Environ. Sci. Technol.* 44: 8082–8088.

Siebers, J. N. and McChesney, M. M. 1988. Measurement and computer model simulation of the volatilisation flux of molinate and methyl parathion from a flooded rice field. In: *California, Department of Pesticide Regulation, Surface Water Environmental Monitoring Reports.* http://www.cdpr.ca.gov/docs/sw/swemreps.htm.

Siebers, J. N., Binner, R., and Wittich, K. P. 2003. Investigation of downwind short-range transport of pesticides after application in agricultural crops. *Chemosphere* 51: 397–407.

Smit, A. A. M. F. R., Leistra, M., and van der Berg, F. 1998. Estimation method for the volatilization of pesticides from plants. *Environmental Planning Bureau Series 2*. DLO Winand Staring Centre: Wageningen, The Netherlands.

Stewart, R. W. 1979. The Atmospheric Boundary Layer. World Meteorological Organization, No. 523.

Stork, A., Ophoff, H., Smelt, J. H., and Fuhr, F. 1998. Volatilization of pesticides: Measurements under simulated field conditions. Developed from a Symposium at the *213th National Meeting of the American Chemical Society*, pp. 21–39, San Francisco, CA, 3–17 April 1997.

Stull, R. B. 1989. *Boundary Layer Meteorology*. Kluwer Academic Publishers: Holland.

Tomlin, C. 1994. *The Pesticide Manual: A World Compendium*, 10th edn. Crop Protection Publications; British Crop Protection Council, Farnham, Survey GU9 7PH, United Kingdom.

Tomlin, C. 1997. *The Pesticide Manual*, 11th edn. Crop Protection Publications; British Crop Protection Council, Farnham, Survey GU9 7PH, United Kingdom.

Tomlin, C. 2003. *The Pesticide Manual*, 13th edn. Crop Protection Publications; British Crop Protection Council, Farnham, Survey GU9 7PH, United Kingdom.

Tremback, C. J., Lyons, W. A., Thorson, W. P., and Walko, R. L. 1993. An emergency response and local weather forecasting software system. In: *Proceedings of the 20th ITM on Air Pollution and Its Application*, 18th edn., pp. 423–429. Plenum Press: New York.

Trevisan, M., Montepiani, C., Ragozza, L., Bartoletti, C., Ionnilli, E., and Del Re, A. A. M. 1993. Pesticides in rainfall and air in Italy. *Environ. Pollut.* 80: 31–39.

Tsal, W. and Cohen, Y. 1991. Dynamic partitioning of semivolatile organics in gas/particle/rain phases during rain scavenging. *Environ. Sci. Technol.* 25: 2012–2023.

Unsworth, J. B., Wauchope, R. D., Klein, A. W., Dorn, E., Zeeh, B., Yeh, S. M., Akerblim, M., Racke, K. D., and Rubin, B. 1999. Significance of long range transport of pesticides in the atmosphere. *Pure Appl. Chem.* 71: 1359–1383.

USDA Pesticide Properties DataBase. http://www.ars.usda.gov/services/docs.htm?docid=14199. Last Modified: 11/06/2009.

Van den Berg, F., Kubiak, R., Benjey, W. G., Majewski, M. S., Yates, S. R., Reeves, G. L., Smelt, J. H., and Van Der Linden, A. M. A. 1999. Emission of pesticides into the air. *Water Air Soil Pollut.* 115: 195–218.

Van Pul, W. A. J., Bidleman, T. F., Brorström-Lundén, E., Builtjes, J. H., Dutchak, S., Duyzer, J. H., Gryning, S., Jones, K. C., Vand Dijk, H. F. G., and Van Jaarsveland, H. A. 1999. Atmospheric transport and deposition of pesticides: An assessment of current knowledge. *Water Air Soil Pollut.* 115: 245–256.

Vanclooster, M., Boesten, J. J. T. I., Trevisan, M., Brown, C. D., Capri, E., Eklo, O. M., Gottesbüren, B., Gouy, V., and van der Linden, A. M. A. 2000. A European test of pesticide-leaching models: Methodology and major recommendations. *Agric. Water Manag.* 44: 1–19.

Venier, M. and Hites, R. A. 2010. Time trend analysis of atmospheric POPs concentrations in the Great Lakes region since 1990. *Environ. Sci. Technol.* 44: 8050–8055.

Vera, T., Muñoz, A., Mellouki, A., Rodenas, M., and Vazquez, M. 2007. The use of EUPHORE facility for studying the atmospheric fate of pesticides. In: *XIII Symposium Pesticide Chemistry – Environmental Fate and Human Health*, Piacenza, Italy.

Vera, T., Muñoz, A., Ródenas, M., Vázquez, M., Mellouki, A., Treacy, J., Al Mulla, I., and Sidebottom, H. 2010. Photolysis of trichloronitromethane (chloropicrin) under atmospheric conditions. *Z. Phys. Chem.* 224: 1039–1057.

Vogel, J. R., Majewski, M. S., and Capel, P. D. 2008. Pesticides in rain in four agricultural watersheds in the United States. *J. Environ. Qual.* 37: 1101–1115.

Wania, F., Axelman, J., and Broman, D. 1998. A review of processess involved in the exchange of persistent organic pollutants across the air sea interface. *Environ. Pollut.* 102: 3–24.

Wania, S. L., Lei, Y. D., Teixeira, C., Muir, D. C. G., and Bidleman, T. 2005. Atmospheric distribution and long-range transport behaviour of organochlorine pesticides in North America. *Environ. Sci. Technol.* 39: 409–420.

Wauchope, R. D., Buttler, T. M., Hornsby, A. G., Augustijn-Beckers, P. W. M., and Burt, J. P. 1991. SCS/ARS/CES pesticide properties database for environmental decision making. *Rev. Environ. Contam. Toxicol.* 123: 1–35.

Wilson, N. K., Chuang, J. C., Morgan, M. K., Lordo, R. A., and Sheldon, L. S. 2007. An observational study of the potential exposures of preschool children to pentachlorophenol, biphenol-A, and nonylphenol at home and daycare. *Environ. Res.* 103: 9–20.

Wolters, A., Leistra, M., Linnemann, V., Klein, M., Schäffer, A., and Vereecken, H. 2004. Pesticide volatilization from plants: Improvement of the PEC model PELMO based on a boundary-layer concept. *Environ. Sci. Technol.* 38: 2885–2893.

Woodrow, J. E. and Seiber, J. N. 1997. Correlation techniques for estimating pesticide volatilization flux and downwind concentrations. *Environ. Sci. Technol.* 31: 523–529.

Woodrow, J. E., Crosby, D. G., and Seiber, J. N. 1983. Vapor-phase photochemistry of pesticides. *Residue Rev.* 85: 111–125.

Yalkowsky, S. H. and Dannenfelser, R. M. 1990. *AQUASOL Database of Aqueous Solubility*, 5th edn. University of Arizona, College of Pharmacy: Tucson, AZ.

Yao, Y., Galarneau, E., Blanchard, P., Alexandrou, N., Brice, D. A., and Li, Y. 2007. Atmospheric atrazine at Canadian IADN sites. *Environ. Sci. Technol.* 41: 7639–7644.

Yao, Y., Harner, T., Blanchard, P., Tuduri, L., Waite, D., Poissant, L., Murphy, C., Belzer, W., Aulagnier, F., and Sverki, E. 2008. Pesticides in the atmosphere across Canadian agricultural regions. *Environ. Sci. Technol.* 42: 5931–5937.

Yusà, V., Coscollà, C., Mellouki, W., Pastor, A., and de la Guardia, M. 2009. Sampling and analysis of pesticides in ambient air. *J. Chromatogr.* A 1216: 2972–2983.

8

Response of Soil Microflora to Pesticides

Mariusz Cycoń and Zofia Piotrowska-Seget

CONTENTS

8.1 Introduction

In modern agriculture, large quantities of pesticides have been used to control pests and weeds and thus increase food production. However, their wide and extensive application and a potential risk that they pose to the soil ecosystem raise a number of environmental concerns. One of the problems that should be addressed is the impact of pesticides on nontarget soil microorganisms, resulting in the perceptible changes of soil properties and the alterations of soil equilibrium for shorter or longer periods. It is well recognized that balance in the soil environment largely depends on the activity of the microorganisms, since they play an essential role in many soil biological processes, including nitrogen transformation, organic matter decomposition, and nutrient release and their availability, as well as stabilization of the soil structure (Edwards and Bater 1990).

Pesticides may affect the microorganisms by reducing their numbers, their biochemical activities, and the diversity of microbial communities (Martínez-Toledo et al. 1998; Chen et al. 2001a; Araújo et al. 2003; Lupwayi et al. 2009a; Cycoń et al. 2010a). The observed changes depend on the types of pesticides, their spectrum of activities, persistence, and dosages applied. Moreover, active substances of pesticide preparations may be used as a source of energy and nutrients by some microorganisms and may be degraded with variable intensities, which may result in the increase of microbial populations (Johnsen et al. 2001; Cycoń et al. 2010b). Certain agrochemicals, which are not utilizable by the soil, might be degraded in the soil by the microorganisms through cometabolism (Dejonghe et al. 2003). Moreover, metabolites, which are the products of pesticide degradation, may be more toxic than the parent substances, resulting in the inhibition of the activities of microbial groups responsible for the degradation of pesticides introduced into the soil (Vonk 1991; Matsushita et al. 2003).

A large number of different assays including measurements of chemical, physical, and biological soil parameters have been used in the studies on pesticide toxicity to soil ecosystem. Chemical and physical soil properties (e.g., organic matter, nutrient status, and soil texture) change very slowly, and therefore a long period of time is needed to observe significant changes. On the contrary, the microbiological indicators may rapidly reflect even small changes that occur in the soil, providing accurate data about the health and quality of the soil (Filip 2002; Schloter et al. 2003). Due to their fast response to contaminants, soil microorganisms are suitable to act as a "biomarker," reflecting the negative effects of pesticide treatment and are commonly used in ecotoxicological tests to evaluate the influence of chemicals on soil systems (Pascual et al. 2000; Filip 2002). Based on the definition of Domsch et al. (1983), a delay in the restitution of normal microbial population or functions within 31–60 days can be considered as having "tolerable" effects, while for more than 60 days indicates "critical" effects.

The literature on the effect of pesticides on soil microorganisms and processes is extremely diverse, ranging from reports of the effects of chemicals on individual species of microorganisms to those on populations of microorganisms and individual biological systems, using a great diversity of testing methods in both field and laboratory. The most common are measurements of global parameters such as microbial respiration, organic matter turnover, microbial biomass, and more specific indicators based on particular microbial activities such as nitrogen fixation, nitrification, and denitrification, as well as soil enzyme activities (Araújo et al. 2003; Chen et al. 2003; Singh and Singh 2005; Cycoń and Piotrowska-Seget 2009). Since these methods do not give insight into all the microbial communities, methods based on molecular techniques have been applied for the studies on the structure of soil microbial communities. These approaches involve the analysis of nucleic acids (DNA and/or RNA) and fatty acids (phospholipids fatty acid (PLFA) and/ or fatty acid methyl ester (FAME)) isolated directly from environmental samples (Seghers et al. 2003; Lin et al. 2008; Hua et al. 2009; Zhang et al. 2010a). In this chapter, both culture-dependent and culture-independent methods and parameters used for the assessment of ecological risk related to pesticide usage are presented and discussed, with regard to their potential successful application for the estimation of the response of soil microorganisms to pesticide treatment.

8.2 Effect of Pesticides on Soil Respiration and Microbial Biomass

The substrate-induced respiration (SIR) method is commonly accepted for the estimation of potential perturbations in microbe-mediated degradation of organic matter in the soil treated with pesticides (Domsch et al. 1983). Responses of soil microorganisms to pesticide application measured by SIR and microbial biomass amount are contrasting. This is largely because of the differences in the soil type and the class and dosage of the pesticide, which results in effects on the cellular metabolism of soil microorganisms. For example, Moreno et al. (2007) found that atrazine applied at concentrations ranging from 0.2 to 250 mg/kg soil did not affect the soil microbial respiration (SMR), whereas when applied at higher dosages (500 and 1000 mg/kg soil), it significantly increased the SMR values. At higher treatments, authors also observed an increase in the microbial biomass with the incubation time; however, at the end of the experiment (45 days), the biomass was lower as compared with that on day 16. To study the side effects of atrazine on microbial activity, they calculated the metabolic quotient (qCO_2), which represents the amount of CO_2–C evolved per unit of microbial carbon per hour. In their study, the values of qCO_2 were significantly higher in the soils treated with the highest herbicide dosages than those in the control soil at each sampling time (Moreno et al. 2007). Increase in the qCO_2 values was also observed by Jones and Ananyeva (2001) as a result of metalaxyl application, showing that fungicide treatment led to a change in the ecophysiological status of the soil microbial community. Metabolic quotient is a parameter used to evaluate the microbial stress in the soil and its increase reflects the harmful effect of a pesticide on microorganisms, which have to use a great part of their energy to survive under unfavorable conditions, resulting in less organic C incorporation into the microbial biomass (Chander and Brookes 1991).

Dose-dependent changes in the values of the basal soil respiration (BSR) in response to napropamide were observed by Hua et al. (2009). In the soil treated with a herbicide at 2–80 mg/kg soil, the increase in BSR was observed 7 days after napropamide application. During the experiment, the values of BSR decreased, and after 56 days in the soils treated with higher dosages (20–80 mg/kg soil), BSR was significantly lower in comparison to the control. This phenomenon was also observed for microbial biomass. On day 56, the high herbicide input (40–80 mg/kg soil) significantly decreased the microbial biomass, whereas the low napropramide dosages (2–20 mg/kg soil) stimulated it (Hua et al. 2009). Dinelli et al. (1998) observed a similar transient increase in SIR in the soils amended with other herbicides, for example, triasulfuron, primisulfuron methyl, and rimsulfuron. Accinelli et al. (2002), studying the short-term effects of six agrochemicals on the microbial activity found that the sulfonylurea herbicides applied up to 20 μg/g soil stimulated soil respiration, whereas at the agricultural rate it did not exert any significant impact on the soil microbial activity. Soil respiration was also stimulated by hexazinone over a 3-week period, when it was applied at a level up to 100 times higher than the recommended field rate (Vienneau et al. 2004).

Changes in the SIR value caused by other herbicides, metazachlor and dinoterb, were observed by Buelke and Malkomes (2001). They found that dinoterb more strongly inhibited SIR as compared with metazachlor and the side effect of dinoterb was more pronounced at 30°C. Busse et al. (2001) observed that glyphosate had no significant effect on microbial respiration in the soil treated with the expected concentrations following field application (5–50 mg/kg soil). However, soil respiration was strongly stimulated by the addition of glyphosate at concentrations of 500 and 5000 mg/kg soil. This herbicide displayed stimulatory effect on the microbial respiration also in forest acidic soils (Stratton and Steward 1992) and agricultural soils, with herbicide application for several years and without history of glyphosate treatment (Araújo et al. 2003). Similarly, methamidophos at both low (0.5 mg/g soil) and high (5.0 mg/g soil) inputs did not display negative effects on soil respiration rate, but in contrast, the respiration rate of the soil microbial biomass C (μg CO_2/g C_{mic} h) was significantly higher (Wang et al. 2008). No effect on the microbial biomass was found in the soil treated with the fungicide vinclozolin and the insecticide λ-cyhalothrin (Lupwayi et al. 2009b). In turn, another insecticide, chlorpyrifos, significantly increased the microbial biomass in the soils treated with both low (0.5 mg/g soil) and high (50 mg/g soil) inputs. Moreover, chlorpyrifos applied at the field rate had no harmful effect on the basal microbial respiration (Dutta et al. 2010).

Smith et al. (2000) studying the impact of the fungicide benomyl on the microorganism activity observed marked increase in the respiration rate, while in the case of the highest dosage of the fungicide, they found a reverse effect. In turn, Chen et al. (2001b) observed that fungicides benomyl, captan, and chlorothalonil suppressed the peak soil respiration by 30%–50%. On the other hand, the studies of Martikainen et al. (1998) showed that the pesticides dimethoate and benomyl had no effect on the soil respiration intensity. Černohlávková et al. (2009) also showed that mancozeb increased the soil respiration at concentrations of 25.6 and 256 mg/kg soil. As indicated by many studies, the changes in the microbial respiration in the soils treated with fungicides (e.g., benomyl, captan, or tebuconazole) have been closely related to the dosages of the pesticides used (Martikainen et al. 1998; Smith et al. 2000; Cycoń et al. 2006).

Many studies have been carried out to evaluate the relationship between microbial biomass, microbial respiration, and degradation rate of the pesticides in different soils. For example, Bolan and Baskaran (1996) and Voos and Groffman (1997) reported the positive correlation between the microbial biomass and the degradation rate of 2,4-dichlorophenoxyacetic acid (2,4-D) and dicamba. Regression analysis of the pesticide transformation rate and soil respiration activity revealed a positive correlation between SIR and the transformation rate constant for metalaxyl and propachlor (Jones and Ananyeva 2001). By contrast, Entry et al. (1994) did not find such correlation studying 2,4-D and atrazine degradation in pasture and forest soils. Similarly, no apparent relationship between the soil microbial biomass and the degradation rates of carbofuran was observed by Karpouzas et al. (2001).

8.3 Effect of Pesticides on Soil Enzyme Activities

Enzyme activities are often proposed as an early and sensitive indicator of microbial response to stress in both natural ecosystems and agroecosystems (Sannino and Gianfreda 2001; Gil-Sotres et al. 2005). Studies on enzyme activities in soil are important as they indicate the potential of soil to support the biochemical processes, which are essential to the maintenance of soil fertility. Any management practice that influences microbial communities and their biochemical activities in soil may be expected to generate changes in the soil enzyme activity level. Kandeler et al. (1996) have also emphasized that the composition of a microbial community determines the potential of that community for enzyme synthesis, and thus, any modification of the microbial community due to environmental factors should be reflected on the level of the soil enzymatic activities.

8.3.1 Dehydrogenase Activity

Since dehydrogenase activity (DHA) reflects the physiological state of microorganisms, DHA has been proposed as the accurate measurement of the potential microbial activity in the soil treated with agrochemicals (Rossel and Tarradellas 1991). However, some authors have criticized this approach as the enzyme is also affected by different soil features such as soil type or pH (Tabatabai 1994). Dehydrogenases occur intracellularly in all living microbial cells and they are linked with microbial respiration processes. They are considered to play an essential role in the initial stages of the oxidation of soil organic matter by transferring hydrogen and electrons from the substrates to the acceptors (Nannipieri et al. 1990).

The increased sensitivity of DHA to insecticides has been reported in some studies (Martínez-Toledo et al. 1998; Pandey and Singh 2006). However, the stimulation of enzyme activity in response to soil amendment with diazinon and other insecticides has also been observed (Gundi et al. 2007; Singh and Singh 2005). In turn, Cycoń et al. (2010c) found no effect of diazinon applied at the recommended field rate on DHA in sandy soils in a 28-day experiment, while the addition of the highest dosage of this insecticide (700 mg/kg soil) resulted in strong DHA inhibition during the whole experimental period. Application of higher rates (5–12.5 kg/ha) of two organophosphorus insecticides (monocrotophos and quinalphos) and two synthetic pyrethroids (cypermethrin and fenvalerate) were found to be either innocuous or toxic to the activities of the dehydrogenase in the soil (Rangaswamy et al. 1993). By contrast, Yao et al. (2006) did not find any effect till 14 days or even with the stimulation of DHA after this time after the addition of the neonicotinoid insecticide acetamiprid.

Varying impacts of herbicides on DHA are also associated with the variations in the soil organic matter content, type of herbicide and its dosage, and farming history (Zabaloy et al. 2008a,b). A positive effect of glyphosate applied up to 200 mg/kg soil on DHA has been reported by some authors (Accinelli et al. 2002; Araújo et al. 2003). Also, the stimulatory effect on DHA has been found for atrazine application at different levels (Moreno et al. 2007). In turn, Grenni et al. (2009) found no changes in DHA between the control and the soils treated with the phenylurea herbicide linuron at the agricultural rate. However, the application of linuron at higher dosages (20 and 400 mg/kg soil) resulted in the decrease of DHA in sandy soils (Cycoń et al. 2010a).

As indicated by the data from various studies, the different levels of some fungicides in the soil have been found inhibitory to DHA. A strong inhibition of DHA was observed as a microbial response to soil amendment with all the dosages of two fungicides, mefenoxam (1–1000 mg/kg soil) and metalaxyl (2–1000 mg/kg soil), during 90-day experimental period (Monkiedje et al. 2002). Similarly, Chen et al. (2001a) observed a negative effect of the fungicides benomyl and captan on DHA. However, this effect was found only at the beginning of the experiment. Inhibitory effect was also found for other fungicides such as azoxystrobin, tebuconazole, and chlorothalonil (Bending et al. 2007).

A low DHA in the soils treated with pesticides may be associated with the death of a part of microbial fraction sensitive to pesticides, and the dehydrogenases released from dead cells do not accumulate in soil since they are rapidly degraded. Free enzymes normally have a short-lived activity because they can be rapidly denatured, degraded, or irreversibly inhibited (Marx et al. 2005). However, a certain proportion of free enzymes may undergo stabilization through adsorption into humic materials, which, despite affecting their catalytic potential, may enable enzyme activity to persist in soil (Badine et al. 2001).

8.3.2 Fluorescein Diacetate Hydrolyzing Activity

Fluorescein diacetate hydrolyzing activity (FDHA) is widely accepted as a valuable and simple method for measuring lipase, protease, and esterase activities in soils (Adam and Duncan 2001). As FDHA is a sensitive and nonspecific test able to depict the hydrolytic activity of soil microorganisms, it is used as a suitable tool for measuring an early detrimental effect of pesticides on soil microbial biomass (Perucci et al. 2000). Depending on the class of pesticides, application rate, and soil type, a decrease, no consistent changes, or an increase in FDHA has been reported. For example, Dutta et al. (2010) reported that the organophosphorus insecticide chlorpyrifos applied at the field rate (0.5 mg/kg soil) into clay loam soil did not adversely affect FDHA and even a stimulation at the beginning of the experiment was observed. By contrast, there was a sharp decrease in FDHA in the soil treated with a higher concentration (100 mg/kg soil) over a 90-day incubation period. A negative effect of chlorpyrifos and endosulfan was also observed in tea garden soil by Bishnu et al. (2008). As revealed by Zabaloy et al. (2008b), the application of 2,4-D into sandy clay loam soil reduced FDHA by 11% in comparison with the control, but this effect was observed only 1 week after herbicide treatment. In turn, this herbicide did not affect FDHA in sandy loam soil over a 21-day experimental period. Authors also found a transient reduction in the hydrolyzing of FDA (fluorescein diacetate) after application of other herbicides, metsulfuron-methyl and glyphosate (Zabaloy et al. 2008b). A transient inhibition of FDHA at the beginning of the experiment (up to day 15) in two soils was also observed by Pal et al. (2005), in the case of the phenylurea fungicide pencycuron.

8.3.3 Phosphatase Activity

Phosphatases are the extracellular enzymes that catalyze the hydrolysis of organic phosphorus compounds into inorganic phosphorus (Pant and Warman 2000). Generally, it is stated that acid and alkaline phosphatases (PHOS-H and PHOS-OH) are insensitive to the toxic effect of pesticides (Boyd and Mortland 1990). For example, Cycoń et al. (2010b) found no effect of diazinon applied at concentrations of 7, 35, and 700 mg/kg soil on PHOS-OH in sandy loam soil over a 28-day incubation period. By contrast, the activity of PHOS-H was significantly stimulated by the applied insecticide during the whole experimental period. Similarly, Megharaj et al. (1999) revealed that the application of another insecticide (fenamiphos) into loamy sand soil did not negatively affect PHOS-OH, and even stimulation up to 20% at a concentration of 100 mg/kg soil was observed. By contrast, the inhibitory effect of insecticides on PHOS-OH was found in the case of acetamiprid applied at higher dosages (Yao et al. 2006).

Sannino and Gianfreda (2001) found that the herbicides glyphosate, paraquat, and atrazine strongly affected the phosphatases in 21 soils. The greatest sensitivity of phosphatase was observed in the glyphosate-treated soils. Authors also suggested that the inhibitory effect of this herbicide may be associated with the presence of phosphate group in its molecules. It has been demonstrated that soil phosphatases are strongly inhibited by inorganic phosphate and phosphate fertilizers (Speir and Ross 1978). Also, Perucci et al. (2000) observed detrimental effects of the sulfonylurea herbicide rimsulfuron and the imidazoline herbicide imazethapyr applied at field and 10-fold field rates on both the phosphatases. A decrease in the activities of both the phosphatases was also found in the soil treated with other herbicides, trifluralin (Wyszkowska and Kucharski 2004), glyphosate and paraquat (Sannino and Gianfreda 2001), or alachlor (Pozo et al. 1994). However, for alachlor, PHOS-H and PHOS-OH decreased significantly shortly after the introduction to soil at concentrations of 5.0–10.0 kg/ha soil, but with time, the enzyme activities reached levels similar to that found in the control (Pozo et al. 1994).

As indicated by the data from various studies, the different levels of some fungicides in soil have been found as inhibitory or stimulatory to phosphatase activities. For example, application of the fungicides mefenoxam (1–1000 mg/kg) and metalaxyl (2–2000 mg/kg) into sandy loam soil resulted in no effect or stimulation of PHOS-H (Monkiedje et al. 2002).

However, the authors found that all fungicide dosages significantly inhibited PHOS-OH over a 90-day incubation period. Also, Chen et al. (2001a) observed a significant decrease in PHOS-H in silt loam soil after treatment with captan (125 mg/kg soil) or benomyl (51 mg/kg soil), and this effect was more pronounced with time. Inhibition of the acitivities of both the phosphatases in sandy loam soil amended with captan during 98 days of incubation was also ascertained by Piotrowska-Seget et al. (2008). Also, Cycoń et al. (2010d) found that the fungicidal mixture of mancozeb and dimethomorph negatively affected both the phosphatases in sandy soils. In addition, the higher the concentration of the fungicides added into the soil, the higher is the decrease in phosphatase activities. However, the authors observed a greater inhibition of this activity in the soil with lower contents of clay and organic matter. Being extracellular enzymes, phosphatases are immobilized by soil colloids, which protect them from degradation in polluted soils (Boyd and Mortland 1990). This fact could explain a greater sensitivity of both the phosphatases to fungicides in sandy soils than clay soils. Moreover, the higher decrease in alkaline phosphatase than acid phosphatase in the soils treated with fungicides may have been associated with the decreased growth of the indigenous fungi that are known as soil producers of alkaline phosphatase (Nannipieri et al. 1990).

8.3.4 Urease Activity

The activity of urease, an extracellular enzyme that catalyzes the hydrolysis of urea to ammonia, may also reflect the response of microorganisms to pesticide application. Depending on the type of herbicide, dosage, application rate, and soil type, the response of microbial community expressed as urease activity (URE) showed a high variability. For example, Ingram et al. (2005) found a significant short-term inhibitory effect of diazinon on the microbial urease–producing community. Similarly, Jung et al. (1995) found that diazinon decreased URE by 50% at concentrations greatly exceeding those found in soils during agricultural practices. A great sensitivity of urease to diazinon addition into soils has also been reported by Cycoń et al. (2010b). However, this effect was transient and URE in the soil with higher clay and organic matter contents recovered to the level as in the control after 28 days of incubation. Previously, Lethbridge and Burns (1975) also observed 40%–50% urease inhibition 60 days after applying other insecticides, malathion, accothion, or thimet, into sandy clay loam. By contrast, the application of the insecticide acetamiprid did not affect URE over a 35-day incubation period (Yao et al. 2006).

A variable activity of urease in soils treated with different herbicides was observed by several authors; however, no effect or even stimulation of URE was generally observed. For example, Moreno et al. (2007) found that application of atrazine into clay loam soil at a wide range of concentrations (0.2–500 mg/kg soil) had no effect at the beginning of the experiment or stimulated URE at two highest dosages on day 45. This phenomenon was also reported by Jorge et al. (2007). Also, Sannino and Gianfreda (2001) observed a significant increase in URE in soils treated with other herbicides, paraquat and glyphosate, over an incubation period. By contrast, Hua et al. (2009) reported that napropamide applied at all used concentrations (2–80 mg/kg soil) inhibited URE in the soil over a 56-day experimental period, and additionally, this effect was more pronounced with time. Similarly, the sulfonylurea herbicide metsulfuron-methyl negatively affected URE in soils (Ismail et al. 1998).

Different levels of some fungicides in the soil have also been found inhibitory or stimulatory to URE. For example, application of the phenylamide fungicide metalaxyl resulted in the gradual decrease in URE in clay loam soil from 10 to 60 days (Sukul 2006).

In turn, Cycoń et al. (2010d) reported that urease activity was affected to a little extent after the application of the fungicidal mixture of mancozeb and dimethomorph into sandy soils. However, shortly after fungicide application at all dosages (15, 75, and 1500 mg/kg soil), URE was significantly lower in both the soils than that in the corresponding controls. Chen et al. (2001a) observed a significant increase of URE in silt loam soil after treatment with captan (125 mg/kg soil) or benomyl (51 mg/kg soil), and this effect was more seen with time.

It is well known that urease is strongly bonded to the organic matter and mineral compounds of soil, and thus, it is protected from degradation and denaturation. The soil-enzyme complexes of high molecular weight are more stable and resistant to degradation than the complexes of low molecular weight (Gianfreda et al. 1994). This fact may support the results obtained by some authors, who observed a higher stability and activity of urease in soils containing higher amounts of organic matter as compared with soils with lower contents of this fraction (Sannino and Gianfreda 2001; Cycoń et al. 2010d).

8.3.5 Other Enzymes

It has been reported that the application of pesticides may also affect the activities of other enzymes, for example, amylase, arylosulfatase, catalase, cellulase, β-glucosidase, nitrate reductase, and proteinase. Depending on the type of pesticide, dosage, application rate, and soil texture, a decrease, no consistent changes, or an increase of these activities has been found. Gundi et al. (2007) observed that the organophosphorus insecticides monocrotophos and quinalphos as well as the synthetic pyrethroid cypermethrin applied at different levels (5, 10, and 25 mg/kg soil) into the soil significantly increased the amylase and cellulase activities over an incubation period. As it was revealed by Yao et al. (2006), acetamiprid at normal field concentration (0.5 mg/kg soil) did not affect the catalase and proteinase activities over a 50-day experimental period. However, the authors found an inhibitory effect of two higher dosages (5 and 50 mg/kg soil) of this insecticide on proteinase activity lasting from 14 days to the end of the incubation period. An inhibitory effect of acetamiprid was also found in the case of other enzymes. Singh and Kumar (2008) observed that the arginine deaminase and nitrate reductase activities were declined by 22% and 41%, respectively, after the first treatment of acetamiprid; however, these activities were recovered at the end of crop season. In turn, Hua et al. (2009) reported that the activity of catalase was enhanced during the initial 7 days of the herbicide napropamide application, but this effect was transient and the enzyme activity was recovered to the control level. Results obtained by Sukul (2006) indicated that the application of the fungicide metalaxyl at the highest dosage (1.6 mg/kg soil) has been shown to be inhibitory to arylsulfatase activity in clay loam soil, and additionally, this effect was more pronounced over time. The author also observed an increase in β-glucosidase activity at all used dosages of metalaxyl at the beginning of the 60-day incubation period. In turn, a strong inhibition of β-glucosidase activity after metalaxyl application (2–1000 mg/kg soil) was found by Monkiedje et al. (2002). However, at lower fungicide treatment levels, this activity was recovered to the control level at the end of the experiment. The authors also found a varied (generally negative) effect of another fungicide, mefenoxam, applied at different concentrations (1–1000 mg/kg soil) on β-glucosidase activity over a 90-day incubation period. Moreover, the effects of these fungicides on β-glucosidase activity were also dependent on their formulations (Monkiedje et al. 2007). By contrast, no effects of the insecticide fenamiphos applied at the recommended field rate (5 mg/kg soil) on β-glucosidase activity were reported by Megharaj et al. (1999).

8.4 Effect of Pesticides on Soil Nitrogen Transformation Rates

With regard to the great importance of nitrogen cycling for soil fertility and plant development, nitrification and ammonification processes are commonly used as the indicators of microbial activity in soils amended with pesticides (Monkiedje et al. 2002; Kara et al. 2004). Generally, insecticides have been reported as compounds stimulating N mineralization and nitrification processes after incorporation into soil. For example, Das and Mukherjee (1994) found that insecticides such as 1,2,3,4,5,6-hexachlorocyclohexane (BHC), phorate, carbofuran, and fenvalerate, at their recommended doses, affected nitrogen turnover by changing the availability of ammonium and nitrogen fixation in the rhizosphere of rice. The application of BHC and phorate resulted in the lowest retention of total N, compared with other insecticides. The greater mineralization of N following the insecticide application brought a significant increase in ammonium and nitrate concentrations in the soil, and this remained more prominent up to the 30th day, followed by a steady decline up to the 60th day, except with BHC and phorate, both of which caused a progressive increase in nitrate concentration until the 45th day. A similar phenomenon was also confirmed by Jana et al. (1998). In turn, Cycoń et al. (2006) found slight negative effects (on days 1 and 7) of two higher treatments (1 and 20 mg/kg soil) of the insecticide λ-cyhalothrin on the nitrate concentrations in sandy loam soil. By contrast, this insecticide did not have an effect on ammonium concentrations as compared with the control. Nitrate concentrations also decreased in the soil samples treated with diazinon; however, this effect was especially seen for a dosage of 700 mg/kg soil (Cycoń et al. 2010b).

Marsh and Davies (1981) observed the long-term strong inhibition of nitrification in dichlorprop- and mecoprop-treated soils. Moreover, nitrification was suppressed significantly by bensulfuron-methyl applied at concentrations of 0.1 and 1.0 µg/g soil (El-Ghamry et al. 2002). Cycoń et al. (2010a) also found that the phenylurea herbicide linuron at the highest dosage (400 mg/kg soil) caused a significant decrease in nitrate concentrations in sandy soils; however, this effect was observed shortly after its application. Kara et al. (2004) reported that the herbicide topogard significantly stimulated ammonification in neutral and alkaline soils, whereas acid soils showed significantly lower ammonium contents in comparison with the controls. Ammonification was also initially inhibited by the triazine herbicide hexazinone; however, ammonium contents were similar or greater in all treatments compared with the controls after 4 weeks (Vienneau et al. 2004). Studies on the successive applications of different pesticides in field conditions over 2 years, carried out by Schuster and Schröder (1990), showed that herbicides had only slight and transient side effects on the ammonification process.

As indicated by the data from some studies, the different levels of some fungicides in soil have been found inhibitory to nitrification. Evidently, the chemical nature of fungicides has decided their effects on this process, albeit soil type has also been of great value (Monkiedje and Spiteller 2002; Kinney et al. 2005). Martens and Bremner (1997) ascertained up to 90% inhibition of nitrification in the soil amended with 50 mg/kg soil of mancozeb. Similarly, Černohlávková et al. (2009) revealed that the nitrification process was very sensitive to mancozeb and its application caused significant decrease in nitrification to 11.2% and 5.6% in arable soil and grassland soil, respectively. An inhibitory effect of mancozeb on the nitrification process was also observed by other authors (Pozo et al. 1994; Kinney et al. 2005; Cycoń et al. 2010d) and may be linked with its fate in soil. The fungicide transformation products such as ethylenethiourea and ethyleneurea, persisting for weeks in soil, were reported as strong inhibitors of soil nitrification (Vonk 1991).

By contrast, being suitable substrates for some microorganisms, these mancozeb intermediates may have stimulated ammonification (Černohlávková et al. 2009). The study performed by Man and Zucong (2009) revealed that the field application rate of chlorothalonil (5.5 mg/kg soil) inhibited the nitrification activity in two soils to a less degree but had no effect on the other four soils. By contrast, chlorothalonil, at the levels of 110 and 220 mg/kg soil, almost completely inhibited nitrification in soils throughout the 14-day incubation period. Also, Monkiedje et al. (2002) observed that the fungicides metalaxyl and mefenoxam at a concentration of 1000 µg/g soil severely inhibited the rate of nitrification in sandy loam soil over a 90-day experimental period. In turn, the ammonification process was strongly stimulated by the applied fungicides. Similar findings were also observed by Burrows and Edwards (2002), who studied the impact of the fungicide carbendazim on the nitrogen transformation processes in soil. They found a significant increase in ammonium concentration in the soil samples treated with the tested pesticide, whereas in the case of nitrification process, a reverse effect was observed. By contrast, Man and Zucong (2009) observed that the inhibitory effect of carbendazim applied even at the highest concentration (220 mg/kg soil) on nitrification was very weak or absent. Also, the study performed by Cycoń et al. (2006) showed that tebuconazole had a marked effect on the nitrogen transformation. This fungicide applied at the highest dosage (270 mg/kg soil) decreased the nitrate concentration in sandy loam soil on day 1, whereas a stimulating effect of the fungicide on nitrate production was found with time. The data obtained for various fungicides could partially be interpreted as the result of mineralization of the fungi killed by these compounds. However, fungal biomass is known to be a poor source of nitrogen as compared with bacteria. Therefore, other chemicals present in the fungicide formulation could stimulate the ammonifying bacteria, resulting in an enhanced production of ammonium. An increase in the ammonium concentrations in the soils treated with benomyl or captan has been reported in some studies (Chen et al. 2001a; Piotrowska-Seget et al. 2008). By contrast, the application of other fungicides to soils has stimulated nitrification as indicated by increased nitrate contents (Chen et al. 2001b; Monkiedje and Spiteller 2002). However, it is difficult to explain why some fungicides could be tolerated by nitrifiers or could even stimulate them. Presumably, soils differ in their nitrifying capacity and those that can maintain a complex population of active nitrifiers are less sensitive to pesticides (Hicks et al. 1990).

Results of some studies have shown that some pesticides indirectly stimulate ammonification probably by killing a part of the microorganisms, which is mineralized, resulting in increased concentrations of ammonium (Das and Mukherjee 2000). Generally, nitrification is more sensitive to pesticides than ammonification, which is conducted by a diverse microbial population (Hicks et al. 1990).

8.5 Effect of Pesticides on Soil Microbial Numbers

The culturable bacteria usually represent an ecologically important fraction of the soil bacterial community. Although only a small fraction of the total bacterial population (0.1%–10% of microorganisms occurring in soil) can grow on agar plates (Littlefield-Wyer et al. 2008), they constitute a large proportion of the total bacterial biomass and play the most important role in nutrient cycles (Ellis et al. 2003). Therefore, quantitative changes in the composition of soil-culturable bacterial communities may serve as important and sensitive indicators of both short- and long-term changes in soil quality. That is the reason why the cultivation-dependent approach is still a useful method for the detection of variable microbial response to toxic substances.

Impact of pesticides on soil microflora is changeable and results not only from the reaction of microorganisms to active substances and formulation additives but also from the development of specific groups of microorganisms. The ability of some microorganisms to grow in the presence of pesticides may result in the compensation of their adverse effects by the increased activity of the remaining part of the soil microbial community. Microorganisms not sensitive to pesticides utilize released nutrients from dead cells, which may result in the increase in their numbers in the soil (Das and Mukherjee 2000).

8.5.1 Heterotrophic Bacteria

It has been reported that some pesticides affect the numbers of soil microflora. For example, Das and Mukherjee (2000) found that the population sizes of bacteria were greatest 30 days following the application of hexachlorocyclohexane (HCH), phorate, and fenvalerate. Earlier, an increase in the numbers of soil bacteria was revealed during the biodegradation of chlorinated hydrocarbons, organophosphates, and synthetic pyrethroid insecticides (Rache and Coats 1988; Das et al. 1995). Significant increases in the numbers of culturable bacteria in the soil amended with diazinon was found by Singh and Singh (2005) and Cycoń et al. (2010b). Das et al. (2005) also found increased numbers of culturable bacteria in the phorate-treated soil. Also, for λ-cyhalothrin applied at different concentrations, the stimulatory effect on bacterial populations in soil was observed by Cycoń et al. (2006). Generally, insecticides are not toxic to bacteria, as these organisms do not have sensitive targets (Topp et al. 1997), and, in addition, these chemicals have been degraded by several bacterial species (Finkelstein et al. 2001; Grant et al. 2002; Cycoń et al. 2009). However, in some studies, decrease in the total numbers of heterotrophic bacteria in the soil treated with insecticides was observed. For example, Cycoń and Piotrowska-Seget (2009) found that fenitrothion significantly decreased the numbers of these bacteria in loamy sand soil after 1 day of experiment at all used dosages (2, 10, and 200 mg/kg soil). However, this effect in lower treatments was transient; the highest dosage negatively affected the numbers of bacteria over a 28-day experimental period. This effect was probably caused by high toxicity of the intermediate metabolites of fenitrothion biodegradation, supporting the results by Matsushita et al. (2003). Also, Das et al. (2005) found that the total bacterial population was reduced due the application of the carbamate insecticide carbofuran at its recommended field rate (1 kg/ha soil).

Studies on the effect of various herbicides on the microbial numbers in soils revealed that these compounds are generally well tolerated by soil bacteria. Cycoń and Piotrowska-Seget (2009) found that after the application of diuron into loamy sand soil, higher counts of heterotrophic bacteria were enumerated. A similar stimulatory effect on the total numbers of these bacteria in soil was also observed by Cycoń et al. (2010a) after the application of linuron, representing the same class of phenylurea herbicides. Generally, these compounds can be degraded by different bacteria and might be used as a source of carbon and nitrogen, resulting in a significant increase in bacterial counts (Dejonghe et al. 2003). Recently, Grenni et al. (2009) indicated the lack of linuron effect on the numbers of viable bacteria in the treated soil because of their adaptation to the presence of the herbicide. Zabaloy et al. (2008b) also observed increased numbers of culturable bacteria in different soils treated with metsulfuron-methyl and glyphosate. In addition, Araújo et al. (2003) reported increased counts of soil heterotrophic bacteria in response to the applied glyphosate. Ratcliff et al. (2006) did not find changes in the culturable bacterial numbers between the control and the glyphosate-treated soil, when the herbicide was used at a dosage of 50 mg/kg soil, whereas a 100-fold higher dosage increased both the heterotrophic bacterial counts and the bacteria:fungi ratio. On the contrary, Busse et al. (2001) reported that glyphosate application decreased the total number of heterotrophic bacteria.

Increased viable bacterial counts in fungicide-treated soil have been found in many studies. For example, Martínez-Toledo et al. (1998) found that after treatment with captan at concentrations of 2, 3.5, and 5 kg/ha soil, bacterial populations did not increase significantly until 14 days, whereas at concentration of 10 kg/ha soil, the populations were significantly higher than that in the control for days 7, 14, and 30. A similar stimulatory effect on the numbers of bacteria was also observed in soils after the application of mefenoxam, metalaxyl (Monkiedje et al. 2002), and tebuconazole (Cycoń et al. 2006), as well as mancozeb and dimethomorph (Cycoń et al. 2010d). Increased bacterial counts in the fungicide-treated soils may be associated with increased levels of nutrients and energy sources released from dead fungal hyphae. Furthermore, soil bacteria can be relieved from competition with indigenous fungi or antagonistic inhibition by the metabolites synthesized by them (Chen et al. 2001b).

8.5.2 Fungal Populations

Apart from bacteria, fungi are responsible for the mineralization of organic matter in soil and the release of available carbon; hence, any disturbance in their counts due to various pesticides may have a harmful impact on soil quality. It has been reported that not all insecticides negatively affect the soil fungi, and even their growth was stimulated in treated soils. For example, Martínez-Toledo et al. (1993) found that the fungal population in an agricultural loam soil was not inhibited in the presence of the chlorinated hydrocarbon insecticide lindane at concentrations of 3.5–15 kg/ha soil. In turn, Das et al. (2005) found a marked increase in the total numbers of fungi in the soil treated with phorate and carbofuran. Similarly, Cycoń et al. (2006) revealed that the addition of λ-cyhalothrin into loamy sand soil at concentrations of 1 and 20 mg/kg soil resulted in higher numbers of fungi compared with that in the control on day 14, but this effect was not seen thereafter. However, some studies reported that organophosphorus insecticides profenofos (Abdell-Malek et al. 1994) as well as chlorpyrifos and quinalphos (Pandey and Singh 2004) decreased the fungal counts in soil. Also, Abd El-Mongy and Abd El-Ghany (2009) indicated that the insecticide chlorpyrifos-methyl affected the numbers of fungi in treated soil; the numbers of colonies decreased after 1–7 days of insecticide application but then increased over time.

Several studies indicated that herbicides appear to be pesticides affecting indigenous soil fungi to a little extent. For example, linuron treatment, even at the highest concentration (400 mg/kg soil), did not significantly change the fungal numbers in sandy soils as compared with the corresponding controls (Cycoń et al. 2010a). It would appear that culturable fungi can tolerate linuron concentrations in soils. Several soil fungi, including different species of *Cunninghamella*, *Mortierella*, *Talaromyces*, *Rhizopus*, *Rhizoctonia*, and *Aspergillus* are able to degrade linuron and other phenylurea herbicides (Sørensen et al. 2003). Also, Araújo et al. (2003) found that the numbers of fungi have increased significantly in the soils treated with glyphosate (2.16 mg/kg soil) after 32 days of incubation. The pronounced effect of glyphosate as well as other herbicides (2,4-D and picloram) on the fungal activity was also confirmed in a previous study (Wardle and Parkinson 1990).

Among the numerous applied compounds, fungicides appear to be pesticides affecting indigenous soil fungi to the most extent. A negative impact of the fungicides on the fungal population size was observed in the case of captan and benomyl (Martínez-Toledo et al. 1998; Busse et al. 2001; Piotrowska-Seget et al. 2008). The numbers of fungi also decreased significantly in the sandy soils treated with the fungicidal mixture (mancozeb and dimethomorph), especially applied at the 100-fold recommended field rate, during a 28-day incubation period (Cycoń et al. 2010d). The results obtained by Cycoń et al. (2006) also indicated that the negative impact of another fungicide (tebuconazole) on the soil fungi was transient (14 days), though the fungicide, with a soil half-life of 49 days, has been suggested as a persistent agrochemical (Strickland et al. 2004).

8.5.3 Bacteria Involved in the Nitrogen Cycling

Among the numerous groups of soil microorganisms, the bacteria involved in the nitrogen transformation processes such as ammonification, nitrification, denitrification, and N_2-fixation are very important. Due to the significant role of these bacteria in soil fertility and total nitrogen cycling, the evaluation of changes in their numbers was often performed in many ecotoxicological studies. For example, Cycoń et al. (2010a) found that the nitrifying bacterial counts decreased in the sandy soils treated with linuron at 400 mg/kg soil; however, this effect was found only on days 1 and 14. Similarly, a negative effect of another herbicide (simazine) applied at 50–300 µg/g soil on the nitrifying bacteria was observed by Martínez-Toledo et al. (1996). In addition, the synergistic action of two herbicides (terbutryn and terbuthylazine) inhibited the activity of the nitrifying bacteria in acidic and neutral soils (Kara et al. 2004). Also, Martínez-Toledo et al. (1998) found that the numbers of this bacterial group significantly decreased in agricultural soils after captan treatment at dose rates of 2–10 kg/ha soil. Application of diazinon at concentrations of 35 and 700 mg/kg soil also resulted in the decline in the numbers of nitrifying bacteria in loamy sand and sandy loam soils over a 28-day incubation period (Cycoń et al. 2010b). A strong inhibition of the nitrifying bacterial activity was also observed by Cycoń et al. (2010d) in the sandy soils treated with the fungicidal mixture of mancozeb and dimethomorph, even at the recommended field rate (15 mg/kg soil). Conversely, the nitrifying bacteria were not affected upon addition of the herbicide alachlor to an agricultural soil (Pozo et al. 1994). Also, Cycoń et al. (2006) found that the application of the fungicide tebuconazole, even at the highest concentration (270 mg/kg soil), resulted in higher numbers of nitrifiers in loamy sand soil on day 28. However, these dosages negatively affected the activity of the nitrifying bacteria immediately after fungicide treatment. A decrease in the numbers of nitrogen-fixing bacteria has often been found in the pesticide-treated soils (Martínez-Toledo et al. 1998; Cycoń et al. 2006; Lin et al. 2008). This bacterial group is regarded as a sensitive microbial indicator of environmental hazard in soils exposed to elevated concentrations of pesticides and other chemicals; however, the effects of these compounds on nitrogen fixation appear to be species specific (Hicks et al. 1990). Gadagi et al. (2004) reported that some insecticides (chlorpyrifos, fenvalerate, and quinalphos) may inhibit the growth and nitrogenase activity of *Azospirillum* sp. at all concentrations used, whereas other insecticides (endosulfan and monocrotophos) may be tolerated only at the recommended dose. By contrast, a stimulation of proliferation of nonsymbiotic N_2-fixing bacteria was found in a laterite soil amended with carbofuran and phorate (Das and Mukherjee 2000). A decrease in the numbers of N_2-fixing bacteria exposed to fungicides such as captan (Martínez-Toledo et al. 1998) and tebuconazole (Cycoń et al. 2006) has also been found. As indicated by several studies, denitrifiers may be regarded as the group more tolerant toward pesticides compared with N_2-fixing bacteria and nitrifiers. It has been reported that different pesticides did not negatively affect the soil denitrifiers and even their growth was stimulated in treated soil (Martínez-Toledo et al. 1998; Cycoń et al. 2006). For example, Cycoń et al. (2010a) found that the numbers of denitrifying bacteria were not affected by linuron in sandy soils over a 28-day experiment. By contrast, the application of glyphosate to grass resulted in a 20–30-fold increase in denitrification compared with untreated soil (Tenuta and Beauchamp 1996). Also, Yeomans and Bremner (1985) found that denitrifying bacteria were either unaffected or even stimulated in the soils treated with different insecticides (lindane, fenitrothion, fonofos, malathion, phorate, terbufos, and carbofuran) and fungicides (mancozeb, maneb, thiram, benomyl, captan, and terrazole). A transient decrease in the numbers of denitrifying bacteria has also been reported in the soil amended with pyrethroid λ-cyhalothrin (Cycoń et al. 2006).

8.6 Effect of Pesticides on Structural and Functional Diversity of Soil Microbial Communities

A negative effect of pesticide application may be also related to the changes in the biodiversity of microbial communities. These communities may be significantly changed even if the total numbers of microorganisms and cycling of nutrient elements such as carbon, nitrogen, sulfur, and phosphoros appear to be unaffected. It happens when the growth of some microorganisms is inhibited and others proliferate in the vacant ecological niches. This replacement may finally lead to altered biological activities of soils at a later point in time. Therefore, the pesticide risk assessment should also involve an examination of the microbial diversity, reflecting both the changes caused by a direct toxicity and the long-term effect resulting from the successions in the microbial community (Johnsen et al. 2001).

The culturable microbial community can be characterized in ecophysiological terms by the r/K strategy concept derived from evolutionary ecology, indicating that there are genetic differences between organisms in their ability to exploit and survive in different environments (De Leij et al. 1993). In soil microbiology, an interpretation of this concept is that the r-fraction represents active, viable bacteria, whereas the K-fraction corresponds to dormant and slow-growing bacteria. Typically, the r-fraction constitutes 5%–20% of SIR in the rhizosphere soil (Stenström et al. 2001). In some studies, the analysis of the bacterial growth dynamics revealed that the K-strategists generally dominated in the soils after treatment with pesticides (e.g., diuron, diazinon, thiram) in comparison with the controls. In addition, the application of pesticides, especially at the highest dosages, resulted in an uneven distribution of the bacterial ecotypes with the domination of one class representing the K-strategists, which was indicated by the decreased ecophysiological index. However, higher abundance of r-strategists was ascertained in the treated soils at the end of the incubation period (Cycoń and Piotrowska-Seget 2009; Cycoń et al. 2010b).

Since the culturable fraction of soil bacteria represent a small fraction of soil microorganisms, the determination of quantitative changes in the composition of the entire microbial communities as a consequence of pesticide application needs the culture-independent methods. In recent years, various methods based on cell molecular signature markers such as DNA or/and fatty acids isolated directly from the soil samples have been shown to be useful for detecting pesticide-induced changes in the compositions of the microbial communities (Ratcliff et al. 2006; Zhang et al. 2009; Crouzet et al. 2010).

8.6.1 PCR–DGGE Profiling

One of the most popular methods based on DNA profiling is denaturing gradient gel electrophoresis (DGGE), which enables rapid and reproducible separation of PCR-amplified 16S rDNA fragments in a mixed PCR. DNA fragments of the same length but different nucleotide sequences are separated based on the difference in the mobilities of the molecules in polyacrylamide gels with a linear denaturant gradient (Muyzer and Smalla 1998). Differences in the electrophoretic profiles between the samples reflect the differences and abundance in a community composition (Kent and Triplett 2002).

Changes in the microbial diversity as a result of herbicide application was found by Hua et al. (2009), who studied the impact of different dosages (0–80 mg/kg soil) of napropamide on the microbial community structure. Napropamide apparently increased the number of bands that are considered to represent the dominant microbial populations, on days 7 and 14 after application. The cluster analysis of 16S rDNA showed that similarities among the different banding patterns obtained for control and pesticide-treated samples were more than 90% on day 7; however, these similarities decreased to 40% over the next week. These results suggested that some particular bacteria may be adapted to the pesticide applied and dominated in the DGGE patterns obtained. Seghers et al. (2003) used PCR–DGGE profiling to study the chronic effect of the herbicides atrazine and metolachlor on the community structure in the soil of a maize monoculture. For the extended analysis, they used the primers for both entire bacterial community and specific group of bacteria such as *Acidobacterium*, the actinomycetes, the ammonium oxidizers, and the methanotrophs. This approach allowed them to observe that the physiological groups of bacteria differed in their reaction to pesticide application. The biggest differences between the control and the pesticide-treated soils were found in the case of methanotrophs. In their DGGE profiles from herbicide-treated soil, some bands were absent or much weaker when compared with the patterns obtained for the control soil. The alterations in the methanotrophic community observed at the first sampling time were also seen in the next year. However, cluster analysis showed that the long-term use of atrazine and metolachlor did not negatively affect the community structures of the tested microbial groups (Seghers et al. 2003).

Similarly, bensulfuron-methyl applied at the field-recommended and 10-fold field-recommended dosages did not affect the bacterial community over 8 weeks in a model paddy soil. PCR–DGGE analysis revealed the same banding patterns in all pesticide-treated and control soils on all sampling days (Saeki and Toyota 2004). By contrast, Ros et al. (2006) analyzing PCR–DGGE banding patterns obtained for the soils treated with atrazine found that regardless of the sampling time, herbicides at concentrations of 10, 100, and 1000 mg/kg soil affected the bacterial community, compared with the control and at a dosage of 1 mg/kg soil. This effect was observed even on day 45 when atrazine had been degraded. The significant changes in both abundance and diversity of ammonia oxidizers group under atrazine and other herbicide (dicamba, fluometuron, metolachlor, and sulfentrazone) exposure over an experimental period were also observed by Chang et al. (2001).

A response of the microbial community to another herbicide was studied by Chen et al. (2009). Application of butachlor at 1- and 10-fold field-recommended dosages shifted the nitrogen-fixation bacterial community and the effect was related to the stages of rice. Changes among this bacterial group were much higher than those observed among the total bacterial community in soils, which suggested that the diazotrophic community might be more susceptible to the addition of butachlor as compared with other groups of microorganisms.

Ferreira et al. (2009) used PCR–DGGE to study the effects of three insecticides (aldicarb, chlorpyrifos, and deltamethrin) and two fungicides (tebuconazole and a mixture of meta-laxyl and mancozeb) on the culturable bacterial communities of three soils with different kinds of agroecological management. They analyzed 16S rDNA profiles of culturable bacteria from soils cultivated with potato and treated with pesticides. The PCR–DGGE patterns of culturable bacterial communities revealed that regardless of the type of the pesticide used, the community structure of soil bacteria was disturbed and the similarity values varied from 5% to 95% in comparison to the control. However, the biggest differences between microbial communities were observed in the first two harvests and, over time, had a tendency to recover. The lowest percentages of similarity of soil bacterial communities were found 32 days after the pesticide application when aldicarb, chlorpyrifos, and tebuconazole had a higher impact on the microorganisms than deltamethrin and fungicidal mixture in comparison to the control. Statistical analysis of the data showed that the response of the bacteria depended on the type of soil and sampling time (Ferreira et al. 2009).

A response of soil microorganism to another fungicide, iprodione, was tested by Wang et al. (2004). They estimated the impact of fungicide by counting the 16S rDNA bands in the DGGE patterns. The differences between the DGGE profiles of 16S rDNA fragments were observed on successive sampling days. The numbers of bands increased at the first sampling times (3–7 days) and decreased after 16 days. No significant differences in the microbial community structures between the control and the soil treated with iprodione at a concentration of 5 µg/g soil was found. By contrast, the bacterial community did not recover till that day in the soil with 50 µg fungicide per gram soil. An ability of the microbial community to recover during 126 days after perturbations caused by carbendazim as revealed by the comparison of the patterns of PCR-amplified 16S rDNA fragments using DGGE obtained for fungicide and control samples was reported by Wang et al. (2009). However, using the same method, they observed that the changes in microbial diversity as a response to pencycuron application (100 mg/kg soil) were seen even on day 120, pointing that the microbial community could not recover to its previous structure.

8.6.2 Phospholipid Fatty Acid Analysis

Phospholipid fatty acids (PLFAs) extracted from soils provide a means for direct in situ measurements of the bacterial and fungal community structures and biomasses, and therefore, they are used to assess the impact of pesticides on soil microbial diversity. Changes in the microbial community structure in response to any environmental stressor are monitored by comparison of relative abundance of signature fatty acids that are specific for groups of microorganisms such as fungi, actinomycetes, Gram-negative (GN), and Gram-positive bacteria (GN) (Zelles 1997; White et al. 1996). Recently, PLFA analysis proved to be a useful tool for the assessment of changes in the community structure of the soil microorganisms exposed to pesticides (Ratcliff et al. 2006; Wang et al. 2008).

Zhang et al. (2009) used this method to study the effect of foliar application of cypermethrin on biomass and the structure of the pepper phylosphere microbial community. PLFA patterns obtained from pepper leaves showed that pesticide application significantly increased the total and bacterial biomasses as compared with the control during the 21-day experiment. By contrast, the amount of fungal fatty acids significantly decreased in the cypermethrin-treated samples. Analysis of the distribution of individual fatty acids revealed that pesticide treatment also significantly decreased the ratio of GP to GN bacteria. Authors found that it was associated with the increase in the amount of four PLFAs

(16:1ω7*t*, 18:1ω7*t*, *cy*17:0) regarded as an indicator of GN bacteria and 19:0 and the decrease in fatty acids (14:0, *i*15:0 and *a*17:0) known as a specification for GP bacteria (Zhang et al. 2009).

In another study, Zhang et al. (2010a) used PLFA approach to assess the community structure of microorganisms in soils differing in their fertility and organic matter contents and treated with the 2,4-D butyl ester at dosages of 10, 100, and 1000 µg/g soil. PLFA profiling showed that biomass expressed as the total PLFA abundance was correlated with herbicide concentration and soil type. The total bacterial and fungal PLFAs decreased with the increasing herbicide concentration, reaching the minimum level at 1000 µg/g in soil with higher organic matter content. In soils of low organic matter amount, the highest PLFA concentration was found in the soil treated with 2,4-D butyl ester at dosage of 100 µg/g. The distribution of indicator fatty acids in PLFA profiles revealed that the herbicide significantly decreased the amount of GN bacteria and the rate of this decrease was correlated with the concentration of the herbicide. In turn, a decrease in GP bacteria biomass was observed only in the soils treated with the highest herbicide concentration (1000 µg/g soil), indicating that GN bacteria were more sensitive to 2,4-D butyl ester than GP bacteria (Zhang et al. 2010a). Another herbicide, imazethapyr, applied into two agricultural soils and incubated for 120 days also changed the structure of the soil microbial community (Zhang et al. 2010b). A decrease in the total amount of PLFAs and lower ratios of GN/GP and fungi/bacteria in the imazethapyr-treated soils in comparison to the control were observed. Additionally, the authors calculated the level of stress expressed as the ratio (cy17:0 + cy19:0)/(16:1ω7*t* + 18:1ω7*t*) and observed that the high pesticide input (1 and 10 mg/kg soil) caused the highest stress in soil with low organic matter content and worse nutrient status. The observed shift in soil microbial community was transient and it recovered after 60 days.

Structural changes in soil microbial communities may be also estimated by the whole-cell FAME profiles isolated from soil. Using this method, Lancaster et al. (2010) showed that the successive glyphosate applications altered the bacterial diversity much more than its single application. They also observed that FAMEs regarded as markers of GN bacteria were present in higher concentrations following five applications in comparison to 1, 2, 3, or 4 applications during first 2 weeks of the experiment.

8.6.3 Community Level Physiological Profile

As microbial response to toxicant involves the study of both the structure and the function of an ecosystem, assessment of microbial communities should consider not only the abundance of bacteria but also the functional diversity and redundancy present in the microbial community (Kent and Triplett 2002). Recently, the most common method used to establish the functional diversity of microbial assemblages is the measurement of community-level physiological profiling (CLPP) based on the ability of the community to utilize a wide range of carbon substrates (Garland and Mills 1991). This technique has shown its value in the field of ecotoxicology, for discriminating between pesticide-treated and nontreated microcosms and field sites (de Lipthay et al. 2004; Mijangos et al. 2009).

The CLPP approach was used by Zabaloy et al. (2008b) for estimating the effect of the herbicides glyphosate, 2,4-D, and metsulfuron-methyl on the soil functional richness. They found that the pesticides used only slightly affected the catabolic potential of the soil microorganisms and this effect was related to the type of soil. Also, no effect of glyphosate on the catabolic richness in the soil exposed to this herbicide for a long period was reported by Busse et al. (2001), who measured the catabolic response of microflora by comparison of SIR rates following the addition of individual C compounds. Similarly, Weaver et al. (2007) reported that glyphosate had only small and transient effects on the soil microbial community even when applied at a dosage much higher than the recommended rate. Mijangos et al. (2009) used the Biolog Ecoplates system for estimating the catabolic potential of the rhizosphere microbial communities of triticale or triticale and pea under glyphosate and 2,4-D exposures. They observed significant differences in the carbon substrate utilization patterns between the rhizosphere microbial communities in the glyphosate-treated (50 and 500 mg/kg soil) and control soil samples on the 15th day after the pesticide application. On the same day, herbicide treatment resulted in a stimulatory effect on the ability of the rhizospheric communities to metabolize the substrates such as phenylethylamine,D-galacturonic acid, Tween 40, Tween 80, and hydroxybutyric acid. By contrast, after 30 days of glyphosate treatment, the microbial assemblages were characterized by a lower capacity of usage of a considerable percentage of the carbon substrates included in the Ecoplates. Interestingly, the response of the microbial communities was different in triticale *versus* triticale and pea combination. In contrast to glyphosate, the application of 2,4-D at a dosage of 5 mg/kg soil increased the functional richness of the microorganisms (Mijangos et al. 2009). A structural shift in the microbial community structure under 2,4-D exposure was earlier proved by Macur et al. (2007). They found that in the soil treated with agriculturally relevant doses of 2,4-D dominated metabolically versatile genera such as *Burkholderia, Bradyrhizobium, Variovorax,* and *Arthrobacter,* whereas in the control soil only *Variovorax*-like isolates were identified. In turn, the addition of metsulfuron-methyl caused transient changes in the percentage of the carbon substrates utilized by the microbial community. A slight increase in the functional richness in the pesticide-treated soil was observed only a week after herbicide application.

Based on the results of the field studies with other herbicides such as metribuzin, imazamox/imazethapyr, glufosinate ammonium metribuzin, triasulfuron, and metsulfuron-methyl, Lupwayi et al. (2004) concluded that the herbicides applied once at recommended rates did not have significant or consistent effects on the microbial functional diversity. The CLPP method was also used to establish changes in the catabolic potential of the microbial communities in soils treated with atrazine at concentrations of 1, 10, 100, and 1000 mg/kg soil (Ros et al. 2006). The authors found that higher pesticide concentrations negatively affected the potential catabolic activity. Values of the diversity index (H') calculated for each class of carbon compounds showed that some carbon groups in the high atrazine-treated soils were degraded at lower rates as compared with the control. Interestingly, a negative effect was also observed after 45 days of atrazine application when the pesticide was totally degraded (Ros et al. 2006).

In turn, Wang et al. (2008) showed in the pot experiments that the organophosphorus insecticide methamidophos markedly changed the functional and genetic diversity of the soil microorganisms. Using the polyphasic approaches including ARDRA (amplified rDNA restriction analysis), PLFA, and CLPP, they revealed that methamidophos decreased the genetic biodiversity and differentially affected the components of the soil microbial community. High input of pesticides (64.48 mg/g soil) decreased the fungal biomass but increased the biomass of the GN bacteria and enhanced their catabolic activity (Wang et al. 2008). By contrast, a minimal impact of the pyrethroid insecticide cypermethrin applied at a concentration of 10 mg/kg soil on the functional diversity was reported by Xie et al. (2009). Similarly, no significant effects of another pyrethroid insecticide (λ-cyhalothrin) on the soil functional bacterial diversity were observed; however, the microbial catabolic abilities in the insecticide-treated soil were altered (Lupwayi et al. 2009b). A transient effect on the soil microbial functional diversity was caused by chlorpyrifos applied at concentrations of 4, 8, and 12 mg/kg soil (Hua et al. 2009). Values of the average well color development (AWCD) were significantly lower in comparison with the control, only within the first 2 weeks of incubation and recovered till the end of the experiment (35 days), reaching the same or extended values of AWCD as in the control. Also, no significant changes were observed for all the values of the diversity index (H') (Hua et al. 2009).

An alteration of the catabolic profiles of the soil bacterial communities was also observed in the soils exposed to the fungicide vinclozolin (Lupwayi et al. 2009b). The authors also observed that combined fungicide and insecticide treatment had a different bacterial community structure compared with the control and single pesticide application. Analysis of the Biolog data revealed that some substrates were utilized more extensively as compared with others. The authors underlined that the measured bacterial functional potential showed shifts caused by fungicide and insecticide application that were not observed by estimating such parameters as bacterial diversity or microbial biomass. Moreover, these shifts can lead to successions in the microbial communities that could have long-term effects on soil activity (Lupwayi et al. 2009b).

8.7 Conclusions

The varied results on the impact of pesticides on microbial number, activity, and diversity indicated that it is difficult to precisely assess the overall influence of pesticides on soil systems and microbe-mediated processes. Interactions between pesticides, soil, and soil microorganisms are complex as many biotic and abiotic factors influence the final value of the measured parameters. The impact of the applied pesticides on the microbial activity is strongly correlated with the soil characteristics, pesticide type and its dose, exposure time, and interactions between these factors. The field-rate pesticide dosages do not usually have long-term adverse effects on the microbial activity. These effects are usually transient and detected immediately after pesticide application. However, higher doses of pesticides very often display a negative effect on the microbial activity, resulting in the alteration of the biological balance of the soil processes.

Due to the high complexity and biodiversity of soil ecosystem, single-process tests do not provide adequate data to predict the overall potential environmental hazards of the pesticides. It seems that polyphasic approaches including measurements of numerous parameters reflecting the functional and structural status of both culturable and nonculturable fractions of the microbial community should be used to assess the response of the soil microorganisms to the pesticides.

References

Abd El-Mongy, M. and Abd El-Ghany, T. M. 2009. Field and laboratory studies for evaluating the toxicity of the insecticide Reldan on soil fungi. *Int. Biodeterior. Biodegradation* 63: 383–388.

Abdell-Malek, A. Y., Moharram, A. M., Abder-Kader, M. I., and Omar, S. A. 1994. Effect of soil treatment with the organophosphorus insecticide Profenofos on the fungal flora and some microbial activities. *Microbiol. Res.* 149: 167–171.

Accinelli, C., Screpanti, C., Dinelli, G., and Vicari, A. 2002. Short-time effects of pure and formulated herbicides on soil microbial activity and biomass. *Int. J. Environ. Anal. Chem.* 82: 519–527.

Adam, G. and Duncan, H. 2001. Development of a sensitive and rapid method for the measurement of total microbial activity using fluorescein diacetate (FDA) in a range of soils. *Soil. Biol. Biochem.* 33: 943–951.

Araújo, A. S. F., Monteiro, R. T. R., and Abarkeli, R. B. 2003. Effect of glyphosate on the microbial activity of two Brazilian soils. *Chemosphere* 52: 799–804.

Badine, N. N. Y., Chotte, J. L., Pate, E., Masse, D., and Rouland, C. 2001. Use of soil enzyme activities to monitor soil quality in natural and improved fallows in semi-arid tropical regions. *Appl. Soil Ecol.* 18: 229–238.

Bending, G. D., Rodríguez-Cruz, S. D., and Lincoln, S. D. 2007. Fungicide impacts on microbial communities in soils with contrasting management histories. *Chemosphere* 69: 82–88.

Bishnu, A., Saha, T., Mazumdar, D., Chakrabarti, K., and Chakraborty, A. 2008. Assessment of the impact of pesticides residues on microbiological and biochemical parameters of a tea garden soil in India. *J. Environ. Sci. Health B* 43: 723–731.

Bolan, N. and Baskaran, S. 1996. Biodegradation of 2,4-D herbicide as affected by its adsorption-desorption behavior and microbial activity of soils. *Aust. J. Soil Res.* 34: 1041–1053.

Boyd, S. A. and Mortland, M. M. 1990. Enzyme interactions with clays and clay–organic matter complexes. In: Bollag, J. M. and Stotzky, G. (eds), *Soil Biochemistry*, vol. 6, pp. 1–28. Marcel Dekker, Inc: New York.

Buelke, S. and Malkomes, H.-P. 2001. Effects of the herbicides metazachlor and dinoterb on the soil microflora and the degradation and sorption of metazachlor under different environmental conditions. *Biol. Fertil. Soils* 33: 467–471.

Burrows, L. A. and Edwards, C. A. 2002. The use of integrated soil microcosms to predict effects of pesticides on soil ecosystems. *Eur. J. Soil Biol.* 38: 245–249.

Busse, M. D., Ratcliff, A. W., Shestak, C. J., and Powers, R. F. 2001. Glyphosate toxicity and the effects of long-term vegetation control on soil microbial communities. *Soil Biol. Biochem.* 33: 1777–1789.

Černohlávková, J., Jarkovský, J., and Hofman, J. 2009. Effects of fungicides mancozeb and dinocap on carbon and nitrogen mineralization in soils. *Ecotoxicol. Environ. Safe.* 72: 80–85.

Chander, K. and Brookes, P. C. 1991. Microbial biomass dynamics during the decomposition of glucose and maize in metal-contaminated soils. *Soil Biol. Biochem.* 23: 917–925.

Chang, Y. J., Hussaiu, A., Stephen, J. R., Mullen, M. D., White, D. C., and Peacock, A. 2001. Impact of herbicides on the abundance and structure of indigenous beta-subgroup ammonia-oxidizer communities in soil microcosms. *Environ. Toxicol. Chem.* 20: 2462–2468.

Chen, S. K., Edwards, C. A., and Subler, S. 2001a. A microcosm approach for evaluating the effects of the fungicides benomyl and captan on soil ecological processes and plant growth. *Appl. Soil Ecol.* 18: 69–82.

Chen, S. K., Edwards, C. A., and Subler, S. 2001b. Effects of the fungicides benomyl, captan and chlorothalonil on soil microbial activity and nitrogen dynamics in laboratory incubations. *Soil Biol. Biochem.* 33: 1971–1980.

Chen, S. K., Edwards, C. A., and Subler, S. 2003. The influence of two agricultural biostimulants on nitrogen transformations, microbial activity, and plant growth in soil microcosms. *Soil Biol. Biochem.* 35: 9–19.

Chen, W.-C., Yen, J.-H., Chang, C.-S., and Wang, Y.-S. 2009. Effects of herbicide butachlor on soil microorganisms and on nitrogen-fixing abilities in paddy soil. *Ecotoxicol. Environ. Safe.* 72: 120–127.

Crouzet, O., Batisson, I., Besse-Hoggan, P., et al. 2010. Response of soil microbial communities to the herbicide mesotrione: A dose–effect microcosm approach. *Soil Biol. Biochem.* 42: 193–202.

Cycoń, M. and Piotrowska-Seget, Z. 2009. Changes in bacterial diversity and community structure following pesticides addition to soil estimated by cultivation technique. *Ecotoxicology* 18: 632–642.

Cycoń, M., Piotrowska-Seget, Z., Kaczyńska, A., and Kozdrój, J. 2006. Microbiological characteristics of a sandy loam exposed to tebuconazole and λ-cyhalothrin under laboratory conditions. *Ecotoxicology* 15: 639–646.

Cycoń, M., Wójcik, M., and Piotrowska-Seget, Z. 2009. Biodegradation of the organophosphorus insecticide diazinon by *Serratia* sp. and *Pseudomonas* sp. and their use in bioremediation of contaminated soil. *Chemosphere* 76: 494–501.

Cycoń, M., Piotrowska-Seget, Z., and Kozdrój, J. 2010a. Linuron effects on microbiological characteristics of sandy soils as determined in a pot study. *Ann. Microbiol.* 60: 439–449.

Cycoń, M., Piotrowska-Seget, Z., and Kozdrój, J. 2010b. Microbial characteristics of sandy soils exposed to diazinon under laboratory conditions. *World J. Microbiol. Biotechnol.* 26: 409–418.

Cycoń, M., Piotrowska-Seget Z., and Kozdrój, J. 2010c. Dehydrogenase activity as an indicator of different microbial responses to pesticide-treated soils. *Chem. Ecol.* 26: 243–250.

Cycoń, M., Piotrowska-Seget, Z., and Kozdrój, J. 2010d. Responses of indigenous microorganisms to a fungicidal mixture of mancozeb and dimethomorph added to sandy soils. *Int. Biodeterior. Biodegradation* 64: 86–96.

Das, A. C. and Mukherjee, D. 1994. Effect of insecticides on the availability of nutrients, nitrogen fixation, and phosphate solubility in the rhizosphere soil of rice. *Biol. Fertil. Soils* 18: 37–41.

Das, A. C. and Mukherjee, D. 2000. Soil application of insecticides influences microorganisms and plant nutrients. *Appl. Soil Ecol.* 14: 55–62.

Das, A. C., Chakrabarty, A., Sukul, P., and Mukherjee, D. 1995. Insecticides: Their effect on microorganisms and persistence in rice soil. *Microbiol. Res.* 150: 187–194.

Das, A. C., Chakravarty, A., Sen, G., Sukul, P., and Mukherjee, D. 2005. A comparative study on the dissipation and microbial metabolism of organophosphate and carbamate insecticides in orchaqualf and fluvaquent soils of West Bengal. *Chemosphere* 58: 579–584.

Dejonghe, W., Berteloot, E., Goris, J., et al. 2003. Synergistic degradation of linuron by a bacterial consortium and isolation of a single linuron-degrading *Variovorax* strain. *Appl. Environ. Microbiol.* 69: 1532–1541.

De Leij, F. A. A. M., Whipps, J. M., and Lynch, J. M. 1993. The use of colony development for the characterization of bacterial communities in soil and roots. *FEMS Microbiol. Ecol.* 27: 81–97.

de Lipthay, J. R., Johnsen, K., Albrechtsen, H. J., Rosenberg, P., and Aamand, J. 2004. Bacterial diversity and community structure of a sub-surface aquifer exposed to realistic low herbicide concentrations. *FEMS Microbiol. Ecol.* 49: 59–69.

Dinelli, G., Vicari, A., and Accinelli, C. 1998. Degradation and side effects of three sulfonylurea herbicides in soil. *J. Environ. Qual.* 27: 1459–1464.

Domsch, K. H, Jabnow, G., and Anderson, T. H. 1983. An ecological concept for the assessment of side-effects of agrochemicals on soil microorganisms. *Res. Rev.* 86: 65–105.

Dutta, M., Sardar, D., Pal, R., and Kole, R. K. 2010. Effect of chlorpyrifos on microbial biomass and activities in tropical clay loam soil. *Environ. Monit. Assess.* 160: 385–391.

Edwards, C. A. and Bater, J. E. 1990. An evaluation of laboratory and field studies for the assessment of the environmental effects of pesticides. In *Proceedings of the Brighton Crop Protection Conference on Pests and Diseases*, Brighton, England. pp. 963–968.

El-Ghamry, A. M., Xu, J. M., Huang, C. Y., and Gan, J. 2002. Microbial response to bensulfuron-methyl treatment in soil. *J. Agric. Food Chem.* 50: 136–139.

Ellis, R. J, Morgan, P., Weightman, A. J, and Fry, J. C. 2003. Cultivation dependent and -independent approaches for determining bacterial diversity in heavy-metal-contaminated soil. *Appl. Environ. Microbiol.* 69: 3223–3230.

Entry, J. A., Donnelly, P. K., and Emmingham, W. H. 1994. Microbial mineralization of atrazine and 2,4-dichlorophenoxyacetic acid in riparian pasture and forest soils. *Biol. Fertil. Soils* 18: 89–94.

Ferreira, E. P. B., Dusi, A. N., Costa, J. R. C., Xavier, G. R., and Rumjanek, N. R. 2009. Assessing insecticide and fungicide effects on the culturable soil bacterial community by analyses of variance of their DGGE fingerprinting data. *Eur. J. Soil Biol.* 45: 466–472.

Filip, Z. 2002. International approach to assessing soil quality by ecologically-related biological parameters. *Agric. Ecosyst. Environ.* 88: 169–174.

Finkelstein, Z. I., Baskunov, B. P., Rietjens, I. M., Boersma, M. G., Vervoort, J., and Golovleva, J. A. 2001. Transformation of the insecticide teflubenzuron by microorganisms. *J. Environ. Sci. Health B* 36: 559–567.

Gadagi, R, S., Tongmin, S., and Chung, J. B. 2004. Chemical insecticide effects on growth and nitrogenase activity of *Azospirillum* sp. OAD-2. *Commun. Soil Sci. Plant Anal.* 35: 495–503.

Garland, J. L. and Mills, A. L. 1991. Classification and characterisation of heterotrophic microbial communities on the basis of pattern of community-level sole-carbon-source utilization. *Appl. Environ. Microbiol.* 57: 2351–2359.

Gianfreda, L., Sanninio, F., Ortega, N., and Nannipieri, P. 1994. Activity of free and immobilized urease in soil: Effects of pesticides. *Soil Biol. Biochem.* 26: 777–784.

Gil-Sotres, F., Trasar-Capeda, C., Leirós, M. C., and Seoane, S. 2005. Different approaches to evaluating soil quality using biochemical properties. *Soil Biol. Biochem.* 37: 877–887.

Grant, R. J., Daniell, T. J., and Betts, W. B. 2002. Isolation and identification of synthetic pyrethroid-degrading bacteria. *J. Appl. Microbiol.* 92: 534–540.

Grenni, P., Caracciolo, A. B., Rodríguez-Cruz, M. S., and Sánchez-Martín, M. J. 2009. Changes in the microbial activity in a soil amended with oak and pine residues and treated with linuron herbicide. *Appl. Soil Ecol.* 41: 2–7.

Gundi, V. A. K. B., Viswanath, B., Chandra, M. S., Kumar, V. N., and Reddy, B. R. 2007. Activities of cellulase and amylase in soils as influenced by insecticide interactions. *Ecotoxicol. Environ. Safe.* 68: 278–285.

Hicks, R. J., Stotzky, G., and van Voris, P. 1990. Review and evaluation of the effects of xenobiotics chemicals on microorganisms in soil. In: Neidelman, S. L. and Laskin, A. I. *Advances in Applied Microbiology* vol. 35, pp. 195–253. San Diego: Academic Press.

Hua, G., Guofeng, C., Zhaoping, L. V., and Hong, Y. 2009. Alteration of microbial properties and community structure in soils exposed to napropamide. *J. Environ. Sci.* 21: 494–502.

Ingram, C. W., Coyne, M. S., and Williams, D. W. 2005. Effects of commercial diazinon and imidacloprid on microbial urease activity in soil and sod. *J. Environ. Qual.* 34: 1573–1580.

Ismail, B. S., Yapp, K. F., and Omar, O. 1998. Effects of metsulfuron-methyl on amylase, urease, and protease activities in two soils. *Aust. J. Soil Res.* 36: 449–456.

Jana, T. K., Debnath, N. C., and Basak, R. K. 1998. Effect of insecticides on the composition of organic matter, ammonification and nitrification in a Fluventic Ustochrept. *J. Int. Soc. Soil Sci.* 46: 133–134.

Johnsen, K., Jacobsen, C. S., Torsvik, V., and Sørensen, J. 2001. Pesticide effects on bacterial diversity in agricultural soils – A review. *Biol. Fertil. Soils* 33: 443–453.

Jones, W. J. and Ananyeva, N. D. 2001. Correlations between pesticide transformation rate and microbial respiration activity in soil of different ecosystems. *Biol. Fertil. Soils* 33: 477–483.

Jorge, M., Martin, A., Carballas, T., and Diaz-Ravina, M. 2007. Atrazine degradation and enzyme activities in an agricultural soil under two tillage systems. *Sci. Total Environ.* 378: 187–194.

Jung, K., Bitton, G., and Koopman, B. 1995. Assessment of urease inhibition assays for measuring toxicity of environmental samples. *Water Res.* 29: 1929–1933.

Kandeler, E., Kampichler, C., and Horak, O. 1996. Influence of heavy metals on the functional diversity of soil communities. *Biol. Fertil. Soils* 23: 299–306.

Kara, E. E, Arli, M., and Uygur, V. 2004. Effects of the herbicide Topogard on soil respiration, nitrification, and denitrification in potato-cultivated soils differing in pH. *Biol. Fertil. Soils* 39: 474–478.

Karpouzas, D. G, Walker, A., Drennan, D. S., and Froud-Williams, R. J. 2001. The effect of initial concentration of carbofuran on the development and stability of its enhanced biodegradation in top-soil and sub-soil. *Pest Manag. Sci.* 57: 72–81.

Kent, A. D. and Triplett, E. W. 2002. Microbial communities and their interactions in soil and rhizosphere ecosystem. *Annu. Rev. Microbiol.* 56: 211–236.

Kinney, C. A., Mandernack, K. W., and Mosier, A. R. 2005. Laboratory investigations into the effects of the pesticides mancozeb, chlorothalonil, and prosulfuron on nitrous oxide and nitric oxide production in fertilized soil. *Soil Biol. Biochem.* 37: 837–850.

Lancaster, S. H., Hollister, E. B., Senseman, S. A., and Gentry, T. J. 2010. Effects of repeated glyphosate applications on soil microbial community composition and the mineralization of glyphosate. *Pest Manag. Sci.* 66: 59–64.

Lethbridge, G. and Burns, R. G. 1975. Inhibition of soil urease by organophosphorus insecticides. *Soil Biol. Biochem.* 8: 99–102.

Lin, X., Zhao, Y., Fu, Q., Umashankara, M. L., and Feng, Z. 2008. Analysis of culturable and uncultur-
able microbial community in bensulfuron-methyl contaminated paddy soils. *J. Environ. Sci.* 20:
1494–1500.
Littlefield-Wyer, J. G., Brooks, P., and Katouli, M. 2008. Application of biochemical fingerprinting and
fatty acid methyl ester profiling to assess the effect of the pesticide atradex on aquatic microbial
communities. *Environ. Pollut.* 153: 393–400.
Lupwayi, N. Z., Harker, K. N., Clayton, G. W., Turkington, T. K., Rice, W. A., and O'Donovan, J. T.
2004. Soil microbial biomass and diversity after herbicide application. *Can. J. Plant Sci.* 84:
677–685.
Lupwayi, N. Z., Harker K. N., Clayton, G. W., O'Donovan, J. T., and Blackshaw R. E. 2009a. Soil
microbial response to herbicides applied to glyphosate-resistant canola. *Agric. Ecosyst. Environ.*
129: 171–176.
Lupwayi, N. Z., Harker K. N., Dosdall, L. M., et al. 2009b. Changes in functional structure of soil
bacterial communities due to fungicide and insecticide application in canola. *Agric. Ecosyst.
Environ.* 130: 109–114.
Macur, R. E., Wheeler, J. T., Burr, M. D., and Inskeep, W. P. 2007. Impacts of 2,4-D application on
soil microbial community structure and on populations associated with 2,4-D degradation.
Microbiol. Res. 162: 37–45.
Man, L. and Zucong, C. 2009. Effects of chlorothalonil and carbendazim on nitrification and denitri-
fication in soils. *J. Environ. Sci.* 21: 458–467.
Marsh, J. A. P. and Davies, H. A. 1981. Effects of dichlorprop and mecoprop on respiration and trans-
formation of nitrogen in two soils. *Bull. Environ. Contam. Toxicol.* 26: 108–115.
Martens, D. A. and Bremner, J. M. 1997. Inhibitory effects of fungicides on hydrolysis of urea and
nitrification of urea nitrogen in soil. *Pestic. Sci.* 49: 344–352.
Martikainen, E., Haimi, J., and Ahtiainen, J. 1998. Effects of dimethoate and benomyl on soil organ-
isms and soil processes – A microcosm study. *Appl. Soil Ecol.* 9: 381–387.
Martínez-Toledo, M. V., Salmerón, V., Rodelas, B., Pozo, C., and González-López, J. 1993. Studies
on the effects of a chlorinated hydrocarbon insecticide, lindane, on soil microorganisms.
Chemosphere 27: 2261–2270.
Martínez-Toledo, M. Y., Salmerón, Y., Rodelas, B., Pozo, C., and González-López, J. 1996. Studies of
the effects of the herbicide simazine on microflora of four agricultural soils. *Environ. Toxicol.
Chem.* 15: 1115–1118.
Martínez-Toledo, M. Y., Salmerón, Y., Rodelas, B., Pozo, C., and González-López, J. 1998. Effects of
the fungicide Captan on some functional groups of soil microflora. *Appl. Soil Ecol.* 7: 245–255.
Marx, M.-C., Kandeler, E., Wood, M., Wermbter, N., and Jarvis, S. C. 2005. Exploring the enzymatic
landscape: Distribution and kinetics of hydrolytic enzymes in soil particle-size fraction. *Soil
Biol. Biochem.* 37: 35–48.
Matsushita, T., Matsui, Y., Ikeba, K., and Inoue, T. 2003. Contribution of metabolites to mutagenicity
during anaerobic biodegradation of fenitrothion. *Chemosphere* 50: 275–282.
Megharaj, M., Singleton, I., Kookana, R., and Naidu, R. 1999. Persistence and effects of fenamiphos
on native algal populations and enzymatic activities in soil. *Soil Biol. Biochem.* 31: 1549–1553.
Mijangos, I., Becerril, J. M., Albizu, I., Epelde, L., and Garibsu, C. 2009. Effects of glyphosate on
rhizosphere soil microbial communities under two different plant compositions by cultivation-
dependent and -independent methodologies. *Soil Biol. Biochem.* 41: 505–513.
Monkiedje, A. and Spiteller, M. 2002. Effects of the phenylamide fungicides, mefenoxam and metalaxyl, on the microbiological properties of a sandy loam and a sandy clay soil. *Biol. Fertil. Soils*
35: 393–398.
Monkiedje, A., Ilori, M. O., and Spiteller, M. 2002. Soil quality changes resulting from the application of the fungicides mefenoxam and metalaxyl to a sandy loam soil. *Soil Biol. Biochem.* 34:
1939–1948.
Monkiedje, A., Spiteller, M., Maniepi, S. J. N., and Sukul, P. 2007. Influence of metalaxyl- and
mefenoxam-based fungicides on chemical and biochemical attributes of soil quality under field
conditions in a southern humid forest zone of Cameroon. *Soil Biol. Biochem.* 39: 836–842.

Moreno, J. L., Aliaga, A., Navarro, S., Hernández, T., and García, C. 2007. Effects of atrazine on microbial activity in semiarid soil. *Appl. Soil Ecol.* 35: 120–127.

Muyzer, G. and Smalla, K. 1998. Application of denaturing gradient gel electrophoresis (DGGE) and temperature gradient gel electrophoresis (TGGE) in microbial ecology. *Ant. Leeuw.* 73: 127–141.

Nannipieri, P., Grego, S., and Ceccanti, B. 1990. Ecological significance of the biological activity in soil. In: Bollag, J. M. and Stotzky, G. (eds), *Soil Biochemistry*, vol. 6, pp. 293–356. Marcel Dekker, Inc: New York.

Pal, R., Chakrabarti, K., Chakraborty, A., and Chowdhury, A. 2005. Pencycuron application to soils: Degradation and effect on microbiological parameters. *Chemosphere* 60: 1513–1522.

Pandey, S. and Singh, D. K. 2004. Total bacterial and fungal population after chlorpyrifos and quinalphos treatments in groundnut (*Arachis hypogaea* L.) soil. *Chemosphere* 55: 197–205.

Pandey, S. and Singh, D. K. 2006. Soil dehydrogenase, phosphomonoesterase and arginine deaminase activities in an insecticide treated groundnut (*Arachis hypogaea* L.) field. *Chemosphere* 63: 869–880.

Pant, H. K. and Warman, P. R. 2000. Enzymatic hydrolysis of soil organic phosphorus by immobilized phosphatases. *Biol. Fertil. Soils* 30: 306–311.

Pascual, J. A., García, G., Hernández, T., Moreno, J. L., and Ros, M. 2000. Soil microbial activity as a biomarker of degradation and remediation processes. *Soil Biol. Biochem.* 32: 1877–1883.

Perucci, P., Dumontet, S., Bufo, S. A., Mazzatura, A., and Casucci, C. 2000. Effects of organic amendment and herbicide treatment on soil microbial biomass. *Biol. Fertil. Soils* 32: 17–23.

Piotrowska-Seget, Z., Engel, R., Nowak, E., and Kozdrój, J. 2008. Successive soil treatment with captan or oxytetracycline affects non-target microorganisms. *World J. Microbiol. Biotechnol.* 24: 2843–2848.

Pozo, C., Salmerón, Y., Rodelas, B., Martínez-Toledo, M. Y., and González-López, J. 1994. Effects of the herbicide alachlor on soil microbial activities. *Ecotoxicology* 3: 4–10.

Rache K. D. and Coats, J. 1988. Comparative biodegradation of organophosphorus insecticides in soil. Specificity of enhanced microbial biodegradation. *J. Agric. Food Chem.* 36: 193–199.

Rangaswamy, V. and Venkateswarlu, K. 1993. Ammonification and nitrification in soil, and nitrogen fixation by *Azospirillum* sp. as influenced by cypermethrin and fenvalerate. *Agric. Ecosyst. Environ.* 45: 311–317.

Ratcliff, A. W., Busse, M. D., and Shestak, C. J. 2006. Changes in microbial community structure following herbicide (glyphosate) addition to forest soils. *Appl. Soil Ecol.* 34: 114–124.

Ros, M., Goberna, M., Moreno, J. L., et al. 2006. Molecular and physiological bacteria diversity of a semi-arid soil contaminated with different levels of formulated atrazine. *Appl. Soil Ecol.* 34: 93–102.

Rossel, D. and Tarradellas, J. 1991. Dehydrogenase activity of soil microflora: Significance in ecotoxicological tests. *Environ. Toxicol. Water Qual.* 6: 17–33.

Saeki, M. and Toyota, K. 2004. Effect of bensulfuron-methyl (a sulfonylurea herbicide) on the soil bacterial community of a paddy soil microcosm. *Biol. Fertil. Soils* 40: 110–118.

Sannino, F. and Gianfreda, L. 2001. Pesticide influence on soil enzymatic activities. *Chemosphere* 45: 417–425.

Schloter, M., Dilly, O., and Munch, J. C. 2003. Indicators for evaluating soil quality. *Agric. Ecosyst. Environ.* 98: 255–262.

Schuster, E. and Schröder, D. 1990. Side-effects of sequentially-applied pesticides on non-target soil microorganisms: Field experiments. *Soil Biol. Biochem.* 22: 367–373.

Seghers, D., Verthé, K., Reheul, D., et al. 2003. Effect of long-term herbicide applications on the bacterial community structure and function in an agricultural soil. *FEMS Microbiol. Ecol.* 46: 139–146.

Singh, D. K. and Kumar, S. 2008. Nitrate reductase, arginine deaminase, urease and dehydrogenase activities in natural soil (ridges with forest) and in cotton soil after acetamiprid treatments. *Chemosphere* 71: 412–418.

Singh, J. and Singh, D. K. 2005. Bacterial, azotobacter, actinomycetes, and fungal population in soil after diazinon, imidacloprid, and lindane treatments in groundnut (*Arachis hypogea* L.) fields. *J. Environ. Sci. Health B* 40: 785–800.

Smith, M. D., Hartnett, D. C., and Rice, C. W. 2000. Effects of long-term fungicide applications on microbial properties in tallgrass soil. *Soil Biol. Biochem.* 32: 935–946.

Sørensen, S. R., Bending, G. D., Jacobsen, C. S., Walker, A., and Aamand, J. 2003. Microbial degradation of isoproturon and related phenylurea herbicides in and below agricultural fields. *FEMS Microbiol. Ecol.* 45: 1–11.

Speir, T. W. and Ross, D. F. 1978. Soil phosphatase and sulphatase. In: Burns, R. G. (ed.), *Soil Enzymes*, pp. 197–250. Academic Press: London.

Stenström, J., Svensson, K., and Johansson, M. 2001. Reversible transition between active and dormant microbial states in soil. *FEMS Microbiol. Ecol.* 36: 93–104.

Stratton, G. W. and Stewart, K. E. 1992. Glyphosate effects on microbial biomass in a coniferous forest soil. *Environ. Toxicol. Water Qual.* 17: 229–234.

Strickland, T. C., Potter, T. L., and Joo, H. 2004. Tebuconazole dissipation and metabolism in Tifton loamy sand during laboratory incubation. *Pest Manag. Sci.* 60: 703–709.

Sukul, P. 2006. Enzymatic activities and microbial biomass in soil as influenced by metalaxyl residues. *Soil Biol. Biochem.* 38: 320–326.

Tabatabai, M. A. 1994. Soil enzymes. In: Weaver, R. M., et al. (eds), *Methods of Soil Analysis, Part 2: Microbiological and Biochemical Properties*, pp. 775–833. Soil Science Society of America: Wisconsin.

Tenuta, M. and Beauchamp, E. G. 1996. Denitrification following herbicide application to a grass sward. *Can. J. Soil Sci.* 76: 15–22.

Topp, E., Vallaeys T., and Soulas, G. 1997. Pesticides: Microbial degradation and effects on microorganisms. In: van Elsas, J. D., Trevors, J. T., Wellington, E. M. H. (eds), *Modern Soil Microbiology*, pp. 547–575. Marcel Dekker, Inc: New York.

Vienneau, D. M., Sullivan, C. A, House, S. K., and Stratton, G. W. 2004. Effect of the herbicide hexazinone on nutrient cycling in a low-pH blueberry soil. *Environ. Toxicol.* 19: 115–122.

Vonk, J. W. 1991. Testing of pesticides for side-effects on nitrogen conversions in soil. *Toxicol. Environ. Chem.* 30: 241–248.

Voos, G. and Groffman, P. M. 1997. Relationships between microbial biomass and dissipation of 2,4-D and dicamba in soil. *Biol. Fertil. Soils* 24: 106–110.

Wang, Y.-S., Wen, C.-Y., Chiu, T.-C., and Yen, J.-H. 2004. Effect of fungicide iprodione on soil bacterial community. *Ecotoxicol. Environ. Safe.* 59: 127–132.

Wang, M.-C., Liu, Y.-H., Wang, Q., et al. 2008. Impacts of methamidophos on the biochemical, catabolic, and genetic characteristics of soil microbial communities. *Soil Biol. Biochem.* 40: 778–788.

Wang, Y.-S., Huang, Y.-J., Chen, W.-C., and Yen, J.-H. 2009. Effect of carbendazim and pencycuron on soil bacterial community. *J. Hazard. Mater.* 172: 84–91.

Wardle, D. A. and Parkinson, D. 1990. Effects of three herbicides on soil microbial biomass and activity. *Plant Soil* 122: 21–28.

Weaver, M. A., Krutz, L. J., Zablotowicz, R. M., and Reddy, K. N. 2007. Effects of glyphosate on soil microbial communities and its mineralization in a Mississippi soil. *Pest Manag. Sci.* 63: 388–393.

White, D. C., Stair, J. O., and Ringelberg, D. B. 1996. Quantitative comparisons of in situ microbial biodiversity by signature biomarker analysis. *J. Ind. Microbiol.* 17: 185–196.

Wyszkowska, J. and Kucharski, J. 2004. Biochemical and physicochemical properties of soil contaminated with herbicide Triflurotox 250 EC. *Pol. J. Environ. Stud.* 13: 223–232.

Xie, W., Zhou, J., and Wang, H., et al. 2009. Short-term effects of copper, cadmium and cypermethrin on dehydrogenase activity and microbial functional diversity in soils after long-term mineral or organic fertilization. *Agric. Ecosyst. Environ.* 129: 450–456.

Yao, X.-h., Min, H., Lü, Z.-h., and Yuan, H.-P. 2006. Influence of acetamiprid on soil enzymatic activities and respiration. *Eur. J. Soil Biol.* 42: 120–126.

Yeomans, J. C. and Bremner, J. M. 1985. Denitrification in soil: Effects of insecticides and fungicides. *Soil Biol. Biochem.* 17: 453–456.

Zabaloy, M. C., Garland, J. L., and Gómez, M. A. 2008a. Microbial respiration in soils of the Argentina pampas after metsulfuron-methyl, 2,4-D and glyphosate treatments. *Commun. Soil Sci. Plant Anal.* 39: 370–385.

Zabaloy, M. C., Garland, J. L, and Gómez, M. A. 2008b. An integrated approach to evaluate the impacts of the herbicides glyphosate, 2,4-D and metsulfuron-methyl on soil microbial communities in the Pampas region, Argentina. *Appl. Soil Ecol.* 40: 1–12.

Zelles, L. 1997. Phospholipid fatty acid profiles in selected members of soil microbial communities in soil. *Chemosphere* 35: 275–294.

Zhang, B., Bai, Z., and Hoefel, D., et al. 2009. The impacts of cypermethrin pesticide application on the non-target microbial community of the pepper plant phyllosphere. *Sci. Total Environ.* 407: 1915–1922.

Zhang, C., Liu, X., Dong, F., Xu, J., Zheng, Y., and Li, J. 2010a. Soil microbial communities response to herbicide 2,4-dichlorophenoxyacetic acid butyl ester. *Eur. J. Soil Biol.* 46: 175–180.

Zhang, C., Xu, J., Liu, X., et al. 2010b. Impact of imazethapyr on the microbial community structure in agricultural soils. *Chemosphere* 81: 800–806.

9

Pesticide Residues in Surface Water and Groundwater

Tatjana Vasiljević, Svetlana Grujić, Marina Radišić,
Nikolina Dujaković, and Mila Laušević

CONTENTS

9.1 Sources of Pesticide Residues in Water...259
    9.1.1 Pesticide Entry Routes into Water 261
9.2 Fate of Pesticides in Water 263
9.3 Regulatory Issues and Levels of Pesticide Residues in Surface Water and
    Groundwater 268
    9.3.1 Regulatory Issues Concerning Pesticide Residues in Surface Water and
        Groundwater 268
    9.3.2 Levels of Pesticide Residues Found in Surface Water and Groundwater......270
9.4 Removal of Pesticide Residues During Drinking Water Production 272
    9.4.1 Water Treatment Technology 273
9.5 Methods for Monitoring Pesticide Residues in Water 276
    9.5.1 Sample Preparation 276
        9.5.1.1 Solid-Phase Extraction...277
        9.5.1.2 Recent Trends 282
    9.5.2 Analyte Detection 282
        9.5.2.1 GC Detection Methods...282
        9.5.2.2 LC Detection Methods 283
        9.5.2.3 Comparison Between GC and LC Methods......................285
        9.5.2.4 Recent Trends 286
Acknowledgments..287
References...287

9.1 Sources of Pesticide Residues in Water

Agricultural food production accounts for about 70%–80% of the total pesticide use. Application of pesticides has greatly increased crop yields and reduced losses. The majority of agricultural pesticides are herbicides. Their widespread use has reduced the extent of mechanical cultivation of soil in the control of weeds. However, with respect to human exposure through food, insecticides and fungicides are the most important pesticides because they are applied shortly before or after harvesting. Herbicide residues are the most frequently detected residues in environmental waters (Manahan 2001). The temporal and spatial changes of pesticide concentration in surface water are related to the timing of

259

its seasonal application. It is more frequent and concentrated during late spring and early summer months, following the spring application, and diminishes significantly in fall and winter (Pereira and Hostettler 1993; Bradford et al. 2010). For instance, it was determined that almost 80% of the annual herbicide load to the Gulf of Mexico occurred during the growing season, that is, from May to August (Clark and Goolsby 2000).

Although agricultural pesticide inputs are dominant, it was shown that pesticide contributions from urban sources to water contamination should not be neglected, sometimes being similar to that of agriculture, especially on a local scale (Bucheli et al. 1998; Kolpin et al. 1998; Blanchoud et al. 2007; Wittmer et al. 2010). There are a number of nonagricultural pesticide sources, originating from urban pesticide use and nonapproved pesticide use or misuse. Pesticides are commonly used in landscaping, gardening, food distribution, and home pest control (Manahan 2001). Outdoor herbicides account for about 85% of pesticides used in urban areas (Braman et al. 1997). They are prone to be flushed into the sewage during rainfall. It was shown that effluents from urban wastewater treatment plants (WWTPs) contribute to a great extent to herbicide pollution of surface water (Nitschke and Schüssler 1998; Kolpin et al. 2006). Farmyard activities, such as pesticide spillage, tank and sprayer filling, use of faulty equipment, washing or waste disposal, etc., can also significantly contribute to pesticide loading to surface water and groundwater. It was estimated that farmyard activities can cause less than 20% of pesticide load into surface waters, originating from WWTPs or direct input of pesticides into surface waters (Gerecke et al. 2002; Neumann et al. 2002; Leu et al. 2004a). Many pesticides are not significantly eliminated in WWTPs (Nitschke et al. 1999), and therefore their residues end up in natural waters after WWTP effluents are discharged. Pollution can be reduced by good agricultural practices and the installation of systems for farmyard wastewater decontamination, which can provide a decontamination level greater than 90% (Fait et al. 2007). In urban areas, some pesticides are used for the protection of construction materials. They are added to outdoor building materials to prevent biological deterioration. For instance, the herbicide mecoprop is used as a protection agent in flat-roof sealing. It was determined that mecoprop loads from flat roofs enter surface waters up to 65% through WWTP effluents (Bucheli et al. 1998; Gerecke et al. 2002). Another nonagricultural source of pesticides is their use in the maintenance of the golf courses. Artificially created wetlands on a golf course have the potential for accepting and storing pesticide-polluted water from within the golf course (Kohler et al. 2004).

Pesticide pollution of surface water and groundwater can arise from point (localized) and nonpoint (diffuse) sources (Carter 2000; Gerecke et al. 2002; Holvoet et al. 2007). In agricultural areas, there are two main sources of pesticide losses to surface waters: farmyards acting as point sources and agricultural fields acting as diffuse sources (Müller et al. 2002; Leu et al. 2004a). Point sources of pollution are largely the result of pesticide-handling procedures, such as spills during filling of the spraying equipment, cleaning of the equipment, usage of faulty equipment, processing of spray waste, etc. Several studies have shown that the major risk of point source pollution is a poor operator handling within the farmyard area where sprayers are filled (Kreuger 1998; Müller et al. 2002; Neumann et al. 2002; Fait et al. 2007). Water contamination can be caused by accidental pesticide spillage as well as inadequate pesticide storage and disposal (Ricking and Schwarzbauer 2008; Frische et al. 2010). Sometimes, pesticide waste originating from agricultural or industrial activities can be approved for discharge into surface waters (Carter 2000). It was determined that for contamination of surface waters, WWTPs (for the treatment of farmyard and urban wastewater) are the most important point source of pesticides, constituting 20%–90% of the total pesticide water pollution in different catchments, especially during dry weather periods (Kreuger 1998; Müller et al. 2002; Neumann et al. 2002, 2003; Leu et al. 2004a; Holvoet et al. 2005).

Whereas the point sources can be controlled, the diffuse sources are much more difficult to handle because they are influenced by various interacting factors including soil type, weather, pesticide properties, and agricultural management practices. Watersheds can be significantly impacted by nonpoint source agricultural inputs. Nonpoint sources are related to the movement of pesticides from the field where they are applied to surface water and groundwater via leaching, surface runoff, spray drift, volatilization and atmospheric deposition, drain flow, groundwater flow, etc. (Carter 2000; Neumann et al. 2002). Owing to their dispersed nature of distribution in the environment, transport from nonpoint sources is difficult to predict, especially in large watersheds.

9.1.1 Pesticide Entry Routes into Water

There are numerous entry routes of pesticides into water. They can enter water either directly, in applications such as mosquito control, or indirectly, such as from drainage of agricultural lands. From agricultural applications, pesticides can contaminate natural waters via agricultural runoff to surface waters, leaching into groundwater, or spray drift. In urban applications, contamination can be caused by leaching from chemically treated surfaces or seepage from accidental spills and leaks, runoff from roads and paved areas, atmospheric fallout, and domestic sewage effluents. Also, in industrial areas, pesticides can reach waters by direct industrial spillages, industrial wastewater effluents, and leaching from industrial landfill sites. The predominant routes of entry arising from diffuse applications of pesticide include surface runoff, spray drift, and field drainage. Less significant routes to surface water include seepage to groundwater (leaching), subsurface lateral flow (interflow), and wet or dry deposition following long-range atmospheric transport.

Leaching or percolation through the soil is a process of pesticide transport directly to underlying groundwater, and subsequently surface water (Fait et al. 2010; Gonzalez et al. 2010; Leistra and Boesten 2010). It is assumed to be negligible as pesticides occur in groundwater in generally low concentrations. Moreover, the slow movement of groundwater favors sorption of pesticides to soil and their degradation, thus further decreasing pesticide concentration (Röpke et al. 2004). Losses of applied pesticide by leaching are typically less than 1% (Carter 2000). Residues that are strongly adsorbed to soil do not leach, but they can be transported in a suspended form in leachate. Leaching is greatly affected by soil irrigation and rainfall regime (Fait et al. 2010; Gonzalez et al. 2010).

When the infiltration capacity of the agricultural soil is exceeded during rainfall, dissolved pesticides and their residues adsorbed to soil particles can move across the treated agricultural soil as surface runoff or overland flow. It is considered to be the most important and major pathway of pesticides from agricultural fields as diffuse sources to surface water (Senseman et al. 1997; Huber et al. 1998, 2000; Schulz et al. 1998; Liess et al. 1999; Bach et al. 2001, 2010). Pesticides that are strongly adsorbed to soil particles are essentially transported via soil erosion (Wu et al. 2004). The abrasive power of surface runoff and the impact of raindrops detach soil particles and cause soil erosion. The partitioning of a pesticide between the solution and the soil solid phase is influenced by factors such as organic carbon and clay content of the soil (Holvoet et al. 2007). For water-soluble pesticides, losses via runoff are considered far more important, because the amount of eroded soil lost from a field is usually small compared with the runoff volume. Pesticide losses in agricultural runoff have been quantified to assess the contamination potential. It was estimated that they are typically less than 1.5% of the applied pesticide (Pereira and Hostettler 1993; Schottler et al. 1994) and sometimes significantly lower, less than 0.1% (Liess et al. 1999; Wu et al. 2004). Although pesticide losses seem small compared to the original concentration applied, their concentrations in water can reach levels detrimental to plant and fish species in water. Numerous factors, such as rainfall, watershed soil type, land use, landscape, slope, etc., affect runoff and, therefore, pesticide concentrations in surface water (Dabrowski et al. 2002; Leu et al. 2004b; Holvoet et al. 2007). For instance, soils situated on steeper slopes are particularly susceptible to surface runoff. However, these factors are regarded as a constant for a given watershed. The temporal and spatial changes of pesticide concentrations in surface water are primarily dependent on precipitation and pesticide use as the two major environmental variables dictating the dynamics of pesticide transport via runoff into surface water in a watershed (Müller et al. 2003; Guo et al. 2004). Precipitation, as the amount of daily rainfall, determines the total amount of runoff water, and pesticide use, as daily amount of used pesticide, represents the source of contamination. Concentrations of pesticides in environmental waters are the highest in late spring and early summer, following the spring application, coinciding with the greatest agricultural runoff from crop lands (Pereira and Hostettler 1993).

Another entry route of pesticides into water is through atmospheric long-range transport from agricultural to urban areas (Foreman et al. 2000; Grynkiewicz et al. 2001; Blanchoud et al. 2002; De Rossi et al. 2003) or to remote mountain areas (Kirchner et al. 2009; Tremolada et al. 2009; Bradford et al. 2010). Pesticides can be transported in the gas phase by volatilization from the crops on which they were applied, dissolved in water droplets by spray drift during pesticide application by spraying and adsorbed on solid particles (as suspended particulate matter) by wind erosion from the soil surface (Dörfler and Scheunert 1997; Bach et al. 2001; Grynkiewicz et al. 2001). Once airborne, the scale of pesticide transport through the atmosphere will depend on a variety of meteorological conditions as well as pesticide physicochemical properties. From the atmosphere, pesticides can be removed by photochemically driven reactions or physical depositional mechanisms, such as dry and wet deposition. Pesticides that are susceptible to photochemical reactions are usually transported through short distances from the source. Pesticides that are less susceptible to these chemical removal processes can be transported through greater distances (Foreman et al. 2000). Pesticides in the gas phase are usually removed by wet deposition and dissolution in rainfall. Particle-bound pesticides can be washed out by precipitation (wet deposition) or removed by dry particle deposition (Sauret et al. 2009; Foreman et al. 2000; Majewski et al. 2000). The concentration of pesticides in precipitation samples revealed seasonal fluctuations, with higher concentrations observed during the application period (Dörfler and Scheunert 1997; Grynkiewicz et al. 2001; De Rossi et al. 2003).

For volatile pesticides, losses after application due to volatilization can be as high as 90% (Carter 2000; Grynkiewicz et al. 2001; Bedos et al. 2002). This is especially important for short-range transport from application areas (Siebers et al. 2003). However, the impact of their later precipitation is negligible compared to that from their direct agricultural application (Dörfler and Scheunert 1997; Carter 2000). The volatilization rate depends on the vapor pressure of pesticide, weather conditions, humidity of the soil, type of the plant surface, etc. (Reichman et al. 2000; Grynkiewicz et al. 2001).

When a pesticide is applied as a spray near a water body, a key route to surface water is the spray drift. Drifting pesticide spray is a complex problem in which equipment design, application parameters, spray physical properties and formulation, and meteorological conditions interact and influence the pesticide loss (Gil and Sinfort 2005). The contribution of pesticide spray drift to surface water pollution is thought to be rather small (Kreuger 1998; Huber et al. 2000; Neumann et al. 2002; Röpke et al. 2004). The amount of pesticide that drifts to water 1 m from its application to crops can be ranging from 0.3% to 3.5% (Carter 2000). The total airborne spray drift is usually 0.8% of the applied dose (Wolters et al. 2008). In general, the amount of spray drift entering the water body increases with decreasing distance from the water body. Deposition is also influenced by landscape, as bankside vegetation and morphology, wind speed and direction, and other weather variables. For instance, the presence of trees at the edge of agricultural field, in front of the water body, and in the direction of wind can almost completely reduce the spray drift (Vischetti et al. 2008). It is estimated that spray droplets smaller than 10 μm can be transported over several kilometers even at low wind speeds during application. Their transport is favored by a continuous decrease in the droplet size due to evaporation (Hüskes and Levsen 1997). Agricultural pesticides can be transported by air through much larger distances and cause contamination in remote areas (Blanchoud et al. 2002; Bradford et al. 2010). It was determined that a large fraction of abundantly applied volatile pesticides reached the remote lakes up to 80 km downwind from the pesticide source with a delay time of 1–2 weeks. Dry deposition to the lake surface was presumably involved as an entry route because the concentration pattern in lake water followed the application rate pattern regardless of rainfall events (Bradford et al. 2010).

In order to remove excess water from slowly permeable soils or those with shallow water tables, land drainage systems are designed. This artificial drainage is responsible for drain flow, the transport of dissolved pesticides or pesticides adsorbed on sediment particles when heavy rainfall and subsequent drainage occur shortly after application, thus limiting the contact of these solutes with the soil. Surface waters receive nonpoint input of pesticides via subsurface field drainage pipes (Kladivko et al. 1999; Gentry et al. 2000). Pesticide losses in this way can be up to 1.9% of the amount applied to the soil but are generally less than 0.1% (Kladivko et al. 1999). Research findings have shown that drain flow is caused by preferential flow through soil macropores in tile-drained structured soils, especially in low-permeable clay soils (Novak et al. 2001; Leu et al. 2004a; Gärdenäs et al. 2006; Köhne et al. 2009). For soils with high clay content, high pesticide concentrations and losses are expected, because such soils are more structured and more prone to pesticide transport by preferential flow. In addition to clay content of the soil, important factors that influence pesticide losses in drain flow are pesticide properties and weather conditions. For pesticides with stronger sorption to soils and shorter half-life, losses were found to be smaller. With the increase of the time between pesticide application and first rainfall and drainage, pesticide losses are expected to decrease (Brown and van Beinum 2009).

Interflow is the lateral movement of water below the soil surface, instead of seeping vertically to the groundwater, which occurs naturally in the absence of artificial drainage (Kahl et al. 2008). Interflow can enter surface water via bankside seepage and will generally contain less pesticide residues than drain flow, as it has passed through the soil matrix having the opportunity for sorption and degradation of the pesticide (Carter 2000). It was found that the antecedent soil water content had a strong effect on triggering of interflow. Namely, as little as 0.1 mm of the rainfall (the "new" water) triggered the pesticide transport, and even two months after application, pesticide traces were detected in water (Kahl et al. 2008). This was explained by mixing small amounts of "new" water with larger amounts of "old" (antecedent) soil water, creating the lateral interflow. Studies have revealed that some of the leading factors influencing the lateral interflow and pesticide transport are antecedent soil moisture and antecedent rainfall (Ng et al. 1995).

9.2 Fate of Pesticides in Water

Pesticides by their nature are designed to destroy unwanted organisms. The key to their selective toxic effect is that they act against certain organisms without adversely affecting others. However, the absolute selectivity is difficult to achieve, and most pesticides create some risk to human health. This results in them being potentially harmful to nontarget organisms, and, therefore, they can be serious pollutants even in low concentrations. The major factors that determine if a pesticide can be approved for use are rapid biodegradation and minimal environmental toxicity. That is why persistent organochlorine pesticides were gradually withdrawn. They were replaced by organophosphorus (OPPs) and organonitrogen (ONPs) pesticides, as they are less stable in the environment than organochlorine compounds (Tankiewicz et al. 2010). The OPPs as a group vary considerably in chemical and environmental properties (Larson et al. 1997). Most of the phosphorus-containing organic compounds act as insecticides. Structurally, they are usually esters and decompose very quickly (Tankiewicz et al. 2010). Because of their low persistence and high effectiveness, these compounds are widely used as systematic insecticides for plant, animal, and soil treatments (Tadeo et al. 2008). In 2001, 33,000 tons of OPPs were sprayed onto crops in the United States (Kiely et al. 2004). ONP is the term covering a large number of different compounds. When plant protection products are considered, this term is usually used for carbamates and triazines. Carbamates are esters of carbamic acid, used worldwide as insecticides, herbicides, fungicides, and plant regulators. They are normally decomposed by soil microorganisms in 3–5 weeks. Triazine pesticides have been synthesized to control broad-leaved weeds in a variety of crops. They are effective at low dosages for destroying broad-leaved weeds in corn and other crops, and they can be used in high dosage as soil sterilants. These compounds have substantial persistence in soil (Tadeo et al. 2008).

World pesticide amount used exceeded 4.1 million tons in 2000 and 2001. Herbicides accounted for the largest portion of total use, followed by insecticides and fungicides. Total world pesticide amount used decreased in 2001 for all pesticide types. Pesticide amount used in the United States in both 2000 and 2001 exceeded 0.5 million tons, in proportions similar to those of world pesticide use. Ten most frequently used pesticides in the United States are presented in Table 9.1. Glyphosate was the most frequently used pesticide in 2001, replacing atrazine which was the most frequently used agricultural pesticide for a number of years (Kiely et al. 2004).

TABLE 9.1

The Most Frequently Used Pesticides in the US Agricultural Crop Production in the Years 2001, 1999, and 1997

Pesticide	Activity	2001 (×1000 t)	1999 (×1000 t)	1997 (×1000 t)
Glyphosate	Herbicide	36–41	30–33	15–17
Atrazine	Herbicide	34–36	34–36	34–37
Metam sodium	Fumigant	26–28	27–28	24–26
Acetochlor	Herbicide	14–16	14–16	14–16
2,4-D	Herbicide	13–15	13–15	13–15
Malathion	Insecticide	9–11	13–15	NA
Methyl bromide	Fumigant	9–11	13–15	17–20
Dichloropropene	Fumigant	9–11	7–9	15–17
Metolachlor-S	Herbicide	9–11	7–9	—
Metolachlor	Herbicide	7–10	12–14	29–31

Source: From Kiely, T., Donaldson, D., and Grube, A. 2004. *Pesticides Industry Sales and Usage. 2000 and 2001 Market Estimates.* US Environmental Protection Agency: Washington, DC. http://www.epa.gov/pesticides/pestsales/01pestsales/market_estimates2001.pdf (accessed 28 January 2011).

The widespread use of pesticides inevitably leads to their presence in surface waters and groundwater. Pesticides can be transformed in the environment into a large number of degradation products, commonly defined as transformation products (TPs), although other terms, such as metabolites or pesticide derivatives can be used. Pesticide transformation is any process in which change in the molecular structure of the pesticide takes place. Most of the applied pesticides are ultimately degraded into universally present compounds, such as carbon dioxide, ammonia, water, mineral salts, and humic substances (Somasundaram and Coats 1991). The rates at which transformation reactions occur in aquatic systems vary considerably for individual chemicals and under different environmental conditions. Some compounds undergo the transformation immediately after or even before application, whereas complete mineralization of some pesticides, such as dichlorodiphenyltrichloroethane (DDT) or mirex, can take decades (Larson et al. 1997). From the same parent compound, different TPs can be created by various transformation processes. Once a pesticide is introduced in the aquatic environment, it can be subjected to a variety of processes: physical (accumulation, deposition, dilution and diffusion); chemical (hydrolysis and oxidation); photochemical (photolysis and photodegradation), and biochemical (biodegradation, biotransformation and bioaccumulation) processes. Photolysis and hydrolysis are main abiotic processes that take place in the aquatic environment (Barceló and Hennion 1997). In photolysis, the light energy and intensity, as well as duration of the sunlight, affect the rate of pesticide degradation. Many environmental parameters, such as pH, temperature, water type, and the presence of humic substances can catalyze or hinder the hydrolysis or photolysis of pesticides (Lartiges and Garrigues 1995; Bachman and Patterson 1999; Iesce et al. 2006).

Many modern-day pesticides are degraded by microbial or chemical processes into nontoxic products. In some instances, transformation products can be more toxic and pose a greater risk to the environment than the parent compound (Belfroid et al. 1998). Moreover, TPs can have different properties that enable them to occur in areas not reached by the parent compounds. The process of degradation generally increases the water solubility and the polarity of the compound. The increase in solubility is caused by the loss of carbon, the incorporation of oxygen, and the addition of carboxylic acid functional groups. For example, atrazine degrades through a combination of physical, chemical, and biochemical processes. The three major degradation products of atrazine are deethylatrazine (DEA), deisopropylatrazine (DIA), and hydroxyl-atrazine (HA) (Kruger et al. 1993). Atrazine's maximum solubility is 33 mg/L, while DEA (loss of two carbon atoms) and DIA (loss of three carbon atoms) have solubilities of 670 mg/L and 3200 mg/L, respectively (Thurman and Meyer 1996). Due to their mobility in soil and water environment, TPs can reach groundwater more easily than parent compounds (Martínez Vidal et al. 2009). Table 9.2 shows a brief summary of TPs that have been studied in the recent years as well as their concentrations found in studied waters.

Two of the most widely used classes of agricultural herbicides are triazines and chloroacetamides (Table 9.1). The difference between these two classes of herbicides is that a substantially larger fraction of chloroacetamides undergoes transformation in soils than in the case of triazines (Hladik et al. 2005). Thus, triazine pesticide TPs are found in lower concentrations than parent compounds, while the opposite is true for chloroacetamide pesticide TPs. Chloroacetamide TPs are readily formed, but once they are transported in an oligotrophic environment, they are likely to be persistent. The two most commonly investigated classes of chloroacetamide TPs, ionic ethanesulfonic acid (ESA) and oxanilic acid (OA), are not likely to present a risk to human health (Heydens et al. 1996, 2000). However, some neutral chloroacetamide derivatives were found to

TABLE 9.2

Transformation Products of Pesticides and Their Concentrations Found in Water Samples

Matrices	Parent Compound (ng/L)	Transformation Products	TP (ng/L)	References
Groundwater and surface water, Iowa, United States	Acetochlor	Acetochlor ESA Acetochlor OA	Groundwater: sum for six TPs 1200, parent compounds <500	Kalkhoff et al. (1998)
	Alachlor	Alachlor ESA Alachlor OA		
	Metolachlor	Metolachlor ESA Metolachlor OA		
			Surface waters: sum for six TPs 6400, parent compounds 130	
Groundwater, Midwestern, United States			Max found concentrations:	Kolpin et al. (1996)
	Alachlor	Alachlor ESA 2,6–diethylanaline	8630 20	
	Atrazine	DEA DIA	2200 1170	
	Cyanazine	Cyanazine amide Deethylcyanazine amide Deethylcyanazine DIA	550 ND ND ND	
	Dacthal	DCPA acid metabolite	2220	
	DDT	DDE	30	
Groundwater, Iowa, United States	Acetochlor (770)	Acetochlor ESA Acetochlor OA	11500 800	Kolpin et al. (2000)
	Alachlor (630)	Alachlor ESA Alachlor OA	8500 33400	
	Atrazine (2100)	DEA DIA HA	590 1100 1300	
	Cyanazine (510)	Cyanazine amide	640	
	Metolachlor (11300)	Metolachlor ESA Metolachlor OA	8600 15300	
Groundwater, Portugal	Atrazine (15–42000)	DEA	140–34000	Goncalves et al. (2007)
	Endosulfan (3–4200)	Endosulfan sulfate	2–920	
Groundwater, Northern Italy	Terbuthylazine (ND–3430)	DET TH DETH	ND–260 ND ND	Guzzella et al. (2006)
	Atrazine (ND–530)	DEA DIA HA DEDIA DEAH DIAH DEDIAH	ND–280 ND–70 ND ND–110 ND–70 ND ND	
Surface water, Iowa, United States	Acetochlor (ND–210)	Acetochlor ESA Acetochlor OA	ND–1600 ND–1400	Kalkhoff et al. (2003)

TABLE 9.2 (Continued)

Transformation Products of Pesticides and Their Concentrations Found in Water Samples

Matrices	Parent Compound (ng/L)	Transformation Products	TP (ng/L)	References
	Alachlor (ND)	Alachlor ESA	ND–3500	
		Alachlor OA	ND–540	
	Atrazine (ND–1500)	DEA	ND–390	
		DIA	ND–360	
		HA	ND–8800	
	Cyanazine (ND–640)	Cyanazine amide	ND–1200	
	Metolachlor (ND–420)	Metolachlor ESA	ND–6700	
		Metolachlor OA	ND–1300	
Surface water, Switzerland	Metolachlor	Metolachlor ESA	Lake Greifensee: 34–55 River Aa Uster: 5–25 River Aa Möuchalfart: 50–280 Parent compound: Lake Greifensee: 2–10 River Aa Uster: <LOQ (2 ng/L) River Aa Möuchalfart: 2–200	Hantscha et al. (2008)
Groundwater and surface water, Valencia, Mediterranean region	Terbuthylazine (ND–422)	DET	ND–936	Hernández et al. (2008b)
		TH	ND–142	
		DETH	<50ᵃ–1257	
	Simazine (ND–611)	2–hydroxy–simazine	ND–236	
		DIA	ND–815	
	Terbumeton (ND–950)	Desethylterbumeton	ND–2223	
		Bentazone-methyl	ND–71	
	Carbendazim (ND–125)	2–aminobenzimidazole	ND–280	
Surface water (Rhône river), France	MCPA (4-chloro-2-methylphenoxy)-acetic acid (ND–4240)	4-chloro-2-methylphenol	ND–1120	Chiron et al. (2009)
		4-chloro-2-methyl-6-nitrophenol	ND–450	
Surface water (Nzoia river), Kenya	Alachlor (ND–350) Metolachlor (ND)	2,6-diethylaniline 2-ethyl-6-methylaniline	sum for two TPs: ND–23000	Osano et al. (2003)

[a] Limit of detection.

ESA, ethanesulfonic acid; OA, oxanilic acid; DEA, deethylatrazine; DIA, deisopropylatrazine; HA, hydroxylatrazine; DET, deethyl-terbuthylazine; TH, 2-hydroxy-terbuthylazine; DETH, deethyl-2-hydroxy-terbuthylazine; DEDIA, deethyl-deisopropyl-atrazine; DEAH, deethyl-2-hydroxy-atrazine; DIAH, deisopropyl-2-hydroxy-atrazine; DEDIAH, deethyl-deisopropyl-2-hydroxy-atrazine; DET, deethyl-terbuthylazine; DDT, dichlorodiphenyltrichloroethane; DDE, dichlorodiphenyldichloroethylene; MCPA, 2-methyl-4-chlorophenoxyacetic acid.

be toxic. For instance, 2,6-diethylaniline and 2-ethyl-6-methyaniline are promutagens, more teratogenic than their parent compounds, metolachlor and alachlor (Kimmel et al. 1986; Osano et al. 2002).

There is considerable information on the occurrence and sources of triazine and chloro-acetamide TPs in surface water and groundwater. However, information on the occurrence and faith of other pesticide classes are rather scarce. Degradation of carbamate and OPPs pesticides was studied in the laboratory conditions, but data on the occurrence of their TPs in the environment is insufficient. Some of the reasons for avoiding the studying of pesti-cide TPs in environmental samples are as follows: many of these compounds are still not well known; they are usually more polar than the parent compounds and their isolation from the water matrix is more difficult; and the commercial availability of reference stan-dards is rather limited. Also, if TPs are included in water analysis, the number of potential analytes to be investigated in water would drastically increase. Another problem could be the selection of relevant pesticide TPs for monitoring studies. For instance, depending on the properties of the matrix, the same pesticide can form different TPs. In order to have a clear picture on water contamination by pesticides, research should include a variety of TPs and their parent compounds from different pesticide classes.

9.3 Regulatory Issues and Levels of Pesticide Residues in Surface Water and Groundwater

The wide usage of pesticides in agriculture not only causes contamination of agricultural crops but also affects the whole environment. Over the last two decades, the extensive con-tamination of aquatic and soil ecosystems has forced the adoption of restrictive legislative measures to protect the environment against pollution.

9.3.1 Regulatory Issues Concerning Pesticide Residues in Surface Water and Groundwater

In the European Union (EU), the presence of pesticides in the aquatic environment is regu-lated through various directives that set restrictive limits for individual and total pesti-cides and demand regular monitoring (Council of the European Communities 1975, 1979, 1980; Council of the European Union 1998, 2000, 2001). The first legislation adopted was the Directive concerning the quality required of surface water intended for the abstraction of drinking water in the Member States (Council of the European Communities 1975). In 1980, the first official act regarding the quality of drinking water was accepted, relating to the quality of water intended for human consumption (Council of the European Communities 1980). The EU adopted the Directive establishing a framework for Community action in the field of water policy, with the objective to improve, protect, and prevent further dete-rioration of water quality across Europe (Council of the European Union 2000). Decision 2455/2001/EC (Council of the European Union 2001) has established a list of 33 priority substances in the field of water protection, the third of which are pesticides. Directive 2008/105/EC (Council of the European Union 2008) set environmental quality standards (EQS) on the concentration of 41 dangerous chemical substances in surface waters, includ-ing 33 priority substances and 8 other pollutants that pose a particular risk to animal and plant life in the aquatic environment and to human health. In that way, EQSs were set for

TABLE 9.3

Environmental Quality Standards (EQS) for Pesticides Set by Directive 2008/105/EC

Pesticide	Chemical Class	Activity	AA-EQS Inland Surface Waters (µg/L)	AA-EQS Other Surface Waters (µg/L)	MAC-EQS Inland Surface Waters (µg/L)	MAC-EQS Other Surface Waters (µg/L)
Alachlor	Chloroacetanilide	Herbicide	0.3	0.3	0.7	0.7
Atrazine	Triazine	Herbicide	0.6	0.6	2.0	2.0
Chlorfenvinphos	Organophosphorus	Insecticide, acaricide	0.1	0.1	0.3	0.3
Chlorpyrifos	Organophosphorus	Insecticide, acaricide	0.03	0.03	0.1	0.1
Aldrin Dieldrin Endrin Isodrin	Cyclodiene		$\Sigma = 0.01$	$\Sigma = 0.005$	n.a.	n.a.
DDT total para-para-DDT	Organochlorine	Insecticide, acaricide	0.025 0.01	0.025 0.01	n.a.	n.a.
Diuron	Phenylurea	Herbicide	0.2	0.2	1.8	1.8
Endosulfan	Organochlorine	Insecticide, acaricide	0.005	0.0005	0.001	0.004
Isoproturon	Phenylurea	Herbicide	0.3	0.3	1.0	1.0
Simazine	Triazine	Herbicide	1	1	4	4
Trifluralin	Dinitroaniline	Herbicide	0.03	0.03	n.a.	n.a.

Source: Council of the European Union. 2008. Directive 2008/105/EC of the European Parliament and of the Council of 16 December 2008 on environmental quality standards in the field of water policy, amending and subsequently repealing Council Directives 82/176/EEC, 83/513/EEC, 84/156/EEC, 84/491/EEC, 86/280/EEC and amending Directive 2000/60/EC of the European Parliament and of the Council. *Off. J. Eur. Union* L348: 84–97.

AA, annual average; MAC, maximum allowable concentration; n.a., not applicable.

a number of pesticides (Table 9.3). Awareness of aquifers' vulnerability caused the adoption of Directive 2006/118/EC (Council of the European Union 2006), on the protection of groundwater against pollution and deterioration. This legislation has set a maximum of 0.1 µg/L for individual pesticides and 0.5 µg/L for total pesticides, including the active substances and their relevant metabolites and degradation and reaction products. These limits coincide with those required by Directive 98/83/EC (Council of the European Union 1998). Restrictive legislations from the EU, concerning the permitted level of pesticide residues in waters, induced the development of novel analytical techniques and the improvement of existing ones, so that the largest possible number of compounds at low levels can be determined in small-volume samples. Which methodology is the most suitable in a given context depends on the type and the complexity of the matrix and on the chemical structure of the target pesticides (Tankiewicz et al. 2010).

In 1972, the United States Environmental Protection Agency (US EPA) enacted and the US Congress ratified the Federal Water Pollution Control Act, commonly referred to as the Clean Water Act, which focused attention to preventing point and nonpoint pollution sources, including pesticides. This Act was amended several times, the last time being in 1987 (US EPA 1987). Another federal law related to pesticides is the Safe Drinking Water Act, initiated in 1974 and amended in 1986 and 1996 (US EPA 1996).

9.3.2 Levels of Pesticide Residues Found in Surface Water and Groundwater

Different researchers investigated the presence of pesticides in surface waters and groundwater and reported ng/L concentrations of these compounds in the European region and on the American continent. The most frequently detected pesticides in surface water and groundwater are shown in Table 9.4. The commonly detected pesticides were those heavily used in the past as well as in the present. Triazine herbicides (atrazine and simazine) were detected most often and at the highest concentrations in surface water. These herbicides are commonly used for weed control in the crop protection. Owing to their toxicity, water solubility, adsorptivity, and persistence in the environment, they are regarded as detrimental water contaminants (Shen and Lee 2003). Phenylureas are reported to be among

TABLE 9.4

Concentration of Pesticides Found in Surface Water and Groundwater

Pesticide	Concentration (ng/L)	Matrix	References
Bentazone	30510	Surface water (Spain)	Kuster et al. (2008)
MCPA	1877		
Simazine	45		
Molinate	222		
Atrazine	314		
Propanil	969		
2,4 D	24		
Mecoprop	8		
Metolachlor	27		
Diuron	15		
MCPA	150	Surface water (Australia)	Jordan et al. (2009)
Atrazine and Acetochlor	200–1000	Surface water (Hungary)	Maloschik et al. (2007)
Prometryn, terbutryn, and diazinon	100–400		
Atrazine	126	Surface water (Spain)	Palma et al. (2009)
Simazine	31		
Diuron	24		
Terbuthylazine	49		
Bentazone	93	Surface water (France)	Gervais et al. (2008)
Diuron	33		
Isoproturon	379		
Triclopyr	66		
Isoproturon	4	Surface water (Germany)	Asperger et al. (2002)
Diuron	9		
Chlortoluron	1		
Simazine	12		
Atrazine	5		
Terbutylazine	1		
Prometryn	4		
Chlorfenvinphos	2		
Atrazine	31	Groundwater (Spain)	Postigo et al. (2010)
Simazine	37		
Diuron	9		
DDT	150–190	Groundwater (India)	Shukla et al. (2006)
α-Endosulfan	1340–2140		
β-Endosulfan	210–870		
Lindane	680–1380		

TABLE 9.4 (Continued)

Concentration of Pesticides Found in Surface Water and Groundwater

Pesticide	Concentration (ng/L)	Matrix	References
Atrazine	380	Groundwater (United	Kolpin et al. (2004)
Metolachlor	3200	States)	
Atrazine	253	Groundwater (Europe)	Loos et al. (2010)
Simazine	127		
Terbutylazine	716		
Bentazone	11		
Propazine	25		
Diuron	279		
Endosulfan	1–12	Groundwater (Morocco)	El Bakouri et al. (2008)
Atrazine	The most	Groundwater (United	Tesoriero et al. (2007)
Metolachlor	commonly	States)	
Alachlor	detected		
Chloroturon	160–1550	Surface water and	Carabias-Martínez
Atrazine	180–1500	groundwater (Spain)	et al. (2003)
Simazine	520	Surface water and	Marín et al. (2006)
Carbendazim	370	groundwater (Spain)	
Diuron	370		
Bromacil	570		
Atrazine	30–280	Surface water and	Hildebrandt et al.
Simazine	10–140	groundwater (Spain)	(2008)
Atrazine	56–73	Surface water and	Carvalho et al. (2008)
Terbuthylazine	27–132	groundwater (Portugal)	
Diuron	56		
Carbendazim	2273	Surface water and	Belmonte Vega et al.
Simazine	918	groundwater (Spain)	(2005)
Diuron	1233		
Carbendazim	8–22	Surface water and	Dujaković et al. (2010)
Dimethoate	23	groundwater (Serbia)	
Carbofuran	25		
Propazine	6–18		

the most widely used herbicides in agriculture today (Barbash et al. 2001). These pesticides have received particular attention in the recent years because of their toxicity and possible carcinogenic properties and considerable time required for their removal from the environment. Diuron, as phenylurea pesticide, was among the most frequently detected pesticides in surface water. As for groundwater, atrazine, simazine, and diuron were the most commonly detected pesticides due to their moderate sorption to soil and relatively high persistence. These compounds were found at levels below the EU limits in European surface waters. Concentrations of these pesticides in European groundwater (Table 9.4) were sometimes above the maximum of 0.1 μg/L for individual pesticides. The residues of carbendazim, a benzimidazole fungicide with an extensive and widespread use in agriculture on fruit, sunflower, sugar beet, and wheat, were reported in several surface water and groundwater studies (Dujaković et al. 2010). Bentazone, an acidic pesticide used for weed control in rice crops in Europe (Kuster et al. 2008), was also reported in a few studies. Additionally, the most frequently detected pesticides (simazine, atrazine, and diuron) and the occasionally detected ones (isoproturon and endosulfan) are considered priority hazardous substances by the Directive 2000/60/EC (Council of the European Union 2000). They are also included in Decision 2455/2001/EC, which establishes a list of priority substances in the field of water policy (Council of the European Union 2001).

9.4 Removal of Pesticide Residues During Drinking Water Production

Drinking water comes from two main sources: surface water and groundwater. Profound groundwater from restrained aquifers is usually free of direct contamination with pesticides. However, shallow aquifers are accessible to contamination from agricultural wastes. The treatment used to produce drinking water from surface water and groundwater must guarantee the removal of pesticides or at least reduce their concentration below the established limits. Pesticide detection rates in treated groundwater were generally much lower than detection rates in treated surface water, but the gap between them steadily decreased in the last 20 years (McInnis 2010).

Intensive studies have been carried out in an attempt to determine what levels of pesticides are acceptable in water supplies. The first effort was made in 1969 when permissible limits for 10 pesticides in public water supplies in the United States were established and used as guidelines (Edwards 1973). Since then, standards for drinking water have been revised several times. As an example, the recommended maximum value for aldrin in drinking water has been reduced, through a number of revisions during the last 50 years, from 17 µg/L (Edwards 1973) to 0.03 µg/L (World Health Organization 2008). A lot of progress in pesticide legislative has been made but for most pesticides, drinking water standards have yet to be set.

The US EPA has the authority to develop nationwide standards, and some of the states are setting local standards as well. Organic pesticides are covered by a guideline limiting the concentration of any single organic chemical to no more than 50 mg/L and the combined concentration of all organics found to no higher than 100 mg/L. Under the Safe Drinking Water Act (US EPA 1996), EPA establishes maximum contaminant levels (MCLs) for pesticides in water delivered to users of a public water system. However, drinking water standards for all pesticides found in water have not been established. EPA has established MCLs for only 21 pesticides (Table 9.5), 10 of which are no longer approved for use (Hetrick et al. 2000).

In the EU, the so-called Drinking Water Directive, the Directive 98/83/EC (Council of the European Union 1998), aims to protect human health from the adverse effects of contamination of water intended for human consumption by ensuring that it is "wholesome and clean." This Directive has set limits for pesticides in water intended for human consumption as 0.1 µg/L for individual pesticides and 0.5 µg/L for the sum of all pesticides. Moreover, depending on the method used for water treatment, water intended for drinking water production is also subjected to a maximum limit that varies between 1 and 5 µg/L for total pesticides (Council of the European Communities 1975).

The complicating factors in setting standards for the individual chemicals is that it is unknown how a given compound might interact with other chemicals to affect human health. Health studies have been conducted on people drinking contaminated water supplies, but these studies are limited by the fact that many health problems are difficult to trace to a specific cause as well as by the fact that the number of such studies is limited.

A survey conducted across India definitely pointed toward high pesticide contamination (DDT, β-endosulfan, α-endosulfan, and lindane) in drinking water (Shukla et al. 2006).

TABLE 9.5

Maximum Contaminant Levels for Pesticides in Drinking Water Set by EPA

Pesticide	Chemical Class	Activity	MCLG (µg/L)	MCL or TT (µg/L)
Alachlor	Chloroacetanilide	Herbicide	0	2
Atrazine	Triazine	Herbicide	3	3
Carbofuran	Carbamate	Insecticide	40	40
Chlordane	Organochlorine	Insecticide	0	2
2,4-D	Phenoxy	Herbicide	70	70
Dalapon	Organochlorine	Herbicide	200	200
1,2-Dibromo-3-chloropropane	—	Nematocide	0	0.2
Dinoseb	Dinitrophenol	Herbicide	7	7
Diquat	Desiccant	Herbicide	20	20
Endothall	Dicarboxylic acid	Herbicide	100	100
Endrin	Organochlorine	Insecticide	2	2
Glyphosate	Glycine derivative	Herbicide	700	700
Heptachlor	Organochlorine	Insecticide	0	0.4
Heptachlor epoxide	Organochlorine	Insecticide	0	0.2
Lindane	Organochlorine	Insecticide	0.2	0.2
Methoxychlor	—	Insecticide	40	40
Oxamyl (Vydate)	Carbamate	Insecticide	200	200
Picloram	Pyridinecarboxylic acid	Herbicide	500	500
Simazine	Triazine	Herbicide	4	4
Toxaphene	—	Insecticide	0	3
2,4,5-TP (Silvex)	—	Herbicide	50	50

Source: US EPA. 2002. National Primary Drinking Water Regulations. http://water.epa.gov/drink/contaminants/index.cfm#Organic (accessed 28 January 2011.)

MCLG, maximum contaminant level goal; MCL, maximum contaminant level; TT, treatment technique (a required process intended to reduce the level of a contaminant in drinking water).

In contrast, the levels of pesticides found in drinking water in Canada were quite low. For example, in the period from 1986 to 2006, from 675 drinking water systems in Canada, 16,166 water samples were tested for 104 pesticides and pesticide degradation products. Only in four samples the concentrations of atrazine or terbufos exceeded the guideline values of the Ontario Drinking Water Quality Standards (McInnis 2010).

Because of the complexity of water treatment technology associated with local water quality conditions and natural treatment systems such as bank filtration (Maeng et al. 2010), it is difficult to establish the standard criteria for defining pesticide-secure water treatment. In order to achieve an adequate control of pesticides in drinking water, it is also necessary to know details of their behavior under drinking water treatment conditions. Water treatment data can be derived from studies providing information on the removal efficiency of the pesticide and the identification of transformation by-products obtained in the laboratories, pilot plant and actual water treatment plant monitoring studies.

9.4.1 Water Treatment Technology

Water treatment involves the deliberate addition of numerous chemicals to improve the safety and quality of the drinking water produced for consumers. The conventional chemicals and materials used in water treatment plants and their effect on pesticide removal

TABLE 9.6

Removal of Some Pesticides, for Which Guideline Values Have Been Established, in Water Purification Processes

Pesticide	Chlorination		Coagulation		Activated Carbon Adsorption		Ozonization	
	R (%)	C (μg/L)	R (%)	C (μg/L)	R (%)	C (μg/L)	R (%)	C (μg/L)
Alachlor					>80	<1	>50	
Aldicarb	>80	<1			>80	<1	>80	
Aldrin/dieldrin			>50		>80	<0.02	>80	<0.02
Atrazine			Limited		>80	<0.1	>50	
Carbofuran	Limited				>80	<1		
Chlordane					>80	<0.1	>80	<0.1
Chlorotoluron					>80	<0.1	>80	<0.1
Cyanazine					>80	<0.1	Limited	
2,4-Dichlorophen-oxyacetic acid			Limited		>80	<1	>80	<1
1,2-DCP					>80	<1		
Dimethoate	>80	<1			>50		>50	
Endrin			Limited		>80	<0.2		
Isoproturon	>50				>80	<0.1	>80	<0.1
Lindane					>80	<0.1	>50	
MCPA					>80	<0.1	>80	<0.1
Mecoprop					>80	<0.1	>80	<0.1
Methoxychlor			>50		>80	<0.1		
Metolachlor					>80	<0.1		
Simazine	Limited		>80	<0. 1	>80	<0.1	>80	<0.1
Terbuthylazine			>80	<0.1				

Source: World Health Organization. 2008. Guidelines for Drinking-water Quality, 3rd ed., Vol. 1, Recommendations. http://www.who.int/water_sanitation_health/dwq/gdwq3rev/en/ (accessed 28 January 2011).

R, removal rate of pesticide in percentage; C,concentration of pesticide that should be achievable; a blank entry indicates that the process is completely ineffective or that data on the effectiveness are not available.

are presented in Table 9.6, adapted from Guidelines for Drinking-water Quality (World Health Organization 2008) where only some pesticides for which guideline values have been established are listed.

Chlorination is employed primarily for microbial disinfection. However, chlorine also acts as an oxidant and can remove or assist in the removal of some easily oxidized pesticides such as aldicarb, dimethoate, and some OPPs pesticides (Acero et al. 2008; Duirk et al. 2010). It is not particularly effective in pesticide removal and has the disadvantage of producing halogenated products. Chlorine reacts mainly with aromatic systems and double bonds, leading to pesticide transformation and, in some cases, the formation of halogenated organic compounds, some of which exhibit a potentially carcinogenic activity (Singer and Reckhow 1999).

Coagulation is the least efficient treatment of those pesticides mentioned in Table 9.6, being efficient for only a few pesticides, regardless of the amount of coagulant employed. Most pesticides are not removed or transformed during physicochemical treatments such

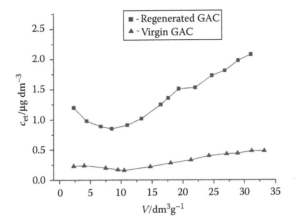

FIGURE 9.1
Cumulative breakthrough curves for 10 different chlorinated pesticides on virgin and regenerated granular activated carbon (GAC). Total pesticide concentration in the influent is 20 μg/L. (From Ninković, M. B., Petrović, R. D., and Laušević, M. D. 2010. Removal of organochlorine pesticides from water using virgin and regenerated granular activated carbon. *J. Serb. Chem. Soc.* 75, 565–573. With permission.)

as coagulation, flocculation, or sedimentation (Pehkonen and Zhang 2002). Therefore, more efficient processes, such as activated carbon adsorption or ozonization, are needed to remove pesticides during drinking water production (Ormad et al. 2008).

Activated carbon adsorption process seems to be the most feasible process for removing pesticides. It has adsorption affinity for contaminants that are hydrophobic. Therefore, a majority of pesticides that exhibit low solubility in water are efficiently removed on fixed-bed adsorbers with carbon, used in water purification plants. Many studies concerning the adsorption capacities of selected types of pesticides on commercially activated carbons have been published (Schreiber et al. 2005; Daneshvar et al. 2007). The adsorption capacity of activated carbon to remove pesticides is affected by competition from other contaminants or natural organic matter that blocks the active seats (Pelaekani and Snoeyink 2000). After saturation, used activated carbon is regenerated and reused in order to reduce the cost of water production and minimize waste. The adsorption capacity of regenerated activated carbon is considerably lower compared to the virgin activated carbon (Figure 9.1). In addition, rinsing of the pesticides after the saturation point is a far more efficient process on regenerated carbon (Ninković et al. 2010).

Consequently, there is a growing interest in developing new carbon adsorbents, which are more effective for the removal of micropollutants, including pesticides (Baćić-Vukčević et al. 2006). Hence, activated carbon fibers (Martin-Gullon and Font 2001), carbon cloth (Vasiljević et al. 2004b, 2006; Ayranci and Hoda 2004), and other alternative carbon materials (Vukčević et al. 2008) with higher adsorption capacity and kinetics have been investigated as a possible replacement for activated carbon in water purification plants. Among all the recently developed nanoadsorbents for drinking water purification, the chemistry of noble metal nanoparticles is truly unique. Synergistic effects could be driven by the use of silver nanoparticles loaded on activated carbon substrate. Both components work synergistically to remove a variety of pesticides from drinking water (Anshup 2009).

Ozone is primarily used as a disinfectant. However, as a powerful oxidant, ozone has many uses in water treatment, including the oxidation of pesticide residues. The intensive treatment by ozone is very effective because it yields very high degradation of pesticides and induces far less toxic products than chlorination. The combination of activated carbon adsorption and ozonization was found to be the most efficient method for degrading the majority of pesticides in the water purification process (Ormad et al. 2008).

9.5 Methods for Monitoring Pesticide Residues in Water

For the protection of human health and the environment, pesticide residues are routinely monitored in food, water, soil, and tissue samples. The survey on pesticide residues in the environment is faced with the identification and quantification of hundreds of compounds with widely different physicochemical properties. Furthermore, ultrasensitive analytical methods are mandatory since the water tolerable limits for most of the pesticides are in low levels of μg/L and, in some cases, ng/L (Council of the European Union 1998, 2008). To be detectable and quantified, the substances present in trace levels must be extracted and concentrated prior to chemical analysis.

Pesticide residues are isolated and concentrated from water samples using different extraction procedures, such as liquid–liquid extraction (LLE) or solid-phase extraction (SPE), usually followed by a solvent evaporation step (Gan and Bondarenko 2008). Gas chromatography (GC) or liquid chromatography (LC) coupled with conventional detectors, such as flame ionization detector (FID) (Pico et al. 1994), electron capture detector (ECD) (Lipinski 2000), nitrogen-phosphorus detector (NPD) (Lipinski 2000), ultraviolet-visible (UV/VIS) (Chiron et al. 1995; Curini et al. 2001), photodiode array (PDA) (Carabias-Martínez et al. 2005), or fluorescence detector (FLD) (Chiron et al. 1995), has been traditionally applied for pesticide residue analysis. Currently, the determination of pesticide residues requires the use of the chromatographic techniques hyphenated to mass spectrometry (MS) as the detection system. In general, MS is widely applied in trace analysis due to its selectivity, sensitivity, and confirmation capability (Gonçalves and Alpendurada 2004).

Different strategies could be applied for monitoring pesticides in water, depending on the objectives pursued (target or nontarget analysis). Both types of analysis have the need for different analytical schemes and may require different instrumentation. An example of target analysis is the inspection of pesticide EQS in surface water. The relevant analytes are preselected by the residue definition given in the EQS regulation (Council of the European Union 2008). In contrast, the EU regulation on residues in drinking water does not contain detailed residue definitions (Council of the European Union 1998). Besides, monitoring and identifying pesticide degradation products, which is considered to be crucial for complete environmental risk assessment, also requires a nontarget analysis. For nontarget screening, instruments must be able to generate sufficient information for the elucidation of residues, such as accurate mass, from which empirical formulae can be deduced.

9.5.1 Sample Preparation

Due to the low levels of pesticides detected in the environment and the complexity of the environmental matrices, preconcentration and cleanup of the samples, prior to chemical analysis, is usually required. The preconcentration is achieved through phase transfer, most often by the use of LLE or SPE (Gan and Bondarenko 2008). SPE is the most widely used procedure for cleanup, as well.

LLE is one of the earliest methods used for analyzing pesticides in water samples. It relies on the partition of analytes between two immiscible liquids, usually an aqueous solution and an organic solvent. Because of its simplicity and also its inclusion in many standard methods, LLE used to be the most popular sample preparation technique for pesticide analysis in many matrices. Its advantages include relatively minimal equipment requirements and low demand on the analyst skills, compatibility with a broad range of pesticides, and reliability. But, there are a number of drawbacks regarding the standard LLE. The most distinguished one is the consumption of large quantities of organic solvents, which makes LLE methods less environment friendly. In addition, polar pesticides, as well as their degradation products, cannot be extracted using this method (Kuster et al. 2006). LLE is generally labor-intensive, time-consuming, and physically demanding. For the reasons mentioned above, LLE has been largely replaced by SPE (Petrovic et al. 2010) in the last decade.

9.5.1.1 Solid-Phase Extraction

The ideal sample preparation method should be fast, accurate, precise, and consuming low amounts of solvent. Furthermore, it should be easily adapted for field work and employ less costly materials. Capability of automation is highly favored as well. SPE is the isolation technique capable of meeting these expectations (Picó et al. 2007). As seen in Tables 9.7 and 9.8, SPE is the technique most widely used for sample preparation prior to chromatographic analysis of pesticide residues in water.

In SPE, the analyte is transferred from the aqueous phase onto a sorbent phase, and then recovered for analysis. A typical SPE sequence includes the activation of the sorbent bed (conditioning), application of the sample, removal of interferences (cleanup), elution of the sorbed analytes, and reconstitution of the extract (Liška 2000). Exact conditions are usually specified by the manufacturer and may vary significantly in the types and volumes of solvents used for conditioning and elution.

Sorbents available in standard SPE include the common sorbents used in LC, such as bonded silica phases and polymers. The most popular phases used in pesticide analysis are octadecyl- (C18) and octyl-silica (C8) and styrene-divinylbenzene copolymers (Tables 9.7 and 9.8). Pesticides are often eluted with methanol, ethyl acetate, dichloromethane, or their mixtures. The preferred sorbent–eluent combination is primarily determined by the polarity of the target analytes and the nature of the sample matrix.

Compared to conventional LLE methods, SPE has several distinctive advantages. SPE generally reduces the analysis time, consumes lower amounts of organic solvents, and is less costly than LLE (Gan and Bondarenko 2008). Furthermore, compatibility of reversed-phase LC systems with aqueous samples allows on-line coupling of SPE with the analytical system. On-line SPE systems consist basically of a binary pump, one or more high-pressure six-port valves, and a clamping system for automated exchange of the SPE columns (Hogenboom et al. 2000). The sample is passed through the cartridge, and subsequent elution with the LC-gradient drives the retained compounds into the chromatographic column. SPE-LC with tandem mass spectrometry detection appears to be the method of choice to control the quality standards for individual pesticides and the sum of all pesticides including their relevant metabolites, degradation and reaction products set up in the Directive 2006/118/EC (Lepom et al. 2009). The disadvantage of SPE is that suspended solids and salts can cause the blockage of SPE cartridges.

TABLE 9.7

Overview of Methods Reported for the Water Analysis of Pesticides Using Gas Chromatography

Pesticides	Samples	Sample Preparation	Gas Chromatography Column	Detector	LOD (ng/L)	References
62, multiclass	SW	SPE (Oasis HLB)	DB-5 MS (30 m × 0.25 mm i.d. and 0.25 μm film)	EI-IT	1–12	Hladik et al. (2008)
14, organochlorine and organophosphorus	SW	LLE (PE and DCM)	DB-5.625	EI-IT	5–50	Tahboub et al. (2005)
8, organochlorine	SW, GW, WW	SPE (Strata-C-18)	HP-5 MS (30 m × 0.25 mm i.d. and 0.25 μm film)	EI-Q	0.5–61	Baugros et al. (2008)
17, organochlorine	SW (sea water)	SPE (ENVI-18) HS-SPME (PDMSPA, CAR, DVB)	HP-5 (30 m × 0.25 mm i.d. and 0.25 μm film)	ECD	0.0018–0.027	Qiu and Cai (2010)
9, organochlorine	DW	on-line SPE (C-18)	Croma-5 (30 m × 0.25 mm i.d. and 0.25 μm film)	EI-Q	10–50	Brondi et al. (2005)
34, multiclass	SW	SPE (Strata-X)	GC × GC TRB-5 MS (30 m × 0.25 mm i.d. and 0.25 μm film) and TRB-50HT (2 m × 0.1 mm i.d. and 0.1 μm film)	TOF	0.5–51	Matamoros et al. (2010)
6, triazines and 3 degradation products	SW	SPE (Bond Elut-ENV)	ZB5 (30 m × 0.25 mm i.d. and 0.25 μm film)	EI-Q	3–115 (LOQ)	Nevado et al. (2007)
18, organochlorine	SW (sea water)	SBSE (PDMS)	HP-5 MS (30 m × 0.25 mm i.d. and 0.25 μm film)	EI-Q	0.014–37.5	Sánchez-Avila et al. (2010)
46, multiclass	SW, DW	On-line SPME (PDMS/DVB)	VF-5 MS (30 m × 0.25 mm i.d. and 0.25 μm film)	EI-IT	4–32	Beceiro-González et al. (2007)
3, organophosphorus	SW, GW	MLPME (n-undecane)	HP-5 MS (30 m × 0.25 mm i.d. and 0.25 μm film)	EI-Q	15–80	Berhanu et al. (2008)
8, multiclass and 3 metabolites	RW	On-line SPME (PA)	DB-5 MS (30 m × 0.32 mm i.d. and 0.25 μm film)	EI-IT	50–50000	Sauret-Szczepanski et al. (2006)

Analyte (number, class)	Water type	Extraction technique	GC column	Detector	LOD/LOQ (ng/L)	Reference
13, organophosphorus	Spiked SW	LLE (DCM)	DB-5 SP-608	ECD NPD	3–29	Tse et al. (2004)
9, organochlorine and organophosphorus	SW, GW	LLE (DCM, hexane)	PE-17 (30 m × 0.32 mm i.d. and 0.25 μm film)	ECD	8–40	Sankararama-krishnan et al. (2005)
16, multiclass	SW, GW	On-line SPME (PA)	Restek Rtx®-1 MS (30 m × 0.25 mm i.d. and 0.25 μm film)	EI-Q	20–300	Filho et al. (2010)
20, multiclass	SW	SPE (C-18)	DB-608 (30 m × 0.32 mm i.d. and 1 μm film)	ECD NPD	0.8–75	Lyytikäinen et al. (2003)
35, halogenated	Spiked SW	On-line multiple SPME (PDMS, PDMS/DVB)	ZB5 (30 m × 0.25 mm i.d. and 0.25 μm film)	ECD NPD	1–100	Lipinski (2000)
5, triazines and chloroacetamides	Farmland run-off	On-line SPME (PDMS)	CP5860 (30 m × 0.25 mm i.d. and 0.25 μm film)	EI-IT	250–500	Rocha et al. (2008)
8, carbamates and organophosphorus	SW	SDME (hexane, iso-octane)	MDN-5S (30 m × 0.25 mm i.d. and 0.25 μm film)	NPD	200–5000	López-Blanco et al. (2005)
44, multiclass	Spiked samples	SPME (PDMS/DVB)	CPSil-8 CB (30 m × 0.25 mm i.d. and 0.25 μm film)	EI-IT	1–12000	Gonçalves and Alpendurada (2004).
11, multiclass	SW	SBSE (PDMS)	HP-5 MS (30 m × 0.25 mm i.d. and 0.25 μm film)	EI-Q	10–240	Penálver et al. (2003)
10, triazines	SW, GW	On-line SPME (PDMS–DVB)	DB-WAX (60 m × 0.25 mm i.d. and 0.5 μm film)	EI-IT	2–17	Frías et al (2003)
8, organophosphorus	SW, DW	SPE (RP-C$_{18}$)	HP-5 (30 m × 0.25 mm i.d. and 0.25 μm film)	NPD FID	50–130 4500–11700	Ballesteros and Parrado (2004)
7, organophosphorus	SW	HS-SPME (PA, PDMS)	DB-1 (30 m × 0.32 mm i.d.) DB-5 MS (30 m × 0.25 mm i.d. and 0.25 μm film)	FTD EI-Q	10–30 10–40	Lambropoulou and Albanis (2001)

LOD, limit of detection; LOQ, limit of quantification; SW, surface water; GW, groundwater; WW, wastewater; RW, rain water; DW, drinking water; PDMS, polydimethylsiloxane; PA, polyamide; CAR, carboxen; DVB, divinylbenzene; DCM, dichloromethane; PE, petroleum ether; FTD, flame thermionic detector.

TABLE 9.8

Overview of Methods Reported for the Water Analysis of Pesticides Using Liquid Chromatography

Pesticides	Samples	Sample Preparation	Liquid Chromatography			LOD (ng/L)	References
			Column	Mobile Phase	Detector		
12, triazines + their metabolites	SW	SPE (propazine-MIP, LiChrolut EN)	Spherisorb S5 ODS2 (250 × 40 mm i.d. and 5 μm particle size)	Water (5 mM phosphate buffer), ACN	PDA	30–100	Carabias-Martínez et al. (2005)
14, multiclass	SW, GW	SPE (Oasis HLB)	Zorbax Eclipse® XDB-C18 (75 × 4.6 mm i.d. and 3.5 μm particle size)	Water (0.1% HAc), MeOH	ESI-IT	0.4–5.5	Dujaković et al. (2010)
31, multiclass	SW	SPE (Oasis HLB)	Acquity BEH C18 (100 × 2.1 mm i.d. and 1.7 μm particle size)	Water (0.1% HCOOH), ACN	ESI-QqQ	3–20	Gervais et al. (2008)
9, multiclass	GW	SPE (Oasis HLB)	Acquity BEH C18 (100 × 2.1 mm i.d. and 1.7 μm particle size)	Water (0.1% HCOOH), ACN	ESI-QqQ	0.1–20	Mezcua et al. (2006)
7, multiclass	Spiked samples	LLE (EtAc) SPE (Oasis HLB)	Phenomenex aqua (150 × 3 mm)	Water (0.1% HAc)–MeOH (0.1% HAc)	ESI-QqQ	<50	Liu et al. (2006)
101, multiclass	SW	SPE (Sep-pak C18)	Zorbax Eclipse XDB-C8 (150 × 4.6 mm i.d. and 5 μm particle size)	Water (0.1 % HCOOH), ACN	ESI-TOF	—	Ferrer and Thurman (2007)
12, multiclass	SW, GW, WW	SPE (Strata-C18)	Uptisphere C_{18} 3HDO (100 × 2.0 mm i.d and 3 μm particle size)	Water (10 mM aqueous NH_4Ac), MeOH	ESI-QqQ	0.2–88.9	Baugros et al. (2008)
14, multiclass	SW, GW, DW	SPE (Oasis HLB)	Atlantis dC18 (150 × 2 mm i.d. and 5 μm particle size)	Water/MeOH (5 mM NH_4Ac)	ESI-QqQ	4.1–43.0	Rodrigues et al. (2007)
6, quaternary ammonium	DW	On-line SPE (Hysphere Resin GP)	Kromasil C8 (200 × 2.1 mm i.d. and 5 μm particle size)	Water (100 mM formic buffer + 20 mM HFBA), ACN	TIS-QqQ ESI-Q-TOF	0.1–2 3–20	Núñez et al. (2004)

Analytes	Matrix	Sample preparation	Column	Mobile phase	Detection	Range	Reference
12, carbamates + their degradation products	SW	SPE (Bond Elut Jr. C18)	X-Terra (250 × 4.6 mm i.d. and 5 μm particle size)	Water–ACN	ESI-QqQ	100–500	El Atrache et al. (2005)
TPs of 4 triazines	Spiked SW	Direct injection	X-Terra C18 (250 × 2.1 mm i.d. and 5 μm particle size)	Water (0.01% HCOOH), MeOH (0.01% HCOOH)	ESI-Q-TOF	—	Ibáñez et al. (2004)
15, multiclass + metabolites	SW	SPE (Oasis HLB)	Nucleodur C18 gravity (125 × 2 mm i.d. and 3 μm particle size) For acidic pesticides: Gromsil C18 (150 × 2 mm, 3 μm)	Water (0.1% HCOOH), MeOH (0.1% HCOOH) (for acidic pesticides 0.6 % HCOOH)	ESI-QqQ	0.15–3	Gomides Freitas et al. (2004)
47, multiclass + TPs		On-line SPE (PRP-1)	Kromasil C18 (125 × 2.0 mm i.d. and 5 μm particle size)	Water (0.01% HCOOH), ACN (0.01% HCOOH)	ESI-QqQ ESI-Q-TOF	—	Hernández et al. (2004)
31, multiclass + some metabolites	WW, DW	Direct injection	Zorbax Eclipse XDB-C18 (50 × 4.6 mm i.d. and 1.8 μm particle size)	Water – MeOH (5 mM ammonium formate)	ESI-QqQ	2–15	Díaz et al. (2008)
20, multiclass	SW, GW	On-line SPE (PLRP-s) for acidic pesticides: (Hysphere Resin GP)	Purospher STAR-RP-18e (125 × 2 mm i.d. and 5 μm particle size)	Water - ACN	ESI-QqQ	0.004–2.8	Kampioti et al. (2005)
5, triazines and phenylureas	SW, GW, DW	SPE (LiChrolut RP-18)	LiChrospher 100 RP-18 (250 × 4 mm i.d. and 5 μm particle size)	Water, ACN	APCI-Q	1.61–10.95	Frías et al. (2004)

SW, surface water; GW, groundwater; WW, wastewater; DW, drinking water; CAN, acetonitrile; HAc, acetic acid; MeOH, methanol; HCOOH, formic acid; NH$_4$Ac, ammonium acetate; TIS, turbo ion spray.

Samples compatible with SPE must be relatively clean (e.g., groundwater). When surface water samples are analyzed, filtration prior to extraction is necessary to remove the suspended solids. This may not be desirable for hydrophobic compounds, because a significant fraction of the analyte may be associated with the suspended solids.

9.5.1.2 Recent Trends

Modern trends in chemical analysis are aimed at the simplification and miniaturization of the sample preparation as well as the minimization of the organic solvent used. In view of this aspect, there is growth in significance of solventless extraction and microextraction techniques.

Solventless extraction techniques, such as solid-phase microextraction (SPME) and stir bar sorptive extraction (SBSE), are used in many applications (Wardencki et al. 2007; Sánchez-Rojas et al. 2009). SPME is performed by immersing a silica fiber coated with a sorbent in an aqueous sample or in the headspace (HS-SPME). Then, the analytes are thermally desorbed into the GC column or eluted with the mobile phase in the mode of LC analysis. SBSE is carried out by stirring the sample with a stir bar covered with a sorbent (polydimethylsiloxane) for a given time. Wardencki et al. (2007) and Lambropoulou and Albanis (2007) published reviews describing recent developments and trends in this area.

Liquid-phase microextraction (LPME) can also be a promising tool to improve the performance of methods used in the control of pesticides in water, particularly in the analysis of nonpolar pesticides (Lambropoulou and Albanis 2007; Trtić-Petrović et al. 2010). LPME-based techniques are simple, rapid, and inexpensive compared to other microextraction techniques (Pinto et al. 2010). In LPME, the solvent can be a single microdrop suspended from a needle (single-drop microextraction, SDME) or can be present in pores of a hydrophobic membrane or separated from the donor phase by a membrane interface (membrane liquid-phase microextraction, MLPME). Recently, a detailed review of the developed LPME techniques and their application in analysis of the pesticide residues in water has been published (Pinto et al. 2010).

Finally, the application of new materials, such as carbon nanotubes or molecularly imprinted polymers, as sorbent materials in SPE has been one of the hot research topics in the last years (Petrovic et al. 2010). Regarding the extraction of pesticides, carbon nanotubes have been shown to outperform many popular solid-phase extractants (Ravelo-Pérez et al. 2010).

9.5.2 Analyte Detection

Pesticides belong to more than 100 substance classes, having widely different chemical and physical properties. For example, some pesticides contain halogens, while others contain phosphorus, sulfur, or nitrogen, and these heteroatoms may have relevance when choosing an appropriate detector for their analysis. Furthermore, a number of compounds are very volatile, but some of them do not evaporate at all. This diversity causes serious problems in the development of a universal residue analytical method, which should have the widest scope possible.

9.5.2.1 GC Detection Methods

Since the early 1970s to the early 1990s, most routine pesticide residue analyses were performed by GC in combination with conventional detectors, such as ECD, NPD, or FID (Gan and Bondarenko 2008). The sensitivity of these detectors is highly specific to the chemical classes of pesticides being analyzed. While ECD is highly sensitive for halogenated pesticides, FID is more universal, but less sensitive, and NPD lies somewhere in between, for N- or P-containing pesticides. As the conventional GC detectors are not universal, they are not appropriate for multiresidue analysis. More importantly, these detectors do not provide any qualitative information on the compounds being analyzed. Confirmation is often required, for example, using a second column of a different polarity.

Capillary GC coupled with MS (GC-MS) revolutionized environmental organic compound analysis in the 1980s (Richardson 2001), particularly with the advent of bench-top instruments. In the 1990s, GC-MS became one of the most attractive and powerful techniques for routine analysis of volatile organic pollutants due to its good sensitivity, high selectivity, and versatility. Nowadays, GC-MS has been widely applied for pesticide monitoring because of its high sensitivity and specificity and for the potential for multiresidue analysis. By the use of GC-MS, simultaneous determination and confirmation of pesticide residues can be obtained with one instrument in one analytical run. A great advantage of GC-MS in identification is that it gives a mass spectrum that can be easily and quickly compared online with a library of more than 200000 mass spectra (Benfenati et al. 2006). Specific pesticide libraries are available with hundreds of compounds.

In GC-MS, ionization of pesticides can be achieved by electron impact ionization (EI) or positive or negative chemical ionization (PCI or NCI). Also, instruments with different mass analyzers, for example, magnetic sectors, quadrupoles (Q), quadrupole ion traps (QIT), and time-of-flight analyzers (TOF) have been used for coupling to GC. However, most of the published studies on residue analysis by GC-MS report on results obtained by quadrupole instruments and EI, as shown in Table 9.7. Chemical ionization (CI) is rarely used. Positive or negative CI–MS gives better selectivity for several pesticides compared to EI, but the signal intensities of different pesticides vary much more compared to EI ionization. Furthermore, CI is not a universal ionization technique, and therefore it is not suitable for multiresidue methods. Finally, mass spectra produced by CI usually contain a smaller number of fragments, thus, offering less information (Vasiljevic et al. 2000).

In general, commonly used low resolution mass analyzers, such as Q and QIT, can suffer from false positives due to low mass resolution (Muir and Sverko 2006). Therefore, experienced analysts are needed to interpret results using confirmatory information, such as full-scan analyses, fragmentation patterns, and ion ratios. A very useful tool for confirmation purposes is tandem mass spectrometry (MS/MS). It uses two stages of mass analysis: first to preselect an ion and the second to analyze fragments induced by the collision of an ion with an inert gas, such as argon or helium. Compared to single-stage MS modes, MS/MS offers a higher degree of selectivity and sensitivity. MS/MS is proved to be a method of choice for the quantification of low levels of target compounds in the presence of high sample matrix background, such as pesticide residues in environmental samples (Gonçalves and Alpendurada 2004).

Major limitations in the application of GC-MS depend on GC limits. Very polar and thermally labile pesticides, not suitable for GC analysis, cannot be analyzed by GC-MS. Other limitations are the cost of the apparatus and the relative complexity of the method, compared with the simpler GC detectors. However, these limitations are gradually becoming less important as GC-MS instruments become simpler and cheaper.

9.5.2.2 LC Detection Methods

In response to environmental problems, arising mainly from the bioaccumulation and global transport of volatile nonpolar pesticides, the use of pesticides has changed through the years, going from persistent, not easily degraded pesticides (e.g., organochlorine pesticides) to more polar, readily degradable pesticides (e.g., N-methylcarbamates). Polar pesticides are not directly amenable for GC analysis. Since the introduction of reversed-phase LC equipped with a UV or FLD detector, which occurred around 1980, LC became adopted as a viable technique complementary to GC for the determination of polar pesticides. The wide application range, easy use, low cost, and improved selectivity (by the use of PDA detection) made UV detectors very popular and most widely used in LC. However, confirmation was difficult for pesticides of the same class due to the high degree of similarity between UV spectra. FLDs are distinctly more selective and sensitive than UV detectors, but without a derivatization step, their applicability is limited to a small number of pesticides having fluorophores.

Until two decades ago, when analyzing pesticide residues, methods based on LC were applied less frequently than GC, because traditional LC detectors are less sensitive than the various GC detection methods (Alder et al. 2006). The breakthrough came in the early 1990s, with the development of atmospheric pressure ionization (API) methods, which made possible the coupling of LC with MS. Compared with traditional detectors, MS detectors have increased the sensitivity of LC detection by several orders of magnitude (Alder et al. 2006). Furthermore, MS has an advantage over conventional detectors because it can provide information for unambiguous analyte identification even with poor LC separation. Otherwise, when using conventional detectors, optimization of LC separation is tedious and time consuming, even with the support of the computer-assisted retention modeling (Onjia et al. 2002; Vasiljevic et al. 2004a).

The most commonly used ionization techniques in LC-MS instruments are electrospray ionization (ESI) and atmospheric pressure chemical ionization (APCI). The selection of the most appropriate ionization source and ionization mode for pesticide analysis, for example, positive ionization (PI) or negative ionization (NI) mode, depends on the pesticide classes investigated. According to Thurman et al. (2001) and Baglio et al. (1999), who evaluated the response of a number of pesticides using either APCI or ESI in LC-MS, neutral and basic pesticides (carbamates, phenylureas, triazines) are more sensitive using APCI (especially in PI mode), whereas cationic and anionic herbicides (bipyridylium ions, sulfonic acids, phenoxy acids, nitrophenols, bentazone) are best ionized with ESI (especially in NI mode). However, as seen in Table 9.8, ESI has found much wider application than APCI, probably due to a wider range (in terms of polarity and molecular weight) of compounds than can be measured with this interface (Thurman et al. 2001). Another general observation is that the NI mode is more selective and less prone to adduct formation than the PI mode (Reemtsma 2003).

Single quadrupole was the predominant configuration of LC-MS in the early 1990s. A disadvantage of single-quadrupole LC-MS is the high intensity of background signals produced from sample matrix and LC solvent clusters (Alder et al. 2006). The chemical background can be significantly reduced if tandem MS in combination with selected reaction monitoring (SRM) is applied (Alder et al. 2006). LC coupled with MS/MS (LC-MS/MS) is capable of differentiation between analyte and matrix signal, as well as between the analytes that coelute, therefore permitting quantification of pesticide traces in very complex matrices. In addition, the use of MS/MS detection allows analysis without the complete chromatographic separation between the analytes, thus shortening the chromatographic run time. Nowadays, LC-MS/MS is one of the most powerful techniques for the analysis of pesticides in a variety of complex matrices (Radišić et al. 2009). Basically, all recently published LC-MS methods for pesticide determination in environmental samples rely on the use of MS/MS detection (Baugros et al. 2008; Petrovic et al. 2010).

The most commonly used mass spectrometers that allow MS/MS experiments are triple quadrupole (TQ) and QIT. This is mainly due to their easier operating performance, better robustness for routine analysis, and relatively low cost compared to TOF or Fourier Transform-Ion Cyclotron Resonance (FT-ICR) instruments. The TQ is the most frequently applied mass analyzer in routine pesticide residue analysis. It displays high sensitivity when working in multiple reaction monitoring (MRM) mode; therefore, it is best suited for achieving the strict MCLs regulated for various toxic compounds in different matrices. The QIT is a small, low-cost, easy-to-use, fast, sensitive, and versatile mass analyzer. Its unique feature is the ability to perform multiple stages of MS (Plomley et al. 2000, 2001). These characteristics make QIT an attractive option for the detection of pesticides and other water contaminants, as many studies have shown (Jeannot et al. 2000; Grujic et al. 2008, 2009, 2010). When performing MS/MS, QIT instruments are generally less sensitive than TQ analyzers, but they have the advantage of working in product-ion scan without losses in sensitivity and the possibility of performing multiple-stage fragmentation (MS^n). The disadvantages of QIT are low resolution, interfering side reactions (because all reactions occur in the same space), and a limited dynamic range.

The procedures for chromatographic separation are very similar and most of them rely on the use of reversed-phase columns. In most cases, a C18-modified silica stationary phase has been applied, while methanol and acetonitrile are commonly used as organic solvents in the mobile phase (Table 9.8). In several methods, acetic acid, formic acid, or ammonium acetate has been added to the mobile phase to improve the separation and enhance the ionization.

One of the major problems in the quantitative analysis using LC-MS is that matrix components coeluting with the analytes from the LC column can interfere with the ionization process in the electrospray, causing ionization suppression or enhancement. This effect is very common when working with complex matrices, such as environmental samples, and is known as the matrix effect (ME) (Niessen et al. 2006). ME can cause significant deterioration of the analytical method precision and accuracy, although the complex environmental matrices, such as soil and sludge, are more prone to ME than water samples. To overcome ME when quantifying, several approaches are available. The use of isotopically labeled target compounds as internal standards is the best option if they are available (Niessen et al. 2006). This approach allows signal suppression (or enhancement) to be corrected, as both labeled and native compounds will undergo the same suppression effect. However, the availability of isotopically labeled analogs as internal standards is frequently limited. Quantification by standard addition is another possibility to correct ME, but this method is not convenient when a high number of samples must be analyzed. Matrix-matched calibration is an additional way to compensate the ME, and it was successfully used in some recently published papers (Grujic et al. 2005; Martínez et al. 2007).

9.5.2.3 Comparison Between GC and LC Methods

In a recent review paper, GC-MS versus LC-MS/MS have been evaluated for the determination of 500 high-priority pesticides (Alder et al. 2006). For each of the selected pesticides, applicability and sensitivity of both methods were compared. LC-MS/MS was shown to be better or exclusively suited for sulfonyl or benzoyl ureas, carbamates, and triazines than GC-MS. Furthermore, LC-MS/MS was found to have a wider scope. Only 49 compounds out of 500 exhibited no response when LC-MS/MS was used (mostly organochlorine compounds) compared to 135 pesticides and metabolites that could not be analyzed by GC-MS. Both GC-MS- and LC-MS-based methods revealed a significant variation in

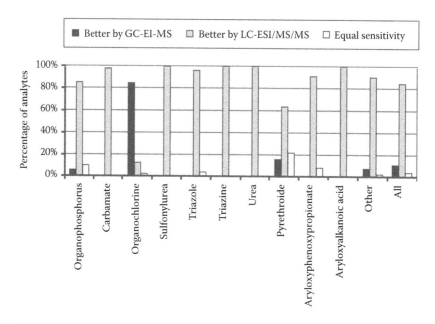

FIGURE 9.2

Comparison of GC-MS sensitivity versus LC–MS/MS sensitivity for different pesticide classes. (From Alder, L., Greulich, K., Kempe, G., and Vieth, B. Residue analysis of 500 high priority pesticides: Better by GC-MS or LC-MS/MS? *Mass Spectrom. Rev.*, 2006, 25, 838–865. Copyright Wiley-VCH Verlag GmbH & Co. KGaA. Reproduced with permission.)

sensitivity, covering a range of 3–4 orders of magnitude, depending on the pesticide analyzed. However, a comparison of the median limits of quantification (LOQs) clearly showed that only for one class of pesticides, the organochlorine compounds, GC-MS achieved better performance (Figure 9.2). For all other classes, higher sensitivity was attained using LC-MS with multiple-stage analysis. Most analytes may be quantified reliably by LC-MS/MS at concentrations between 0.1 and 1 µg/L. In contrast, the median LOQs observed by GC-MS are distinctly higher, up to 100 µg/L. The better performance of LC-MS/MS is probably determined by a larger injection volume and a lower amount of fragmentation during ionization.

9.5.2.4 Recent Trends

The fastest growing chromatography trend is the use of ultra performance liquid chromatography (UPLC). UPLC is an LC technique that uses small-diameter particles (typically 1.7 µm) in the stationary phase and short columns that allow higher pressures and ultimately narrower LC peaks. UPLC provides higher peak capacity, greater resolution, increased sensitivity, and higher speed of analysis compared to the conventional high-performance liquid chromatography (HPLC) (Mezcua et al. 2006; Richardson 2009). UPLC has been successfully applied to pesticide residue analyses in water samples (Mezcua et al. 2006; Gervais et al. 2008).

The other significant chromatography trend remains to be the use of two-dimensional GC (GC × GC). The main benefits of GC × GC are an increased chromatographic resolution, improved analyte detectability due to the cryofocusing in the thermal modulator, and chemical class ordering in the contour plots (Matamoros et al. 2010). The potential of the GC × GC coupling to MS has been shown in the simultaneous determination of 97 environmental contaminants, including pesticides at ng/L levels (Matamoros et al. 2010).

Recently, one of the hottest trends in water contaminant analysis has been the use of LC with full scan and high-resolution MS (HRMS) to identify unknown contaminants (nontarget analysis) or provide confirmation data (Richardson 2009). This identification relies on the mass accuracy obtained, with mass errors below 2 mDa or 5 ppm normally being accepted for a positive identification (Hernández et al. 2004). Nowadays, the most commonly used mass analyzer in HRMS instruments is TOF. For example, it has been shown that the higher mass resolution of TOF allows the detection of some pesticides in river water, even when they are accompanied by isobaric compounds (Hogenboom et al. 1999). Furthermore, LC-TOF has been the selected technique for the determination of pesticide degradation products in environmental, biological, and food matrices (Martínez Vidal et al. 2009). Even more useful in terms of confirmatory analysis is a quadrupole-TOF hybrid instrument (Q-TOF) as it allows MS^2 experiments to be performed, thus improving selectivity (Hernández et al. 2004; Barceló and Petrovic 2007). LC-Q-TOF has been successfully applied in the nontarget analysis of pesticides (Meng et al. 2010) and their metabolites (Hernández et al. 2008a) as well as disinfection by-products (Brix et al. 2009).

Acknowledgments

The authors greatly appreciate the financial support from the Ministry of Science and Technological Development of the Republic of Serbia (Project No. 172007).

References

Acero, J. L., Benitez, F. J., Real, F. J., and Gonzalez, M. 2008. Chlorination of organophosphorus pesticides in natural waters. *J. Hazard. Mater.* 153: 320–328.

Alder, L., Greulich, K., Kempe, G., and Vieth, B. 2006. Residue analysis of 500 high priority pesticides: Better by GC–MS or LC–MS/MS? *Mass Spectrom. Rev.* 25: 838–865.

Anshup, T. P. 2009. Noble metal nanoparticles for water purification: A critical review. *Thin Solid Films* 517: 6441–6478.

Asperger, A., Efer, J., Koal, T., and Engewald, W. 2002. Trace determination of priority pesticides in water by means of high-speed on-line solid-phase extraction–liquid chromatography–tandem mass spectrometry using turbulent-flow chromatography columns for enrichment and a short monolithic column for fast liquid chromatographic separation. *J. Chromatogr. A* 960: 109–119.

Ayranci, E. and Hoda, N. 2004. Studies on removal of metribuzin, bromacil, 2,4-D and atrazine from water by adsorption on high area carbon cloth. *J. Hazard. Mater.* 112: 163–168.

Bach, M., Huber, A., and Frede, H.-G. 2001. Input pathways and river load of pesticides in Germany – A national scale modeling assessment. *Water Sci. Technol.* 43: 261–268.

Bach, M., Letzel, M., and Kaul, U., et al. 2010. Measurement and modeling of bentazone in the river Main (Germany) originating from point and non-point sources. *Water Res.* 44: 3725–3733.

Bachman, J. and Patterson, H. H. 1999. Photodecomposition of the carbamate pesticide carbofuran: Kinetics and the influence of dissolved organic matter. *Environ. Sci. Technol.* 33: 874–881.

Bačić-Vukčević, M., Udovičić, A., Laušević, Z., Perić-Grujić, A., and Laušević, M. 2006. Surface characteristics and modification of different carbon materials. *Mater. Sci. Forum* 518: 217–222.

Baglio, D., Kotzias, D., and Larsen, B. R. 1999. Atmospheric pressure ionisation multiple mass spectrometric analysis of pesticides. *J. Chromatogr. A* 854: 207–220.

Ballesteros, E. and Parrado, M. J. 2004. Continuous solid-phase extraction and gas chromatographic determination of organophosphorus pesticides in natural and drinking waters. *J. Chromatogr. A* 1029: 267–273.

Barbash, J. E., Thelin, G. P., Kolpin, D. W., and Gilliom, R. J. 2001. Major herbicides in ground water: Results from the national waterquality assessment. *J. Environ. Qual.* 30: 831–845.

Barceló, D. and Petrovic, M. 2007. Challenges and achievements of LC-MS in environmental analysis: 25 years on. *TrAC Trends Anal. Chem.* 26: 2–11.

Barceló, D. and Hennion, M. C. 1997. *Trace Determination of Pesticides and their Degradation Products in Water.* Elsevier Science B.V: Amsterdam.

Baugros, J. B., Giroud, B., Dessalces, G., Grenier-Loustalot, M. F., and Cren-Olivè, C. 2008. Multiresidue analytical methods for the ultra-trace quantification of 33 priority substances present in the list of REACH in real water samples. *Anal. Chim. Acta* 607: 191–203.

Beceiro-González, E., Concha-Graña, E., Guimaraes, A., Gonçalves, C., Muniategui-Lorenzo, S., and Alpendurada, M. F. 2007. Optimisation and validation of a solid-phase microextraction method for simultaneous determination of different types of pesticides in water by gas chromatography–mass spectrometry. *J. Chromatogr. A* 1141: 165–173.

Bedos, C., Rousseau-Djabri, M.-F., Flura, D., Masson, S., Barriuso, E., and Cellier, P. 2002. Rate of pesticide volatilization from soil: An experimental approach with a wind tunnel system applied to trifluralin. *Atmos. Environ.* 36: 5917–5925.

Belfroid, A. C., van Drunen, M., Beek, M. A., Schrap, S. M., van Gestel, C. A. M., and van Hattum, B. 1998. Relative risks of transformation products of pesticides for aquatic ecosystems. *Sci. Total Environ.* 222: 167–183.

Belmonte Vega, A., Garrido Frenich, A., and Martínez Vidal, J. L. 2005. Monitoring of pesticides in agricultural water and soil samples from Andalusia by liquid chromatography coupled to mass spectrometry. *Anal. Chim. Acta* 538: 117–127.

Benfenati, E., Natangelo, M., and Tavazzi, S. 2006. Gas chromatography/mass spectrometry methods in pesticide analysis. In: Meyers, R. A. (ed.), *Encyclopedia of Analytical Chemistry.* John Wiley & Sons, Inc. http://onlinelibrary.wiley.com/doi/10.1002/9780470027318.a1709/abstract (accessed 28 January 2011).

Berhanu, T., Megersa, N., Solomon, T., and Jönsson, J. Å. 2008. A novel equilibrium extraction technique employing hollow fiber liquid phase microextraction for trace enrichment of freely dissolved organophosphorus pesticides in environmental waters. *Int. J. Environ. Anal. Chem.* 88: 933–945.

Blanchoud, H., Garban, B., Ollivon, D., and Chevreuil, M. 2002. Herbicides and nitrogen in precipitation: Progression from west to east and contribution to the Marne River (France). *Chemosphere* 47: 1025–1031.

Blanchoud, H., Moreau-Guigon, E., Farrugia, F., Chevreuil, M., and Mouchel, J. M. 2007. Contribution by urban and agricultural pesticide uses to water contamination at the scale of the Marne watershed. *Sci. Total Environ.* 375: 168–179.

Bradford, D. F., Heithmar, E. M., Tallent-Halsell, N. G., et al. 2010. Temporal patterns and sources of atmospherically deposited pesticides in Alpine Lakes of the Sierra Nevada, California, U.S.A. *Environ. Sci. Technol.* 44: 4609–4614.

Braman, S. K., Oetting, R. D., and Florkowski, W. 1997. Assessment of pesticide use by commercial landscape maintenance and lawn care firms in Georgia. *J. Entomol. Sci.* 32: 403–411.

Brix, R., Bahi, N., Lopez De Alda, M. J., Farré, M., Fernandez, J.-M., and Barceló, D. 2009. Identification of disinfection by-products of selected triazines in drinking water by LCQ–ToF–MS/MS and evaluation of their toxicity. *J. Mass Spectrom.* 44: 330–337.

Brondi, S. H. G., Spoljaric, F. C., and Lanças, F. M. 2005. Ultratraces analysis of organochlorine pesticides in drinking water by solid phase extraction coupled with large volume injection/gas chromatography/mass spectrometry. *J. Sep. Sci.* 28: 2243–2246.

Brown, C. D. and van Beinum, W. 2009. Pesticide transport via sub-surface drains in Europe. Review. *Environ. Pollut.* 157: 3314–3324.

Bucheli, T. D., Müller, S. R., Voegelin, A., and Schwarzenbach, R. P. 1998. Bituminous roof sealing membranes as major sources of the herbicide (*R,S*)-mecoprop in roof runoff waters: Potential contamination of groundwater and surface waters. *Environ. Sci. Technol.* 32: 3465–3471.

Carabias-Martínez, R., Rodríguez-Gonzalo, E., and Herrero-Hernández, E. 2005. Determination of triazines and dealkylated and hydroxylated metabolites in river water using a propazine-imprinted polymer. *J. Chromatogr. A* 1085: 199–206.

Carabias-Martínez, R., Rodríguez-Gonzalo, E., Fernández-Laespada, M. E., Calvo-Seronero, L., and Sánchez-San Román, F. J. 2003. Evolution over time of the agricultural pollution of waters in an area of Salamanca and Zamora (Spain). *Water Res.* 37: 928–938.

Carter, A. D. 2000. How pesticides get into water – And proposed reduction measures. *Pestic. Outlook* 11: 149–156.

Carvalho, J. J., Jerónimo, P. C. A., Gonçalves, C., and Alpendurada, M. F. 2008. Evaluation of a multiresidue method for measuring fourteen chemical groups of pesticides in water by use of LC-MS-MS. *Anal. Bioanal. Chem.* 392: 955–968.

Chiron, S., Valverde, A., Fernandez-Alba, A., and Barceló, D. 1995. Automated sample preparation for monitoring groundwater pollution by carbamate insecticides and their transformation products. *J. AOAC Int.* 78: 1346–1352.

Chiron, S., Comoretto, L., Rinaldi, E., Maurino, V., Minero, C., and Vione, D. 2009. Pesticide by-products in the Rhône delta (Southern France). The case of 4-chloro-2-methylphenol and of its nitroderivative. *Chemosphere* 74: 599–604.

Clark, G. M. and Goolsby, D. A. 2000. Occurrence and load of selected herbicides and metabolites in the lower Mississippi River. *Sci. Total Environ.* 248: 101–113.

Council of the European Communities. 1975. Council Directive of 16 June 1975 concerning the quality required of surface water intended for the abstraction of drinking water in the Member States (75/440/EEC). *Off. J. Eur. Commun.* L194: 26–36.

Council of the European Communities. 1979. Council Directive of 9 October 1979 concerning the methods of measurement and frequencies of sampling and analysis of surface water intended for the abstraction of drinking water in the Member States (79/869/EEC). *Off. J. Eur. Commun.* L271: 1–14.

Council of the European Communities. 1980. Council Directive of 15 July 1980 relating to the quality of the water intended for human consumption (80/778/EEC). *Off. J. Eur. Commun.* L229: 11–29.

Council of the European Union. 1998. Council Directive 98/83/EC of 3 November 1998 on the quality of water intended for human consumption. *Off. J. Eur. Commun.* L330: 32–54.

Council of the European Union. 2000. Directive 2000/60/EC of the European Parliament and of the Council of 23 October 2000 establishing a framework for Community action in the field of water policy. *Off. J. Eur. Commun.* L327: 1–72.

Council of the European Union. 2001. Decision 2455/2001/EC of the European Parliament and of the Council of 20 November 2001 establishing the list of priority substances in the field of water policy and amending Directive 2000/60/EC. *Off. J. Eur. Commun.* L331: 1–5.

Council of the European Union. 2006. Directive 2006/118/EC of the European Parliament and of the Council of 12 December 2006 on the protection of groundwater against pollution and deterioration. *Off. J. Eur. Union* L372: 19–31.

Council of the European Union. 2008. Directive 2008/105/EC of the European Parliament and of the Council of 16 December 2008 on environmental quality standards in the field of water policy, amending and subsequently repealing Council Directives 82/176/EEC, 83/513/EEC, 84/156/EEC, 84/491/EEC, 86/280/EEC and amending Directive 2000/60/EC of the European Parliament and of the Council. *Off. J. Eur. Union* L348: 84–97.

Curini, R., Gentili, A., Marchese, S., Perret, D., Arone, L., and Monteleone, A. 2001. Monitoring of pesticides in surface water: Off-line SPE followed by HPLC with UV detection and confirmatory analysis by mass spectrometry. *Chromatographia* 53: 244–250.

Dabrowski, J. M., Peall, S. K. C., van Niekerk, A., Reinecke, A. J., Day, J. A., and Schulz, R. 2002. Predicting runoff-induced pesticide input in agricultural sub-catchment surface waters: Linking catchment variables and contamination. *Water Res.* 36: 4975–4984.

Daneshvar, N., Aber, S., Khani, A., and Khataee, A. R. 2007. Study of imidaclopride removal from aqueous solution by adsorption onto granular activated carbon using an on-line spectrophoto-metric analysis system. *J. Hazard. Mater.* 144: 47–51.

De Rossi, C., Bierl, R., and Riefstahl, J. 2003. Organic pollutants in precipitation: Monitoring of pesticides and polycyclic aromatic hydrocarbons in the region of Trier (Germany). *Phys. Chem. Earth* 28: 307–314.

Díaz, L., Llorca-Pórcel, J., and Valor, I. 2008. Ultra trace determination of 31 pesticides in water samples by direct injection–rapid resolution liquid chromatography–electrospray tandem mass spectrometry. *Anal. Chim. Acta* 624: 90–96.

Dörfler, U. and Scheunert, I. 1997. S-triazine herbicides in rainwater with special reference to the situation in Germany. *Chemosphere* 35: 77–85.

Duirk, S. E., Desetto, L. M., Davis, G. M., Lindell, C., and Cornelison, C. T. 2010. Chloramination of organophosphorus pesticides found in drinking water sources. *Water Res.* 44: 761–768.

Dujaković, N., Grujić, S., Radišić, M., Vasiljević, T., and Laušević, M. 2010. Determination of pesticides in surface and ground waters by liquid chromatography–electrospray–tandem mass spectrometry. *Anal. Chim. Acta* 678: 63–72.

Edwards, C. A. 1973. Pesticide residues in soil and water. In: Edwards, C. A. (ed), *Environmental Pollution by Pesticides*, pp. 409–458. Plenum Press: London.

El Atrache, L. L., Sabbah, S., and Morizur, J. P. 2005. Identification of phenyl-N-methylcarbamates and their transformation products in Tunisian surface water by solid-phase extraction liquid chromatography–tandem mass spectrometry. *Talanta* 65: 603–612.

El Bakouri, H., Ouassini, A., Morillo, J., and Usero, J. 2008. Pesticides in ground water beneath Loukkos perimeter, Northwest Morocco. *J. Hydrol.* 348: 270–278.

Fait, G., Balderacchi, M., Ferrari, F., Ungaro, F., Capri, E., and Trevisan, M. 2010. A field study of the impact of different irrigation practices on herbicide leaching. *Eur. J. Agron.* 32: 280–287.

Fait, G., Nicelli, M., Fragoulis, G., Trevisan, M., and Capri, E. 2007. Reduction of point contamination sources of pesticide from a vineyard farm. *Environ. Sci. Technol.* 41: 3302–3308.

Ferrer, I. and Thurman, E. M. 2007. Multi-residue method for the analysis of 101 pesticides and their degradates in food and water samples by liquid chromatography/time-of-flight mass spectrometry. *J. Chromatogr. A* 1175: 24–37.

Filho, A. M., dos Santos, F. N., and Pereira, P. A. D. P. 2010. Development, validation and application of a method based on DI-SPME and GC–MS for determination of pesticides of different chemical groups in surface and groundwater samples. *Microchem. J.* 96: 139–145.

Foreman, W. T., Majewski, M. S., Goolsby, D. A., Wiebe, F. W., and Coupe, R. H. 2000. Pesticides in the atmosphere of the Mississippi River Valley, part II-air. *Sci. Total Environ.* 248: 213–226.

Frías, S., Rodríguez, M. A., Conde, J. E., and Pérez-Trujillo, J. P. 2003. Optimisation of a solid-phase microextraction procedure for the determination of triazines in water with gas chromatography–mass spectrometry detection. *J. Chromatogr. A* 1007: 127–135.

Frische, K., Schwarzbauer, J., and Ricking, M. 2010. Structural diversity of organochlorine compounds in groundwater affected by an industrial point source. *Chemosphere* 81: 500–508.

Gan, J. and Bondarenko, S. 2008. Determination of pesticides in water. In: Tadeo, J. L. (ed), *Analysis of Pesticides in Food and Environmental Samples*, pp. 231–256. CRC Press: Boca Raton, FL.

Gärdenäs, A. I., Šimůnek, J., Jarvis, N., and van Genuchten, M. Th. 2006. Two-dimensional modelling of preferential water flow and pesticide transport from a tile-drained field. *J. Hydrol.* 329: 647–660.

Gentry, L. E., David, M. B., Smith-Starks, K. M., and Kovacic, D. A. 2000. Nitrogen fertilizer and herbicide transport from tile drained fields. *J. Environ. Qual.* 29: 232–240.

Gerecke, A. C., Schärer, M., Singer, H. P., et al. 2002. Sources of pesticides in surface waters in Switzerland: Pesticide load through waste water treatment plants – Current situation and reduction potential. *Chemosphere* 48: 307–315.

Gervais, G., Brosillon, S., Laplanche, A., and Helen, C. 2008. Ultra-pressure liquid chromatography–electrospray tandem mass spectrometry for multiresidue determination of pesticides in water. *J. Chromatogr. A* 1202: 163–172.

Gil, Y. and Sinfort, C. 2005. Emission of pesticides to the air during sprayer application: A bibliographic review. *Atmos. Environ.* 39: 5183–5193.

Gomides Freitas, L., Götz, C. W., Ruff, M., Singer, H. P., and Müller, S. R. 2004. Quantification of the new triketone herbicides, sulcotrione and mesotrione, and other important herbicides and metabolites, at the ng/l level in surface waters using liquid chromatography–tandem mass spectrometry. *J. Chromatogr. A* 1028: 277–286.

Gonçalves, C. and Alpendurada, M. F. 2004. Solid-phase micro-extraction–gas chromatography–(tandem) mass spectrometry as a tool for pesticide residue analysis in water samples at high sensitivity and selectivity with confirmation capabilities. *J. Chromatogr. A* 1026: 239–250.

Goncalves, C. M., Esteves da Silva, J. C. G., and Alpendurada, M. F. 2007. Evaluation of the pesticide contamination of groundwater sampled over two years from vulnerable zone in Portugal. *J. Agric. Food Chem.* 55: 6227–6235.

Gonzalez, M., Miglioranza, K. S. B., Aizpún, J. E., Isla, F. I., and Peña, A. 2010. Assessing pesticide leaching and desorption in soils with different agricultural activities from Argentina (Pampa and Patagonia). *Chemosphere* 81: 351–358.

Grujic, S., Radisic, M., Vasiljevic, T., and Lausevic, M. 2005. Determination of carbendazim residues in fruit juices by liquid chromatography–tandem mass spectrometry. *Food Addit. Contam.* 22: 1132–1137.

Grujic, S., Vasiljevic, T., Lausevic, M., and Ast, T. 2008. Study on the formation of an amoxicillin adduct with methanol using electrospray ion trap tandem mass spectrometry. *Rapid Commun. Mass Spectrom.* 22: 67–74.

Grujic, S., Vasiljević, T., and Laušević, M. 2009. Determination of multiple pharmaceutical classes in surface and ground waters by liquid chromatography–ion trap–tandem mass spectrometry. *J. Chromatogr. A* 1216: 4989–5000.

Grujic, S., Vasiljevic, T., Radisic, M., and Lausevic, M. 2010. Determination of pesticides by matrix solid-phase dispersion and liquid chromatography–tandem mass spectrometry. In: Nollet, L. M. L. and Rathore, H. S. (eds), *Handbook of Pesticides: Methods of Pesticide Residues Analysis*, pp. 141–164. CRC Press: Boca Raton, FL.

Grynkiewicz, M., Polkowska, Z., Górecki, T., and Namieśnik, J. 2001. Pesticides in precipitation in the Gdańsk region (Poland). *Chemosphere* 43: 303–312.

Guo, L., Nordmark, C. E., and Spurlock, F. C., et al. 2004. Characterizing dependence of pesticide load in surface water on precipitation and pesticide use for the Sacramento River watershed. *Environ. Sci. Technol.* 38: 3842–3852.

Guzzella, L., Pozzoni, F., and Giuliano, G. 2006. Herbicide contamination of surficial groundwater in Northern Italy. *Environ. Pollut.* 142: 344–353.

Hantscha, S., Singer, H., Canonica, S., Schwarzenbach, R. P., and Fenner, K. 2008. Input dynamics and fate of the herbicide metolachlor and his highly mobile transformation product metolachlor ESA. *Environ. Sci. Technol.* 42: 5507–5513.

Hernández, F., Ibáñez, M., Sancho, J. V., and Pozo, Ó. J. 2004. Comparison of different mass spectrometric techniques combined with liquid chromatography for confirmation of pesticides in environmental water based on the use of identification points. *Anal. Chem.* 76: 4349–4357.

Hernández, F., Sancho, J. V., Ibáñez, M., and Grimalt, S. 2008a. Investigation of pesticide metabolites in food and water by LC–TOF–MS. *TrAC Trends Anal. Chem.* 27: 862–872.

Hernández, F., Ibáñez, M., Pozo, Ó., and Sancho, J. V. 2008b. Investigating the presence of pesticide transformation products in water by using liquid chromatography-mass spectrometry with different mass analyzers. *J. Mass Spectrom.* 43: 173–184.

Hetrick, J., Parker, R., Pisigan, J. R., and Thurman, N. 2000. Progress report on estimating pesticide concentrations in drinking water and assessing water treatment effects on pesticide removal and transformation. Briefing Document for a Presentation to the FIFRA Scientific Advisory Panel (SAP). http://www.epa.gov/scipoly/sap/meetings/2000/…/sept00_sap_dw_0907.pdf (accessed 28 January 2011).

Heydens, W. F., Siglin, J. C., Holson, J. F., and Stegeman, S. D. 1996. Subchronic, developmental, and genotoxic studies with the ethane sulfonate metabolite of alachlor. *Fundam. Appl. Toxicol.* 33: 173–181.

Heydens, W. F., Wilson, A. G. E., Kraus, L. J., Hopkins, W. E., and Hotz, K. J. 2000. Ethane sulfonate metabolite of alachlor: Assessment of oncogenic potential based on metabolic and mechanistic considerations. *Toxicol. Sci.* 55: 36–43.

Hildebrandt, A., Guillamón, M., Lacorte, S., Tauler, R., and Barceló, D. 2008. Impact of pesticides used in agriculture and vineyards to surface and groundwater quality (North Spain). *Water Res.* 42: 3315–3326.

Hladik, M. L., Hsiao, J. J., and Roberts, L. 2005. Are neutral chloroacetamide herbicide degradets of potential environmental concern? Analysis and occurrence in the Upper Chesapeake Bay. *Environ. Sci. Technol.* 39: 6561–6574.

Hladik, M. L., Smalling, K. L., and Kuivila, K. M. 2008. A multi-residue method for the analysis of pesticides and pesticide degradates in water using HLB solid-phase extraction and gas chromatography–ion trap mass spectrometry. *Bull. Environ. Contam. Toxicol.* 80: 139–144.

Hogenboom, A. C., Niessen, W. M. A., Little, D., and Brinkman, U. A. Th. 1999. Accurate mass determinations for the confirmation and identification of organic microcontaminants in surface water using on-line solid-phase extraction liquid chromatography electrospray orthogonal-acceleration time-of-flight mass spectrometry. *Rapid Commun. Mass Spectrom.* 13: 125–133.

Hogenboom, A. C., Hofman, M. P., Jolly, D. A., and Niessen, W. M. A., Brinkman, U. A. T. 2000. On-line dual-precolumn-based trace enrichment for the determination of polar and acidic microcontaminants in river water by liquid chromatography with diode-array UV and tandem mass spectrometric detection. *J. Chromatogr. A* 885: 377–388.

Holvoet, K., van Griensven, A., Seuntjens, P., and Vanrolleghem, P. A. 2005. Sensitivity analysis for hydrology and pesticide supply towards the river in SWAT. *Phys. Chem. Earth* 30: 518–526.

Holvoet, K. M. A., Seuntjens, P., and Vanrolleghem, P. A. 2007. Monitoring and modeling pesticide fate in surface waters at the catchment scale. *Ecol. Model.* 209: 53–64.

Huber, A., Bach, M., and Frede, H. G. 1998. Modeling pesticide losses with surface runoff in Germany. *Sci. Total Environ.* 223: 177–191.

Huber, A., Bach, M., and Frede, H. G. 2000. Pollution of surface waters with pesticides in Germany: Modeling non-point source inputs. *Agric. Ecosyst. Environ.* 80: 191–204.

Hüskes, R. and Levsen, K. 1997. Pesticides in rain. *Chemosphere* 35: 3013–3024.

Ibáñez, M., Sancho, J. V., Pozo,Ó. J., and Hernández, F. 2004. Use of quadrupole time-of-flight mass spectrometry in environmental analysis: Elucidation of transformation products of triazine herbicides in water after UV exposure. *Anal. Chem.* 76: 1328–1335.

Iesce, M. R., Greca, M., Cermola, F., Rubino, M., Isidori, M., and Pascarella, L. 2006. Transformation and ecotoxicity of carbamic pesticides in water. *Environ. Sci. Pollut. Res.* 13: 105–109.

Jeannot, R., Sabik, H., Sauvard, E., and Genin, E. 2000. Application of liquid chromatography with mass spectrometry combined with photodiode array detection and tandem mass spectrometry for monitoring pesticides in surface waters. *J. Chromatogr. A* 879: 51–71.

Jordan, T. B., Nichols, D. S., and Kerr, N. I. 2009. Selection of SPE cartridge for automated solid-phase extraction of pesticides from water followed by liquid chromatography-tandem mass spectrometry. *Anal. Bioanal. Chem.* 394: 2257–2266.

Kahl, G., Ingwersen, J., and Nutniyom, P., et al. 2008. Loss of pesticides from a litchi orchard to an adjacent stream in northern Thailand. *Eur. J. Soil Sci.* 59: 71–81.

Kalkhoff, S. J., Kolpin, D. W., Thruman, E. M., Ferrer, I., and Barcelo, D. 1998. Degradation of chloroacetanilide herbicides: The prevalence of sulfonic and oxanilic acid metabolites in Iowa groundwater and surface waters. *Environ. Sci. Technol.* 32: 1738–1740.

Kalkhoff, S. J., Lee, K. E., Porter, S. D., Terrio, P. J., and Thurman, E. M. 2003. Herbicides and herbicide degradation products in upper Midwest agricultural streams during august base-flow conditions. *J. Environ. Qual.* 32: 1025–1035.

Kampioti, A. A., Borba da Cunha, A. C., López de Alda, M., and Barceló, D. 2005. Fully automated multianalyte determination of different classes of pesticides, at picogram per litre levels in water, by on-line solid-phase extraction–liquid chromatography–electrospray–tandem mass spectrometry. *Anal. Bioanal. Chem.* 382: 1815–1825.

Kiely, T., Donaldson, D., and Grube, A. 2004. *Pesticides Industry Sales and Usage. 2000 and 2001 Market Estimates.* US Environmental Protection Agency: Washington, DC. http://www.epa.gov/pesticides/pestsales/01pestsales/market_estimates2001.pdf (accessed 28 January 2011).

Kimmel, E. C., Casida, J. E., and Ruzo, L. O. 1986. Formamidine insecticides and chloroacetanilide herbicides: Disubstituted anilines and nitrosobenzenes as mammalian metabolites and bacterial mutagens. *J. Agric. Food Chem.* 34: 157–161.

Kirchner, M., Faus-Kessler, T., Jakobi, G., et al. 2009. Vertical distribution of organochlorine pesticides in humus along Alpine altitudinal profiles in relation to ambiental parameters. *Environ. Pollut.* 157: 3238–3247.

Kladivko, E. J., Grochulska, J., Turco, R. F., Van Scoyoc, G. E., and Eigel, J. D. 1999. Pesticide and nitrate transport into subsurface tile drains of different spacings. *J. Environ. Qual.* 28: 997–1004.

Kohler, E. A., Poole, V. L., Reicher, Z. J., and Turco, R. F. 2004. Nutrient, metal, and pesticide removal during storm and nonstorm events by a constructed wetland on an urban golf course. *Ecol. Eng.* 23: 285–298.

Köhne, J. M., Köhne, S., and Šimůnek, J. 2009. A review of model applications for structured soils: b) Pesticide transport. *J. Contam. Hydrol.* 104: 36–60.

Kolpin, D. W., Thurman, E. M., and Goolsby, D. A. 1996. Occurrence of selected pesticides and their metabolites in near-surface aquiferis of Midwesern United States. *Environ. Sci. Technol.* 30: 335–340.

Kolpin, D. W., Barbash, J. E., and Gilliom, R. J. 1998. Occurrence of pesticides in shallow groundwater of the United States: Initial results from the national water-quality assessment program. *Environ. Sci. Technol.* 32: 558–566.

Kolpin, D. W., Thurman, E. M., and Linhart, S. M. 2000. Finding minimal herbicide concentrations in ground water? Try looking their degradets. *Sci. Total Environ.* 248: 115–122.

Kolpin, D. W., Schnoebelen, D. J., and Thurman, E. M. 2004. Degradates provide insight to spatial and temporal trends of herbicides in ground water. *Ground Water* 42: 601–608.

Kolpin, D. W., Thurman, E. M., Lee, E. A., Meyer, M. T., Furlong, E. T., and Glassmeyer, S. T. 2006. Urban contributions of glyphosate and its degradate AMPA to streams in the United States. *Sci. Total Environ.* 354: 191–197.

Kreuger, J. 1998. Pesticides in stream water within an agricultural catchment in southern Sweden, 1990–1996. *Sci. Total Environ.* 216: 227–251.

Kruger, E. L., Somasundaram, L., Coats, J. R., and Kanwar, R. S. 1993. Persistence and degradation of [14C]atrazine and [14C]deisopropylatrazine as affected by soil depth and moisture conditions. *Environ. Toxicol. Chem.* 12: 1959–1967.

Kuster, M., López de Alda, M., and Barceló, D. 2006. Analysis of pesticides in water by liquid chromatography–tandem mass spectrometric techniques. *Mass Spectrom. Rev.* 25: 900–916.

Kuster, M., López de Alda, M. J., Barata, C., Raldùa, D., and Barceló, D. 2008. Analysis of 17 polar to semi-polar pesticides in the Ebro river delta during the main growing season of rice by automated on-line solid-phase extraction-liquid chromatography–tandem mass spectrometry. *Talanta* 75: 390–401.

Lambropoulou, D. A. and Albanis, T. A. 2001. Optimization of headspace solid-phase microextraction conditions for the determination of organophosphorus insecticides in natural waters. *J. Chromatogr. A* 922: 243–255.

Lambropoulou, D. A. and Albanis, T. A. 2007. Liquid-phase micro-extraction techniques in pesticide residue analysis. *J. Biochem. Biophys. Methods* 70: 195–228.

Larson, S. J., Capel, P. D., and Majewski, M. S. 1997. *Pesticides in Surface Waters: Distribution, Trends, and Governing Factors.* Ann Arbor Press: Chelsea, MI.

Lartiges, S. B. and Garrigues, P. P. 1995. Degradation kinetics of organophosphorus and organoni-trogen pesticides in different waters under various environmental conditions. *Environ. Sci. Technol.* 29: 1246–1254.

Leistra, M. and Boesten, J. J. T. I. 2010. Pesticide leaching from agricultural fields with ridges and fur-rows. *Water Air Soil Pollut.* 213: 341–352.

Lepom, P., Brown, B., Hanke, G., Loos, R., Quevauviller, P., and Wollgast, J. 2009. Needs for reliable analytical methods for monitoring chemical pollutants in surface water under the European Water Framework Directive. *J. Chromatogr. A* 1216: 302–315.

Leu, C., Singer, H., Stamm, C., Müler, S. R., and Schwarzenbach, R. P. 2004a. Simultaneous assess-ment of sources, processes, and factors influencing herbicide losses to surface waters in a small agricultural catchment. *Environ. Sci. Technol.* 38: 3827–3834.

Leu, C., Singer, H., Stamm, C., Müler, S. R., and Schwarzenbach, R. P. 2004b. Variability of herbicide losses from 13 fields to surface water within a small catchment after a controlled herbicide application. *Environ. Sci. Technol.* 38: 3835–3841.

Liess, M., Schulz, R., Liess, M. H.-D., Rother, B., and Kreuzig, R. 1999. Determination of insecticide contamination in agricultural headwater streams. *Water Res.* 33: 239–247.

Lipinski, J. 2000. Automated multiple solid phase micro extraction. An approach to enhance the limit of detection for the determination of pesticides in Water. *Fresen. J. Anal. Chem.* 367: 445–449.

Liška, I. 2000. Fifty years of solid-phase extraction in water analysis-historical development and overview. *J. Chromatogr. A* 885: 3–16.

Liu, F., Bischoff, G., Pestemer, W., Xu, W., and Kofoet, A. 2006. Multi-residue analysis of some polar pesticides in water samples with SPE and LC–MS–MS. *Chromatographia* 63: 233–237.

Loos, R., Locoro, G., Comero, S., et al. 2010. Pan-European survey on the occurrence of selected polar organic persistent pollutants in ground water. *Water Res.* 44: 4115–4126.

López-Blanco, C., Gómez-Álvarez, S., Rey-Garrote, M., Cancho-Grande, B., and Simal-Gándara, J. 2005. Determination of carbamates and organophosphorus pesticides by SDME–GC in natural water. *Anal. Bioanal. Chem.* 383: 557–561.

Lyytikäinen, M., Kukkonen, J. V. K., and Lydy, M. J. 2003. Analysis of pesticides in water and sedi-ment under different storage conditions using gas chromatography. *Arch. Environ. Contam. Toxicol.* 44: 437–444.

Maeng, S. K., Ameda, E., Sharma, S. K., Grutzmacher, G., and Amy, G. L. 2010. Organic micropollut-ant removal from wastewater effluent-impacted drinking water sources during bank filtration and artificial recharge. *Water Res.* 44: 4003–4014.

Majewski, M. S., Foreman, W. T., and Goolsby, D. A. 2000. Pesticides in the atmosphere of the Mississippi River Valley, part I – rain. *Sci. Total Environ.* 248: 201–212.

Maloschik, E., Ernst, A., Hegedűs, G., Darvas, B., and Székács, A. 2007. Monitoring water-polluting pesticides in Hungary. *Microchem. J.* 85: 88–97.

Manahan, S. E. 2001. *Fundamentals of Environmental Chemistry.* CRC Press LLC: Boca Raton, FL.

Marín, J. M., Sancho, J. V., Pozo, O. J., López, F. J., and Hernández, F. 2006. Quantification and confirma-tion of anionic, cationic and neutral pesticides and transformation products in water by on-line solid phase extraction–liquid chromatography–tandem mass spectrometry. *J. Chromatogr. A* 1133: 204–214.

Martínez, D. B., Galera, M. M., Vázquez, P. P., and García, M. D. G. 2007. Simple and rapid determi-nation of benzoylphenylurea pesticides in river water and vegetables by LC–ESI–MS. 2007. *Chromatographia* 66: 533–538.

Martínez Vidal, J. L., Plaza-Bolaños, P., Romero-González, R., Garrido Frenich, A. 2009. Determination of pesticide transformation products: A review of extraction and detection methods. *J. Chromatogr. A* 1216: 6767–6788.

Martin-Gullon, J., and Font, R. 2001. Dynamic pesticide removal with activated carbon fibers. *Water Res.* 35: 516–520.

Matamoros, V., Jover, E., and Bayona, J. M. 2010. Part-per-trillion determination of pharmaceuticals, pesticides, and related organic contaminants in river water by solid-phase extraction followed by comprehensive two-dimensional gas chromatography time-of-flight mass spectrometry. *Anal. Chem.* 82: 699–706.

McInnis, P. 2010. Pesticides in Ontario's Treated Municipal Drinking Water 1986–2006. Ontario Ministry of the Environment. http://www.ene.gov.on.ca/stdprodconsume/groups/lr/@ene/@resources/documents/resource/std01_079845.pdf (accessed 28 January 2011).

Meng, C. K., Zweigenbaum, J., Furst, P., and Blanke, E. 2010. Finding and confirming nontargeted pesticides using GC/MS, LC/quadrupole-time-of-flight MS, and databases. *J. AOAC Int.* 93: 703–711.

Mezcua, M., Agüera, A., Lliberia, J. L., Cortés, M. A., Bagó, B., and Fernández-Alba, A. R. 2006. Application of ultra performance liquid chromatography–tandem mass spectrometry to the analysis of priority pesticides in groundwater. *J. Chromatogr. A* 1109: 222–227.

Muir, D. and Sverko, E. 2006. Analytical methods for PCBs and organochlorine pesticides in environmental monitoring and surveillance: A critical appraisal. *Anal. Bioanal. Chem.* 386: 769–789.

Müller, K., Bach, M., Hartmann, H., Spiteller, M., and Frede, H. G. 2002. Point- and nonpoint-source pesticide contamination in the Zwester Ohm catchment, Germany. *J. Environ. Qual.* 31: 309–318.

Müller, K., Deurer, M., Hartmann, H., Bach, M., Spiteller, M., and Frede, H.-G. 2003. Hydrological characterisation of pesticide loads using hydrograph separation at different scales in a German catchment. *J. Hydrol.* 273: 1–17.

Neumann, M., Schulz, R., Schäfer, K., Müller, W., Mannheller, W., and Liess, M. 2002. The significance of entry routes as point and non-point sources of pesticides in small streams. *Water Res.* 36: 835–842.

Neumann, M., Liess, M., and Schulz, R. 2003. A qualitative sampling method for monitoring water quality in temporary channels or point sources and its application to pesticide contamination. *Chemosphere* 51: 509–513.

Nevado, J. J. B., Cabanillas, C. G., Llerena, M. J. V., and Robledo, V. R. 2007. Sensitive SPE–GC–MS–SIM screening of endocrine-disrupting herbicides and related degradation products in natural surface waters and robustness study. *Microchem. J.* 87: 62–71.

Ng, H. Y. F., Gaynor, J. D., Tan, C. S., and Drury, C. F. 1995. Dissipation and loss of atrazine and metolachlor in surface and subsurface drain water: A case study. *Water Res.* 29: 2309–2317.

Niessen, W. M. A., Manini, P., and Andreoli, R. 2006. Matrix effects in quantitative pesticide analysis using liquid chromatography–mass spectrometry. *Mass Spectrom. Rev.* 25: 881–899.

Ninković, M. B., Petrović, R. D., and Laušević, M. D. 2010. Removal of organochlorine pesticides from water using virgin and regenerated granular activated carbon. *J. Serb. Chem. Soc.* 75: 565–573.

Nitschke, L. and Schüssler, W. 1998. Surface water pollution by herbicides from effluents of waste water treatment plants. *Chemosphere* 36: 35–41.

Nitschke, L., Wilk, A., Schossler, W., Metzner, G., and Lind, G. 1999. Biodegradation in laboratory activated sludge plants and aquatic toxicity of herbicides. *Chemosphere* 39: 2313–2323.

Novak, S. M., Portal, J.-M., and Schiavon, M. 2001. Effects of soil type upon metolachlor losses in subsurface drainage. *Chemosphere* 42: 235–244.

Núñez, O., Moyano, E., and Galceran, M. T. 2004. Time-of-flight high resolution versus triple quadrupole tandem mass spectrometry for the analysis of quaternary ammonium herbicides in drinking water. *Anal. Chim. Acta* 525: 183–190.

Onjia, A., Vasiljevic, T., Cokesa, D., and Lausevic, M. 2002. Factorial design in isocratic high-performance liquid chromatography of phenolic compounds. *J. Serb. Chem. Soc.* 67: 745–751.

Ormad, M. P., Miguel, N., Claver, A., Matesanz, J. M., and Ovelleiro, J. L. 2008. Pesticides removal in the process of drinking water production. *Chemosphere* 71: 97–106.

Osano, O., Admiral, W., and Otieno, D. 2002. Developmental disorders in embryos of the frog *Xenopus laevis* induced by chloroacetanilide herbicides and their degradation products. *Environ. Toxicol. Chem.* 21: 375–379.

Osano, O., Nzyuko, D., Tole, M., and Admiral, W. 2003. The fate of chloroacetanilide herbicides and their degradation products in the Nzoia Basin, Kenya. *Ambio* 32: 424–427.

Palma, P., Kuster, M., Alvarenga, P., et al. 2009. Risk assessment of representative and priority pesticides, in surface water of the Alqueva reservoir (South of Portugal) using on-line solid phase extraction-liquid chromatography-tandem mass spectrometry. *Environ. Int.* 35: 545–551.

Pehkonen, S. O. and Zhang, Q. 2002. The degradation of organophosphorus pesticides in natural waters: A critical review. *Crit. Rev. Environ. Sci. Technol.* 32: 17–72.

Pelaekani, C. and Snoeyink, V. L. 2000. Competitive adsorption between atrazine and methylene blue on activated carbon: The importance of pore size distribution. *Carbon* 38: 1423–1436.

Penálver, A., García, V., Pocurull, E., Borrull, F., and Marcé, R. M. 2003. Stir bar sorptive extraction and large volume injection gas chromatography to determine a group of endocrine disrupters in water samples. *J. Chromatogr. A* 1007: 1–9.

Pereira, W. E. and Hostettler, F. D. 1993. Nonpoint source contamination of the Mississippi River and its tributaries by herbicides. *Environ. Sci. Technol.* 27: 1542–1552.

Petrovic, M., Farré, M., Lopez de Alda, M., Perez, S., Postigo, C., Köck, M., Radjenovic, J., Gros, M., and Barcelo, D. 2010. Recent trends in the liquid chromatography–mass spectrometry analysis of organic contaminants in environmental samples. *J. Chromatogr. A* 1217: 4004–4017.

Pico, Y., Louter, A. J. H., Vreuls, J. J., and Brinkman, U. A. Th. 1994. On-line trace level enrichment gas chromatography of triazine herbicides, organophosphorus pesticides, and organosulfur compounds from drinking and surface waters. *Analyst* 119: 2025–2031.

Picó, Y., Fernández, M., Ruiz, M. J., and Font, G. 2007. Current trends in solid-phase-based extraction techniques for the determination of pesticides in food and environment. *J. Biochem. Biophys. Methods* 70: 117–131.

Pinto, M. I., Sontag, G., Bernardino, R. J., and Noronha, J. P. 2010. Pesticides in water and the performance of the liquid-phase microextraction based techniques. *Microchem. J.* 96: 225–237.

Plomley, J. B., Lausevic, M., and March, R. E. 2000. Determination of dioxins/furans and PCBs by quadrupole ion-trap gas chromatography–mass spectrometry. *Mass Spectrom. Rev.* 19: 305–365.

Plomley, J. B., Lausevic, M., and March, R. E. 2001. Analysis of dioxins and polychlorinated biphenyls by quadrupole ion-trap gas chromatography-mass spectrometry. In: Niessen, W. M. A. (ed.), *Current Practice of Gas Chromatography-Mass Spectrometry*, pp. 95–116. Marcel Dekker: New York.

Postigo, C., López de Alda, M. J., Barceló, D., Ginebreda, A., Garrido, T., and Fraile, J. 2010. Analysis and occurrence of selected medium to highly polar pesticides in groundwater of Catalonia (NE Spain): An approach based on on-line solid phase extraction–liquid chromatography–electro-spray-tandem mass spectrometry detection. *J. Hydrol.* 383: 83–92.

Qiu, C. and Cai, M. 2010. Ultra trace analysis of 17 organochlorine pesticides in water samples from the Arctic based on the combination of solid-phase extraction and headspace solid-phase microextraction–gas chromatography-electron-capture detector. *J. Chromatogr. A* 1217: 1191–1202.

Radišić, M., Grujić, S., Vasiljević, T., and Laušević, M. 2009. Determination of selected pesticides in fruit juices by matrix solid-phase dispersion and liquid chromatography–tandem mass spectrometry. *Food Chem.* 113: 712–719.

Ravelo-Pérez, L. M., Herrera-Herrera, A. V., Hernández-Borges, J., and Rodríguez-Delgado, M.Ă. 2010. Carbon nanotubes: Solid-phase extraction. *J. Chromatogr. A* 1217: 2618–2641.

Reemtsma, T. 2003. Liquid chromatography-mass spectrometry and strategies for trace-level analysis of polar organic pollutants. *J. Chromatogr. A* 1000: 477–501.

Reichman, R., Mahrer, Y., and Wallach, R. 2000. A combined soil-atmosphere model for evaluating the fate of surface-applied pesticides. 2. The effect of varying environmental conditions. *Environ. Sci. Technol.* 34: 1321–1330.

Richardson, S. D. 2001. Mass spectrometry in environmental sciences. *Chem. Rev.* 101: 211–254.

Richardson, S. D. 2009. Water analysis: Emerging contaminants and current issues. *Anal. Chem.* 81: 4645–4677.

Ricking, M. and Schwarzbauer, J. 2008. HCH residues in point-source contaminated samples of the Teltow Canal in Berlin, Germany. *Environ. Chem. Lett.* 6: 83–89.

Rocha, C., Pappas, E. A., and Huang, C. 2008. Determination of trace triazine and chloroacetamide herbicides in tile-fed drainage ditch water using solid-phase microextraction coupled with GC–MS. *Environ. Pollut.* 152: 239–244.

Rodrigues, A. M., Ferreira, V., Cardoso, V. V., Ferreira, E., and Benoliel, M. J. 2007. Determination of several pesticides in water by solid-phase extraction, liquid chromatography and electrospray tandem mass spectrometry. *J. Chromatogr. A* 1150: 267–278.

Rodriguez-Mozaz, S., López de Alda, M. J., and Barceló, D. 2004. Monitoring of estrogens, pesticides and bisphenol A in natural waters and drinking water treatment plants by solid-phase extraction–liquid chromatography–mass spectrometry. *J. Chromatogr. A* 1045: 85–92.

Röpke, B., Bach, M., and Frede, H.-G. 2004. DRIPS – A DSS for estimating the input quantity of pesticides for German river basins. *Environ. Model. Software* 19: 1021–1028.

Sánchez-Avila, J., Quintana, J., Ventura, F., Tauler, R., Duarte, C. M., and Lacorte, S. 2010. Stir bar sorptive extraction-thermal desorption–gas chromatography–mass spectrometry: An effective tool for determining persistent organic pollutants and nonylphenol in coastal waters in compliance with existing Directives. *Mar. Pollut. Bull.* 60: 103–112.

Sánchez-Rojas, F., Bosch-Ojeda, C., and Cano-Pavón, J. M. 2009. A review of stir bar sorptive extraction. *Chromatographia* 69: 79–94.

Sankararamakrishnan, N., Sharma, A. K., and Sanghi, R. 2005. Organochlorine and organophosphorous pesticide residues in ground water and surface waters of Kanpur, Uttar Pradesh, India. *Environ. Int.* 31: 113–120.

Sauret, N., Wortham, H., Strekowski, R., Herckès, P., and Nieto, L. I. 2009. Comparison of annual dry and wet deposition fluxes of selected pesticides in Strasbourg, France. *Environ. Pollut.* 157: 303–312.

Sauret-Szczepanski, N., Mirabel, P., and Wortham, H. 2006. Development of an SPME–GC–MS/MS method for the determination of pesticides in rainwater: Laboratory and field experiments. *Environ. Pollut.* 139: 133–142.

Schottler, S. P., Eisenreich, S. J., and Capel, P. D. 1994. Atrazine, alachlor, and cyanazine in a large agricultural river system. *Environ. Sci. Technol.* 28: 1079–1089.

Schreiber, B., Brinkmann, T., Schmalz, V., and Worch, E. 2005. Adsorption of dissolved organic matter onto activated carbon-the influence of temperature, absorption wavelength, and molecular size. *Water Res.* 39: 3449–3456.

Schulz, R., Hauschild, M., Ebeling, M., Nanko-Drees, J., Wogram, J., and Liess, M. 1998. A qualitative field method for monitoring pesticides in the edge-of-field runoff. *Chemosphere* 36: 3071–3082.

Senseman, S. A., Lavy, T. L., Mattice, J. D., Gbur, E. E., and Skulman, B. W. 1997. Trace level pesticide detections in Arkansas surface waters. *Environ. Sci. Technol.* 31: 395–401.

Shen, G. and Lee, H. K. 2003. Determination of triazines in soil by microwave-assisted extraction followed by solid-phase microextraction and gas chromatography–mass spectrometry. *J. Chromatogr. A* 985: 167–174.

Shukla, G., Kumar, A., Bhanti, M., Joseph, P. E., and Taneja, A. 2006. Organochlorine pesticide contamination of ground water in the city of Hyderabad. *Environ. Int.* 32: 244–247.

Siebers, J., Binner, R., and Wittich, K.-P. 2003. Investigation on downwind short-range transport of pesticides after application in agricultural crops. *Chemosphere* 51: 397–407.

Singer, P. C. and Reckhow, D. A. 1999. Chemical oxidation. In: Letterman, R. D. (ed), *Water Quality and Treatment*, pp. 12.1–12.51. McGraw-Hill, Inc.: New York.

Somasundaram, L. and Coats, J. L. 1991. Pesticide transformation products in the environment. In: Somasundaram, L. and Coats, J. L. (eds), *Pesticide Transformation Products: Fate and Significance in the Environment*, pp. 2–9. American Chemical Society: Washington, DC.

Tadeo, J. L., Sánchez-Brunete, C., and González, L. 2008. Pesticides: Classification and properties. In: Tadeo, J. L. (ed), *Analysis of Pesticides in Food and Environmental Samples*, pp. 1–34. CRC Press: Boca Raton, FL.

Tahboub, Y. R., Zaater, M. F., and Al-Talla, Z. A. 2005. Determination of the limits of identification and quantitation of selected organochlorine and organophosphorous pesticide residues in surface water by full-scan gas chromatography/mass spectrometry. *J. Chromatogr. A* 1098: 150–155.

Tankiewicz, M., Fenik, J., and Biziuk, M. 2010. Determination of organophosphorus and organonitrogen pesticides in water samples. *TrAC Trends Anal. Chem.* 29: 1050–1063.

Tesoriero, A. J., Saad, D. A., Burow, K. R., Frick, E. A., Puckett, L. J., and Barbash, J. E. 2007. Linking ground-water age and chemistry data along flow paths: Implications for trends and transformations of nitrate and pesticides. *J. Contam. Hydrol.* 94: 139–155.

Thurman, E. M. and Meyer, M. T. 1996. Herbicide metabolites in surface water and groundwater: Introduction and overview. In: Meyer, M. T. and Thurman, E. M. (ed), *Herbicide Metabolites in Surface Water and Groundwater*, pp. 1–15. American Chemical Society: Washington, DC.

Thurman, E. M., Ferrer, I., and Barceló, D. 2001. Choosing between atmospheric pressure chemical ionization and electrospray ionization interfaces for the HPLC/MS analysis of pesticides. *Anal. Chem.* 73: 5441–5449.

Tremolada, P., Parolini, M., Binelli, A., Ballabio, C., Comolli, R., and Provini, A. 2009. Preferential retention of POPs on the northern aspect of mountains. *Environ. Pollut.* 157: 3298–3307.

Trtić-Petrović, T., Đorđević, J., Dujaković, N., Kumrić, K., Vasiljević, T., and Laušević, M. 2010. Determination of selected pesticides in environmental water by employing liquid-phase micro-extraction and liquid chromatography–tandem mass spectrometry. *Anal. Bioanal. Chem.* 397: 2233–2243.

Tse, H., Comba, M., and Alaee, M. 2004. Method for the determination of organophosphate insecticides in water, sediment and biota. *Chemosphere* 54: 41–47.

United States Environmental Protection Agency (US EPA). 1987. Clean Water Act. http://www.epa.gov/oecaagct/lcwa.html (accessed 28 January 2011).

US EPA. 1996. Safe Drinking Water Act. http://water.epa.gov/lawsregs/rulesregs/sdwa/index.cfm (accessed 28 January 2011).

US EPA. 2002. National Primary Drinking Water Regulations. http://water.epa.gov/drink/contaminants/index.cfm#Organic (accessed 28 January 2011).

Vasiljevic, T., Lausevic, M., and March, R. E. 2000. Mass spectrometry analysis of polychlorinated biphenyls: Chemical ionization and selected ion chemical ionization using methane as a reagent gas. *J. Serb. Chem. Soc.* 65: 431–438.

Vasiljevic, T., Onjia, A., Cokesa, D., and Lausevic, M. 2004a. Optimization of artificial neural network for retention modeling in high-performance liquid chromatography. *Talanta* 64: 785–790.

Vasiljević, T., Baćić, M., Laušević, M., and Onjia, A. 2004b. Surface composition and adsorption properties of activated carbon cloth. *Mater. Sci. Forum* 453–454: 163–168.

Vasiljević, T., Spasojević, J., Baćić, M., Laušević, M., and Onjia, A. 2006. Adsorption of phenol and 2,4-dinitrophenol on activated carbon cloth: The influence of sorbent surface acidity and pH. *Sep. Sci. Technol.* 41: 1061–1075.

Vischetti, C., Cardinali, A., and Monaci, E., et al. 2008. Measures to reduce pesticide spray drift in a small aquatic ecosystem in vineyard estate. *Sci. Total Environ.* 389: 497–502.

Vukčević, M., Kalijadis, A., Dimitrijević-Branković, S., Laušević, Z., and Laušević, M. 2008. Surface characteristics and antibacterial activity of a silver-doped carbon monolith. *Sci. Technol. Adv. Mater.* 9: 015006 (7pp).

Wardencki, W., Curyło, J., and Namieśnik, J. 2007. Trends in solventless sample preparation techniques for environmental analysis. *J. Biochem. Biophys. Methods* 70: 275–288.

Wittmer, I. K., Bader, H.-P., Scheidegger, R., et al. 2010. Significance of urban and agricultural land use for biocide and pesticide dynamics in surface waters. *Water Res.* 44: 2850–2862.

Wolters, A., Linnemann, V., van de Zande, J. C., and Vereecken, H. 2008. Field experiment on spray drift: Deposition and airborne drift during application to a winter wheat crop. *Sci. Total Environ.* 405: 269–277.

World Health Organization. 2008. Guidelines for Drinking-water Quality, 3rd Ed., Volume 1, Recommendations. http://www.who.int/water_sanitation_health/dwq/gdwq3rev/en/ (accessed 28 January 2011).

Wu, Q., Riise, G., Lundekvam, H., Mulder, J., and Haugen, L. E. 2004. Influences of suspended particles on the runoff of pesticides from an agricultural field at Askim, SE-Norway. *Environ. Geochem. Health* 26: 295–302.

Section IV

Pesticides Residues in Vegetation

10

Pesticide Residues in Vegetation

Antonia Garrido Frenich, Isabel Mª Campoy Jiménez,
Mª del Mar Bayo Montoya, and José Luis Martínez Vidal

CONTENTS

10.1 Introduction

Pesticides have been used worldwide for pest control in crops, allowing the increase in crop productivity and improved product quality. Their use has been considered a revolution in agriculture, but, due to their high toxicity, many of them were included in the Stockholm Convention (1972) and declared as persistent organic pollutants (POPs). From 1970s, many of these pesticides have been forbidden in most countries; however, in 1992, approximately 2.5 million tons of pesticides were used around the world (Pimmentel et al. 1992). Although pesticide use has been controlled gradually, there may be more than 20,000 tons of obsolete pesticides in developing countries, and stocks are growing because a lot of these countries do not have environmental policies in this regard (FAO 1997).

Pesticides are present in all environmental compartments; they can be transported through great distances in the atmosphere before removal by deposition or degradation, depending on their residence time. Transport in ocean and atmosphere and fluxes between these compartments are essential to understand global cycling of the contaminants (Stemmler and Lammel 2009). In the troposphere, they can be removed from the air phase via atmospheric deposition to water and soil, and significant amounts of pesticides, mainly persistent compounds, can be found in soils, water, groundwater, sediments, biota, and vegetation (Barber et al. 2005).

In the aquatic ecosystems, there is a continuous exchange of pesticides among land, sediment, sediment–water interface, interstitial waters, aquatic organisms, and air–water interface. Absorption of lipophilic compounds by sediments plays an important role in pesticide absorption by aquatic organisms, and this explains to a large degree why the toxicity of pesticides to wildlife is predominantly an aquatic problem in comparison with terrestrial ecosystems (Cooper 1991).

In terrestrial ecosystems, soil pollutants can be taken up by plants and ultimately passed into the human food chain, where they can be very persistent and have a high bioaccumulation potential along the food chain (Hayes and Laws 1991).

There are a number of different pathways by which pesticides may enter vegetation. Deposition from the bulk atmosphere is the main process of movement of the pesticides from the troposphere to the terrestrial surface, where they are likely stored in terrestrial vegetation and soil (Syversen and Beckam 2004). In addition, they can be taken up by roots, may be volatilized from the soil followed directly by foliar uptake from the gaseous phase, or resuspended in the soil particles by varied mechanisms such as the actions of wind and rain, which may result in their subsequent capture by vegetation (Smith and Jones 2000).

Only a small portion of POPs deposited by rain reaches the stream water due to the vegetation buffer effects. The dominating retention process in the surface runoff buffer zones is the trapping of sediments and sediment-bound nutrients (Syversen and Beckam 2004).

The levels of pesticides in/on vegetation are influenced by different factors, which can be divided into three groups: first, the characteristics of the pesticide, mainly related to its stability in general and to its metabolites; second, environmental aspects, primarily temperature, precipitation, humidity, and wind, whose influence is lower; and third, the plant species implicated, structure of its cuticle, and its stage and rate of growth. Of all these factors, the stability of the pesticide, which depends on its physical and chemical characteristics, and the plant species and the rate of growth are the most important (Edwards 1975). However, for a specific pesticide, its adsorption, mobility, and degradation in a determinate plant species are highly influenced by climatic conditions.

In general, low temperature and high water content may reduce the degradation of pesticides in plants (Syversen and Beckam 2004). In that sense, there are several studies that demonstrate that in high mountain environments and in cold places, in general, the accumulation and deposition of persistent organic compounds, such as pesticides, are higher. Several studies have shown that regions with cold climates are susceptible to be enriched with POPs (e.g., Vighi 2006; Wang et al. 2006). Compounds prone to enrichment in cold environments are typically those with subcooled liquid vapor pressures between 0.01 and 1.0 Pa at 25°C (including many pesticides, highly volatiles). It is hypothesized that cold temperatures promote the deposition and accumulation of pesticides and other POPs, which tend to evaporate in warmer areas and be transported by air currents and deposited in colder alpine and arctic areas. Simonich and Hites (1995) have shown that the distribution of relatively volatile organochlorine compounds (such as hexachlorobenzene (HCB)) is dependent on the latitude and demonstrates the global distillation effect. By contrast, less volatile compounds (such as endosulfan) are not as effectively distilled and tend to remain in the region of use (Simonich and Hites 1995).

Natural vegetation has an important role in determining the general environmental fate of these chemicals, especially the levels of air concentrations, and there are many studies that use vegetation as a passive sampler to monitor remote areas (e.g., Calamari et al. 1991; Jensen et al. 1992). In many researches, plant samples have been used to qualitatively indicate the atmospheric contamination levels, since plants accumulate and circulate the residues of the pollutants, so they can offer indirect evidence of the atmospheric transport of POPs to remote areas (Wang et al. 2006). Also, they can be used to discover the sources of organic pollutants, to find out contamination within different places, such as cities or countries, and to determine the global contamination of organic pollutants (Holoubek et al. 1999). For instance, in Paris, leaves from plane trees (*Platanus vulgaris*) have been used as bioindicators of the pollution by the organochlorines in air. This study demonstrated the high power of bioconcentration of this kind of compounds in plants, as compared with their concentration in air (Granier and Chevreuil 1992).

In recent years, several studies have described the importance of vegetation as a pollutant sink (Holoubek et al. 1999). Plants can adsorb as particles from soil and water and as volatile compounds from air and suffer atmospheric deposition in their leaves, because of which plant biomass plays an important role as an indicator of environmental pollution (Barriada-Pereira et al. 2005).

According to a study about the estimation of the half-lives of pesticides, the degradation of these compounds in plants is more efficient than in soil but slower than in air. Moreover, the concentration of pesticides decreases vertically from the cuticle and their concentration in the cells of the subcuticle is often zero. It is because plants normally metabolize pesticides, which become water-soluble compounds and other residues that are less toxic or nontoxic for the plants (Juraske et al. 2008). Because of this, vegetation has been widely used to mitigate or reduce the negative effects produced by the contaminants in a determinate zone; this process is known as phytoremediation.

Numerous laboratory and field studies have established the importance of aquatic macrophytes and their associated phytoremediation processes as a mitigating and mobilizing factor of the organic contaminants in the environment. More concretely, the roles of those plant communities have been widely studied for the creation of natural buffer zones designed to mitigate the impacts of the pesticides associated with waters following runoff and spray drift from agricultural areas. Moreover, rooted aquatic vegetation can aid in remobilizing the pollutants from contaminated sediments. To sum up, phytoremediation processes can mitigate the contaminants from soil, sediment, and water (Armitage et al. 2008).

Most of the compiled studies are focused on the POPs, and experimental studies are predominant, which are carried out using manipulated plant samples in the laboratory. There are a few studies about the presence of pesticides in natural vegetation, and those performed studies are basically focused on the organochlorine pesticides, since they are the most toxic and persistent ones.

10.2 Methodology

Many studies about pesticides in natural plants worldwide have been compiled, independent of the date, sampling method, or analytical method used for the extraction and determination of pesticide residues. The studies that were carried out in vegetation from cultivated areas or the experimental studies in which the analyses were realized on the samples from the plants previously treated with pesticides are not considered.

10.3 Results and Discussion

The data from a total of 636 pesticide concentrations in vegetation samples from 38 countries around the world have been compiled (Table 10.1). The organochlorine pesticides are the most studied, particularly dichloro-diphenyl-trichloroethane (DDT), considering

TABLE 10.1

List of Pesticides and Vegetation Species Studied in Different Countries

Country	Reference	Pesticides	Species
Antarctica	Bacci et al. 1986	α-HCH, γ-HCH, p,p´-DDE, p,p´-DDT, HCB	Lichen, moss
Argentina	Calamari et al. 1991	HCB, α-HCH, γ-HCH, p,p´-DDT, o,p´-DDT, p,p´-DDE	Lichen, moss
Austria	Calamari et al. 1994	α-HCH, HCB, γ-HCH, o,p´-DDE, p,p´-DDE, o,p´-DDT, p,p´-DDT	Pine
Benin-Burkina-Faso	Bacci et al. 1988	α-HCH, γ-HCH, p,p´-DDE, p,p´-DDD, o,p´-DDT, p,p´-DDT	Mango (*Magnifera indica*)
Bolivia	Calamari et al. 1991	HCB, α-HCH, γ-HCH, p,p´-DDT, o,p´-DDT, p,p´-DDE	Lichen
Brazil (Amazon)	Calamari et al. 1991	HCB, α-HCH, γ-HCH, p,p´-DDT, o,p´-DDT, p,p´-DDE	Lichen, moss
Canada	Davidson et al. 2003	α-HCH, HCB, p,p´-DDE, β-Endosulfan	Pine
Cape Town	Calamari et al. 1991	HCB, α-HCH, γ-HCH, p,p´-DDT, o,p´-DDT, p,p´-DDE	Lichen, moss
China	Davidson et al. 2003	HCB	Pine
Croatia	Krauthacker et al. 2001	HCB, α-HCH, β-HCH, γ-HCH, p,p´-DDE, p,p´-DDD, p,p´-DDT, o,p´-DDT	*Larix decidu, Pinus nigra, Tilia platyphyllos, Picea abies, Thuja occidentalis, Abies alba*
Czech Republic	Calamari et al. 1994, Holoubeck et al. 2000	α-HCH, HCB, γ-HCH, o,p´-DDE, p,p´-DDE, o,p´-DDT, p,p´-DDT, HCB	Moss (*Hypnum cuprresiforme*), *Pinus sylvestris*
Denmark	Jensen et al. 1992	DDE, DDT, DDT+DDE, α-HCH, γ-HCH, PCP, HCB	*Pinus sylvestris*
Estonia	Hellström et al. 2004	HCB	Pine
Finland	Calamari et al. 1994	HCB, α-HCH, γ-HCH, p,p´-DDT, o,p´-DDT, p,p´-DDE, o,p´-DDE	Pine
France	Jensen et al. 1992	DDE, DDT, DDT+DDE, α-HCH, γ-HCH, PCP, HCB	*Pinus sylvestris*
Germany	Jensen et al. 1992, Hellström et al. 2004	DDE, DDT, DDT+DDE, α-HCH, γ-HCH, PCP, HCB	*Pinus sylvestris*
Ghana	Bacci et al. 1988	HCB, α-HCH, γ-HCH, p,p´-DDE, p,p´-DDD, o,p´-DDT, p-p´-DDT	Mango (*Magnifera indica*)
Greece	Calamari et al. 1994	α-HCH, HCB, γ-HCH, o,p´-DDE, p,p´-DDE, o,p´-DDT, p,p´-DDT	Pine
Guatemala	Calamari et al. 1991	HCB, α-HCH, γ-HCH, p,p´-DDT, o,p´-DDT, p,p´-DDE	Lichen, moss
Guinea	Bacci et al. 1988	α-HCH, γ-HCH, p,p´-DDE, p,p´-DDD, o,p´-DDT, p,p´-DDT, HCB	Mango (*Magnifera indica*)
Holland	Calamari et al. 1994	α-HCH, HCB, γ-HCH, o,p´-DDE, p,p´-DDE, o,p´-DDT, p,p´-DDT	Pine
Iceland	Calamari et al. 1991	HCB, α-HCH, γ-HCH, p,p´-DDT, o,p´-DDT, p,p´-DDE	Lichen, moss
India	Calamari et al. 1991	HCB, α-HCH, γ-HCH, p,p´-DDT, o,p´-DDT, p,p´-DDE	Mango (*Magnifera indica*)
Indonesia	Calamari et al. 1991	HCB, α-HCH, γ-HCH, p,p´-DDT, o,p´-DDT, p,p´-DDE	Lichen, moss

TABLE 10.1 (Continued)

List of Pesticides and Vegetation Species Studied in Different Countries

Country	Reference	Pesticides	Species
Italy	Calamari et al. 1994, Bacci et al. 1988	HCB, α-HCH, γ-HCH, o,p′-DDE, p,p′-DDT, o,p′-DDT, p,p′-DDE	*Pinus sylvestris, Pinus pinea,* lichen
Ivory Coast	Calamari et al. 1991	HCB, α-HCH, γ-HCH, p,p′-DDT, o,p′-DDT, p,p′-DDE	Mango (*Magnifera indica*)
Kenya	Calamari et al. 1991	HCB, α-HCH, γ-HCH, p,p′-DDT, o,p′-DDT, p,p′-DDE	Mango (*Magnifera indica*)
Mali	Bacci et al. 1988	α-HCH, γ-HCH, p,p′-DDE, p,p′-DDD, o,p′-DDT, p,p′-DDT, HCB	Mango (*Magnifera indica*)
Nepal	Calamari et al. 1991, Wang et al. 2007	HCB, α-HCH, γ-HCH, p,p′-DDT, o,p′-DDT, p,p′-DDE, o,p′-DDD, α-Endosulfan	Lichen, moss, grass, *Picea abies*
Norway	Calamari et al. 1991, Jensen et al. 1992, Oeckenden et al. 2007	DDE, DDT, DDT+DDE, α-HCH, γ-HCH, PCP, HCB, p,p′-DDT, o,p′-DDT, p,p′-DDE	*Pinus sylvestris,* lichen
Poland	Jensen et al. 1992	DDE, DDT, DDT+DDE, α-HCH, γ-HCH, PCP, HCB	*Pinus sylvestris*
Slovenia	Weiss 2001	HCB	*Picea abies*
Spain	Barriada-Pereira 2001, Villa et al. 2003, Perez-Ruzafa et al. 2001	HCB, α-HCH, β-HCH, γ-HCH, δ-HCH, p,p′-DDT, p,p′-DDD, p,p′-DDE, o,p′-DDE, o,p′-DDD, o,p′-DDT, endosulfan, endrín aldehído, aldrín, endrín, dieldrín, heptacloro	*Pinus sylvestris, Ruppia cirrhosa, Caulerpa prolifera, Cymodocea nodosa, Enteromorpha flexuosa, Chaetomorpha linum, Laurencia obtusa, Hypnea cervicornis, Acetabularia calyculus*
Sweden	Jensen et al. 1992, Villeneuve et al. 1985, Knulst et al. 1995	HCB, α-HCH, γ-HCH, p,p′-DDE, p,p′-DDD, p,p′-DDT, DDE, DDT, DDT+DDE, PCP	Moss (*Hylocomium splendens* and *Pleurozium schreberi*), lichen, *Pinus sylvestris*
Switzerland	Jensen et al. 1992	DDE, DDT, DDT+DDE, α-HCH, γ-HCH, PCP	*Pinus sylvestris*
Tanzania	Mahugija-Marco and Kisimba 2007	p,p′-DDT, o,p′-DDT, p,p′-DDE, o,p′-DDE, p,p′-DDD, o,p′-DDD	Mango (*Magnifera indica*)
Tristan da Cunha	Calamari et al. 1991	HCB, α-HCH, γ-HCH, p,p′-DDT, o,p′-DDT, o,p′-DDE	Mango (*Magnifera indica*)
Venezuela	Calamari et al. 1991	HCB, α-HCH, γ-HCH, p,p′-DDT, o,p′-DDT, p,p′-DDE	Lichen, moss

its isomers o,p′-DDT and p,p′-DDT, and metabolites, such as dichloro-diphenyl-ethylene (DDE), its isomers (o,p′-DDE and p,p′-DDE), and dichloro-diphenyl-dichloroethane (DDD), its isomers (o,p′-DDD and p,p′-DDD), whose data were collected from over 250 different plant species concentrations in a large number of countries.

Hexachlorocyclohexane (α-, β-, and γ-HCH isomers) has also been widely studied; data from 160 different concentrations in plant samples have been compiled. HCB pesticides are in the third place (80 data have been compiled).

About other pesticides, such as endosulfan and pentachlorophenol (PCP), 12 studies have been carried out, and 8 results have been compiled about aldrin, dieldrin, endrin aldehyde, and heptachlor in different species.

10.3.1 Spatial Distribution

Away from polluted areas, deposition from the bulk atmosphere is the main way by which persistent compounds arrive to vegetation. Therefore, the concentrations of these pollutants in plants are a reflection of their atmospheric concentrations in the area. This theory has been confirmed by a group of studies that used vegetation as a passive sampler for monitoring pollution in contaminated areas.

DDT has been analyzed in natural vegetation samples in a total of 34 countries. Most of the studies have been carried out in the Northern Hemisphere; a total of 141 have been performed in Europe, 37 in Africa, and 24 in Asia. The rest have been carried out in America (principally in the South) and in Antarctica (Figure 10.1).

In most performed analysis of DDTs (considering the DDT, DDE, o,p′-DDT, p,p′-DDT, o,p′-DDE, p,p′-DDE, o,p′-DDD, and p,p′-DDD congeners), positive results have been observed (about 90% of the compiled studies), although in 48% of the studies, the mean concentration is below 1 µg/kg (data in dry weight). However, in 25% of the analyses, average concentrations between 1 and 5 µg/kg were found. Only, in 11% of the cases, the average concentrations of DDT were above 20 µg/kg. The highest concentrations obtained were for p,p′-DDT, whereas the lowest concentrations were for the isomer o,p′-DDE.

Figure 10.1 shows the average concentrations of DDTs found in different species of *Pinus* spp., *Mangifera indica*, and lichens/mosses, which constitute most of the studied species (on a global scale) around the world. The high concentrations obtained in Africa (Kenia, Ghana, Mali), India, and the Amazonas are outstanding. The highest concentration, found in the bibliography, occurs in Tanzania, in *M. indica* leaves: a concentration of 485 µg/kg (concretely the p,p′-DDT; Mahugija-Marco and Kisimba 2007). In Europe, the high concentrations in

FIGURE 10.1
Distribution of DDTs (mean concentrations) in *Pinus* spp. (triangle), *Magnifera indica* (circle) and lichen/moss (square) around the world. The size of the symbol indicates different mean concentration of DDT (see legend).

Italy and Czech Republic (average concentrations above 10 μg/kg) are outstanding. In general, as Calamari et al. showed in 1991, DDT contamination is more important in tropical zones, contrary to what happens to other compounds, such as HCB. The HCH compound has been studied in a total of 35 countries around the world. Most of the zones where HCH has been studied in vegetation samples are in the Northern Hemisphere, mainly in the European continent, where 68% of the total studies have been performed.

In most of the studies, the analyzed congeners were α- and γ-HCH (lindane), although, in some cases, the congeners β- and δ-HCH have been also included (Barriada-Pereira et al. 2005; Krauthacker et al. 2001). In 49% of the analyzed samples, the average concentrations of this contaminant were below 1 μg/kg (in dry weight of the analyzed species). In 33% of the cases, the detected concentrations were between 1 and 10 μg/kg. Only in 18% of the cases, average concentrations of HCH above 10 μg/kg were obtained.

In Europe, high average concentrations, in general, of HCH were found in *Pinus* spp. and lichens/mosses (Figure 10.2). Also, the average concentrations obtained in the areas of the Amazon, India, and Nepal are outstanding. The highest value of the concentration of HCH, concretely of α-HCH, was found in New Delhi (India), in mango (*M. indica*) leaves: a concentration of 106.9 μg/kg (Krauthacker et al. 2001). Significant concentrations of HCH have been found in samples from high mountains characterized by low temperatures, in areas such as Antarctica, Iceland, Nepal and Kenya. That confirms the processes of large-scale transport of this contaminant and its higher accumulation in cold places. In that sense, high mountain areas, such as cold zones, act like condensers for HCH (Barber et al. 2005; Calamari et al. 1991).

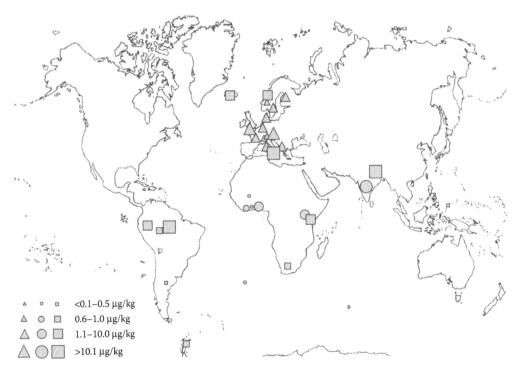

△ ○ ▫ <0.1–0.5 μg/kg
△ ○ ☐ 0.6–1.0 μg/kg
△ ◎ ☐ 1.1–10.0 μg/kg
△ ● ☐ >10.1 μg/kg

FIGURE 10.2
Distribution of HCH (mean concentrations) in *Pinus* spp. (triangle), *Magnifera indica* (circle) and lichen/moss (square) around the world. The size of the symbol indicates different mean concentration of HCH (see legend).

In the studies carried out on other species, high concentrations have been obtained, for example, 97.3 µg/kg of HCH was found in *Chaetomorpha linum* in Murcia (Spain). In addition, especially high concentrations of HCH, with values above 20 mg/kg for the congeners α- and β-HCH were obtained in a study performed on the species *Cytisus striatus*, *Avena sativa*, *Vicia sativa*, *Solanum nigra*, and *Chenopodium vulgare* in a polluted area in Galicia (Spain) (Barriada-Pereira et al. 2005).

The HCB pesticide in plant samples in a total of 33 countries, most of them in the Northern Hemisphere, has been studied. By considering the average values, in 51% of the samples the concentrations were below 0.5 µg/kg, and in 12% the concentrations were between 0.5 and 1 µg/kg. Figure 10.3 shows the average concentration of HCB in vegetation samples obtained from different countries. The highest concentrations are found in the Northern Hemisphere, principally Iceland, Norway, Italy, and Czech Republic; the maximum value of concentration, 4.8 µg/kg, was detected in pines in Estonia and Germany (Hellström et al. 2004). In the Southern Hemisphere, only the results obtained in Antarctica and Kenya stand out.

In general, higher concentrations were found in the Northern Hemisphere because most part of the emissions took place there. Moreover, the values are very homogeneous within the countries because the use of HCB was banned in most of them and, therefore, the emission sources have been reduced.

These results agree with the data reported by Barber et al. (2005) about the concentration of this pollutant in air on a global scale. Furthermore, as in the case of HCH, HCB is accumulated in cold places and high mountain areas.

FIGURE 10.3
Distribution of HCB (mean concentrations) in *Pinus* spp. (triangle), *Magnifera indica* (circle) and lichen/moss (square) around the world. The size of the symbol indicates different mean concentration of HCB (see legend).

The presence of endosulfan has been studied in a less number of occasions; only three studies were compiled. In Spain, it was analyzed in eight different species of aquatic plants, although contamination by this compound was found in only four of the target species (Pérez-Ruzafa et al. 2000). The rest of the studies were carried out in high mountain areas, in Canada (Davidson et al. 2003) and in the Mt. Qomolangma (Himalayas) (Wang et al. 2007), and the presence of endosulfan was found in both the zones.

Regarding the endrin aldehyde, aldrin, dieldrin, and heptachlor pesticides, they have been analyzed in a study performed on eight species of aquatic plants in Spain. Endrin contamination was not found in this zone, but for the rest of the target compounds, the results ranged between 3.40 μg/kg for aldrin and 1474.9 μg/kg for endrin aldehyde.

The PCP compound has also been studied in different zones of the European continent (Sweden, Norway, Denmark, Germany, France, Switzerland, and Poland), in *Pinus sylvestris* (Jensen et al. 1992). Values of the concentrations of this pollutant, which ranged between 0.29 μg/kg in Germany and 0.90 μg/kg in Sweden and Norway, have been recorded.

10.3.2 Temporal Evolution

The insecticide DDT was implemented in the late 1940s and was used until the 1970s in large quantities in agricultural areas around the world. Although emissions of this compound ceased in 1990, due to its persistence, DDT continues circulating among different environmental compartments (Stemmler and Lammel 2009).

Concentrations of DDTs in plant samples have declined along the years, from 1988 to 2007 (Figure 10.4a), except for high concentration values found by Mahugija-Marco and

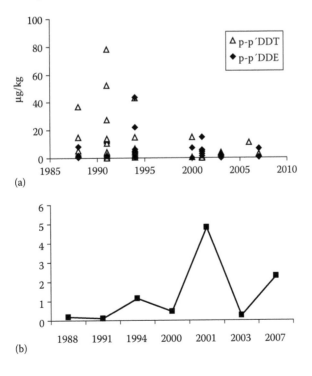

(a)

(b)

FIGURE 10.4
(a) Temporal evolution of concentratrions of p-p′DDT and p-p′DDE in vegetation samples from 34 countries around the world. (b) Temporal evolution of p-p′DDE / p-p′DDT ratio.

Kisimba (2007) in Tanzania and by Barriada-Pereira et al. (2005) in Spain (values not shown, outliers). It can be said that the temporal evolution of DDT continues the pattern of applications, which agrees with the ideas published by Stemmler and Lammel (2009).

The relationship between p,p'-DDE and p,p'-DDT may be an indicator of the age of DDT in the environment, since the p,p'-DDE is a breakdown product (Rapport and Eisenreich 1986). Figure 10.4b shows a general increase in this ratio, indicating that DDT sources have decreased along the years; but there are still sources of emission of these organochlorine compounds in some areas, as Wang et al. (2007) and Stemmler and Lammel (2009) indicated, in India and many Asian and African countries, where dicofol is used as a miticide, a product that contains a high concentration of p,p'-DDT and o,p'-DDT.

HCH concentrations in vegetation samples have been studied during the period between 1991 and 2006. Throughout this period, there has been a reduction in the presence of α-HCH congeners and γ-HCH (the most studied) in vegetation. The highest concentrations were recorded in 1991 and 1994 (Figure 10.5a). In 2005 and 2010, high concentration levels of HCH were found in contaminated areas of Tanzania (Mahugija-Marco and Kisimba 2007) and Spain (Barriada-Pereira et al. 2005), which have not been included in the studies of temporal evolution, because they were considered outliers.

According to Atlas and Giam (1988), the proportion of these isomers (α-/γ-HCH ratio) is an indicator of the age of lindane in air, so that the higher the ratio, the higher is the seniority. The relationship between the concentrations of both the isomers in plant samples throughout the years (Figure 10.5b) has been studied, and there has been a substantial reduction from 1991 to 1994 and then an increase such that in 2003, the values obtained were around 1.5/1. The results suggest the widespread use of γ-HCH in the middle of the 1990s and then a gradual decrease since its use was banned in many countries, although it is still used in some areas as Wang et al. (2007) noted after detecting high levels of HCH in the Himalayas.

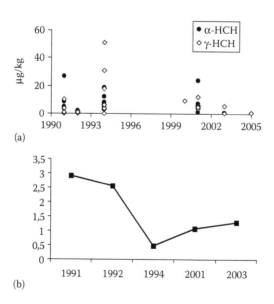

FIGURE 10.5
(a) Temporal evolution of concentrations of α-HCH y γ-HCH in vegetation samples from 35 countries around the world. (b) Temporal evolution of α-HCH / γ-HCH ratio.

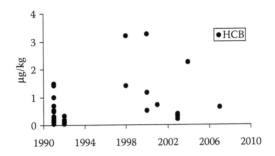

FIGURE 10.6
Temporal evolution of concentration of HCB in vegetation samples from 33 countries around the world.

Figure 10.6 shows the temporal evolution of HCB concentration in different plant species collected from around the world. The high values found in Estonia and Germany in 2004 by Hellström are not represented in the figure because they were considered outliers. There is a slight tendency of the concentrations of the compound to decrease over the years. This agrees with the data reported by Barber et al. (2005) about emissions and global distribution of this pollutant. The use of this pesticide was banned for the first time in Argentina and Hungary in the 1960s; in the following decades, between 1970 and 2000, the use of HCB had been banned in most countries. Currently, its use is only allowed in China and Russia (Barber et al. 2005).

10.3.3 Distribution of Pesticides in Vegetation Species

The most studied plant species to determine the presence of pesticides in natural plants have been the genus *Pinus* (*P. sylvestris* and *P. canariensis* mainly), lichens/mosses, and mangoes (*M. indica*). Together, the *Pinus* species account for approximately 75% of total studies on vegetation. In general, it can be said that the *Pinus* spp. were used as an indicator of pesticide contamination in the Northern Hemisphere, while in the Southern Hemisphere, the mango (*M. indica*) plants were used. Lichens and mosses have been studied in several areas throughout the world due to their worldwide distribution. The rest of the performed works were focused on different herbaceous species and other aquatic plants. Figures 10.1 through 10.3 show a worldwide distribution of the studies on different species. The results obtained in these species have been discussed in section "Geographical Distribution".

DDT has been studied mainly in tree species, highlighting the genus *Pinus* (*P. sylvestris* and *P. canariensis*) and mangoes (*M. indica*). Also, studies have been carried out in other species such as *Larix decidua, P. nigra, Tilia platyphyllos, Picea abies, Thuja occidentalis,* and spruce (*Abies alba*). DDT was also analyzed in lichens and mosses and in a variety of herbaceous plants (*Cy. striatus, A. sativa, V. sativa, S. nigra,* and *Ch. vulgare*), as well as in aquatic species (*Ruppia cirrhosa, Cymodocea nodosa, Caulerpa prolifera, Enteromorpha flexuosa, C. linum, Laurencia obtusa, Hypnea cervicornis,* and *Acetabularia calyculus*). The maximum concentration of DDT, concretely p,p′-DDT, was found in mangoes (*M. indica*) in Tanzania (Mahugija-Marco and Kisimba 2007).

HCH has been analyzed in a variety of plant species, particularly lichens, herbaceous plants, aquatic plants, and marine and woody vegetation. The species of these three groups of plants are the cited ones for DDT. Results obtained from herbaceous species were particularly high, ranging from 10 mg/kg of δ-HCH in *S. nigra* to 40.26 mg/kg of β-HCH in *Cy. striatus* (Barriada-Pereira et al. 2005), but it should be noted that the study was conducted in a contaminated area. In the study of aquatic plants (species named above for DDT), the results ranged from 0.005 mg/kg in *E. flexuosa* in Spain to 97.3 mg/kg in *C. linum* in the same place. Finally, HCH concentrations obtained in different tree species of pines and mangoes ranged between 0.9 mg/kg of β-HCH in *Pic. abies* (spruce) in Croatia and 12.0 mg/kg of γ-HCH in *T. platyphyllos* in the same country (Krauthacker et al. 2001).

HCB has been detected, especially in the species of *Pinus* spp. and lichens and mosses, since, as noted above, most of these studies have been conducted in the Northern Hemisphere. Some studies in mangoes (*M. indica*) and other species, such as *L. decidua*, *P. nigra*, *T. platyphyllos*, *Pic. abies*, *Th. occidentalis*, *Ab. alba*, and grasses have also been conducted. The maximum value of HCB was found in pines in Milan (Italy) (Calamari et al. 1994).

The aldrin, dieldrin, endrin, endrin aldehyde, heptachlor, and endosulfan pesticides were studied in the following aquatic species: *R. cirrhosa*, *Cym. nodosa*, *Ca. prolifera*, *E. flexuosa*, *C. linum*, *La. obtus*, *H. cervicornis*, and *Ac. calyculus*, in Spain. The maximum concentration value was obtained for endrin aldehyde in *C. linum*. Endosulfan in pines, spruces, and grasses in the mountainous regions of Alberta (Canada) and the Mt Qomolangma (Himalaya) was also studied.

However, the PCP compound has only been compiled in a study on *P. sylvestris* in seven European countries (Sweden, Norway, Denmark, Germany, France, Switzerland, and Poland).

10.3.4 Distribution with Elevation

The relationship between the presence of POPs and the altitude has been extensively studied worldwide in high mountain regions (e.g., Davidson et al. 2003; Villa et al. 2003). These studies have established gradients of correlation between both the parameters for different locations. According to the studies on grasses, there is a positive correlation between the presence of HCH and HCB and elevation, while in the case of o,p′-DDT, p,p′-DDT, and aldrin, there is an inverse correlation between the concentrations of these compounds and elevation (Wang et al. 2007). In general, it can be said that high elevation areas may act as condensers of POPs (Wang et al. 2007). This is due mainly to low temperatures in these areas.

10.4 Conclusions

The number of studies on the pesticides in natural vegetation are scarce, compared with those on the pesticides in other environmental compartments such as air, soil, or water (Barber et al. 2005). Most studies have been conducted in Europe, considering only organochlorine pesticides, mainly DDT and lindane, so the discussed results can be very slant.

However, vegetation is at the base of the food chain and plays a key role in the nutrient cycles in nature, so it is important to include vegetation in monitoring programs of environmental pollution.

The concentrations of pesticides in different plant species depend largely on the chemical properties of the compounds and the meteorological conditions in the areas, although also on the characteristics of each plant species. Therefore, the choice of the species to be used to monitor pollution by POPs is of great importance. In general, there should be plant species with a wide distribution in the study area to obtain representative results. However, it must be considered that vegetation is a problematic tool for monitoring pollutants, since there may be differences in the accumulation of the compounds between plant species, for example, between conifers, deciduous trees, and herbaceous species (Vighi 2006). To consider the use of vegetation as an environmental indicator, the collection, extraction, and analysis of the samples should be standardized.

It is important to study the presence of pesticides in the vegetation in areas located far from the sources of pollution (reference areas) when conducting studies on regional and global scales, to know the levels of "background pollution."

The HCH and HCB organochlorine pesticides, which are more volatile, are concentrated mainly in cold areas and high mountain areas around the world (Barber et al. 2005; Krauthacker et al. 2001), although mainly in the Northern Hemisphere. However, DDT has higher levels of pollution in tropical areas, confirming the results obtained in the previous studies (Calamari et al. 1991).

Although the results show the general trends of decreasing levels of DDT along the years in plant samples, the concentrations of other organochlorine pesticides, such as lindane, show that people are still using these compounds in some places around the world, despite them being banned in most countries (Wang et al. 2007).

It would be interesting to carry out studies about the presence of other persistent pesticides, such as chlorpyrifos, chlorfenvinphos, and others, which are widely used and which appear in significant concentrations in freshwater, marine water, wastewater, sediments, and soils.

Acknowledgment

The authors gratefully acknowledge the Consejería de Innovación Ciencia y Empresa de la Junta de Andalucía-FEDER (P05-FQM-0202) for the financial support.

References

Armitage, J. M., Franco, A., Gomez, S., and Cousins, I. T. 2008. Modelling the potential influence of particle deposition on the accumulation of organic contaminants by submerged aquatic vegetation. *Environ. Sci. Technol.* 42: 4052–4059.

Atlas, E. L. and Giam, C. S. 1988. Ambient concentration and precipitation scavenging of atmospheric organic pollutants. *Water Air Soil Pollut.* 38: 19–36.

Bacci, E., Calamari, D., Gaggi, C., Biney, C., Focardi, S., and Morosini, M. 1988. Organochlorine pesticides and PCB residues in plant foliage (*Mangifera indica*) from west Africa. *Chemosphere* 17: 693–702.

Barber, J. L., Sweetman, A. J., Van Wijk, D., and Jones, K. C. 2005. Hexachlorobenzene in the global environment: Emissions, levels, distribution, trends and processes. *Sci. Total Environ.* 349: 1–44.

Barriada-Pereira, M., Gozález-Castro, M. J., Muniategui-Lorenzo, S. et al. 2005. Organochlorine pesticides accumulation and degradation products in samples of contaminated area in Galicia (NW Spain). *Chemosphere* 58(11): 1571–1578.

Calamari, D., Bacci, E., Focardi, S., Gaggi, C., Morosini, M., and Vighi, M. 1991. Role of plant biomass in the global environmental partitioning of chlorinated hydrocarbons. *Environ. Sci. Technol.* 25: 1489–1495.

Calamari, D., Tremolada, P., Guardo, A. D., and Vighi, M. 1994. Chlorinated hydrocarbons in pine needles in Europe: Fingerprint for the past and recent use. *Environ. Sci. Technol.* 28: 429–434.

Cooper, K. 1991. Effects of pesticides on wildlife. In Hayes, W. J., Jr. and Laws, E. R., Jr. (eds), *Handbook of Pesticide Toxicology, Vol 1: General Principles*, pp. 463–496. Academic Press: New York.

Davidson, D. A., Wilkinson, A. C., and Blais, J. M. 2003. Orographic cold-trapping of persistent organic pollutants by vegetation in mountains of Western Canada. *Environ. Sci. Technol.* 37: 209–215.

Edwards, C. A. 1975. Factors that affect the persistence of pesticides in plant and soils. *Pure Appl. Chem.* 42: 39–56.

(FAO) Food and Agriculture Organization of the United Nations. 1997. Prevention and disposal of obsolete and unwanted pesticide stocks in Africa and Near East. FAO Pesticide disposal Series no. 5.

Granier, L. and Chevreuil, M. 1992. On the use of tree leaves as bioindicators of the contamination of air by organochlorines in France. *Water Air Soil Pollut.* 64: 575–584.

Hayes, W. J. and Laws, E. R. 1991. *Handbook of Pesticide Toxicology*. Academic Press: San Diego, CA.

Hellström, A., Kylin, H., Strachan, W. M. J., and Jensen, S. 2004. Distribution of some organo chlorine compounds in pine needles from Central and Northern Europe. *Environ. Pollut.* 12: 829–848.

Holoubek, I., Korinek, P., Seda, Z., et al. 2000. The use of mosses and pine needles to detect persistent organic pollutants at local and regional scales. *Environ. Pollut.* 109: 283–292.

Jensen, S., Ericsson, G., Kylin, H., and Strachan, W. M. J. 1992. Atmospheric pollution by persistent organic compounds: monitoring with pine needles. *Chemosphere* 24: 229–245.

Juraske, R., Antón, A., and Castells, F. 2008. Estimating half-lives in/on vegetation for use in multi-media fate and exposure models. *Chemosphere* 70: 1748–1755.

Knulst, J. C., Westling, H. O., and Brorström-Lundén, E. 1995. Airbone organic micropollutant in mosses and humus as indicators for local versus long range sources. *Environ. Monitor. Assess.* 36: 75–91.

Krauthacker, B., Romanic, S. H., and Reiner, E. 2001. Polychlorinated biphenyls and organochlorine pesticides in vegetation samples collected in Croatia. *Bull. Contam. Toxicol.* 66: 334–341.

Mahugija-Marco, J. A. and Kisimba, M. A. 2007. Organochlorine pesticides and metabolites in young leaves of *Mangifera indica* from sites near a point source in coast region, Tanzania. *Chemosphere* 48: 832–837.

Pérez-Ruzafa, A., Navarro, S., Barba, A., et al. 2000. Presence of pesticides throughout trophic compartments on the food web in the Mar Menor lagoon (SE Spain). *Marine Pollut. Bull.* 40: 140–151.

Pimmentel, D., Acquay, H., Biltonen, M., et al. 1992. Environmental and economic cost of pesticide use. *BioScience* 542: 750–760.

Rapport, R. A. and Eisenreich, S. J. 1986. Atmospheric deposition of toxaphene to eastern North America derived from peat accumulation. *Atmos. Environ.* 20: 2367–2379.

Simonich, S. L. and Hites, R. A. 1995. Global distribution of persistent organochlorine compounds. *Science* 269: 1851–1854.

Smith, K. E. C. and Jones, K. C. 2000. Particles and vegetation: Implications for the transfer of particle-bound organic contaminants to vegetation. *Sci. Total Environ.* 246: 207–236.

Stemmler, I. and Lammel, G. 2009. Cycling of DDT in the global environment 1950–2002: World ocean returns the pollutant. *Geophys. Res. Lett.* 36: 1–5.

Syversen, N. and Bechmann, M. 2004. Vegetative buffer zones as pesticide filters for simulated surface runoff. *Ecol. Eng.* 22: 175–184.

Vighi, M. 2006. The role of high mountains in the global transport of persistent organic pollutants. *Ecotoxicol. Environ. Safety* 63: 108–112.

Villa, S., Finizio, A., Díaz-Díaz, R., and Vighi, M. (2003). Distribution of organochlorine pesticides in pine needles of an oceanic island: The case of Tenerife (Canary Islands, Spain). *Water Air Soil Pollut.* 146: 335–349.

Villeneuve, J. P., Holm, E., and Cattini, C. 1985. Transfer of chlorinated hydrocarbons in the food chain lichen-reindeer-man. *Chemosphere* 14(11/12): 1651–1658.

Wang, X.-P., Yao, T.-D., Cong, Z.-Y., Yan, X.-L., Kang, S.-C., and Zhang, Y. 2006. Gradient distribution of persistent organic contaminants along northern slope of central-Himalayas, China. *Sci. Total Environ.* 372: 193–202.

Wang, X.-P., Yao, T.-D., Cong, Z.-Y., Yan, X.-L., Kang, S.-C., and Zhang, Y. 2007. Distribution of persistent organic pollutants in soil and grasses around Mt.Qomolangma, China. *Arch. Environ. Contam. Toxicol.* 5: 153–162.

Weiss, P. 2001. Organic pollutants at remote forest sites of Slovenia and Carinthia. Umweltbundesamt/Federal Environmental Agency, Austria Report No. BE-195.

Section V

Pesticides Residues in Animals

11

Pesticide Residues in Aquatic Invertebrates

**Sarun Keithmaleesatti, Wattasit Siriwong, Marija Borjan,
Kristen Bartlett, and Mark Robson**

CONTENTS

Aquatic invertebrates are defined as invertebrate animals that depend on aquatic ecosystems or moist environments for at least a portion of their lifecycle. Occasionally, taxonomists also include the semiaquatic invertebrates, which inhabit shores and vegetation surrounding aquatic environments. Habitats for aquatic invertebrates include wetlands, lakes, streams, rivers, oceans, and other waters (Pechenik 1996). The term aquatic invertebrate refers to swimming nekton, floating plankton, bottom-dwelling benthos, and surface-dwelling neuston. Aquatic invertebrates are diverse species that range widely in size, habitat, behavior, characteristics, food preference, and evolutionary relationships. In the food web, aquatic invertebrates are often divided into herbivores, detritivores, carnivores, and parasites. However, many scientists classify aquatic organisms based on feeding method (shredders, grazers, and suspension feeders) rather than food type. For the purpose of this book, aquatic invertebrates are classified into three major phyla: Rotifera, Mollusca, and Arthropoda.

Phylum Rotifera is a group of microscopic (50–500 μm in length), free-living organisms made up of over 2000 described species. Although this group contains a diverse array of species, including some that are capable of withstanding extreme conditions, 95% are found in freshwater ecosystems and 5% are found in marine ecosystems. Rotifers often make up an important component of the plankton in both freshwater and marine ecosystems. Normally, 40–500 organisms are found per liter of freshwater lakes or ponds. Rotifers are an important source of food for fish; therefore, they play a significant role in determining the structure of the aquatic community and in mediating the flow of energy throughout the freshwater ecosystem (Pechenik 1996).

Phylum Mollusca is the second largest animal phylum in regard to numbers of described species. There are approximately 50,000–120,000 described species of living mollusks as well as over 70,000 described fossil species. The majority of these species belong to the classes Cephalopoda, Gastropoda, and Bivalvia. About 75%–80% of all living mollusks belong to the class Gastropoda. Gastropods, which include snails, slugs, conchs, sea slugs, and their relatives, are found in freshwater, marine, and terrestrial ecosystems and are considered carnivorous, herbivorous, and detritivorous in the food chain. Class Bivalvia, which includes clams, oysters, and mussels, consists of 7000 species, 10%–15% of which are found in freshwater ecosystems. Both gastropods and bivalves are primarily benthic organisms, although many mollusks spend their larval stages as plankton to aid in dispersal (Pechenik 1996). Mollusks are very diverse, not only in size and superficial structure but also in behavior and habitat. A unique characteristic of the organisms of this phylum is their shell, which is secreted by a thin membrane called the mantle. Although many cephalopods, such as octopi and squid, do not have shells, they are classified as mollusks due to the presence of the mantle and mantle cavity.

Organisms in the phylum Arthropoda have an exoskeleton (external skeleton), segmented body, and jointed appendages (Triplehorn and Johnson 2005). Although new species are continually being described, it is estimated that there are over 5 million species of arthropods worldwide. Common aquatic arthropods that humans are most familiar with include lobsters, crabs, and shrimps. However, aquatic insects constitute the majority of described species. Although there are 13 different arthropod classes, the majority of aquatic arthropods belong to the classes Insecta (mayflies, dragonflies, mosquitoes, etc.), Arachnida (water mites, spiders, etc.), Malacostraca (shrimp, crayfish, and crab), Branchiopoda (fairy shrimp, clam shrimp, water fleas, etc.), and Maxillopoda (barnacles, copepods, etc.). Two major groups of aquatic arthropods, which are the focus of this chapter, include class Insecta and subphylum Crustacean (classes Malacostraca and Maxillopoda), which are abundant and important members of the aquatic and terrestrial ecosystems.

The class Insecta is the most diverse class of animals (over 750,000 described species), which has been found in every ecosystem on the planet. Because of their importance in both agriculture and human health, they are often a major target of pesticide applications. For example, mosquitoes, which inhabit practically all parts of the earth except the polar regions, can carry disease-causing organisms, such as the malarial parasite, Yellow Fever virus, Dengue virus, or filarial worm. As a result, persistent pesticide compounds were widely used to control potential mosquito vectors. Aquatic insects are made up of benthic, nekton, neuston, and plankton species, which vary in their feeding strategies. In addition, aquatic insects may have varying degrees of pollution tolerance, creating unique ecological niches for many species. Because insects are both numerous and diverse in aquatic ecosystems, they have received a lot of attention for their use in monitoring water quality.

11.1 History of Pesticide

Synthetic pesticides were first introduced in the twentieth century. Organochlorine pesticides (OCPs), organophosphate pesticides (OPPs), carbamates (CMs), and pyrethroids have been used to control insects around the world since the 1940s. Dichlorodiphenyltrichloroethane (DDT), the first OCP, was developed in 1939 by Paul Muller and was used to control malaria during World War II. The development of DDT prompted the search for other OCPs such as hexachlorocyclohexane (HCH), chlordane, and aldrin. At that time, these became the most widely used pesticides for agricultural and public health purposes due to their broad spectrum, high persistence, and low cost (Perry et al. 1998). OCPs were heavily used worldwide from 1940 to 1970. Starting in the 1950s, there were increasing reports of the adverse health effects due to OCPs, including effects on the reproductive system of nontarget organisms. The chemical properties of OCPs (hydrophobic, lipophilic, and low biodegradation potential) help facilitate the bioaccumulation in organisms, which can lead to adverse health effects. Accordingly, OCPs were banned in many countries including the United States, Japan, and Sweden since the 1970s (Keithmaleesatti et al. 2009).

OPPs were discovered and synthesized by German scientists in the 1930s. Parathion was one of the earliest OP insecticides used. OP compounds were used as warfare agents by Nazi Germany in World War II. After World War II, many OPPs, such as malathion, were produced and used on a worldwide scale. Today, more than 100 OPPs are used in countries such as Thailand, Vietnam, and India for public health purposes (Raghavendra et al. 2010) and to protect crops (Gupta 2006).

CMs, derivatives of carbamic acid, are most commonly used for gardening and agriculture. The first successful CM produced was carbaryl, which was developed in the 1950s. The mode of action of CMs and OPPs is similar; however, CM is lower in toxicity than OPPs (Gupta 2006).

The last group of pesticides, the pyrethroids, is currently the most widely used group of pesticides. Pyrethroids are synthetic chemicals, which were developed to enhance the insecticidal properties of natural pyrethrins. The active component of natural pyrethrin is extracted from the East African chrysanthemum flower, which was long known to have insecticidal properties. Natural pyrethrin products pose far less risks to mammals than OPPs and CM insecticides but have no residual or long-term control qualities as they are rapidly degraded by sunlight. However, they are extremely toxic to insects and fish. Examples of common pyrethroids include permethrin, cypermethrin, deltamethrin, allethrin, and fenvalerate. Synthetic pyrethroids are classified into two types: type I and type II. Type I pyrethroid is a non-α-cyanopyrethroid including natural pyrethrins, allethrin, tetramethrin, permethrin, kadethrin, etc. Type II pyrethroid is α-cyanopyrethroid including cypermethrin, deltamethrin, and fenvalerate (Perry et al. 1998). Modes of Action of pesticides are summarized in Table 11.1.

11.1.1 Pesticide Exposure in the Aquatic Invertebrate Food Web

Pesticide exposure can occur directly or indirectly. Direct exposure is the deliberate application of pesticides to control weeds and pests or the vectors of disease-causing organisms. Indirect exposure is the result of pesticide runoff from riparian land or spray-drift deposition. Some banned pesticides, such as DDT and its derivatives, persist in the sediment where aquatic invertebrates at the base of the food web feed. The food web is a network of consumer–resource interactions among a group of organisms, populations, or aggregate trophic units

TABLE 11.1

Summary of Pesticide Mechanisms of Action on Target Organisms

Pesticide Groups	General Toxic Effect	Specific Site of Action
Organochlorine	Nervous system inhibition	GABA receptor
Organophosphate	Nervous system inhibition	Acetylcholinesterase
Carbamate	Nervous system inhibition	Acetylcholinesterase
Synthetic pyrethroid	Nervous system inhibition	GABA receptor

Source: From Delorenzo, M. E., Scott, G. I., and Ross, P. E. 2001. Toxicity of pesti-
cides to aquatic microorganisms: A review. *Environ. Toxicol. Chem.* 20: 84–98;
Perry, A. S., Yamamato, I., Ishaaya, I., and Perry, R. 1998. *Insecticides in
Agriculture and Environment: Restrospects and Prospects.* Narosa: New Delhi.

(Winemiller and Polis 1996). Aquatic invertebrates play an important role as nutrient sources of the aquatic food web and food pyramid. Aquatic invertebrates, for example, zooplankton such as rotifers and copepods, are food resources for both mega aquatic invertebrates and fish. Since energy transfer between trophic levels is generally inefficient, only 10%–20% of energy is transferred between trophic levels, the food web is important to the ecosystem.

Bioaccumulation, which is the net accumulation of a contaminant, such as pesticides, in and on an organism from multiple sources in the environment, may result upon exposure. Once in the food web, pesticide concentrations increase from one trophic level to the next higher trophic level throughout the food web (Figure 11.1). Biomagnification

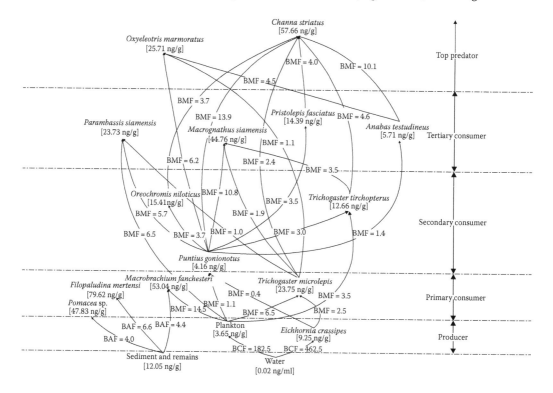

FIGURE 11.1
The bioconcentration and biomagnification of DDT and its derivatives in the aquatic food web of lower Chao Phraya River Basin, Central Thailand. (From Robson, M., Hamilton, G., and Siriwong, W. 2010. Pest control and pesticides. In: Frumkin, H. (ed.), *Environmental Health*, 2nd edn, p. 595.)

will result when the chemical becomes more concentrated as it moves up the food chain, resulting in higher concentrations in the predators than in the prey (Moriarty 1983; Newman 2001).

11.1.2 Dose Response (Acute and Chronic)

Acute toxicity, as a result of a single pesticide exposure, is most commonly measured using median lethal dose or lethal concentration (LD_{50} or LC_{50}, respectively). Lethal dose (LD_{50}) can be calculated as the known amount of toxicant per amount of body weight (milligrams toxicant/kilogram body weight) or as the amount of toxicant per animal (milligrams toxicant/animal). In contrast, the LC_{50} is based on the amount of toxicant in an environmental medium (milligrams toxicant/liter of media), where the amount of toxicant that enters the organism is not known. Because of the subtle differences in uncertainties associated between the two metrics, confusion persists even among toxicologists about the differences between lethal dose and lethal concentration.

In contrast, chronic toxicity is usually monitored after continuous exposure to sublethal dose or concentration over a given period of time. This is calculated using no observable effect concentration (NOEC) or level (NOEL), which is the highest concentration or level where no effect is observed (Stark and Banks 2003). The end points of interest for chronic toxicities are life span, weight gain, reproduction, cancer, and birth defects.

11.1.3 Pesticides in Aquatic Ecosystem

Although many pesticides are used to control agricultural pests in farm crops, organisms in aquatic ecosystems can be indirectly affected by the toxicity of pesticides. Aquatic ecosystems are contaminated by five major routes: (1) surface runoff and sediment transported from treated soil, (2) industrial waste discharged from factory effluents, (3) direct application as aerial spray or granules to control pests inhabiting water, (4) spray drift from normal agricultural operation, and (5) municipal waste discharge (Ray and Ghosh 2006). Many pesticides that contaminate aquatic ecosystems can be hazardous to aquatic invertebrates. Broad-spectrum pesticides are designed to kill a wide variety of pests. As a result, broad-spectrum pesticides will not only kill the targeted pest (snail, worm, and beetle), but might kill nontargeted organisms as well (rotifer, copepods, and beneficial insects). Moreover, some persistent pesticides, such as DDT, can be taken up in the food web where toxins can be transferred from the aquatic ecosystem to the terrestrial ecosystem.

11.2 Effect of Pesticides on Aquatic Invertebrates

11.2.1 Effect of Pesticides on the Phylum Rotifera

In the food web, rotifers include predators, competitors, and prey. Due to their small size and biology, they are often a nontarget organism of pesticide usage. Rotifers are an important part of the nutrient cycle, and biological activities including decomposition and mineralization in rice fields and aquatic resources. Furthermore, they have the potential to affect the aquatic ecosystem by increasing soil pH (Kikuchi et al. 1977), changing the soil microbial population (Wavre and Brinkhurst 1971; Kikuchi and Kurihara 1977), and providing a food source for aquaculture species (Aston et al. 1982; Marian and Pandian 1984).

11.2.1.1 Effects of Organochlorine on Rotifers

OCPs are toxic in both vertebrates and invertebrates. Many studies have shown that OCPs have reduced the reproductive success of raptors such as the Bald eagle *Haliaeetus leucocephalus* (Bowerman et al. 1995; Clark et al. 1998) and disrupted the sex hormones in reptiles such as the American alligator *Alligator mississippiensis* (Guillette et al. 1994) and common snapping turtle *Chelydra serpentina* (de Solla et al. 1998). Furthermore, OCPs have disrupted estrogenic hormones in mammals. These reproductive complications may be due to the bioaccumulation of OCPs in high trophic level animals. Invertebrate organisms are at the base of the food chain and take up and transfer OCPs into the aquatic food chain. Studies have shown that OCPs, such as aldrin, can significantly affect the duration of embryonic development of the rotifer *Brachionus calyciflorus pallas* at concentrations ranging from 0.04 to 1.28 mg/L (Huang et al. 2007). In addition, the OCPs DDT, dicofol, endosulfan, and lindane have been shown to significantly decrease the population growth rate and reproduction of *B. calyciflorus* (Xi et al. 2007).

11.2.1.2 Effects of Organophosphorus on Rotifers

OPPs are known to adversely affect rotifers. Ke et al. (2009) reported that OPPs such as dichlorvos, triazophos, and chlorpyrifos have a significant effect on the population growth rate of the freshwater rotifer *B. calyciflorus* in China. Studies showed that rotifers died after being exposed to 10,000 µg/L of chlorpyrifos in a 24-h time period. Reproductive effects from diazinon and fenitrothion were tested using *B. plicatilis*. The results showed that diazinon and fenitrothion were highly acutely toxic with a 24-h LC_{50} of 88.39 µM and 229.76 µM, respectively (Marcial et al. 2005).

11.2.2 Effect of Pesticides on Phylum Mollusca

Mollusks, such as apple snails, *Pomacea canaliculata*, are major pests in rice fields. Apple snails damage crops, particularly rice, by feeding on germinating seeds and young seedlings. Crops in many countries in Southeast Asia such as Thailand, Vietnam, and Cambodia are affected by these species. Farmers usually use molluscicides to control these pests. Molluscicides, known as snail bait and snail pellets, are used in agriculture or gardening to control gastropod pests such as slugs and snails that can damage crops. For example, endosulfan is a common chlorinated molluscicide used in agriculture to kill mollusks in Thailand.

11.2.2.1 Effect of Organochlorine on Mollusks

OCPs are known to be potential endocrine disruptors. Residues from DDT and its derivatives affect gonad development and cause reproductive abnormalities in mollusks (Binelli et al. 2004). Long-term exposures to OCPs can also affect the freshwater snail,

Filopaludina martensi, by disrupting the endocrine system. Single-cell gel electrophoresis (SCGE) assay showed that DDT and its derivatives are genotoxins, causing damage to DNA in *Dreissena polymorpha* (Binelli et al. 2008). Freshwater mussels such as unionid mussels, *Elliptio complanata*, have been used as sentinel animals to detect OCP residues in freshwater ecosystems in North America (Renaud et al. 2004; Won et al. 2005) and used as a bioindicator in Lake Maggiore, Italy (Riva et al. 2010). Reports have shown that OCPs have been detected in mollusks in several countries including Southeast Asia. Siriwong (2006) and Siriwong et al. (2008) reported that low concentrations of banned OCPs, such as HCH, heptachlor and heptachlor epoxide, DDT and its derivatives, and endosulfan (α-, β-, and -sulfate) are still present in apple snail *Pomacea* sp. Furthermore, in Canada, the Rideau River near Ottawa was used to detect OCPs in *D. polymorpha* and *E. complanata* (Renaud et al. 2004) and also OCPs presented in *D. polymorpha* in Maggiore Lake, Italy (Binelli et al. 2004; Binelli et al. 2007).

11.2.2.2 Effect of Organophosphate on Mollusks

OPPs such as malathion, thiometon, and disulfoton were used to control insects and protect crops. In 1986, the Rhine River was accidently contaminated with several tons of pesticides, mainly thiometon and disulfoton. After three and a half months, high concentrations of pesticides were detected in sediment and the Zebra mussel *D. polymorpha* (a freshwater bivalve), which is an important part of the diving duck's diet. Significant mortality was seen in those ducks that had fed on the mussels (Dauberschmidt 1995). In vitro, it was found that 88% and 93% of *D. polymorpha* died when they were administered 50 mg/L of thiometon and 30 mg/L of disulfoton, respectively (Dauberschmidt et al. 1996). Additionally, Dauberschmidt et al. (1997) reported that organophosphates, including 10 mg/L of disulfoton and 6 mg/L of thiometon, can create subacute exposures (10 day) on *D. polymorpha*. Chlorpyrifos, a broad-spectrum organophosphate insecticide, has been shown to reduce cholinesterase activity in the Asian clam, *Corbicula fluminea*, and reduced their capacity to burrow into the substrate (Cooper and Bidwell 2006).

11.2.2.3 Effect of Carbamate on Mollusks

Carbamate pesticides such as carbaryl (1-naphthyl methyl carbamate) have been used for pest control in crops and domestic animals. CMs are not persistent in the ecosystem. However, CM residues have been detected in surface water. The toxicity of CMs is similar to OPs. However, there is some documentation reporting that CMs are higher in toxicity and exhibit a greater sensitivity to the inhibition of AChE activity than OPs. In vivo, the gastropod *Biomphalaria glabrata* was exposed to 1 mg/L of carbaryl, which significantly inhibited enzyme activity (Kristoff et al. 2010).

11.2.2.4 Effect of Pyrethroid on Mollusks

Synthetic pyrethroids were used for agricultural and public health purposes after OCPs were banned. Pyrethroids are very toxic to insects and crustaceans, but are less toxic to mammals (Coats et al. 1989). However, both acute and lethal effects on nontarget invertebrates are of interest. A study in India administered cypermethrin and alphamethrin in the freshwater snail *Lymnaea acuminata*. This resulted in a significant increase in the number of eggs and egg masses, but the survivability of hatchlings was reduced (Tripathi and Singh 2004).

11.2.3 Effect of Pesticides on Phylum Arthropoda

Arthropods are advantageous in that they are a source of food for other aquatic organisms and are natural predators. However, many are also pests in the aquatic ecosystem. For example, mosquito larvae are controlled by a variety of pesticides with various modes of action. These include biorational pesticides (*Bacillus thuringiensis israelensis*), organophosphates (temephos), and growth regulators (methoprene). OPPs affect the nervous system of organisms (Perry et al. 1998). Organism mortality and behavior are affected by pesticide concentration. Exposure to higher doses can lead to death, while lower doses can influence organism behavior (Haynes 1988).

11.2.3.1 Effect of OCP on Arthropods

OCPs interfere with axonal and synaptic transmission. For example, DDT delays the inward sodium current and increases the depolarizing action potential, eventually resulting in insect death (Perry et al. 1998).

In the class Insecta, OCP contamination is linked to mortality rate, survival rate, and reduction rate of a population (Bartrons et al. 2007; Schulz and Liess 1995). Broad-spectrum OCPs can affect both the target and the nontarget aquatic insects. Although OCPs may be used to kill insect larvae (mosquitoes, chironomids), other aquatic insects can be killed indirectly (e.g., coleopterans, hemipterans; Simpson and Roger 1995).

In the subphylum Crustacea, aquatic organisms can also be indirectly affected by OPPs. Standard methods for aquatic ecotoxicity assessment recommend that the *Daphnia* spp. be used as a model because of its sensitivity to chemical environmental stressors. *Daphnia* is a herbivore and an important source of food for fish in the aquatic ecosystem. *Daphnia magna* is often used as a sentinel for ecotoxicity testing. It was shown that chlordane (at concentrations greater than 2.9 µg/L) affected the survival of *D. magna*. The same study showed that concentrations of 0.7 µg/L or greater reduced the length of adults and affected reproduction (number of offspring per adult and brood size; Manar et al. 2009). Acute toxicity of endosulfan in the freshwater shrimp *Caridina laevis* in Indonesia was observed at 2.38 µg/L (24 h LC_{50}) and 1.02 µg/L (96 h LC_{50}) (Sucahyo et al. 2008). The LC_{50} after 48 h of exposure to chlordane, DDT, and lindane in Juvenile shrimp *Penaeus vannamei* from Mexico was 0.0632, 0.0087, and 0.0039 mg/L, respectively. The concentrations of OCPs at LC_{50} can reduce the percentage of protein more than 25% (Reyes et al. 1996a). Chlordane (0.27 µg/L), DDT (0.13 µg/L), and lindane (0.19 µg/L) can decrease glycogen synthesis of the white leg shrimp (Reyes et al. 1996b). Additionally, crab embryos are an effective sink for DDT and its degradates (Smalling et al. 2010).

11.2.3.2 Effect of OPPs on Arthropods

OP exposure can result in intermediate syndrome (IMS), OP-induced delayed neuropathy (OPIDN), and chronic OP-induced neuropsychiatric disorder (COPIND) in arthropods. OPs are also known to cause tumorigenic effects in arthropods (Ray and Ghosh 2006). Chlorpyrifos toxicity is believed to result primarily from metabolic activation of the parent compound to the oxon metabolite. Acetylcholinesterase (AChE), at the neural junction, is deactivated by the oxon metabolite. This results in overstimulation of the peripheral and central nervous systems (Buchwalter et al. 2004).

Many of the OPPs are used to target insects, and therefore, they may have greater impacts on nontarget organisms in the class Insecta. *Chironomus riparius* larvae were exposed in vitro during the second to fourth instar. The first sign of toxicity was knockdown in all larvae. Tremors were monitored after 1 h of exposure. No movement or response to physical stimuli was observed. At high concentrations, AChE assays exposed that homogenates from *C. riparius* larvae were refractory to chlorpyrifos; in opposition, chlorpyrifos-oxon in a concentration-dependent manner inhibited homogenates from *C. riparius* larvae (Buchwalter et al. 2004).

OPs are also toxic to aquatic organisms in the subphylum Crustacea. Chlorpyrifos is highly toxic to *Daphnia magna* juveniles, 24-h LC_{50} is 1.28 nM, and it is more toxic to *Daphnia* than malathion and CM (Moore et al. 1998; Kikuchi et al. 2000). In the case of freshwater shrimp, there was decreased survival of juvenile *Caridina laevis*, at 24- and 96-h LC_{50} when exposed to diazinon at 0.76 and 0.59 µg/L, respectively (Sucahyo et al. 2008). The question is why. Organisms have a tendency to be more sensitive to pesticides in the younger stages. This may be due to younger organisms having a greater surface area to volume ratio, and their metabolic capacity may be different (Kefford et al. 2004; Sucahyo et al. 2008).

11.2.3.3 Effect of Carbamate Pesticides on Arthropods

Carbamate pesticides are known to inhibit type "B" esterases, including cholinesterase and carboxylesterase, by binding to the active site and phosphorylating the enzyme. However, the action in aquatic organisms is less clear, such as how the transfer of oxygen occurs (Barata et al. 2004). Carbofuran is the most toxic carbamate pesticide. Carbofuran derivatives, *N*-nitroso, and its toxic metabolites (3-hydroxycarbofuran and 3-ketocarbofuran) increased the likelihood of mutagenicity in *Salmonella typhimurium* strains TA98 and TA100 (Nelson et al. 1981). Carbofuran has been linked to cytochrome P450 activities in organisms (Ray and Ghosh 2006) and has been monitored in insects since the 1960s (Snyder 2000). Furthermore, Carbaryl affected the predator–prey interactions within a zooplankton community (Chang et al. 2005).

Many CMs are highly toxic to the class Insecta. Furadan is the commercial name of carbofuran, which is used to control insects in rice fields in Senegal (Mullié et al. 1991). Carbofuran is hazardous to fish and invertebrates (Eisler 1985). Acute toxicity of carbofuran was studied in Senegal rice fields in 1985. The study found that carbofuran may be highly toxic to aquatic fauna in rice fields. Additionally, the number of aquatic macroinsects was significantly reduced after carbofuran exposure (Mullié et al. 1991).

CMs can have deleterious effects on organisms in the subphylum Crustacea. CMs have the potential to change the dominant species in aquatic ecosystems. Low concentrations of carbaryl influenced *Daphnia* spp.; moreover, *Daphnia* was the most sensitive organism to carbaryl (Hanazato 1998). The acute toxicity of carbofuran was monitored in freshwater shrimp *Caridina laevis*. The study found the 24-h LC_{50} to be 1.70 mg/L. and 0.95 mg/L. for the 96-h LC_{50} (Sucahyo et al. 2008). *Daphnia magna*, which is widely used to test toxicity, were exposed to carbofuran to assess acetylcholinesterase (AchE) and carboxylesterase (CbE) inhibition. The LC_{50} of carbofuran on *D. magna* was 762.93 nM (Barata et al. 2004). Additionally, Dobsikova (2003) reported a 24-h LC_{50} and 48-h LC_{50} of 0.0447 and 0.0187 mg/L, respectively, in *D. Magna*.

11.2.3.4 Effect of Pyrethroid on Arthropods

Synthetic pyrethroids are the most commonly used pesticides for controlling insects because of their low toxicity to mammals (Friberg-Jensen et al. 2003). However, many studies have shown that cypermethrin, a synthetic pyrethroid used in crop protection, is extremely toxic to aquatic invertebrates and may cause a reduction in feeding activity, survival, and reproduction (Day and Kaushik 1987a,b; Kjølholt et al. 1991).

Flies in the class, Insecta, including mosquitoes, blackflies, and chironomids had an LC_{50} less than 1 µg/L when exposed to pyrethroids (Anderson 1989). The 24-h LC_{50} of cypermethrin, cis-*cypermethrin*, deltamethrin, permethrin and its cis isomer, and fenvalerate for *Culex* larvae ranged from 0.02–7 µg/L (Mulla et al. 1978). Aquatic insects including the black fly *Simulium vittatum*, caddisfly *Hydropsyche* and *Cheumatopsyche* spp., mayfly Family Heptageniidae, damselfly *Enallagma* and *Ischnura* spp., and water scavenger beetle *Hydrophilus* spp. were exposed to three synthetic pyrethroids including permethrin, cypermethrin, and bifenthrin. The 24-h LC_{50} of this experiment presented that the synthetic pyrethroid was highly toxic to aquatic insects. Bifenthrin is highly toxic to both mayfly and damselfly (Siegfried 1993).

Pyrethroids have been scrutinized to determine their effects on organisms in the subphylum Crustacea. Many reports have shown that pyrethroid pesticides are highly toxic to fish and aquatic invertebrates (Christensen et al. 2005; Clark et al. 1989; Hill 1989). The toxicity of pyrethroids is due to the chemical impact on the sodium channels in the nervous system. Exposure to pyrethroids can produce sublethal effects on aquatic invertebrates. Feeding rate and swimming ability of aquatic invertebrates are also impacted after exposure (Christensen et al. 2005). Crustacean zooplankton are important organisms in the aquatic ecosystem. They aid in the regeneration of nutrients from primary organic matter and are an important component in the food web (Valiela 1991 in Barnes and Mann 1991). Synthetic pyrethroids can directly and indirectly affect crustacean zooplankton. They directly affect the crustacean zooplankton by reducing the grazer control of phytoplankton, protozoans, and rotifers, which results in an altered biomass and species composition in the food web. An indirect effect can occur through impacting the other trophic levels in the aquatic food web (Friberg-Jensen et al. 2003).

Cypermethrin is a popular synthetic pyrethroid used to control mosquitoes. In toxicity testing, *Daphnia magna* was treated with cypermethrin at concentrations of 0.05, 0.1, 0.2, 0.3, 0.6, and 1.0 µg/L. The results show that 0.1 µg/L of cypermethrin reduced the content of chlorophyll pigment, affected swimming ability, and increased mortality of *D. magna* (Christensen et al. 2005). Additionally, Friberg-Jensen et al. (2003) found that *Daphnia* species rapidly decreased by more than 10% in three treatments (0.47, 1.70, and 6.10 µg/L) in contrast to the control.

Red swamp crayfish *Cyprinus carpio* was used to monitor the acute toxicity of three synthetic pyrethroids: cypermethrin, deltamethrin, and cyfluthrin. The results found that *C. carpio* was highly sensitive to all three pyrethroids. The 24-h LC_{50} of cypermethrin, deltamethrin, and cyfluthrin was 0.14, 0.17, and 0.22 µg/L, respectively (Morolli et al. 2006).

Lambda-cyhalothrin, which is a synthetic pyrethroid, is highly toxic to aquatic invertebrates. For example, *Caridina laevis* is very sensitive to lambda-cyhalothrin. Sucahyo et al. (2008) found the concentrations for 24-h LC_{50} and 96-h LC_{50} to be 0.87 µg/L and 0.33 µg/L, respectively. Similarly, Wang et al. (2007) reported that the 96-h LC_{50} for lambda-cyhalothrin on *Macrobrachium nipponense* was 0.04 µg/L. The 48-h LC_{50} for lambda-cyhalothrin on *Daphnia*

TABLE 11.2

Summary List of Toxicity of Organochlorine on Aquatic Invertebrates

Pesticide Name	Species	Taxonomy Classification of Species	Test End Point	References
DDT, Dicofol, endosulfan, and lindane	*Brachionus calyciflorus*	Phylum Rotifera	Reproduction	Xi et al. (2007)
Aldrin	*Brachionus calyciflorus*	Phylum Rotifera	Embryonic development	Huang et al. (2007)
Hexachlorobenzene, α-, β-, and γ-HCHs (hexachlorocyclohexanes), DDT and derivatives	*Dreissena polymorpha*	Phylum Mollusca	Enzyme activities and DNA	Faria et al. (2010)
Hexachlorocyclohexane, DDE, DDD, and DDT	*Potamopyrgus antipodarum*	Phylum Mollusca	Reproductive ability	Mazurova et al. (2008)
DDT	*Dreissena polymorpha*	Phylum Mollusca	Reproductive ability	Binelli et al. (2004)
DDT	*Metapenaeus monoceros*	Phylum Arthropoda	Behavioral and enzyme activity	Tu et al. (2010)
OCPs	*Hemigrapsus oregonensis* and *Pachygrapsus crassipes*	Phylum Arthropoda	Embryos	Smalling et al. (2010)
Chlordane	*Daphnia magna*	Phylum Arthropoda	24-, 48-h LC_{50} 21 day NOEC Reproduction	Manar et al. (2009)
Endosulfan	*Caridina laevis*	Phylum Arthropoda	24-, 96-h LC_{50}	Sucahyo et al. (2008)
Chlordane, DDT, and lindane	*Penaeus vannamei*	Phylum Arthropoda	48-h LC_{50} and glycogen	Reyes et al. (1996a,b)
Lindane	*Limnephilus lunatus*	Phylum Arthropoda	96-h LC_{50} and chronic	Schulz and Liess (1995)

magna was 0.39 μg/L (Barata et al. 2006) and 0.04 μg/L for *D. galeata* (Schroer et al. 2004). This indicates that synthetic pyrethroids are hazardous to aquatic crustaceans. Tables 11.2 through 11.5 are summaries of organochlorine, organophosphorus, CM, and synthetic pyrethroid pesticides, respectively.

11.3 Conclusion

Although most aquatic invertebrates are nontarget organisms, studies have shown that aquatic invertebrates are highly sensitive to pesticides. While OCPs have been banned in many countries, OCP residues are still detected in aquatic invertebrates. An important characteristic of OCPs is biological magnification in the food chain. In biomagnification, aquatic invertebrates are not directly affected by pesticides. However, other nontarget organisms are exposed to pesticides through the food web. Both organophosphates and

TABLE 11.3

Summary List of Toxicity of Organophosphorus on Aquatic Invertebrates

Pesticide Name	Species	Taxonomy Classification of Species	Test End Point	References
Dichlorvos, triazophos, and chlorpyrifoss	*Brachionus calyciflorus*	Phylum Rotifera	Population growth and sexual reproduction	Ke et al. (2009)
Diazinon and fenitrothion	*Brachionus plicatilis*	Phylum Rotifera	LC_{50} and reproduction	Marcial et al. (2005)
Chlorpyrifos	*Corbicula fluminea*	Phylum Mollusca	24-, 48-, 72-, and 96-h LC_{50} and behavior	Cooper and Bidwell (2006)
Malathion, demeton-S-methyl, thiometon, and disulfoton	*Dreissena polymorpha*	Phylum Mollusca	96-h LC_{50}, 10-day subacute organ concentration, and enzyme activities	Dauberschmidt et al. (1996, 1997)
Chlorpyrifos and malathion	*Hemigrapsus oregonensis* and *Pachygrapsus crassipes*	Phylum Arthropoda	Embryos	Smalling et al. (2010)
Diazinon and profenofos	*Caridina laevis*	Phylum Arthropoda	24- and 96-h LC_{50}	Sucahyo et al. (2008)
Malathion and chlorpyrifos	*Daphnia magna*	Phylum Arthropoda	24-h LC_{50}, acetylcholinesterase (AChE) and carboxylesterase (CbE) activities	Barata et al. (2004)
Chlorpyrifos	*Chironomus riparius*	Phylum Arthropoda	AChE	Buchwalter et al. (2004)
Parathion	*Limnephilus bipunctatus*	Phylum Arthropoda	96-h LC_{50} and chronic	Schulz and Liess (1995)

TABLE 11.4

Summary List of Toxicity of Carbamate on Aquatic Invertebrates

Pesticide Name	Species	Taxonomy of Species	Test End Point	References
Carbaryl	*Biomphalaria glabrata*	Phylum Mollusca	Cholinesterase (ChE) and carboxylesterase (CbE) activities	Kristoff et al. (2010)
Carbofuran	*Caridina laevis*	Phylum Arthropoda	24- and 96-h LC_{50}	Sucahyo et al. (2008)
Carbaryl	*Mesocyclops pehpeiensis*	Phylum Arthropoda	Predator–prey relationship	Chang et al. (2005)
Carbofuran	*Daphnia magna*	Phylum Arthropoda	24-h LC_{50}, acetylcholinesterase (AChE) and carboxylesterase (CbE) activities	Barata et al. (2004)
Carbofuran	*Daphnia magna*	Phylum Arthropoda	24-, 48-h LC_{50}	Dobsikova (2003)
Carbaryl	*Daphnia galeata*	Phylum Arthropoda	Population dynamics	Hanazato (1998)

TABLE 11.5

Summary List of Toxicity of Synthetic Pyrethroid on Aquatic Invertebrates

Pesticide Name	Species	Taxonomy of Species	Test End Point	References
Cypermethrin, alphamethrin	*Lymnaea acuminata*	Phylum Mollusca	Reproduction and oxidative metabolism	Tripathi and Singh (2004)
Bifenthrin, cyfluthrin, and permethrin	*Hemigrapsus oregonensis and Pachygrapsus crassipes*	Phylum Arthropoda	Embryos	Smalling et al. (2010)
Lambda-cyhalothrin	*Caridina laevis*	Phylum Arthropoda	24- and 96-h LC_{50}	Sucahyo et al. (2008)
Lambda-cyhalothrin	*Macrobrachium nipponense*	Phylum Arthropoda	96-h LC_{50}	Wang et al. (2007)
Deltamethrin, cyfluthrin, and cypermethrin	*Procambarus clarkii*	Phylum Arthropoda	24- and 48-h LC_{50}	Morolli et al. (2006)
Lambda-cyhalothrin	*Daphnia magna*	Phylum Arthropoda	48-h LC_{50}	Barata et al. (2006)
Cypermethrin	*Daphnia magna*	Phylum Arthropoda	Behavior (feeding and swimming)	Christensen et al. (2005)
Bifenthrin, cyfluthrin, cypermethrin, deltamethrin, esfenvalerate, lambda-cyhalothrin, and permethrin	*Hyalella azteca*	Phylum Arthropoda	LC_{50}	Weston et al. (2005)
Lambda-cyhalothrin	*Daphnia galeata*	Phylum Arthropoda	48-h LC_{50}	Schroer et al. (2004)
Cypermethrin	*Daphnia* spp.	Phylum Arthropoda	Population, no effect concentration (NEC), and median effect concentration (EC_{50})	Friberg-Jensen et al. (2003)
Bifenthrin, cypermethrin, permethrin	*Simulium vittatum, Hydropsyche* and *Cheumatopsyche* spp., Heptageniidae, *Enallagma* and *Ischnura* spp., *Hydrophilus* spp.	Phylum Arthropoda	24-h LC_{50}	Siegfried (1993)
Cypermethrin, *cis*-cypermethrin, deltamethrin, and fenvalerate	*Culex* spp. and *Aedes* spp.	Phylum Arthropoda	LC_{50}	Anderson (1989)
Fenvalerate	*Culex pipiens* (larvae)	Phylum Arthropoda	LC_{50}	Coats et al. (1989)

CMs affect the nervous system of aquatic organisms and have similar target sites in aquatic invertebrates. Due to their high toxicity, the mortality rate is often the first noticeable effect in aquatic organisms. In contrast, synthetic pyrethroids have a low toxicity to mammals. However, many aquatic organisms are highly sensitive to synthetic pyrethroids, and even low concentrations of synthetic pyrethroids, in vitro and in vivo, are capable of killing aquatic invertebrates at ppb levels.

Acknowledgment

We wish to acknowledge the support of NIH Fogarty International Center ITREOH D43 TW007849.

References

Anderson, R. L. 1989. Toxicity of synthetic pyrethroids to freshwater invertebrates. *Environ. Toxicol. Chem.* 8: 403–410.

Aston, R. J., Sadler, K., and Milner, A. G. P. 1982. The effects of temperature on the culture of *Brunchiura sowerbyi* (Oligochaeta, Tubificidae) in activated sludge. *Aquaculture* 29: 137–145.

Barata, C., Solayan, A., and Porte, C. 2004. Role of B-esterases in assessing toxicity of organophosphorus (chlorpyrifos, malathion) and carbamate (carbofuran) pesticides to *Daphnia magna*. *Aquat. Toxicol.* 66: 125–139.

Barata, C., Baird, D. J., Nogueira, A. J. A., Soares, A., and Riva, M. C. 2006. Toxicity of binary mixtures of metals and pyrethroid insecticides to *Daphnia magna* Straus. Implications for multi-substance risks assessment. *Aquat. Toxicol.* 78: 1–14.

Bartrons, M., Grimalt, J. O., and Catalan, J. 2007. Concentration of organochlorine compounds and polybromodiphenyl ethers during metamorphosis of aquatic insect. *Environ. Sci. Technol.* 41: 6137–6141.

Binelli, A., Bacchetta, R., Mantecca, P., Ricciardi, F., Provini. A., and Vailati, G. 2004. DDT in Zebra mussel from Lake Maggiore (N. Italy): level of contamination and endocrine disruptions. *Aquat. Toxicol.* 69: 175–188.

Binelli, A., Riva, C., and Provini, A. 2007. Biomarkers in Zebra mussel for monitoring and quality assessment of Lake Maggiore (Italy). *Biomarkers* 12: 349–368.

Binelli, A., Riva, C., Cogni, D., and Provini, A. 2008. Genotoxic effects of p,p′-DDT (1,1,1-trichloro-2,2-bis-(chlorophenyl)ethane) and its metabolites in Zebra mussel (*D. polymorpha*) by SCGE assay and micronucleus test. *Environ. Mol. Mutagen.* 49(5): 406–415.

Bowerman, W. W., Giesy, J. P., Best, D. A., and Kramer, V. J. 1995. A review of factors affecting productivity of bald eagles in the Great Lakes region: Implications for recovery. *Environ. Health Perspect.* 103: 51–59.

Buchwalter, D. D., Sandahl, F., Jenkins, J. J., and Curtis, L. R. 2004. Roles of uptake, biotransformation, and target site sensitivity in determining the differential toxicity of chlorpyrifos to second to fourth instar *Chironomous riparius* (Meigen). *Aquat. Toxicol.* 66: 149–157.

Chang, K. H., Sakamoto, M., and Hanazato, T. 2005. Impact of pesticide application on zooplankton communities with different densities of invertebrate predators: An experimental analysis using small-scale mesocosms. *Aquat. Toxicol.* 72: 373–382.

Christensen, B. T., Lauridsen, T. L., Ravn, H. W., and Bayley, M. 2005. A comparison of feeding efficiency and swimming ability of *Daphnia magna* exposed to cypermethrin. *Aquat. Toxicol.* 73: 210–220.

Clark, J. R., Goodman, L. R., Borthwick, P. W., Patrick, J. M., Cripe, G. M., Moody, P.M., Moore J. C., and Lores, E. M. 1989. Toxicity of pyrethroids to marine invertebrates and fish: A literature review and test results with sediment-sorbed chemicals. *Environ. Toxicol. Chem.* 8: 411–416.

Clark, K. E., Niles, L. J., and Stansley, W. 1998. Environmental Contaminants Associated with Reproductive Failure in Bald Eagle (*Haliaeetus leucocephalus*) Eggs in New Jersey. *Bull. Environ. Contam. Toxicol.* 61: 247–254.

Coats, J. R., Symonik, D. M., Bradbury, S. P., Dyer, S. D., Timson, L. K., and Atchison, G. J. 1989. Toxicology of synthetic pyrethroids in aquatic organisms: An overview. *Environ. Toxicol. Chem.* 8: 671–679.

Cooper, N. L. and Bidwell, J. R. 2006. Cholinesterase inhibition and impacts on behavior of the Asian clam, *Corbicula flumines*, after exposure to an organophosphate insecticide. *Aquat. Toxicol.* 76: 258–267.

Dauberschmidt, C. 1995. Organophosphorus pesticides in the fresh mollusc *Dreissena polymorpha*. Doctoral dissertation, Swiss Federal Institute of Technology.

Dauberschmidt, C., Dietrich, D. R., and Schlatter, C. 1996. Toxicity of organophosphorus insecticides in the zebra mussel *Dreissena polymorpha* P. *Arch. Environ. Contam. Toxicol.* 30: 373–378.

Dauberschmidt, C., Dietrich, D. R., and Schlatter, C. 1997. Organophosphorus in the zebra mussel *Dreissena polymorpha*: Subacute Exposur, Body Burdens, and Organ Concentrations. *Arch. Environ. Contam. Toxicol.* 33: 42–46.

Day, K. E. and Kaushik, N. K. 1987a. Short-term exposure of zooplankton to the synthetic pyrethroid, fenvalerate, and its effects on rates of filtration and assimilation of the alga, *Chlamydomonas reinhardii*. *Arch. Environ. Contam. Toxicol.* 16: 423–432.

Day, K. E. and Kaushik, N. K. 1987b. An assessment of the chronic toxicity of the synthetic pyrethroid, fenvalerate, to *Daphnia galeata mendotae*, using life tables. *Environ. Pollut.* 44: 13–26.

Delorenzo, M. E., Scott, G. I., and Ross, P. E. 2001. Toxicity of pesticides to aquatic microorganisms: A review. *Environ. Toxicol. Chem.* 20: 84–98.

de Solla, S. R., Bishop, C. A., Van Der Kraak, G., and Brooks, R. J. 1998. Impact of organochlorine contamination on levels of sex hormones and external morphology of common snapping turtles *Chelydra serpentina serpentina* in Ontario, Canada. *Environ. Health Perspect.* 106: 253–260.

Dobsikova, R. 2003. Acute toxicity of carbofuran to selected species of aquatic and terrestrial organisms. *Plant Protect. Sci.* 39: 103–108.

Eisler, R. 1985. Carbofuran hazard to fish, wildlife and invertebrates: A synoptic review. *US Fish Wildl. Serv. Biol. Rep.* 85(1.3): 36 pp.

Faria, M., Huertas, D., Soto, D. X., Grimalt, J. O., Catalan, J., Riva, M. C., and Barata, C. 2010. Contaminant accumulation and multi-biomarker responses in field collected zebra mussels (*Dreissena polymorpha*) and crayfish (*Procambarus clarkii*), to evaluate toxicological effects of industrial hazardous dumps in the Ebro river (NE Spain). *Chemosphere* 78: 232–240.

Friberg-Jensen, U., Wendt-Rasch, L., Woin, P., and Christoffersen, K. 2003. Effects of the pyrethroid insecticide, cypermethrinon a freshwater community studied under field conditions. I: Direct and indirect effects on abundance measures of organisms at different trophic levels. *Aquat. Toxicol.* 63: 357–371.

Guillette, L. J., Gross, T. S., Masson, G. R., Matter, J. M., Percival, H. F., and Woodward, A. R. 1994. Developmental abnormalities of the gonad and abnormal sex hormone concentrations in juvenile alligators from contaminated and control lakes in Florida. *Environ. Health Perspect.* 102: 680–688.

Gupta, R. C. 2006. Classification and uses of organophosphates and carbamates. In: Gupta, R. C. (ed.), *Toxicity of Organophosphates and Carbamate Pesticides*, pp. 5–24. Academic Press (Elsevier): San Diego, CA.

Hanazato, T. 1998. Response of a zooplankton community to insecticide application in experimental ponds: A review and the implications of the effects of chemicals on the structure and functioning of freshwater communities. *Environ. Pollut.* 101: 361–373.

Haynes, K. F. 1998. Sublethal effects of neurotoxic insecticides on insect behavior. *Ann. Rev. Entomol.* 33: 149–168.

Hill, I. R. 1989. Aquatic organisms and pyrethroids. *Pestic. Sci.* 27: 429–465.

Huang, L., Xi, Y. L., Zha, C. W., and Zhao, L. L. 2007. Effect of aldrin on life history characteristics of rotifer *Brachionus calyciflorus* Pallas. *Bull. Environ. Contam. Toxicol.* 79: 524–528.

Ke, L. X., Xi, Y. L., Zha, C. W., and Dong, L. L. 2009. Effects of three organophosphorus pesticides on population growth and sexual reproduction of rotifer *Brachionus calyciflorus* Pallas. *Acta Ecol. Sin.* 29: 182–185.

Kefford, B. J., Dalton, A., Palmer, C. G., and Nugegoda, D. 2004. The salinity tolerance of eggs and hatchlings of selected aquatic macroinvertebrates in southeast Australia and South Africa. *Hydrobiologia* 517: 179–192.

Keithmaleesatti, S., Varanusupakul, P., Siriwong, W., Thirakhupt, K., Robson, M., and Kitana, N. 2009. Contamination of organochlorine pesticides in nest soil, egg, and blood of the snail-eating turtle (*Malayemys macrocephala*) from the Chao Phraya River Basin, Thailand. *Proc. World Acad. Sci. Eng. Technol.* 40: 459–464.

Kikuchi, E., Fumsaka, C., and Kurihara, Y. 1977. Effects of tubificids (*Branchiura sowerbyi* and *Limnodrilus socialis*) on the nature of a submerged soil ecosystem. *Jpn. J. Ecol.* 27: 163–170.

Kikuchi, E. and Kurihara, Y. 1977. In vitro studies on the effects of tubificids on the biological chemical and physical characteristics of submerged ricefield soil and overlying water. *Oikos* 29: 348–356.

Kikuchi, M., Sasaki, Y., and Wakabayashi, M. 2000. Screening of organophosphate insecticide pollution in water by using *Daphnia magna*. *Ecotox. Environ. Safe.* 47: 239–245.

Kjølholt, J., Schou, L., and Hald, C. 1991. Økotoksikologisk vurdering af cypermethrin (ICI). COWIconsult, Danish Environmental Protection Agency, Ministry of Environment and Energy, Denmark, Document no. 20754-01.

Kristoff, G., Guerrero, N. R. V., and Cochón, A. C. 2010. Inhibition of cholinesterases and carboxylesterases of two invertebrate species, *Biomphalaria glabrata* and *Lumbriculus variegatus*, by the carbamate pesticide carbaryl. *Aquat. Toxicol.* 96: 115–123.

Manar, R., Bessi, H., and Vasseur, P. 2009. Reproduvtive effects and bioaccumulation of chlordane in *Daphnia magna*. *Environ. Toxicol. Chem.* 28: 2150–2159.

Marcial, H. S., Hagiwara, A., and Snell, T. W. 2005. Effect of some pesticides on reproduction of rotifer *Brachionus plicatilis*. *Hydrobiologia* 546: 569–575.

Marian, M. P. and Pandian, Y. J. 1984. Culture and harvesting technique for *Tubifex tubifex*. *Aquaculture* 42: 303–315.

Mazurova, E., Hilscherova, K., Jalova, V., Kohler, H. R., Triebskorn, R., Giesy, J. P., and Blaha, L. 2008. Endocrine effects of contaminated sediments on the freshwater snail *Potamopyrgus antipodarum in vivo* and in the cell bioassays *in vitro*. *Aquat. Toxicol.* 89: 172–179.

Moore, M. T., Huggett, D. B., Gillespie, W. B., Rodgers, J. H., and Cooper, C. M. 1998. Comparative toxicity of chlordane, chlorpyrifos, and aldicarb to four aquatic testing organisms. *Arch. Environ. Contam. Toxicol.* 34: 152–157.

Moriarty, F. 1983. *Ecotoxiciology. The Study of pollutants in Ecosystems*. Academic Press: London.

Morolli, C., Quaglio, F., Rocca G. D., Malvisi, J., and Salvo, A. D. 2006. Evaluation of the toxicity of synthetic pyrethroids to red swamp crayfish (*Procambarus clarkii*, Girard 1852) and common carp (*Cyprinus carpio*, L. 1758). *Bull. Français Pêche Piscicult.* 380–381: 1381–1394.

Mulla, M. S., Navvb-Gojrati, H. A., and Darwazeh, H. A. 1978. Biological activity and longevity is new synthetic pyrethroids against mosquitoes and some nontarget insect. *Mosquito News* 38: 90–96.

Mullié, W. C., Verwey, P. J., Berends, A. G., Sène, F., Koeman, J. H., and Everts, J. W. 1991. The impact of Furadan 3G (carbofuran) applications on aquatic macroinvertebrates in irrigated rice in Senegal. *Arch. Environ. Contam. Toxicol.* 20: 177–182.

Newman, M. C. 2001. *Fundamentals of Ecotoxicoolgy*. CRC Press: Boca Raton, FL.

Nelson, J., MacKinnon, E. A., Mower, H. F., and Wong, L. 1981. Mutagenicity of *N*-nitroso derivatives of carbofuran and its toxic metabolites. *J. Toxicol. Environ. Health* 7: 519–531.

Pechenik, J. A. 1996. *Biology of the Invertebrates*. 3rd edn. Wm.C. Brown: Dubuque, IA.

Perry, A. S., Yamamato, I., Ishaaya, I., and Perry, R. 1998. *Insecticides in Agriculture and Environment: Restrospects and Prospects*. Narosa: New Delhi.

Raghavendra, K., Verma, V., Srivastava, H. C., Gunasekaran, K., Sreehari, U., and Dash, A. P. 2010. Persistence of DDT, malathion & deltamethrin resistance in *Anopheles culicifacies* after their sequential withdrawal from indoor residual spraying in Surat district, India. *Indian J. Med. Res.* 132: 260–264.

Ray, A. K. and Ghosh, M. C. 2006. Aquatic toxicity of organophosphates and carbamates. In: Gupta, R. C. (ed.), *Toxicity of Organophosphates and Carbamate Pesticides*, pp. 657–672. Academic Press (Elsevier): San Diego, CA.

Renaud, C. B., Martel, A. L., Kaiser, K. L. E., and Comba, M. E. 2004. A comparison of organochlorine contaminant levels in the Zebra Mussel, Dreissena polymorpha, versus its unionid attachment, Elliptio complanata, in the Rideau river, Ontario. *Water Qual. Res. J. Canada* 39: 83–92.

Reyes, J. G. G., Jasso, A. M., and Lizarraga, C. V. 1996a. Toxic effects of organochlorine pesticides on *Penaeus vannamei* shrimps in Sinaloa, Mexico. *Chemosphere* 33: 567–575.

Reyes, J. G. G., Medina, J. A., and Villagrana, L. C. 1996b. Physiological and biochemical changes in shrimps larvae *Penaeus vannamei* intoxicated with organochlorine pesticides. *Mar. Pollut. Bull.* 32: 872–875.

Riva, C., Binelli A., Parolini, M., and Provini, A. 2010. The Case of Pollution of Lake Maggiore: a 12-Year Study with the Bioindicator Mussel *Dreissena polymorpha*. *Water Air Soil Pollut.* 210: 75–86.

Robson, M., Hamilton, G., and Siriwong, W. 2010. Pest control and pesticides. In: Frumkin, H. (ed.), *Environ. Health: From Global to Local*, 2nd edn., Chapter 17, pp. 591–634. Jossey Bass Wiley Publishers.

Schroer, A. F. W., Belgers, J. D. M., Brock, T. C. M., Matser, A. M., Maund, S. J., and Van den Brink, P. J. 2004. Comparison of laboratory single species and field population-level effects of the pyrethroid insecticide lambda-cyhalothrin on freshwater invertebrates. *Arch. Environ. Contam. Toxicol.* 46: 324–335.

Schulz, R. and Liess, M. 1995. Chronic effects of low insecticide concentrations on freshwater caddisfly larvae. *Hydrobiologia* 29: 103–113.

Siegfried, B. D. 1993. Comparative toxicity of pyrethroid insecticide to terrestrial and aquatic insect. *Environ. Toxicol. Chem.* 12: 1683–1689.

Simpson, I. C. and Roger, P. A. 1995. The impact of pesticides on non-target aquatic invertebrates in wetland rice fields: A review. In: Pingali, P. L. and Roger, P.A. (eds), *Impact of Pesticides and Farmer Health and the Rice Environment*, IRRI Editions, pp. 249–270. Los Banos: Philippines.

Siriwong, W. 2006. *Organochlorine Pesticide Residues in Aquatic Ecosystem and Health Risk Assessment of Local Agricultural Community*. Doctoral dissertation, Graduate School, Chulalongkorn University.

Siriwong, W., Thirakhupt, K., Sitticharoenchai, D., Rohitrattana, J., Borjan, M., and Robson, M. 2008. DDT and derivatives in indicator species of the aquatic food web of Rangsit agricultural area, Central Thailand. *Ecol. Indicators* 9: 878–882.

Smalling, K. L., Morgan, S., and Kuivila, K. K. 2010. Accumulation of current-use and organochlorine pesticides in crab embryos from northern California, USA. *Environ. Toxicol. Chem.* 29: 2593–2599.

Snyder, M. 2000. Cytochrome P450 enzymes in aquatic invertebrates: Recent advance and future directions. *Aquat. Toxicol.* 48: 529–547.

Stark, J. D. and Banks, J. E. 2003. Population-level effects of pesticides and other toxicants on arthropods. *Ann. Rev. Entomol.* 48: 505–519.

Sucahyo, D., van Straalen, N. M., Krave, A., and van Gestel, C. A. M. 2008. Acute toxicity of pesticides to the tropical freshwater shrimp *Caridina laevis*. *Ecotoxicol. Environ. Safe.* 69: 421–427.

Tripathi, P. K. and Singh, A. 2004. Toxic effects of cypermethrin and alphamethrin on reproduction and oxidative metabolism of the freshwater snail, *Lymnaea acuminata*. *Ecotoxicol. Environ. Safe.* 58: 227–235.

Triplehorn, C. A. and Johnson, N. F. 2005. *Borror and Delong's Introduction to the Study of Insects*, 7th edn. Brooks/Cole: Pacific Grove, CA.

Tu, H. T., Silvestre, F., Phuong, N. T., and Kestrmont, P. 2010. Effects of pesticides and antibiotics on penaeid shrimp with special emphases on behavioral and biomarker responses. *Environ. Toxicol. Chem.* 29: 929–938.

Valiela, I. 1991. Organisms and ecosystem. In: Barnes, R. S. K. and Mann, K. H. (eds), *Fundamentals of Aquatic Ecology*, pp. 3–26. Cambridge University Press: Cambridge.

Wang, W., Cai. D. J., Shan, Z. J., Chen, W. L., Poletika, N., and Gao, X. W. 2007. Comparison of the acute toxicity of gamma-cyhalothrin and lambda-cyhalothrin to zebra fish and shrimp. *Regul. Toxicol. Pharmacol.* 47: 184–188.

Wavre, M. and Brinkhurst, R. O. 1971. Interaction between some tubificid oligochaetes and bacteria found in the sediments of Toronto Harbour, Ontario. *J. Fish. Res. Board Can.* 28: 335–341.

Weston, D. P., Holmes, R. W., You, J., and Lydy, M. J. 2005. Aquatic toxicity due to residential use of pyrethroid insecitices. *Environ. Sci. Technol.* 39: 9778–9784.

Winemiller, K. O. and Polis, G. A. 1996. Food webs: What can they tell us about the world. In: Polis, G. A. and Winemiller, K. O. (eds), *Food Webs: Integration of Patterns and Dynamics*, pp. 1–22. Chapman & Hall: New York.

Won, S. J., Novill, A., Custodia, N., Rie, M. T., Fitzgerald, K., Osada, M., and Callard, I. P. 2005. The freshwater mussel *Elliptio complanata* as a sentinel species: Vitellogenin and steroid receptors. *Integr. Comp. Biol.* 45: 72–80.

Xi, Y. L., Chu, Z. X., and Xu, X. P. 2007. Effect of four organochlorine pesticides on the reproduction of freshwater rotifer *Brachionus calyciflorus* Pallas. *Environ. Toxicol. Chem.* 26: 1695–1699.

12

Pesticide Residues in Soil Invertebrates

Prem Dureja and Rajendra Singh Tanwar

CONTENTS

12.1 Introduction

In nature, the soil is one of the key elements that enable life on earth. It plays a central role in all terrestrial ecosystems and functions as the habitat for many organisms and as a filter and buffer, allowing clean groundwater storage. Important parts of the natural cycle of carbon, nitrogen, phosphorus, and sulfur take place in the soil. The main ecological functions of soil are those related to organic matter breakdown and nutrient mineralization by soil invertebrates and microbes. The soil-dwelling organisms play a crucial role in the ecosystem by mediating the geochemical cycling of elements and nutrient supplies to plants. They are also very beneficial to soil structure and structural stability. The widespread introduction of chemical pesticides to control pests (microbial pathogens, weeds, nematodes, snails, insects, and rodents) has significantly increased grain yields. However, because pesticides are often unspecific, they have the potential to profoundly modify the soil communities. It is important to understand how pesticide use affects the ecology of the ecosystem and to consider the implication of these changes.

12.2 Soil Invertebrates

Soil invertebrates are some of the smallest and most important invertebrates on the planet and provide a wide range of ecosystem services. They are the backbone of many ecologically important processes such as nutrient cycling, soil formation, and decomposition as well as composing the vital second, third, and fourth trophic levels of food webs. Species of invertebrates inhabiting soils have been estimated to constitute 23% of the total diversity of all living organisms (Decaëns et al. 2006). They include such animals as earthworms, slugs, land snails, ants, and other insects. These animals carry out the early stages of the physical and chemical decomposition of all types of organic debris in or on the soil. Most soil invertebrates also act as carriers of microbial propagules (e.g., seeds, spores), and so they inoculate the organic matter as it is passed through their bodies. The final stages of biochemical decomposition are accomplished by microbes, thus recycling nutrients, forming humus, and fostering soil particle aggregation (Dindal 1980).

These invertebrates provide essential regulatory services (decreasing greenhouse gas emissions, carbon sequestration, flood control, and more). The size of soil invertebrates is diverse, ranging from the smallest microfauna such as nematodes and protozoa (<200 µm) to the macrofauna such as mollusks and annelids. The invertebrates inhabiting soils are very sensitive to disturbance as the soil is their habitat and the place from which they can extract all the resources they need. Microarthropods, *Enchytraeidae*, and the many groups of the mesofauna (0.2–2 mm) live in the air-filled soil pores. The largest arthropods, Mollusca, Annelida, and Crustacea constitute the macrofauna that live in the surface litter or in nests and burrows that they create in the soil (Lavelle et al. 1997). In some places, vertebrates of the megafauna may become conspicuous elements of the soil fauna. Soil invertebrates and the many important functions they provide have resulted in the use of their diversity and abundance as a measure of soil quality.

A greater proportion (>80%) of biomass of terrestrial invertebrates is represented by earthworms, which play an important role in structuring and increasing the nutrient content of the soil. Therefore, they can be suitable bioindicators of chemical contamination of the soil in terrestrial ecosystems, providing an early warning of deterioration in soil quality (Culy and Berry 1995; Sorour and Larink 2001; Bustos-Obregón and Goicochea 2002). This is important for protecting the health of natural environments and is of increasing interest in the context of protecting human health (Beeby 2001) as well as other terrestrial vertebrates that prey upon earthworms (Dell'Omo et al. 1999). The suitability of earthworms as bioindicators of soil toxicity is largely due to the fact that they ingest large quantities of the decomposed litter, manure, and other organic matter deposited on soil, helping to convert them into rich topsoil (Reinecke and Reinecke 1999; Sandoval et al. 2001). Moreover, studies have shown that earthworm skin is a significant route of contaminant uptake (Lord et al. 1980), and thus the investigation of earthworm biomarkers in the ecological risk assessment (ERA) can be helpful (Sanchez-Hernandez 2006).

The technological advances in agriculture have led to an increased production and emission of chemical substances, which end up in the soil. The soil constituents, such as clay and organic matter, have a great capacity to retain chemicals. Therefore, the soil is a net sink for all kinds of chemicals, and their concentrations are often considerably higher than in any other environmental compartment (Verhoef and Van Gestel 1995).

12.3 Pesticides in Soil

Pesticides reach soil as a result of direct application to it or by drift during dusting and spraying to foliage or by being washed down by rains after application. Soil is the main accumulator of pesticides, and it is the main source of secondary contamination of the components of the biosphere. Pesticides pervade subsoil waters and thus reach the reservoirs. From the soil, some pesticides are transferred into plants through their root systems. When soil treatment takes place, pesticides are lifted together with the soil dust into the atmospheric air and, assuredly, to the working atmospheric air. In the soil, pesticides reach all of the soil inhabitants, especially the soil invertebrates.

Once a pesticide is in the soil, it will most likely follow one of three pathways. It will move through the soil with water, attach to soil particles, or be metabolized by organisms in the soil. Soil texture (percent sand, silt, and clay) and structure play a large role in the transport processes of pesticides. Soils that are very sandy allow water to move through them quickly, do not attach easily to pesticides, and generally do not contain a large population of soil organisms relative to other soil types. Soils that are high in clays and organic matter slow the movement of water, attach easily to many pesticides, and generally have a higher diversity and population of soil organisms that can metabolize the pesticide.

Some of the factors that influence the persistence of pesticides in soils are as follows: firstly, the factors include characteristics of the pesticide, including its overall stability either as a parent compound or as a metabolite, volatility, solubility, and formulation, and the method and site of application. Secondly, these also include the environmental factors, particularly temperature, precipitation (and humidity), and air movement (wind). Of these factors, the most important ones seem to be the chemical stability and physical characteristics of the pesticide, its stability exerting the greatest influence, otherwise volatility being more important in soil. The granular formulations persist longer than emulsions, which again persist longer than wettable powders. Heavy clayey soil retains pesticides longer than light sandy ones, and both organochlorines and organophosphates persist longer in acid than in alkaline soils. Higher the organic matter and clay mineral content in a soil, longer is the persistence of a chemical. Higher the temperature and soil moisture, lesser is the persistence. The breakdown of insecticides is more in soils having microorganisms than in sterile soils.

From soil, the persistent chemicals enter into soil invertebrates or water or are broken down by microorganisms and physical factors. Such chemicals are mostly DDT and dieldrin (aldrin also breaks down in soil to dieldrin). Both DDT and dieldrin are reported to be more prominent soil contaminants in Western Countries because of large-scale and indiscriminate uses.

12.3.1 Pesticides in Soil Invertebrates

The effects of pesticides on soil fauna are a highly complex issue, and researchers have had difficulty making generalizations. Variables include (1) the abundance of biocidal compounds from various chemical families, (2) great differences in persistence of pesticide compounds in the environment, (3) the diversity of invertebrate organisms in different soil communities, (4) metabolic products of different organisms that ingest pesticides, (5) the many chemical and physical varieties of different agricultural soil ecosystems, and (6) the psychological, cultural, and traditional agricultural practices of people who use pesticides (Dindal 1980). Where effects of pesticides have been observed and analyzed, the biotic responses are equally variable: (1) soil fauna may exhibit either a direct response to pesticides or more often an indirect secondary response; (2) only certain organisms are affected in a detrimental fashion, because some populations actually increase; (3) certain pesticide residues accumulate in tissues of some soil organisms with no apparent ill effects; and (4) certain sensitive species are killed from acute or chronic exposure to biocides. In almost all cases, the structures and functions of soil communities are modified by pesticide use (Dindal 1980).

Soil invertebrates are often directly exposed to pesticides, as they live in a number of habitats that are deliberately sprayed to control insect pests, fungi, or weeds or to protect human beings from disease vectors. Others are directly exposed by the deposition of insecticide sprays that miss their target, for example, soil-dwelling invertebrates in sprayed forests. In either case, exposure may be through contact or ingestion. Other invertebrates may be indirectly affected through the removal or reduction of food sources, be they vegetable, fungal, or animal. Insecticides are designed specifically to kill insects, and thus most invertebrates are sensitive to these chemicals. Sensitivity to other pesticides varies, but some herbicides and fungicides are also directly and highly toxic to this group of organisms.

Thompson and Edwards 1974, discussed the effects of pesticides on nontarget animals. Brown 1978, reviewed the effects of insecticides, herbicides, and fungicides on soil invertebrates. The pesticide-dependent population reduction has been documented (House et al. 1987; Reddy 1989; Reddy et al. 1996).

The impact of agrochemicals on soil fauna diversity and soil functions has become an issue of great concern. Agrochemicals are applied in the environment to fulfill a specific purpose but, at the same time, may cause damage to the soil biota, decreasing its diversity, growth, or reproduction, and consequently decreasing organic matter decomposition and soil fertility. Therefore, there is an increasing need for appropriate methods to assess the side effects of these chemicals on soil ecosystem.

12.3.2 Pesticide Residues in Earthworms

Earthworms are very useful and important members of the soil invertebrate fauna. They help in soil formation by feeding on the decaying organic matter. Their burrowing and feeding activities help in the continued maintenance of soil structure, aeration, drainage, and fertility. They erect pebble-free top layers by moving large amounts of soil from deeper strata up to the surface. They enhance the soil microbial fauna and hence the microbial degradation of various substances. Aristotle called them "the intestines of the earth" and Charles Darwin defined the extreme importance of earthworms in the breaking down of dead plants and animal matter that reach the soil and in the continued maintenance of soil structure (Darwin 1881). Their contribution to complex processes such as litter decomposition, nutrient cycling, and soil formation is very important, hence the presence of earthworms is beneficial to agroecosystems (Edwards and Fletcher 1988; Lavelle 1988; Lavelle et al. 1997; Edwards 1998; Eriksen-Hamel and Whalen 2007). As a major component of the soil ecosystem, they are susceptible to xenobiotics such as pesticides, and effects could be seen at the species, population, and even the community level (Edwards and Bohlen 1992). In fact, it was Kennel (1972, 1990) who first reported that reduced litter decomposition in orchards could indicate pesticide effects on earthworms. Since then, the use of earthworms in ecotoxicological studies is common, and a large database on pesticide effects on earthworms exists (Spurgeon et al. 2003; Frampton et al. 2006), varying from biomarkers (VanGestel and Weeks 2004; Rodriguez-Castellanos and Sanchez-Hernandez 2007) to field effects (Förster et al. 2006; Reinecke and Reinecke 2007; Casabé et al. 2007). Earthworm casts and burrow walls exhibit higher concentrations of total and plant-available elements than surrounding soil, and it has been recognized that surface-feeding species horizontally and vertically disseminate microorganisms, spores, pollen, and seeds and can reduce plant pathogens through the digestion of fungal spores. Therefore, practices that reduce earthworm populations in the soil can lead to a reduction in soil health.

Earthworms are generally found in the top 12" to 18" of the soil because this is where food is most abundant. The worm ingests soil and organic matter, which is swallowed and ground in the gizzard. The ejected material, or casting, is used to line the burrow or deposited at the entrance. Earthworm activity depends directly on the soil moisture and temperature. They become active when the soil thaws in the spring and move deeper in late summer as the soil dries.

The use of pesticides that may reduce the numbers or activity of earthworms can result in adverse effects on soil fertility, which in turn may influence crop yields. Second, reductions in the populations of earthworms may lead to a decrease in food supply for their predators. Third, the presence of pesticide residues in earthworms may lead to the poisoning of predators after consumption. Bioaccumulation may occur at secondary levels, with earthworm predators such as birds, or at tertiary levels, for example, when a fox preys upon a bird that has consumed earthworms containing pesticide residues.

DDT residues in soil and earthworms from 50 sites in Delhi were monitored. DDT was detected in all but two samples each of soil and earthworms. Among DDT residues, p,p'-DDE was most common and was found in 48 samples each of soil and earthworms; p,p'-DDT was detected in only 43 soil samples and 46 earthworm samples. p,p'-TDE and o,p'-DDT were also present in smaller concentrations in 29 and 15 soil samples and in 43 and 25 earthworm samples, respectively (Pillai 1986). A maximum total DDT concentration of 2.6 ppm was detected in the soil from Durga Nagar in the vicinity of a DDT factory. The highest concentration of 37.7 ppm total DDT in earthworms was also obtained from the same site. The maximum concentration factor found in the earthworms was 551. The total DDT concentration in the earthworms and soil showed significant correlation (Yadav et al. 1981).

The most important consequences of pesticide contamination of earthworms were seen in the 1960s and 1970s on species at higher trophic levels in food chains, when adverse effects on predators of earthworms were extensively reported. In addition to the example of local declines in the population of American robins noted earlier, poisoning of birds after eating DDT-contaminated earthworms has been reported from several countries. In the UK, several studies carried out in orchards in the early 1970s reported deaths of blackbirds (*Turdus merula*), song and mistle thrushes (*Turdus philomelos* and *T. viscivorus*), gray partridge (*Perdix perdix*), red-legged partridge (*Alectoris rufa*), pheasant (*Phasianus colchicus*), and tawny owl (*Strix aluco*). Blackbirds and song thrushes found dead in a British apple orchard contained 81–128 mg/kg DDE in breast muscle, and the authors suggested that, based on the residues detected in other live birds, most thrushes in the orchard were carrying near lethal amounts of DDT (Bailey et al. 1974). However, in this study, the concentrations in the earthworms were much lower than the threshold value that Beyer and Gish (1980) suggested: a concentration of 32 mg/kg of total DDT in earthworms could be a hazard to reproduction in some bird species, implying the lower risk to the earthworms in most of the leisure areas in Beijing in this present condition. Although DDTs are of little risk to the worms, it was probably virulent for the higher trophic organisms in the urban environment because of the bioaccumulation of DDTs through the food chain such as soil-earthworm-bird.

Work carried out in a Norfolk orchard during the 1960s showed that thrushes could accumulate DDT in their fat deposits from eating worms and other invertebrates. The death of birds was associated with the release of DDT into the bloodstream, following the mobilization of fat reserves during periods of stress (Edwards 1973; McEwan and Stephenson 1979).

DDT passed to them by the consumption of contaminated earthworms may also affect vertebrates other than birds. Frogs of the species *Rana temporaria* may be affected by DDT arising from the consumption of earthworms and slugs containing residues of this compound. Moles (*Talpa europaea*) also eat earthworms. Although moles trapped in a Norfolk orchard, where DDT had been used, contained high levels of this compound, they appeared to be unaffected (McEwan and Stephenson 1979).

Organochlorines other than DDT may be transmitted through earthworms to predators. Aldrin and dieldrin have been implicated in several instances of bird deaths in the United Kingdom from consumption of contaminated earthworms, but these reports were over-shadowed by the principal effects on birds of these compounds when used as seed dress-ings. Apart from the effects of contamination by the organochlorine insecticides, a few instances of harm to earthworms or their predators from pesticide applications have been reported. A number of deaths of black-headed gulls (*Larus ridibundus*), common gulls (*Larus canus*), and lapwings (*Vanellus vanellus*) resulted from the consumption of granules (which had not been incorporated as recommended into soil) and possibly of earthworms con-taining the carbamate insecticide aldicarb shortly after the introduction of this compound in the United Kingdom in the mid-1970s (McEwan and Stephenson 1979). Experiments indicate that vertebrates do not accumulate residues of pyrethroid insecticides that may be taken up by earthworms. The first soil-active pyrethroid insecticide, tefluthrin, proved to have a relatively high toxicity to earthworms (at 2 mg/kg soil) in laboratory studies.

Six species of earthworms from an arable soil were analyzed for residues of aldrin, dieldrin, DDT, and γ-BHC. The ratios between the concentrations of γ-BHC, aldrin + diel-drin, and p,p′-DDT + p,p′-DDE in the earthworms and in the soil (concentration factors) were similar, but the residue concentrations were consistently higher in the smaller, more shallow-living species *Allolobophora caliginosa*, *A. chlorotica*, and *A. rosea* than in the larger, deeper-living species *Lumbricus terrestris*, *A. longa*, and *Octolasion cyaneum*. In arable field plots, dieldrin residues in *A. longa* and *A. chlorotica* increased with increasing concentra-tions in the soil, but the concentration factors decreased. The concentrations of residues in earthworms (W) appear to be related to those in the soil (S) by an equation of the form $W = aS^b$, where a and b are constants, the latter being about 0.79 for a wide range of resi-due data embracing the uptake of aldrin-dieldrin, DDT components, and γ-BHC by earth-worms (Wheatley and Hardman 1968).

12.3.3 Effect of Pesticides on Earthworm

Earthworms, in the light of their importance in maintaining soil structure and fertility, are considered to be one of the most important indicator species in the soil. They incor-porate decaying organic matter into the soil and improve soil aeration, drainage, and water-holding capacity. The use of pesticides that may reduce the numbers or activity of earthworms can result in adverse effects on soil fertility, which in turn may influence crop yields. Second, reductions in populations of earthworms may lead to a decrease in food supply for their predators. Third, the presence of pesticide residues in earthworms may lead to the poisoning of predators after consumption. Bioaccumulation may occur at secondary levels, with earthworm predators such as birds, or at tertiary levels, for exam-ple, when a fox preys upon a bird that has consumed earthworms containing pesticide residues.

Risk assessments of toxicity to earthworms are required for the registration of pesti-cides in many countries, and considerable efforts have been made to devise appropriate procedures for evaluating the toxicity of pesticides to this key group of soil-inhabiting invertebrates. A standard method involves mixing a range of concentrations of a chemical into a loam soil containing a population of the earthworm species *Eisenia fetida*, which is easily cultured in the laboratory. The lethal concentration to kill 50% of earthworms (LC_{50}) and no observed effect levels (NOELs) may then be measured. If the NOEL is greater than the expected environmental concentration, then it is unlikely that the pesticide will have appreciable effects under field conditions.

The effects of pesticides on earthworms depend on the type of pesticide and its rate of application, earthworm species and age, and environmental conditions. Earthworm metabolizes pesticides as well as concentrates them from soil. Aldrin, for example, is rapidly epoxidized to dieldrin by microsomal oxidases located in the intestine of earthworms. Dieldrin is relatively stable in earthworms.

A single application of pesticides such as diazinon, isazophos, benomyl, carbaryl, and bendiocarb in soil caused significant short-term reduction in earthworm numbers (Potter et al. 1990). The most common action of agrochemicals on earthworms has been found to be coiling of the body and longitudinal muscle contraction, after which the body becomes rigid and sometimes swellings appear on the body surface. The swelling often bursts, creating bleeding sores. Pesticides such as propoxur, methidathion, endosulfan, triazophos, carbofuran, terbufos, and methamidophos cause such symptoms.

Zoren et al. (1986) reported the teratogenic effect of benomyl, that is, posterior segment regeneration of the earthworm, including an increased frequency of segmental groove anomalies and a variety of monstrosities, including two tails. Anton et al. (1990) reported the occurrence of various malformations due to the fungicide captan. Many carbamates have been reported to produce tumors and swellings along the earthworm's body (Stenersen 1979a). Hans et al. (1990) reported that different chemicals produced different symptoms in earthworms for chronic toxicity. For example, coiling and curling was caused by aldrin, whereas lifting of the body and extrusion of coelomic fluid was caused by heptachlor. Similarly, excretion of the mucus, glandular swelling, segmental construction and white banding, and hyperactivity were caused by endosulfan, lindane, and imazalil, respectively. Low dosages of phosphamidon caused hyperactivity of *L. mauritii*. Stenersen (1979b) observed immobility and rigidity in all tested species of earthworm caused by carbaryl and carbofuran.

One group of fungicides—the benzimidazoles—is highly toxic to earthworms, and studies have shown reductions of the activity in soils to which compounds such as carbendazim have been applied. These fungicides, which inhibit microtubule assembly in annelid and some other invertebrate groups as well as sensitive fungi, have been used to control earthworms on ornamental lawns and golf greens.

Studies carried out by Haque and Ebling (1983) by testing the toxicity of eight fungicides, five herbicides, and ten insecticides showed that, despite the diverse chemical structure of the compounds tested, there was species specificity in the earthworm toxicity response to different pesticides. Behavioral disturbances of earthworms caused by the insecticides were more severe than those caused by the herbicides or fungicides. There was also a difference in the sensitivity of worms of the same species to different pesticides.

Earthworms, *Aporrectodea caliginosa*, showed a 70%–80% inhibition of acetylcholine esterase (Booth et al. 2000a) when exposed to 28 mg/g of chlorpyrifos in the soil. However, there were no effects on the enzyme efficiency when exposed to recommended doses in the field (O'Halloran et al. 1999). Chlorpyrifos added to soil was toxic to six species of earthworms with 14-day no observed effect concentrations for survival of 46–875 mg/g (Ma and Bodt 1993).

Many of these pesticides have drastic effects on the nervous system of the earthworms. When earthworms were exposed to benomyl, they showed sublethal neurotoxic effects (Drews et al. 1987; Drews and Callahan 1988).

There was impairment of locomotory reflexes at the dosage, which were approximately two orders of magnitude less than LC_{50}. Both carbamates and organophosphorus pesticides inhibit acetylcholine activity (Edwards and Fisher 1991). Carbamates are described as "extremely toxic to earthworms in comparison to organophosphorus" (Roberts and Dorough 1984). Symptoms of intoxication to earthworms, which are caused by carbofuran, include rigidity, immobility, coiling, sores, and segmental swelling, whereas only immobility was observed in the case of organophosphorus pesticides. Aldicarb, endosulfan, benomyl, and calcium cyanide caused constrictions of the body.

Phorate has been found to be extremely toxic to earthworms and has almost eliminated them from many soils even when applied at normal agricultural doses. Dimethoate affected the perinuclear position of the Golgi apparatus in the neurons of the nerve cord, and the Golgi apparatus was fragmented into short elements and granules (El-Banhaway et al. 1982). In a laboratory study, the fungicide pyrazophos showed the greatest reduction in population number and biomass dynamics of the earthworm *Dendrobaena veneta* (Karaman et al. 1996).

The earthworm *Pheretima posthuma* is reported to bioaccumulate the nonextractable soil-bound residues of DDT and HCH (Verma and Pillai 1991) and also found to facilitate the steady mobilization of soil-bound residues of the insecticide to readily bioavailable labile forms. The pesticides also affect the growth of fast-growing *Eisenia fetida*. There is loss of weight or slowing of growth in some of the species.

Sublethal effects of terbuthylazine and carbofuran on the growth and reproduction of *Eisenia andrei* were investigated by Brunninger et al. (1994) over a period of three generations by measuring the cocoon production of worms. Inhibition of cocoon production was found in the parental generation.

Hatchlings raised from cocoons to provide the F1 generation grew more rapidly when treated with terbuthylazine. This increase in vitality was also observed in cocoon production. Groups treated with terbuthylazine produced more cocoons than control. The F2 generation was raised from hatchings of the F1 generation, and here, the terbuthylazine treatments increased earthworms' growth, but not cocoon production. Exposure of carbofuran decreased cocoon production in all generations. Growth of the F1 generation was not influenced by low carbofuran concentration.

The sublethal effect of dieldrin was studied on *Eudrilus eugeniae*, on the ultrastructure of spermatozoa. Dieldrin at a relatively low concentration caused structural damage, especially to the nucleus of the sperm (Reinecke et al. 1995). The burrowing ability and mortality in earthworm species native to North Canadian forest soils were affected at relatively low concentrations of organophosphorus pesticides such as fenitrothion. *Dendrobaena octaedra*, the forest litter–dwelling species, was approximately eight times more sensitive to the chemical than *E. fetida*. However, when *D. octaedra* was exposed to the pesticide in litter, the earthworm was considerably less susceptible to fenitrothion (Addison and Holmes 1995). Diflubenzuron, an insect growth regulator, showed little effect on mortality and gross anatomical changes of earthworms. This is because earthworms have no chitin in their body (Maurya and Chattoraj 1994).

Fumigants such as dichloropropane and dichloropropene applied to the soil for the control of pathogens and nematodes permeate the soil as vapors and kill most of the earthworms, even those that live in deep burrows. Chloropicrin is also very toxic to all earthworms.

Xiao et al. (2006) suggested that growth can be regarded as a sensitive parameter to evaluate the toxicity of acetochlor on earthworms. Helling et al. (2000) tested in laboratory the effect of copper oxychloride, while Yasmin and D'Souza (2007) investigated the impact of carbendazim, glyphosate, and dimethoate on *E. fetida* and found a significant reduction in the earthworm growth in a dose-dependent manner. According to Van Gestel et al. (1992), parathion affects the growth of *Eisenia andrei*. Booth et al. (2000b) studied the effects of two organophosphates, chlorpyrifos and diazinon, while Mosleh et al. (2003a) investigated the toxicity of aldicarb, cypermethrin, profenofos, chlorfluazuron, atrazine, and metalaxyl in the earthworm *Aporrectodea caliginosa* and observed a reduction in growth rate in all pesticide-treated worms. Mosleh et al. (2002, 2003b) studied the effects of endosulfan and aldicarb on *Lumbricus terrestris* and have suggested growth rate as an important biomarker for contamination by endosulfan and aldicarb. Zhou et al. (2007) assessed and found that chlorpyrifos had an adverse effect on growth in earthworms exposed to 5 mg/kg chlorpyrifos after eight weeks. Some studies have shown that the growth of earthworms appeared to be more severely affected at the juvenile stage.

Toxicity of avermectins to soil invertebrates in soil and in feces from recently treated sheep was studied. Abamectin was more toxic than doramectin. In soil, earthworms (*Eisenia andrei*) were most affected with LC_{50}s of 18 and 228 mg/kg dry soil, respectively, while LC_{50}s were 67–111 and >300 mg/kg for springtails (*Folsomia candida*), isopods (*Porcellio scaber*), and enchytraeids (*Enchytraeus crypticus*), respectively. EC_{50}s for the effect on reproduction of springtails and enchytraeids were 13 and 38 mg/kg, respectively, for abamectin, and 42 and 170 mg/kg, respectively, for doramectin. For earthworms, NOEC was 10 and 8.4 mg/kg for abamectin and doramectin effects, respectively, on body weight. Earthworm reproduction was not affected. This study indicates a potential risk of avermectins for soil invertebrates colonizing feces from recently treated sheep (Lucija et al. 2008). Malathion affects the reproduction as well as morphology of *Eisenia foetida*. It also causes skin rupture (Reddy and Rao 2008).

Bharati and Subba Rao (1986) observed that the sublethal concentrations of phosphamidon, monocrotophos, and dichlorvos reduce carbohydrate and glycogen content and increase phosphorylase "a" and "b" activity in the muscle and blood of the earthworm, *Lampito mauritii* (Kingberg). Fenitrothion produces similar effects on carbohydrate metabolism of the body wall muscle of *Octochaetona pattoni* (Varadaraj 1986). Reddy et al. (1983) have noted an increase in the blood sugar level under fenitrothion treatment. Although persistence of insecticides in soil and their effects on density of nontarget organisms is minimal at a normal agricultural dose, the effect is obvious at population metabolism with changes in physiological and biochemical responses.

Soils that contain significant copper residues have been observed to have few earthworms (Van Rhee 1967; Van Zwieten et al. 2004), reduced surface activity (fewer castings visible at the soil surface), and greater litter build-up (Ma 1984). Copper has been shown to reduce the burrowing activity of earthworms, which in turn led to increased soil bulk density in a vineyard (Eijsackers et al. 2005). Likewise, Gaw et al. (2003) described the lack of pesticide breakdown in soils where copper was a co-contaminant. There is clear evidence that soil organisms and thus soil functions can be affected by pesticides. At sites in a study on avocado orchards in northern NSW (New South Wales) (Merrington et al. 2002), an absence of earthworms in areas of copper contamination was accompanied by a thick layer of organic matter (ca. 10–30 cm deep) that was clearly stratified on the soil surface, with little evidence of breakdown and incorporation into the subsurface layers. This phenomenon is not seen in other subtropical horticultural systems where copper fungicides are not used. A strong correlation has been observed between the soil-copper concentration and the level of copper in earthworm tissues (Ma et al. 1983). It has been noted that earthworms exhibit chronic toxic responses at relatively low concentrations of copper (<4–16 mg/kg) (Helling et al. 2000). The enchytraeid worm *Cognettia sphagnetorum* was shown to actively avoid copper-contaminated soil (Salminen and Haimi 2001). Copper oxychloride has recently been shown to reduce populations of the earthworm, *Aporrectodea caliginosa*, in field trials six months following the application of the fungicide (Maboeta et al. 2003). Elevated copper concentrations have been shown to reduce beneficial mycorrhizal associations (Liao et al. 2003), reduce microbial activity and function (Bogomolov et al. 1996), and impact a range of mesofauna (Böckl et al. 1998). Direct tests have shown that concentrations of copper in the soil ranging between 100 and 150 ppm are in most cases sufficient to produce a consistent decrease in earthworm numbers. Thus, it is worrisome that higher concentrations of copper are found in the soil in many orchards around the world.

Potter et al. (1990) found that a single application of the fungicide benomyl or the insecticides ethoprop, carbaryl, or bendiocarb at labeled rates reduced earthworm populations in turfgrass by 60%–90%, with significant effects lasting for 20 weeks. Other insecticides including diazinon and chlorpyrifos caused less severe, but significant earthworm mortality in some tests (Potter et al. 1990). Similarly, Choo and Baker (1998) reported that the nematicide fenamiphos and the insecticide endosulfan reduced the growth and inhibited the reproduction of the earthworm *Aporrectodea trapezoides* (Lumbricidae) when applied at recommended rates to an Australian pasture soil. The nematicide carbofuran is also well documented as having an adverse effect on earthworm growth and activity (Parmelee et al. 1990; Reddy 1999).

Fungicides are generally highly toxic to earthworms, especially copper and zinc residues from copper sulfate and carbamates, respectively. Soil fumigants, nematicides, and fungicides such as D–D mixture (dichloropropane:dichloropropene), metham-sodium, and methyl bromide are highly toxic to earthworms. The majority of fumigants and contact nematicides are toxic to earthworms as well (Edwards and Bohlen 1992). Carbamate fungicides such as benomyl and carbendazim are also highly toxic to earthworms. For example, it has been reported that about 1.8 kg/ha per year of benomyl may destroy all the *Lumbricus terrestris* and most of the *Allolobophora* spp. present in an apple orchard in England (Brown 1978).

One group of fungicides—the benzimidazoles—is highly toxic to earthworms, and studies have shown reductions of the activity in soils to which compounds such as carbendazim have been applied. These fungicides, which inhibit microtubule assembly in annelids and some other invertebrate groups as well as sensitive fungi, have been used to control earthworms on ornamental lawns and golf greens.

12.3.4 Effect of Pesticides on Soil Nematode Communities

Nematodes are the most abundant metazoans on the Earth and also the most abundant invertebrates in the soil ecosystem where they perform many ecological functions. and particularly bacterivorous nematodes play an important role in the nutrient cycling (Wood 1988). They interact closely with other soil organisms and their activity affects primary production, decomposition, energy flows, and nutrient cycling, especially nitrogen cycle (Sohlenius 1980; Freckman and Baldwin 1990). Soil nematodes belong to microfauna living in the pore-water of soil top layer, where they can be exposed substantially to the soil contaminants that are usually accumulated here. The thin cuticle covering their body is water-permeable, which makes them very sensitive to the uptake of the dissolved fraction of contaminants. After exposure and effects of toxicants, the ability of nematodes to play their ecological roles may be impaired with possible deleterious effects at the ecosystem level. In ecological studies, change in nematode community structure is used as a sensitive marker of environmental stress as well as of pollution (Yeates 2003). Nematodes possess many characteristics that make them ideal as bioindicators for soil health assessment (Bongers and Ferris 1999), and their fauna composition, together with its ecological indices, has emerged as a useful monitor of environmental conditions and soil ecosystem function (Neher 2001).

There is a growing concern over the impacts of such pesticides on nontarget free-living nematodes. Yardim and Edwards (1998) studied the responses of nematode communities to mixtures of insecticides (carbaryl, endosulfan, and esfenvalerate), fungicides (chlorothalonil), and herbicides (trifluralin and paraquat) in a tomato field and found that the temporal community dynamics of different nematode trophic groups differed with the pesticide applications. Yeates et al. (1999) reported that the diversity of the nematode community was increased and bacterivores were consistently low in herbicide-treated plots over a 7-year period. Chen et al. (2003) monitored the effect of acetochlor on the nematode community structure in a soybean field. The results showed that the soil nematode community structure could be greatly influenced by acetochlor, and the numbers of total nematode and trophic groups were reduced, and species richness (SR) was effective in distinguishing the differences between the acetochlor plots and the control plots. Pen-Mouratov and Steinberger (2005) investigated the effect of two pesticides, Nemacur and Edigan, on soil nematodes in a desert system, and the results indicated that the numbers of total nematodes, fungivores, and bacterivores were decreased in the pesticide treatments. Wada and Toyota (2008) conducted a pot test to evaluate the toxicity of imicyafos and fosthiazate to nontarget organisms and reported that the two organophosphorus nematicides effectively suppressed a plant-parasitic nematode *Pratylenchus penetrans* but had little impact on free-living nematodes and the soil microbial community. Smith et al. (2000) showed that long-term application of benomyl in a tall grass prairie had no effect on herbivorous nematodes but reduced populations of certain groups of fungal-feeding and predatory nematodes. Wardle et al. (1995) have suggested that evidence for direct negative effects of herbicides on nematode populations were more likely to be indirect effects from changes arising in the quantity and quality of plant inputs (e.g., dead organic matter from weeds) to the soil.

12.3.5 Effect of Pesticides on Microarthropods

Soil arthropods play important roles in breaking down plant remains and controlling crop pests by parasitism and predation and are food sources for larger animals. For sprays, the concentration of pesticides will be greatest at the soil surface immediately after application, and the species most vulnerable to potential adverse effects are those that live on the soil surface. Examples of such species include spiders as well as ground and rove beetles belonging to the Carabidae and Staphylinidae families that are predators of aphids and that may contribute to the control of these insect pests. Many regulatory authorities require an assessment of the effects of pesticides on nontarget terrestrial arthropods, particularly if the compound is intended for application to soil (Oomen 1998).

Soil fauna including microarthropods (collembolans, mites) and microfauna (protozoa) may be affected by soil-applied pesticides with possible flow-on consequences for organic matter mineralization. *Collembola* are among the most abundant soil arthropods and play an important role in decomposer food webs (Petersen 2002). *Collembola* are known to be food generalists (Scheu and Folger 2004). The diet of most species is composed of a mixture of detritus, algae, bacteria, and fungi and varies with season (Wolters 1985). Due to their feeding activity, Collembola affect decomposition processes and the microstructure of the soil (Cragg and Bardgett 2001). The modification of decomposition processes results in changes in nutrient mineralization and may ultimately affect plant growth (Gange 2000). Bardgett and Chan (1999) and Scheu et al. (1999) showed that the presence of Collembola resulted in changes in plant growth and plant shoot N contents. The effects of Collembola on decomposition processes and nutrient mineralization may depend on species composition and dominance structure of collembolan communities, since different species of Collembola have different feeding preferences (Cole et al. 2004).

Rajagopal et al. (1990) reported that soil fauna that included mainly mites and collembolans were affected by insecticide application in the field. The effect of pesticides on the soil mesofauna is complex because of their action on both predacious and nonpredacious groups. Therefore, several researchers have reported increased populations of collembolan and mites after the use of pesticides, mainly because of mortality of predacious mites, which exercise a great check on other mites and collembolans (Veeresh and Rajagopal 1989). Epstein et al. (2000) reported the effect of broad-spectrum insecticides on epigeal arthropod biodiversity in Pacific Northwest apple orchards. They reported that the highly mobile invertebrates were strongly susceptible to application of broad-spectrum neural active insecticides.

The effect of dimethoate was found to be species specific. Folker-Hansen et al. (1996) reported the effect of dimethoate on the body growth of representatives of the soil mesofauna and it was proved to be sex and species specific. All the collembolan species displayed a sexual dimorphism for body length, and females grew longer than the males in the case of treatment with dimethoate. Diversity, a measure of community structure, was higher in control farms, while dimethoate application reduced the soil fauna diversity. In station one, *Trips* spp. disappeared in the treated farms, whereas they were present in the control farms. In station two, *Collembola* spp. and seed corn maggots disappeared in the treated farms, whereas they were present in the control farms. On the other hand, *Collembola* spp., *Symphyla* spp., and beetle larvae were absent in the treated farms, whereas at the same time they were present in the control farms in station number three.

Everts et al. (1989) reported that fields, which were mechanically treated against weeds, had higher species diversity than fields chemically treated by deltamethrin, fenitrothion, and bromophos-ethyl. The present results are supported by the report of Stevenson et al. (2002); they reported that pesticide-free cornfields had soil invertebrates' densities that were significantly higher than in the treated cornfields. Entry of pesticides into the ecosystem may cause imbalances in the ecological equilibrium. Rogers and Potter (2003) studied the effect of spring imidacloprid application for white grub control on parasitism of Japanese beetle (Coleoptera: *Scarabaeidae*) by *Tiphia vernalis* (Hymenoptera: *Tiphiidae*). They indicated that applying imidacloprid in early spring can interfere with biological control by *T. vernalis*, whereas postponing preventive grub treatments until June or July, after the wasps' flight period, will help to conserve *T. vernalis* population. These deleterious effects greatly modify some biological functions, such as soil organic matter decomposition and nutrient availability in the soil by reduction of the diversity of soil biota (Ferraro and David 2000). From the outcomes of this study, we can conclude that dimethoate residues in the soil of Zendan Valley are beyond the TTLC (total threshold limit concentrations) for total dimethoate in soil. Soil microarthropod density and diversity are low in the contaminated soil with dimethoate and omethoate; this will lead to the reduction of agriculture production for such areas and thus lower the economy of the country.

Foliar and soil insecticides were found to be lethal to Collembola, and insecticide applications resulted in a strong decline in the density of total Collembola. The application of chlorpyrifos reduced collembolan density to a greater extent than dimethoate; the effect of the combined application on total collembolan numbers was similar to that of chlorpyrifos only. The foliar insecticide, dimethoate, significantly decreased the density of epigeic *Collembola* (Christiansen 1964), including the genera *Isotoma* and Lepidocyrtus, and also tended to decrease the density of *Entomobrya* and *Sminthuridae*. In contrast, the density of endogeic genera remained little affected. Probably, the effect of the application of dimethoate, which functions as a contact pesticide, was restricted to surface-living species and its effect penetrated little into the soil. Joy and Chakravorty (1991) reported that dimethoate is lethal to many species of Collembola; however, its effect in the field is likely to vary between species depending on exposure. In addition to reducing overall collembolan abundance, the application of insecticides changed the dominant structure of the collembolan community.

Overall, the soil insecticide caused such a strong change in the composition of the collembolan community that the additional application of foliar insecticide had no further effect. In fact, the soil insecticide application virtually eradicated most species of Collembola (Endlweber et al. 2006). The nematicide, carbofuran, for example, has been shown in many studies to have a negative impact on soil microarthropods, particularly Collembola and predatory mites (Prostigmata and Mesostigmata), with populations taking up to a year to recover (Reddy 1999). A similar effect has been noted with the insecticide dimethoate (Martikainen et al. 1998). Long-term use of the fungicide mancozeb in vineyards has also been reported to reduce populations of the predatory mite *Typhlodromus pyri* (Auger et al. 2004). Herbicides, however, have generally been found to have negligible effects on soil microarthropod populations (Martikainen et al. 1998; Cortet et al. 2002), although some early reports have shown that atrazine, simazine, glyphosate, and paraquat can cause temporary reductions in microarthropod activity (Edwards 1989). More recently, Rebecchi et al. (2000) reported that the sulfonylurea herbicide triasulfuron caused a decrease in some collembolan species in an agricultural soil. Effects of insecticides and fungicides on protozoan and nematode populations are also more pronounced than those of herbicides (Gupta and Yeates 1997). For example, Petz and Foissner (1989) found that mancozeb altered the community structure but not the absolute number of protozoans in a field soil, while Ekelund (1999) demonstrated that various groups of protozoa (notably flagellates) were reduced by field application rates of the fungicide fenpropimorph.

A 5-month durational period of the recolonization of soil arthropods was monitored on endosulfan-treated soil from April to August 2007, within the 5–10 cm depth at varying concentrations of the pesticide (Iloba and Ekrakene 2010). The result showed that there was consistent decrease in the mean numbers of soil arthropod sampled from April to June and the decrease was more as the concentration of applied endosulfan increased. However, July to August witnessed a very remarkable increase in the mean soil arthropod sampled compared to the control stations, an indication of recolonization. Among the seven groups of soil microarthropod sampled, the mean number from groups were statistically different ($p < 0.05$) at all concentrations compared to the control with Acarina, Collembola, Coleoptera, and Myriapoda being the most abundant while Hymenoptera, Isoptera, and Crustacea being the least in abundance. The most recovered group was the Acarina while the least was Crustacea with recolonization ability being greatest in stations with higher concentrations.

The application of the insecticide dimethoate, particularly in the autumn, has led to reductions in the numbers of ground-dwelling beetles in cereals and other crops.

Reductions in populations of up to 90% were observed in some trials. In some cases, this decrease in predatory beetles was associated with the increased incidence of aphids, one of their major food sources (Al-Haifi et al. 2006).

The significance of predators, such as spiders, in controlling pests in agricultural eco-systems (Marc et al. 1999) and as an important food source for farmland birds has been recognized (Wilson et al. 1999). In addition, the importance of Collembola for increasing fungal decomposition (Cragg and Bardgett 2001) and maintaining spider numbers when pest populations are low (Marcussen et al. 1999; Bilde et al. 2000; Harwood et al. 2001, 2003, 2004; Agusti et al. 2003) has highlighted the need for more studies investigating mechanisms for soil invertebrate population regulation. Therefore, the effects of the insecticide, chlorpyrifos, on spider and Collembola communities were studied. Initially, twelve species of Collembola were identified from the insecticide-treated and control plots. Species diversity, richness, and evenness were all reduced in the chlorpyrifos-treated plots, although the total number of Collembola species increased tenfold despite the abundance of some spider species being reduced. The dominant collembolan in the insecticide-treated plots was *Ceratophysella denticulata*, accounting for over 95% of the population. Forty-three species of spider were identified. There were a reduced number of spiders in insecticide-treated plots due mainly to a lower number of the linyphiid *Tiso vagans*. Immature Collembola and spiders were either absent or in low numbers on the insecticide plots, respectively. Juveniles are often more sensitive to xenobiotics than adults (Fountain and Hopkin 2001), and it is possible that adults are better at detecting and avoiding insecticides (Fabian and Petersen 1994). It emphasizes the importance of understanding the effects of soil management practices on soil biodiversity, which is under increasing pressure from land. Chlorpyrifos is toxic to spiders (Pekár 2002) and also reduces other beneficial predators such as *Staphylinidae* (Wang et al. 2001) and Carabidae (Curtis and Horne 1995).

12.3.6 Effect of Pesticides on Beneficial Soil Microorganisms

One spoonful of healthy soil has millions of tiny organisms including fungi, bacteria, and a host of others. These microorganisms play a key role in helping plants utilize the soil nutrients needed to grow and thrive. Microorganisms also help the soil store water and nutrients, regulate water flow, and filter pollutants (Pell et al. 1998). The heavy treatment of soil with pesticides can cause populations of beneficial soil microorganisms to decline. According to soil scientist Dr. Elaine Ingham, "If we lose both bacteria and fungi, then the soil degrades. Overuse of chemical fertilizers and pesticides have effects on the soil organisms that are similar to human overuse of antibiotics. Indiscriminate use of chemicals might work for a few years, but after awhile, there aren't enough beneficial soil organisms to hold onto the nutrients" (Santos and Flores 1995). For example, plants depend on a variety of soil microorganisms to transform atmospheric nitrogen into nitrates that plants can use. Common landscape herbicides disrupt this process: triclopyr inhibits soil bacteria that transform ammonia into nitrite (Moorman et al. 1992); glyphosate reduces the growth and activity of both free-living nitrogen-fixing bacteria in soil (Fabra et al. 1997) and those that live in nodules on plant roots (Arias et al. 1993); and 2,4-D reduces nitrogen fixation by the bacteria that live on the roots of bean plants (Singh and Singh 1989; Tözüm-Çalgan and Sivaci-Güner 1993), reduces the growth and activity of nitrogen-fixing blue-green algae (Martens and Bremner 1993; Frankenberger and Tabatabai 1991), and inhibits the transformation by soil bacteria of ammonia into nitrates. Mycorrhizal fungi grow with the roots of many plants and aid in nutrient uptake. These fungi can also be damaged by herbicides in the soil. One study found that oryzalin and trifluralin both inhibited the growth of certain species of mycorrhizal fungi (Kelley and South 1978). Roundup has been shown to be toxic to mycorrhizal fungi in laboratory studies, and some damaging effects were seen at concentrations lower than those found in soil following typical applications (Estok et al. 1989; Chakravarty and Sidhu 1987). Triclopyr was also found to be toxic to several species of mycorrhizal fungi (Moorman 1989), and oxadiazon reduced the number of mycorrhizal fungal spores.

In South Africa, the feeding activity of soil organisms was higher in soil from organic vineyards than from conventionally treated sites (Reinecke et al. 2008). The number of earthworms was 1.3–3.2 times higher in organic compared to conventional plots, and the length of plant roots colonized by mycorrhizae was 40% higher in organic than in conventional systems (Mäder et al. 2002). Triclopyr, a herbicide, caused a major reduction in the growth of mycorrhizae at elevated soil levels (Chakravarty 1987).

The sulfonylurea herbicides metsulfuron and (to a lesser extent) chlorsulfuron caused a reduction in the growth of *Pseudomonas* soil bacteria (Boldt and Jacobsen 2006). In laboratory tests, a combination of two sulfonylurea herbicides, bensulfuron-methyl (B) and metsulfuron-methyl, caused a considerable reduction in soil microbial biomass over the first 15 days (El-Ghamry et al. 2001). In bacterial communities in soil, bromoxynil (a nitrile herbicide) caused major changes in species composition and diversity. Bromoxynil inhibited the growth of bacteria capable of degrading chemicals in the soil (Baxter and Cummings 2008). Also, captan (a fungicide) and the herbicide glyphosate caused a shift among species in bacterial communities in the soil (Widenfalk et al. 2008). Certain organophosphate insecticides (e.g., dimethoate) can decrease the activity and biomass of soil microorganisms, while others (such as fosthiazate) may actually result in an increase in microbial biomass (Eisenhauer et al. 2009). How pesticides affect long-term soil fertility is not well understood as this depends on many factors.

Although pesticides are designed to control pest species, the extent of their selectivity for pests in some cases is not great, and other organisms are injured, including soil microorganisms. Inhibitions of microbial activity are most pronounced from fungicides and fumigants and the suppression may remain for long periods. The impact may be so great that the natural balance among the resident soil microbial populations is upset and new organisms, such as plant disease vectors, become prominent. Moreover, certain nutrient cycles regulated by microorganisms are inhibited by fungicides and fumigants in such a way that significant adverse effects on plant growth and nutrition become evident. The lack of widespread concern for these antimicrobial agents is not because of their lack of toxicity but rather because they are not as widely used as the other two major classes of pesticides (Alexander 1981).

Insecticides have received a lot of attention in the past. These compounds may be applied directly to soil for the control of soil-borne insects, or they may reach the soil from drifting sprays or when treated plant remains fall to the ground or are mixed with the soil during normal farming practices. Inhibition of some microbial processes or suppressions of individual populations of bacteria, fungi, or actinomycetes occur. On the other hand, the toxicity is generally not marked, and the beneficial effects of the insecticides in controlling insect pests argue for their use. The US regulatory agencies have not acted on the basis of possible long-term harm insecticides might have on microbial processes, but a few instances of major suppression of microbial activities in the field have been noted, so that a change in policy in regard to their use does not appear warranted (Alexander 1980).

12.4 Conclusion

Pesticide residues in soil, in addition to eliminating or reducing parasitic microbes, are also toxic to the nonparasitic and ecologically useful soil microbial and vertebrate population. Pesticides may reduce certain microorganism populations while they stimulate the growth of others, especially the saprophytic and spore-forming types. Soil chemical properties may also be altered by accumulation of residual pesticides and their metabolites. These processes may disrupt the ecological balance in the soil microenvironment, first by simplifying the microbial population and possibly by reducing soil fertility and its ability to support life. Nontarget or residual pesticide toxicity would also disrupt the population of some of the valuable soil invertebrates such as earthworms, predatory mites, centipedes, and carabid beetles. The accumulation of pesticides in resistant or tolerant species may provoke episodes of toxicity to organisms higher in the food chain. The chlorinated hydrocarbons are likely to be most ecotoxic. The use of such pesticides ought to be restricted. Efforts to find alternatives to pesticides, especially research into biological control, should be intensified.

References

Addison, J. A. and Holmes, S. B. 1995. Comparison of forest soil microcosm and acute toxicity studies for determining effects of fenitrothion on earthworms. *Ecotoxicol. Environ. Safe.* 30: 127–133.

Agusti, N., Shayler, S. P., Harwood, J. D., Vaughan, I. P., Sunderland, K. D., and Symondson, W. O. C., 2003. Collembola as alternative prey sustaining spiders in arable ecosystems: Prey detection within predators using molecular markers. *Mol. Ecol.* 12: 3467–3475.

Al-Haifi, M. A., Khan, M. Z., Abdullah Murshed, V., and Ghole, S. 2006. Effect of Dimethoate Residues on Soil Micro-arthropods Population in the Valley of Zendan,Yemen. *J. Appl. Sci. Environ. Manag.* 10(2): 37–41.

Alexander, M. 1980. How agricultural technologies affect productivity of croplands and rangelands by affecting microbial activity in soil. OTA, Background Paper.

Alexander, M. 1981. Biodegradation of chemicals of environmental concern. *Science* 211: 132–138.

Anton, E., Lahorda, E., and Laborda, P. 1990. Acute toxicity of the fungicide captan to the earthworm *Eisenia foetida* (Savingny). *Bull. Environ. Contam. Toxicol.* 45: 82–87.

Arias, R. N. and Fabra de Peretti, A. 1993. Effects of 2,4-dichlorophenoxyacetic acid on *Rhizobium* sp. growth and characterization of its transport. *Toxicol. Lett.* 68: 267–273.

Auger, P., Kreiter, S., Mattioda, H., and Duriatti, A. 2004. Side effects of mancozeb on *Typhlodromus pyri* (Acari: Phytoseiidae) in vineyards: Results of multi-year field trials and a laboratory study. *Exp. Appl. Acarol.* 33: 203–213.

Bardgett, R. D. and Chan, K. F. 1999. Experimental evidence that soil fauna enhance nutrient mineralization and plant nutrient uptake in montane grassland ecosystems. *Soil Biol. Biochem.* 31: 1007–1014.

Bailey, S., Bunyan, P. J., Jennings, D. M., Norris, J. D., Stanley, P. I., and Williams, J. H. 1974. Hazards to wildlife from the use of DDT in orchards:II a further study. *Agro-Ecosys* 1: 323–338.

Baxter, J. and Cummings, S. P. 2008. The degradation of the herbicide bromoxynil and its impact on bacterial diversity in a top soil. *J. Appl. Microbiol.* 104(6): 1605–1616.

Beeby, A. 2001. What do sentinels stand for? *Environ. Pollut.* 112(2): 285–298.

Beyer, W. N. and Gish, C. D. 1980. Persistence in earthworms and potential hazards to birds of soil applied DDT, dieldrin and heptachlor. *J. Appl. Ecol.* 17: 295–307.

Bharati, Ch. and Subba Rao, B. V. S. S. R. 1986. Effect of sub-lethal and lethal concentrations of organo-phosphate insecticides on some biochemical constituents of the earthworm, *Lampito mauritii* (Kingberg). In: Dash, M. C., Senapati, B. K., and Mishra, P. C. (eds), *Verms and Vermicomposting*, pp. 82–92. Five Star Printing Press: Burla, India

Bilde, T., Axelsen, J. A., and Toft, S. 2000. The value of Collembola from agricultural soils as food for a generalist predator. *J. Anim. Ecol.* 37: 672–683.

Böckl, M., Blay, K., Fischer, K., Mommertz, S., and Filser, J. 1998. Colonisation of a copper-decontaminated by micro- and mesofauna. *Appl. Soil Ecol.* 9 (1-3 Special Issue SI): 489–494.

Bogomolov, D. M., Chen, S. K., Parmelee, R. W., Subler, S., and Edwards, C. A. 1996. An ecosystem approach to soil toxicity testing: A study of copper contamination in laboratory soil micro-cosms. *Appl. Soil Ecol.* 4: 95–105.

Boldt, T. S. and Jacobsen, C. S. 2006. Different toxic effects of the sulfonylurea herbicides metsulfuron methyl, chlorsulfuron and thifensulfuron methyl on fluorescent pseudomonads isolated from an agricultural soil. *FEMS Microbiol. Lett.* 161(1): 29–35.

Bongers, T. and Ferris, H. 1999. Nematode community structure as a bioindicator in environmental monitoring. *Trends Ecol. Evol.* 14: 224–228.

Booth, L. H., Heppelthwaite, V. J., and O'Halloran, K. 2000a. Growth, development and fecundity of the earthworm *Aporrectodea caliginosa* after exposure to two organophosphates. *NZ Plant Protect.* 53: 221–225.

Booth, L. H., Hodge, S., and O'Halloran, K. 2000b. Use of cholinesterase in *Aporrectodea caliginosa* (Oligochaeta; Lumbricidae) to detect organophosphate contamination: Comparison of labora-tory tests, mesocosms and field trials. *Environ. Toxicol. Chem.* 19: 417–422.

Brown, A. W. A. 1978. *Ecology of Pesticides.* John Wiley & Sons: New York.

Brunninger, B., Viswanathan, R., and Beese, F. 1994. Terbuthylazine and carbofuran effects on growth and reproduction within three generations of *Eisenia andrei* (Oligochaeta). *Biol. Fertil. Soil* 18: 83–88.

Bustos-Obregón, E. and Goicochea, R. I. 2002. Pesticide soil contamination mainly affects earthworm male reproductive parameters. *Asian J. Androl.* 4(3): 195–199.

Casabé, N., Piola, L., Fuchs, J., Oneto, M. L., Pamparato, L., Basack, S., Giménez, R., Massaro, R., Papa, J. C., and Kesten, E. 2007. Ecotoxicological assessment of the effects of glyphosate and chlorpyrifos in an Argentine soya field. *J. Soils Sediments* 7: 232–239.

Chakravarty, P. and Sidhu, S. S. 1987. Effects of glyphosate, hexazinone and triclopyron in vitro growth of five species of ectomycorrhizal fungi. *Eur. J. Forest Pathol.* 17: 204–210.

Chen, L. J., Li, Q., and Liang, W. J. 2003. Effects of agrochemicals on nematode community structure in a soybean field. *Bull. Environ. Contam. Toxicol.* 71: 755–760.

Choo, L. P. D. and Baker, G. H. 1998. Influence of four commonly used pesticides on the survival, growth and reproduction of the earthworm Aporrectodea trapezoides. *Aust. J. Agric. Res.* 49: 1297–1303.

Christiansen, K. 1964. Bionomics of Collembola. *Annu. Rev. Entomol.* 9: 147–178.

Cole, L., Staddon, P. L., Sleep, D., and Bardgett, R. D. 2004. Soil animals influence microbial abun-dance, but not plant–microbial competition for soil organic nitrogen. *Funct. Ecol.* 18: 631–640.

Cortet, J., Gillon, D., Joffre, R., Ourcival, J. M., and Poinsot-Balaguer, N. 2002. Effects of pesticides on organic matter recycling and microarthropds in a maize field: Use and discussion of the litter bag methodology. *Eur. J. Soil Biol.* 38: 261–265.

Cragg, R. and Bardgett, R. D. 2001. How changes in soil faunal diversity and composition within a trophic group influence decomposition processes. *Soil Biol. Biochem.* 33: 2073–2081.

Culy, M. D. and Berry, E. C. 1995. Toxicity of soil-applied granular insecticides to earthworm popula-tions in cornfields. *Down to Earth* 50: 20–25.

Curtis, J. E. and Horne, P. A. 1995. Effect of chlorpyrifos and cypermethrin applications on nontarget invertebrates in a conservation-tillage crop. *J. Aust. Entomol. Soc.* 34: 229–231.

Darwin, C. R. 1881. The formation of vegetable mould, through the action of worms, with observa-tions on their habits. John Murray, London.

Decaëns, T., Jiménez, J. J., Gioia, C., Measey, G. J., and Lavelle, P. 2006. The values of soil animals for conservation biology. *Eur. J. Soil Biol.* 42: 523–538.

Dell'Omo, G., Turk, A., and Shore, R. F. 1999. Secondary poisoning in the common shrew (*Sorex araneus*) fed earthworms exposed to an organophosphate pesticide. *Environ. Toxicol. Chem.*. 18(2): 237–240.

Dindal, D. L. (ed.). 1980. Soil Biology as Related to Land Use Practices. *Proc. 7th Int. Soil Zool. Colloq.*, Syracuse, N.Y. EPA-560/13-80-038. EPA, Wash.

Drews, C. D., Zoren, M. J., and Callahan, C. A. 1987. Sublethal neurotoxic effects of the fungicide benomyl on earthworms (*Eisenia fetida*). *Pestic. Sci.* 19: 197–208.

Drews, C. D. and Callahan, C. A. 1988. Electrophysiological detection of sublethal neurotoxic effects in intact earthworms. In: Edwards, C. A. and Neuhauser, E. F. (eds), *Earthworms in Waste and Environmental Management*, pp. 355–356. SPB Academic Publ.: *The Netherlands*.

Edwards, C. A. 1973. *Environmental Pollution by Pesticides*. Plenum Press: NewYork.

Edwards, C. A. 1989. Impact of herbicides on soil microcosms. CRC Critical Reviews in *Plant Science*. 8: 221–298.

Edwards, C. A. 1998. *Earthworm Ecology*. St. Lucie Press: New York.

Edwards, C. A. and Bohlen, P. J. 1992. The effect of toxic chemicals on earthworm. *Rev. Environ. Contam. Toxicol.* 125: 23–29.

Edwards, C. A. and Fisher, S. W. 1991. The use of cholinesterase measurements in assessing the impacts of pesticides on terrestrial and aquatic invertebrates. In: Mineau, P. (ed.), *Cholinesterase Inhibiting Insecticides: Impact on Wildlife and the Environment*, pp. 255–276. Elsevier, The Hague: *The Netherlands*.

Edwards, C. A. and Fletcher, K. E. 1988. Interactions between earthworms and microorganisms in organic matter breakdown. *Agric. Ecosyst. Environ.* 20(3): 235–249.

Eijsackers, H., Beneke, P., Maboeta, M., Louw, J. P. E., and Reinecke, A. J. 2005. The implications of copper fungicide usage in vineyards for earthworm activity and resulting sustainable soil quality. *Ecotoxicol. Environ. Safe.* 62: 99–111.

Eisenhauer, N., et al. 2009. No interactive effects of pesticides and plant diversity on soil microbial biomass and respiration. *Appl. Soil Ecol.* 42(1): 31–36. http://dx.doi.org/10.1016/j.apsoil.2009.01.005 (abstract).

Ekelund, F. 1999. The impact of the fungicide fenpropimorph (Corbel®) on bacterivorous and fungivorous protozoa in soil. *J. Appl. Ecol.* 36: 233–243.

El-Banhaway, M. A., El-Ganzuri, M. A., and El-Akkad, M. M. 1982. Morphological changes of the golgi apparatus of the nerve and intestinal cells of the earthworm *Allolobophora caliginosa* living on insecticide contaminated soil. *Pak. J. Zool.* 18: 1–7.

El-Ghamry, A., et al. 2001. Combined effects of two sulfonylurea herbicides on soil microbial biomass and Nmineralization. *J. Environ. Sci.* 13(3): 1878–7320.

Endlweber, K., Schaedler, M., and Scheu, S. 2006. Effects of foliar and soil insecticide applications on the collembolan community of an early set-aside arable field. *Appl. Soil Ecol.* 31: 136–146.

Epstein, D. L., Zack, R. S., Brunner, J. F., Gut, L., and Brown, J. J. 2000. Effects of broad-spectrum insecticides on epigeal arthropods biodiversity in Pacific Northwest apple orchards. *J. Environ. Entomol.* 29(2): 340–348.

Eriksen-Hamel, N. S. and Whalen, J. K. 2007. Competitive interactions affect the growth of *Aporrectodea caliginosa* and *Lumbricus terrestris* (Oligochaeta: Lumbricidae) in single and mixed-species laboratory cultures. *Eur. J. Soil Biol.* 43: 142–150.

Estok, D., Freedman, B., and Boyle, D. 1989. Effects of the herbicides 2,4-D, glyphosate, hexazinone, and triclopyr on the growth of three species of ectomycorrhizal fungi. *Bull. Environ. Contam. Toxicol.* 42: 835–839.

Everts, J. W., Aukema, B., Hengeveld, R., and Koeman, J. H. 1989. Side effects of pesticides on ground-dwelling predatory arthropods in arable ecosystems. *Environ. Pollut.* 59(3): 203–225.

Fabian, M. and Petersen, H. 1994. Short-term effects of the insecticide dimethoate on activity and spatial distribution of a soil-inhabiting collembolan Folsomia fimetaria Linne' (Collembola: Isotomidae). *Pedobiologia* 38: 289–302.

Fabra, A., Duffard, R., and Evangelista de Duffard, A. 1997. Toxicity of 2,4-dichlorophenoxyacetic acid in pure culture. *Bull. Environ. Contam. Toxicol.* 59: 645–652.

Ferraro, D. O. and David, P. 2000. Pesticides in agroecosystems and their ecological effect on the structure and function of soil faunal population. *J. Rachel Carson Council* 2(2): 79–91.

Folker-Hansen, P., Krogh, P. H., and Holmstrup, M. 1996. Effect of dimethoate on body growth of representatives of the soil living mesofauna. *Ecotoxicol. Environ. Safe.* 33: 207–216.

Förster, B., Garcia, M., Francimari, O., and Römbke, J. 2006. Effects of carbendazim and lambda-cyhalothrin on soil invertebrates and leaf litter decomposition in semi-field and field tests under tropical conditions (Amazonia, Brazil). *Eur. J. Soil Biol.* 42: 171–179.

Fountain, M. T. and Hopkin, S. P. 2001. Continuous monitoring of *Folsomia candida* (Insecta: Collembola) in a metal exposure test. *Ecotoxicol. Environ. Safe.* 48: 275–286.

Frampton, G. K., Jänsch, S., Scott-Fordsman, J. J., Römbke, J., and Van den Brink, P. J. 2006. Effects of pesticides on soil invertebrates in laboratory studies: A review and analysis using species sensitivity distributions. *Environ. Toxicol. Chem.* 25: 2480–2489.

Frankenberger, W. T., Jr. and Tabatabai, M. A. 1991. Factors affecting L-asparaginase activity in soils. *Biol. Fert. Soils* 11: 1–5.

Freckman, D. W., and Baldwin, J. G. 1990. Nematoda. In: Dindal, D. I. (ed.), *Soil Biology Guide*, pp. 155–200. John Wiley and Sons: New York.

Gange, A. 2000. Arbuscular mycorrhizal fungi, Collembola and plant growth. *Trends Ecol. Evol.* 15: 369–372.

Gaw, S. K., Palmer, G., Kim, N. D., and Wilkins, A. L. 2003. Preliminary evidence that copper inhibits the degradation of DDT to DDE in pip and stonefruit orchard soils in the Auckland region, New Zealand. *Environ. Pollut.* 122: 1–5.

Gupta, V. V. S. R. and Yeates, G. W. 1997. Soil microfauna as bioindicators of soil health. In: Pankhurst, C. E., Doube, B. M. and Gupta, V. V. S. R. (eds), *Biological Indicators of Soil Health*, pp. 201–233. CAB International: Wallingford.

Hans, R. K., Gupta, R. C., and Beg, M. V. 1990. Toxicity assessment of four insecticides to earthworm *Pheretima posthuma. Bull. Environ. Contam. Toxicol.* 45: 358–364.

Haque, A. and Ebling, W. 1983. Toxicity determination of pesticides to earthworms in the soil substrate. *Zeit. Pflanzenschutz.* 90: 395–408.

Harwood, J. D., Sunderland, K. D., and Symondson, W. O. C., 2001. Living where the food is: Web location by linyphiid spiders in relation to prey availability in winter wheat. *J. Appl. Ecol.* 38: 88–99.

Harwood, J. D., Sunderland, K. D., and Symondson, W. O. C. 2003. Web-location by linyphiid spiders: Prey-specific aggregation and foraging strategies. *J. Anim. Ecol.* 72: 745–756.

Harwood, J. D., Sunderland, K. D., and Symondson, W. O. C. 2004. Prey selection by linyphiid spiders: Molecular tracking of the effects of alternative prey on rates of aphid consumption in the field. *Mol. Ecol.* 13: 3549–3560.

Helling, B., Reinecke, S. A., and Reinecke, A. J. 2000. Effects of the fungicide copper oxychloride on the growth and reproduction of *Eisenia fetida* (Oligochaeta). *Ecotoxicol. Environ. Safe.* 46(1): 108–116.

House, G. J., Worshman, A. D., Sheets, T. J., and Stinner, R. E. 1987. Herbicide effects on soil arthropod dynamics and wheat straw decomposition in a north Carolina no-tillage agroecosystem. *Biol. Fertil. Soils* 4: 109–114.

Iloba, B. N. and Ekrakene, T. 2010. Soil arthropods recovery rates from 5–10 cm depth within 5 months period following endosulfan (an Organochlorine pesticide) treatment in designated plots in Benin City. *Nigeria J. Entomol. Res.* 34 (2):1–6.

Joy, V. C. and Chakravorty, P. P. 1991. Impact of pesticides on nontarget microarthropod fauna in agricultural soil. *Ecotoxicol. Environ. Safe.* 22: 8–16.

Karaman, S., Ognjanovic, R., Stojanovic, M., and Nikolic, T. 1996. Pesticide effect on number and biomass dynamics of the earthworm *Dendrobaena veneta. Megadrilogic.* 6: 83–86.

Kelley, W. D. and South, D. B. 1978. In vitro effects of selected herbicides on growth and mycorrhizal fungi. *Weed Science Society of America Meeting*, pp. 38. Auburn University: Auburn, AL.

Kennel, W. 1972. Schadpilze als Objekte integrierte Pflanzens chutzmassnahmen im Obstbau. Z. Pflanzenkrankh. Pflanzenpath. Pflanzenschutz 79: 400–406.

Kennel, W. 1990. The role of the earthworm *Lumbricus terrestris* in integrated fruit production. *Acta Hortic.* 285: 149–156.

Lavelle, P. 1988. Earthworm activities and the soil system. *Biol. Fertil. Soils* 6: 237–251.

Lavelle, P., Bignell, D., and Lepage, M. 1997. Soil function in a changing world: The role of invertebrate ecosystem engineers. *Eur. J. Soil Biol.* 33: 159–193.

Liao, J. P., Lin, X. G., Cao, Z. H., Shi, Y. Q., and Wong, M. H. 2003. Interactions between arbuscular mycorrhizae and heavy metals under sand culture experiment. *Chemosphere* 50: 847–853.

Lord, K. A., Briggs, G. G., Neale, M. C., and Manlove, R. 1980. Uptake of pesticides from water and soil by earthworms. *Pestic. Sci.* 11(4): 401–408.

Lucija, K., Eržen, N. K., Hogerwerf, L., and van Gestel, C. A. M. 2008. Toxicity of abamectin and doramectin to soil invertebrates. *Environ. Pollut.* 151: 182–189.

Ma, W. C. 1984. Sub lethal toxic effects of copper on growth, reproduction and litter breakdown activity in the earthworm *Lumbricus rubellus*, with observations on the influence of temperature and soil pH. *Environ. Pollut.* 33: 207–219.

Ma, W.-A. and Bodt, J. A. 1993. Toxicity of the insecticide chlorpyrifos to six species of earthworms (Oligochaeta, Lumbricidae) in standardized soil tests. *Bull. Environ. Contam. Toxicol.* 50: 864–870.

Ma, W-C., Edelman, T., Van Beersum, I., and Jans, T. 1983. Uptake of cadmium, zinc, lead and copper by earthworms near a zinc smelting complex: Influence of pH and organic matter. *Bull. Environ. Contam. Toxicol.* 30: 424–427.

Maboeta, M. S., Reinecke, S. A., and Reinecke, A. J. 2003. Linking lysosomal biomarker and population responses in a field population of Aporrectodea caliginosa (Oligochaeta) exposed to the fungicide copper oxychloride. *Ecotoxicol. Environ. Saf.* 56(3): 280–287.

Mäder, P., et al. 2002. Soil fertility and biodiversity in organic farming. *Science* 296(5573): 1694–1697.

Marc, P., Canard, A., and Ysnel, F. 1999. Spiders (Araneae) useful for pest limitation and bioindication. *Agric. Ecosyst. Environ.* 74: 229–273.

Marcussen, B. M., Axelsen, J. A., and Toft, S. 1999. The value of two Collembola species as food for a linyphiid spider. *Entomol. Exp. Appl.* 92: 29–36.

Martens, D. A. and Bremner, J. M. 1993. Influence of herbicides on transformations of urea nitrogen in soil. *J. Environ. Sci. Health B.* 28: 377–395.

Martikainen, E., Haimi, J., and Ahtiainen, J. 1998. Effects of dimethoate and benomyl on soil organisms and soil processes – A microcosm study. *Appl. Soil Ecol.* 9: 381–387.

Maurya, N. and Chattoraj, A. N. 1994. Insecticidal interaction with a non-target soil organism earthworm. In: Prasad, D. and Gaur, H. S. (eds.), *Soil Environment and Pesticides*, 201–231. Venus Publishing House: New Delhi, India.

McEwan, F. L. and Stephenson, G. R. 1979. *The Use and Significance of Pesticides in the Environment.* New York: John Wiley.

Merrington, G., Rogers, S. L., and Van Zwieten, L. 2002. The potential impact of long-term copper fungicide usage on soil microbial biomass and microbial activity in an avocado orchard. *Aust. J. Soil Res.* 40: 749–759.

Moorman, T. B. 1989. A review of pesticide effects on microorganisms and microbial processes related to soil fertility. *J. Prod. Agric.* 2(1): 14–23.

Moorman, T. B., et al. 1992. Production of hydrobenzoic acids by Bradyrhizobium japonicum strains after treatment with glyphosate. *J. Agric. Food Chem.* 40: 289–293.

Mosleh, Y. Y., Paris-Palacios, S., Couderchet, M., and Vernet, G. 2002. Biological effects of two insecticides on earthworms (*Lumbricus terrestris* L.) under laboratory conditions. *Mededelingen Rijksuniversiteit te Gent. Fakulteit van de Landbouwkundige en Toegepaste Biologische Wetenschappen.* 67(2): 59–68.

Mosleh, Y. Y., Ismail, S. M. M., Ahmed, M. T., and Ahmed, Y. M. 2003a. Comparative toxicity and biochemical responses of certain pesticides to the mature earthworm *Aporrectodea caliginosa* under laboratory conditions. *Environ. Toxicol.* (5): 338–346.

Mosleh,Y. Y, Paris-Palacios, S., Couderchet, M., and Vernet, G. 2003b. Acute and sublethal effects of two insecticides on earthworms (*Lumbricus terrestris* L.) under laboratory conditions. *Environ. Toxicol.* 18(1): 1–8.

Neher, D. A. 2001. Role of nematodes in soil health and their use as indicators. *J. Nematol.* 33: 161–168.

O'Halloran, K., Booth, L. H., Hodge, S., Thomsen, S., and Wratten, S. D., 1999. Biomarker responses of the earthworm *Aporrectodea caliginosa* to organophosphates: Hierarchical tests. *Pedobiologia* 43: 646–651.

Oomen, P. A. 1998. Risk assessment and risk management of pesticide effects on non-target arthropods in Europe. *Proceedings of the Brighton Crop Protection Conference, Pests and Diseases*, pp. 591–598. BCPC: Farnham (GB).

Parmelee, R. W., Beare, M. H., Cheng, W., Hendrix, P. F., Rider, S. J., Crossley, D. A., and Coleman, D. C. 1990. Earthworms and enchytraeids in conventional and no tillage agroecosystems: A biocide approach to assess their role in organic matter breakdown. *Biol. Fertil. Soils* 10: 1–10

Pekár, S. 2002. Susceptibility of the spider Theridion impressum to 17 pesticides. *J. Pest Sci.* 75: 51.

Pell, M., Stenberg, B., and Torstensson, L. 1998. Potential denitrification and nitrification tests for evaluation of pesticide effects in soil. *Ambio* 27: 24–28.

Pen-Mouratov, S. and Steinberger, Y. 2005. Responses of nematode community structure to pesticide treatments in an arid ecosystem of the Negev Desert. *Nematology* 7: 179–191.

Petersen, H. 2002. General aspects of collembolan ecology at the turn of the millennium. *Pedobiologia.* 46: 246–260.

Petz, W. and Foissner, W. 1989. The effects of mancozeb and lindane on the soil microfauna of a spruce forest: A field study using a completely randomized block design. *Biol. Fertil. Soils* 7: 225–231.

Pillai, M. K. 1986. Pesticide pollution of soil, water and air in Delhi area, India. *Sci. Total Environ.* 55: 321–327.

Potter, D. A., Powell, A. L., and Shith, K. M. S. 1990. Degradation of turfgrass thatch by earthworms (Oligochaeta: Lumbricidae) and other soil invertebrates. *J. Econ. Entomol.* 83: 203–211.

Rajgaopal, D., Kumar, P., and Gowda, G. 1990. Effect of newer granular insecticides on soil fauna in groundnut cropping system. *J. Soil Biol. Ecol.* 10(1): 36–40.

Rebecchi, L., Sabatini, M. A., Cappi, C., Grazioso, P., Vicari, A., Dinelli, G., and Bertolini, D. 2000. Effects of a sulphonylurea herbicide on soil microarthropods. *Biol. Fertil. Soils.* 30: 312–317.

Reddy, M. V. 1989. Soil pollution and soil arthropod population. In: Mishra, P. C. (ed.), *Soil Pollution and Soil Organisms*, pp. 209–253. Ashish Publishing House: New Delhi.

Reddy, M. V. 1999. Soil management and beneficial soil meso and macrofauna with particular reference to semi-arid tropical alfisol. In: Reddy, V. J. (ed.), *Management of Tropical Agroecosystems and the Beneficial Soil Biota*, pp. 223–271. Science Publishers, Inc.: USA.

Reddy, N. C. and Rao, V. 2008. Biological response of earthworm, *Eisenia foetida* (savigny) to an organophosphorous pesticide, profenofos. *J. Ecotoxicol. Environ. Safe.* 71(2): 574–582.

Reddy, D. R., Purshotam, K. R., and Ramamurthi, R. 1983. Influence of fenitrothion on carbohydrate constituents and some enzyme levels in the earthworm *Lampito mauritii*. *Experentia* 30: 1109–1110.

Reddy, V. S., Reddy, M. V., Lee, K. K., Rao, K. P. C., and Srinivas, S. T. 1996. Response of some soil meso- and macro-fauna populations to soil management during crop and fallow periods on a semi-arid tropical alfisol (India). *Eur. J. Soil Biol.* 32: 123–129.

Reinecke, A. J., et al. 2008. The effects of organic and conventional management practices on feeding activity of soil organisms in vineyards. *African Zool.* 43(1): 66–74.

Reinecke, S. A. and Reinecke, A. J. 1999. Lysosomal response of earthworm coelomocytes induced by longterm experimental exposure to heavymetals. *Pedobiologia* 43(6): 585–593.

Reinecke, S. A. and Reinecke, A. J. 2007. The impact of organophosphate pesticides in orchards on earthworms in the Western Cape, South Africa. *Ecotoxicol. Environ. Saf.* 66: 244–251.

Reinecke, S. A., Reinecke, A. J., and Froneman, M. L. 1995. The effects of dieldrin on the sperm ultrastructure of the earthworm *Eudrilus eugeniae* (Oligochaeta). *Environ. Toxicol. Chem.* 14: 961–965.

Roberts, B. L. and Dorough, H. W. 1984. Relative toxicities of chemicals to the earthworms *Bisenia foetida*. *Environ. Toxicol. Biochem.* 3: 67–78.

Rodriguez-Castellanos, L. and Sanchez-Hernandez, J. C. 2007. Earthworm biomarkers of pesticide contamination: Current status and perspectives. *J. Pestic. Sci.* 32: 360–371.

Rogers, M. E. and Potter, D. A. 2003. Effect of spring imidacloprid application for white grub control on parasitism of Japanese beetle (Coleoptera: Scarabaeidae) by Tiphia vernalis (Hymenoptera: Tiphiidae). *J. Econ. Entomol.* 96(5): 1412–1419.

Salminen, J. and Haimi, J. 2001. Life history and spatial distribution of the enchytraeid worm *Cognettia sphagnetorum* (Oligochaeta) in metal-polluted soil: Below-ground sink-source population dynamics? *Environ. Toxicol. Chem.* 20: 1993–1999.

Sanchez-Hernandez, J. C. 2006. Earthworm biomarkers in ecological risk assessment. *Rev. Environ. Contam. Toxicol.* 188: 85–126.

Sandoval, M. C., Veiga, M., Hinton, J., and Klein, B. 2001. Review of biological indicators for metal mining effluents: A proposed protocol using earthworms. *Proceedings of the 25th Annual British Columbia Reclamation Symposium*, pp. 67–79.

Santos, A. and Flores, M. 1995. Effects of glyphosate on nitrogen fixation of free-living heterotrophic bacteria. *Lett. Appl. Microbiol.* 20: 349–352.

Scheu, S. and Folger, M. 2004. Single and mixed diets in Collembola: Effects on reproduction and stable isotope fractionation. *Funct. Ecol.* 18: 94–102.

Scheu, S., Theenhaus, A., and Jones, T. H. 1999. Links between the detritivore and the herbivore system: Effects of earthworms and Collembola on plant growth and aphid development. *Oecologia* 199: 541–551.

Singh, J. B. and Singh, S. 1989. Effect of 2,4-dichlorophenoxyacetic acid and maleic hydrazide on growth of bluegreen algae (cyanobacteria) Anabaena doliolum and Anacystis nidulans. *Sci. Cult.* 55: 459–460.

Smith, M. D., Hartnett, D. C., and Rice, C. W. 2000. Effects of long-term fungicide applications on microbial properties in tallgrass prairie soil. *Soil Biol. Biochem.* 32: 935–946.

Sohlenius, B. 1980. Abundance, biomass and contribution to energy flow by soil nematodes in terrestrial ecosystems. *J. Nematol.* 11: 86–196.

Sorour, J. and Larink, O. 2001. Toxic effects of benomyl on the ultra structure during spermatogenesis of the earthworm *Eisenia fetida. Ecotoxicol. Environ. Safe.* 50(3): 180–188.

Spurgeon, D. J., Weeks, J. M., and van Gestel, C. A. M. 2003. A summary of eleven years progress in earthworm ecotoxicology. *Pedobiologia* 47: 588–606.

Stenersen, J. 1979a. Action of pesticides on earthworm's part 1: The toxicity of cholinesterase inhibiting insecticides to earthworms as evaluated by laboratory tests. *Pestic. Sci.* 10: 66–74.

Stenersen, J. 1979b. Action of pesticides on earthworm's part 3: Inhibition and reaction of cholinesterase in *Eisenia foetida* after treatment with cholinesterase inhibiting insecticides. *Pestic. Sci.* 10: 113–122.

Stevenson, K., Anderson, R. V., and Vigue, G. 2002. The density and diversity of soil invertebrates in conventional and pesticide free corn. *Trans. Ill. State Acad. Sci.* 95(1): 1–9.

Thompson, A. R. and Edwards, C. A. 1974. Effects of pesticides on non-target invertebrates in freshwater and soil. In: Guenzi, W. D. (ed.), *Pesticides in Soil and Water*, pp. 341–386, Soil Sci. Soc. of America Inc.: Madison, WI.

Tözüm-Çalgan, S. R. D., and Sivaci-Güner, S. 1993. Effects of 2,4-D and methylparathion on growth and nitrogen fixation in cyanobacterium, Gloeocapsa. *Int. J. Environ. Stud.* 23: 307–311.

Van Gestel, C. A. M., Dirven-Van Breemen, E. M., and Baerselman, R. 1992. Comparison of sub lethal and lethal criteria for nine different chemicals in standardized toxicity tests using the earthworm *Eisenia andrei. Ecotoxicol. Environ. Safe.* 23(2): 206–220.

Van Gestel, C. and Weeks, J. 2004. Recommendations of the 3rd International Workshop on Earthworm Ecotoxicology, Aarhus, Denmark, August 2001. *Ecotoxicol. Environ. Saf.* 57: 100–105.

Van Rhee, J. A. 1967. Development of earthworm populations in orchard soils. In: Graff, O. and Satchell, J. (eds), *Progress in Soil Biology*, pp. 360–371. North Holland Publishing Company: Amsterdam, Netherlands.

Van Zwieten, L., Rust, J., Kingston, T., Merrington, G., and Morris, S. 2004. Influence of copper fungicide residues on occurrence of earthworms in avocado orchard soils. *Sci. Total Environ.* 329: 29–41.

Varadaraj, G. 1986. Effect of sublethal concentration of Sumithion on carbohydrate metabolism in the earthworm *Octochaetona pattoni*. In: Dash, M. C., Senapati, B. K., and Mishra, P. C. (eds), *Verms and Vermicomposting*, pp. 93–96. Five Star Printing Press: Burla, India.

Veeresh, G. K. and Rajagopal, D. 1989. Pesticides and soil fauna. In: Veeresh, G. K. (ed.), *Applied Soil Biology and Ecology*, pp. 317–343, Oxford and IBH Publishing Co. Pvt. Ltd.

Verhoef, H. A. and Van Gestel, C. A. M. 1995. Methods to assess the effects of chemicals on soils. In: Linthurst, R. A., Bourdeau, P. and Tardiff, R. G. (eds), *Methods to Assess the Effects of Chemicals on Ecosystems*, pp. 223–257. John Wiley and Sons Ltd.: New York.

Verma, A. and Pillai, M. K. K. 1991. Bioavailability of soil-bound residues of DDT and HCH to earthworms. *Current-Science* 61(12): 840–843.

Wada, S. and Toyota, K. 2008. Effect of three organophosphorous nematicides on non-target nematodes and soil microbial community. *Microbes Environ.* 23: 331–336.

Wang, Y., Crocker, R. L., Wilson, L. T., Smart, G., Wei, X., Nailon W. T. Jr., and Cobb, P. P. 2001. Effect of nematode and fungal treatments on nontarget turfgrass-inhabiting arthropod and nematode populations. *Environ. Entomol.* 30:196–203.

Wardle, D. A., Yeates, G. W., Watson, R. N., and Nicholson, K. S. 1995. The detritus food web and the diversity of soil fauna as indicators of disturbance regimes in agroecosystems. *Plant Soil* 170: 35–43.

Wheatley, G. A. and Hardman, J. A. 1968. Organochlorine insecticide residues in earthworms from arable soils. *J. Sci. Food Agric.* 19(4): 219–225.

Widenfalk, A., et al. 2008. Effects of pesticides on community composition and activity of sediment microbes – Responses at various levels of microbial community organization. *Environ. Pollut.* 152(3): 576–584.

Wilson, J. D., Morris, A. J., Arroyo, B. E., Clark, S. C., and Bradbury, R. B. 1999. A review of the abundance and diversity of invertebrate and plant foods of granivorous birds in northern Europe in relation to agricultural change. *Agric. Ecosyst. Environ.* 75: 13–30.

Wolters, V. 1985. Untersuchungen zur Habitatbildung und Nahrungsbiologie der Springschwänze Collembola eines Laubwaldes unter besonderer Berücksichtigung ihrer Funktion in der Zersetzerkette. Ph.D. Dissertation, Universität Göttingen.

Wood, W. B. 1988. The nematode *Caenorhabditis elegans*. Cold Spring Harbor Laboratory: New York.

Xiao, N., Jing, B., Ge, F., and Liu, X. 2006. The fate of herbicide acetochlor and its toxicity to *Eisenia fetida* under laboratory conditions. *Chemosphere* 62(8): 1366–1373.

Yadav, D. V., Mittal, P. K., Agarwal, H. C., and Pillai, M. K. 1981. Organochlorine insecticide residues in soil and earthworms in the Delhi area, India, August–October, 1974. *Pestic. Monit. J.* 15(2): 80–85.

Yardim, E. N. and Edwards, C. A. 1998. The effects of chemical pest, disease and weed management practices on the trophic structure of nematode populations in tomato agroecosystems. *Appl. Soil Ecol.* 7: 137–147.

Yasmin, S. and D'Souza, D. 2007. Effect of pesticides on the reproductive output of *Eisenia fetida*,. *Bull. Environ. Contam. Toxicol.* 79(5): 529–532.

Yeates, G. W. 2003. Nematodes as soil indicators: Functional and biodiversity aspects. *Biol. Fertil. Soils* 37: 199–210.

Yeates, G. W., Wardle, D. A., and Waston, R. N. 1999. Responses of soil nematode populations, community structure, diversity and temporal variability to agricultural intensification over a seven-year period. *Soil Biol. Biochem.* 31: 1721–1733.

Zhang, S. P., Duan, C. Q., Chen, H., Fu, Y. H., Wang, X. H., and Yu, Z. F. 2007. Toxicity assessment for chlorpyrifos-contaminated soil with three different earthworm test methods. *J. Environ. Sci.* 19(7): 854–858.

Zoren, M. J., Heppner, T. J., and Drews, C. D. 1986. Terratogenic effects of the fungicide benomyl on posterior segmental regeneration in the earthworm *Eisenia fetida*. *Pestic. Sci.* 17: 641–652.

13

Pesticide Residues in Fish

Prem Dureja and Hamir Singh Rathore

CONTENTS

13.1 Introduction

The synthetic pesticides had a glorious past and did a yeoman's service to mankind in increasing agricultural production and improving food and health quality. Before World War II, pesticides in common use were predominantly inorganic materials, such as sulfur, lead, copper, arsenic, boron, and mercury, as well as botanical ones, such as nicotine, pyrethrum, and rotenone. The advent of organic insecticides really began with the discovery of DDT's insecticidal properties (1939–1942) by Paul Muller. The use of DDT in controlling mosquito vectors saved millions of lives in the Indian subcontinent by preventing malaria. Its astonishing efficacy led to the development of a variety of other synthetic organic pesticides such as aldrin, dieldrin, toxaphene, and endosulfan followed by organophosphates (malathion, diazinon, and chlorpyrifos), carbamates (carbaryl, carbofuran, and Baygon), and synthetic pyrethroids, the second most significant group after organophosphorus compounds.

13.1.1 Pesticides and the Environment

If the credits of pesticides include enhanced economic potential in terms of increased production of food and fiber and amelioration of vector-borne diseases, then their debits have

TABLE 13.1

Common Side Effects of Pesticides

Environmental Component	Side Effects
Abiotic	Residues in soil, water, and air
Plants	Presence of residues, phytotoxicity, vegetation changes (due to the use of herbicides)
Animals, birds, insects, etc.	Residues, physiological effects, mortality in certain wild life species, mortality of beneficial predators and parasites, insect population changes (outbreak of secondary pests), genetic disorders
Man	Biochemical changes, residues in tissues and organs, effect of occupational exposure, mortality and deformations
Food	Presence of the residue

resulted in serious health implications to man and his environment. There is now overwhelming evidence that some of these chemicals do pose potential risk to humans and other life forms and unwanted side effects to the environment (Table 13.1).

No segment of the population is completely protected against exposure to pesticides, and the potentially serious health effects, though a disproportionate burden, are shouldered by the people of developing countries and by the high-risk groups in each country.

Ideally, a pesticide must be lethal to the targeted pests but not to the nontarget species, including man. Unfortunately, this is not the case. So the controversy of use and abuse of pesticides has surfaced. The rampant use of these chemicals, under the adage, "if little is good, a lot more will be better," has played havoc with human and other life forms.

13.1.2 Adverse Effects of Pesticides

Carson in 1962 warned that organochlorine (OC) compounds could pollute the tissues of virtually every life form on the earth, the air, the lakes, and the oceans, including the fish that live in them and the birds that feed on the fish. Later, the US National Academy of Sciences stated that the DDT metabolite DDE causes eggshell thinning and that the bald eagle population in the United States declined primarily because of exposure to DDT and its metabolites. Organochlorine pesticides (OCPs), which act in a similar way as natural hormones, may lead to disorders of the endocrine system in animals and humans (endocrine disruptors) (Trudeau and Tyler 2007) and reduction in reproduction of animals and humans (Hart et al. 2003; Henny et al. 2008). Gregoraszczuk et al. (2008) reported a stimulatory effect of environmental organic contaminants extracted from the liver oil of Atlantic cod on testosterone, 17β-estradiol, and progesterone secretion by porcine ovarian follicular cells. Golec et al. (2003) observed morphological abnormalities in sperm as well as decreases in sperm count, percentage of viable sperm, and fecundity rate in male humans exposed to pesticides. Endocrine disruptors are being increasingly implicated in infertility, menstrual irregularities, spontaneous abortions, birth defects, endometriosis, and breast cancer (Bhatt 2000). Many studies indicate that OCPs may damage the immune system and cause defects in the male reproductive system, low sperm concentration, and increased incidence of cancer (Porta et al. 1999; Taylor and Harrison 1999; Abell et al. 2000).

TABLE 13.2

Some Important Pesticides and the Nature of Impact

Pesticide	Food Contaminated	Impact
Organochlorines		
Aldrin	Fish, milk, and fat	Accumulation in human tissues
DDT	Fish, milk, fat, and meat	Accumulation in human tissues, low toxicity, induction of liver enzymes
Lindane	Fruits, vegetables, cereals, milk, and fat	Accumulation in human tissues, toxic to bone marrow, probably carcinogenic
Heptachlor	Fish, milk, and fat	Accumulation in human tissues, toxic to bone marrow, probably carcinogenic
Organophosphorus		
Malathion	Fruits	Acute toxicity, neurotoxicity
Parathion	Fruits and vegetables	Acute toxicity, neurotoxicity

The excessive toxic residues of pesticides may accumulate in living organisms and cause acute or chronic toxicity. These residues may affect the soil fertility and lower the drought tolerance of the crops. The nature of pesticide impact through the toxic residues is indicated in Table 13.2.

Generally, the type of pesticide used varies with geographical area. For example, humid countries such as mid-European and the Scandinavian countries use mainly herbicides for weed control problems, while Mediterranean countries use pesticides for agricultural fungus control. Hence, different countries are facing different environmental problems. Overall, during pesticide application, less than 5% of the pesticide applied can reach its intended target because of losses from volatilization or washout, which further reinforces the need for assessing the environmental impact of these compounds. To do this, it is important to know their chemical properties, environmental fate, and impact on nontarget species in order to protect wildlife and humans from pesticide risks.

13.2 Movement of Pesticides in Water

Pesticides enter the water from various sources such as runoff from agricultural lands, direct deposition from spray operations, industrial effluents, spraying of cattle, dust, and rainfall. In water, the residues and their degradation products are distributed between the truly dissolved form and those incorporated into sediments, benthic invertebrates, aquatic plants, plankton, aquatic invertebrates, suspended detritus, and fish. Pesticides can leave aquatic systems by volatilization or codistillation, as residues in fish, which are eaten by humans, birds, and animals, or by degradation, burial in sediments, or overflow. Some of the pesticides also end up in groundwater systems by leaching down through the soil. They easily find their way from the soil surface to the atmosphere, due to their high vapor pressures, and migrate with the air currents to distant areas and then later fall back to land as precipitation, into streams, lakes, and even drinking water. That is why pollution by pesticides occurs even in remote areas, where they have not been used for decades (Wania and Mackay 1996). Figure 13.1 shows the dynamic movement of pesticides in the aquatic environment.

Investigation of the atmosphere over Europe indicated that the highest concentrations of OCPs are observed over the region where OCPs were produced and used in high doses and over urban areas. A higher concentration of lindane was found over the southern and

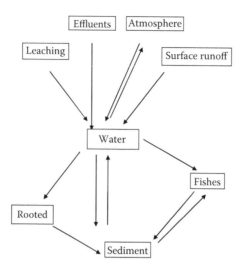

FIGURE 13.1
Dynamic movement of pesticides in the aquatic environment.

eastern parts of Europe (Jaward et al. 2004). As a result of all these pathways, pesticides are widely found in rivers. OCPs cause major damage to aquatic life (Van der Werf 1996). Pimentel et al. (1993) estimated that from 1977 to 1987 in the United States, 6–14 million fish per year were killed by pesticides.

13.2.1 Hazards of Pesticide Residues in Water

The two main hazards of pesticide residues in water are the deposition of chemicals in the bodies of fish, which are effective filters of suspended particulate matter, and in the bodies of aquatic organisms that form the food for the fish, and as a result of both, more chemical gets deposited in the fish than is found free in water. The accumulation of pesticides in fish is of great importance because they form food for humans and birds. OCPs are characterized by high persistence, low polarity, low aqueous solubility, and high lipid solubility (lipophilicity), and as a result, they have a potential to bioaccumulate in fish and in the food chain, posing a great threat to human health and the environment globally (Lars 2000). For example, omega-3 fatty acids offer excellent health benefits, but they come from the fatty tissues of fish, which are also the best part of the fish to store pesticides. Problems occur when people who want to improve their health naturally rely on omega-3 fatty acids but unknowingly ingest high amounts of pesticides. It is known that OCPs in water are taken up very rapidly by living organisms. Holden (1962) and Marth (1965) demonstrated that fish can take up 80%–90% of the DDT from water through gills. It is therefore important to determine the levels of OCP residues in fish samples, considering the adverse health effects OCPs cause in humans.

13.3 Fish Exposed to Pesticides

Fish and aquatic animals are exposed to pesticides in three primary ways: (1) dermally, direct absorption through the skin by swimming in pesticide-contaminated waters, (2) breathing, by direct uptake of pesticides through the gills during respiration, and (3) orally, by drinking pesticide-contaminated water or feeding on pesticide-contaminated prey. Poisoning by consuming another animal that has been poisoned by a pesticide is termed "secondary poisoning." For example, fish feeding on dying insects poisoned by insecticides may themselves be killed if the insects they consume contain large quantities of pesticides or their toxic by-products.

Exposure of fish and other aquatic animals to a pesticide depends on its biological availability (bioavailability), bioconcentration, biomagnification, and persistence in the environment.

13.3.1 Bioavailability

Bioavailability refers to the amount of pesticide in the environment available to fish and wildlife. Some pesticides rapidly undergo degradation after application. Some bind tightly to soil particles suspended in the water column or to stream bottoms, thereby reducing their availability. Some are quickly diluted in water or rapidly volatize into the air and are less available to aquatic life. Bioconcentration is the accumulation of pesticides in animal tissue at levels greater than those in the water or soil to which they were applied. Some fish may concentrate certain pesticides in their body tissues and organs (especially fats) at levels 10 million times greater than in the water.

13.3.2 Biomagnification

Biomagnification is the accumulation of pesticides at each successive level of the food chain. Some pesticides can bioaccumulate (build up) in the food chain. For example, if a pesticide is present in small amounts in water, it can be absorbed by water plants that are, in turn, eaten by insects and minnows. These organisms also become contaminated. At each step in the food chain, the concentration of the pesticide increases. Fish, such as bass or trout, repeatedly consume contaminated animals, and they bioconcentrate high levels in their body fat. Fish can pass these poisons on to humans (Figure 13.2).

In other words, biomagnification occurs when the toxic burden of a large number of organisms at a lower trophic level is accumulated and concentrated by a predator in a higher trophic level. Phytoplankton and bacteria in aquatic ecosystems, for instance, take up heavy metals or toxic organic molecules from water or sediments. Their predators—zooplankton and small fish—collect and retain the toxins from many prey organisms, building up higher concentrations of toxins. The top carnivores in the food chain—game fish, fish-eating birds, and humans—can accumulate such high toxin levels that they suffer adverse health effects. One of the first known examples of bioaccumulation and

DDT in water   DDT in zooplankton   DDT in small fish   DDT in large fish   DDT in fish-eating birds

FIGURE 13.2
Bioaccumulation and biomagnification.

biomagnification was DDT, which accumulated through food chains so that by the 1960s it was shown to be interfering with the reproduction of peregrine falcons, brown pelicans, and other predatory birds at the top of their food chains.

> Organisms lower in the food chain take up and store toxins from the environment. They are eaten up by larger predators, which are eaten, in turn, by even larger predators. The highest members of the food chain can accumulate very high levels of the toxin (Cunningham and Saigo 1999).

13.3.3 Bioaccumulation

The process of bioaccumulation starts when pesticides applied to the agricultural land run off during storms into rivers, streams, and eventually to the oceans. The pesticides become part of the water column, and fish ingest the pesticides, usually through their gills, although sometimes through their fish scales. The pesticides enter their organs and fat tissue and are sequestered there. More and more pesticides are ingested and stored in organs and tissues. These pesticides accumulate up the food chain as big fish eat little fish and eventually as humans eat the fish.

Bioaccumulation is affected by how much of the pollutant gets into each organism, the effectiveness of the elimination process of each organism, organism metabolic processes, organism fat content, and the solubility of the pollutant. To get an accurate picture of threats to humans, this type of information for each organism within the food chain must be obtained and added. The EPA has identified tolerance levels for pollutants (EPA 2006). Bioaccumulation values help to determine tolerance levels in which no harm to human health will take place. Fish advisories for pregnant women and children are more stringent because they have different metabolic processes occurring; the same stringency applies to people with serious illnesses.

Organochlorines, even at very low concentrations, interfere with reproduction, growth, and development. This is one reason why there are many federal agencies that study and regulate pesticides including the EPA, Food and Drug Administration (FDA), and National Institutes of Health.

The FDA produces a total diet study that evaluates the pesticide residues found in fish, meats, dairy, fruits, and vegetables as prepared (cooked) in an average US diet (Egan 2002). In 2007, 15.6% of the total fish and shellfish sample contained pesticide residues but no fish or shellfish had levels in violation of federal standards set to protect human health. The total diet study showed that bioaccumulation of pesticides in cooked fish is not problematic to adults eating an average US diet, as long as no serious health issues are apparent.

13.4 Monitoring Pesticide Residues in Fish and Fish Products

As already stated, fish accumulate chemicals from the environment because contaminants wash into lakes and rivers and eventually reach the ocean. Some get degraded slowly and can spread around the world. Heavy metals such as mercury, pesticide residues, and other compounds move up the food chain from plankton to small fish and larger fish that eat smaller ones. For this reason, nearly all fish have traces of contaminants. However, only a few species and sizes of fish contain relatively large amounts. These include larger older fish, some carnivorous species, and fish from polluted waters.

The concentration of pesticides in fish is the consequence of the pollution of the environment in which they live. In order to ensure food safety, opening of the market to import, and consumer awareness to pesticides or environmental pollutants, a number of studies have been carried out to monitor the amount of pesticides present in fish from various sources and countries.

Fish inhabiting coastal areas have often been proposed as sentinels for monitoring pollutants of land-based origin, because they may concentrate indicative hydrophobic compounds in their tissues, directly from water through respiration and also through the diet. The red mullet (*Mullus barbatus*) is a benthic and territorial fish of commercial interest in the NW Mediterranean region, which has been used in several studies of coastal pollution monitoring (Porte and Albaigés 2002; UNEP 1997). Fish were collected from 10 stations along the Spanish, French, and Corsica and Sardinia coasts, to grant a wide coverage of the area encompassing large pollution gradients. The major pesticides identified in fish tissues were p,p'-DDT, their metabolites p,p'-DDD and p,p'-DDE, and hexachlorobenzene (HCB). The highest concentrations (up to 230 ng/g ww) were found in the area of influence of Barcelona (stations 2–4), whereas the lowest (<1 ng/g ww) were found in Corsica and Sardinia Islands (stations 9 and 10). Tissue concentration ranges may then be considered as representative of pristine, moderately contaminated, and highly contaminated areas, and they are consistent with those found in sediments (Tolosa et al. 1995), supporting the close relationship of the red mullet uptake with the benthic environment.

Regional surveys on the pesticide residues (organochlorines) in fish and fish products were conducted in Southeast Asia (SEAFDEC member countries). A total of eight countries, namely Cambodia, Indonesia, Malaysia, Myanmar, Philippines, Singapore, Thailand, and Vietnam, participated in JTF II (Japanese Trust Fund) project on research and analysis of chemical residues and contamination in fish and fish products. The participating regional laboratories conducted a survey each in their respective countries. Fish and fish products that are of economic and social importance to the participating countries were targeted. Samples of fish and fish products were sent to their local or collaborating laboratories for analysis. Dried products from both marine and freshwater species as well as fish and fish products that are dried or from inland and estuarine waters and rivers were also targeted. A total of 35 fish and fish products were surveyed. The results of the survey were submitted in the database of the Fish and Fish Products Safety Information Network. From the results of the surveys conducted on pesticide residues, only a few samples have exceeded national or international regulatory limits. It can therefore be concluded that the fish and fish products from these countries are safe for both domestic consumption and export (Japanese Trust Fund II Project on Research and Analysis of Chemical Residues and Contamination in Fish and Fish Products 2004–2008).

Fish are a valuable source of high-grade proteins, and they occupy an important position in the socioeconomic condition of the South Asian countries by providing the population not only the nutritious food but also income and employment opportunities. India ranks second in the world in inland fish production. West Bengal is the largest producer of inland fish and consumes 11.67 lakh tons of it annually, the highest among all the states in India. The rivers in West Bengal are the major source of fish. At the same time, these rivers are the sink of the effluents generated from the industries and the usage of households (Chowdhury et al. 1994). As a consequence, there is a likelihood of bioaccumulation of organic and inorganic pollutants such as heavy metals and pesticides in fish and other aquatic organisms, and without proper monitoring program for the safety evaluation to consumers, the situation may rise to an alarming level (Amorastt et al. 1983).

The National Institute of Oceanography, India, conducted several studies on chlorinated hydrocarbons in marine biota from the Indian region. Zooplankton and bottom-feeding fish (four species) from the Bay of Bengal were studied in the vicinity of the Coleroon River mouth and from the northern Bay of Bengal. Total DDT concentrations ranging from 1.31 to 115.9 ng/g w/w in different fish species and 4.0 to 1587.8 ng/g w/wt in zooplankton were found. Aldrin levels were 0.32 to 4.23 ng/g in the fish tissue and ND to 0.78 ng/g in zooplankton (Shailaja and Singhal 1994).

Levels of total DDT in marine fish from different landing centers along the Tamil Nadu and Pondicherry coasts on the southeast coast of India were from ND to 2.38 ng/g (Rajendran et al. 1992). Out of four species of marine fish, the concentration of pesticide residues was highest in black pomfret fish (*Parastromateus niger*), containing 0.0003 mg/kg heptachlor, 0.09 mg/kg aldrin, 0.001 mg/kg dieldrin, 0.004 mg/kg endrin, 0.003 mg/kg p,p'-DDE, 0.002 mg/kg p,p'-DDD, 0.007 mg/kg o,p'-DDT, and 0.042 mg/kg p,p'-DDT (Radhakrishnan and Antony 1989). The concentration of pesticides increases in marine biota in the monsoon season (Shailaja and Nair 1997; Shailaja and Sengupta 1989).

In other studies, DDT was reported from 3 to 128 ng/kg in marine biota (Nigam et al. 1998). DDT and its metabolites, HCH isomers, chlordane compounds, and HCB were determined in river dolphin blubber and prey fish collected during 1993–1996 from the river, Ganges, in India (Senthil Kumar et al. 1999). The DDT metabolites in blubber (21,000–64,000 ng/g ww) were the predominant compounds found in dolphin tissues and fish that constitute the diet of dolphins. The levels of chlordane compounds and HCB in dolphin blubber were 860–1,900, 45–240, and 7.7–19 ng/g ww, respectively. The concentrations in fish were 3–13 ng/g ww for chlorinated compounds, respectively. Compared to the levels of total DDT, chlordane, and HCB observed during 1988–1992 (Kannan et al. 1994), the levels in 1994–1996 were doubled for all the compounds except HCB (Senthil Kumar et al. 1999).

An investigation was conducted from 2001 to 2005 for determining the residual concentration of five pesticides, namely total-HCH, total-DDT, total-endosulfan, dimethoate, and malathion in fish samples collected from various points of the river Ganga. It was found that total-HCH concentration remains above the MRL values for maximum number of times in comparison to four other pesticides. The pesticide contamination to fish may be due to indiscriminate discharge of polluted and untreated sewage sludge to the river. The pesticide contents in some places are alarming (Md. Wasim Aktar et al. 2009).

An analysis of four fish species from a freshwater lake in Jaipur, India, revealed that the residues of aldrin and DDT were low in muscle tissue compared to other tissues (Bakre et al. 1990). The levels of total DDT in the brain averaged 0.82–12.84 µg/g (Bakre et al. 1990).

In 1999, freshwater fish were analyzed for pesticide residues in three ICAR (Indian Council of Agricultural Research) centers in India. Out of 36 samples collected from Kerala, none of the samples contained residues of DDT, whereas all the 37 samples collected from Assam contained DDT. Similarly, none of the 9 samples collected from Andhra Pradesh contained DDT. In Kalyani, West Bengal, 48 samples of the fish Rohu and Catla collected from Calcutta market were examined, and 36 samples (75%) were contaminated with DDT residues; 14 samples (29.2%) exceeded the MRL value. In Coimbatore, out of 12 samples of common small fish, 75% of the samples contained DDT. In Vellayani, sea fish such as Mackerel and Salmon from the market were monitored over a period of 12 months and none of the samples contained any detectable residue (ICAR 2002).

As part of a larger study assessing contamination status of inland wetlands of India, a study was conducted to evaluate OCP residues in fish collected from different inland wetlands in Karnataka, India, and their suitability for human consumption. Among the OCPs tested, isomers of hexachlorocyclohexane (HCH) were the most frequently detected with β- and γ-HCH as the main pollutants. The average concentration of $\sum$HCH and $\sum$DDT ranged from 2.1 to 51.7 µg/kg and below detection level to 12.3 µg/kg, respectively. Other OCPs such as heptachlor epoxide, dieldrin, and endosulfan were found at lower levels. Among various fish species tested, the higher pesticide burden was recorded in *Anguilla bicolor bicolor* (77.9 µg/kg) and the lower was in *Heteropneustes fossilis* (2.1 µg/kg). OCPs detected in this study were well below the tolerance limits recommended for fish. The calculated daily dietary intake of OCPs in all the species examined was lower than the maximum acceptable dietary intake (ADI) limits prescribed for human consumption.

The level and distribution patterns of some OCPs were monitored in fish samples of the Gomti River, India, collected from three sites. In the fish muscles, OCPs ranged between 2.58–22.56 ng/g (mean value: 9.66 ± 5.60 ng/g). Neither spatial nor temporal trends could be observed in the distribution of the OCPs. Aldrin was the predominant OCP, whereas HCB and methoxychlor could not be detected. α-HCH and β-HCH among the isomers of HCH and p,p'-DDE among the metabolites of DDT were the most frequently detected OCPs. The results revealed that the fish of the Gomti River are contaminated with various OCPs (Malik et al. 2007).

OCP residues were determined in 10 species of fish caught at Cochin and Rameshwaram coast and sold in Coimbatore, Tamil Nadu, India. Species were selected on the basis of their regular availability throughout the year and commercial value (Muralidharan et al. 2009). A total of 389 fish were analyzed for organochlorine residues and their suitability for human consumption was evaluated. Results showed varying levels of residues of hexachlorocyclohexane (HCH), DDT, heptachlor epoxide, endosulfan, and dieldrin. Among the 10 species, a high concentration of pesticide residues was recorded in *Sardinella longiceps, Carangoides malabaricus, Chlorophthalmus agassizi, Saurida tumbil*, and *Rastrelliger kanagurta*. The variation in total organochlorine residues among species and between places was not significant (p >0.05). Only five species of fish showed monthly variation in residue levels and there was no significant correlation between the body size and residue levels in the tissue. About 22% of the fish exceeded the maximum residue limits (MRL) of total HCH prescribed by FAO/WHO for fish products. The calculated dietary intake of total HCH through consumption of *C. malabaricus, C. agassizi*, and *S. longiceps* exceeded the maximum acceptable dietary intake (ADI) limits prescribed for human consumption. The present study recommends continuous monitoring of environmental contaminants in marine fish to assess the possible impact on human health.

A number of studies have indicated that the residues of organochlorine pesticides including HCH, DDT, endosulfan, and their metabolites are commonly occurring substances in water of the river and its estuary. Unusual content of the pesticides was reported by Nayak et al. (1995) in the middle stretch (Varanasi) of the river (Table 13.3). Moderate content of HCH compounds was recorded by Kumari et al. (2001). DDT and its analogs were noticed at moderate levels by Ray (1992), and Halder et al. (1989). Ray (1992) also reported moderate content of endosulfan.

Bojakowska and Gliwicz (2005) showed that river sediments in Poland are commonly contaminated with OCPs. They were found in 97.5% of the analyzed sediments from different rivers. The dominant compound was β-HCH in roach (*Rutilus rutilus*) and bream (*Abramis brama*) muscles and endrin in the muscle tissue of ide (*Leuciscus idus*). Mean concentrations of OCPs in the gonads ranged from 0.385 to 0.544 ng/g ww for α-HCH, 0.745 to 0.832 ng/g ww for γ-HCH, 0.479 to 0.576 ng/g ww for dieldrin, and 0.381 to 0.684 ng/g ww

TABLE 13.3

Mean Concentration of Organochlorine Pesticides in the Gonads and Muscles of Fish from the River Course of Oder River

| | Concentration (ng/g ww) | | | | | |
| | Roach, *Rutilus rutilus* | | Bream, *Abramis brama* | | Ide, *Leuciscus idus* | |
Pesticides	Gonads	Muscles	Gonads	Muscles	Gonads	Muscles
α-HCH	0.38 ± 0.17	0.21 ± 0.13	0.42 ± 0.15	0.12 ± 0.09	0.54 ± 0.15	0.33 ± 0.11
β-HCH	0.51 ± 0.14	0.44 ± 0.12	0.64 ± 0.13	0.38 ± 0.08	0.24 ± 0.20	0.13 ± 0.10
γ-HCH	0.75 ± 0.24	0.34 ± 0.15	0.83 ± 0.11	0.32 ± 0.04	0.82 ± 0.04	0.41 ± 0.08
Heptachlor	nd	nd	0.15 ± 0.11	nd	0.14 ± 0.04	nd
Heptachlor epoxide	0.12 ± 0.09	nd	0.15 ± 0.03	nd	nd	nd
Aldrin	0.15 ± 0.03	nd	0.22 ± 0.09	0.15 ± 0.03	0.36 ± 0.08	0.31 ± 0.10
Dieldrin	0.52 ± 0.09	0.16 ± 0.04	0.58 ± 0.08	0.23 ± 0.09	0.48 ± 0.09	0.15 ± 0.03
Endrin	0.38 ± 0.16	0.20 ± 0.11	0.60 ± 0.24	0.20 ± 0.09	0.68 ± 0.22	0.52 ± 0.21

Source: From Tomza-Marciniak, A. and Witczak, A. 2010. Distribution of endocrine-disrupting pesticides in water and fish from the Oder River, Poland. *Acta Ichthyol. Piscatoria* 40(1): 1–9.
nd, not detected.

for endrin. Concentrations of the studied compounds in the water taken from the Oder River followed the order: endrin > γ-HCH > α-HCH > dieldrin > β-HCH > heptachlor ≈ aldrin > heptachlor epoxide. The highest log BCF (Bioaccumulation Concentration Factor) was obtained for fish gonads and ranged from 1.5 (endrin) to 3.4 (heptachlor epoxide). The estimated daily intake (EDI) varied from 0.0014% to 0.097% of the acceptable daily intake (ADI). Although concentrations of the compounds in the muscle were lower than reported by Kannan et al. (1995), Yim et al. (2005), Sun et al. (2006), and Hinck et al. (2008), the fact that the most toxic pesticides occurred in the highest concentrations seems to be alarming. Endrin is highly toxic to all animals, especially fish and other aquatic biota (Anonymous 1991a,b), while β-HCH is the most toxic isomer of hexachlorocyclohexane (Anonymous 2006). The muscle levels of γ-HCH and dieldrin were significantly lower (p ≤ 0.05) than the gonad levels. Similar relationships for these two compounds were reported by Sapozhnikova et al. (2005). In another study, concentrations of γ-HCH and dieldrin in fish muscles ranged from 0.318 to 0.411 ng/g ww and from 0.146 to 0.234 ng/g ww, respectively. Protasowicki et al. (2007) reported higher levels of these pesticides in the muscles of roach and bream harvested from the lower Oder River in the years 1996–2000. Ciereszko and Witczak (2002) also found higher γ-HCH content in the muscles of roach and bream from the Szczecin Lagoon, which averaged within 1.9–0.4 and 2.0–0.3 ng/g ww, respectively.

According to Sun et al. (2006), dieldrin residue levels in fishery products in the years 2001–2003 varied widely from 0.63 to 30.68 ng/g. In contrast, Hinck et al. (2008) reported lower levels of the pesticides in fish from the Savannah River and Pee Dee River (USA) that averaged at 3.2 and 1.73 ng/g ww, respectively. In a study, heptachlor and heptachlor epoxide residues were found in 67% of all gonad samples. None of these two pesticides was found in the muscles of the examined fish. Hinck et al. (2008) reported that heptachlor epoxide residues in the muscles of largemouth bass and common carp from the Savannah River averaged 0.26–0.06 ng/g ww and 0.36–0.07 ng/g ww, respectively, and in fish from the Pee Dee River, its levels were nearly twice as high. Worth noticing is that intersex gonads were identified in 42% of male bass (Hinck et al. 2008) due to exposure to potential endocrine-disrupting compounds. At the same time, relatively high vitellogenin concentrations in male fish occurred, which could have indicated environmental exposure to estrogenic or antiandrogenic chemicals. The study revealed that concentrations of OCPs in fish gonads were significantly higher than in muscle tissue.

The OCP residues in fish sold in markets and fishing sites in Abidjan, Côte d'Ivoire, were investigated. Of the 16 OCPs considered, 11 were present in the samples analyzed at various concentrations ranging from 0.4 to 14.4 µg/kg. Fresh product samples were mostly contaminated by DDD. The catfish, with a total average concentration of 27.2 µg/kg of fresh product, was the most contaminated species. Heads (27.8 µg/kg of fresh product) and viscera (17.5 µg/kg of fresh product) were the most contaminated parts of the fish species analyzed. The fishing port of Vridi was the most contaminated site. The species collected on this site presented a total average concentration of 24.4 µg/kg of fresh product. The comparison of total concentration mean of organochlorine pesticides in the species collected, with the maximum residue limits (MRL) set for the fishery products, suggests that health risks faced by populations in Abidjan through fish consumption are currently low (Biego et al. 2010).

Some organochlorine pesticides may occur in the tissues of aquatic biota at levels even a thousand times higher than in the water. The levels of DDT and HCH in the Qiantang River (China) varied from 0.001 to 0.026 ng/mL and 0.010 to 0.023 ng/mL, respectively (Zhou et al. 2006). Falandysz et al. (1999) reported that similar concentrations of γ-HCH (0.012 ng/mL) and β-HCH (0.012 ng/mL) also were detected in the Vistula River (Poland).

The Oder (Odra) is one of the longest rivers in Europe (741.9 km). In Widuchowa, the Oder River bifurcates, with its eastern and western branches enclosing a wetland area of Międzyodrze, which is one of the largest fens in Europe with unique flora and fauna. A study was conducted to monitor the presence of pesticides in water and fish from the Oder River in Poland (Tomza-Marciniak and Witczak 2010). The concentrations of OCPs in fish gonads were found to be significantly higher than in the muscle tissue. The dominant pesticide in the gonads of roach and bream was γ-HCH, while β-HCH predominated the muscle tissues of these fish. Endrin, on the other hand, was the major pesticide in the muscle tissue of ide. The mean concentrations of OCPs in the gonads ranged from 0.385 to 0.544 ng/g ww for α-HCH, 0.745 to 0.832 ng/g ww for γ-HCH, 0.479 to 0.576 ng/g ww for dieldrin, and 0.381 to 0.684 ng/g ww for endrin (Table 13.3). Concentrations of these pesticides in the water taken from the Oder River followed the order: endrin > γ-HCH > α-HCH > dieldrin > β-HCH > heptachlor ≈ aldrin > heptachlor epoxide. The highest log BCF (bioconcentration factor) was obtained for fish gonads and ranged from 1.5 (endrin) to 3.4 (heptachlor epoxide). Estimated daily intake (EDI) varied from 0.0014% to 0.097% of the acceptable daily intake (ADI) (Table 13.4).

Bioconcentration factors are used for quantitative assessment of bioaccumulation, understood as a process of accumulation of a substance in the body of fish in proportion to its concentration in the aquatic environment (Zhou et al. 2007). Water and sediment samples were also analyzed for thirteen organochlorine pesticides (OCPs) in 18 fish species from the Qiantang River (Zhou et al. 2007). Total concentrations of OCPs in fish muscles ranged from 7.43 to 143.79 ng/g ww, with the highest concentration recorded in the sole fish (*Cynoglossus abbreviatus*), a benthic carnivore. The results indicated that carnivorous fish have higher OCP concentration than other fish with different feeding modes. OCP concentration in fish was in the range of 1.86–5.85, 2.65–133.51, and 1.94–12.48 ng/g for HCHs (α-, β-, γ-, δ-HCH), DDTs (p,p'-DDD, p,p'-DDE, p,p'-DDT, o,p'-DDD), and other OCPs (aldrin, dieldrin, endrin, heptachlor, heptachlor epoxide), respectively. The highest OCP

TABLE 13.4

Estimated Daily Intake of Organochlorine Pesticides Consumed by Humans with the Fish from the Lower Oder River

| Pesticides | MCM (ng/g ww) | EDI | | ADI |
		(ng/kg bw per day)	(% ADI)	(ng/kg bw per day)
α-HCH	0.222	0.056	—	No ADI
β-HCH	0.317	0.079	—	No ADI
γ-HCH	0.335	0.089	0.0018	0–5000[a]
Heptachlor	—	—	—	0–100[b]
Heptachlor epoxide	—	—	—	No ADI
Aldrin	0.232	0.058	0.074	0–100[b]
Dieldrin	0.180	0.045	0.057	
Endrin	0.306	0.077	0.097	0–200[c]

Source: From Tomza-Marciniak, A. and Witczak, A. 2010. Distribution of endocrine-disrupting pesticides in water and fish from the Oder River, Poland. *Acta Ichthyol. Piscatoria* 40(1): 1–9.

MCM, mean concentration in fish muscles; EDI, mean concentration in fish muscles × fish consumption (g per day) per Kg body weight (60 kg); ADI, acceptable daily intake.

[a] ADI recommended in 2002 by JMPR (*Joint FAO/WHO Meeting on Pesticide Residues*).
[b] ADI recommended in 1991; as sum of aldrin and dieldrin.
[c] ADI recommended in 1971.

concentration in fish organs of four big fish species was found in the brain of the silver carp (*Hypophthalmichthys molitrix*), 289.26 ng/g ww, followed by the kidney, liver, heart, and gill. Among the OCPs analyzed, DDE, γ-HCH, and heptachlor were the predominant contaminants in the fish muscle, which indicated that there was recent input of lindane. A significant correlation was observed between concentrations of DDTs and lipid content as well as between OCPs and lipid contents in fish species. Both field water bioconcentration factors (BCF) and sediment BCF showed a positive correlation with octanol–water partition coefficients (K_{ow}) in the sole fish. The values of log BCFs were from 1.5 to 3.4 for fish gonads and from 1.3 to 3.1 for fish muscles. Zhou et al. (2007) reported log BCF values higher than 3 for HCH isomers in the muscles of freshwater fish. Pandit et al. (2006) obtained similar results in studies on bioaccumulation of organochlorine pesticides in the coastal marine environment of Mumbai. Assuming that, in 2007, a statistical Polish consumer ate about 5.4 kg of fish (Anonymous 2008), the total daily intake of the examined pesticides from fish averaged at 0.41 ng/kg body weight. EDI values obtained in this study were lower than that reported in the literature. According to Sun et al. (2006), the EDI for dieldrin was 70.45%–93.56% of ADI. Zhou et al. (2008) reported that estimated daily intake (EDI) of HCH isomers through shellfish by humans was 5.83 ng/kg body wt/day (ten times higher than EDI).

The common paddy fish, *Trichogaster pectoralis* Regan, is a major source of protein for many Malaysian farmers. Daily fish consumption varies between 54 and 205 g, and is a function of the availability of fish (Chen 1982). For example, fish are harvested once per season at Tanjung Piandang, and the daily estimated fish intake at that site is only 54 g. At Sungai Burong and Parit Tanjung Piandang, however, fish are caught twice weekly, and the daily fish consumption is 173 and 205 g, respectively. Farmers from the remaining two sites, Sungai Kota and Jalan Bharu, indicated that fish were collected once every 1–4 weeks and hence fish consumption there is intermediate. Several common pesticides were found in the fish obtained from different sites at varying concentrations. The highest average pesticide residues were observed in fish from Tanjung Piandang, and these concentrations were attributed to the long exposure period. At this collection site, the dieldrin level in fish varied from 18.8 to 40.3 ng/g. At Jalan Bharu, the range for dieldrin was even greater (8.0–63.8 ng/g). Numerous factors contribute to this variability, but spot application of pesticide may be the most important one. It is therefore important to examine many sub-samples from each site in environmental surveys. The average daily intake of pesticide residues was calculated from these data, assuming an average body weight of 60 kg for the consumer. The highest observed daily intakes were as follows: aldrin/dieldrin, 44.4 ng/kg of body weight; chlordane, 55.0 ng/kg of body weight; HCH, β-isomer, 15.6 ng/kg of body weight and γ-isomer, 8.4 ng/kg of body weight; and DDT, 21.8 ng/kg of body weight. These findings were compared with the acceptable daily intake (ADI) values recommended for these compounds by FAO/WHO. This showed that all the observed daily intakes of pesticide residues were below the FAO/WHO ADIs. In fact, the aldrin and dieldrin levels approach the ADI (0.1 ng/kg of body weight) only if the highest observed concentration in fish (24.9 ng/g) is multiplied by the highest consumption rate of 205 g/day.

One hundred samples of fish from Karadj and Latian Lake (two of the reservoirs supplying Tehran with potable water) were analyzed for the detection and determination of pesticide residues. DDT and its metabolites TDE and DDE were the pesticides commonly found. In the Karadj reservoir, the mean concentration of total DDT found in whole fish was 0.049 ± SD 0.0032 ppm while that of the Latian reservoir was 0.065 ± SD 0.007 ppm. The mean concentration of the pesticide in fat was 3.11 ± 0.21 and 4.85 ± 0.13 ppm, respectively. This shows the Latian reservoir to be more contaminated. Extreme concentrations of pesticides in Latian fish were found in *Varicorhinus nikolskii*, while in Karadj fish, the highest level of the pesticide occurred in *Salmo trutta* morpha *fario* and the lowest concentration in *Varicorhinus nikolskii*. Samples collected in summer showed the highest concentrations of the pesticides.

The residues of p,p'-DDE and HCB were detected with a frequency of 100%, in fish samples from the Sene-Gambian, whereas p,p'-DDT was detected in eight, heptachlor epoxide in six, and endosulfan sulfate in five of the nine fish samples (Manirakiza et al. 2002).

Some OCP residues in tilapia fish (*Tilapia zillii*), sediment, and water samples from Lake Bosumtwi (the largest natural lake in Ghana) were determined. p,p'-DDE was the predominant residue in all the samples analyzed and was detected in 82% of water samples, 98% of sediment samples, and 58% of fish samples at concentrations of 0.061 ± 0.03, 8.342 ± 2.96, and 5.232 ± 1.30 ng/g, respectively. DDT was detected in 78%, at a mean concentration 0.012 ± 0.62 ng/g, of water samples analyzed. The mean concentrations of DDT in sediments and fish were 4.41 ± 1.54 and 3.645 ± 1.81 ng/g, respectively. The detection of lower levels of DDT than its metabolite, DDE, in the samples implies that the presence of these contaminants in the lake is a result of past usage of the pesticides.

In six fish species, namely *Heterotis niloticus, Channa obscura, Hepsetus odoe, Tilapia zillii, Clarias gariepinus,* and *Chrysichthys nigrodigitatus*, collected from the sampling towns, Weija and Nsawam along the Densu river basin in the Greater Accra Region of Ghana, the concentrations of OCs ranged from 0.3 to 71.3 μg/kg (correct), which were, however, below the Australian Maximum Residue Limit (MRL) of 50–1000 μg/kg (correct) for freshwater fish. The results of this work indicate that OCP residues are present in the Densu basin; however, the concentrations of the OC residues determined in the fish samples were below the stipulated Australian MRL. Of all fourteen OCPs, seven are among the banned pesticides of the Environmental Protection Agency (EPA) of Ghana. The mean concentrations of the banned OCs detected range from 0.30 to 35.2 μg/kg, with the highest concentration of 35.2 μg/kg recorded for δ-HCH in the *Channa obscura* sp. sampled from Nsawam. The lowest mean concentration of 0.30 μg/kg was recorded for endrin, aldrin, and endrin-aldehyde in the species *Clarias gariepinus*, *Clarias gariepinus*, and *Channa obscura*, respectively. The metabolites, namely p,p'-DDE and endosulfan sulfate, were also detected in six of the eight fish species used for the investigation. From the results of this work, organochlorine pesticide residues in the Densu river basin are likely to have originated from various sources, such as the use of pesticide products for agricultural purposes by farmers along the Densu river basin (Afful et al. 2010).

The distribution and residue levels of BHC, lindane, endrin, o,p'-DDT, p,p'-DDE, and p,p'-DDD in *Mugil* and *Tilapia* species collected from Egyptian Delta lakes were evaluated. Residues of BHC and p,p'-DDT were found in most fish samples obtained from El-Manzala, El-Burullus, Edku, and Maryut lakes. Fish samples from Lake Edku showed the highest residue levels of the studied pesticides. Residues of BHC and p,p'-DDT in *Mugil* and *Tilapia* species are within the range found in fish from the northwest Atlantic but lower than the levels reported for other fish species in the United States. *Mugil* and *Tilapia* species could be used as indicator organisms for monitoring the residue levels of OCPs and their distribution.

A great deal of domestic, agricultural, and industrial wastes is thrown in Bahr El-Bakar drainage canal. This drain finally pours its water in Manzala Lake, north of Egypt, where extensive fishing activities exist. Fish is considered an important source of food to a great portion of the human population. As large as a total of 400 ng/L of chlorinated organic pesticides in water samples and 364 ng/g in the sediment samples were estimated, whereas, the concentration of these compounds in fish samples was as high as 137 ng/g.

A National Pesticide Monitoring Program was initiated by the Biological Research Centre, Baghdad in 1981 to determine the residue levels of persistent compounds in the environment and biota. The purpose of the present work was to find out the OC residue levels in different fish species collected from a polluted region in Diyala River (Rustamiya, Baghdad), where the largest sewage treatment plant in Baghdad discharges treated water. Seventy fish, 1–3 years old, of 11 different species were collected in 30 different collections from October 1983 until July 1984 from the same sampling station on the lower part of Diyala River at Rustamiya. Among the 30 composite samples collected, 12 consisted of only one species (one fish), which was mostly *Chondrostoma regium*, and others consisted of 2–5 species; however, only one composite consisted of 6 species. The numbers of fish caught for each species were as follows: *Chondrostoma regium: 19, Liza abu: 11, Barbus grypus: 11, B. xanthopterus: 6, B. luteus: 6, B. belaweyi: 4, Mystus pelusius: 4, Aspius vorax: 2, Heteropneustes fossilis: 3, Cyprinus carpio: 2,* and *Garra rufa: 2.* All the freshwater fish showed a high percentage of OCs (90–100%). Generally speaking, higher levels of all insecticide residues were detected in low-fat content fish of all species. Chlordane, including cis- and trans-chlordane, and hydroxychlordane showed the highest average residue level of 73.495 ppm (fat basis) in low-fat content group of the species *B. xanthopterus,* and 68.816 ppm (fat basis) in high-fat content group of the same species. The concentration of the pesticides was higher in low-fat content groups of all species than in high-fat content groups.

It was reported that, following endosulfan application for tsetse fly control in Botswana, an increased fatality in freshwater fish was observed and endosulfan residues were detected in the tissues (Mathlessen et al. 1982). This is an example of nontarget organisms being affected by applications of insecticides in vector control. A high concentration of pesticide residues was detected in the liver of the indigenous fish sampled for the study followed by the intestines, gills, brain, and skeletal muscles (Heath 1992). The pesticide residues were rapidly absorbed into the body through the gills of freshwater fish species (Lloyd 1992) and accumulated in the various organs, sometimes bioconcentrated by as

much as 2000-fold over the concentration of the residue in water (Devi et al. 1981). The liver
has been shown to be the major organ for the metabolism of the residues in fish (Peterson
and Bately 1993).

Among the several freshwater fish species caught in two rivers in South Africa, the
African catfish showed highest concentrations of residues (Heath 1992) and can be used as
an indicator species for pollution. Grobler (1994) detected and measured pesticide residues
in the catfish caught in the Olifants River, South Africa, and dams in Zimbabwe, respectively. They found that the catfish contained elevated levels of pesticide residues compared
with other freshwater fish species.

The bioaccumulation of carbofuran and endosulfan in tissues is demonstrated in several
other species of fish (Berbert et al. 1989; Ferrando et al. 1992; Kale et al. 1996; Singh and
Garg 1992). Of the tissues analyzed for residues, the liver contained the highest level of
carbofuran and endosulfan residues, followed by the intestines, gills, brain, and skeletal
muscle. The liver and gills of *Clarias gariepinus* caught in the Olifants River, South Africa,
were also found to accumulate high levels of heavy metals (Du Preez et al. 1997).

The levels of organochlorine pesticides were monitored in 18 fish species from Konya
markets, Turkey. These species were selected on the basis of their importance to local
human fish consumption. DDT and its metabolites and HCH were the predominant contaminants in fish muscles. The mean concentrations of summation operator DDT were in
the range between 0.0008 and 0.0828 µg/g. DDT was the predominant residue in *Sparus
aurata*. Detectable levels of HCH, aldrin, and heptachlor were found in most samples.
However, dieldrin, endrin, β-endosulfan, p,p'-DDT, and p,p'-DDD were not detected in
Salmo trutta. The mean of endrin ranged from 0.0040 µg/g (*Triglia lineate*) to 0.0326 µg/g
(*Trachurus trachurus*) (Leyla et al. 2009).

A study was carried out between May 2002 and August 2003 in Meric Delta, Turkey,
which is located at a site where the Meric River falls into the Aegean Sea. Residues of OCPs
in fish (*Cyprinus carpio*) samples from Meric Delta were analyzed (Belda and Kolankaya
2006). The α- and β-HCH were the predominant HCH isomers in all analyzed fish samples
and ranged between 319.5 and 968.15 ng/g and between 397.5 and 876.4 ng/g, respectively.
The concentration levels of p,p'-DDT (ranging from 2.68 to 52.45 ng/g) were consistently
higher than its metabolite p,p'-DDE, indicating a recent use of this OCP in the area.

Residue levels of organochlorine pesticides in 13 commercially important fish species collected from the NW Arabian Gulf were monitored. While most of the residues
were below the detection limit of 1 tg/kg wet weight, relatively low concentrations of
ΣDDT, endrin, and dieldrin were detected in the edible tissue of these fish. The ΣDDT
residue levels ranged from 2 to 1 tg/kg weight, endrin ranged from none detected (rid) to
45 tg/kg, and dieldrin from ND to 5 tg/g. A definite correlation was established between
total organochlorine pesticide residues and lipid content (r = 0.6) for the NW Arabian
Gulf fish. A comparison with fish from Hor-al-Hammar Lake (an area that used to be
sprayed with pesticides) has shown that the latter contained significantly higher residue
levels. The ΣDDT residue levels ranged from 5 to 45 lig/kg wet weight, endrin from 3 to
83 tg/kg, and dieldrin from ND to 4 tg/kg. Based upon the observation that the original
DDT (p,p'-DDT) was identified in the NW Arabian Gulf fish, it has been concluded that
there was a recent input of DDT to this region. Since DDT application has been banned in
Iraq, consequently, it was assumed that DDT must originate from a more remote source.
Endrin is considered to be the most toxic of all commercial insecticides to fish (Johnson
and Finley 1980), and this was the second most dominant compound in fish from the
NW Arabian Gulf. Endrin residues were detected in approximately 90% of these fish,
with mean values ranging from 1 tg/kg in large-scale sole, croaker, and elongate ilisha

to 28 tg/kg wet weight in silver-banded croaker. Slightly higher concentrations of endrin have been detected in fish from the Hor al-Hammar Lake, with residue levels ranging from 3 tg/kg in carp to 67 pg/kg wet weight in minnows. Endrin has not been detected in the Arabian Gulf before, which may be due to the fact that this insecticide is relatively short-living and has been used in few occasions in this region.

Although dieldrin has been officially banned in Iraq since 1976, its residues are expected to persist owing to its long use for agricultural and public health purposes. However, residue levels of dieldrin in the fish from the NW Arabian Gulf were at or near the detection threshold of 1 tg/kg wet weight. More frequent residues of dieldrin were observed in the Hor al-Hammar Lake. No significant difference ($\alpha = 0.05$) in the residue levels between fish types for lindane, α-endosulfan, p,p'-DDE, p,p'-DDT, and dieldrin was observed. The aldrin levels in Nile perch (*Lates niloticus*) were significantly higher than the levels in Nile tilapia (*Oreochromis niloticus*). No difference was observed in the distribution of residues in the different parts of Nile tilapia, although a difference for p,p'-DDE was observed in the Nile perch. No significant difference was observed in the average fat content of the tissue of Nile perch and Nile tilapia; however, the distribution of fat was significantly different in the different parts of the fish, with the abdominal portion having the highest amount of fat. There was no correlation observed in this study between fat content and organochlorine concentration. Lower p,p'-DDT residue levels as compared with the p,p'-DDE levels observed indicate that DDT is no longer in use. The levels of OCP residues found in fish samples were below the FAO, US FDA, Australian, and German extraneous residue limits and maximum residue limits.

Lake Chicot is located in the alluvial plain of the Mississippi River in eastern Arkansas, an area noted for its agricultural productivity. Because of the intensive agriculture in this region, numerous organochlorine pesticides were applied for several decades to control insect pests. Samples of fish from isolated and flow-through portions of Lake Chicot, Arkansas, were analyzed for residual pesticide concentrations. DDT, and its metabolites, and heptachlor were significantly ($a = 0.05$) higher in spotted gar (*Lepisosteus oculatus*) and yellow bullhead catfish (*Ictalurus natalis*) than in other species examined. Pesticide concentrations did not exceed the acceptable levels established by the US Environmental Protection Agency, although toxaphene levels in one white crappie (*Pomoxis annularis*) and one freshwater drum (*Aplodinotus grunniens*) were as high as 0.01 tg/kg. Bottom-feeding and piscivorous fish had consistently higher concentrations of pesticides than fish belonging to other feeding groups. The main body of the lake, with a large drainage area (930 km^2) had higher concentrations of suspended solids than did the isolated northern basin and produced fish with significantly ($a = 0.05$) higher levels of toxaphene, DDD, and DDT. Insecticide concentrations were consistently greater in viscera with toxaphene, DDT, DDE, and DDD levels significantly ($a = 0.05$) greater than in either whole fish or fish flesh samples. Eight years after the ban, residual pesticides in fish from both basins of Lake Chicot were still significantly ($a = 0.05$) higher during years of increased runoff, indicating the importance of watershed management practices in long-term downstream water resources (Cooper and Knight 1987).

A study of the content of persistent OCPs (DDT and its metabolites, α-, β-, γ-isomers of hexachlorocyclohexane) in the tissues and organs of some fish and molluskan species from the lower reaches of the Tumen River and the contiguous part of Peter the Great Bay (Sivuchya Bay and Zapadnaya Bay of the Furugelm) and Amursky bay showed that the highest total content of HCHs (785.60 ng/g of gross mass) was revealed in the digestive gland of the Japanese scallop *Mizuhopecten yessoensis* from Zapadnaya Bay and in the brain and the liver of the starry flounder *Platichthys stellatus* from the Tumen

River (390.80 and 340.29 ng/g of gross mass). The maximum total content of DDT (270.70 ng/g of gross mass) was recorded in the brain of the dark plaice *Pleuronectes obscurus* from Zapadnaya Bay and in the liver of the far eastern smooth flounder *Pleuronectes pinnifasciatus* caught in Amursky Bay to the west of Skrebtsov Island (212.80 ng/g of gross mass). The level of HCHs and DDT in mollusks and flounders from Zapadnaya Bay was higher than that in the same species from Sivuchya Bay. The concentration of OCPs in *P. pinnifasciatus* from the inner part of Amursky Bay at Skrebtsov Island was higher than in individuals of that species from the open part of the Bay at Peschany Pen. The DDT/DDE and DDT/$\sum$DDT ratios evidenced the recent entry of DDT into the ecosystem of Peter the Great Bay. The southwest part of Peter the Great Bay, from the mouth of the Tumen River up to Furugelm Island, was contaminated by HCHs to a greater extent than Amursky Bay. OCPs accumulated in appreciable quantities in the organs of fish and mollusks of Peter the Great Bay, though their present content does not exceed sanitary and hygienic standards, a subsequent monitoring of their concentrations in biota is necessary (Syasina 2003).

Many research activities have been carried out in China to investigate the OCPs in marine organisms. The results showed that the fish and shellfish species in the Chinese coastal zone are contaminated by OCPs. DDTs were at the top level. A recent report released in Asian countries revealed that the contents of DDTs in the marine benthos along the coastal areas of China were at the top level (Monirith et al. 2003). Results also showed that South China is the most contaminated coastal sea among the major seas in China (Monirith et al. 2003; Chen et al. 1996, 2000, 2002; Fang et al. 2001; Phillips 1985; Tanabe et al. 1987). There are possible reasons behind this observation. The coastal area of Guangdong province in South China has 60,000 fishing ships, which is above 1/5 of the total number in China. It can be estimated that about 30–60 tons of DDT may be introduced to the coastal environment of Guangdong, including the Pearl River Delta. According to Fu et al. (2003), the Pearl River is believed to carry a considerable load of chlorinated pesticides, up to 863 tons per annum, which is the highest among China's rivers. A high ratio of DDT/(DDD+DDE) in sediment (Hong et al. 1995; Zhang 2001; Mai et al. 2002) as well as water (Zhou et al. 2001; Luo et al. 2004) samples indicated the relatively recent releases of DDT (Zhang et al. 2002; Luo et al. 2004; Zhou and Wong 2004; Chen et al. 2006). However, comparatively lower levels of DDTs were found in marine organisms collected from the East coast. But the survey of the National Bureau of Coastal Zone Protection during 1980–1987 showed that the organochlorine flux just carried by Yangtze River (the longest river in China) was 239.3 tons per year, which accounted for 19.8% of the total flux by Chinese river catchments into the marine coastal sites. Bohai Sea, Northern China, collects pollutants from major rivers, namely Yalu, Daliao, Luan, and Yellow, and it showed comparatively intermediate levels of DDT among the recent researches. Wu et al. (1999) noted high concentrations of DDTs in the river sediments from Northern China where a factory with a high manufacturing capacity of DDT is located.

Fish tissues from different fishery types (freshwater farmed, seawater farmed, and seawater wild) collected from the Pearl River Delta, China, showed significantly higher $\sum$OCP levels in seawater-farmed fish than others, and among three freshwater-farmed species, the lowest levels occurred in filter-feeding fish (bighead carp). Liver contained the highest $\sum$OCP levels, while no significant differences were found among other tissues. Among DDT components, p,p'-DDT was abundant in seawater fish, while for freshwater fish, p,p'-DDE was the predominant congener, except for northern snakehead (34% for p,p'-DDE and 30% for p,p'-DDT). The new source of DDTs to freshwater fish ponds was partly attributed to dicofol, whereas sewage discharged from the Pearl River Delta and antifouling paint were likely the DDTs sources to seawater-farmed fish. Occurrence of organochlorine pesticides in fish tissues was examined to assess input sources and modes of bioaccumulation in the Pearl River Delta, China (Guo et al. 2008).

Fishery products are very popular in the Taiwanese cuisine. The major raw materials include fish, crab, shrimp, bivalve, and cephalopod. Pesticide residues in fishery products should be of great concern to consumers. The residue data will enable us to estimate pesticide dietary intakes in citizens and to compare them with acceptable daily intakes (ADIs) of the Food and Agricultural Organization (FAO) as well as with the World Health Organization (WHO), so as to assess potential health hazards.

A study was carried out to find the concentration levels of pesticide residues in edible parts of consumed fishery products in the market. A total of 920 samples of fishery products were analyzed for pesticide residues from the years 2001 to 2003. Overall, 65.40% of fish, 93.55% of shellfish, 84.92% of crustacean, and 98.33% of cephalopod samples contained no detectable residues. There are no detectable residues of carbamate and pyrethroid pesticide in surveyed samples. OCP residues are detected in 144 (34.60%) fish, 4 (6.45%) shellfish, and 28 (15.7%) crustacean samples. Organophosphate (OP) pesticide residues are detected in 69 (11.37%) fish, 2 (1.05%) crustacean, and 1 (1.67%) cephalopod samples. Fish had higher detection rates of OC and OP pesticides. These pesticides, however, were not detected in cephalopod or shellfish. Multipesticide residues were detected only in 22 (3.62%) fish samples. This result is similar to the no detectable rate of FDA domestic seafood in 1998, 1999, and 2000 separately (21.50% in 1998, 28.90% in 1999, 24.60% in 2000) but is higher than that in 2001 (18.4%) (FDA 1999, 2000, 2002, 2003). As for the 91 pesticides, DDT, dieldrin, chlorpyrifos, fenitrothion, fenthion, and prothion have been detected from fishery products. The predominant residues are DDTs (including p,p'-DDD, p,p'-DDE, and p,p'-DDT). DDT and dieldrin were banned in the 1970s in Taiwan for their long persistence in environment. There are no studies on OC pesticide incidence in products of fish farms in Taiwan so far. Chlorpyrifos, following DDTs, is one of the main detectable residues in fishery products, especially in the nationwide supermarkets (in 30 samples, 10.07%) and regional supermarkets (in 29 samples, 14.08 %). The next detectable residue is fenitrothion. Chlorpyrifos and fenitrothion are widely used not only in agriculture but also in sanitation in Taiwan. In fishery product processing, both supermarkets have the highest standards on hygienic conditions. However, in the national supermarkets (nationwide chain store), the service of processing is at sight, whereas in the regional supermarket, the same service is through the central kitchen.

Chlorpyrifos is one of the most widely used active ingredients for pest control in the world. Many studies have been conducted in examining critical aspects of chlorpyrifos products as they relate to health and safety (Dishburger et al. 1997; Johnson et al. 1998; Clegg and Gemert 1999). There are no reports directly pertaining to the contamination of sanitary chemicals on fishery products. In both kinds of supermarkets, higher multipesticide residues have been detected: in 9 samples of national supermarket (3.02%) and 12 samples of regional supermarket (5.83%).

The Indian Ocean coastal environment of Kenya contains critical terrestrial and aquatic habitats, which comprise unique ecosystems and support a rich biological diversity and valuable assortment of natural resources, which may be adversely affected by pesticide residues transported from inland waterways into coastal waters via rivers and estuaries through flooding, deposition, and rainfall (UNEP 1998). Marine fish are regarded as good bioindicator species for evaluating pesticide exposure toxicity levels in marine environments, and their susceptibility is depicted by reduction in species richness (Matin 2002).

OCP residues were analyzed in samples of fish taken from four different locations along the Indian Ocean coast of Kenya. Considering all the four sampling sites in the two sampling seasons, the ranges of concentrations of residues detected in the fish samples (in mg/kg ww) were as follows: lindane 16.1 (in sample from Mombasa)—1445 (Sabaki), aldrin 1.55 (Kilifi)—323 (Kilifi), dieldrin 4.81 (Ramisi)—109 (Sabaki), endosulfan 5.91 (Mombasa)—54.6 (Sabaki), p,p0-DDT 9.11 (Mombasa)—29.3 (Kilifi), p,p0-DDE 1.94 (Kilifi)—97.5 (Sabaki), and p,p'-DDD 1.68 (Mombasa)—98.9 (Kilifi). The concentration ranges obtained in fish in this study indicate that Mombasa Old Town was the least contaminated by organochlorine residues as higher residues were detected near Malindi, at Funzi Lazy lagoon and at Kilifi creek, which were near the confluences of Rivers Sabaki, Ramisi, and Goshi at the Indian Ocean coast, indicating that the sources of the residues in fish of the Indian Ocean coast were through discharge from the rivers. Seasonal variation in the concentration of residues detected in fish samples was clear, with higher residue levels being recorded in the rainy season in May compared with those detected in the dry season in January. This seasonal variation was particularly observable in p,p'-DDT residues, which were only detected in samples taken in the rainy season but not in those taken in the dry season in all the sampling sites. The high DDT/DDE and aldrin/dieldrin concentration ratios also indicated that the sources of these residues were recent. This conclusion was also supported by high BCF values of lindane in fish sampled in Sabaki and of aldrin sampled from Kilifi and Ramisi. Overall, the concentration ranges of some of the residues such as DDT, DDE, and DDD were comparable with those reported earlier for fish samples from Lake Kariba in Zimbabwe, but concentration ranges of lindane and aldrin were much higher than those reported in fish from Lake Kariba. The concentration ranges of the residues of dieldrin, lindane, and DDE were comparable with those obtained in fish from Tana River in Kenya but the other residues of aldrin, endosulfan, DDD, and DDT were much higher than those detected in fish from River Tana (Barasa et al. 2008).

The natural gifts of aquatic resources, fish, and other fishery items have played an important role in the culture and lifestyle of the people of Bangladesh. Fisheries and aquatic resources are economically, ecologically, culturally, and aesthetically important to the nation. In Bangladesh, at present, there are 260 freshwater fish species, 12 species of exotic fish, 475 species of marine fish, and 60 species of prawn and shrimp available (Chandra 2006). Fisheries sector contributes to 5.24% of GDP (Gross Domestic Product), 63% of animal protein supply, and 4.76% of foreign exchange earnings for the nation (Chandra 2006). During the winter season when the natural depression attains shallow water and even dry up, there is huge amount of fish catches from freshwater and marine water. During this time, the Bay of Bengal, the coastal crisscross channels, and other depressions remain calm and quiet and are strengthened as a result fishing activities, and huge fish are harvested during this period than the other seasons.

Long-term preservation of fish by drying is common practice in Bangladesh. This practice is usually carried out in the remote coastal isolated islands and inland depressions, where chilling and freezing facilities are lacking. The finally dried fish products are generally stored in a dump warehouse near coastal towns. In addition to this, the weather is humid, particularly during the monsoon period, and the dry fish absorb moisture rapidly and become suitable for infestation by beetles and mites. The most unexpected cause of infestation is improper drying of the fish by fishermen to prevent a large reduction in their weight, as the fishermen earn more profit selling the dry fish having more weight. For the protection of dry fish from infestation, they use a mixture of OC (DDT and heptachlor) insecticides (Bhuiyan et al. 2008). Some analyses in Bangladesh show alarming pollutants in fish like DDT and heptachlor (BCAS 1990). In Kuakata (a fish-processing zone in

Bangladesh), high levels of DDT powder (locally known as white powder) are used, though Bangladesh banned the "dirty dozen" in 1997 (Barua 2007). Therefore, a study was carried out for the detection and determination of the concentration levels of insecticides (DDT and heptachlor) in dry fish under normal condition (without washing) and after various traditional washing procedures to elucidate the actual concentration level of insecticides taken in through dry fish.

Twenty samples of five different fish species analyzed were found to contain organochlorine insecticides. The concentrations of DDT in the samples of Bombay duck at normal, cold washing, hot washing, and boiling treatment were 97.0195, 76.604, 49.802, and 38.195 ppb, respectively. In the samples of ribbon fish, the concentrations were 36.563, 29.966, 21.565, and 9.918 ppb, respectively. In the samples of shrimp at normal, cold washing, hot washing, and boiling treatment these were 12.617, 7.673, 6.779, and 5.136 ppb, respectively.

In the samples of Chinese pomfret at normal, cold washing, hot washing, and boiling treatment, concentrations were 712.155, 372.119, 256.847, and 197.740 ppb, respectively. In the samples of Indian salmon at normal, cold washing, hot washing, and boiling treatment concentrations were 737.238, 469.106, 142.188, and 135.516 ppb, respectively. The concentrations of heptachlor in the samples of Bombay duck at normal, cold washing, hot washing, and boiling treatment were 1.237, 0.752, 0.456 ppb, and not detected respectively.

In the samples of ribbon fish following different treatments, concentrations were 1.208, 1.991, 0.915, and 0.562 ppb respectively. In the samples of shrimp at normal, cold washing, hot washing, and boiling treatment, these were 44.806, 29.863, 26.665, and 16.868 ppb, respectively. In the samples of Chinese pomfret at normal, cold washing, hot washing, and boiling treatment, concentrations were 5.318, 4.186, 2.811, and 2.139 ppb, respectively. In the samples of Indian salmon at normal, cold washing, hot washing, and boiling treatment, concentrations were 4.834, 4.117, 2.208, and 1.272 ppb, respectively.

From the above results, it was found that all the samples of dry fish contained DDT and heptachlor, and these insecticides were not removed 100% by any type of washing, even after washing was followed by 10 min boiling, (except in Bombay duck boiling treatment sample, in which heptachlor was not detected).

In the Republic of Bénin, aquatic ecosystems are subject to poisoning risks due to the inappropriate use of pesticides, such as washing of empty bottles in rivers and using pesticides to catch fish. In some areas, cotton fields are located near riverbanks, increasing the probability of pesticide emission to the river. To assess contamination levels in the Ouémé River catchment area, different fish species were collected from different geographical areas along the river. DDT, its metabolites, and its isomers were the most frequently identified pesticides in fish flesh; α-endosulfan, β-endosulfan, dieldrin, telodrin, lindane, and octachlorostyrene were also detected. Concentrations of pesticide residues in fish ranged from 0 to 1364 ng/g lipid. A preliminary risk assessment indicated that the daily intake of chlorinated pesticides by people consuming fish from the Ouémé River is still rather low and does not present an immediate risk.

The OCP residues were measured in three species of fish: *Tilapia zillii* (redbelly tilapia), *Ethmalosa fimbriata* (Bonga shad), and *Chrysichthys nigrodigitatus* (catfish). These fish species are a significant part of the diet of residents of Lagos, Nigeria. The mean concentration of OCPs ranged from 0.01 to 8.92 ppm. The concentrations of the OCPs (except for HCHs) in fish samples in this study were below the extraneous residue limit of 5 ppm, set by the Codex Alimentarius Commission of FAO-WHO-1997. However, the concentrations were higher than those detected during previous studies of fish samples from Lake

Victoria, Uganda, in Africa (Kasozi et al. 2005). Also, the levels were quite high when compared with the allowable Federal Environmental Protection Agency (FEPA) (now Federal Ministry of Environment) limit and can be harmful if the trend is not checked. The study also showed that the concentrations of OCPs were higher in adult than in juvenile fish in most cases, and there was no correlation observed between the fat content and the total concentration of OCPs (Adeyemi et al. 2008).

13.5 Effects of Pesticide Residues on Fish Population

Fish species are sensitive to enzymic and hormone disruptors. Chronic exposure to low levels of pesticides may have a more significant effect on fish populations than acute poisoning. Doses of pesticides that are not high enough to kill fish are associated with subtle changes in behavior and physiology that impair both survival and reproduction (Kegley et al. 1999). Biochemical changes induced by pesticidal stress lead to metabolic disturbances, inhibition of important enzymes, retardation of growth, and reduction in the fecundity and longevity of the organism (Murty 1986). Liver, kidney, brain, and gills are the most vulnerable organs of a fish exposed to the medium containing any type of toxicant (Jana and Bandyopadhyaya 1987). The fish show restlessness, rapid body movement, convulsions, difficulty in respiration, excess mucus secretion, change in color, and loss of balance when exposed to pesticides. Similar changes in behavior are also observed in several fish exposed to different pesticides (Haider and Inbaraj 1986).

Singh and Singh (2008) reported that the ovary is an important organ for the bioaccumulation of OCPs. Their study indicated that during the reproductive phase, OCPs are transferred to the ovary from the liver, which may cause reproductive disorders. They observed a decrease in gonadosomatic index (GSI) and plasma levels of testosterone and 17β-estradiol in female fish captured from polluted rivers (Ganga and Gomti River) compared to the same species from unpolluted ponds. The ratio of pesticide levels in fish and water equaled 100:1.

The Great Lakes fish are contaminated with chlorinated pesticides such as DDT, DDE, mirex, and dieldrin as well as trace amounts of metals such as lead and mercury (Susan et al. 2001). Lake trout, which became extinct in the Great Lakes in the 1950s, has been shown to be very sensitive to dioxins and polychlorinated biphenyls (PCBs) when exposed as embryos. Several species of salmon introduced into the Great Lakes have severely enlarged thyroid glands, which is a strong evidence of hormone disruption. Salmon in the Lake Erie shows a variety of reproductive and developmental problems, for example, early sexual development and a loss of the typical male secondary sexual characteristics, such as heavy protruding jaws and red coloration on the flanks.

Some pesticides can indirectly affect fish by interfering with their food supply or altering the aquatic habitat, even when the concentrations are too low to affect the fish directly. Other agricultural chemicals are capable of killing salmon and other aquatic animals directly and within a short period of time. For example, in 1996, the herbicide acrolein was responsible for the death of approximately 92,000 steel-head, 114 juvenile Coho salmon, 19 resident rainbow trout, and thousands of nongame fish in the Bear Creek, a tributary of the Rogue River. Several laboratory experiments show that sublethal concentrations of agrochemicals can affect many aspects of salmon biology, including a number of behavioral effects.

Some pesticides such as organochlorines, organophosphates, and carbamates are known to cause morphological damage to the fish testis. They also affect female fish in the same way. They cause delayed oocyte development and inhibition of steroid hormone synthesis.

Experimental exposure of fish to pesticides has shown to depress protein values in the brain, gills, muscle, kidney, and liver. In the kidney and the liver, there is evidence of significant decrease in the protein content due to stress in elimination and also in metabolism (Tilak et al. 1991).

Organophosphorus pesticides vary in their toxicity to different species. Interference with endocrine hormones affects reproduction, immune function, development, and neurological functions in several species of wild animals. In fish, endocrine disruptors interrupt normal development and cause male fish to have female characteristics. These outward symptoms of developmental disruption are accompanied by reduced fertility and even sterility in adults as well as lower hatching rates and viability of offspring. Many studies show a direct relationship between concentrations of pesticides and related chemicals in fish tissues and depressed hormone concentrations. Disruption of the balance of endocrine hormones during the development of young fish can also cause defects of the skeletal system, resulting in deformities and stunted growth (Goodbred 1997). The common pesticide synergist piperonyl butoxide increases carbaryl toxicity (carbaryl is a neurotoxic carbamate pesticide). In fish, acute toxicity of a carbaryl–piperonyl butoxide mixture was over 100 times that of carbaryl alone (Singh and Agarwal 1989). In addition, carbaryl increases the acute toxicity of the phenoxy herbicide 2,4-D, the insecticides rotenone and dieldrin (an organochlorine), and the wood preservative pentachlorophenol (Statham and Lech 1975). Sublethal effects of the organophosphate insecticide phenthoate are also synergized by carbaryl in fish, resulting in AChE inhibition (Rao and Rao 1989) and both morphological and behavioral changes (Rao and Rao 1987). While the toxicity of combinations of chemicals is rarely studied, the ability of carbaryl to interact with a large number of chemical classes is striking.

Polychlorinated pesticides are endocrine-disrupting chemicals (EDCs). They may be harmful, as they disturb the endocrine system and reproduction in animals and humans (Fossi et al. 1999; Taylor and Harrison 1999). These changes are most visible in aquatic animals, as over 100,000 endocrine-active chemicals are discharged directly into freshwater and marine ecosystems (Trudeau and Tyler 2007). Their presence in the environment poses a threat to numerous populations and reduces biodiversity (Fossi et al. 2001).

OCPs have been extensively used for agriculture and vector control purposes in Tanzania. The pesticides applied on land eventually find their way to the aquatic environment, thus contaminating it. The pesticides are transported to aquatic bodies by rain runoff, rivers, and streams and associate with biotic and abiotic macroparticles (Colombo et al. 1990). They are removed from the surface to the benthic layers by settling of the particles into the water column (Allan 1986). The lipophilic nature, hydrophobicity, and low chemical and biological degradation rates of OCPs have led to their accumulation in biological tissues and the subsequent magnification of their concentrations in organisms progressing up the food chain (Swackhamer and Hites 1988; Vassilopoulou and Georgakopoulous-Gregoriades 1993). Consumption of biota from a contaminated aquatic body is considered to be an important route of exposure to persistent OC compounds (Johansen et al. 1996). Humans, being the final link in the food chain, are mostly affected, and consequently, the general public has become increasingly concerned about the potential risk to human health from the consumption of such polluted biota. Furthermore, OCPs in the aquatic environment have been reported to cause reproductive depression in aquatic biota (Helle et al. 1976).

13.5.1 Pesticide Residues in Edible Parts of Fish

Residues of OCPs in edible parts of fish are another problem (Darko et al. 2008; Li et al. 2008). Among food products, fish are considered the main source of these compounds in human diet. Fish products sometimes contain significant amounts of these compounds that pose a health risk to consumers (Binelli and Provini 2004; Sun et al. 2006). As OCPs have the ability to induce endocrine, metabolic, and reproductive disorders, their concentrations in food products, especially in edible fish species, should be monitored. Fish have been also used in numerous studies as the most effective bioindicators in the environmental monitoring of aquatic ecosystems (Kannan et al. 1995; Van der Oost et al. 2003; Hinck et al. 2008).

High bioaccumulation factors of these compounds in the gonads indicate a potential threat to populations of many animal species—especially aquatic ones, known to accumulate higher amounts of lipophilic organic compounds. In aquatic systems, OCPs easily join food webs, and their concentrations increase with each trophic level (Sun et al. 2006). In the case of fish, the accumulation of these compounds in the gonads may result in a reduced reproduction potential as well as in a decrease in fry number and developmental disorders.

Fish act as nonpolar media that can adsorb hydrophobic organic chemicals within the water column. Since birds and humans consume fish, this makes fish good biomonitors for xenobiotic pollutants. The ingestion of foods contaminated with persistent lipophilic pesticides can result in the accumulation of these pesticides in humans. The potential for pesticide residues to cross the placental barrier (Waliszewski et al. 2000), even if they were in trace concentrations, may cause serious damage in the newborns and therefore raises great concern. OCPs have already been implicated in a broad range of adverse human health and environmental effects, including reproductive failure and birth defects (Edwards 1987), immune system dysfunction, endocrine disruption, and cancers (Paasivirta 1991; World Wildlife Fund 1999). Because these chemicals are toxic to living organisms, an increased accumulation in the food chain may pose serious health hazards to the general populace (Jayashree and Vasudevan 2007). The pesticides can cause toxic symptoms similar to those caused by dioxin-like compounds, including developmental abnormalities and growth suppression, disruption of the endocrine system, impairment of immune function, and cancer promotion (GESAMP 1993). In order to avert any environmental or health disaster, it is necessary to put up mechanisms for monitoring residue levels in the ecosystem as well as the food chain (Shingare et al. 2009).

13.5.1.1 *Fish and Endocrine Disruption*

A study (Zaheer Khan and Law 2005) of sex hormones in carp indicates that pesticides may be affecting the ratio of estrogen to testosterone in both male and female fish. At stream sites with the highest concentrations of pesticides, the hormone ratio in the carp is significantly lower, indicating potential abnormalities in the endocrine system. The authors of the study conclude, "Reconnaissance assessment of sex steroid hormones in carp from United States streams indicates that fish in some streams within all regions studied may be experiencing some degree of endocrine disruption." According to the US Fish and Wildlife Service (FWS), "Endocrine disruption has the potential to compromise proper development in organisms, leading to reproductive, behavioral, immune system, and neurological problems, as well as the development of cancer. Effects often do not show up until later in life."

Pesticides are useful tools in agriculture and forestry, but their contribution to the gradual degradation of the aquatic ecosystem cannot be ignored (Konar 1975; Basak and Konar 1976; 1977). The aquatic ecosystem as a greater part of the natural environment is also faced with the threat of a shrinking genetic base and biodiversity. *Anabas testudineus, Channa punctatus,* and other indigenous small fish use paddy fields as breeding and nursery grounds. *Barbodes gonionotus* is an important species for integrated rice-fish farming. Pesticides at high concentrations are known to reduce the survival, growth, and reproduction of fish (Mckim et al. 1976) and produce many visible effects on fish (Johnson 1968). Due to the residual effects of pesticides, important organs such as the kidney, liver, gills, stomach, brain, muscles, and genital organs are damaged. Until the use of pesticides in crop farming is replaced by other means of pest control such as integrated pest management, less toxic pesticides at lowest possible doses need to be recommended.

13.6 Conclusion

Fish and shellfish raised by farming are likely to concentrate environmental contaminants and aquaculture drugs such as pesticide residues, antibiotic residues, hormone residues, etc. All these residues pose various kinds of health risks to consumers of these fish and fishery products. Thus, with various kinds of human activity, the occurrence of pathogens, pollutants, toxicants, and other undesirable compounds in fishery products is fast becoming a common phenomenon in developed and developing countries.

In spite of all these health hazards, fish and shellfish continue to be in great demand as food material in the developed world on account of the better taste, nutritional quality, and medicinal properties. To avoid public health problems in using fish and shellfish as food for mass consumption, several quality assurance programs were developed and enforced from time to time. Today, various kinds of quality standards like Codex standards, US FDA standards, EU Norms, BIS (Bureau of Indian Standards) standards, etc. are in operation at national and international levels. To achieve these standards, several quality assurance programs have also been developed and practised in different parts of the world. The HACCP system of the United States, the European Council directives, the QMP (Quality Management Program) of Canada, and TQM (Total Quality Management) of Japan are such quality assurance programs aimed to ensure safety and quality of fish and fishery products consumed in these countries.

With change of time, all these quality standards and quality assurance programs became more and more stringent and mandatory and posed severe challenges to the developing countries, which used to export a major share of their fish and fishery products to these developed countries. As a result, many developing countries including India had to face trade ban on fish exports. The imposition of HACCP in the early 1990s by US FDA, the EU ban of fishery products from India in 1997 on account of sanitation and hygiene, and the recent rejection of several consignments on account of antibacterial substances, antibiotic residues, heavy metal residues, muddy moldy flavor, etc. by the United States, EU, and Japan played havoc in Indian seafood industry. Even though TQM of Japan aimed total quality management, it failed to ensure both safety and quality, probably due to certain lacunae like calibration, good laboratory practice, good personnel policy, etc. There were several attempts to improve the TQM concept of Japan to make it suitable to tackle all problems of safety and quality of a given product.

References

Abell, A., Ernst, E., and Bonde, J. 2000. Semen quality and sexual hormones in greenhouse workers. *Scan. J. Work Environ. Health* 26(6): 492–500.

Adeyemi, D., Ukpo, G., Anyakora, C., and Unyimadu, J. P. 2008. Organochlorine pesticide residues in fish samples from Lagos Lagoon, Nigeria. *Am. J. Environ. Sci.* 4(6): 649–653.

Afful, S., Anim, A. K., and Serfor-Armah, Y. 2010. Spectrum of organochlorine pesticide residues in fish samples from Densu Basin. *Res. J. Environ. Earth Sci.* 2(3): 133–138.

Allan, R. J. 1986. The role of particulate matter in the transport and burial of contaminants in aquatic ecosystems. In: Hart, B. T. (ed.) *The Role of Particulate Matter in the Transport and Fate of Pollutants, Water Studies Center, Chisholm*. Institute of Technology: Melbourne, Australia.

Amorastt, M., Varothae, S., and Chunaoowatanakul, S. 1983. Determination of pesticides and some of the heavy metal residues in fish and water samples. *Proc. Fresh Water Fishes Epidemic* 1982–1983 Research Affairs, (pp. 135–152). Chulalongkorn University, Bangkok (Thailand) and Assessment Programme (GEMS/Food) in collaboration with Codex Committee on Pesticide Residues.

Anonymous. 1991a. Heptachlor. Evaluation for acceptable daily intake. IMPR (Joint FAO/WHOMeetingon Pesticide Residues) http://www.inchem.org//jmpr/jmpmono/v91pr13.htm.

Anonymous. 1991b. Endrin. Health and safety guide. 60. IPCS International Programme On Chemical Safety. http://www.inchem.org.

Anonymous. 2006. Assessment of lindane and other HCH isomers. United States Environmental Protection Agency (EPA). Regulation No. EPA-HQ-OPP-2006-0034-0002. http://www.epa.gov.

Anonymous. 2008. Mały rocznik statystyczny Polski. [Concise statistical yearbook of Poland.] Głowny Urząd Statystyczny, Warszawa.

Bakre, P. P., Misra, V., and Bhatnagar, P. 1990. Residue of organochlorine insecticides in fish from Mahla water reservoir, Jaipur, India. *Bull. Environ. Contam. Toxicol.* 45: 394–398.

Barasa, M. W., Lalah, J. O., and Wandiga, S. O. 2008. Seasonal variability of persistent organochlorine pesticide residues in marine fish along the Indian Ocean coast of Kenya. *Toxicol. Environ. Chem.* 90(3): 535–547.

Barua, S. 2007. Escaping from organic chemical pollutants. The new nation. Internet edition. November 19, 2007; http://nation.ittefaq.com/rss.xml.

Basak, P. K. and Konar, S. K. 1976. Toxicity of six insecticides to fish. *Geobios* 3: 209–210.

BCAS. Bangladesh Env. News Letter; No. 6, Vol. 2, Dhaka, Bangladesh. 1990; http://www.sos-arsenic.net/english/export/1.html.

Belda, E. and Kolankaya, D. 2006. Determination of organochlorine pesticide residues in water, sediment, and fish samples from the Meriç Delta, Turkey. *Int. J. Environ. Anal. Chem.* 86(1–2): 161–169.

Berbert, P. R. T., Abreu, J. M., Gradvohl Andj, M. P., and De-Abre, M. 1989. Toxicity of endosulfan to native and exotic fishes and crustaceans from the south of Bahia. *Agrotropica* 1: 144–152.

Bhatt, R. V. 2000. Environmental influence on reproductive health. *Int. J. Gynecol. Obstet.* 70(1): 69–75.

Bhuiyan, M. N. H., Bhuiyan, H. R., Rahim, M., Ahmed, K., Haque, K. M. F., Hassan, M. T., and Bhuiyan, M. N. I. 2008. Screening of organochlorine insecticides (DDT and heptachlor) in dry fish available in Bangladesh. *Bangladesh J. Pharmacol.* 3: 114–120.

Biego, G. H. M., Yao, K. D., Ezoua, P., and Kouadio, L. P. 2010. Assessment of organochlorine pesticides residues in fish sold in abidjan markets and fishing sites. *Afr. J. Food Agric. Nutr. Dev.* 10(2): 2152–2165.

Binelli, A. and Provini, A. 2004. Risk for human health of some pops due to fish from Lake Iseo. *Ecotoxicol. Environ. Saf.* 58(1): 139–145.

Bojakowska, I. and Gliwicz, T. 2005. Chloroorganiczne pestycydy i polichlorowane bifenyle w osadach rzek Polski. [Chlorinated pesticides and polychlorinated biphenyls in river sediments of Poland.] *Przegląd Geologiczny* 53: 649–655. [In Polish.]

Carson, R. 1962. *Silent Spring*, Houghton Mifflin Co.: New York.

Chandra, K. J. 2006. Fish parasitological studies in Bangladesh: A review. *J. Agric. Rural Dev.* 4: 9–18.

Chen, D. F. K. 1982. Analysis of organochlorine pesticide residues in Malaysian fish and the health risk to human consumption. Ann Arbor, University of Michigan, 1982 (doctoral thesis). Chlorinated insecticide in fishes from the Bay of Bengal. *Mar. Pollut. Bull.* 24: 567–570.

Chen, W., Zhang, L., Xu, L., Wang, X., and Hong, H. 1996. Concentrations and distributions of HCHs, DDTs and PCBs in surface sediments of sea areas between Xiamen and Jinmen. *J. Xaimen Univ. (Natural Science).* 35: 936–940 (in Chinese).

Chen, W., Hong, H., Zhang, L., Xu, L., Wang, X., and Hong, L. 2000. Residue levels and distribution features of organochlorine pollutants in the surface sediments of the sea areas between Minjiang Estuary and Mazu. *Mar. Sci. Bull.* 19: 53–58 (in Chinese).

Chen, W., Zhang, L., Xu, L., Wang, X., Hong, L., and Hong, H. 2002. Residue levels of HCHs, DDTs and PCBs in shellfish from coastal areas of east Xiamen Island and Minjiang Estuary, China. *Mar. Pollut. Bull.* 45(1–12): 385–390.

Chen, L. G., Mai, B. X., Bi, X. H., Ran, Y., Luo, X. J., Chen, S. J., Sheng, G. Y., Fu, J. M., and Zeng, E. Y. 2006. Concentration levels, compositional profiles, and gas-particle partitioning of polybrominated diphenyl ethers in the atmosphere of an urban city in South China. *Environ. Sci. Technol.* 40: 709–714.

Chowdhury, A., Raha, P., Guha, P., Kole, R., Banerjee, H., and Das, M. K. 1994. Effect of pesticides on the ulcerative disease syndrome of fish – a case study. *Pollut. Res.* 13: 161–167.

Ciereszko, W. and Witczak, A. 2002. Concentration of pcbs and selected pesticides in bottom sediments, zebra mussel and in some more important fish species of the Szczecin Lagoon. *Acta Ichthyologica et Piscatoria* 32(1): 35–40.

Clegg, D. J. and Gemert, M. V. 1999. Expert panel report of human studies on chlorpyrifos and/or other organophosphate exposures. *J. Toxicol. Environ. Health. Part B: Crit. Rev.* 2: 257–279.

Colombo, J. G., Khalil, M. F., Horth, A. C., and Catoggio, J. A. 1990. Distribution of chlorinated pesticides and individual polychlorinated biphenyls in biotic and abiotic compartments of the Rio de La Plata. *Environ. Sci. Technol.* 24: 498–505.

Cooper, C. M. and Knight, S. S. 1987. Residual Pesticides In fishes from Lake chicot, Arkansas. *Proc. Ark. Acad. Sci.* 41: 26–28.

Cunningham, W. P. and Saigo, B. W. 1999. *Environmental Science*, Custom edition. pp. 193–194. Mcgraw-Hill College: Boston.

Darko, G., Akoto, O., and Oppong, C. 2008. Persistent organochlorine pesticide residues in fish, sediments and water from Lake Bosomtwi, Ghana. *Chemosphere* 72(1): 21–24.

Devi, A. P., Rato, D. M. R., Tuak, K. S., and Murty, A. S. 1981. Relative toxicities of technical materials, isomers, and formulations of endosulfan to *Channa punctatus*. *Bull. Environ. Contam. Toxirol.* 27: 239–243.

Dishburger, H. J., McKellar, R. L., Pennington, J. Y., and Rice, J. R. 1997. Determination of residues of chlorpyrifos [insecticide], its oxygen analogue, and 3,5,6-trichloro-2-pyridinol in tissues of cattle fed. *J. Agric. Food Chem.* 25: 1325–1329.

Du Preez, H. H., Van Der Merwe, M., and Van Vuren, J. H. J. 1997. Bioaccumulation of selected metals in Mrican sharptooth catfish, *Clarias gariepinus* from the lower Olifants river, Mpumalanga, South Mrica. *Koedoe* 40: 77–90.

Edwards, C. A. 1987. The environmental impact of pesticides. *Parasitis* 86: 309–329.

Egan, K. 2002. FDA's Total Diet Study: Monitoring U.S. Food Supply Safety. Reproduced from Food safety magazine June/July 2002 with permission of the publisher. ©2002 by The Target Group.

EPA. 2006. The Evaluation of Methods for Creating Defensible, Repeatable, Objective and Accurate Tolerance Values for Aquatic Taxa. EPA 600/R-06/045 May 2006.

Falandysz, J., Brudnowska, B., Iwata, H., and Tanabe, S. 1999. Pestycydy chloroorganiczne i polichlorowane bifenyle w wodzie wiślanej. [Organochlorine pesticides and polychlorinated biphenyls in water of the Vistula River.] *Roczniki Państowego Zakładu Higieny* 50(2): 123–130. [In Polish.]

Fang, Z. Q., Cheung, R. Y. H., and Wong, M. H. 2001. Concentrations and distribution of organochlorinated pesticides and PCBs in green-lipped mussels, *Perna viridis* collected from the Pearl River estuarine zone. *Acta Scientiae Circumstantiae* 21(1): 113–116 (in Chinese).

FDA. 1999. Food and Drug Administration Pesticide Program: Residue Monitoring 1998, US Food and Drug Administration Center for Food Safety and Applied Nutrition.

FDA. 2000. Food and Drug Administration Pesticide Program: Residue Monitoring 1999, US Food and Drug Administration Center for Food Safety and Applied Nutrition.

FDA. 2002. Food and Drug Administration Pesticide Program: Residue Monitoring 2000, US Food and Drug Administration Center for Food Safety and Applied Nutrition.

FDA. 2003. Food and Drug Administration Pesticide Program: Residue Monitoring 2001, US Food and Drug Administration Center for Food Safety and Applied Nutrition.

Ferrando, M. D., Gamo, M., and Andreu, E. 1992. Accumulation and distribution of pesticides in *Anguilla anguilla* from Albiferra lake (Spain). *J. Environ. Biol.* 13: 75–82.

Fossi, M. C., Casini, S., and Marsili, L. 1999. Nondestructive biomarkers of exposure to endocrine disrupting chemicals in endangered species of wildlife. *Chemosphere* 39 (8): 1273–1285.

Fossi, M. C., Casini, S., Ancora, S., Moscatelli, A., Ausili, A., and Notarbartolo di Sciara G. 2001. Do endocrine disrupting chemicals threaten Mediterranean swordfish? Preliminary results of vitellogenin and Zonaradiata proteins in *Xiphiasgladius. Mar. Environ. Res.* 52 (5): 477–483.

Fu, J. M., Mai, B. X., Sheng, G. Y., Zhang, G., Wang, X. M., Peng, P. A., Xiao, X. M., Ran, R., Cheng, F. Z., Peng, X. Z., Wang, Z. S., and Tang, U. W. 2003. Persistent organic pollutants in environment of the Pearl River Delta, China: An overview. *Chemosphere* 52: 1411–1422.

GESAMP. 1993. Impact of Oil and Related Chemicals and Wastes on the marine Environment. Rep. Stu. GESAMP 50: 180 p.

Golec, J., Hanke, W., and Dąbrowski, S. 2003. Ryzyko zaburzeń płodności u osob zawodowo eksponowanych na pestycydy. [The risk of infertility in people occupationally exposed to pesticides.] *Med. Pr.* 54(5): 465–472.

Goodbred, S. L., Gilliom, R. J., Gross, T. S., Denslow, P., Bryant, W. L., and Schoeb, T. R. 1997. Reconnaissance of 17B-estradiol, 11-ketotestosterone, vitellognin, and gonad histopathology incommon carp of United States streams: Potential for contaminant-induced endocrine disruption: U.S. Geological Survey, Open-File Report, pp. 96–627.

Gregoraszczuk, E. L., Milczarek, K., Wójtowicz, A. K., Berg, V., Skaare, J. U., and Ropstad, E. 2008. Steroid secretion following exposure of ovarian follicular cells to three different natural mixtures of persistent organic pollutants (pops). *Reprod. Toxicol.* 25(1): 58–66.

Grobler, D. F. 1994. A note on PCBs and chlorinated hydrocarbon pesticide residues in water, fish and sediment from Olifants River, Eastern Transvaal, South Africa. *Water SA* 20(3): 187–194.

Guo, Y., Meng, X.-Z., Tang, H.-L., and Zeng, E. Y. 2008. Tissue distribution of organochlorine pesticides in fish collected from the Pearl River Delta, China: Implications for fishery input source and bioaccumulation. *Environ. Pollut.* 155(1): 150–156.

Haider, S. and Inbaraj, M. 1986. Relative toxicity of technical material and commercial formulation of malathion and endosulfan to a freshwater fish, *Channa punctatus* (Bloch). *Ecotoxicol. Environ. Saf.* 11: 347–351.

Halder, P., Raha, P., Bhatiacharya, A., Choudhury, A., and Adityachoudhury, N. 1989. Studies on the residues of DDT and endosulfan occurring in Ganga water. *Indian J. Environ. Health.* 31: 156–161.

Hart, C. A., Nisbet, I. C. T., Kennedy, S. W., and Hahn, M. E. 2003. Gonadal feminization and halogenated environmental contaminants in common terns (*Sterna hiduro*): Evidence that ovotestes in
male embryos do not persist to the prefledgling stage. *Ecotoxicology* 12(1–4): 125–140.

Heath, R. G. M. 1992. The levels of organochlorine pesticides in indigenous fish from two rivers that
flow through the Kruger ational Park, South Africa. *Water Supply* 10: 177–185.

Helle, E., Olsson, M., and Jensen, S. 1976. DDT and PCB levels and reproduction in ringed seal from
the Bothnian Bay. *Ambio* 5: 188–189.

Henny, C. J., Grove, R. A., and Kaiser, J. L. 2008. Osprey distribution, abundance, reproductive
success and contaminant burdens along lower Columbia River, 1997/1998 versus 2004. *Arch.
Environ. Contam. Toxicol.* 54(3): 525–534.

Hinck, J. E., Balzer, V. S., Denslow, N. D., Echols, K. R., Gale, R. W., Wieser, C., May, T. W., Ellersieckm,
M., Coyle, J. J., and Tillitt, D. E. 2008. Chemical contaminants, health indicators, and reproductive biomarker responses in fish from rivers in the Southeastern United States. *Sci. Total Environ.*
390(2–3): 538–557.

Holden, A. V. 1962. A study of the absorption of C14-1abeled DDT from water by fish. *Ann. Appl.
Biol.* 50: 467.

Hong, H., Xu, L., Zhang, L., Chem, J. C., Wong, Y. S., and Wan, T. S. M. 1995. Environmental fate and
chemistry of organic pollutants in the sediment of Xiamen and Victoria Harbours. *Mar. Pollut.
Bull.* 31: 229–236.

ICAR. 2002. All India coordinated Research Project on Pesticide residue ICAR.

Jana, S. and Bandyopadhyaya, S. 1987. Effect of heavy metals on some biochemical parameters in the
freshwater fish *Channa punctatus*. *Environ. Ecol.* 5: 488–493.

Japanese Trust Fund II Project on Research and Analysis of Chemical Residues and Contamination
in Fish and Fish Products, 2004–2008.

Jaward, F. M., Farrar, N. J., Harner, T., Sweetman, A. J., and Jones, K. C. 2004. Passive air sampling
of PCBs, PBDEs and organochlorine pesticides across Europe. *Environ. Sci. Technol.* 38: 34–41.

Jayashree, R. and Vasudevan, N. 2007. Effect of Tween 80 added to the soil on the degradation of
endosulphan by Pseudomonas aeruginosa. *Int. J. Environ. Sci. Technol.* 4(2): 203–210.

Johansen, H. R., Alexander, J., Rossland, O. J., Planting, S., Lovik, M., Gaarder, P. I., Gdynia, W., Bjerve,
K. S., and Becher, G. 1996. pcdds, pcdfs and pcbs in human blood in relation to consumption of
crabs from a contaminated Fjord area in Norway. *Environ. Health Perspect.* 7: 756–764.

Johnson, D. W. 1968. Pesticide and fishes, a review of selected literature, *Trans. Am Fish Soc.* 97: 398

Johnson, W. L. and Finley, M. T. 1980. *Handbook of Acute Toxicity of Chemicals to Fish and Aquatic
Invertebrates*, U.S. Fish and Wildlife Service, Washington, DC, Resour. Publ. 137, 98 pp.

Johnson, D. E., Seidler, F. J., and Slotkin, T. A. 1998. Early biochemical detection of delayed neurotoxicity resulting from developmental exposure to chlorpyrifos. *Brain Res. Bull.* 45: 143–147.

Kale, S. P., Sarma, G., Goswami, U. C., and Ragh, K. 1996. Uptake and distribution of 14C carbofuran
and 14C-HCH in catfish. *Chemosphere* 33: 449–451.

Kannan, K., Tanabe, S., Tatsukawa, R., and Sinha, R. K. 1994. Biodegradation capacity and residue
pattern of organochlorines in Ganges river Dolphins from India. *Toxicol. Environ. Chem.* 42:
249–261.

Kannan, K., Tanabe, S., and Tatsukawa, S. 1995. Geographical distribution and accumulation features
of organochlorine residues in fish in tropical Asia and Oceania. *Environ. Sci. Technol.* 29(10):
2673–2683.

Kasozi, G. N., Kiremire, B. T., Bugenyi, F. W. B., Kirsch, N. H., and Nkedi-Kizza, P. 2005. Organochlorine
residues in fish and water samples from Lake Victoria, Uganda. *J. Environ. Qual.* 35(2): 584–589.

Kegley, S., Neumeister, L., and Martin, T. 1999. *Ecological Impacts of Pesticides in California*, p. 99.
Pesticide Action Network: San Francisco, CA.

Konar, S. K. 1975. Pesticides and aquatic ecosystems. *Indian J. Fish.* 22(1–2): 80–85.

Kumari, A., Sinha, R. K., Gopal, K., and Prasad, K. 2001. Dietary intake of persistent organochlorine
residue through Gangetic fishes in India. *Int. J. Ecol. Environ. Sci.* 27: 117–120.

Lars, H. 2000. Environmental exposure to persistent organohalogen and health risks. In: Lennart, M.
(ed.), *Environmental Medicine*. Ch: 12, Retrieved from: www.envimed.com.

Leyla, K., İhsan A., and Abdurrahman, A. 2009. Some organochlorine pesticide residues in fish species in Konya, Turkey. *Chemosphere* 74(7): 885–889.

Li, X., Gan, Y., Yang, X., Zhou, J., Dai, J., and Xu, M. 2008. Human health risk of organochlorine pesticides (ocps) and polychlorinated biphenyls (pcbs) in edible fish from Huairou Reservoir and Gaobeidian Lake in Beijing, China. *Food Chem.* 109(2): 348–354.

Lloyd, R. 1992. What is pollution? Definition and effects. In: *Pollution and Freshwater Fish*, pp. 12–23. Hartnolls: Bodmin, Cornwall.

Luo, X. J., Mai, B. X., Yang, Q. S., Fu, J. M., Sheng, G. Y., and Wang, Z. S. 2004. Polycyclic aromatic hydrocarbons (PAHs) and organochlorine pesticides in water columns from the Pearl River and the Macao harbor in the Pearl River Delta in South China. *Mar. Pollut. Bull.* 48: 1102–1115.

Machiwa, J. F. 2000. Heavy metals and organic pollutants in sediments of Dar es Salaam harbor prior dredging in 1999. *Tanz. J. Sci.* 26: 29–45.

Mai, B. X., Fu, J. M., Sheng, G. Y., Kang, Y. H., Lin, Z., Zhang, G., Min, Y. S., and Zeng, E. Y. 2002. Chlorinated and polycyclic aromatic hydrocarbons in riverine and estuarine sediments from Pearl River Delta, China. *Environ. Pollut.* 117: 457–474.

Malik, A., Kunwar P. Singh, and Ojha P. 2007. Residues of Organochlorine Pesticides in Fish from the Gomti River, India. *Bull. Environ. Contam. Toxicol.* 78: 335–340.

Manirakiza, P., Akimbamijo, O., Covaci, A., Adediran, S. A., Cisse, I., Fallc, S. T., and Schepens, P. 2002. Persistent chlorinated pesticides in fish and cattle fat and their implications for human serum concentrations from the Sene-Gambian region. *J. Environ. Monit.* 4: 609–617.

Marth, E. H. 1965. Residues and some effects of chlorinated hydrocarbon insecticides in biological materials. *Res. Rev.* 9: 1–89.

Mathlessen, P., Fox, P. J., Douthwaite, R. J., and Woo, A. B. 1982. Accumulation of endosulfan residues in fish and their predators after aerial spraying for the control of tsetse fly in Botswana. *Pestic. Sci.* 13: 39–48.

Matin, M. A. 2002. Pesticides in Bangladesh. In: Taylor, M. D., Klaine, S. J., Carvahlo, F. P., Barcello, D., and Everaarts, J. M. (eds), Pesticide residues in coastal tropical ecosystems. Distribution, fate, & effects, pp. 137–58. Taylor and Francis: London.

McKim, J. M., Olson, G. F., Holcombe, G. W., and Hunt, E. P. 1976. Long-term effects of methylmercuric chloride on three generations of brook trout (Salvelinus fontinalis): toxicity, accumulation, distribution and elimination. *J. Fish. Res. Bd. Can.* 33: 2726–2739.

Md. Wasim Aktar, Paramasivam, M., Sengupta, D., Purkait, S., Ganguly, M., and Banerjee, S. 2009. Impact assessment of pesticide residues in fish of Ganga river around Kolkata in West Bengal. *Environ. Monit. Assess.* 157: 97–104.

Monirith, I., Ueno, D., Takahashi, S., Nakata, H., Sudaryanto, A. and Subramanian, A. 2003. Asia-Pacific Mussel Watch: monitoring contamination of persistent organochlorine compounds in coastal waters of Asian countries. *Mar. Pollut. Bull.* 46: 281–300.

Muralidharan, S., Dhananjayan, V., and Jayanthi, P. 2009. Organochlorine pesticides in commercial marine fishes of Coimbatore, India and their suitability for human consumption. *Environ. Res.* 109(1): 15–21.

Murty, A. S. 1986. *Toxicity of Pesticides to fish*, Vols. I and II, 483 pp, 355 pp. CRC Press Inc: Boca Raton, FL.

Nayak, A. K., Raha, R., and Das, A. K. 1995. Organochlorine pesticides residues in middle stream of the Ganga River, India. *Bull. Environ. Contam. Toxicol.* 54: 68–75.

Nigam, U., Hans, R. K., Prakash, S., Seth, T. D., and Siddiqui, M. K. J. 1998. Biomagnification of organochlorine pesticides and metals in biota of an Indian lake. *Poll. Res.* 17(1): 83–86.

Paasivirta, J. 1991. *Chemical Ecotoxicology*. Lewis Publishers: Chelsea, MI.

Pandit, G. G., Sahu, S. K., Sharma, S., and Puranik, V. D. 2006. Distribution and fate of persistent organochlorine pesticides in coastal marine environment of Mumbai. *Environ. Int.* 32(2): 240–243.

Peterson, S. M. and Bately, G. E. 1993. The fate of endosulfan in aquatic ecosystem. *Environ. Pollut.* 82: 143–152.

Phillips, D. J. H. 1985. Organochlorine and trace metals in green-lipped mussel's *perna viridis* from HongKong waters: a test of indicator ability. *Mar. Ecol. Progr Series* 21: 252–258.

Pimentel, D., Acquay, H., Biltonen, M., Rice, P., Silva, M., Nelson, J., Lipner, V., Giordano, S., Horowitz, A., and D'Amore, M. 1993. Assessment of environmental and economic costs of pesticide use. In: Pimentel, D. and Lehman, H. (eds), *The Pesticide Question: Environment, Economics and Ethics,* pp. 47–84. Routledge, Chapman and Hall: New York.

Porta, M., Malats, N., Jariod, M., Grimalt, J. O., Rifà, J., Carrato, A., Guarner, L., Salas, A., Santoiago-Silva, M., Corominas, J. M., Andreu, M., and Real, F. X. 1999. Serum concentration of organochlorine compounds and K-ras mutation in exocrine pancreatic cancer. *Lancet* 354(9196): 2125–2129.

Porte, C. and Albaigés, J. 2002. Residues of pesticides in aquatic organisms. *Revue Méd. Vét.* 153(5): 345–350.

Protasowicki, M., Ciereszko, W., Perkowska, A., Ciemniak, A., Bochenek, I., Brucka-Jastrzębska, E., and Błachuta, J. 2007. Metale ciężkie i chlorowane węglowodory w niektorych gatunkach ryb z rzeki Odry. [Heavy metals and polychlorinated biphenyls in certain species of fish from the Oder River.] *Rocznik Ochrona Środowiska* 9: 95–105.

Radhakrishnan, A. G. and Antony, P. D. 1989. Pesticide residues in marine fishes. *Fishery Technol.* 26: 60–61.

Rajendran, R. B., Karunagaran, V. M., Babu, S., and Subramanian, A. N. 1992. Levels of chlorinated insecticide in fishes from the Bay of Bengal. *Mar. Pollut. Bull.* 24: 567–570.

Rao, K. R. S. S. and Rao, J. C. 1987. Independent and combined action of carbaryl and phenthoate on adverse effects of pesticides and chemicals on snake head, *Channa punctatus* (Bloch). *Curr. Sci.* 56: 331–332.

Rao, K. R. S. S. and Rao, K. V. R. 1989. Combined action of carbaryl and phenthoate on the sensitivity of the acetylcholinesterase system of the fish. *Ecotoxicol. Environ. Saf.* 17: 12–15.

Ray, P. K. 1992. Measurement of Ganga River quality-heavy metals and pesticides. Project report. Industrial Toxicology Research Centre, Lucknow, India.

Sapozhnikova, Y., Zubcov, N., Hungerford, S., Roy, L. A., Boicenco, N., Zubcov, E., and Schlenk, D. 2005. Evaluation of pesticides and metals in fish of the Dniester River, Moldova. *Chemosphere* 60(2): 196–205.

Senthil Kumar, K., Kannan, K., Sinha, R. K., Tanabe, S., and Giesy, J. P. 1999. Bioaccumulation profiles of polychlorinated biphenyl congeners and organochlorine pesticides in Ganges river dolphins. *Environ. Toxicol. Chem.* 18: 1511–1520.

Shailaja, M. S. and Sengupta, R. 1989. DDT residues in fishes from the eastern Arabian Sea. *Mar. Pollut. Bull.* 20: 629–630.

Shailaja, M. S. and Singhal, S. Y. S. 1994. Organochlorine pesticides compounds in organism from the Bay of Bengal. *Estuar. Coast. Shelf Sci.* 39: 219–226.

Shailaja, M. S. and Nair, M. 1997. Seasonal difference in organochlorine pesticide concentrations of zooplankton and fish in the Arabian Sea. *Mar. Environ. Res.* 44: 253–274.

Shingare, A. M., Gaikwad, M. V., Pathan, T. S., Thete, P. B. and Khillare, Y. K. 2009. Study on bioaccumulation of lindane in various tissues of *Channa gachua* from Aurangabad district (MS) India. *World J. Zool.* 4(2): 148–152.

Singh, D. K. and Agarwal, R. A. 1989. Toxicity of piperonyl butoxide-carbaryl synergism on the snail *Lymnaea acuminata. Int. Revue ges. Hydrobiol.* 74: 689–699.

Singh, B. and Garg, A. K. 1992. Observation on the accumulation of 14C-labelled carbofuran in different organs of fish. *Nat. Acad. Sci. Lett.* 15: 343–344.

Singh, P. and Singh, V. 2008. Pesticide bioaccumulation and plasma sex steroids in fishes during breeding phase from north India. *Environ. Toxicol. Pharmacol.* 25(3): 342–350.

Statham, C. N. and Lech, J. J. 1975. Potentiation of the acute toxicity of several pesticides and herbicides in trout by carbaryl. *Toxicol. Appl. Pharmacol.* 34: 83–87.

Sun, F., Wong, S. S., Li, G. C., and Chen, S. N. 2006. A preliminary assessment of consumer's exposure to pesticide residues in fisheries products. *Chemosphere* 62(4): 674–680.

Susan, L., Schantz, S. L., Gasior, D. M., Polverejan, E., McCaffrey, R. J., Sweeney, A. M., Humphrey, H. E. B., and Gardiner, J. C. 2001. Impairments of memory and learning in older adults exposed to polychlorinated biphenyls via consumption of Great Lakes fish. *Environ. Health Perspect.* 109: 605–611.

Swackhamer, D. and Hites, R. A. 1988. Occurrence and bioaccumulation of organochlorine compounds in fish from Siskiwit Lake, Isle Royale, Lake Superior. *Environ. Sci. Technol.* 22: 543–548.

Syasina, I. G. 2003. Organochlorine pesticides in fishes and mollusks from lower reaches of the Tumen River and of the contiguous part of Peter the Great Bay (Sea of Japan). *Russian J. Mar. Biol.* 29(1): 23–30.

Tanabe, S., Tatsukawa, R., and Phillips, D. J. H. 1987. Mussels as bioindicators of PCB pollution: A case study of uptake and release of PCB isomers and congeners in green-lipped mussels (*perna viridis*) in Hong Kong waters. *Environ. Pollut.* 47: 41–48.

Taylor, M. R. and Harrison, P. T. C. 1999. Ecological effect of endocrine disruption: Current evidence and research priorities. *Chemosphere* 39(8): 1237–1248.

Tilak, K. S., Janardhana Rao, N. H., and Lakshmi, J. 1991. Effect of pesticides mixed in different ratios to the freshwater *Lareo rohita*. *J. Ecotoxicol. Environ. Onit.* 1: 49–52.

Tolosa, I., Bayona, J. M., and Albaigés, J. 1995. Spatial and temporal distribution, fluxes, and budgets of organochlorinated compounds in Northwest Mediterranean sediments. *Environ. Sci. Technol.* 29: 2519–2522.

Tomza-Marciniak, A. and Witczak, A. 2010. Distribution of endocrine-disrupting pesticides in water and fish from the Oder River, Poland. *Acta Ichthyologica Et Piscatoria* 40(1): 1–9.

Trudeau, V. and Tyler, C. H. 2007. Endocrine disruption. *Gen. Comp. Endocrinol.* 153(1–3): 13–14.

UNEP. 1997. Report of the meeting of experts to review the MEDPOL biomonitoring programme. Athens. Greece. Document UNEP-(OCA)/MED WG. 132/7.

UNEP. 1998. Eastern African Action Plan – Kenya. Strategic action plan for land-based sources and activities affecting the marine, coastal and associated freshwater environment in the Eastern African region, UNEP, Nairobi.

Van der Oost, R., Beyer, J., and Vermeulen, N. P. E. 2003. Fish bioaccumulation and biomarkers in environmental risk assessment: A review. *Environ. Toxicol. Pharmacol.* 13(3): 57–149.

Van der Werf, H. M. G. 1996. Assessing the impact of pesticides on the environment. *Agric. Ecosyst. Environ.* 60(2–3): 81–96.

Vassilopoulou, V. and Georgakopoulous-Gregoriades, E. 1993. Factors influencing the uptake of organochlorines in red mullet (*Mullus barbatus*) from a gulf of Central Greece. *Mar. Poll. Bull.* 26: 285–287.

Waliszewski, S. M., Aguirre, A. A., Infanzon, R. M., and Siliceo, J. 2000. Carry-over of persistent organochlorine pesticides through placenta to fetus. *Salud Publica Mex.* 42: 384–90.

Wania, F. and Mackay, D. 1996. Tracking the distribution of persistent organic pollutants. *Environ. Sci. Technol/News* 30(9): 390A–396A.

World Wildlife Fund, 1999. Hazards and exposures associated with DDI and synthetic pyrethroids used vector control. World Wildlife Fund, Washington, DC.

Wu, S. J., Chang, Y. H., Fang, C. W., and Pan, W. H. 1999. Food sources of weight, calories, and three macro-nutrients- NAHSIT 1993–1996. *Nutr. Sci. J.* 24: 41–58.

Yim, U. H., Hong, S. H., Shim, W. J., and Oh, J. R. 2005. Levels of persistent organochlorine contaminants in fish from Korea and their potential health risk. *Arch. Environ. Contam. Toxicol.* 48(3): 358–366.

Zaheer Khan, M. and Law, F. C. P. 2005. Adverse effects of pesticides and related chemicals on enzyme and hormone systems of fish, amphibians and reptiles: A review. *Proc. Pakistan Acad. Sci.* 42(4): 315–323.

Zhang, Y. L. 2001. *Ph.D. Thesis*, Xiamen University.

Zhang, G., Parker, A., House, A., Mai, B., Li, X., Kang, Y., and Wang, Z. 2002. Sedimentary records of DDT and HCH in the Pearl River Delta, South China. *Environ. Sci. Technol.* 36: 3671–3677.

Zhou, H. Y. and Wong, M. H. 2004. Screening of organochlorines in freshwater fish collected from the Pearl River Delta, People's Republic of China. *Arch. Environ. Contam. Toxicol.* 46: 106–113.

Zhou, J. L., Maskaoui, K., Qiu, Y. W., Hong, H. S., and Wang, Z. D. 2001. Polychlorinated biphenyl congeners and organochlorine insecticides in the water column and sediments of Daya Bay, China. *Environ. Pollut.* 113: 373–384.

Zhou, R., Zhu, L., Yang, K., and Chen, Y. 2006. Distribution of organochlorine pesticides in surface water and sediments from Qiantang River, East China. *J. Hazard. Mater.* 137(1): 68–75.

Zhou, R., Zhu, L., and Kong, Q. 2007. Persistent chlorinated pesticides in fish species fromqiantang River in East China. *Chemosphere* 68(5): 838–847.

Zhou, R., Zhu, L., and Kong, Q. 2008. Levels and distribution of organochlorine pesticides in shellfish from Qiantang River, China. *J. Hazard. Mater.* 152(3): 1192–1120.

14

Pesticide Residues in Birds and Mammals

Raimon Guitart

CONTENTS

14.1 Introduction

Pesticides have been used extensively since the end of World War II and are, at present, integral to modern agriculture, forestry, animal production, and public health across the world. Roughly, some 3000 million kg of pesticides are applied worldwide per year, 500 million kg corresponding to the United States. In spite of the unquestionable benefits, many of these pest-control chemicals have also the potential or measurable adverse effects on nontarget vertebrate domestic animals and wildlife (Kendall and Smith 2003; Poppenga 2007; Rattner 2009; Poppenga and Oehme 2010). By design, manufacture, and use, pesticides are biologically active and, in most cases, toxic substances. In the twentieth and twenty-first centuries, the fate of pesticide residues in the environment has become a critical issue in both developed and developing countries.

Although human health continues to be the main concern for the regulators, as a part of environmental risk assessment, it is important to balance economic and health benefits to the general public against acceptability of the adverse effects on nontarget fauna when pesticides are used (Burger 1997). Some of these negative aspects are drastic and easily assessable by means of sign observation, necropsy findings, and/or analytical chemistry results, but others are less obvious and more difficult to ascertain due to interspecific variations in susceptibility among animals, especially in the wild and when long-term effects or multiple exposures are considered (Hoffman et al. 1990; Blus and Henny 1997; Kendall and Smith 2003; Walker 2003; Plumlee 2004; Brakes and Smith 2005; Mineau 2005; Berny 2007; Poppenga 2007; Smith et al. 2007; Martínez-Haro et al. 2008; Aktar et al. 2009; Sonne 2010).

Vertebrate wildlife exists in a much less well controlled environment than companion animals, livestock, and poultry and is therefore exposed to a greater variety of different pesticides through contaminated air, water, soil, or food (Newton 1998; Smith et al. 2007). Even when no adverse health effects can be verified or confirmed with the current premortem and postmortem techniques and methodologies available, the measurement of contaminant levels in wildlife can be used as an index of the environmental quality of the ecosystems (Peakall 1992).

Pets can play the same role in domestic environments (O'Brien et al. 1993; Backer et al. 2001; Kunisue et al. 2005; Schmidt 2009). Moreover, monitoring the pesticide levels in the samples of animal foods, including meat, milk, and eggs, which historically preceded those of the residues in wildlife, started when concerns arose about the safety of human foods and is, nowadays, routinely carried out in many countries (Keith 1996; Fontcuberta et al. 2008; Glynn et al. 2009; Lutze et al. 2009).

Given the breadth of the topic, this chapter is not an exhaustive treatise, but will hopefully provide an updated overview and the state of the art of major avian and mammalian pesticide residue exposure data. This review is focused (but is not restricted) to a selection of information published in the last 10 years, from 2000 up to the end of 2010, at least for the most typically monitored pesticide pollutants. Table 14.1 gives the common names, the Chemical Abstracts Service (CAS) registry numbers, and the oral acute toxicities (LD_{50}, μg/g BW) of some selected pesticides in birds (usually, mallard ducks, quails, or chickens) and mammals (rat), which will be discussed in depth next.

14.1.1 Addressing the Terminology

Definitions of terms and concepts differ among authors and need to be addressed for the purpose of this review, in order to proceed further. First of all, the negative effects of pesticides in individuals or populations of birds and mammals can be classified basically into two main groups: direct and indirect (Boatman et al. 2004; Berny 2007). Direct effects require exposure to the pesticides and are complex processes that have some different variants and types, not necessarily exclusive among them. We have, for example, lethal and sublethal, accidental and intentional, primary and secondary, and short- (acute) and long-term (chronic) exposures, which will be discussed next with some examples. This group of direct effects is, by far, the most studied one.

TABLE 14.1

Oral Acute Toxicity (LD$_{50}$, µg/g of Body Weight) of Some Selected Pesticides in Birds (Usually Chicken, Mallard duck, or Quail) and Mammals (Rat)

Common Name	CAS Registry Number	LD$_{50}$ Birds	LD$_{50}$ Mammals
Aldicarb	116-06-3	0.5	2
Aldrin	309-00-2	42.1	39
Brodifacoum	56073-10-0	2	0.16
Bromadiolone	28772-56-7	1,600	0.49
Carbofuran	1563-66-2	3.16	5
Chlordane	57-74-9	1,200	200
Chlorophacinone	3691-35-8	100	2.1
Chlorpyrifos	2921-88-2	13.3	82
Coumaphos	56-72-4	13.3	13
p,p'-DDD	72-54-8	2,000	113
p,p'-DDE	72-55-9	—	880
p,p'-DDT	50-29-3	595	87
Diazinon	333-41-5	4.21	66
Dichlorvos	62-73-7	22.01	17
Dicofol	115-32-2	4,365	575
Dieldrin	60-57-1	10.78	38.3
Difenacoum	56073-07-5	50	0.68
Difethialone	104653-34-1	0.87	0.55
Endosulfan	115-29-7	33	18
Famphur	52-85-7	1.8	28
Fenthion	55-38-9	11	180
Flocoumafen	90035-08-8	300	0.25
Fonofos	944-22-9	12	3
Glyphosate	1071-83-6	4,640	4,873
Hexachlorobenzene (HCB)	118-74-1	6,400	10,000
α-Hexachlorocyclohexane	319-84-6	—	177
β-Hexachlorocyclohexane	319-85-7	—	6,000
γ-Hexachlorocyclohexane (Lindane)	58-89-9	2,000	76
Malathion	121-75-5	1,485	290
Methomyl	16752-77-5	23.7	14.7
Mirex	2385-85-5	2,400	235
Monocrotophos	6923-22-4	4	8
Monosodium methanearsonate (MSMA)	2163-80-6	—	700
Paraquat	4685-14-7	262	100
Pentachlorophenol (PCP)	87-86-5	380	27
Permethrin	52645-53-1	13,500	383
Strychnine	57-24-9	23	2.35
Toxaphene	8001-35-2	31	50
Warfarin	81-81-2	942	1.6
Zinc phosphide	1314-84-7	13.5	12

The CAS registry numbers are also provided for unequivocal identification. Data were compiled from the ChemIDplus Advanced of the United States National Library of Medicine (http://chem.sis.nlm.nih.gov/chemidplus); Canadian Council of Ministers of the Environment. 1999. Canadian tissue residue guidelines for the protection of wildlife consumers of aquatic biota: DDT (total). In: *Canadian Environmental Quality Guidelines*. Canadian Council of Ministers of the Environment: Winnipeg; and Martínez-Haro, M., Mateo, R., Guitart, R., Soler-Rodríguez, F., Pérez-López, M., Maria-Mojica, P., and García-Fernández, A. J. 2008. Relationship of the toxicity of pesticide formulations and their commercial restrictions with the frequency of animal poisonings. *Ecotoxicol. Environ. Safe.* 69: 396–402.

On the other hand, the indirect effects are subtler, because the pesticides exert their effects without the necessity of exposure of the affected individuals (lethal effects can occur, but not related to direct pesticide poisoning), and take place when the use of certain chemicals has effects on plants or prey and thus on the diet of their natural consumers (Freemark and Boutin 1995; Mañosa et al. 2001; Sullivan and Sullivan 2003; Boatman et al. 2004; Morris et al. 2005).

14.1.1.1 Indirect Effects

It is well known that the density and reproduction of specialist carnivorous species are intimately linked to the population dynamics of their prey, affecting productivity or survival. Indirect effects have been demonstrated, for example, in the treatments with rodenticides in England (Brakes and Smith 2005), intended to eradicate rats (*Rattus* spp.), which also affected the populations of nontarget small mammals such as European wood mice (*Apodemus sylvaticus*), bank voles (*Clethrionomys glareolus*), and field voles (*Microtus agrestis*). These animals are important in the diet of many species such as barn owls (*Tyto alba*), long-eared owls (*Asio otus*), short-eared owls (*Asio flammeus*), tawny owls (*Strix aluco*), hen harriers (*Circus cyaneus*), European kestrels (*Falco tinnunculus*), stouts (*Mustela erminea*), and weasels (*Mustela nivalis*), species that were in turn indirectly affected by the use of the pesticides through the reduction of available food.

A more recent example of indirect effects was reported in France (Poulin et al. 2010) and was related to the use of the microbial insecticide *Bacillus thuringiensis* var. *israelensis* (Bti), considered the most selective and least toxic pesticide currently available worldwide to control mosquitoes. House martins (*Delichon urbicum*) in Bti-treated zones were adversely affected in their clutch size (2.3 vs. 3.2 chicks produced per nest) and in the number of young that actually fledged, both smaller than in the control zones. The explanation was related to the diet: small prey (<2.5 mm), such as flying ants, were consumed in greater proportion in treated areas by house martins, while large prey (<7.5 mm), such as midges, mosquitoes, dragonflies, and spiders, dominated in the control ones.

14.1.1.2 Direct Effects

Lethal poisoning in nontarget birds and mammals due to significant exposures to pesticides are not uncommon (Fleischli et al. 2004; Kwon et al. 2004; Wang et al. 2007; Berny et al. 2010; Guitart et al. 2010a,b). However, most of the reported mortalities are due to intentional poisonings, which are frequently a result of acute exposure to a restricted number of well-known and, sometimes, easily available poisonous pesticides, mainly organophosphorus (OP) and carbamate (CB) insecticides, strychnine, and anticoagulant rodenticides (ARs) (de Snoo et al. 1999; Berny 2007; Wang et al. 2007; Muzinic 2007; Hernández and Margalida 2008; Martínez-Haro et al. 2008). When pesticides are used in a deliberate or illegal attempt to poison animals, secondary poisonings, whereby a predator or scavenger consumes the flesh of primarily exposed species, can easily occur because of the high doses of pesticides usually employed (Wobeser et al. 2004; Martínez-Haro et al. 2008).

Accidental lethal poisonings are a less common observable event if pesticides are used in accordance with the manufacturer instructions and governmental regulations, and most episodes are related to the use of treated seed, bait, or wood preservatives and the spraying of grasslands (Berny et al. 1997; de Snoo et al. 1999; Fleischli et al. 2004; Pain et al. 2004). However, when pesticides are misused or when registered products are not applied to the approved target and at the proper application rate indicated on the product label, they can also represent a hazard, especially under some circumstances (Swiggart et al. 1972; Guitart et al. 1996b; Goldstein et al. 1999; Berny 2007; Martínez-Haro et al. 2008; Anadón et al. 2009). For example, in a retrospective study of the US Geological Survey National Wildlife Health Center (NWHC) mortality database from 1980 to 2000, in which 335 mortality events attributed to anticholinesterase poisoning involving 8877 carcasses of 103 different avian species were analyzed, 14% of the mortality events were associated with pesticide use, 10% with misuse, and 76% with undetermined use (Fleischli et al. 2004). In another survey carried out in Spain in the period 1990–2006 with the cinereous vulture (*Aegypius monachus*), an endangered species worldwide that inhabits Mediterranean forested areas, in which 241 incidents affecting 464 vultures were examined, the approved use of pesticides was responsible for only 1.3% of the incidents, whereas up to 98% of them were considered intentional poisonings (Hernández and Margalida 2008).

Lethal and sublethal exposures of animals to pesticides may result either from primary consumption of the toxic bait (or solid or liquid formulations of pesticides) or, less commonly, from direct inhalation or dermal contact with the substance, or through secondary contamination or poisoning (Vermeer et al. 1974; Newton 1998; de Snoo et al. 1999; Smith et al. 2007; Vyas et al. 2007). Tertiary (and quaternary) expositions are also possible: a great horned owl (*Bubo virginianus*) was found dead in Oregon under a tree after it fed on a red-tailed hawk (*Buteo jamaicencis*) that died following ingestion of a black-billed magpie (*Pica pica*) that obtained the insecticide famphur from the hair of dusted cattle (Henny et al. 1987).

Exposure to environmental pesticide contaminants can also be through transfer from gravid females into the eggs in birds and through transplacental and lactational pathways in mammals (Wolkers et al. 2004; Smith et al. 2007; Haraguchi et al. 2009; Miranda Filho et al. 2009; Van den Steen et al. 2009b; Mwevura et al. 2010).

Short-term exposure is more often associated with acute or single ingestion, inhalation, or topical contact with the pesticide. Secondary exposure can also occur with fast-acting toxicants and thus also with short-term exposures, but is usually related to highly or moderately persistent compounds, such as the organochlorine (OC) fungicide, hexachlorobenzene (HCB), and several OC insecticides, or the ARs, and consequently, with long-term or chronic exposures (Brakes and Smith 2005). As many OC pesticides are now banned worldwide and their environmental levels are declining (Bignert et al. 1995; Addison and Stobo 2001; Braune et al. 2001, 2005; Damstra et al. 2002; Herkert 2004; Aguilar and Borrell 2005; Verreault et al. 2005; Borrell and Aguilar 2007; Bustnes et al. 2007; Helgason et al. 2008; Walker et al. 2008a; Leonel et al. 2010; Rigét et al. 2010), there is growing concern about the risk of secondary poisoning to valued wildlife populations due to ARs, and this is manifested by an increasing number of reported studies in the scientific literature of the last 15–20 years (Newton et al. 1990; Eason and Spurr 1995; Eason et al. 2002; Mineau and Tucker 2002; Fournier-Chambrillon et al. 2004; Hoare and Hare 2006; Dowding et al. 2010).

On the other hand, there are some cases that pose problems in the distinction between primary and secondary poisonings. Zinc phosphide, for example, is a rodenticide that is hydrolyzed in the acidic environment of the stomach, liberating phosphine gas, which is responsible for the inhibition of oxidative phosphorylation and the production of reactive-oxygen free radicals (Plumlee 2004). As undigested zinc phosphide residues have been shown to be localized almost entirely in the gastrointestinal tract of the target small rodents, predators, and scavengers, consuming such poisoned animals can be a risk in fact of primary hazard (Srterner and Mauldin 1995; Mauldin et al. 1996). Even the attending veterinarians of poisoned animals can be at risk if phosphine is inhaled during nasogastric intubation (Drolet et al. 1996), and care should be taken also during necropsy procedures.

14.1.2 Birds, Mammals, Pesticides, and Biomarkers

There are about 2,000 genera and 10,000 species of birds in the Aves class, and approximately 1,200 genera and 5,500 species of mammals in the Mammalia class. With such numbers, it is not surprising that they possess a wide array of life history strategies, and both classes inhabit the ecosystems across the globe, including the oceans and the Arctic and Antarctic regions. Raptorial and fish-eating birds and cetaceans, pinnipeds, and mustelids are the most classic groups used as environmental sentinels for monitoring pesticides and other contaminant residues (Sheffield 1997; Newton 1998; Ross 2000; Burger and Gochfeld 2004; Hollamby et al. 2006; Helander et al. 2008; Grove et al. 2009), although in a few cases with some unnecessary controversy (Basu et al. 2007, 2009; Bowman and Schulte-Hostedde 2009). Most of these animals have long lifespans, so pollutant burdens may be integrated in some complex way over time. In addition, the conservation of biodiversity is becoming an integral part of the sustainable management of ecosystems, so almost any bird or mammal species can also be used as primary or surrogate receptor species in the environmental risk assessments by themselves (Strause et al. 2007).

The most persistent (and most analyzed) pesticides are ubiquitous and may appear in environments such as the Arctic, Antarctic, and other remote and cold areas such as high altitude mountains that had never been treated with these chemicals (Cipro et al. 2010; Rigét et al. 2010). This is especially true for the so-called persistent organic pollutants (POP), a group of carbon-based chemicals characterized by their low solubility in water, semivolatility, and resistance to photolytic, chemical, and biological degradation (Jones and de Voogt 1999; El-Shahawi et al. 2010). The combination of these chemical and physical properties facilitates long half-lives in the environment and their long-range transport processes, notably by the "global-distillation" mechanism, in which these compounds evaporate in warmer regions (where they were or are produced or used) and are transported by the atmosphere to cooler regions of the world (by higher altitude or latitude), where POPs condense and therefore enter the food web (Braune et al. 2005; Bhatt et al. 2009). By contrast, for most polar POPs, transport by ocean currents may also play an important role (Lohmann et al. 2007). All these pollutants are also known for their ability to bioconcentrate (or bioaccumulate) and biomagnify (or bioamplify) under environmental conditions due to their lipophilic structures and resistance to metabolism, thereby potentially achieving toxicologically relevant concentrations in some animal groups, especially those situated at the top of the food web via trophic transfer (Jones and de Voogt 1999; Braune et al. 2005).

The "classical" POPs (informally called the "dirty dozen") were 12 chemicals addressed by the Stockholm Convention on Persistent Organic Pollutants of 2001 and include eight intentionally produced OC insecticides (aldrin, chlordane, 1,1,1-trichloro-2,2-di(4-chlorophenyl)ethane (p,p′-DDT), dieldrin, endrin, heptachlor, mirex, and toxaphene) and the polychlorinated biphenyls (PCBs), the two unintentionally produced substances polychlorinated dibenzo-p-dioxins (PCDD) and polychlorinated dibenzofurans (PCDF), and HCB, which had both origins (intentional as a fungicide and nonintentional as a byproduct in industrial processes) (Zitko 2003a,b; Lohmann et al. 2007; El-Shahawi et al. 2010). This international treaty, which aims to eliminate or restrict the production and use of these POPs, can be traced in its origins to the Rachel Carson's classic 1963 book Silent Spring, which alerted the public to the toxic side effects of OC insecticides such as DDT and its derivatives, with innovative and even provocative new concepts (Burger 1997; Pollock 2001; Rattner 2009). However, the use of DDT is still "tolerated" in a few countries in Africa, Asia, and Latin America for the control of malaria (Turusov et al. 2002), as a total ban has been demonstrated in an increase of this vector-transmitted illness (Attaran and Maharaj 2000).

Despite the partial or total regulation of most OC pesticides, they are still ubiquitously present in the environment. All these recalcitrant POPs are organohalogen compounds (as also are some newly added pesticides to the list of new or potentially new POPs such as chlordecone, endosulfan, and α-, β-hexachlorocyclohexane (HCH) and lindane (γ-HCH)) (Lohmann et al. 2007; Bhatt et al. 2009; Muir et al. 2010), which allow a sensitive detection by means of the electron capture detector (ECD). Coupled to a gas chromatograph (GC), the GC/ECD instrumentation played an important role in demonstrating the extent of the contamination with these substances since mid-1960s (Lovelock 1991). Resolution of analytes later improved with the introduction of the open tubular columns ("capillary columns") and selectivity with the substitution of the ECD by a single-quadrupole mass spectrometer (MS). In the last few years, more sophisticated and powerful detection techniques, such as tandem MS (MS^2) or MS-time of flight (MS-TOF), and at present, some of the current methods of choice to detect and quantify POPs and other pesticide residue contaminants and/or their transformations products became available, and the combination with liquid chromatography (LC) allows even a larger-scale screening, identification, and quantitation of pesticides in environmental samples (Barceló and Petrovic 2007; Margariti et al. 2007; Muir and de Wit 2010; El-Shahawi et al. 2010; Petrovic et al. 2010).

Analytical chemistry and therefore chemical monitoring were and continue to be a cornerstone for the assessment of wildlife exposure to POPs and other pesticides using residue determination in the tissues and other biological material such as eggs, milk, or feces (Peakall 1992; Keith 1996). However, biomarkers of response or toxic effect (Walker et al. 1996; Timbrell 1998) and thus biological monitoring are increasingly used in environmental sciences to identify the impact of exposure to, and effects caused by, pesticides (Hoffman et al. 1990; Walker 1995). A classic example was the work by Ratcliffe (1967), in which he described decreases in eggshell thickness in peregrine falcons (*Falco peregrinus*) and Eurasian sparrowhawks (*Accipiter nisus*) in Britain from 1946 onward. Eggshell thinning related to DDT, and specifically its metabolite 1,1-bis-(4-chlorophenyl)-2,2-dichloroethene (p,p'-DDE), is now well documented and has been considered the cause of reproductive failure in raptors and fish-eating birds in the past (Hickey and Anderson 1968; Blus et al. 1972; Bignert et al. 1995; Blus 1996, 2003; Keith 1996; Newton 1998; Gervais et al. 2000; Giesy et al. 2003; Kendall and Smith 2003; Henny et al. 2008b; Rattner 2009). Nevertheless, studies on Canadian peregrine falcon, for example, suggest that eggshell thinning due to the high DDE content in the eggs continued to be a problem until recently (Johnstone et al. 1996).

Acetylcholinesterase (AChE) (EC 3.1.1.7) activity is another frequently used biomarker to verify, in this case, OP and CB exposure and effects (Newton 1998; Walker 1995; Mineau and Tucker 2002; Berny 2007). These pesticides, mainly applied as insecticides, have been widely used in agriculture since the ban on OCs in the 1970s. They are neurotoxicants that inactivate AChE, which is the enzyme responsible for the breakdown of acetylcholine into acetate and choline, through the inhibition by phosphorylation or carbamylation. The neurotransmitter is present at the cholinergic nerve endings and myoneural junctions, and accumulation in the synaptic cleft leads to cholinergic toxicity, characterized by tremors, convulsions, and eventually death from respiratory failure. As chemical detection requires GC or HPLC equipment with an appropriate and selective (but also expensive) detector such as an MS, usually exposure to anticholinesterase agents determining the parent compounds or their metabolites (there is the additional problem that some OPs and CBs are rapidly metabolized) is substituted in many laboratories by the measurement of the AChE activity in brain or the cholinesterase activity in the blood, plasma, or serum (Peakall 1992; Keith 1996; Maul and Farris 2005; Berny 2007). Although the toxic effects of these compounds are associated with the inhibition of AChE activities in the central nervous system

(CNS), depressed AChE and other cholinesterases may also occur in the plasma of birds and in the red blood cells of mammals (Mineau and Tucker 2002). Research in this field is intense, and other biomarkers have been proposed as alternatives, such as carboxylesterases (CbEs) (EC 3.1.1.1), which are also susceptible to phosphorylation or carbamylation (Fossi et al. 1992; Walker 1995). Recently, Sogorb et al. (2007) determined the activity of CbEs in the plasma of 106 healthy individuals belonging to 11 different wild bird species, including Eurasian eagle owls (*Bubo bubo*), Montagu's harriers (*Circus pygargus*), and griffon vultures (*Gyps fulvus*), for baseline purposes in future uses.

14.2 Herbicides

In general, herbicides are thought to have low toxicity to vertebrate animals when used at the recommended doses, and direct damage to wildlife or domestic animals has scarcely been reported in the literature (Swiggart et al. 1972; Freemark and Boutin 1995; Berny et al. 2010; Guitart et al. 2010a), although paraquat and other bipyridinium compounds constitute a remarkable exception. Moreover, the most modern herbicides have short half-lives in the environment and bioconcentration, biomagnification, or secondary exposures are initially not expected to occur (Berny 2007). In spite of this general rule, recently, the herbicide dimethyl tetrachloroterephthalate (dacthal or DCPA) and its isomer dimethyl tetrachlorophthalate (diMe-TCP) have been consistently found, though at low concentrations (2.0–10.3 pg/g WW for DCPA, 6.9–85.5 pg/g WW for diMe-TCP), in the eggs of osprey (*Pandion haliaetus*) collected in the state of Washington, United States, indicating that these contaminants can be bioaccumulated in an aquatic food web and persist in a high trophic level species such as osprey (Chu et al. 2007). Also, in the tissues of common bottlenose dolphins (*Tursiops truncatus*) collected in Israel, the herbicides prometryn, simazine, simetryn, and trifluralin were detected (Shoham-Frider et al. 2009).

14.2.1 Arsenic-Containing Herbicides

A lethal poisoning episode was reported by Swiggart et al. (1972) with the older pesticide arsenic acid, when it was applied by air erroneously in a big area in Tennessee, United States, killing at least 11 white-tailed deers (*Odocoileus virginianus*) of the zone. Analyses revealed average levels of 18.96, 17.78, and 22.50 µg/g WW in the liver, kidney, and rumen contents, respectively, indicating acute exposure and poisoning.

Monosodium methanearsonate (MSMA) is an organic arsenical used usually for weed control, but, in Canada, it has only been registered as a cheap way in the forest industry to combat natural infestations of both mountain pine beetle (*Dendroctonus ponderosae*) and spruce beetle (*D. rufipennis*) outbreaks and for tree thinning. Blood samples from 32 individuals of three species of woodpeckers (hairy woodpeckers (*Picoides villosus*), red-naped sapsuckers (*Sphyrapicus nuchalis*), and three-toed woodpeckers (*Picoides dorsalis*)) breeding within 1 linear km of MSMA treatment area were analyzed in British Columbia for arsenic content, and the results were found to range widely from 0.05 to 2.14 µg/g DW with a geometric mean of 0.16 µg/g, which were interpreted as follows: there is significant accumulation and transfer of organic arsenic within the food chain at levels that may have toxicological consequences to avian wildlife (Morrissey et al. 2007).

14.2.2 Paraquat

Paraquat is a contact-action herbicide, used to control a broad range of unwanted plants in cereal, oilseed, fruit, and vegetable cultivations. Due to its polar and ionic nature, it is unlikely to bioaccumulate in food chains. Paraquat is toxic to animals and man, with mammals generally being more sensitive than birds based on the LD_{50}.

Deliberate or accidental poisoning has been described in the literature for several domestic mammal species (Bischoff et al. 1998; Philbey and Morton 2001; Berny et al. 2010; Guitart et al. 2010a). Paraquat has also been implicated in the decline of the European brown hare (*Lepus europaeus*) populations in Europe, and although a retrospective study of reported incidents in France and the United Kingdom revealed that the residues of the herbicide were detected in the stomach, urine, liver, and kidney contents of some animals, Edwards et al. (2000) concluded that this is confirmation of exposure but not of death due to paraquat.

14.2.3 Glyphosate

Glyphosate, currently one of the most widely used herbicides in a huge variety of agricultural, lawn and garden, aquatic, and forestry situations, is an OP herbicide without anticholinesterase activity. It is considered poorly toxic, and only a few poisoning cases have been described in the literature (Burgat et al. 1998; Berny et al. 2010). Most of the adverse effects of glyphosate use to wildlife are related to indirect effects on birds and mammals (Sullivan and Sullivan 2003; Aktar et al. 2009), such as when treatment in some zones caused dramatic decreases in the populations of native birds (MacKinnon and Freedman 1993).

Residues of glyphosate can be detected in the viscera of herbivorous animals in amounts that remained above 1 µg/g for the first 2 weeks after treatment, then decreased to almost the minimum detection limit (0.10 µg/g) by day 55, with the levels in carnivorous small mammals being distinctly lower (Newton et al. 1984), suggesting that biomagnification does not occur.

14.3 Fungicides

The only fungicide for which extensive data on the residues in birds and mammals is available is HCB, a classical POP with a strong tendency to accumulate in food chains and the lipid-rich tissues of animals (Zitko 2003b).

Older inorganic fungicides that are mercury-, cadmium-, or chromium-based are considered bioaccumulative as well as very toxic and nonselective products and are mostly prohibited today. Less problematic copper-based fungicides are still in use in many countries, although chronic copper poisoning is one of the main diseases associated with the sudden death of sheep across the world (Plumlee 2004; Roubies et al. 2008; Guitart et al. 2010a).

Organomercurial compounds were used in agriculture as fungicidal seed dressings. In Sweden, a widespread contamination of the environment and serious secondary poisonings in some mammalian predators such as red foxes (*Vulpes vulpes*), martens (*Martes martes*), and polecats (*Mustela putorius*) during the late 1950s and early 1960s because of the use of these compounds in agriculture and also in the wood-pulp industry were recorded (Wren 1986; Rattner 2009). Pheasants and other seed-eating birds collected in Alberta, Canada, where mercury seed dressings were extensively used, contained up to 5.9 μg/g WW of mercury in the liver, which resulted in the closure of pheasant and partridge hunting areas in Alberta in 1969 (Fimreite 1970). Jefferies and French (1976) analyzed the specimens of the European wood mouse taken from a wheat field that had been drilled 2 months previously with wheat dressed with dieldrin and mercury, and found that the whole body mercury concentrations were much higher on average (0.83 ± 0.44 μg/g WW) than those found before and 13 days after drilling (0.04 ± 0.005 and 0.39 ± 0.04 μg/g WW, respectively).

14.3.1 Hexachlorobenzene

Due to its low acute toxicity (see Table 14.1), HCB was used as a fungicide for seed treatment, but most countries banned its use during the last four decades (Barber et al. 2005). The International Agency for Research on Cancer (IARC) classified this chlorinated compound as possibly carcinogenic to humans (group 2B) (IARC 2010).

Currently, HCB enters the environment mostly and unintentionally as a byproduct and waste material from the manufacture of chlorinated solvents and other chlorinated substances and from old disposal sites (Zitko 2003b). Usually, it is considered that the environmental levels have been globally declining in the last few decades (Mañosa et al. 2003; Braune 2007; Borrell and Aguilar 2007; Helgason et al. 2008; Vorkamp et al. 2009; Rigét et al. 2010). Although temporal trends vary depending on the time period measured, the media studied, and the study location, Barber et al. (2005) estimated an average half-life of approximately 9 years, from all the published studies.

14.3.1.1 Birds

Birds have been proposed as ideal candidates for environmental monitoring and surveillance purposes since they integrate the POP levels along the food web (Galassi et al. 2002; Sanpera et al. 2003; Dong et al. 2004; Dell'Omo et al. 2008; Helgason et al. 2008; van Drooge et al. 2008; Dhananjayan and Muralidharan 2010). Although chemical monitoring using, for example, the blood (or plasma) and tissues from living and dead birds, respectively, have been reported (Herrera et al. 2000; Choi et al. 2001; Guruge et al. 2001; Alleva et al. 2006; Sakellarides et al. 2006; Cesh et al. 2008; van Drooge et al. 2008; Elliot et al. 2009; Martínez-López et al. 2009; Dhananjayan and Muralidharan 2010; Rajaei et al. 2010a,b; Skarphedinsdottir et al. 2010; Dhananjayan et al. 2011), for practical and even ethical or legal reasons, the eggs are the material of choice in the majority of studies (Jaspers et al. 2005; Van den Steen et al. 2006, 2009a,b; García-Fernández et al. 2008).

Eggs have been considered a suitable material in the breeding areas, given that their chemical composition mainly reflects that of the female diet during the period of egg formation (Braune et al. 2001; Verreault et al. 2006; Van den Steen et al. 2009b), the contaminant levels found in the eggs may reveal local pollution (Aurigi et al. 2000; Galassi et al. 2002; Chu et al. 2007; Dell'Omo et al. 2008; Van den Steen et al. 2010), and they are easy to collect and the lipid content is high (Van den Steen et al. 2009a). Compared with the liver and other tissues of adult females, eggs tend to reflect more long-term contaminant exposure (Braune et al. 2001; Vorkamp et al. 2004). Although, usually it is considered, for practical purposes, that a single egg could represent statistically the entire clutch (Berny et al. 2002; Muñoz-Cifuentes et al. 2003; Dong et al. 2004; Van den Steen et al. 2006), for many species, such as the Audouin's gull (*Larus audouinii*), there is a laying order effect on egg

HCB and other POP contents (Pastor et al. 1995). The question of the laying influence on POPs has been recently reviewed by Van den Steen et al. (2009b), but although a decline in POP concentrations is usually observed in relation to the laying order, within the clutch the variation was larger than that among clutches (Van den Steen et al. 2009a). Finally, another potential drawback of the use of eggs is that, for many protected species, only unhatched, addled, sterile, and/or broken eggs are collected and analyzed (Hernández et al. 2008; Clark et al. 2009; Vorkamp et al. 2009; Best et al. 2010); so the data are potentially biased due to the worst situation.

Tables 14.2 and 14.3 show the concentrations of HCB in nonraptorial and raptorial birds' eggs from selected papers published in the last 10 years. A comparison of the levels among species is difficult to carry out, since most studies have been done on whole eggs (yolk and albumen) while others have used only yolk; units are reported sometimes on wet and sometimes on dry or even lipid weight basis; and many authors give arithmetic means, while some others give geometric ones, medians, or simply a range. However, for practical purposes and as a sufficient approximation for basic evaluation, some recalculations can be made taking into account the values of lipid content and moisture in eggs reported in the recent literature. Lipid content usually ranges from a minimum of 3.8% to a maximum of 18% (Mora et al. 2008; Henny et al. 2009), but typically values around 7%–12% are reported for all species (Elliot et al. 2001; Vorkamp et al. 2004; Jaspers et al. 2005; Braune 2007; Bustnes et al. 2007; Henny et al. 2008a; Helgason et al. 2008; Lavoie et al. 2010). Moisture ranges from a minimum of 63.9% to a maximum of 81.7% (Henny et al. 2008a; Mora et al. 2008), but most values are around 70%–78% (Elliot et al. 2001; Mañosa et al. 2003; Albanis et al. 2003; Braune 2007; Lavoie et al. 2010).

Accordingly, assuming round percentages of 10% for lipid and 25% for dry matter in eggs, no great differences for HCB can be observed, as concentrations ranged from nondetectable (below the limit of detection or quantitation) to the highest value of ≈66.5 ng/g WW in Atlantic puffins (*Fratercula arctica*) from Northern Norway (Helgason et al. 2008). Values ≥30 ng/g WW were also found in the species of the Arctic or sub-Arctic regions, such as black guillemots (*Cepphus grylle*) from Greenland (Vorkamp et al. 2004), northern fulmars (*Fulmarus glacialis*) and thick-billed murres (*Uria lomvia*) from Canada (Braune et al. 2001; Braune 2007), and herring gulls (*Larus argentatus*), black-legged kittiwakes (*Rissa tridactyla*), and common guillemots (*Uria aalge*) from Norway (Helgason et al. 2008).

In a study of the subcutaneous fat of 61 individuals belonging to eight albatross species (*Diomedea* spp. and *Phoebetria palpebrata*) from the North Pacific and Southern Ocean, Guruge et al. (2001) found no marked differences in the content of HCB among the species or between the hemispheres, in contrast to other POPs for which differences of at least one order of magnitude higher was detected in North Pacific individuals. This was attributed by the authors to the dispersive nature of HCB though long-range atmospheric transport, which contributes to uniform levels on a planetary scale, including the Antarctic region (Goerke et al. 2004).

14.3.1.2 Mammals

HCB, like other POPs, is highly lipophilic and thus accumulates in the blubber, a dense vascularized fatty tissue beneath the skin, of marine mammals. Blubber has a lipid content of 60%–90% and may constitute up to 50% of the body mass of some species at certain life stages (Colborn and Smolen 1996; Dietz et al. 2000). A selection of the concentration values of HCB in the blubber of different marine mammals (cetaceans and pinnipeds), published in the last 10 years, is presented in Table 14.4. Although not fully aquatic and lacking "true"

TABLE 14.2

Organochlorine Residues in the Eggs (Y = Yolk, A = Albumen) of Different Species of Nonraptorial Birds of the World

Species	Country	N	HCB	o,p'-DDD	p,p'-DDD	p,p'-DDE	o,p'-DDT	p,p'-DDT	ΣDDTs	Sample	Units	Source
Anas platyrhynchos	Italy	6	—	17	7	2,760	7	304	3,095	Y + A	AM ng/g DW	Galassi et al. (2002)
Anhinga rufa	South Africa	14	4.1 (1.9–7.8)	—	—	260 (56–870)	—	1.26 (0.80–2.7)	260 (57–870)	Y + A	AM ng/g WW	Bouwman et al. (2008)
Alca torda	Canada	20	234	—	—	2,563	—	—	2,719	Y + A	AM ng/g LW	Lavoie et al. (2010)
Ardea alba	United States	13	—	—	7 (3–33)	1,085 (369–4,701)	—	3 (1–105)	—	Y + A	GM ng/g WW	Wainwright et al. (2001)
Ardea alba	United States	12	NC	—	1.23	560 (400–860)	—	2.49	—	Y + A	GM ng/g WW	Henny et al. (2008a)
Ardea herodias	Canada	10	5	—	0.9	125	—	0.9	—	Y + A	AM ng/g WW	Harris et al. (2003)
Ardea herodias	Canada	10	7	—	1	359	—	4	—	Y + A	AM ng/g WW	Harris et al. (2003)
Ardea herodias	Canada	18	—	—	—	496 (328–850)	—	—	—	Y + A	AM ng/g WW	Champoux et al. (2010)
Ardea purpurea	Spain	14	5.5	—	—	—	—	—	477.3	Y + A	AM ng/g WW	Barata et al. (2010)
Ardeola bacchus	China	17	—	—	30.15 (7.91–251.9)	2,745 (613–7,401)	—	22.26 (6.13–206.5)	2,812 (631–7,705)	Y + A	GM ng/g DW	Dong et al. (2004)
Bubulcus ibis	China	13	—	—	8.56 (1.50–37.39)	670 (210–3,002)	—	12.56 (1.84–121.5)	700 (215.5–3,038)	Y + A	GM ng/g DW	Dong et al. (2004)

Species	Location	n									Units	Reference
Bubulcus ibis	South Africa	20	0.78 (0.29–2.5)	—	—	24 (3.2–94)	—	—	24 (3.2–95)	Y + A	AM ng/g WW	Bouwman et al. (2008)
Bubulcus ibis	Pakistan	10	—	—	126.2 (ND–372.0)	64.2 (ND–188.2)	36.3 (14.1–71.7)	27.0 (ND–75.1)	—	Y + A	AM ng/g WW	Malik et al. (2011)
Bubulcus ibis	Pakistan	10	—	—	165.6 (ND–317.2)	58.2 (ND–188.2)	28.7 (ND–95.5)	41.4 (ND–152.3)	—	Y + A	AM ng/g WW	Malik et al. (2011)
Cepphus grylle	Greenland	20	37 (21–50)	—	—	—	—	—	120 (70–190)	Y + A	AM ng/g WW	Vorkamp et al. (2004)
Cepphus grylle	Canada	20	85.6	—	—	1,244	—	—	1,250	Y + A	AM ng/g LW	Lavoie et al. (2010)
Egretta garzetta	France	37	—	—	—	86 (13–2,020)	—	—	—	Y + A	GM ng/g WW	Berry et al. (2002)
Egretta garzetta	Hong Kong	9	—	—	—	1,000 (530–1,700)	—	—	1,200 (560–2,200)	Y + A	AM ng/g WW	Connell et al. (2003)
Egretta garzetta	Pakistan	8	16.3 (/−25.0)	—	—	—	—	—	728 (ND–9,982)	Y + A	GM ng/g DW	Sanpera et al. (2003)
Egretta garzetta	Pakistan	14	10.3 (/31.4)	—	—	—	—	—	2,665.9 (ND–5,332.8)	Y + A	GM ng/g DW	Sanpera et al. (2003)
Egretta garzetta	China	36	—	—	40.71 (5.80–357.7)	1,980 (390.8–11,886)	—	196.03 (13.70–1,279.60)	2,061 (390.8–13,474)	Y + A	GM ng/g DW	Dong et al. (2004)
Egretta garzetta	Thailand	12	—	—	—	69.5 (33.4–116)	—	—	—	Y	AM ng/g WW	Keithmaleesatti et al. (2007)
Egretta garzetta	Spain	4	11.8	—	—	—	—	—	550.2	Y + A	AM ng/g WW	Barata et al. (2010)

(continued)

TABLE 14.2 (Continued)

Organochlorine Residues in the Eggs (Y = Yolk, A = Albumen) of Different Species of Nonraptorial Birds of the World

Species	Country	N	HCB	o,p'-DDD	p,p'-DDD	p,p'-DDE	o,p'-DDT	p,p'-DDT	ΣDDTs	Sample	Units	Source
Egretta intermedia	Pakistan	20	4.23 (2.21–62.71)	—	—	165.5 (30.0–18.840)	—	8.52 (3.36–86.47)	—	Y + A	GM ng/g WW	Sampera et al. (2002)
Fratercula arctica	Norway	5	665	—	—	1,206	—	270	1,476	Y + A	AM ng/g LW	Helgason et al. (2008)
Fulmarus glacialis	Canada	15	32.8	—	2.4	192	—	14.4	209	Y + A	AM ng/g WW	Braune et al. (2001) and Braune (2007)
Fulmarus glacialis	Canada	15	13.2	—	1.0	112	—	10.9	124	Y + A	AM ng/g WW	Braune (2007)
Gallus gallus	Belgium	40	—	—	—	—	—	—	862 (ND–21,390)	Y + A	AM ng/g LW	Van Overmeire et al. (2009)
Gallus gallus	China	24	—	0.053	0.711	1.190	0.059	0.394	2.42	Y + A	AM ng/g WW	Tao et al. (2009)
Gallus gallus	Jordan	134	ND	ND	ND	31 (5–200)	ND	142 (10–600)	72 (5–600)	Y + A	AM ng/g LW	Ahmad et al. (2010)
Himantopus mexicanus	United States	12	1.75	—	0.82	550 (190–1,700)	—	1.40	—	Y + A	GM ng/g WW	Henny et al. (2008a)
Lanius ludovicianus	United States	21	—	—	—	660 (ND–4,040)	—	40 (ND–390)	—	Y + A	AM ng/g WW	Herkert (2004)
Larus argentatus	Norway	6	308	—	—	4,044	—	140	4,184	Y + A	AM ng/g LW	Helgason et al. (2008)
Larus argentatus	Canada	18	87.5	—	—	2,691	—	—	2,818	Y + A	AM ng/g LW	Lavoie et al. (2010)

Species	Country	n								Age	Statistic / units	Reference
Larus audouinii	Greece	9	—	1,022 (77–2,833)	—	1 (ND-3)	3 (ND-6)	27 (ND-49)	—	Y + A	AM ng/g DW	Goutner et al. (2001)
Larus audouinii	Greece	15	—	1,265 (339–2,503)	—	22 (ND-291)	4 (ND-16)	35 (16-75)	—	Y + A	AM ng/g DW	Goutner et al. (2001)
Larus cachinnans	Greece	15	—	235 (32.3–556)	ND	0.90 (0-3.06)	0.49 (0-1.05)	2.65 (0-5.91)	—	Y + A	AM ng/g WW	Albanis et al. (2003)
Larus cachinnans	Greece	12	—	162.2 (13.3–288)	0.22 (0-2.60)	1.24 (0-3.86)	0.35 (0-1.39)	2.85 (0-10.36)	—	Y + A	AM ng/g WW	Albanis et al. (2003)
Larus crassirostris	Japan	10	74	—	—	—	—	—	1,522	Y	AM ng/g LW	Choi et al. (2001)
Larus dominicanus	Chile	12	11.8	—	—	—	—	—	39.8	Y + A	AM ng/g WW	Muñoz-Cifuentes et al. (2003)
Larus maculipennis	Chile	10	9.2	—	—	—	—	—	866.4	Y + A	AM ng/g WW	Muñoz-Cifuentes et al. (2003)
Larus marinus	Canada	20	123	—	—	5,438	—	—	5,496	Y + A	AM ng/g LW	Lavoie et al. (2010)
Larus melanocephalus	Greece	15	—	112 (11–735)	(ND-1)	2 (ND-7)	1 (ND-4)	2 (ND-11)	—	Y + A	AM ng/g WW	Goutner et al. (2005)
Larus melanocephalus	Greece	15	—	186 (9–711)	ND	1 (ND-7)	1 (ND-3)	1 (ND-9)	—	Y + A	AM ng/g WW	Goutner et al. (2005)
Nycticorax nycticorax	United States	9	—	—	7 (1-16)	1,480 (562–4,876)	—	4 (1-48)	—	Y + A	GM ng/g WW	Wainwright et al. (2001)
Nycticorax nycticorax	Hong Kong	9	—	—	—	590 (200–1,200)	—	—	600 (210–1,200)	Y + A	AM ng/g WW	Connell et al. (2003)

(continued)

TABLE 14.2 (Continued)

Organochlorine Residues in the Eggs (Y = Yolk, A = Albumen) of Different Species of Nonraptorial Birds of the World

Species	Country	N	HCB	o,p'-DDD	p,p'-DDD	p,p'-DDE	o,p'-DDT	p,p'-DDT	ΣDDTs	Sample	Units	Source
Nycticorax nycticorax	China	65	—	—	81.01 (6.06–1,204)	5,464 (998–28,656)	—	22.84 (5.38–310.9)	5,627 (1,063–8,793)	Y + A	GM ng/g DW	Dong et al. (2004)
Nycticorax nycticorax	United States	11	0.73	—	0.97	560 (140–11,000)	—	2.74	—	Y + A	GM ng/g WW	Henny et al. (2008a)
Parus major	Belgium	38	9.0 (2.9–15)	—	—	562 (258–1,367)	—	39 (12–126)	—	Y + A	AM ng/g LW	Van den Steen et al. (2006)
Pelecanus erythrorhynchos	United States	30	—	—	46 (10–170)	520 (150–1,200)	—	NC (10–30)	—	Y + A	GM ng/g WW	Wiemeyer et al. (2005)
Phalacrocorax brasilianus	Chile	10	3.1	—	—	—	—	—	107.9	Y + A	AM ng/g WW	Muñoz-Cifuentes et al. (2003)
Phalacrocorax carbo	Romania	10	5.54	—	—	—	—	—	12,400	Y	AM ng/g DW	Aurigi et al. (2000)
Phalacrocorax carbo	Greece	38	—	111.16 (8.58–336.11)	0.31 (0–9.28)	0.42 (0–4.02)	0.29 (0–5.18)	2.98 (0–9.05)	—	Y + A	AM ng/g WW	Konstantinou et al. (2000)
Phalacrocorax carbo	Greece	15	—	115.58 (45.18–438.87)	0.43 (0–5.84)	0.00 (0–0.04)	1.00 (0–8.18)	3.33 (0.37–5.25)	—	Y + A	AM ng/g WW	Konstantinou et al. (2000)
Phalacrocorax pygmaeus	Romania	4	30.15	—	—	—	—	—	10,417	Y	AM ng/g DW	Aurigi et al. (2000)
Phoenicopterus roseus	Spain	53	—	—	—	721 (145–6,600)	—	—	—	Y + A	GM ng/g WW	Guitart et al. (2005)
Pygoscelis adéliae	Antarctica	9	7.63	—	—	18	—	—	23	Y + A	AM ng/g WW	Schiavone et al. (2009)

Species	Location	n								Age	Stat (units)	Reference
Plegadis chihi	United States	16	NA (ND–10)	—	NA (ND–80)	2,140 (130–14,000)	—	NA (ND–200)	—	Y + A	GM ng/g WW	King et al. (2003)
Plegadis chihi	United States	10	—	—	—	3,630 (NA–8,000)	—	—	—	Y + A	GM ng/g WW	Yates et al. (2010)
Podiceps cristatus	Italy	6	—	40	5	23,033	70	220	23,368	Y + A	AM ng/g DW	Galassi et al. (2002)
Podiceps cristatus	Italy	4	—	—	—	42,440	—	—	—	Y + A	AM ng/g DW	Cortinovis et al. (2008)
Podiceps cristatus	Italy	9	—	—	—	4,650	—	—	—	Y + A	AM ng/g DW	Cortinovis et al. (2008)
Pygoscelis antarctica	Antarctica	26	18.91 (4.99–39.1)	—	—	—	—	—	15.8 (2.67–38.0)	Y + A	AM ng/g WW	Cipro et al. (2010)
Pygoscelis pappua	Antarctica	9	16.2 (14.2–19.3)	—	—	—	—	—	5.47 (3.10–9.95)	Y + A	AM ng/g WW	Cipro et al. (2010)
Rissa tridactyla	Canada	15	25.3	—	—	60.2	—	—	60.2	Y + A	AM ng/g WW	Braune et al. (2001) and Braune (2007)
Rissa tridactyla	Canada	15	15.7	—	—	43.5	—	—	43.5	Y + A	AM ng/g WW	Braune (2007)
Rissa tridactyla	Norway	6	488	—	—	1,489	—	73.4	1,562	Y + A	AM ng/g LW	Helgason et al. (2008)
Rissa tridactyla	Canada	20	267	—	—	860	—	—	930	Y + A	AM ng/g LW	Lavoie et al. (2010)
Recurvirostra avosetta	Greece	20	—	252 (1–694)	3 (1–11)	1 (ND–5)	1 (ND–6)	5 (ND–44)	—	Y + A	AM ng/g WW	Goutner et al. (2005)

(continued)

TABLE 14.2 (Continued)

Organochlorine Residues in the Eggs (Y = Yolk, A = Albumen) of Different Species of Nonraptorial Birds of the World

Species	Country	N	HCB	o,p'-DDD	p,p'-DDD	p,p'-DDE	o,p'-DDT	p,p'-DDT	ΣDDTs	Sample	Units	Source
Somateria mollissima	United States	20	7.47 (3.60–11.0)	ND	ND	8.02 (2.80–17.0)	—	ND	—	Y + A	AM ng/g WW	Franson et al. (2004)
Somateria mollissima	Canada	18	18.8	—	—	229	—	—	229	Y + A	AM ng/g LW	Lavoie et al. (2010)
Sterna hirundo	Spain	13	18.3 (ND-221)	—	69 (ND-1,342)	1,184 (337–5,749)	—	—	—	Y	GM ng/g WW	Mateo et al. (2004)
Sterna hirundo	Spain	34	15.4 (ND-221)	—	170 (ND-1,011)	684 (134–5,063)	—	—	—	Y	GM ng/g WW	Mateo et al. (2004)
Sterna hirundo	Greece	13	—	342 (105–700)	1 (ND-8)	3 (ND-7)	1 (ND-2)	4 (ND-7)	—	Y + A	AM ng/g WW	Goutner et al. (2005)
Sula leucogaster	Mexico	10	—	—	—	49.0	—	—	—	Y + A	AM ng/g WW	Mellink et al. (2009)
Uria aalge	Norway	7	467	—	—	937	—	25.2	962	Y + A	AM ng/g LW	Helgason et al. (2008)
Uria lomvia	Canada	15	46.6	—	—	99.6	—	—	99.6	Y + A	AM ng/g WW	Braune et al. (2001) and Braune (2007)
Uria lomvia	Canada	15	27.6	—	—	103	—	—	103	Y + A	AM ng/g WW	Braune (2007)

AM, arithmetic mean; GM, geometric mean, range between parentheses; DW, dry weight; LW, lipid weight; WW, wet (or fresh) weight; NA, not available; NC, not calculated; ND, not detected.

TABLE 14.3

Organochlorine Residues in the Eggs (Y = Yolk, A = Albumen) of Different Species of Birds of Prey of the World

Species	Country	N	HCB	o,p'-DDD	p,p'-DDD	p,p'-DDE	o,p'-DDT	p,p'-DDT	ΣDDTs	Sample	Units	Source
Accipiter gentilis	Spain	24	14.0 (3.3–286.1)	—	—	292.5 (38.5–3,660)	—	—	—	Y + A	GM ng/g WW	Mañosa et al. (2003)
Accipiter gentilis	Spain	8	ND	—	ND	154	—	ND	—	Y + A	AM ng/g WW	Martínez-López et al. (2007)
Aquila adalberti	Spain	102	—	—	—	1,163 (NA–18,001)	—	—	—	Y + A	GM ng/g WW	Hernández et al. (2008)
Aquila chrysaetos	United Kingdom	126	—	—	—	44	—	—	—	Y + A	GM ng/g WW	Shore et al. (2006b)
Aquila chrysaetos	Sweden	14	—	—	—	—	—	—	3,700	Y + A	AM ng/g LW	Helander et al. (2008)
Asio otus	Russia	10	NC (ND–410)	—	ND	410 (60–4,580)	—	ND	—	Y + A	GM ng/g WW	Henny et al. (2003a)
Athene cunicularia	United States	9	—	—	—	7,520 (1,500–33,000)	—	—	—	Y + A	GM ng/g WW	Gervais et al. (2000)
Athene noctua	Belgium	39	180 (35–1,110)	—	—	—	—	—	1,400 (200–7,300)	Y + A	AM ng/g LW	Jaspers et al. (2005)
Bubo virginianus	United States	7	—	—	—	—	—	—	1,305 (618–2,013)	Y + A	GM ng/g WW	Strause et al. (2007)
Buteo buteo	Spain	8	15.5 (4.5–34.7)	—	—	113.0 (15.9–909.6)	—	—	—	Y + A	GM ng/g WW	Mañosa et al. (2003)

(continued)

TABLE 14.3 (Continued)

Organochlorine Residues in the Eggs (Y = Yolk, A = Albumen) of Different Species of Birds of Prey of the World

Species	Country	N	HCB	o,p'-DDD	p,p'-DDD	p,p'-DDE	o,p'-DDT	p,p'-DDT	ΣDDTs	Sample	Units	Source
Falco columbarius	United Kingdom	493	—	—	—	3,050	—	—	—	Y + A	GM ng/g WW	Shore et al. (2006b)
Falco femoralis	Mexico	26	7.7 (2–48)	—	—	3,487 (648–21487)	—	—	—	Y + A	GM ng/g WW	Mora et al. (2008)
Falco femoralis	Mexico	7	8.8 (5–16)	—	—	2,055 (753–8005)	—	—	—	Y + A	GM ng/g WW	Mora et al. (2008)
Falco peregrinus	United States	60	10 (1–45)	9.2 (0.4–140.4)	24 (1–121)	4,260 (620–31,870)	10 (0–102)	7.9 (0.3–356.5)	—	Y + A	GM ng/g WW	Clark et al. (2009)
Falco peregrinus	Greenland	36	170	—	—	38,800	—	—	40,000	Y + A	Med ng/g LW	Vorkamp et al. (2009)
Falco tinnunculus	Spain	14	—	—	—	17,900 (530–44,000)	—	—	—	Y + A	AM ng/g DW	Mateo et al. (2000)
Falco tinnunculus	Russia	8	ND	—	ND	20 (ND–390)	—	ND	—	Y + A	GM ng/g WW	Henny et al. (2003a)
Falco tinnunculus	Italy	27	15.46 (ND–84.46)	—	12.33 (ND–12.33)	395.9 (16.18–2,313)	—	ND	—	Y + A	AM ng/g WW	Dell'Omo et al. (2008)
Falco vespertinus	Russia	8	ND	—	ND	50 (20–550)	—	ND	—	Y + A	GM ng/g WW	Henny et al. (2003a)
Gyps bengalensis	India	2	—	—	2,150 (1,960–2,340)	2,960 (2,460–3,460)	—	1,000 (690–1,310)	—	Y + A	AM ng/g WW	Muralidharan et al. (2008)

Species	Country	n									Units	Reference
Haliaeetus albicilla	Sweden	6	—	—	—	14,000	—	—	—	Y + A	GM ng/g LW	Helander et al. (2002)
Haliaeetus albicilla	Sweden	54	—	—	—	—	—	—	120,000	Y + A	AM ng/g LW	Helander et al. (2008)
Haliaeetus albicilla	Sweden	21	—	—	—	—	—	—	17,000	Y + A	AM ng/g LW	Helander et al. (2008)
Haliaeetus leucocephalus	United States	25	12 (<17–27)	2 (<0.06–6)	273 (149–498)	5,630 (2,890–12,500)	NC	7 (<0.06–34)	—	Y + A	GM ng/g WW	Buck et al. (2005)
Haliaeetus leucocephalus	United States	21	—	—	—	1,239 (449–2,794)	—	—	—	Y + A	GM ng/g WW	Anthony et al. (2007)
Haliaeetus leucocephalus	United States	23	—	—	—	673 (228–1,917)	—	—	—	Y + A	GM ng/g WW	Anthony et al. (2007)
Haliaeetus leucocephalus	United States	24	—	—	—	576 (274–1,011)	—	—	—	Y + A	GM ng/g WW	Anthony et al. (2007)
Haliaeetus leucocephalus	United States	111	—	—	—	5,600	—	—	5,900	Y + A	AM ng/g WW	Best et al. (2010)
Haliaeetus leucocephalus	United States	86	—	—	—	1,390	—	—	1,400	Y + A	AM ng/g WW	Best et al. (2010)
Hieraaetus pennatus	Spain	14	1.0	—	2.0	936	—	ND	—	Y + A	AM ng/g WW	Martinez-López et al. (2007)
Milvus milvus	Spain	5	—	—	—	1,574 (882–2,071)	—	15.4 (10–20)	—	Y + A	AM ng/g LW	Jiménez et al. (2007)

(continued)

TABLE 14.3 (Continued)

Organochlorine Residues in the Eggs (Y = Yolk, A = Albumen) of Different Species of Birds of Prey of the World

Species	Country	N	HCB	o,p'-DDD	p,p'-DDD	p,p'-DDE	o,p'-DDT	p,p'-DDT	ΣDDTs	Sample	Units	Source
Milvus milvus	Spain	12	—	—	0,11 (0.03–1)	3,654 (152–33,549)	—	103 (3–608)	3,795 (177–34,158)	Y + A	GM ng/g WW	Gómara et al. (2008)
Pandion haliaetus	United States	17	—	—	110 (30–370)	930 (310–1,700)	—	10 (ND–20)	—	Y + A	GM ng/g WW	Clark et al. (2001)
Pandion haliaetus	United States	10	3.8 (2.2–11.1)	—	98.5 (11.0–706.4)	2,347 (241–12,054)	—	25.3 (ND–2,811.7)	—	Y + A	GM ng/g WW	Henny et al. (2003b)
Pandion haliaetus	United States	29	3.66 (1.0–17.1)	—	198.61 (38.7–1,213.2)	4,872 (977–18,377)	—	19.77 (0.2–161.4)	—	Y + A	GM ng/g WW	Henny et al. (2004)
Pandion haliaetus	Spain	9	—	—	—	521 (57–1,016)	—	21.1 (9–50)	—	Y + A	AM ng/g LW	Jiménez et al. (2007)
Pandion haliaetus	United States	40	1.73 (0.68–184)	—	69.7 (23.9–281)	1,521 (985–2,294)	—	4.92 (0.74–43.9)	—	Y + A	GM ng/g WW	Henny et al. (2008b)
Pandion haliaetus	United States	10	0.70	—	5.43	210	—	1.40	—	Y + A	GM ng/g WW	Henny et al. (2009)
Strix aluco	Norway	139	111.7 (18.0–1,264)	—	—	2,743 (236–43,288)	—	—	—	Y + A	AM ng/g LW	Bustnes et al. (2007)

AM, arithmetic mean; GM, geometric mean; Med, median, range between parentheses; DW, dry weight; LW, lipid weight; WW, wet (or fresh) weight; NA, not available; NC, not calculated; ND, not detected.

TABLE 14.4

Comparison of Organochlorine Residue Levels (HCB and DDTs) in the Blubber of Cetaceans and Pinnipeds from Various Waters and in Fat from Polar Bears

Species	Country	N and sex	HCB	o,p'-DDD	p,p'-DDD	p,p'-DDE	o,p'-DDT	p,p'-DDT	ΣDDTs	Units	Source
Arctocephalus australis	Brazil	3M + 5F	1.1 (0.22–4.1)	—	—	—	—	—	660 (20–2,800)	AM ng/g WW	Fillmann et al. (2007)
Balaenoptera acutorostrata	Korea	11M	100 (50–180)	—	—	—	—	—	4,500 (1,800–11,000)	AM ng/g LW	Moon et al. (2010)
Balaenoptera musculus	Canada	38M	225.8	—	551.9	1,978.9	—	887.6	3,418.4	AM ng/g LW	Metcalfe et al. (2004)
Balaenoptera musculus	Canada	27F	90.0	—	215.2	832.4	—	306.6	1354.3	AM ng/g LW	Metcalfe et al. (2004)
Callorhinus ursinus	Japan	5F	0.95 (0.39–2.1)	—	—	—	—	—	1,300 (620–2,700)	AM ng/g LW	Kajiwara et al. (2004)
Delphinapterus leucas	Canada	34M	184 (40.3–845)	385 (<1–3,960)	973 (131–16,900)	4,590 (358–138,000)	641 (74.0–21,700)	944 (111–30,200)	9,440 (1,340–201,000)	GM ng/g WW	Hobbs et al. (2003)
Delphinapterus leucas	Canada	10F	126 (11.7–1,070)	13.1 (<1–4,840)	469 (17.7–2,580)	1,420 (53.0–15,200)	232 (3.38–1,800)	478 (7.23–4,610)	3,560 (86.6–24,400)	GM ng/g WW	Hobbs et al. (2003)
Delphinapterus leucas	Canada	18M	508 (366–706)	—	—	2,007 (1,365–2,952)	—	671 (328–1374)	3,492 (2,289–5,329)	GM ng/g WW	Stern et al. (2005)
Delphinapterus leucas	Canada	12M	246 (151–402)	—	—	396 (247–637)	—	314 (190–517)	1,184 (727–1929)	GM ng/g WW	Stern et al. (2005)
Delphinus capensis	Korea	12M	110 (20–230)	—	—	—	—	—	14,000 (800–23,000)	AM ng/g LW	Moon et al. (2010)

(continued)

TABLE 14.4 (Continued)

Comparison of Organochlorine Residue Levels (HCB and DDTs) in the Blubber of Cetaceans and Pinnipeds from Various Waters and in Fat from Polar Bears

Species	Country	N and sex	HCB	o,p'-DDD	p,p'-DDD	p,p'-DDE	o,p'-DDT	p,p'-DDT	ΣDDTs	Units	Source
Delphinus capensis	Korea	10F	100 (20–170)	—	—	—	—	—	13,000 (730–22,000)	AM ng/g LW	Moon et al. (2010)
Delphinus delphis	Spain	11M + 10F + 5U	—	—	2,100 (300–6,500)	23,300 (2,100–130,400)	2,300 (200–13,500)	3,600 (300–14,900)	31,000 (3,000–165,000)	AM ng/g LW	Borrell and Aguilar (2005)
Megaptera novaeangliae	Canada	12F	153.0	—	225.6	812.2	—	84.4	1,122.2	AM ng/g LW	Metcalfe et al. (2004)
Mirounga leonina	Antarctica	7F	12.6	6.39	0.56	108.5	1.82	5.09	126.53	AM ng/g LW	Miranda Filho et al. (2009)
Monachus monachus	Sahara	8M + 13F + 5U	57 (12–187)	—	11 (1–107)	391 (40–1,064)	—	48 (8–149)	445 (48–1,198)	AM ng/g LW	Borrell et al. (2007)
Monachus monachus	Greece	2M + 4F + 5U	242 (43–732)	—	422 (17–780)	34,803 (213–173,982)	—	1,010 (954–3,574)	36,229 (259–178,253)	AM ng/g LW	Borrell et al. (2007)
Monodon monoceros	Greenland	33M + 17F	1,001	—	—	—	—	—	4,191	AM ng/g LW	Dietz et al. (2004a)
Neophocaena phocaenoides	Hong Kong	6M	152 (87–250)	—	—	—	—	—	89,200 (26,000–260,000)	GM ng/g LW	Ramu et al. (2006)
Neophocaena phocaenoides	Korea	7M	130 (70–180)	—	—	—	—	—	11,000 (2,600–24,000)	AM ng/g LW	Park et al. (2010)
Neophocaena phocaenoides	Korea	10F	40 (10–130)	—	—	—	—	—	2,300 (480–5,000)	AM ng/g LW	Park et al. (2010)

Species	Location	n								Units	Reference
Odobenus rosmarus	Greenland	8M	1.52 (1.11–2.16)	—	213 (122–265)	2,108 (1,013–3,777)	74.4 (38.3–125)	917 (541–1,623)	3,406 (1,924–5,854)	AM ng/g WW	Muir et al. (2000)
Odobenus rosmarus	Norway	17M	0.3 (0.1–3.0)			123.8 (4.6–5,065)				GM ng/g LW	Wolkers et al. (2006)
Orcaella brevirostris	India	2M + 3F	7.4 (3.6–17)			—			2,220 (310–10,000)	GM ng/g LW	Kannan et al. (2005)
Orcinus orca	Japan	5F	930 (370–1,500)			—			66,000 (28,000–120,000)	AM ng/g LW	Kajiwara et al. (2006)
Orcinus orca	United States	8M	653 (480–1,600)			—			72,066 (24,000–160,000)	GM ng/g LW	Krahn et al. (2007)
Orcinus orca	Japan	1M + 8F	1,860 (640–6,240)			89,372 (36,300–237,000)			97,087 (40,200–240,000)	GM ng/g LW	Haraguchi et al. (2009)
Otaria flavescens	Argentina	25M	—			—			854	GM ng/g LW	Borrell et al. (2010)
Otaria flavescens	Argentina	24F	—			—			511	GM ng/g LW	Borrell et al. (2010)
Phocoena phocoena	Black Sea	11M	575 (487–926)			—			77,329 (54,993–156,750)	Med ng/g LW	Weijs et al. (2010)
Pontoporia blainvillei	Brazil	20M + 6F	31.1 (4.83–114)			—			836 (219–3,447)	GM ng/g LW	Leonel et al. (2010)
Pontoporia blainvillei	Brazil	4M	54 (22–108)			—			3,080 (1,180–6,890)	AM ng/g LW	Yogui et al. (2010)
Pontoporia blainvillei	Brazil	4F	13 (10–15)			—			1,100 (304–2,280)	AM ng/g LW	Yogui et al. (2010)

(continued)

TABLE 14.4 (Continued)

Comparison of Organochlorine Residue Levels (HCB and DDTs) in the Blubber of Cetaceans and Pinnipeds from Various Waters and in Fat from Polar Bears

Species	Country	N and sex	HCB	o,p'-DDD	p,p'-DDD	p,p'-DDE	o,p'-DDT	p,p'-DDT	ΣDDTs	Units	Source
Pusa caspica	Russia	12F	8.0 (2.0–18)	—	—	—	—	—	9,200 (780–24,000)	AM ng/g LW	Tanabe et al. (2003)
Pusa hispida	Norway	6M	6.4	—	—	1,318.1	—	—	—	GM ng/g LW	Wolkers et al. (2004)
Pusa hispida	Norway	4F	2.1	—	—	188.3	—	—	—	GM ng/g LW	Wolkers et al. (2004)
Pusa hispida	Greenland	20	—	—	—	—	—	—	190	Med ng/g LW	Vorkamp et al. (2008)
Pusa hispida	Greenland	9M + 6F	—	—	—	350	—	—	—	AM ng/g LW	Letcher et al. (2009)
Pusa sibirica	Russia	5F	<0.005	—	—	—	—	—	4,000 (3,200–4,600)	AM ng/g LW	Tanabe et al. (2003)
Stenella coeruleoalba	Spain	5	—	—	3,050	45,600	1,850	4,590	55,090	AM ng/g LW	Aguilar and Borrell (2005)
Stenella coeruleoalba	Spain	15M + 5F + 7U	—	—	5,500 (1,000–13,100)	58,800 (6,000–116,400)	5,100 (100–13,100)	9,500 (200–25,000)	79,000 (7,000–189,000)	AM ng/g LW	Borrell and Aguilar (2005)
Stenella coeruleoalba	Japan	5M + 1F	321 (53–620)	—	—	—	—	—	37,186 (3,500–88,000)	GM ng/g LW	Isobe et al. (2009)

Species	Location	n							Value	Stat/Units	Reference
Stenella coeruleoalba	Spain	28	—	—	—	—	—	—	73,000 (8,000–258,000)	AM ng/g LW	Castrillon et al. (2010)
Stenella longirostris	Tanzania	11M + 7F	42.06 (19–85)	—	—	—	—	—	8,883 (1,700–76,000)	GM ng/g LW	Mwevura et al. (2010)
Tursiops aduncus	Tanzania	10M + 8F	64.76 (6–380)	—	—	—	—	—	10,324 (500–93,000)	GM ng/g LW	Mwevura et al. (2010)
Tursiops truncatus	Spain	2M + 3F + 2U	—	—	—	—	—	—	52,820	AM ng/g LW	Borrell et al. (2006)
Tursiops truncatus	Portugal	4M + 3F	—	—	—	—	—	—	31,030	AM ng/g LW	Borrell et al. (2006)
Tursiops truncatus	Spain	2M + 2F + 5U	639 (ND–1,200)	—	—	—	—	—	52,250 (ND–153,900)	GM ng/g LW	Borrell and Aguilar (2007)
Tursiops truncatus	Canary Islands	7M	43.2 (ND–98)	—	—	—	—	—	9,553 (950–21,050)	AM ng/g WW	Carballo et al. (2008)
Tursiops truncatus	Israel	2M + 3F	—	—	217 (5–2050)	9,628 (715–135,000)	—	316 (10–4,200)	—	GM ng/g WW	Shoham-Frider et al. (2009)
Ursus maritimus	Greenland	25F	—	—	—	—	—	—	462	AM ng/g LW	Dietz et al. (2004b)
Ursus maritimus	Greenland	16M	—	—	—	—	—	—	559	AM ng/g LW	Dietz et al. (2004b)
Ursus maritimus	Greenland	7M + 12F	—	—	—	467 (67–1,019)	—	—	496 (73–1,113)	AM ng/g LW	Sandala et al. (2004)

(continued)

TABLE 14.4 (Continued)

Comparison of Organochlorine Residue Levels (HCB and DDTs) in the Blubber of Cetaceans and Pinnipeds from Various Waters and in Fat from Polar Bears

Species	Country	N and sex	HCB	o,p'-DDD	p,p'-DDD	p,p'-DDE	o,p'-DDT	p,p'-DDT	ΣDDTs	Units	Source
Ursus maritimus	Canada	14F	75.3 (39.3–229)	—	—	200 (94.5–349)	—	—	210 (94.5–380)	GM ng/g LW	Verreault et al. (2005)
Ursus maritimus	Canada	16F	87.1 (33.2–249)	—	—	112 (63.9–280)	—	—	126 (77.6–329)	GM ng/g LW	Verreault et al. (2005)
Ursus maritimus	Greenland	16F	60.0 (25.5–311)	—	—	268 (82.8–830)	—	—	309 (95.4–960)	GM ng/g LW	Verreault et al. (2005)
Ursus maritimus	Greenland	8M	235 (21–795)	—	—	268 (84–638)	—	—	—	AM ng/g LW	Sonne et al. (2009)
Ursus maritimus	Greenland	3F	59 (26–89)	—	—	128 (118–143)	—	—	—	AM ng/g LW	Sonne et al. (2009)

AM, arithmetic mean; GM, geometric mean; Med, median, range between parentheses; LW, lipid weight; WW, wet (or fresh) weight; ND, not detected; M, male; F, female; U, unknown sex.

blubber, the polar bear (*Ursus maritimus*) is often regarded as a marine mammal because it lives on sea ice for most of the year (Rigét et al. 2010) and is included in the list when pesticide determinations in fat are done. Reported levels varied greatly, being highest in killer whales (*Orcinus orca*) from the United States and Japan (Kajiwara et al. 2006; Krahn et al. 2007; Haraguchi et al. 2009) and lowest in most pinnipeds, which are thought to possess an exceptional capacity to eliminate HCB by biotransformation (Goerke et al. 2004).

Temporal trends in marine mammals have been published. Hobbs et al. (2001) found that the levels of HCB in blubber of fin whales (*Balaenoptera physalus*) of the coast of Eastern Canada fell from 218–288 ng/g LW to 96.3 between 1971–1972 and 1991, and a significant decline in franciscana dolphins (*Pontoporia blainvillei*) in Southern Brazil over a 10-year period, 1994–2004 (Leonel et al. 2010) and a decline in common bottlenose dolphins by a factor of 4.9 in Western Mediterranean between 1987 and 2002 (Borrell and Aguilar 2007) were noted. Vorkamp et al. (2008) reported a significant decrease in HCB levels between 1994 and 2006 in the blubber of ringed seals (*Pusa hispida*, formerly *Phoca hispida*) from West Greenland. However, there are several studies that show the levels of HCB remaining unchanged in marine mammals. In Northern fur seals (*Callorhinus ursinus*) collected in Japan, HCB levels were found to be rather constant since the early 1970s (Kajiwara et al. 2004), and in gray seals (*Halichoerus grypus*) collected near Sable Island, Canada, the levels did not change between 1984 and 1994 (Addison and Stobo 2001). Similarly, Muir et al. (2000) found no significant trend in HCB concentration in the Atlantic walrus (*Odobenus rosmarus*) between 1978 and 1988.

HCB levels and trends have also been reported in other mammalian species. For example, in Sweden, concentrations have fallen by a rate of 7.5% per year in bovine adipose tissue sampled between 1991 and 2004, from medians of 3.4–3.9 to 1.3–2.7 ng/g LW (Glynn et al. 2009). In butter samples from many countries and regions of the world, concentrations of HCB showed a much narrower range (i.e., a more even distribution) than that observed for other OC pesticides, with concentrations varying "only" between 0.34 and 6.2 ng/g LW (Kalantzi et al. 2001).

14.3.2 Pentachlorophenol

Pentachlorophenol (PCP) and its derivatives such as sodium pentachlorophenate (NaPCP) have been widely used until recently for agricultural, industrial, and public health purposes, as wood preservatives, termite deterrents, bactericides, weed killers, and/or molluscicides. Following extensive application of NaPCP as a molluscicide in rice fields in Surinam, Vermeer et al. (1974) detected residues of PCP in the liver samples from cattle egrets (*Bubulcus ibis*), common egrets (*Ardea alba*), and snowy egrets (*Egretta thula*) residing in the vicinity of PCP-treated rice fields, at levels of 0.07–0.19 µg/g WW. Egrets, herons, and jacanas were found sick or dead during the period of pesticide application, and this also includes 50 snail kites (*Rostrhamus sociabilis*), birds of prey in which high PCP residues were found in the brain (artithmetic mean ± SE, 11.25 ± 1.11 µg/g WW) and liver (45.56 ± 2.18 µg/g WW) of 17 corpses analyzed, suggesting a probable death after contaminated snail ingestion.

More recently, PCP was detected at average levels of 51 (range 5–343) ng/g LW in the whole blood of polar bears collected in 1999–2001 in central East Greenland (Sandala et al. 2004). In the blubber of male and female ringed seals from East Greenland, Letcher et al. (2009) reported concentrations of 1.0 ± 0.4 ng/g LW and hypothesized a possible origin in polar bear samples of PCP through the diet (they consume large quantities of blubber of seals) or also by metabolization of HCB.

14.4 Insecticides

Insecticides are chemicals developed, produced, and used to control the populations of insects. Together with the group of rodenticides, they can be considered the most valuable pesticides used in agriculture, animal and public health, industry, and households. However, some representatives of the OP, CB, and pyrethroid groups are also the most acutely toxic ones of the pesticides for nontarget vertebrates (Table 14.1), and the medium-to long-term problems of contamination associated with the massive use of the OC group are still one of the most active and productive areas of research in Environmental Toxicology today.

14.4.1 Organochlorine Insecticides

Although a long time has passed since the banning of OC insecticides in most parts of the world, the residues are still present in any kind of biotic or abiotic sample (provided that a sensitive analytical method is used to detect them). Their bioaccumulatory properties and capacity to biomagnify and their potential adverse effects on the reproduction and survival of organisms have stressed the necessity of monitoring their environmental levels since mid-1960s (Keith 1996; Fisk et al. 2005). The undesirable effects of some of these chemicals are linked to the occurrence of immunological, pathological, reproductive, and teratogenic dysfunctions in various vertebrate species. Moreover, some of these compounds, such as chlordane, chlordecone, p,p′-DDT, HCHs, mirex, or toxaphene, are classified by the IARC as possibly carcinogenic to humans (group 2B) (IARC 2010).

The interference with the endocrine systems of wild animals has received particular attention in recent times (Vos et al. 2000; Tanabe 2002; Giesy et al. 2003; Porte et al. 2006; Fossi et al. 2007; Bernanke and Köhler 2009). The International Programme on Chemical Safety (IPCS) defines endocrine disruptors (EDs) as exogenous substances or mixtures that alter the function(s) of the endocrine system and consequently cause adverse health effects in an intact organism, or its progeny, or (sub)populations (Damstra et al. 2002). Such pollutants can modulate or disrupt the endocrine system via several mechanisms, including interfering with hormone synthesis, transport, metabolism, ligand binding, and gene expression. Reproductive failure, teratogenic effects, and eggshell thinning in animals exposed to EDs during embryonic development have been discussed extensively (Vos et al. 2000; Damstra et al. 2002; Tanabe 2002; Giesy et al. 2003; Porte et al. 2006; Hotchkiss et al. 2008; Bernanke and Köhler 2009; Sonne 2010).

Due to the biomagnification capacity of recalcitrant POPs, the trophic position has commonly been considered the main factor explaining the pollutant concentrations in bird and mammalian tissues (Dietz et al. 2000; Fisk et al. 2001; Goerke et al. 2004; Alleva et al. 2006; Corsolini 2009; Dhananjayan and Muralidharan 2010; Skarphedinsdottir et al. 2010). This explains in part the vast use of carnivorous (or piscivorous) species such as birds of prey, seabirds, or marine mammals for monitoring purposes in environmental toxicology studies.

Over the past two decades, stable isotope analysis (SIA) has become an important technique in environmental toxicology and ecotoxicology studies. Although many stable isotopes can be used, nitrogen-15 ($^{15}N/^{14}N$ expressed relative to a standard, or $\delta^{15}N$) and carbon-13 ($^{13}C/^{12}C$, or $\delta^{13}C$) isotopic signatures have increasingly gained more attention for unraveling the structures of food webs (Fisk et al. 2001; Elliott 2005; Elliott et al. 2009; Lavoie et al. 2010). Whereas $\delta^{15}N$ value is used as an indicator to assess the trophic position occupied by a population and the food web transfer of POPs, increasing approximately 3%–5% between trophic levels, $\delta^{13}C$ provides some information about the within-ecosystem compartment that the animal group exploited, such as their particular foraging site, with terrestrial communities usually being depleted of ^{13}C when compared to marine communities and benthic/littoral communities being enriched with ^{13}C when compared to pelagic communities (Fisk et al. 2001; Dietz et al. 2004a; Elliott 2005; Borrell et al. 2006; Krahn et al. 2007; Elliott et al. 2009; Ramírez et al. 2009). Similarly, the use of the fatty acid profile, and especially the ratio of n-3/n-6 ($\omega3/\omega6$) polyunsaturated fatty acids (PUFA), can also provide significant information on the diet and thus on POP exposure (Guitart et al. 1996a; Iverson et al. 1997; Mateo et al. 2004; Wolkers et al. 2006).

The trophic position certainly plays an important role, but several other factors might confound with the correct interpretation of spatial variations or temporal trends of insecticide POPs. Age, gender, pregnancy, and nutritional and health statuses of the populations studied are some of them, as is metabolism, in general, which has also emerged as significant recently (Fisk et al. 2001; Guruge et al. 2001; Dietz et al. 2004b; Goerke et al. 2004; Andersen et al. 2006; Shore et al. 2006b; Cid et al. 2007; Gouteux et al. 2008; Vorkamp et al. 2008; Lavoie et al. 2010). Even small differences in the diet may contribute significantly to variations in the contaminant levels in the samples from the same species collected in different subzones of the same area or between years (Goutner et al. 2001; Galassi et al. 2002; Mateo et al. 2004; Christensen et al. 2005; Borrell et al. 2006; Anthony et al. 2007; Elliott et al. 2009). Mañosa et al. (2003) showed that a shift in the diet in northern goshawk (*Accipiter gentilis*) from European rabbits (*Oryctolagus cuniculus*) to birds (mainly Passeriformes), due to a severe reduction of the population of the first type of preys in Spain, probably masked any temporal decline in OC levels. In addition, the role of migration to more contaminated latitudes or zones may, or may not, introduce unexpected changes in the pollutant levels (Braune et al. 2001; Guruge et al. 2001; Aguilar et al. 2002; Hario et al. 2004; Bustnes et al. 2006; Helgason et al. 2008; van Drooge et al. 2008; Corsolini 2009; Grove et al. 2009; Rajaei et al. 2010a). Last but not the least, a study carried out with glaucous gulls (*Larus hyperboreus*) breeding in the Norwegian Arctic suggested that climate variability, and thus the predicted climate change, may also influence POP concentrations (Bustnes et al. 2010). In fact, Geisz et al. (2008) found glacier melting to be a potential source for the movement of ΣDDT to the Antarctic marine food web as levels have not declined in Adélie penguins (*Pygoscelis adeliae*) from the Western Antarctic Peninsula for more than 30 years.

In the samples either from bird (Wainwright et al. 2001; Henny et al. 2003b, 2008a; Sanpera et al. 2003; Guitart et al. 2005; Van den Steen et al. 2006; Bustnes et al. 2007; Martínez-López et al. 2007; Keithmaleesatti et al. 2007; Bouwman et al. 2008; Helgason et al. 2008; Malik et al. 2011) or mammal origin (Hobbs et al. 2003; Tanabe et al. 2003; Kajiwara et al. 2004; Ramu et al. 2006; Fillmann et al. 2007; Vorkamp et al. 2008; Lemarchand et al. 2010; Moon et al. 2010; Mwevura et al. 2010; Park et al. 2010), usually the residues of the DDT family (o,p'-DDD, p,p'-DDD, o,p'-DDE, p,p'-DDE, o,p'-DDT and p,p'-DDT) have dominated, among OC insecticides, in the majority of the reports during the last 10 years. A selection of published values determined in different species and areas of the world are shown in Tables 14.2 and 14.3 for birds and in Table 14.4 for marine mammals. Total DDT (ΣDDT) concentrations are the sum of all the DDT family compounds. In this regard, it should be mentioned that the ratio p,p'-DDT/p,p'-DDE (or sometimes ΣDDT/p,p'-DDE) has been used as an estimate of the time elapsed since the use of the insecticide, with the ratio increasing over time as the p,p'-DDT degrades.

Other important OC insecticides are the chlordane, cyclodiene, HCH, and toxaphene families and endosulfan and mirex. Chlordane is a mixture of structurally related OC compounds whose major constituents include cis- and trans-chlordane, cis- and trans-nonachlor and heptachlor, with heptachlor epoxide being the main product of transformation of the latter and one of the most common cyclodienes found in animal samples, while oxychlordane is a metabolite of both cis- and trans-chlordane (Wiemeyer 1996). Toxaphene is a term used to refer to a complex mixture of chemicals that consists primarily of chlorinated bornanes and chlorinated bornenes. The inconsistencies in the analytical methods (usually a GC but coupled to various detectors) and quantification standards of varying compositions hinder the comparison of chlordane and toxaphene concentrations among laboratories, making spatial or trend analyses difficult. Vorkamp et al. (2004) considered that the levels of toxaphene and chlordane-related compounds were comparable to that of DDT, making them equally important for monitoring programs and studies on POPs in the Arctic. Typically, for all Arctic birds, the general order of egg (or tissue) concentrations of OC pesticides is ΣDDT$\approx\Sigma$chlordanes$\approx\Sigma$toxaphene>Σchlorobenzenes$\approx\Sigma$HCH, while for all marine mammals (and similar to polar bears), the general order of tissue concentrations is ΣDDT$\approx\Sigma$chlordanes>Σchlorobenzenes$\approx\Sigma$HCH$\approx\Sigma$toxaphene (Letcher et al. 2010). In Africa, the residues in the fauna were the greatest for DDTs, followed by those of dieldrin, HCH, endosulfan, and endrin, with small amounts of aldrin and toxaphene being found in some animals (Wiktelius and Edwards 1997). Although not with as much reported levels as DDTs and other OC pesticides, chlordane and toxaphene concentrations have been determined in different vertebrate animal samples in recent times (Muir et al. 2000; Braune et al. 2001, 2005; Guruge et al. 2001; Hobbs et al. 2001; Wainwright et al. 2001; Connell et al. 2003; Harris et al. 2003; Henny et al. 2003b; Tanabe et al. 2003; Vorkamp et al. 2004, 2008; Wiemeyer et al. 2005; Andersen et al. 2006; Wolkers et al. 2006; Anthony et al. 2007; Braune 2007; Cid et al. 2007; Bustnes et al. 2007; Bouwman et al. 2008; Carballo et al. 2008; Gouteux et al. 2008; Grove and Henny 2008; Helgason et al. 2008; Clark et al. 2009; Haraguchi et al. 2009; Letcher et al. 2009, 2010; Best et al. 2010; Champoux et al. 2010; Lavoie et al. 2010; Moon et al. 2010; Park et al. 2010; Rigét et al. 2010). Maruya et al. (2005) found levels of toxaphene ranging from 12 to 260 ng/g WW in the livers of adult swallows (*Petrochelidon* spp.) collected in Texas and concluded that the liver profiles in these insectivorous birds appear to be intermediate in complexity between those in invertebrates and fish (more complex) and mammals (less complex).

The cyclodienes, and particularly aldrin, dieldrin, and endrin, were implicated in numerous wildlife incident mortalities when still in use (Newton et al. 1992; Peakall 1996). In very recent times, their concentrations in bird or mammal tissues are scarcely reported as the environmental levels found are usually well below many other OC pesticides (Konstantinou et al. 2000; Dong et al. 2004; Buck et al. 2005; Braune 2007; Dell'Omo et al. 2008; Grove and Henny 2008; Muralidharan et al. 2008; Clark et al. 2009; Best et al. 2010; Champoux et al. 2010; Cipro et al. 2010; Dhananjayan and Muralidharan 2010; Leonel et al. 2010; Mwevura et al. 2010; Dhananjayan et al. 2011; Malik et al. 2011). However, in the United States, it is considered that some peregrine falcons might still be at risk for dieldrin residue effects (Clark et al. 2009), and, in Argentina, an individual of the great kiskadee (*Pitangus sulphuratus*), a passerine bird, was found with 5,245 ng/g LW in the carcass (Cid et al. 2007). The levels of dieldrin in the subcutaneous adipose tissues of polar bears from East Greenland (range: 43–640 ng/g LW) were also considered to be of concern (Sonne et al. 2009),

HCH levels are the highest in India (Kalantzi et al. 2001; Kannan et al. 2005; Muralidharan et al. 2008; Aktar et al. 2009; Dhananjayan and Muralidharan 2010; Dhananjayan et al. 2011) and China (Dong et al. 2004). The predominance of β-HCH reflects its bioaccumulative nature and resistance to enzymatic degradation in the animal body, while lower proportions of α- and γ-isomers suggest metabolic degradation (Tanabe et al. 1997). In fact, β-HCH differs largely in its physicochemical properties from all the rest of HCH isomers (Bhatt et al. 2009). The presence of an unusually high proportion of lindane could suggest a recent use of this pesticide, as has been observed in birds from France (Berny et al. 2002), Spain (Herrera et al. 2000; Mañosa et al. 2003; Guitart et al. 2005; van Drooge et al. 2008; Martínez-López et al. 2009), Greece (Sakellarides et al. 2006), Argentina (Cid et al. 2007), India (Dhananjayan and Muralidharan 2010), Iran (Rajaei et al. 2010b), and Pakistan (Malik et al. 2011); marine mammals from Tanzania (Mwevura et al. 2010); ring-tailed lemurs (*Lemur catta*) from Madagascar (Rainwater et al. 2009); mustelids from France (Lemarchand et al. 2010); and bovine milk from Iran (Bayat et al. 2010).

On the other hand, endosulfan is a special case and a candidate to be included in a new group of POPs (Weber et al. 2010). Contrary to many other OC pesticides, by the year 2010, this insecticide had only been banned in approximately 60 countries, and India and China are the biggest users due to its inexpensive economical cost (Lubick 2010). Commercial endosulfan is a mixture of two isomers, namely α- and β-endosulfan, and in the environment, the cyclic sulfate group of endosulfan can be oxidized to the corresponding endosulfan sulfate, which appears to be the most persistent. Residues of endosulfan, although normally at much lower concentrations than other OC pesticides in the biota, have been reported, and many data on the trends point to an increase in the environmental levels, and there is a growing concern about their impact especially on the marine environment (Li and Macdonald 2005; Muir and de Wit 2010; Weber et al. 2010).

Interestingly, mirex levels in the blood of south polar skuas (*Catharacta maccormicki*) from Antarctica were among the highest reported in birds, the mean level being 3 and 26 times higher than those in comparable species of the Arctic glaucous gull and great black-backed gull (*Larus marinus*): 20.7 vs. 6.9 and 0.8 ng/g WW, respectively, exceeding even p,p'-DDE (Bustnes et al. 2006). Mirex was also found in the eggs of African darter, cattle egret, and other species from South Africa, where this insecticide (but also a flame retardant) has never been registered as a pesticide (Bouwman et al. 2008). In franciscana dolphins from Southern Brazil, Leonel et al. (2010) found that the mirex levels did not differ over the period from 1994 to 2004 or even show a slight increase; and in polar bears from the Hudson Bay, Canada, McKinney et al. (2010) found no significant changes in the contaminant levels between 1991 and 2007.

14.4.1.1 Birds

Several studies have reported OC concentrations in eggs from different areas of the world in order to evaluate both temporal and spatial variations in the pollutant status of particular ecosystems and to determine the potential adverse effects on animal populations (Mañosa et al. 2003; Muñoz-Cifuentes et al. 2003; Herkert 2004; Mateo et al. 2004; Vorkamp et al. 2004; Buck et al. 2005; Braune 2007; Bustnes et al. 2007; Strause et al. 2007; Henny et al. 2009; Best et al. 2010; Van den Steen et al. 2010). In birds, POP insecticides may impair reproduction (reduced egg production, eggshell thinning, decreased fertility and hatchability, and others), affect the ability to compete for food and habitat, impair the capacity to avoid predation, and contribute to reduced parental attentiveness (Peakall 1996; Berny et al. 2002; Giesy et al. 2003; Walker 2003; Fisk et al. 2005; Mineau 2005; Bernanke and Köhler 2009; Grove et al. 2009). For example, during the late 1960s and into the 1970s, instances of reproductive failure, developmental anomalies, and population declines were reported for the species of raptorial and fish-eating colonial water birds from the North American Great Lakes ecosystem (Hoffman et al. 1990; Bernanke and Köhler 2009; Best et al. 2010).

More surprisingly, deaths due to OC insecticides of adult birds are still being diagnosed even in the United States: in New York, potentially lethal levels of chlordane, dieldrin, and DDTs were detected in the brains of 165 birds representing 15 species, mainly blue jays (*Cyanocitta cristata*), American crows (*Corvus brachyrhynchos*), and Cooper's hawks (*Accipiter cooperii*) (Okoniewski et al. 2006).

It is known that most biotransformation processes of p,p'-DDT in vertebrates end up as p,p'-DDE. This is the most persistent metabolite and the major representative of the insecticide DDT-group in almost all published data (Mateo et al. 2000; Muñoz-Cifuentes et al. 2003; Van den Steen et al. 2006, 2010; Jiménez et al. 2007; Keithmaleesatti et al. 2007; Bouwman et al. 2008; Cortinovis et al. 2008; Gómara et al. 2008; Helgason et al. 2008; Henny et al. 2008a; Clark et al. 2009; Vorkamp et al. 2009; Ahmad et al. 2010; Champoux et al. 2010), Greece being an intriguing exception with the highest proportion of o,p'-DDD (Konstantinou et al. 2000; Goutner et al. 2001, 2005; Albanis et al. 2003) and Pakistan with DDD, mostly the p,p'-isomer (Malik et al. 2011).

The notorious and well-documented effect of eggshell thinning caused by DDE, resulting in crushed eggs and breeding failure, showed extreme interspecific differences in sensitivities (Bernanke and Köhler 2009). Blus (1996) proposed as calculated no-effect level for DDE in eggs in a range between 100 ng/g WW for the brown pelican (*Pelecanus occidentalis*) and 2000 ng/g WW for the peregrine falcon, while Mañosa et al. (2003) suggested values between 1000 and 2000 ng/g WW for the goshawk. However, the effect on eggshell thickness is an important but no exclusive factor related to reproductive success in wild birds. Elliott et al. (2001) found that the highest concentration of DDE in osprey eggs that hatched was 9200 ng/g WW, whereas the highest concentration in eggs that did not hatch was 22,900, and proposed that the levels of DDE that affect hatching success ranged between 6,000 and 10,000 ng/g WW. In brown pelicans, 3000 ng/g WW is associated with near total reproductive failure (Blus 2003). Recently, Best et al. (2010) suggested a critical value associated with a decrease in nest success below 0.7 young/occupied territory at 4500 ng/g WW DDE for bald eagles (*Haliaeetus leucocephalus*). Using two Ardeid species, the little egret (*Egretta garzetta*) and the black-crowned night heron (*Nycticorax nycticorax*), Connell et al. (2003) established a threshold level of 1000 ng/g WW at which there was a significant level of reduction in the survival of the young.

With these (and many others estimated or calculated) critical levels, DDTs were still considered to pose a hazard for many avian species. Yates et al. (2010), for example, indicate that white-faced ibis (*Plegadis chihi*) nesting at Carson Lake, Nevada, might represent the worst-case scenario for continuing DDE problems, due to the extreme sensitivity to DDE in the eggs of this species. Similar problems were described for white-faced ibises from Cibola Lake, Arizona (King et al. 2003). However, other nonraptorial species such as the red-legged partridge (*Alectoris rufa*) in Spain (Herrera et al. 2000); the purple heron (*Ardea purpurea*) in Greece (Sakellarides et al. 2006); the African darter (*Anhinga rufa*) in South Africa (Bouwman et al. 2008); the green heron (*Butorides virescens*) in the United States (Wainwright et al. 2001); the little egret in Romania (Aurigi et al. 2000), Hong Kong (Connell et al. 2003), and Greece (Sakellarides et al. 2006); the intermediate egret (*Egretta intermedia*) in Pakistan (Sanpera et al. 2002); the lesser black-backed gull (*Larus fuscus*) in Finland (Hario et al. 2004); the brown-hooded gull (*Larus maculipennis*) in Chile (Muñoz-Cifuentes et al. 2003); the black-crowned night heron in Hong Kong (Connell et al. 2003), China (Dong et al. 2004), and Greece (Sakellarides et al. 2006); the greater flamingo (*Phoenicopterus ruber*) in Greece (Sakellarides et al. 2006); and the great crested grebe (*Podiceps cristatus*) in Italy (Galassi et al. 2002; Cortinovis et al. 2008) have also been considered at risk due to the high DDE (or ΣDDTs) residue concentrations detected in recent times.

Among raptors, because of their predatory lifestyle, often at the top of the food chain, the situation of the Eurasian sparrowhawk in Spain (van Drooge et al. 2008); the Spanish imperial eagle (*Aquila adalberti*) in Spain (Hernández et al. 2008); the northern aplomado falcon (*Falco femoralis septentrionalis*) in Mexico (Mora et al. 2008); the European kestrel in the Canary Islands, Spain (Mateo et al. 2000); the Eurasian hobby (*Falco subbuteo*) in Spain (van Drooge et al. 2008); the white-backed vulture (*Gyps bengalensis*) in India (Muralidharan et al. 2008); the white-tailed sea eagles (*Haliaeetus albicilla*) in Sweden (Helander et al. 2002) and in the United Kingdom (Walker et al. 2010); the bald eagle in the United States (Buck et al. 2005; Cesh et al. 2008); the booted eagle (*Hieraaetus pennatus*) in Spain (Martínez-López et al. 2007); the red kite (*Milvus milvus*) in Spain (Gómara et al. 2008); and the osprey in the United States (Henny et al. 2003b) have also been considered to be of concern.

These are some of the wild bird species that were judged, at least, in "at risk" situation for DDT residues, in scientific publications of the last 10 years. But the lack of data, results, or conclusions for this period about other species, zones, regions, or countries should not be, obviously, interpreted at any cost as free of problems derived from DDTs and/or other recalcitrant POP insecticides and/or other pollutants in the environment.

Moreover, there are human health implications in some cases. In poultry (*Gallus gallus*), the levels of OC pesticides were measured in home-produced eggs in Belgium, and it is interesting to note that 17% of the egg samples had levels above the norm for ΣDDT, established at 500 ng/g LW (Van Overmeire et al. 2009).

14.4.1.2 Mammals

Since the early works by Koeman and van Genderen (1966) or Holden and Marsden (1967), among others, the OC pesticide levels in marine mammals have centered a lot of the attention of scientists (Aguilar et al. 2002; Rattner 2009). The literature on these compounds in the blubber and other tissues of marine mammals across the world, especially from the Arctic and more recently, the Antarctic regions, has been extensively and comprehensively reviewed in the last 15 years (Norstrom and Muir 1994; Tanabe et al. 1994; Muir and Norstrom 2000; Aguilar et al. 2002; Goerke et al. 2004; Riget et al. 2004, 2010; Braune et al. 2005; Fisk et al. 2005; Houde et al. 2005; Li and Macdonald 2005; Corsolini 2009; Sonne et al. 2009; Hoferkamp et al. 2010; Letcher et al. 2010; Sonne 2010).

Marine mammals are especially vulnerable to OC pesticides because these chemicals accumulate efficiently in their thick blubber layers. They are long-lived animals, and cetaceans and pinnipeds are thought to have a reduced capacity to metabolize DDTs and other POPs when compared to terrestrial mammals, allowing a higher biomagnification through the food web (Norstrom and Muir 1994; Tanabe et al. 1994; Houde et al. 2005). Polar bears are an exception among marine mammals, as they have a superior OC biotransformation capacity and have the highest levels of OC metabolites, although some of them have also demonstrated ED activity (Braune et al. 2005; Letcher et al. 2009; Sonne 2010). Seals have been considered particularly sensitive, since the natural stress in extreme cold environments is often more severe than in temperate regions, rendering this group of animals more susceptible to the effects of anthropogenic pollutants (Schiavone et al. 2009).

In wild marine mammal populations, elevated POP levels have been associated with pathological changes and abnormalities of various forms (skin lesions, skeletal deformities, and liver and kidney damage), biochemical modifications (especially on hormonal and vitamin A status), reproductive impairment (some of them related to ED activities), increased tumor incidence and depression of the immune system, and, as a consequence of the latter, the occurrence of large die-offs among seal and cetacean species (Nyman et al. 2002;

Borrell and Aguilar 2005; Fisk et al. 2005; Fossi et al. 2007; Sonne et al. 2008, 2009; Bernanke and Köhler 2009; Letcher et al. 2010; Routti et al. 2010; Sonne 2010). Since the 1980s, mass mortality events of aquatic mammals occurred all over the world (Gulland and Hall 2007). Immunological impairment caused by POPs was considered relevant in the extension of the epizootic, which started in the summer of 1990 and continued through 1991 and 1992 in the striped dolphins (*Stenella coeruleoalba*) in the Mediterranean Sea, an infection produced by a morbillivirus (Domingo et al. 1990, 1992), when it was speculated that the total population declined to just a third of its initial level (Aguilar and Raga 1993; Guitart et al. 1996a). The much less important epizootic event that occurred in 2007–2008 in the Mediterranean striped dolphins was also associated with lower levels of OCs, with mean levels for ΣDDTs of 73,000 ng/g LW in the 2007–2008 outburst compared to 392,000 ng/g LW in 1990, and the authors considered the extension of the epizootic not to have enhanced this time (Castrillon et al. 2010).

Until recently, the levels of DDTs in marine mammals were among the highest ever recorded in vertebrates; and, in some cases, they still remain very high. In Table 14.4, we can find seven species that contain concentrations above 20,000 ng/g LW of this group of chemicals: short-beaked common dolphin (*Delphinus delphis*), Mediterranean monk seal (*Monachus monachus*), finless porpoise (*Neophocaena phocaenoides*), harbor porpoise (*Phocoena phocoena*), killer whale, striped dolphin, and common bottlenose dolphin from different parts of the world (Borrell and Aguilar 2005; Kajiwara et al. 2006; Ramu et al. 2006; Borrell et al. 2006, 2007; Krahn et al. 2007; Haraguchi et al. 2009; Isobe et al. 2009; Castrillon et al. 2010; Weijs et al. 2010). As a matter of fact, marine mammals from the Arctic and some regions of Europe, North America, and Asia have been found to be particularly contaminated with OC pesticides (Guitart et al. 1996a; Aguilar et al. 2002; Houde et al. 2005; Kajiwara et al. 2006; Ramu et al. 2006; Rigét et al. 2010). Overall, Houde et al. (2005) point out that the highest concentrations of POPs were found in delphinoid cetaceans from industrialized areas of the Northern hemisphere compared to the Southern hemisphere. But regional differences not related to big latitude changes also occur: one recent example is the Mediterranean monk seal, for which extremely significant differences were found between two separate populations collected in the 1995–1999 period from Greece, in the Mediterranean Sea, and the Western Sahara, in the Atlantic Ocean (Borrell et al. 2007).

The usually high or very high levels of DDTs and other OC pesticides found in these animals and the quantity of tissues available, for example, from stranded animal episodes have led some authors to investigate the tissue distribution of these pollutants (Guitart et al. 1996a; Kannan et al. 2005; Ramu et al. 2005; Fillmann et al. 2007; Carballo et al. 2008; Isobe et al. 2009; Schiavone et al. 2009; Shoham-Frider et al. 2009; Moon et al. 2010; Weijs et al. 2010). In striped dolphins, for example, studies indicate that the distribution of POP insecticides in the body (muscle, liver, blubber, etc.) depends on the lipid content, as usually lipid normalized concentrations showed no significant differences, with the brain being an exception: lower levels in this organ are probably due to the existence of the blood–brain barrier (Guitart et al. 1996a; Isobe et al. 2009).

Less obvious or studied in birds or terrestrial mammals, in marine mammals there are many works reporting differences of OC pesticide concentrations between males and females with growth (Guitart et al. 1996a; Hobbs et al. 2001, 2003; Dietz et al. 2004a; Metcalfe et al. 2004; Wolkers et al. 2004; Stern et al. 2005; Carballo et al. 2008; Borrell et al. 2010; Mwevura et al. 2010; Park et al. 2010). These sex- and sexual maturity–related variations are generally attributed to the transfer of pollutants from the mother to the calf or pup during lactation, resulting in lower OC levels in the mother, although variation in diet, body size, metabolic ability, and other biological traits may partially account for sex-related differences in pollutant loads (Borrell and Aguilar 2005; Kajiwara et al. 2006; Haraguchi et al. 2009; Leonel et al. 2010). Therefore, the first pregnancy marks the start of the decrease phase in females. It is considered that the transfer during lactation is much higher than that occurring through deposition in the tissues of the calf or pup during gestation (Wolkers et al. 2004). The magnitude of the reproductive transfer is extremely variable, ranging from 7% to 100%, depending on the species and the compound (Houde et al. 2005).

Reports of OC pesticide residues devoted to other wild mammal groups are scarcer. Livers of male river otters (*Lontra canadensis*) collected from Oregon and Washington, in 1994–1999, were found, in general, with low OC insecticide levels, although some individuals exceeded 500,000 ng/g DW for p,p'-DDE and p,p'-DDD (Grove and Henny 2008). In Eurasian otters (*Lutra lutra*) from France, the level of ΣDDTs was 2060 (200–2600) and that of lindane was 450 (ND-900) ng/g LW in the liver samples (Lemarchand et al. 2010).

Bats have been subject to several studies, because world populations have declined dramatically (Aktar et al. 2009). Clark (2001), using the samples collected in 1930, 1956, 1965, 1973, and 1988, provided data supporting the hypothesis that DDT played a major role in the severe population decline of Brazilian free-tailed bats (*Tadarida brasiliensis mexicana*) at Carlsbad Cavern, New Mexico. More recently, Allinson et al. (2006) reported, in the carcass samples of southern bent-wing bat (*Miniopterus schreibersii bassanii*) at Bat Cave, Australia, levels of p,p'-DDE ranging from 11,000 to 59,000 ng/g LW, p,p'-DDT ranging from 110 to 1600 ng/g LW, and p,p'-DDD ranging from 35 to 620 ng/g LW.

Two populations of British Columbia grizzly bears (*Ursus arctos horribilis*) with different feeding habits were studied by Christensen et al. (2005), and it was found that, overall, maritime bears were more contaminated with many POPs than the interior bears, with chlordane levels ranging from 213 to 27,606 ng/g LW and ΣDDTs ranging from 28 to 20,277 ng/g LW in fat. In China, captive giant pandas (*Ailuropoda melanoleuca*) and red pandas (*Ailurus fulgens*) found dead from disease showed 16.3–83.2 and 14.6–56.2 ng/g LW of ΣDDT in the livers, respectively (Hu et al. 2008).

In Madagascar, the levels of OC insecticides in the blood of ring-tailed lemurs were considered to be the background concentrations and were characterized within DDTs by a dominance of p,p'-DDT and the absence of p,p'-DDE (Rainwater et al. 2009).

In pet dogs, the levels of DDTs usually detected are low, with the concentrations of HCHs and chlordanes being even slightly higher than those of DDTs in Japan (Kunisue et al. 2005). However, cats have higher values of DDTs: 99.8 (77–756) ng/g LW in the adipose tissues, in Italy (n = 84); and 99–140 (29–400) ng/g LW in the genital organs, in Japan (Kunisue et al. 2005; Storelli et al. 2009).

OC insecticides can get accumulated in fat-rich foods. Food is therefore one of the main sources of human exposure and is the reason why animal-origin fats need to be preferentially monitored in order to assess the potential health risks for consumers. Data for milk and butter are available today for almost every country. Highest concentrations of OC insecticide residues have been found in India, China, Iran, Mexico, Colombia, and Spain, with the HCH and DDT families and endolsulfan being especially problematic in some regions or countries, sometimes reflecting the current use at the time of sampling (Badia-Vila et al. 2000; Kalantzi et al. 2001; Battu et al. 2004; Nag and Raikwar 2008; Bayat et al. 2010; Castilla-Pinedo et al. 2010). In a comparative study of butter from several countries, Kalantzi et al. (2001) found levels between 0.410 (United States) and 248.9 (India) ng/g LW for ΣDDTs and between ND (Australia, Denmark, Italy and United States) and 222.8 (India) ng/g LW for ΣHCHs.

14.4.2 Organophosphorus and Carbamate Insecticides

Cholinesterase-inhibiting compounds such as OP and CB insecticides (some representatives also with acaricide and nematicide activities) have largely replaced the OC group. They are often considered nonpersistent products in the environment; thus, they have a low potential for bioaccumulation and biomagnification but a high potential for acute toxicity (Poppenga 2007). Moreover, due to their extensive use in agriculture for crop protection and in many other areas and situations, such as golf courses, forestry, gardens, and in veterinary medicine for parasite control on livestock and because some products remain quite toxic for days or weeks after application, continuous unwanted exposure to OPs and CBs occur.

Among the most vulnerable of nontarget species are birds, for which lethal episodes are frequently reported in the literature (Henny et al. 1987; Newton 1998; de Snoo et al. 1999; Goldstein et al. 1999; Fleischli et al. 2004; Kwon et al. 2004; Pain et al. 2004; Wobeser et al. 2004; Muzinic 2007; Berny and Gaillet 2008; Elliott et al. 2008; Hernández and Margalida 2008; Guitart et al. 2010b; Otieno et al. 2010; Novotný et al. 2011). One of the worst and latest episodes occurred in the pampas (grasslands) of Argentina between 1995 and 1996, when the inadequate use of monocrotophos was responsible for the deaths of an estimated 5095 Swainson's hawks (*Buteo swainsoni*), which represented approximately 1% of the world population of this raptor species at the time (Goldstein et al. 1999). In the United States, the lethal risk to birds has generally declined over the period between 1991 and 2003 in most crops, which was attributed to a replacement of the more toxic insecticides, largely OP and CB products (Mineau and Whiteside 2006). Of 24 different pesticides identified in avian mortalities in the United States, the most frequent were famphur (18%), carbofuran (15%), diazinon (12%), and fenthion (5.1%), with Falconiformes reportedly being killed most frequently (49% of all die-offs) but Anseriformes found dead in the greatest numbers (64% of 8877 recovered carcasses) (Fleischli et al. 2004). In Spain, a review of the analytical data pooled from various institutions indicated that 58 wild birds, including 10 cinereous vultures and one Spanish imperial eagle, were poisoned by different pesticides between 1990 and 2005, especially by aldicarb and carbofuran (27.59% each) and methomyl (8.62%) (Martínez-Haro et al. 2008). Moreover, there is growing concern that exposure to AChE-inhibiting insecticides may also have negative sublethal effects on avian populations by affecting the physiological and behavioral responses, including the loss of orientation in migratory birds (Strum et al. 2010).

Lethal cases in mammals, some primary and other secondary poisonings, have also been reported (Goldstein et al. 1999; Wobeser et al. 2004; Wang et al. 2007; Martínez-Haro et al. 2008; Berny et al. 2010; Guitart et al. 2010a,b; Novotný et al. 2011). The illegal use of anticholinesterase pesticides in Canada between 1967 and 2002 was reviewed by Wobeser et al. (2004), and coyotes (*Canis latrans*), red foxes, and striped skunks (*Mephitis mephitis*) were implicated in many of the episodes. Novotný et al. (2011) reported cases in Eurasian otters with carbofuran and cases in wild boars (*Sus scrofa*) and bisons (*Bison* spp.) with methomyl in the Czech Republic.

The best diagnosis of poisoning with OPs and CBs is the measurement of AChE activity in the brain, which is best correlated with morbidity and mortality than chlolinesterase inhibition in the plasma or blood. As a general rule, it is considered that a 20% decrease in activity is indicative of abnormal exposure, while lethal intoxication is usually associated with a 50% decline, although these figures can be lower or higher depending on the situation (Goldstein et al. 1999; Mineau and Tucker 2002; Wobeser et al. 2004). In particular, species, age, sex, reproductive condition, nutritional status, and sample handling and laboratory techniques have been demonstrated to introduce some variability (Mayack and Martin 2003; Maul and Farris 2004). At least for Northern cardinals (*Cardinalis cardinalis*), Maul and Farris (2005) concluded that monitoring of the plasma cholinesterase reactivation with pyridine-2-aldoxime methochloride (2-PAM) appeared to be a more sensitive indicator of exposure than the diagnostic threshold. Using this biomarker of toxic effect, it is not easy to distinguish between OPs and CBs, although some differences in the reactivation of the phosphorylated and carbamylated enzymes exist and can be used (Smith et al. 1995; Mineau and Tucker 2002; Wobeser et al. 2004).

Chemical monitoring is rarely used for these insecticides and metabolites, and usually only for forensic purposes and analyzing preferentially the gastrointestinal contents (Goldstein et al. 1999; Fleischli et al. 2004; Kwon et al. 2004; Pain et al. 2004; Wobeser et al. 2004; Muzinic 2007; Wang et al. 2007; Berny and Gaillet 2008; Elliott et al. 2008; Otieno et al. 2010). Determination of the AChE activity in the plasma, as a nondestructive biomarker, can be extensively (and nonexpensively) used to monitor "normal" exposition in different bird groups, related to environmental stress from typical agricultural pesticide application regimes (Wilson et al. 2001; Maul and Farris 2005; Martínez-Haro et al. 2007; Vergara et al. 2008; Strum et al. 2010) or in "special" circumstances as during locust control operations of large areas (Fildes et al. 2006).

Many lethal episodes in birds are related to granular formulations, which present a serious hazard to birds. Microgranules and granules are always used for the most toxic pesticides, such as the OPs, chlorpyrifos, disulphoton, ethoprop, fonofos, and phorate, and the CBs, aldicarb and carbofuran (Kendall and Smith 2003; Elliott et al. 2008; Martínez-Haro et al. 2008; Otieno et al. 2010). Birds eat granules exposed on the soil surface, mistaking them for food or grit. Elliott et al. (2008) found in Canada 211 waterfowl corpses, dead from primary poisoning (average AChE inhibition in brain being 74%, range 69%–78%), and 15 raptors, mainly bald eagles, dead from secondary poisoning, after granular application of fonofos. In Canada, the use of granular insecticides in agriculture was found linked to population declines of several bird species, including the American robin (*Turdus migratorius*), the horned lark (*Eremophila alpestris*), the house sparrow (*Passer domesticus*), and the mourning dove (*Zenaida macroura*) (Mineau et al. 2005).

Due to the wide and extensive use, it is not unsuspected that the residues of these insecticide compounds may appear in the milk in dairy cattle. In Mexico, of 96 milk samples examined by GC for 13 OP insecticides, eight (12%) contained residues exceeding established maximum residue limits (MRL) for dichlorvos (n = 5 samples), phorate, chlorpyrifos, and chlorfenvinphos (n = 1, each) (Salas et al. 2003). In Italy and Spain, none of the 298 and 242 milk samples examined, respectively, contained detectable levels of OPs above the MRL fixed by the European Commission, although, for example, residues of chlorpyriphos, dichlorvos, or coumaphos were found (Gazzotti et al. 2009; Melgar et al. 2010). Sengupta et al. (2010) reported levels of OP pesticides, mainly dimethoate and malathion, in the range of 11–98 ng/g WW in meat samples of cow, goat, and chicken collected from five locations in India.

14.4.3 Pyrethroid Insecticides

Synthetic pyrethroids have their origins in the naturally occurring insecticides in pyrethrum extracts, collectively called pyrethrins, which are obtained from the flowers of the chrysanthemum species (*Chrysanthemum cinerariaefolium*). Pyrethroids were developed to increase the photostability and residence time in environmental conditions of pyrethrins while retaining some other interesting characteristics, such as a rapid neurotoxicant insecticidal activity and a relatively low acute mammalian toxicity (Anadón et al. 2009). Thus, pyrethroids are considered relatively persistent compounds, which show very low water solubility. They are used in agriculture on a variety of fruits and vegetables, pet and livestock protection, and in suburban and urban areas, and gained popularity among the general public as the more common household insecticide products available today for both indoor and outdoor applications.

Most papers have dealt with pyrethroid determination in different food samples of vegetable origin, but animal matrices such tissues and milk have been less studied (Di Muccio et al. 1997; Sun et al. 2003; Akre and MacNeil 2006; Fernández-Álvarez et al. 2008; Khay et al. 2009). Usually, routine detection of pyrethroid residues in tissues and milk from domestic animals give negative or under MRL results (Battu et al. 2004; Fernández-Álvarez et al. 2008; Khay et al. 2009; Lutze et al. 2009; Sereda et al. 2009). However, in some regions of Africa, aerial and ground sprayings to control the tsetse flies (*Glossina* spp.) and also the use of dip and pour-on applications of pyrethroids in cattle cause some concerns on the environmental impact of such practices: for example, residues of pyrethroids in the dung of treated animals have been associated with the death of dung beetles (Grant 2001).

Cats are particularly susceptible to the poisoning with permethrin, a pyrethroid widely used in flea control products such as spot-ons, sprays, collars, and shampoos (Anadón et al. 2009; Poppenga and Oehme 2010). Most cases are due to the application of canine permethrin flea products to cats, which is considered one of the most common causes of feline deaths (Boland and Angles 2010).

14.5 Acaricides

Dicofol, an OC compound structurally related to the insecticide DDT, is a nonsystemic acaricide with little insecticide activity, currently banned or severely restricted in many countries. Most of the problems associated with this miticide are related to the manufacture process and the subsequent presence of impurities such as DDTs, especially in the technical product (Di Muccio et al. 1988; Qiu et al. 2005). Dicofol itself has been experimentally found to cause eggshell thinning in several species of birds (Wiemeyer 1996; Wiemeyer et al. 2001), although no adverse effects have been documented in field studies, and rare detections of dicofol in the biota are generally low (Blus and Henny 1997). For example, Hoferkamp et al. (2010), in their review, were unable to find any results for dicofol in the Arctic environmental media.

However, in other parts of the world, the situation may differ. Malik et al. (2011) reported for the first time, residues of this compound in birds from Pakistan, at levels ranging from ND-155 ng/g WW in eggs collected from heronries at the River Ravi and the River Chenab and attributed the detection to its use as a pesticide in agricultural applications. In bovine milk sampled in India, dicofol was positive in 17 out of 325 (5.23%) samples, and in four of these positive samples, it was above the MRL (Nag and Raikwar 2008).

14.6 Rodenticides

Unwanted nontarget domestic animal and wildlife exposures can occur when chemicals are used to control and/or eradicate outbreaks of rodent (rats, mice, voles, and others) pests in agricultural and urban areas or wild lands. The majority of rodenticide-related exposures involved long-acting ARs (Poppenga and Oehme 2010).

Historically, many natural (mineral and toxin) poisonous and nonselective substances have been used for this purpose; fortunately most of them are now prohibited in many countries and/or applications. However, prohibition does not mean that exposition (usually in criminal acts) cannot occur. This is the case of thallium sulfate, once widely used as a rodenticide, for which poisoning cases are still seen (Volmer et al. 2006). But the most striking case is that of strychnine, an extremely poisonous alkaloid obtained from the seeds of the plant *Strychnos nux-vomica*. It is a fast-acting product, so it is unexpected that living vertebrate animals may carry or accumulate residues of strychnine in their bodies. Cases of domestic and wildlife poisonings are described frequently in European Laboratories of Toxicology (Martínez-López et al. 2006; Wang et al. 2007; Hernández and Margalida 2008; Martínez-Haro et al. 2008; Berny et al. 2010; Guitart et al. 2010b). Interestingly, Martínez-López et al. (2006) were able to detect strychnine residues by GC/MS in the carcass of a Bonelli's eagle (*Hieraaetus fasciatus*) found in the field, which had died 20–45 days before.

14.6.1 Zinc Phosphide and Phosphine

Zinc phosphide (Zn_3P_2) acts through the release of phosphine gas (PH_3) (Plumlee 2004). It is a rodenticide considered relatively safe for nontarget animals, as secondary poisoning of scavenger species is unlikely (Sterner and Mauldin 1995). However, direct ingestion of zinc phosphide baits or the consumption of the digestive tracts of poisoned animals poses a risk for many domestic and wildlife species (Mauldin et al. 1996; Drolet et al. 1996; Poppenga et al. 2005). Aluminum phosphide (AlP) is used more as an insecticide for stored cereal grains and can also generate phosphine gas (Proudfoot 2009).

14.6.2 Anticoagulant Rodenticides

ARs are vitamin K antagonists that prevent the synthesis of functionally active blood-clotting factors II, VII, IX, and X, and therefore, kill rodents by predisposing them to fatal hemorrhage (Murphy 2007). These compounds are, at present, some of the preferred methods of controlling rodent infestations worldwide.

Chemically, ARs are classified into two classes: 4-hydroxycoumarin derivatives (brodifacoum, bromadiolone, difenacoum, warfarin, and some others belong to this group), and 1,3-indanedione derivatives (with chlorophacinone and diphacinone being the most prominent examples) (Murphy 2007; Pelfrène 2010). However, from the clinical point of view, both types of chemical classes produce similar basic symptomatology and require comparable treatment, based on the use of vitamin K_1 as antidote (Plumlee 2004; Pelfrène 2010).

More practical and environmentally useful is the classification based on the toxicological properties (Murphy 2007). Warfarin, the first AR introduced in the market shortly after the end of World War II, and others such as chlorophacinone, coumatetralyl, or diphacinone are called first-generation ARs or FGARs. They were so successful and thus so extensively used in the following years that it led to the development of rodent resistance to their effects (Pelz et al. 2005; Ishizuka et al. 2007), the first population of warfarin-resistant brown rats (*Rattus norvegicus*) being discovered in a Scottish farm in 1958 (Boyle 1960). This prompted the introduction of new products in the 1970s and 1980s, such as brodifacoum, bromadiolone, difenacoum, difethialone, or flocoumafen, now defined as second-generation ARs, or SGARs (and sometimes described also as "superwarfarins" or as long-acting anticoagulant rodenticides (LAARs)). The pharmacological target molecule, vitamin K 2,3-epoxide reductase, and the mode of action of SGARs are the same as those of the FGAR type (Pelfrène 2010; Watanabe et al. 2010).

However, there are some important differences between the two groups. FGARs are characterized by moderate toxicity in terms of LD_{50} (see Table 14.1) and often need continuous exposure to the bait to be effective in rodent control. SGARs are much more acutely toxic, generally providing a lethal dose after a single feeding, and tend to have longer half-lives and thus a considerably higher persistence in animal tissues. For example, Vandenbroucke et al. (2008) determined the comparative pharmacokinetics in mice (*Mus musculus*) of eight freely available ARs in the Belgian market after a single oral dose, and the shortest elimination half-lives in the plasma were found for the FGARs coumatetralyl (0.52 days), chlorophacinone (11.7 days), and warfarin (14.9 days), while the longest were for the SGARs brodifacoum (91.7 days), difethialone (38.9 days), and bromadiolone (33.3 days). Moreover, due to the great affinity to the binding sites in the liver, the elimination half-lives in this organ were found to range from 15.8 days for coumatetralyl to 307.4 days for brodifacoum.

Owing to its extensive use, poisoning in domestic and farm animals or wild birds and mammals may occur in cases of direct intentional or accidental ingestion of the poisoned baits (Stone et al. 1999; Mineau and Tucker 2002; Berny et al. 2005, 2010; Guitart et al. 2010b). After an aerial baiting operation with brodifacoum to eradicate rats and mice and reduce rabbit numbers on Motuihe Island, New Zealand, in groups of pukekos (*Porphyrio porphyrio*) and paradise shelducks (*Tadorna variegata*) that were being monitored, mortalities of 48/98 (49%) and 31/52 (60%) were observed, respectively (Dowding et al. 1999). A recent case of mass poisoning occurred in Spain in 2007 when an outbreak of common voles (*Microtus arvalis*) causing potentially severe crop damage was inappropriately treated with large-scale application of wheat grain treated with 0.005% chlorophacinone in the open field, causing the death of dozens of individuals of domestic pigeons (*Columba livia*), a mostly granivorous species in the wild (Sarabia et al. 2008; Olea et al. 2009). The poisoning of game animals such as pigeons or white-tailed deers, which can potentially be exposed to high doses of ARs due to the direct ingestion of poisoned baits, has raised concerns regarding the potential health effects if these animals are hunted and consumed by humans (Stone et al. 1999; Olea et al. 2009).

On the other hand, the delayed action of such compounds, with mortality in rodents occurring days or even weeks after initial bait consumption, together with the high toxicity and long elimination half-lives especially observed with the SGARs, has resulted in secondary exposure and even lethal poisoning of various nontarget animals, mainly affecting the populations of birds of prey and carnivorous mammals worldwide who feed on poisoned rodents (Newton et al. 1990; Berny et al. 1997; Howald et al. 1999; Shore et al. 2003; Murray and Tseng 2008; Dowding et al. 2010). However, any animal, including an invertebrate, which has consumed the rodenticide bait, could potentially serve as a source of secondary exposure for a predator or a scavenger (Eason and Spurr 1995; Howald et al. 1999; Thorsen et al. 2000; Spurr et al. 2005; Hoare and Hare 2006; Dowding et al. 2010).

Massive mortality has been barely described in the literature after secondary poisoning. For example, 28 animals, principally red foxes and buzzards (*Buteo buteo*), were recovered from a single area in France, near the Swiss border, where bromadiolone was applied to carrot baits at a dose of 100 mg/kg for the control of European water voles (*Arvicola terrestris*) (Berny et al. 1997). A similar episode was reported by Kupper et al. (2006), when again an application of bromadiolone baits to control water voles in Northern Switzerland killed at least 40 red foxes. In 2002, hundreds of nontarget wildlife deaths, including some individuals of the saker falcons (*Falco cherrug*), corsac foxes (*Vulpes corsac*), and Pallas' cats (*Otocolobus manul*), were detected in Mongolia, following the application of bromadiolone-treated wheat grains that were spread by aircraft over large areas of steppe grassland to control an outbreak of Brandt's voles (*Microtus brandti*) (Winters et al. 2010).

In recent years, concern has grown also on the potential sublethal effects of AR exposures to animals, due to their widespread and large-scale occurrence. Monitoring of pesticide residues in many countries has shown the wanted or unwanted extension of the problem (see Tables 14.5 and 14.6), and the situation has been particularly well documented in the United Kingdom (Newton et al. 1990; McDonald et al. 1998; Shore et al. 1999, 2003; Dowding et al. 2010; Walker et al. 2010), France (Berny et al. 1997; Fournier-Chambrillon et al. 2004; Lemarchand et al. 2010; Sage et al. 2010), United States (Stone et al. 1999, 2003), Canada (Howald et al. 1999; Albert et al. 2010), and, especially, New Zealand (Murphy et al. 1998; Eason et al. 2002; Spurr et al. 2005; Hoare and Hare 2006). Usually, the pattern of exposure to ARs reflects usage patterns in each region or country considered.

New Zealand is an island country particularly vulnerable, which has now lost over 40% of its prehuman land-bird fauna (Atkinson 2001; Clout and Russell 2006). The reason is that biological invasions of alien mammals have devastating effects principally on insular ecosystems, affecting wildlife populations and biodiversity in different ways (Atkinson 2001; Courchamp et al. 2003; Jones et al. 2008; Gibbs 2009). Warfarin, bromadiolone, and, more recently, brodifacoum, which have been used since at least 1980, have been the AR-type toxicants of choice for the eradication of introduced rats from offshore islands in New Zealand (Towns and Broome 2003; Hoare and Hare 2006; Howald et al. 2007). By the end of 2004, there were 218 successful eradications of 17 different alien mammal species reported from around 90 New Zealand islands (Towns and Broome 2003; Clout and Russell 2006). Pioneering techniques that have been developed, which included the advent of new rodenticide delivery techniques such as aerial broadcast or the use of pulsed rather than continuous baiting strategies, encouraged the application of similar campaigns in islands of other parts of the world (Howald et al. 1999; Taylor et al. 2000; Thorsen et al. 2000; Donlan et al. 2003; Orueta et al. 2005). Invasive rodents have been eradicated from at least 284 islands worldwide, totaling around 476 km^2 (Howald et al. 2007).

14.6.2.1 Birds

Based on necropsy findings and analytical results, cases of confirmed or suspected lethal effects after direct AR exposure through the consumption of poisoned baits in wild birds have been described in the literature, and the species involved included domestic pigeons, crested wood partridges (*Rollulus roulroul*) and other partridges, New Zealand robins (*Petroica australis*), little spotted kiwis (*Apteryx owenii*), western wekas (*Gallirallus australis*), pukekos, saddlebacks (*Philesturnus carunculatus*), New Zealand kakas (*Nestor meridionalis*), paradise shelducks, mallards (*Anas platyrhynchos*), and common ravens (*Corvus corax*) (Dowding et al. 1999; Howald et al. 1999; Taylor et al. 2000; Borst and Counotte 2002; Hoare and Hare 2006; Sarabia et al. 2008; Olea et al. 2009; Guitart et al. 2010b). However, death after secondary poisoning is a more common feature and had been especially diagnosed in birds of prey (Borst and Counotte 2002; Stone et al. 2003; Murray and Tseng 2008; Albert et al. 2010; Guitart et al. 2010b).

TABLE 14.5

Concentrations of Selected First and Second Anticoagulant Rodenticides Detected in the Livers of Wild Birds Across the World

Species	Country	N	N Positive	Brodifacoum	Bromadiolone	Chlorophacinone	Coumatetralyl	Difenacoum	Difethialone	Diphacinone	Warfarin	Source
Accipiter cooperii	United States	50	18	0.10 (0.008–0.22)	0.35 (0.04–0.60)	—	—	—	—	0.10	0.10	Stone et al. (2003)
Anas platyrhynchos	New Zealand	2	2	1.07 (0.90–1.23)	—	—	—	—	—	—	—	Eason et al. (2002)
Anas platyrhynchos	Spain	3	3	—	—	1.34 (0.71–2.17)	—	—	—	—	—	Olea et al. (2009)
Apteryx australis	New Zealand	29	14	0.09 (0.01–0.69)	—	—	—	—	—	—	—	Eason et al. (2002)
Bubo virginianus	United States	13	13	0.33 (0.01–0.73)	0.14	—	—	—	—	—	—	Stone et al. (1999)
Bubo virginianus	United States	53	43	0.21 (0.007–0.97)	0.23 (0.05–1.08)	—	—	—	—	—	0.73	Stone et al. (2003)
Bubo virginianus	Canada	61	43	0.052 (0.001–0.609)	0.042 (0.005.0.571)	0.0029 (0.0025–0.014)	—	—	0.004 (0.0025–0.03)	0.011 (0.008–0.012)	0.0037 (0.0025–0.72)	Albert et al. (2010)
Buteo buteo	France	16	15	—	0.4* (0.2–1.3)	0.3* (0.2–0.5)	—	—	—	—	—	Berny et al. (1997)
Buteo buteo	Spain	3	1	—	—	0.12	—	—	—	—	—	Olea et al. (2009)
Buteo jamaicensis	United States	7	7	0.65 (0.23–1.60)	—	—	—	—	—	—	—	Stone et al. (1999)
Buteo jamaicensis	United States	78	45	0.21 (0.006–1.28)	0.23 (0.08–0.50)	0.18	—	—	—	0.34	—	Stone et al. (2003)
Cathartes aura	Netherlands	2	2	0.190	—	—	—	—	—	—	—	Borst and Counotte (2002)
Circus approximans	New Zealand	2	2	0.64 (0.61–0.66)	—	—	—	—	—	—	—	Eason et al. (2002)
Columba livia	France	4	4	—	—	3.4* (1.7–3.5)	—	—	—	—	—	Berny et al. (1997)

Species	Country										Reference
Columba livia	Spain	66	64	—	—	6.6 (0.55–50.11)	—	—	—	—	Olea et al. (2009)
Corvus corax	Canada	13	13	1.35 (0.98–2.52)	—	—	—	—	—	—	Howald et al. (1999)
Falco tinnunculus	United Kingdom	18	14	0.041 (0.009–0.115)	0.136 (0.010–0.515)	0.029	—	0.157 (0.005–1.074)	—	—	Walker et al. (2010)
Fringilla coelebs	New Zealand	3	3	1.43 (0.12–2.31)	—	—	—	—	—	—	Eason et al. (2002)
Gallirallus australis	New Zealand	7	7	1.08 (0.11–2.30)	—	—	—	—	—	—	Eason et al. (2002)
Gymnorhina tibicen	New Zealand	10	2	0.47 (0.08–0.99)	—	—	—	—	—	—	Eason et al. (2002)
Melanocorypha calandra	Spain	7	2	—	—	1.57 (1.04–2.09)	—	—	—	—	Olea et al. (2009)
Milvus migrans	France	5	5	—	0.4* (0.3–0.6)	—	—	—	—	—	Berny et al. (1997)
Milvus milvus	France	62	27	—	(0.2–5.6)	(0.9–5.2)	—	—	—	—	Berny and Gaillet (2008)
Milvus milvus	United Kingdom	23	17	0.125** (0.071–0.222)	0.073** (0.056–0.094)	—	—	0.052** (0.040–0.067)	—	—	Walker et al. (2008a)
Milvus milvus	United Kingdom	3	2	—	0.085 (0.055–0.115)	—	—	0.084	—	—	Walker et al. (2010)
Nestor meridionalis	New Zealand	3	3	2.87 (1.20–4.10)	—	—	—	—	—	—	Eason et al. (2002)
Ninox novaeseelandiae	New Zealand	3	3	1.84 (0.97–3.44)	—	—	—	—	—	—	Eason et al. (2002)
Otus asio	United States	22	10	1.16 (0.007–0.47)	0.30 (0.05–0.50)	—	—	—	—	—	Stone et al. (2003)
Philesturnus carunculatus	New Zealand	4	2	0.33 (0.05–0.60)	—	—	—	—	—	—	Eason et al. (2002)
Porphyrio porphyrio	New Zealand	9	9	0.86 (0.52–1.35)	—	—	—	—	—	—	Dowding et al. (1999)

(continued)

TABLE 14.5 (Continued)

Concentrations of Selected First and Second Anticoagulant Rodenticides Detected in the Livers of Wild Birds Across the World

Species	Country	N	Positive	Brodifacoum	Bromadiolone	Chlorophacinone	Coumatetralyl	Difenacoum	Difethialone	Diphacinone	Warfarin	Source
Strix varia	Canada	25	23	0.074 (0.001–0.927)	0.084 (0.002–1.012)	0.0043 (0.0025–0.015)	—	—	0.003 (0.0025–0.017)	0.010 (0.010–0.012)	0.0026 (0.0025–0.005)	Albert et al. (2010)
Tadorna variegata	New Zealand	4	4	0.56 (0.24–0.80)	—	—	—	—	—	—	—	Eason et al. (2002)
Tyto alba	United Kingdom	145	15	0.180 (0.019–0.515)	—	—	—	0.036 (0.005–0.106)	—	—	—	Newton et al. (1990)
Tyto alba	Canada	78	48	0.027 (0.01–0.47)	0.020 (0.005–0.72)	—	—	—	0.019 (0.0025–0.72)	0.010 (0.010–0.02)	0.0025 (0.0025–0.008)	Albert et al. (2010)
Tyto alba	United Kingdom	62	39	0.093 (0.004–0.911)	0.044 (0.004–0.396)	0.007 (0.004–0.009)	0.033	0.032 (0.003–0.127)	—	—	—	Walker et al. (2010)
Turdus merula	New Zealand	7	7	0.55 (0.01–1.10)	—	—	—	—	—	—	—	Eason et al. (2002)

Values are arithmetic means (and/or range), otherwise stated (* = median, ** = geometric mean), of positive cases and are expressed as µg/g WW. Animals studied were usually found dead or moribund in the wild, or captured, and, in some cases, a diagnosis of anticoagulant poisoning was made.

TABLE 14.6

Concentrations of Selected First and Second Anticoagulant Rodenticides Detected in Livers of Wild Mammals Across the World

Species	Country	N	N Positive	Brodifacoum	Bromadiolone	Chlorophacinone	Coumatetralyl	Difenacoum	Difethialone	Diphacinone	Warfarin	Source
Capreolus capreolus	France	2	2	—	1.55* (1.2–1.9)	—	—	—	—	—	—	Berny et al. (1997)
Cervus elaphus	New Zealand	33	11	(0.00–0.03)	—	—	—	—	—	—	—	Eason et al. (2002)
Erinaceus europaeus	New Zealand	21	10	0.20 (0.0–1.31)	—	—	—	—	—	—	—	Spurr et al. (2005)
Erinaceus europaeus	United Kingdom	120	33	0.05	0.59	—	—	0.10	—	—	—	Dowding et al. (2010)
Felis catus	New Zealand	71	57	(0.078–1.84)	—	—	—	—	—	—	—	Eason et al. (2002)
Felis catus	New Zealand	10	8	0.41 (0.0–1.25)	—	—	—	—	—	—	—	Spurr et al. (2005)
Lepus capensis	France	15	13	—	1.4* (1.2–1.6)	2.3* (0.2–8.3)	—	—	—	—	—	Berny et al. (1997)
Lepus granatensis	Spain	16	6	4.18 (1.09–9.52)	—	—	—	—	—	—	—	Olea et al. (2009)
Lutra lutra	France	11	3	—	6.6 (6.0–7.1)	5.0	—	—	—	—	—	Fournier-Chambrillon et al. (2004)
Lutra lutra	France	20	2	—	0.63 (0.40–0.85)	—	—	—	—	—	—	Lemarchand et al. (2010)
Mustela erminea	United Kingdom	44	9	0.12	0.197 (0.04–0.38)	—	0.0074 (0.0046–0.0097)	—	—	—	—	McDonald et al. (1998)
Mustela erminea	New Zealand	40	31	0.37	—	—	—	—	—	—	—	Murphy et al. (1998)
Mustela erminea	New Zealand	115	98	(0.008–1.32)	—	—	—	—	—	—	—	Eason et al. (2002)
Mustela erminea	New Zealand	55	39	0.15 (0.0–0.74)	—	—	—	—	—	—	—	Spurr et al. (2005)
Mustela furo	New Zealand	16	9	1.01	—	—	—	—	—	—	—	Murphy et al. (1998)
Mustela furo	New Zealand	6	5	0.82 (0.0–2.43)	—	—	—	—	—	—	—	Spurr et al. (2005)

(continued)

TABLE 14.6

Concentrations of Selected First and Second Anticoagulant Rodenticides Detected in Livers of Wild Mammals Across the World

Species	Country	N	N Positive	Brodifacoum	Bromadiolone	Chlorophacinone	Coumatetralyl	Difenacoum	Difethialone	Diphacinone	Warfarin	Source
Mustela lutreola	France	31	1	—	5.0	—	—	—	—	—	—	Fournier-Chambrillon et al. (2004)
Mustela nivalis	United Kingdom	10	3	—	0.25	—	0.0295 (0.0085–0.06)	—	—	—	—	McDonald et al. (1998)
Mustela nivalis	New Zealand	14	10	1.26	—	—	—	—	—	—	—	Murphy et al. (1998)
Mustela nivalis	New Zealand	18	13	0.33 (0.0–0.93)	—	—	—	—	—	—	—	Spurr et al. (2005)
Mustela putorius	United Kingdom	50	13	0.008	0.079 (0.016–0.217)	—	—	0.106 (0.005–0.321)	—	—	—	Shore et al. (1999)
Mustela putorius	United Kingdom	50	18	0.06 (0.05–0.07)	0.12 (0.03–0.19)	—	—	0.30 (0.02–0.92)	—	—	—	Shore et al. (2003)
Mustela putorius	France	33	5	—	3.4 (0.6–9.0)	—	—	—	—	—	—	Fournier-Chambrillon et al. (2004)
Mustela vison	France	47	7	—	2.7 (1.9–4.2)	5.5 (3.4–8.5)	—	—	—	—	—	Fournier-Chambrillon et al. (2004)
Odocoileus virginianus	United States	6	6	0.29 (0.12–0.41)	—	—	0.5	—	—	0.93	—	Stone et al. (1999)
Oryctolagus cuniculus	France	13	12	—	1.35* (1.3–1.4)	2.9* (1.1–14.3)	—	—	—	—	—	Berry et al. (1997)
Sciurus carolinensis	United States	6	6	1.68 (0.53–4.10)	—	0.62	—	—	—	2.0	0.228	Stone et al. (1999)
Sus scrofa	France	6	6	—	0.6* (0.4–3.6)	1.2* (0.6–1.4)	—	—	—	—	—	Berry et al. (1997)
Sus scrofa	New Zealand	35	21	(0.007–1.38)	—	—	—	—	—	—	—	Eason et al. (2002)
Vulpes vulpes	France	31	31	—	1.5* (0.8–6.9)	0.3* (0.2–0.6)	—	—	—	—	—	Berry et al. (1997)

Values are arithmetic means (and/or range), otherwise stated (* = median), of positive cases and are expressed as μg/g WW. Animals studied were usually found dead or moribund in the wild, or captured, and, in some cases, a diagnosis of anticoagulant poisoning was made.

Recently, Watanabe et al. (2010) explained the apparent contradiction of the reported high resistance (expressed as the LD_{50}) of some bird species (e.g., chickens, ring-necked pheasants (*Phasianus colchicus*), and northern bobwhites (*Colinus virginianus*)) to warfarin when compared with brown rats and the fact that many observed cases of death episodes in the wild occur in raptors. They found that there is a wide interspecies difference among birds in xenobiotic metabolism and sensitivity to ARs based on vitamin K 2,3-epoxide reductase kinetics and inhibition and warfarin metabolic activity.

Table 14.5 shows the levels of FGAR and SGAR residues determined in birds from different periods of time, zones, and sources. Animals studied were usually found dead or moribund in the wild, or captured, and, in a few cases, a veterinary diagnosis of anticoagulant poisoning was eventually made.

In situations in which ARs are commonly used against different rodent pests, exposure of raptors is usually more common than of other groups of birds. For example, in France, a study carried out in the wetlands and marshes of Loire Atlantique's department where ARs are employed to combat the large and semiaquatic rodent coypus (*Myocastor coypus*), it was determined that from 30 raptors (11 common buzzards, 10 barn owls, 5 tawny owls, and 4 European kestrels) and 29 waterbirds (15 mallards, 13 Eurasian coots (*Fulica atra*), and 1 common moorhen (*Gallinula chloropus*)) found dead or moribund in the wild, 22 raptors (73%) and only 4 waterfowl (14%) presented levels ≥ 0.008 µg/g WW of residues (Lambert et al. 2007). In spite of this, even when rodent control campaigns using heavy doses of ARs are performed in the field, an increase in the exposure of raptors is not expected to occur if managed correctly, the results of nontarget exposures depending less on the amount of bait used rather than on the way it is used (Shore et al. 2006a).

The Strigiformes is an order comprising some 200 species of birds of prey, and due to their diet and feeding habits, some of their representatives have been used in many occasions to monitor the exposure to the residues of ARs. Newton et al. (1990) detected brodifacoum and difenacoum at medium–low concentrations (<0.52 and <0.11 µg/g WW, respectively) in the livers of 15 (10.3%) out of 145 barn owls found dead in the United Kingdom and submitted to their laboratory for a survey period of 7 years, between 1983 and 1989. Only one bird was diagnosed postmortem as poisoned with brodifacoum (0.43 µg/g), while in the rest of 144 animals, the cause of death was ascribed to trauma, starvation, or shot. In fact, barn owls have been monitored in the United Kingdom since 1983 by the Predatory Bird Monitoring Scheme (PBMS) (Newton et al. 1990; Walker et al. 2008a). The PBMS in Britain covers a long-term monitoring program that examines the levels of pesticide and other pollutant residues in selected avian wildlife species (Walker et al. 2008a). Analytical methods have varied with the time passed, and currently the number of ARs determined is higher and the limit of quantification (LoQ) is lower than in previous years. But, when data were normalized for the four SGARs, brodifacoum, bromadiolone, difenacoum, and flocoumafen and a common LoQ of 0.01 µg/g WW was applied to each, the percentage of positive individuals for any SGAR plotted against time shows that exposure had increased over time, from an almost undetectable proportion in 1983 to a maximum of 53% in 2003, decreasing again to 30%–35% during the period from 2004 to 2006 (Walker et al. 2010). The increase during this long period was driven by an augment of exposure mainly due to bromadiolone and difenacoum, while brodifacoum and flocoumafen exposures were more erratic and showed no clear trend. However, it should be stressed that when all the AR residues determined were considered and more sensitive methods were used with lower LoQs, the true level of exposure of barn owls in 2006 (n = 62) was 62.9% (6.4% to any FGAR, 62.9% to any SGAR), with a total of 25.8% of the examined individuals exposed to multiple rodenticides (Walker et al. 2010).

Nevertheless, of 172 livers from tawny owls collected in two distant periods of years examined in the United Kingdom for some SGARs, it was found that 33 (19.2%) of the whole samples contained detectable concentrations of one or more of them, the individual occurrence being 11.6%, 5.8%, 4.7%, and 0% for bromadiolone, difenacoum, brodifacoum, and flocoumafen, respectively (Walker et al. 2008b). Although the SGARs have been hypothesized as a potential cause for the decline of British owl populations, there was found no clear variation or trend in exposure between individuals in the 1990–1993 and 2003–2005 examined periods in this work, and the authors concluded that there was no clear evidence to implicate SGARs as a major factor affecting tawny owl numbers.

A recently published survey carried out in British Columbia and the Yukon region of Western Canada between 1988 and 2003 (Albert et al. 2010), in which the livers of 78 barn owls, 61 great horned owls, and 25 barred owls (*Strix varia*) were examined, 70% had residues of at least one AR, and, of these, 41% had more than one detected. Bromadiolone (52%) and brodifacoum (51%) were the most commonly detected ARs in all three species, followed by difethialone (9%), diphacinone (5%), chlorophacinone (4%), and warfarin (3%) residues. Only brodifacoum was found in barred owls to increase its presence over time. Based on necropsy findings and residue levels, a final diagnosis of SGAR poisoning was established for 6 animals (3 barred, 2 barn, and 1 great horned owls), although other animals were considered suspected of AR poisoning based on the liver residues.

The PBMS also carried out systematic AR residue analysis in European kestrels since 1997 (Walker et al. 2010). Using the same considerations for calculations explained above for barn owls (LoQ of 0.01 µg/g WW), the adjusted percentage of birds with any detectable SGAR liver residue was 50%–70% (mainly due to bromadiolone and difenacoum), with no apparent temporal trend. However, individually it was found that brodifacoum progressively increased from about 5% during the period 1997–2000 to a 17% in the period 2005–2006 (Walker et al. 2010).

Another diurnal raptor monitored by the PBMS is the red kite (Walker et al. 2010). Livers of 23 red kites found dead in Great Britain between 1994 and 2005 were analyzed, and 73.9% of them showed one or more residues of SGARs (Walker et al. 2008a). In France, Berny and Gaillet (2008) found that over the period 1992–2002, 27 out of 62 red kites submitted to the Toxicology Laboratory of the College of Veterinary Medicine in Lyon were diagnosed of AR poisoning, with 24 cases due to bromadiolone and 3 due to chlorophacinone (Table 14.5).

14.6.2.2 Mammals

ARs continue to be a major cause for morbidity and mortality for pets such as dogs and cats and can even affect production animals (Robben et al. 1998; Kohn et al. 2003; Berny et al. 2005, 2010; Del Piero and Poppenga 2006). Carnivorous mammals are commonly exposed to ARs usually when they feed on dead and dying poisoned rats and mice (or other rodent species), although a few cases of suspected or confirmed primary poisoning have been described in the literature in herbivorous or omnivorous species such as gray squirrels (*Sciurus carolinensis*), common hares (*Lepus capensis*), Iberian hares (*Lepus granatensis*), European rabbits, raccoons (*Procyon lotor*), wild boars, sheep (*Ovis aries*), white-tailed deers, red deers (*Cervus elaphus*), and roe deers (*Capreolus capreolus*) (Berny et al. 1997, 2005; Stone et al. 1999; Eason et al. 2002; Del Piero and Poppenga 2006; Olea et al. 2009).

The European hedgehog (*Erinaceus europaeus*) is an omnivorous mammal that feeds mainly on a wide range of invertebrates, but preferring insects. Exposure to ARs in this species may be a result of direct ingestion of the poison baits or more habitually through the consumption of contaminated invertebrates that ingest rodent baits, carcasses, feces, or soil-bound residues (Spurr et al. 2005; Dowding et al. 2010). It has been shown that snails or beetles can carry considerable amounts of AR residues (Eason and Spurr 1995; Howald et al. 1999; Thorsen et al. 2000), and other insectivorous mammals such as shrews (*Sorex* spp.) can also be in danger of exposition (Brakes and Smith 2005). The European hedgehog has been investigated recently in the United Kingdom for the exposure to ARs, showing that indirect contamination in carnivorous species, other than those who consume directly poisoned rodents, is possible (Dowding et al. 2010). Results showed that the prevalence of exposure was widespread, with 66.7% of 120 analyzed livers by HPLC/MS from hedgehogs found dead positive for combined residues of FGARs (22.5%) and SGARs (57.5%) (some animals presented more than one residue compound). However, there was no clear pathological evidence found that this was a contributory factor to lethal effects, as necropsy revealed no sign of hemorrhage. Samples were analyzed also by HPLC with fluorescence detection (these are the results show in Table 14.6), and it was found that the prevalence of exposition was clearly higher using HPLC/MS due to its superior sensibility.

Red fox is a well-known carnivorous and the most widespread and abundant species of fox. In France, a study of anticoagulant poisonings of wildlife carried out from 1991 to 1994 yielded 31 confirmed diagnoses for bromadiolone and chlorophacinone lethal toxicoses, based on a liver threshold of 0.2 µg/g and/or lesions consistent with anticoagulant poisoning (Berny et al. 1997). Recently, a noninvasive technique using feces was developed to monitor both recent and previous exposure of foxes to bromadiolone, useful to control baiting strategies and exposure to nontarget carnivores (Sage et al. 2010). The same study, in which animals were sacrificed at the end of the experimental procedure and a complete necropsy was performed in the corpses, also revealed that apparently red foxes were more susceptible to the effects of bromadiolone than dogs (Sage et al. 2010).

Exposure of other predators such as mustelids, some representatives considered pests in various countries, to ARs is well documented (see Table 14.6). A poisoning operation with brodifacoum in New Zealand targeting rabbits also killed stoats and ferrets (*Mustela furo*) (Alterio 1996). Shore et al. (1996) analyzed the livers of 24 polecats in the United Kingdom during 1992–1994, and SGAR residues were detected in 29% (7/24) of the animals, difenacoum being the most prevalent one (6/7). During the period 1994–1997, the proportion was similar (23%, 6/26), but bromadiolone was the residue most detected that time (Shore et al. 1999). In a subsequent study with newly added individuals, 35.1% (13/37) male and 38.5% (5/13) female polecats were positive for difenacoum, bromadiolone, and, to a lesser extent, brodifacoum (Shore et al. 2003). McDonald et al. (1998) showed that ARs were detected in 23% (9/40) of stoats and 30% (3/10) of weasels in England between 1996 and 1997, revealing also that exposure was more prevalent in female stoats than in males. Analysis of polecat feces confirmed rats as the principal prey items, although woodmice and voles were also taken (Brakes and Smith 2005).

In France, Fournier-Chambrillon et al. (2004) estimated the exposure of 122 dead free-ranging riparian mustelids of four species collected between 1990 and 2002 and found residues of bromadiolone or chlorophacinone in 3% (1/31) of European minks (*Mustela lutreola*), 15% (7/47) of American minks (*Mustela vison*), 15% (5/33) of polecats, and 27% (3/11) of European otters of the collected carcasses. Another study with the latter species detected 10% (2/20) individuals with bromadiolone residues (Lemarchand et al. 2010).

14.7 Conclusions

Monitoring of wildlife for pesticide residues provides data that can be used in risk assessments and to reduce sublethal contamination and mortality in nontarget species. Most studies are devoted to OC pesticide residues and, overall, a decline in the environmental levels has been observed, although it was most evident during the 1970s and 1980s and less dramatic thereafter. Endosulfan and HCHs can be an exception, and their temporal trends deserve more attention in the future. In order to compare better results, the use of concentrations based on LW is suggested. Many predators and scavengers are declining in numbers, and exposure to SGARs might therefore be of importance in conservation biology, and more studies are needed. Other areas such as herbicides, fungicides other than HCB, and acaricides are less covered and merit probably more attention by scientists.

References

Addison, R. F. and Stobo, W. T. 2001. Trends in organochlorine residue concentrations and burdens in grey seals (*Halichoerus grypus*) from Sable Is., NS, Canada, between 1974 and 1994. *Environ. Pollut.* 112: 505–513.

Aguilar, A. and Borrell, A. 2005. DDT and PCB reduction in the western Mediterranean from 1987 to 2002, as shown by levels in striped dolphins (*Stenella coeruleoalba*). *Mar. Environ. Res.* 59: 391–404.

Aguilar, A. and Raga, J. A. 1993. The striped dolphin epizootic in the Mediterranean Sea. *Ambio* 22: 524–528.

Aguilar, A., Borrell, A., and Reijnders, P. J. 2002. Geographical and temporal variation in levels of organochlorine contaminants in marine mammals. *Mar. Environ. Res.* 53: 425–452.

Ahmad, R., Salem, N. M., and Estaitieh, H. 2010. Occurrence of organochlorine pesticide residues in eggs, chicken and meat in Jordan. *Chemosphere* 78: 667–671.

Akre, C. J. and MacNeil, J. D. 2006. Determination of eight synthetic pyrethroids in bovine fat by gas chromatography with electron capture detection. *J. AOAC Int.* 89: 1425–1431.

Aktar, M. W., Sengupta, D., and Chowdhury, A. 2009. Impact of pesticides use in agriculture: Their benefits and hazards. *Interdisc. Toxicol.* 2: 1–12.

Albanis, T. A., Goutner, V., Konstantinou, I. K., and Frigis K. 2003. Organochlorine contaminants in eggs of the yellow-legged gull (*Larus cachinnans michahellis*) in the North Eastern Mediterranean: Is this gull a suitable biomonitor for the region? *Environ. Pollut.* 126: 245–255.

Albert, C. A., Wilson, L. K., Mineau, P., Trudeau, S., and Elliott, J. E. 2010. Anticoagulant rodenticide in three owl species from Western Canada, 1988–2003. *Arch. Environ. Contam. Toxicol.* 58: 451–459.

Alleva, E., Francia, N., Pandolfi, M., De Marinis, A. M., Chiarotti, F., and Santucci, D. 2006. Organochlorine and heavy-metal contaminants in wild mammals and birds of Urbino-Pesaro Province, Italy: An analytic overview for potential bioindicators. *Arch. Environ. Contam. Toxicol.* 51: 123–134.

Allinson, G., Mispagel, C., Kajiwara, N., Anan, Y., Hashimoto, J., Laurenson, L., Allinson, M., and Tanabe, S. 2006. Organochlorine and trace metal residues in adult southern bent-wing bat (*Miniopterus schreibersii bassanii*) in southeastern Australia. *Chemosphere* 64: 1464–1471.

Alterio, N. 1996. Secondary poisoning of stoats (*Mustela erminea*), feral ferrets (*Mustela furo*), and feral house cats (*Felis catus*) by the anticoagulant poison, brodifacoum. *N. Z. J. Zool.* 23: 331–338.

Anadón, A., Martínez-Larrañaga, M. R., and Martínez, M. A. 2009. Use and abuse of pyrethrins and synthetic pyrethroids in veterinary medicine. *Vet. J.* 182: 7–20.

Andersen, G., Føreid, S., Skaare, J. U., Jenssen, B. M., Lydersen, C., and Kovacs, K. M. 2006. Levels of toxaphene congeners in white whales (*Delphinapterus leucas*) from Svalbard, Norway. *Sci. Total. Environ.* 357: 128–137.

Anthony, R. G., Miles, A. K., Ricca, M. A., and Estes, J. A. 2007. Environmental contaminants in bald eagle eggs from the Aleutian archipelago. *Environ. Toxicol. Chem.* 26: 1843–1855.

Atkinson, I. A. E. 2001. Introduced mammals and models for restoration. *Biol. Conserv.* 99: 81–96.

Attaran, A. and Maharaj, R. 2000. DDT for malaria control should not be banned. *Br. Med. J.* 321: 1403–1404.

Aurigi, S., Focardi, S., Hulea, D., and Renzoni, A. 2000. Organochlorine contamination in bird's eggs from the Danube Delta. *Environ. Pollut.* 109: 61–67.

Backer, L. C., Grindem, C. B., Corbett, W. T., Cullins, L., and Hunter, J. L. 2001. Pet dogs as sentinels for environmental contamination. *Sci. Total. Environ.* 274: 161–169.

Badia-Vila, M., Ociepa, M., Mateo, R., and Guitart, R. 2000. Comparison of residue levels of persistent organochlorine compounds in butter from Spain and from other European countries. *J. Environ. Sci. Health B* 35: 201–210.

Barata, C., Fabregat, M. C., Cotín, J., Huertas, D., Solé, M., Quirós, L., Sanpera, C., Jover, L., Ruiz, X., Grimalt, J. O., and Piña, B. 2010. Blood biomarkers and contaminant levels in feathers and eggs to assess environmental hazards in heron nestlings from impacted sites in Ebro basin (NE Spain). *Environ. Pollut.* 158: 704–710.

Barber, J. L., Sweetman, A. J., van Wijk, D., and Jones, K. C. 2005. Hexachlorobenzene in the global environment: Emissions, levels, distribution, trends and processes. *Sci. Total. Environ.* 349: 1–44.

Barceló, D. and Petrovic, M. 2007. Challenges and achievements of LC-MS in environmental analysis: 25 years on. *Trends Analyt. Chem.* 26: 2–11.

Basu, N., Scheuhammer, A. M., Bursian, S. J., Elliott, J., Rouvinen-Watt, K., and Chan, H. M. 2007. Mink as a sentinel species in environmental health. *Environ. Res.* 103: 130–144.

Basu, N., Head, J., Scheuhammer, A. M., Bursian, S. J., Rouvinen-Watt, K., and Chan, H. M. 2009. The mink is still a reliable sentinel species in environmental health. *Environ. Res.* 109: 940–941.

Battu, R. S., Singh, B., and Kang, B. K. 2004. Contamination of liquid milk and butter with pesticide residues in the Ludhiana district of Punjab state, India. *Ecotoxicol. Environ. Safe.* 59: 324–331.

Bayat, S., Esmaili Sari, A., Bahramifar, N., Younesi, H., and Dahmarde Behrooz, R. 2010. Survey of organochlorine pesticides and polychlorinated biphenyls in commercial pasteurized milk in Iran. *Environ. Monit. Assess.* DOI 10.1007/s10661-010-1544-y].

Bernanke, J. and Köhler, H. R. 2009. The impact of environmental chemicals on wildlife vertebrates. *Rev. Environ. Contam. Toxicol.* 198: 1–47.

Berny, P. 2007. Pesticides and the intoxication of wild animals. *J. Vet. Pharmacol. Ther.* 30: 93–100.

Berny, P. and Gaillet, J. R. 2008. Acute poisoning of red kites (*Milvus milvus*) in France: Data from the Sagir network. *J. Wildl. Dis.* 44: 417–426.

Berny, P. J., Buronfosse, T., Buronfosse, F., Lamarque, F., and Lorgue, G. 1997. Field evidence of secondary poisoning of foxes (*Vulpes vulpes*) and buzzards (*Buteo buteo*) by bromadiolone, a 4-year survey. *Chemosphere* 35: 1817–1829.

Berny, P., Sadoul, N., Dol, S., Videman, B., Kayser, Y., and Hafner, H. 2002. Impact of local agricultural and industrial practices on organic contamination of little egret (*Egretta garzetta*) eggs in the Rhone Delta, southern France. *Environ. Toxicol. Chem.* 21: 520–526.

Berny, P., Alves, L., Simon, V., and Rossi, S. 2005. Intoxications des ruminants par les raticides anticoagulants: Quelle réalité? *Rev. Méd. Vét.* 156: 449–455.

Berny, P., Caloni, F., Croubels, S., Sachana, M., Vandenbroucke, V., Davanzo, F., and Guitart, R. 2010. Animal poisoning in Europe. Part 2: Companion animals. *Vet. J.* 183: 255–259.

Best, D. A., Elliott, K. H., Bowerman, W. W., Shieldcastle, M., Postupalsky, S., Kubiak, T. J., Tillitt, D. E., and Elliott, J. E. 2010. Productivity, embryo and eggshell characteristics, and contaminants in bald eagles from the Great Lakes, USA, 1986 to 2000. *Environ. Toxicol. Chem.* 29: 1581–1592.

Bhatt, P., Kumar, M. S., and Chakrabarti, T. 2009. Fate and degradation of POP-hexachlorocyclohexane *Crit. Rev. Environ. Sci. Technol.* 39: 655–695.

Bignert, A., Litzén, K., Odsjö, T., Olsson, M., Persson, W., and Reutergårdh, L. 1995. Time-related factors influence the concentrations of SDDT, PCBs and shell parameters in eggs of Baltic guillemot (*Uria aalge*), 1861–1989. *Environ. Pollut.* 89: 27–36.

Bischoff, K., Brizzee-Buxton, B., Gatto, N., Edwards, W. C., Stair, E. L., and Logan, C. 1998. Malicious paraquat poisoning in Oklahoma dogs. *Vet. Hum. Toxicol.* 40: 151–153.

Blus, L. J. 1996. DDT, DDD, and DDE in birds. In: Beyer, W. N., Heinz, G. H., and Redmon-Norwood, A. W. (eds), *Environmental Contaminants in Wildlife: Interpreting Tissue Concentrations*, pp. 49–71. Lewis Publishers: Boca Raton, FL.

Blus, L. J. 2003. Organochlorine pesticides. In: Hoffman, D. J., Rattner B. A., Burton, G. A., and Cairns J. Jr. (eds), *Handbook of Ecotoxicology*, pp. 313–403. Lewis Publishers: Boca Raton, FL.

Blus, L. J. and Henny, C. J. 1997. Field studies on pesticides and birds: Unexpected and unique relations. *Ecol. Appl.* 7: 1125–1132.

Blus, L. J., Gish, C. D., Belisle, A. A., and Prouty, R. M. 1972. Logarithmic relationship of DDE residues to eggshell thinning. *Nature* 235: 376–377.

Boatman, N. D., Brickle, N. W., Hart, J. D., Milsom, T. P., Morris, A. J., Murray, A. W. A., Murray, K. A., and Robertson, P. A. 2004. Evidence for the indirect effect of pesticides on farmland birds. *Ibis* 146(Supplement 2): 131–143.

Boland, L. A. and Angles, J. M. 2010. Feline permethrin toxicity: Retrospective study of 42 cases. *J. Feline Med. Surg.* 12: 61–71.

Borrell, A. and Aguilar, A. 2005. Differences in DDT and PCB residues between common and striped dolphins from the southwestern Mediterranean. *Arch. Environ. Contam. Toxicol.* 48: 501–508.

Borrell, A. and Aguilar, A. 2007. Organochlorine concentrations declined during 1987–2002 in western Mediterranean bottlenose dolphins, a coastal top predator. *Chemosphere* 66: 347–352.

Borrell, A., Aguilar, A., Tornero, V., Sequeira, M., Fernández, G., and Alis, S. 2006. Organochlorine compounds and stable isotopes indicate bottlenose dolphin subpopulation structure around the Iberian Peninsula. *Environ. Int.* 32: 516–523.

Borrell, A., Cantos, G., Aguilar, A., Androukaki, E., and Dendrinos, P. 2007. Concentrations and patterns of organochlorine pesticides and PCBs in Mediterranean monk seals (*Monachus monachus*) from Western Sahara and Greece. *Sci. Total. Environ.* 381: 316–325.

Borrell, A., Garcia-Solà, A., Aguilar, A., García, N. A., and Crespo, E. A. 2010. Organochlorine residues in South American sea lions, *Otaria flavescens* (Shaw, 1800): Bioaccumulation and time trends. *Bull. Environ. Contam. Toxicol.* 84: 731–737.

Borst, G. H. A. and Counotte, G. H. M. 2002. Shortfalls using second-generation anticoagulant rodenticides. *J. Zoo Wildl. Med.* 33: 85.

Bouwman, H., Polder, A., Venter, B., and Skaare, J. U. 2008. Organochlorine contaminants in cormorant, darter, egret, and ibis eggs from South Africa. *Chemosphere* 71: 227–241.

Bowman, J. and Schulte-Hostedde, A. I. 2009. The mink is not a reliable sentinel species. *Environ. Res.* 109: 937–939.

Boyle, C. M. 1960. Case of apparent resistance of *Rattus norvegicus* Berkenhout to anticoagulant poisons. *Nature* 188: 517.

Brakes, C. R. and Smith, R. H. 2005. Exposure of non-target small mammals to rodenticides: Short-term effects, recovery and implications for secondary poisoning. *J. Appl. Ecol.* 42: 118–128.

Braune, B. M. 2007. Temporal trends of organochlorines and mercury in seabird eggs from the Canadian Arctic, 1975–2003. *Environ. Pollut.* 148: 599–613.

Braune, B. M., Donaldson, G. M., and Hobson, K. A. 2001. Contaminant residues in seabird eggs from the Canadian Arctic. Part I. Temporal trends 1975–1998. *Environ. Pollut.* 114: 39–54.

Braune, B. M., Outridge, P. M., Fisk, A. T., Muir, D. C., Helm, P. A., Hobbs, K., Hoekstra, P. F., Kuzyk, Z. A., Kwan, M., Letcher, R. J., Lockhart, W. L., Norstrom, R. J., Stern, G. A., and Stirling, I. 2005. Persistent organic pollutants and mercury in marine biota of the Canadian Arctic: An overview of spatial and temporal trends. *Sci. Total. Environ.* 351–352: 4–56.

Buck, J. A., Anthony, R. G., Schuler, C. A., Isaacs, F. B., and Tillitt, D. E. 2005. Changes in productivity and contaminants in bald eagles nesting along the lower Columbia River, USA. *Environ. Toxicol. Chem.* 24: 1779–1792.

Burgat, V., Keck, G., Guerre, P., Bigorre, V., and Pineau, X. 1998. Glyphosate toxicosis in domestic animals: A survey from the data of the Centre National d'Informations Toxicologiques Veterinaires (CNITV). *Vet. Hum. Toxicol.* 40: 363–367.

Burger, J. 1997. The historical basis for ecological risk assessment. *Ann. N. Y. Acad. Sci.* 837: 360–371.

Burger, J. and Gochfeld, M. 2004. Marine birds as sentinels of environmental pollution. *EcoHealth* 1: 263–274.

Bustnes, J. O., Tveraa, T., Henden, J. A., Varpe, O., Janssen, K., and Skaare, J. U. 2006. Organochlorines in antarctic and arctic avian top predators: A comparison between the South Polar Skua and two species of northern hemisphere gulls. *Environ. Sci. Technol.* 40: 2826–2831.

Bustnes, J. O., Yoccoz, N. G., Bangjord, G., Polder, A., and Skaare, J. U. 2007. Temporal trends (1986–2004) of organochlorines and brominated flame retardants in tawny owl eggs from northern Europe. *Environ. Sci. Technol.* 41: 8491–8497.

Bustnes, J. O., Gabrielsen, G. W., and Verreault, J. 2010. Climate variability and temporal trends of persistent organic pollutants in the arctic: A study of glaucous gulls. *Environ. Sci. Technol.* 44: 3155–3161.

Canadian Council of Ministers of the Environment. 1999. Canadian tissue residue guidelines for the protection of wildlife consumers of aquatic biota: DDT (total). In: *Canadian Environmental Quality Guidelines.* Canadian Council of Ministers of the Environment: Winnipeg.

Carballo, M., Arbelo, M., Esperón, F., Mendez, M., de la Torre, A., and Muñoz, M. J. 2008. Organochlorine residues in the blubber and liver of bottlenose dolphins (*Tursiops truncatus*) stranded in the Canary Islands, North Atlantic Ocean. *Environ. Toxicol.* 23: 200–210.

Castilla-Pinedo, Y., Alvis-Estrada, L., and Alvis-Guzmán, N. 2010. Exposición a órganoclorados por ingesta de leche pasteurizada comercializada en Cartagena, Colombia. *Rev. Salud Pública* 12: 14–26.

Castrillon, J., Gomez-Campos, E., Aguilar, A., Berdié, L., and Borrell, A. 2010. PCB and DDT levels do not appear to have enhanced the mortality of striped dolphins (*Stenella coeruleoalba*) in the 2007 Mediterranean epizootic. *Chemosphere* 81: 459–463.

Cesh, L. S., Williams, T. D., Garcelon, D. K., and Elliott, J. E. 2008. Patterns and trends of chlorinated hydrocarbons in nestling bald eagle (*Haliaeetus leucocephalus*) plasma in British Columbia and Southern California. *Arch. Environ. Contam. Toxicol.* 55: 496–502.

Champoux, L., Moisey, J., and Muir, D. C. 2010. Polybrominated diphenyl ethers, toxaphenes, and other halogenated organic pollutants in great blue heron eggs. *Environ. Toxicol. Chem.* 29: 243–249.

Choi, J. W., Matsuda, M., Kawano, M., Wakimoto, T., Iseki, N., Masunaga, S., Hayama, S., and Watanuki, Y. 2001. Chlorinated persistent organic pollutants in black-tailed gulls (*Larus crassirostris*) from Hokkaido, Japan. *Chemosphere* 44: 1375–1382.

Christensen, J. R., MacDuffee, M., Macdonald, R. W., Whiticar, M., and Ross, P. S. 2005. Persistent organic pollutants in British Columbia grizzly bears: Consequence of divergent diets. *Environ. Sci. Technol.* 39: 6952–6960.

Chu, S., Henny, C. J., Kaiser, J. L., Drouillard, K. G., Haffner, G. D., and Letcher, R. J. 2007. Dacthal and chlorophenoxy herbicides and chlorothalonil fungicide in eggs of osprey (*Pandion haliaetus*) from the Duwamish-Lake Washington-Puget Sound area of Washington state, USA. *Environ. Pollut.* 145: 374–381.

Cid, F. D., Antón, R. I., and Caviedes-Vidal, E. 2007. Organochlorine pesticide contamination in three bird species of the Embalse La Florida water reservoir in the semiarid midwest of Argentina. *Sci. Total. Environ.* 385: 86–96.

Cipro, C. V., Taniguchi, S., and Montone, R. C. 2010. Occurrence of organochlorine compounds in *Euphausia superba* and unhatched eggs of *Pygoscelis* genus penguins from Admiralty Bay (King George Island, Antarctica) and estimation of biomagnification factors. *Chemosphere* 78: 767–771.

Clark, D. R. Jr. 2001. DDT and the decline of free-tailed bats (*Tadarida brasiliensis*) at Carlsbad Cavern, New Mexico. *Arch. Environ. Contam. Toxicol.* 40: 537–543.

Clark, K. E., Stansley, W., and Niles, L. J. 2001. Changes in contaminant levels in New Jersey osprey eggs and prey, 1989 to 1998. *Arch. Environ. Contam. Toxicol.* 40: 277–284.

Clark, K. E., Zhao, Y., and Kane, C. M. 2009. Organochlorine pesticides, PCBs, dioxins, and metals in postterm peregrine falcon (*Falco peregrinus*) eggs from the Mid-Atlantic states, 1993–1999. *Arch. Environ. Contam. Toxicol.* 57: 174–184.

Clout, M. N. and Russell, J. C. 2006. The eradication of mammals from New Zealand islands. In: Koike, F., Clout, M. N., Kawamichi, M., De Poorter, M., and Iwatsuki, K. (eds), *Assessment and Control of Biological Invasion Risks*, pp. 127–141. Shoukadoh Book Sellers and the World Conservation Union (IUCN): Kyoto Gland.

Colborn, T. and Smolen, M. J. 1996. Epidemiological analysis of persistent organochlorine contaminants in cetaceans. *Rev. Environ. Contam. Toxicol.* 146: 92–157.

Connell, D. W., Fung, C. N., Minh, T. B., Tanabe, S., Lam, P. K., Wong, B. S., Lam, M. H., Wong, L. C., Wu, R. S., and Richardson, B. J. 2003. Risk to breeding success of fish-eating Ardeids due to persistent organic contaminants in Hong Kong: Evidence from organochlorine compounds in eggs. *Water Res.* 37: 459–467.

Corsolini, S. 2009. Industrial contaminants in Antarctic biota. *J. Chromatogr. A* 1216: 598–612.

Cortinovis, S., Galassi, S., Melone, G., Saino, N., Porte, C., and Bettinetti, R. 2008. Organochlorine contamination in the Great Crested Grebe (*Podiceps cristatus*): Effects on eggshell thickness and egg steroid levels. *Chemosphere* 73: 320–325.

Courchamp, F., Chapuis, J. L., and Pascal, M. 2003. Mammal invaders on islands: Impact, control and control impact. *Biol. Rev.* 78: 347–383.

Damstra, T., Barlow, S., Bergman, A., Kavlock, R., and van der Kraak, G. (eds). 2002. *Global Assessment of the State-of-the-Science of Endocrine Disruptors*, p. 180. International Programme on Chemical Safety, World Health Organization: Geneva.

Dell'Omo, G., Costantini, D., Wright, J., Casagrande, S., and Shore, R. F. 2008. PCBs in the eggs of Eurasian kestrels indicate exposure to local pollution. *Ambio* 37: 452–456.

Del Piero, F. and Poppenga, R. H. 2006. Chlorophacinone exposure causing an epizootic of acute fatal hemorrhage in lambs. *J. Vet. Diagn. Invest.* 18: 483–485.

de Snoo, G. R., Scheidegger, N. M. I., and De Jong, F. M. W. 1999. Vertebrate wildlife incidents with pesticides: A European survey. *Pest. Sci.* 55: 47–54.

Dhananjayan, V. and Muralidharan, S. 2010. Levels of organochlorine pesticide residues in blood plasma of various species of birds from India. *Bull. Environ. Contam. Toxicol.* 85: 129–136.

Dhananjayan, V., Muralidharan, S., and Jayanthi, P. 2011. Distribution of persistent organochlorine chemical residues in blood plasma of three species of vultures from India. *Environ. Monit. Assess.* 173: 803–811.

Dietz, R., Riget, F., Cleemann, M., Aarkrog, A., Johansen, P., and Hansen, J. C. 2000. Comparison of contaminants from different trophic levels and ecosystems. *Sci. Total. Environ.* 245: 221–231.

Dietz, R., Riget, F., Hobson, K. A., Heide-Jørgensen, M. P., Møller, P., Cleemann, M., de Boer, J., and Glasius, M. 2004a. Regional and inter annual patterns of heavy metals, organochlorines and stable isotopes in narwhals (*Monodon monoceros*) from West Greenland. *Sci. Total. Environ.* 331: 83–105.

Dietz, R., Riget, F. F., Sonne, C., Letcher, R., Born, E. W., and Muir, D. C. 2004b. Seasonal and temporal trends in polychlorinated biphenyls and organochlorine pesticides in East Greenland polar bears (*Ursus maritimus*), 1990–2001. *Sci. Total. Environ.* 331: 107–124.

Di Muccio, A., Camoni, I., Citti, P., and Pontecorvo, D. 1988. Survey of DDT-like compounds in Dicofol formulations. *Ecotoxicol. Environ. Safe.* 16: 129–132.

Di Muccio, A., Pelosi, P., Barbini, D. A., Generali, T., Ausili, A., and Vergori, F. 1997. Selective extraction of pyrethroid pesticide residues from milk by solid-matrix dispersion. *J. Chromatogr. A* 765: 51–60.

Domingo, M., Ferrer, L., Pumarola, M., Marco, A., Plana, J., Kennedy, S., McAliskey, M., and Rima, B. K. 1990. Morbillivirus in dolphins. *Nature* 348: 21.

Domingo, M., Visa, J., Pumarola, M., Marco, A. J., Ferrer, L., Rabanal, R., and Kennedy, S. 1992. Pathological and immunocytochemical studies of morbillivirus infection in striped dolphins (*Stenella coeruleoalba*). *Vet. Pathol.* 29: 1–10.

Dong, Y. H., Wang, H., An, Q., Ruiz, X., Fasola, M., and Zhang, Y. M. 2004. Residues of organochlorinated pesticides in eggs of water birds from Tai Lake in China. *Environ. Geochem. Health* 26: 259–268.

Donlan, C. J., Howald, G. R., Tershy, B. R., and Croll, D. A. 2003. Evaluating alternative rodenticides for island conservation: Roof rat eradication from the San Jorge Islands, Mexico. *Biol. Conserv.* 14: 29–34.

Dowding, C. V., Shore, R. F., Worgan, A., Baker, P. J., and Harris, S. 2010. Accumulation of anticoagulant rodenticides in a non-target insectivore, the European hedgehog (*Erinaceus europaeus*). *Environ. Pollut.* 158: 161–166.

Dowding, J. E., Murphy, E. C., and Veitch, C. R. 1999. Brodifacoum residues in target and non-target species following an aerial poisoning operation on Motuihe Island, Hauraki Gulf, New Zealand. *N. Z. J. Ecol.* 23: 207–214.

Drolet, R., Laverty, S., Braselton, W. E., and Lord, N. 1996. Zinc phosphide poisoning in a horse. *Equine Vet. J.* 28: 161–162.

Eason, C. T. and Spurr, E. B. 1995. Review of the toxicity and impacts of brodifacoum on non-target wildlife in New Zealand. *N. Z. J. Zool.* 22: 371–379.

Eason, C. T., Murphy, E. C., Wright, G. R., and Spurr, E. B. 2002. Assessment of risks of brodifacoum to non-target birds and mammals in New Zealand. *Ecotoxicology* 11: 35–48.

Edwards, P. J., Fletcher, M. R., and Berny, P. 2000. Review of the factors affecting the decline of the European brown hare, *Lepus europaeus* (Pallas, 1778) and the use of wildlife incident data to evaluate the significance of paraquat. *Agric. Ecosyst. Environ.* 79: 95–103.

Elliott, J. E. 2005. Chlorinated hydrocarbon contaminants and s isotope ratios in pelagic seabirds from the North Pacific Ocean. *Arch. Environ. Contam. Toxicol.* 49: 89–96.

Elliott, J. E., Wilson, L. K., Henny, C. J., Trudeau, S. F., Leighton, F. A., Kennedy, S. W., and Cheng, K. M. 2001. Assessment of biological effects of chlorinated hydrocarbons in osprey chicks. *Environ. Toxicol. Chem.* 20: 866–879.

Elliott, J. E., Birmingham, A. L., Wilson, L. K., McAdie, M., Trudeau, S., and Mineau, P. 2008. Fonofos poisons raptors and waterfowl several months after granular application. *Environ. Toxicol. Chem.* 27: 452–460.

Elliott, K. H., Cesh, L. S., Dooley, J. A., Letcher, R. J., and Elliott, J. E. 2009. PCBs and DDE, but not PBDEs, increase with trophic level and marine input in nestling bald eagles. *Sci. Total. Environ.* 407: 3867–3875.

El-Shahawi, M. S., Hamza, A., Bashammakh, A. S., and Al-Saggaf, W. T. 2010. An overview on the accumulation, distribution, transformations, toxicity and analytical methods for the monitoring of persistent organic pollutants. *Talanta* 80: 1587–1597.

Fernández-Álvarez, M., Llompart, M., Lamas, J. P., Lores, M., García-Jares, C., Cela, R., and Dagnac, T. 2008. Development of a solid-phase microextraction gas chromatography with microelectron-capture detection method for a multiresidue analysis of pesticides in bovine milk. *Anal. Chim. Acta* 617: 37–50.

Fildes, K., Astheimer, L. B., Story, P., Buttemer, W. A., and Hooper, M. J. 2006. Cholinesterase response in native birds exposed to fenitrothion during locust control operations in eastern Australia. *Environ. Toxicol. Chem.* 25: 2964–2970.

Fillmann, G., Hermanns, L., Fileman, T. W., and Readman, J. 2007. Accumulation patterns of organochlorines in juveniles of *Arctocephalus australis* found stranded along the coast of Southern Brazil. *Environ. Pollut.* 146: 262–267.

Fimreite, N. 1970. Mercury uses in Canada and their possible hazards as sources of mercury contamination. *Environ. Pollut.* 1: 119–131.

Fisk, A. T., Hobson, K. A., and Norstrom, R. J. 2001. Influence of chemical and biological factors on trophic transfer of persistent organic pollutants in the Northwater Polynya marine food web. *Environ. Sci. Technol.* 35: 732–738.

Fisk, A. T., de Wit, C. A., Wayland, M., Kuzyk, Z. Z., Burgess, N., Letcher, R., Braune, B., Norstrom, R., Blum, S. P., Sandau, C., Lie, E., Larsen, H. J., Skaare, J. U., and Muir, D. C. 2005. An assessment of the toxicological significance of anthropogenic contaminants in Canadian arctic wildlife. *Sci. Total. Environ.* 351–352: 57–93.

Fleischli, M. A., Franson, J. C., Thomas N. J., Finley, D. L., and Riley, W. 2004. Avian mortality events in the United States caused by anticholinesterase pesticides: A retrospective summary of National Wildlife Health Center records from 1980 to 2000. *Arch. Environ. Contam. Toxicol.* 46: 542–550.

Fontcuberta, M., Arqués, J. F., Villalbí, J. R., Martínez, M., Centrich, F., Serrahima, E., Pineda, L., Duran, J., and Casas, C. 2008. Chlorinated organic pesticides in marketed food: Barcelona, 2001–06. *Sci. Total. Environ.* 389: 52–57.

Fossi, M. C., Leonzio, C., Massi, A., Lari, L., and Casini, S. 1992. Serum esterase inhibition in birds: A nondestructive biomarker to assess organophosphorus and carbamate contamination. *Arch. Environ. Contam. Toxicol.* 23: 99–104.

Fossi, M. C., Casini, S., and Marsili, L. 2007. Potential toxicological hazard due to endocrine-disrupting chemicals on Mediterranean top predators: State of art, gender differences and methodological tools. *Environ. Res.* 104: 174–182.

Fournier-Chambrillon, C., Berny, P. J., Coiffier, O., Barbedienne, P., Dassé, B., Delas, G., Galineau, H., Mazet, A., Pouzenc, P., Rosoux, R., and Fournier, P. 2004. Evidence of secondary poisoning of free-ranging riparian mustelids by anticoagulant rodenticides in France: Implications for conservation of European mink (*Mustela lutreola*). *J. Wildl. Dis.* 40: 688–695.

Franson, J. C., Hollmén, T. E., Flint, P. L., Grand, J. B., and Lanctot, R. B. 2004. Contaminants in molting long-tailed ducks and nesting common eiders in the Beaufort Sea. *Mar. Pollut. Bull.* 48: 504–513.

Freemark, K. and Boutin, C. 1995. Impacts of agricultural herbicide use on terrestrial wildlife in temperate landscapes: A review with special reference to North America. *Agric. Ecosyst. Environ.* 52: 67–91.

Galassi, S., Saino, N., Melone, G., and Croce, V. 2002. DDT homologues and PCBs in eggs of great crested grebe (*Podiceps cristatus*) and mallard (*Anas platyrhynchos*) from Lake Maggiore (Italy). *Ecotoxicol. Environ. Safe.* 53: 163–169.

García-Fernández, A. J, Calvo, J. F., Martínez-López, E., María-Mojica, P., and Martínez, J. E. 2008. Raptor ecotoxicology in Spain: A review on persistent environmental contaminants. *Ambio* 37: 432–439.

Gazzotti, T., Sticca, P., Zironi, E., Lugoboni, B., Serraino, A., and Pagliuca, G. 2009. Determination of 15 organophosphorus pesticides in Italian raw milk. *Bull. Environ. Contam. Toxicol.* 82: 251–254.

Geisz, H. N., Dickhut, R. M., Cochran, M. A., Fraser, W. R., and Ducklow, H. W. 2008. Melting glaciers: A probable source of DDT to the Antarctic marine ecosystem. *Environ. Sci. Technol.* 42: 3958–3962.

Gervais, J. A., Rosenberg, D. K., Fry, D. M., Trulio, L., and Sturm, K. K. 2000. Burrowing owls and agricultural pesticides: Evaluation of residues and risks for three populations in California, USA. *Environ. Toxicol. Chem.* 19: 337–343.

Gibbs, G. W. 2009. The end of an 80–million year experiment: A review of evidence describing the impact of introduced rodents on New Zealand's 'mammal-free' invertebrate fauna. *Biol. Invas.* 11: 1587–1593.

Giesy, J. P., Feyk, L. A., Jones, P. D., Kannan, K., and Sanderson, T. 2003. Review of the effects of endocrine-disrupting chemical in birds. *Pure Appl. Chem.* 75: 2287–2303.

Glynn, A., Aune, M., Nilsson, I., Darnerud, P. O., Ankarberg, E. H., Bignert, A., and Nordlander, I. 2009. Declining levels of PCB, HCB and p,p′-DDE in adipose tissue from food producing bovines and swine in Sweden 1991–2004. *Chemosphere* 74: 1457–1462.

Goerke, H., Weber, K., Bornemann, H., Ramdohr, S., and Plötz, J. 2004. Increasing levels and biomagnification of persistent organic pollutants (POPs) in Antarctic biota. *Mar. Pollut. Bull.* 48: 295–302.

Goldstein, M. I., Lacher Jr., T. E., Woodbridge, B., Bechard, M. J., Canavelli, S. B., Zaccagnini, M. E., Cobb, G. P., Scollon, E. J., Tribolet, R., and Hooper, M. J. 1999. Monocrotophos-induced mass mortality of Swainson's hawks in Argentina, 1995–1996. *Ecotoxicology* 8: 201–214.

Gómara, B., González, M. J., Baos, R., Hiraldo, F., Abad, E., Rivera, J., and Jiménez, B. 2008. Unexpected high PCB and total DDT levels in the breeding population of red kite (*Milvus milvus*) from Doñana National Park, south-western Spain. *Environ. Int.* 34: 73–78.

Gouteux, B., Muir, D. C., Backus, S., Born, E. W., Dietz, R., Haug, T., Metcalfe, T., Metcalfe, C., and Øien, N. 2008. Toxaphene in minke whales (*Balaenoptera acutorostrata*) from the North Atlantic. *Environ. Pollut.* 153: 71–83.

Goutner, V., Albanis, T., Konstantinou, I., and Papakonstantinou, K. 2001. PCBs and organochlorine pesticide residues in eggs of Audouin's gull (*Larus audouinii*) in the north-eastern Mediterranean. *Mar. Pollut. Bull.* 42: 377–388.

Goutner, V., Albanis, T. A., and Konstantinou, I. K. 2005. PCBs and organochlorine pesticide residues in eggs of threatened colonial charadriiform species (Aves, Charadriiformes) from wetlands of international importance in northeastern Greece. *Belg. J. Zool.* 135: 157–163.

Grant, I. F. 2001. Insecticides for tsetse and trypanosomiasis control: Is the environmental risk acceptable? *Trends Parasitol.* 17: 10–14.

Grove, R. A. and Henny, C. J. 2008. Environmental contaminants in male river otters from Oregon and Washington, USA, 1994–1999. *Environ. Monit. Assess.* 145: 49–73.

Grove, R. A., Henny, C. J., and Kaiser, J. L. 2009. Osprey: Worldwide sentinel species for assessing and monitoring environmental contamination in rivers, lakes, reservoirs, and estuaries. *J. Toxicol. Environ. Health B Crit. Rev.* 12: 25–44.

Guitart, R., Guerrero, X., Silvestre, A. M., Gutiérrez, J. M., and Mateo, R. 1996a. Organochlorine residues in tissues of striped dolphins affected by the 1990 Mediterranean epizootic: Relationships with the fatty acid composition. *Arch. Environ. Contam. Toxicol.* 30: 79–83.

Guitart, R., Mateo, R., Gutiérrez, J. M., and To-Figueras, J. 1996b. An outbreak of thiram poisoning on Spanish poultry farms. *Vet. Hum. Toxicol.* 38: 287–288.

Guitart, R., Clavero, R., Mateo, R., and Mañez, M. 2005. Levels of persistent organochlorine residues in eggs of Greater Flamingos from the Guadalquivir marshes (Doñana), Spain. *J. Environ. Sci. Health B* 40: 753–760.

Guitart, R., Croubels, S., Caloni, F., Sachana, M., Davanzo, F., Vandenbroucke, V., and Berny, P. 2010a. Animal poisoning in Europe. Part 1: Livestock and poultry. *Vet. J.* 183: 249–254.

Guitart, R., Sachana, M., Caloni, F., Croubels, S., Vandenbroucke, V., and Berny, P. 2010b. Animal poisoning in Europe. Part 3: Wildlife. *Vet. J.* 183: 260–265.

Gulland, F. M. D. and Hall, A. J. 2007. Is marine mammal health deteriorating? Trends in the global reporting of marine mammal disease. *EcoHealth* 4: 135–150.

Guruge, K. S., Watanabe, M., Tanaka, H., and Tanabe, S. 2001. Accumulation status of persistent organochlorines in albatrosses from the North Pacific and the Southern Ocean. *Environ. Pollut.* 114: 389–398.

Haraguchi, K., Hisamichi, Y., and Endo, T. 2009. Accumulation and mother-to-calf transfer of anthropogenic and natural organohalogens in killer whales (*Orcinus orca*) stranded on the Pacific coast of Japan. *Sci. Total. Environ.* 407: 2853–2859.

Hario, M., Hirvi, J. P., Hollmén, T., and Rudbäck, E. 2004. Organochlorine concentrations in diseased vs. healthy gull chicks from the northern Baltic. *Environ. Pollut.* 127: 411–423.

Harris, M. L., Elliott, J. E., Butler, R. W., and Wilson, L. K. 2003. Reproductive success and chlorinated hydrocarbon contamination of resident great blue herons (*Ardea herodias*) from coastal British Columbia, Canada, 1977 to 2000. *Environ. Pollut.* 121: 207–227.

Helander, B., Olsson, A., Bignert, A., Asplund, L., and Litzén, K. 2002. The role of DDE, PCB, coplanar PCB and eggshell parameters for reproduction in the white-tailed sea eagle (*Haliaeetus albicilla*) in Sweden. *Ambio* 31: 386–403.

Helander, B., Bignert, A., and Asplund, L. 2008. Using raptors as environmental sentinels: Monitoring the white-tailed sea eagle *Haliaeetus albicilla* in Sweden. *Ambio* 37: 425–431.

Helgason, L. B., Barrett, R., Lie, E., Polder, A., Skaare, J. U., and Gabrielsen, G. W. 2008. Levels and temporal trends (1983–2003) of persistent organic pollutants (POPs) and mercury (Hg) in seabird eggs from Northern Norway. *Environ. Pollut.* 155: 190–198.

Henny, C. J., Kolbe, E. J., Hill, E. F., and Blus, L. J. 1987. Case histories of bald eagles and other raptors killed by organophosphorus insecticides topically applied to livestock. *J. Wildl. Dis.* 23: 292–295.

Henny, C. J., Galushin, V. M., Khokhlov, A. N., Malovichko, L. V., and Iljukh, M. P. 2003a. Organochlorine pesticides in eggs of birds of prey from the Stavropol Region, Russia. *Bull. Environ. Contam. Toxicol.* 71: 163–169.

Henny, C. J., Kaiser, J. L., Grove, R. A., Bentley, V. R., and Elliott, J. E. 2003b. Biomagnification factors (fish to Osprey eggs from Willamette River, Oregon, U.S.A.) for PCDDs, PCDFs, PCBs and OC pesticides. *Environ. Monit. Assess.* 84: 275–315.

Henny, C. J., Grove, R. A., Kaiser, J. L., and Bentley, V. R. 2004. An evaluation of osprey eggs to determine spatial residue patterns and effects of contaminants along the Lower Columbia River, U.S.A. In: Chancellor, R. D. and Meyburg, B. Y. (eds), *Raptors Worldwide*, pp. 369–388. WWGBP/MME: Budapest.

Henny, C. J. Anderson, T. W., and Crayon, J. J. 2008a. Organochlorine pesticides, polychlorinated biphenyls, metals and trace elements in waterbird eggs, Salton Sea, California, 2004. *Hydrobiologia* 604: 137–149.

Henny, C. J., Grove, R. A., and Kaiser, J. L. 2008b. Osprey distribution, abundance, reproductive success and contaminant burdens along lower Columbia River, 1997/1998 versus 2004. *Arch. Environ. Contam. Toxicol.* 54: 525–534.

Henny, C. J., Kaiser, J. L., and Grove, R. A. 2009. PCDDs, PCDFs, PCBs, OC pesticides and mercury in fish and osprey eggs from Willamette River, Oregon (1993: 2001 and 2006) with calculated biomagnification factors. *Ecotoxicology* 18: 151–173.

Herkert, J. R. 2004. Organochlorine pesticides are not implicated in the decline of the loggerhead shrike. *Condor* 106: 702–705.

Hernández, M. and Margalida, A. 2008. Pesticide abuse in Europe: Effects on the Cinereous vulture (*Aegypius monachus*) population in Spain. *Ecotoxicology* 17: 264–272.

Hernández, M., González, L. M., Oria, J., Sánchez, R., and Arroyo, B. 2008. Influence of contamination by organochlorine pesticides and polychlorinated biphenyls on the breeding of the Spanish imperial eagle (*Aquila adalberti*). *Environ. Toxicol. Chem.* 27: 433–441.

Herrera, A., Ariño, A., Conchello, M. P., Lazaro, R., Bayarri, S., Yagüe, C., Peiro, J. M., Aranda, S., and Simon, M. D. 2000. Red-legged partridges (*Alectoris rufa*) as bioindicators for persistent chlorinated chemicals in Spain. *Arch. Environ. Contam. Toxicol.* 38: 114–120.

Hickey, J. J. and Anderson, D. W. 1968. Chlorinated hydrocarbons and eggshell changes in raptorial and fish-eating birds. *Science* 162: 271–273.

Hoare, J. M. and Hare, K. M. 2006. The impact of brodifacoum on non-target wildlife: Gaps in knowledge. *N. Z. J. Ecol.* 30: 157–167.

Hobbs, K. E., Muir, D. C. G., and Mitchell, E. 2001. Temporal and biogeographic comparisons of PCBs and persistent organochlorine pollutants in the blubber of fin whales from eastern Canada in 1971–1991. *Environ. Pollut.* 114: 243–254.

Hobbs, K. E., Muir, D. C., Michaud, R., Béland, P., Letcher, R. J., and Norstrom, R. J. 2003. PCBs and organochlorine pesticides in blubber biopsies from free-ranging St. Lawrence River Estuary beluga whales (*Delphinapterus leucas*), 1994–1998. *Environ. Pollut.* 122: 291–302.

Hoferkamp, L., Hermanson, M. H., and Muir, D. C. 2010. Current use pesticides in Arctic media; 2000–2007. *Sci. Total. Environ.* 408: 2985–2994.

Hoffman, D. J., Rattner, B. A., and Hall, R. J. 1990. Wildlife toxicology. *Environ. Sci. Technol.* 24: 276–283.

Holden, A. V. and Marsden, K. 1967. Organochlorine pesticides in seals and porpoises. *Nature* 216: 1274–1276.

Hollamby, S., Afema-Azikuru, J., Waigo, S., Cameron, K., Gandolf, A. R., Norris, A., and Sikarskie, J. G. 2006. Suggested guidelines for use of avian species as biomonitors. *Environ. Monit. Assess.* 118: 13–20.

Hotchkiss, A. K., Rider, C. V., Blystone, C. R., Wilson, V. S., Hartig, P. C., Ankley, G. T., Foster, P. M., Gray, C. L., and Gray, L. E. 2008. Fifteen years after "Wingspread"-environmental endocrine disrupters and human and wildlife health: Where we are today and where we need to go. *Toxicol. Sci.* 105: 235–259.

Houde, M., Hoekstra, P. F., Solomon, K. R., and Muir, D. C. 2005. Organohalogen contaminants in delphinoid cetaceans. *Rev. Environ. Contam. Toxicol.* 184: 1–57.

Howald, G. R., Mineau, P., Elliott, J. E., and Cheng, K. M. 1999. Brodifacoum poisoning of avian scavengers during rat control on a seabird colony. *Ecotoxicology* 8: 431–447.

Howald, G., Donlan, C. J., Galván, J. P., Russell, J. C., Parkes, J., Samaniego, A., Wang, Y., Veitch, D., Genovesi, P., Pascal, M., Saunders, A., and Tershy, B. 2007. Invasive rodent eradication on islands. *Conserv. Biol.* 21: 1258–1268.

Hu, G. C., Luo, X. J., Dai, J. Y., Zhang, X. L., Wu, H., Zhang, C. L., Guo, W., Xu, M. Q., Mai, B. X., and Weit, F. W. 2008. Brominated flame retardants, polychlorinated biphenyls, and organochlorine pesticides in captive giant panda (*Ailuropoda melanoleuca*) and red panda (*Ailurus fulgens*) from China. *Environ. Sci. Technol.* 42: 4704–4709.

IARC. 2010. *Agents Classified by the IARC Monographs*, Vols 1–100. Available http://monographs.iarc.fr/ENG/Classification/index.php (accessed December 2010).

Ishizuka, M., Okajima, F., Tanikawa, T., Min, H., Tanaka, K. D., Sakamoto, K. Q., and Fujita, S. 2007. Elevated warfarin metabolism in warfarin-resistant roof rat (*Rattus rattus*) in Tokyo. *Drug Metab. Dispos.* 35: 62–66.

Isobe, T., Ochi, Y., Ramu, K., Yamamoto, T., Tajima, Y., Yamada, T. K., Amano, M., Miyazaki, N., Takahashi, S., and Tanabe, S. 2009. Organohalogen contaminants in striped dolphins (*Stenella coeruleoalba*) from Japan: Present contamination status, body distribution and temporal trends (1978–2003). *Mar. Pollut. Bull.* 58: 396–401.

Iverson, S. J., Frost, K. J., and Lowry, L. F. 1997. Fatty acid signatures reveal fine scale structure of foraging distribution of harbor seals and their prey in Prince William Sound, Alaska. *Mar. Ecol. Prog. Ser.* 151: 255–271.

Jaspers, V., Covaci, A., Maervoet, J., Dauwe, T., Voorspoels, S., Schepens, P., and Eens, M. 2005. Brominated flame retardants and organochlorine pollutants in eggs of little owls (*Athene noctua*) from Belgium. *Environ. Pollut.* 136: 81–88.

Jefferies, D. J. and French, M. C. 1976. Mercury, cadmium, zinc, copper and organochlorine insecticide levels in small mammals trapped in a wheat field. *Environ. Pollut.* 10: 175–182.

Jiménez, B., Merino, R., Abad, E., Rivera, J., and Olie, K. 2007. Evaluation of organochlorine compounds (PCDDs, PCDFs, PCBs and DDTs) in two raptor species inhabiting a Mediterranean island in Spain. *Environ. Sci. Pollut. Res.* 14: 61–68.

Johnstone, R. M., Court, G. S., Fesser, A. C., Bradley, D. M., Oliphant, L. W., and MacNeil, J. D. 1996. Long-term trends and sources of organochlorine contamination in Canadian tundra Peregrine Falcons, *Falco peregrinus tundrius*. *Environ. Pollut.* 93: 109–120.

Jones, K. C. and de Voogt, P. 1999. Persistent organic pollutants (POPs): State of the science. *Environ. Pollut.* 100: 209–221.

Jones, H. P., Tershy, B. R., Zavaleta, E. S., Croll, D. A., Keitt, B. S., Finkelstein, M. E., and Howald, G. R. 2008. Severity of the effects of invasive rats on seabirds: A global review. *Conserv. Biol.* 22: 16–26.

Kajiwara, N., Ueno, D., Takahashi, A., Baba, N., and Tanabe, S. 2004. Polybrominated diphenyl ethers and organochlorines in archived northern fur seal samples from the Pacific coast of Japan, 1972–1998. *Environ. Sci. Technol.* 38: 3804–3809.

Kajiwara, N., Kunisue, T., Kamikawa, S., Ochi, Y., Yano, S., and Tanabe, S. 2006. Organohalogen and organotin compounds in killer whales mass-stranded in the Shiretoko Peninsula, Hokkaido, Japan. *Mar. Pollut. Bull.* 52: 1066–1076.

Kalantzi, O. I., Alcock, R. E., Johnston, P. A., Santillo, D., Stringer, R. L., Thomas, G. O., and Jones, K. C. 2001. The global distribution of PCBs and organochlorine pesticides in butter. *Environ. Sci. Technol.* 35: 1013–1018.

Kannan, K., Ramu, K., Kajiwara, N., Sinha, R. K., and Tanabe, S. 2005. Organochlorine pesticides, polychlorinated biphenyls, and polybrominated diphenyl ethers in Irrawaddy dolphins from India. *Arch. Environ. Contam. Toxicol.* 49: 415–420.

Keith, J. O. 1996. Residue analyses: How they were used to assess the hazards of contaminants to wildlife. In: Beyer, W. N., Heinz, G. H., and Redmon-Norwood, A. W. (eds), *Environmental Contaminants in Wildlife: Interpreting Tissue Concentrations*, pp. 1–47. Lewis Publishers: Boca Raton, FL.

Keithmaleesatti, S., Thirakhupt, K., Pradatsudarasar, A., Varanusupakul, P., Kitana, N., and Robson, M. 2007. Concentration of organochlorine in egg yolk and reproductive success of *Egretta garzetta* (Linnaeus, 1758) at Wat Tan-en non-hunting area, Phra Nakhorn Si Ayuthaya Province, Thailand. *Ecotoxicol. Environ. Safe* 68: 79–83.

Kendall, R. I. and Smith, P. N. 2003. Wildlife toxicology revisited. *Environ. Sci. Technol.* 37: 178A–183A.

Khay, S., Abd El-Aty, A. M., Choi, J. H., Shin, E. H., Shin, H. C., Kim, J. S., Chang, B. J., Lee, C. H., Shin, S. C., Jeong, J. Y., and Shim, J. H. 2009. Simultaneous determination of pyrethroids from pesticide residues in porcine muscle and pasteurized milk using GC. *J. Sep. Sci.* 32: 244–251.

King, K. A., Zaun, B. J., Schotborgh, H. M., and Hurt, C. 2003. DDE-induced eggshell thinning in white-faced ibis: A continuing problem in the western United States. *Southwest Nat.* 48: 356–364.

Koeman, J. H. and van Genderen, H. 1966. Some preliminary note on residues of chlorinated hydrocarbon insecticides in birds and mammals in the Netherlands. *J. Appl. Ecol.* 3: 99–106.

Kohn, B., Weingart, C., and Giger, U. 2003. Haemorrhage in seven cats with suspected anticoagulant rodenticide intoxication. *J. Feline Med. Surg.* 5: 295–304.

Konstantinou, I. K., Goutner, V., and Albanis, T. A. 2000. The incidence of polychlorinated biphenyl and organochlorine pesticide residues in the eggs of the cormorant (*Phalacrocorax carbo sinensis*): An evaluation of the situation in four Greek wetlands of international importance. *Sci. Total. Environ.* 257: 61–79.

Krahn, M. M., Hanson, M. B., Baird, R. W., Boyer, R. H., Burrows, D. G., Emmons, C. K., Ford, J. K., Jones, L. L., Noren, D. P., Ross, P. S., Schorr, G. S., and Collier, T. K. 2007. Persistent organic pollutants and stable isotopes in biopsy samples (2004/2006) from Southern Resident killer whales. *Mar. Pollut. Bull.* 54: 1903–1911.

Kunisue, T., Nakanishi, S., Watanabe, M., Abe, T., Nakatsu, S., Kawauchi, S., Sano, A., Horii, A., Kano, Y., and Tanabe, S. 2005. Contamination status and accumulation features of persistent organochlorines in pet dogs and cats from Japan. *Environ. Pollut.* 136: 465–476.

Kupper, J., Grobosch, T., Kistler, R., Sydler, T., and Naegeli, H. 2006. Bromadiolon-Vergiftung bei Füchsen. *Schweiz. Arch. Tierheilkd.* 148: 405–408.

Kwon, Y. K., Wee, S. H. and Kim, J. H. 2004. Pesticide poisoning events in wild birds in Korea from 1998 to 2002. *J. Wildl. Dis.* 40: 737–740.

Lambert, O., Pouliquen, H., Larhantec, M., Thorin, C., and L'Hostis, M. 2007. Exposure of raptors and waterbirds to anticoagulant rodenticides (difenacoum, bromadiolone, coumatetralyl, coumafen, brodifacoum): Epidemiological survey in Loire Atlantique (France). *Bull. Environ. Contam. Toxicol.* 79: 91–94.

Lavoie, R. A., Champoux, L., Rail, J. F., and Lean, D. R. 2010. Organochlorines, brominated flame retardants and mercury levels in six seabird species from the Gulf of St. Lawrence (Canada): Relationships with feeding ecology, migration and molt. *Environ. Pollut.* 158: 2189–2199.

Lemarchand, C., Rosoux, R., and Berny, P. 2010. Organochlorine pesticides, PCBs, heavy metals and anticoagulant rodenticides in tissues of Eurasian otters (*Lutra lutra*) from upper Loire River catchment (France). *Chemosphere* 80: 1120–1124.

Leonel, J., Sericano, J. L., Fillmann, G., Secchi, E., and Montone, R. C. 2010. Long-term trends of polychlorinated biphenyls and chlorinated pesticides in franciscana dolphin (*Pontoporia blainvillei*) from Southern Brazil. *Mar. Pollut. Bull.* 60: 412–418.

Letcher, R. J., Gebbink, W. A., Sonne, C., Born, E. W., McKinney, M. A., and Dietz, R. 2009. Bioaccumulation and biotransformation of brominated and chlorinated contaminants and their metabolites in ringed seals (*Pusa hispida*) and polar bears (*Ursus maritimus*) from East Greenland. *Environ. Int.* 35: 1118–1124.

Letcher, R. J., Bustnes, J. O., Dietz, R., Jenssen, B. M., Jørgensen, E. H., Sonne, C., Verreault, J., Vijayan, M. M., and Gabrielsen, G. W. 2010. Exposure and effects assessment of persistent organohalogen contaminants in arctic wildlife and fish. *Sci. Total. Environ.* 408: 2995–3043.

Li, Y. F. and Macdonald, R. W. 2005. Sources and pathways of selected organochlorine pesticides to the Arctic and the effect of pathway divergence on HCH trends in biota: A review. *Sci. Total. Environ.* 342: 87–106.

Lohmann, R., Breivik, K., Dachs, J., and Muir, D. 2007. Global fate of POPs: Current and future research directions. *Environ. Pollut.* 150: 150–165.

Lovelock, J. E. 1991. The Electron-Capture Detector and green politics. *LC-GC Int.* 4: 26–35.

Lubick, N. 2010. Environment. Endosulfan's exit: U.S. EPA pesticide review leads to a ban. *Science* 328: 1466.

Lutze, J., Derrick, J., Korth, W., and MacLachlan, D. J. 2009. Monitoring of pesticides and veterinary drugs in Australian cattle: Verification of the residue control system. *Food Addit. Contam. B* 2: 99–111.

MacKinnon, D. S. and Freedman, B. 1993. Effects of silvicultural use of the herbicide glyphosate on breeding birds of regenerating clearcuts in Nova Scotia, Canada. *J. Appl. Ecol.* 30: 395–406.

Malik, R. N., Rauf, S., Mohammad, A., Eqani, S. A., and Ahad, K. 2011. Organochlorine residual concentrations in cattle egret from the Punjab Province, Pakistan. *Environ. Monit. Assess.* 173: 325–341.

Mañosa, S., Mateo, R., and Guitart, R. 2001. A review of the effects of agricultural and industrial contamination on the Ebro Delta biota and wildlife. *Environ. Monit. Assess.* 71: 187–205.

Mañosa, S., Mateo, R., Freixa, C., and Guitart, R. 2003. Persistent organochlorine contaminants in eggs of northern goshawk and Eurasian buzzard from northeastern Spain: Temporal trends related to changes in the diet. *Environ. Pollut.* 122: 351–359.

Margariti, M. G., Tsakalof, A. K., and Tsatsakis, A. M. 2007. Analytical methods of biological monitoring for exposure to pesticides: Recent update. *Ther. Drug Monit.* 29: 150–163.

Martínez-Haro, M., Viñuela, J., and Mateo, R. 2007. Exposure of birds to cholinesterase-inhibiting pesticides following a forest application for tick control. *Environ. Toxicol. Pharmacol.* 23: 347–349.

Martínez-Haro, M., Mateo, R., Guitart, R., Soler-Rodríguez, F., Pérez-López, M., Maria-Mojica, P., and García-Fernández, A. J. 2008. Relationship of the toxicity of pesticide formulations and their commercial restrictions with the frequency of animal poisonings. *Ecotoxicol. Environ. Safe.* 69: 396–402.

Martínez-López, E., Romero, D., María-Mojica, P., Navas, I., Gerique, C., Jiménez, P., and García-Fernández, A. J. 2006. Detection of strychnine by gas chromatography-mass spectrometry in the carcase of a Bonelli's eagle (*Hieraaetus fasciatus*). *Vet. Rec.* 159: 182–183.

Martínez-López, E., Maria-Mojica, P., Martínez, J. E., Calvo, J. F., Wright, J., Shore, R. F., Romero, D., and García-Fernández, A. J. 2007. Organochlorine residues in booted eagle (*Hieraaetus pennatus*) and goshawk (*Accipiter gentilis*) eggs from southeastern Spain. *Environ. Toxicol. Chem.* 26: 2373–2378.

Martínez-López, E., Romero, D., María-Mojica, P., Martínez, J. E., Calvo, J. F., and García-Fernández, A. J. 2009. Changes in blood pesticide levels in booted eagle (*Hieraaetus pennatus*) associated with agricultural land practices. *Ecotoxicol. Environ. Safe.* 72: 45–50.

Maruya, K. A., Smalling, K. L., and Mora, M. A. 2005. Residues of toxaphene in insectivorous birds (*Petrochelidon* spp.) from the Rio Grande, Texas. *Arch. Environ. Contam. Toxicol.* 48: 567–574.

Mateo, R., Carrillo, J., and Guitart, R. 2000. p,p′-DDE residues in eggs of European kestrel *Falco tinnunculus* from Tenerife, Canary Islands, Spain. *Bull. Environ. Contam. Toxicol.* 65: 780–785.

Mateo, R., Gil, C., Badía-Vila, M., Guitart, R., Hernández-Matías, A., Sanpera, C., and Ruiz, X. 2004. Use of fatty acids to explain variability of organochlorine concentrations in eggs and plasma of common terns (*Sterna hirundo*). *Ecotoxicology* 13: 545–554.

Maul, J. D. and Farris, J. L. 2004. The effect of sex on avian plasma cholinesterase enzyme activity: A potential source of variation in an avian biomarker endpoint. *Arch. Environ. Contam. Toxicol.* 47: 253–258.

Maul, J. D. and Farris, J. L. 2005. Monitoring exposure of northern cardinals, *Cardinalis cardinalis*, to cholinesterase-inhibiting pesticides: Enzyme activity, reactivations, and indicators of environmental stress. *Environ. Toxicol. Chem.* 24: 1721–1730.

Mauldin, R. E., Goldade, D. A., Engeman, R. M., Goodall, M. J., Craver, R. K., and Johnston, J. J. 1996. Determination of zinc phosphide residues in the California ground squirrel (*Spermophilus beecheyi*) by Gas Chromatography-Flame Photometric Detection. *J. Agric. Food Chem.* 44: 189–194.

Mayack, D. T. and Martin, T. 2003. Age-dependent changes in plasma and brain cholinesterase activities of house wrens and European starlings. *J. Wildl. Dis.* 39: 627–637.

McDonald, R. A., Harris, S., Turnbull, G., Brown, P., and Fletcher, M. 1998. Anticoagulant rodenticides in stoats (*Mustela erminea*) and weasels (*Mustela nivalis*) in England. *Environ. Pollut.* 103: 17–23.

McKinney, M. A., Stirling, I., Lunn, N. J., Peacock, E., and Letcher, R. J. 2010. The role of diet on long-term concentration and pattern trends of brominated and chlorinated contaminants in western Hudson Bay polar bears, 1991–2007. *Sci. Total. Environ.* 408: 6210–6222.

Melgar, M. J., Santaeufemia, M., and García, M. A. 2010. Organophosphorus pesticide residues in raw milk and infant formulas from Spanish northwest. *J. Environ. Sci. Health B*, 45: 595–600.

Mellink, E., Riojas-López, M. E., and Luévano-Esparza, J. 2009. Organochlorine content and shell thickness in brown booby (*Sula leucogaster*) eggs in the Gulf of California and the southern Pacific coast of Mexico. *Environ. Pollut.* 157: 2184–2188.

Metcalfe, C., Koenig, B., Metcalfe, T., Paterson, G., and Sears, R. 2004. Intra- and inter-species differences in persistent organic contaminants in the blubber of blue whales and humpback whales from the Gulf of St. Lawrence, Canada. *Mar. Environ. Res.* 57: 245–260.

Mineau, P. 2005. A review and analysis of study endpoints relevant to the assessment of "long term" pesticide toxicity in avian and mammalian wildlife. *Ecotoxicology* 14: 775–799.

Mineau, P. and Tucker, K. R. 2002. Improving detection of pesticide poisoning in birds. *J. Wildl. Rehab.* 25: 4–13.

Mineau, P. and Whiteside, M. 2006. Lethal risk to birds from insecticide use in the United States – A spatial and temporal analysis. *Environ. Toxicol. Chem.* 25: 1214–1222.

Mineau, P., Downes, C. M., Kirk, D. A., Byane, E., and Csizy, M. 2005. Patterns of bird species abundance in relation to granular insecticide use in the Canadian prairies. *Ecoscience* 12: 267–278.

Miranda Filho, K. C., Metcalfe, C. D., Metcalfe, T. L., Muelbert, M. M., Robaldo, R. B., Martinez, P. E., Colares, E. P., and Bianchini, A. 2009. Lactational transfer of PCBs and chlorinated pesticides in pups of southern elephant seals (*Mirounga leonina*) from Antarctica. *Chemosphere* 75: 610–616.

Moon, H. B., Kannan, K., Choi, M., Yu, J., Choi, H. G., An, Y. R., Choi, S. G., Park, J. Y., and Kim, Z. G. 2010. Chlorinated and brominated contaminants including PCBs and PBDEs in minke whales and common dolphins from Korean coastal waters. *J. Hazard. Mater.* 179: 735–741.

Mora, M. A., Montoya, A. B., Lee, M. C., Macías-Duarte, A., Rodríguez-Salazar, R., Juergens, P. W., and Lafón-Terrazas, A. 2008. Persistent environmental pollutants in eggs of aplomado falcons from Northern Chihuahua, Mexico, and south Texas, USA. *Environ. Int.* 34: 44–50.

Morris, A. J., Wilson, J. D., Whittingham, M. J., and Bradbury, R. B. 2005. Indirect effects of pesticides on breeding yellowhammer (*Emberiza citrinella*). *Agric. Ecosyst. Environ.* 106: 1–16.

Morrissey, C. A., Albert, C. A., Dods, P. L., Cullen, W. R., Lai, V. W., and Elliott, J. E. 2007. Arsenic accumulation in bark beetles and forest birds occupying mountain pine beetle infested stands treated with monosodium methanearsonate. *Environ. Sci. Technol.* 41: 1494–1500.

Muir, D. C. G. and Norstrom, R. J. 2000. Geographical differences and time trends of persistent organic pollutants in the Arctic. *Toxicol. Lett.* 112–113: 93–101.

Muir, D. C. and de Wit, C. A. 2010. Trends of legacy and new persistent organic pollutants in the circumpolar arctic: Overview, conclusions, and recommendations. *Sci. Total. Environ.* 408: 3044–3051.

Muir, D. C. G., Born, E. W., Koczansky, K., and Stern, G. A. 2000. Temporal and spatial trends of persistent organochlorines in Greenland walrus (*Odobenus rosmarus rosmarus*). *Sci. Total. Environ.* 245: 73–86.

Muñoz-Cifuentes, J., Becker, P. H., Sommer, U., Pacheco, P., and Schlatter, R. 2003. Seabird eggs as bioindicators of chemical contamination in Chile. *Environ. Pollut.* 126: 123–137.

Muralidharan, S., Dhananjayan, V., Risebrough, R., Prakash, V., Jayakumar, R., and Bloom, P. H. 2008. Persistent organochlorine pesticide residues in tissues and eggs of white-backed vulture, *Gyps bengalensis* from different locations in India. *Bull. Environ. Contam. Toxicol.* 81: 561–565.

Murphy, E. C., Clapperton, B. K., Bradfield, P. M. F., and Speed, H. J. 1998. Brodifacoum residues in target and non-target animals following large-scale poison operations in New Zealand podocarp-hardwood forests. *N. Z. J. Zool.* 25: 307–314.

Murphy, M. J. 2007. Anticoagulant rodenticides. In: Gupta, R. C. (ed.), *Veterinary Toxicology*, pp. 525–547. Academic Press: New York.

Murray, M. and Tseng, F. 2008. Diagnosis and treatment of secondary anticoagulant rodenticide toxicosis in a red-tailed hawk (*Buteo jamaicensis*). *J. Avian Med. Surg.* 22: 41–46.

Muzinic, J. 2007. Poisoning of seventeen Eurasian griffons (*Gyps fulvus*) in Croatia. *J. Raptor Res.* 41: 239–242.

Mwevura, H., Amir, O. A., Kishimba, M., Berggren, P., and Kylin, H. 2010. Organohalogen compounds in blubber of Indo-Pacific bottlenose dolphin (*Tursiops aduncus*) and spinner dolphin (*Stenella longirostris*) from Zanzibar, Tanzania. *Environ. Pollut.* 158: 2200–2207.

Nag, S. K. and Raikwar, M. K. 2008. Organochlorine pesticide residues in bovine milk. *Bull. Environ. Contam. Toxicol.* 80: 5–9.

Newton, I. 1998. Pesticides and Pollutants. In: Newton, I. (ed.), *Population Limitation in Birds*, pp. 407–447. Academic Press: London.

Newton, I., Wyllie, I., and Freestone, P. 1990. Rodenticides in British barn owls. *Environ. Pollut.* 68: 101–117.

Newton, I., Wyllie, I., and Asher, A. 1992. Mortality from the pesticide aldrin and dieldrin in British sparrowhawks and kestrels. *Ecotoxicology* 1: 31–44.

Newton, M., Howard, K. M., Kelpsas, B. R., Danhaus, R., Lottman, C. M., and Dubelman, S. 1984. Fate of glyphosate in an Oregon U.S.A. forest ecosystem. *J. Agric. Food Chem.* 32: 1144–1151.

Norstrom, R. J. and Muir, D. C. 1994. Chlorinated hydrocarbon contaminants in arctic marine mammals. *Sci. Total. Environ.* 154: 107–128.

Novotný, L., Misík, J., Honzlová, A., Ondráček, P., Kuča, K., Vávra, O., Rachač, V., and Chloupek, P. 2011. Incidental poisoning of animals by carbamates in the Czech Republic. *J. Appl. Biomed.* 9: 1–5.

Nyman, M., Koistinen, J., Fant, M. L., Vartiainen, T., and Helle, E. 2002. Current levels of DDT, PCB and trace elements in the Baltic ringed seals (*Phoca hispida baltica*) and grey seals (*Halichoerus grypus*). *Environ. Pollut.* 119: 399–412.

O'Brien, D. J., Kaneene, J. B., and Poppenga, R. H. 1993. The use of mammals as sentinels for human exposure to toxic contaminants in the environment. *Environ. Health Perspect.* 99: 351–368.

Okoniewski, J. C., Stone, W. B., and Hynes, K. P. 2006. Continuing organochlorine insecticide mortality in wild birds in New York, 2000–2004. *Bull. Environ. Contam. Toxicol.* 77: 726–731.

Olea, P., Sánchez-Barbudo, I., Viñuela, J., Barja, I., Mateo-Tomás, P., Piñeiro, A., Mateo, R., and Purroy, F. 2009. Lack of scientific evidence and precautionary principle in massive release of rodenticides threatens biodiversity: Old lessons need new reflections. *Environ. Conserv.* 36: 1–4.

Orueta, J. F., Aranda, Y., Gómez, T., Tapia, G. G., and Sánchez-Mármol, L. 2005. Successful eradication of invasive rodents from a small island through pulsed baiting inside covered stations. *Biol. Invas.* 7: 141–147.

Otieno, P. O., Lalah, J. O., Virani, M., Jondiko, I. O., and Schramm, K. W. 2010. Carbofuran and its toxic metabolites provide forensic evidence for furadan exposure in vultures (*Gyps africanus*) in Kenya. *Bull. Environ. Contam. Toxicol.* 84: 536–544.

Pain, D. J., Gargi, R., Cunningham, A. A., Jones, A., and Prakash, V. 2004. Mortality of globally threatened Sarus cranes *Grus antigon* from monocrotophos poisoning in India. *Sci. Total. Environ.* 326: 55–61.

Park, B. K., Park, G. J., An, Y. R., Choi, H. G., Kim, G. B., and Moon, H. B. 2010. Organohalogen contaminants in finless porpoises (*Neophocaena phocaenoides*) from Korean coastal waters: Contamination status, maternal transfer and ecotoxicological implications. *Mar. Pollut. Bull.* 60: 768–774.

Pastor, D., Jover, L., Ruiz, X., and Albaigés, J. 1995. Monitoring organochlorine pollution in Audouin's gull eggs: The relevance of sampling procedures. *Sci. Total. Environ.* 162: 215–223.

Peakall, D. 1992. *Animal Biomarkers as Pollution Indicators*. Chapman & Hall: London.

Peakall, D. 1996. Dieldrin and other cyclodiene pesticides in wildlife. In: Beyer, W. N., Heinz, G. H., and Redmon-Norwood, A. W. (eds), *Environmental Contaminants in Wildlife: Interpreting Tissue Concentrations*, pp. 73–97. Lewis Publishers: Boca Raton, FL.

Pelfrène, A. F. 2010. Rodenticides. In: Krieger, R. (ed.), *Hayes' Handbook of Pesticide Toxicology*, pp. 2153–2217. Academic Press: London.

Pelz, H. J., Rost, S., Hünerberg, M., Fregin, A., Heiberg, A. C., Baert, K., MacNicoll, A. D., Prescott, C. V., Walker, A. S., Oldenburg, J., and Müller, C. R. 2005. The genetic basis of resistance to anticoagulants in rodents. *Genetics* 170: 1839–1847.

Petrovic, M., Farré, M., de Alda, M. L., Pérez, S., Postigo, C., Köck, M., Radjenovic, J., Gros, M., and Barceló, D. 2010. Recent trends in the liquid chromatography-mass spectrometry analysis of organic contaminants in environmental samples. *J. Chromatogr. A* 1217: 4004–4017.

Philbey, A. W. and Morton, A. G. 2001. Paraquat poisoning in sheep from contaminated water. *Aust. Vet. J.* 79: 842–843.

Plumlee, K. H. (ed.). 2004. *Clinical Veterinary Toxicology*. Mosby: St Louis.

Pollock, C. G. 2001. Silent spring revisited: A 21st-century look at the effect of pesticides on wildlife. *J. Avian Med. Surg.* 15: 50–53.

Poppenga, R. H. 2007. Avian toxicology. In: Gupta, R. C. (ed.), *Veterinary Toxicology*, pp. 663–688. Academic Press: New York.

Poppenga, R. H. and Oehme, F. W. 2010. Pesticide use and associated morbidity and mortality in veterinary medicine. In: Krieger, R. (ed.), *Hayes' Handbook of Pesticide Toxicology*, pp. 285–301. Academic Press: London.

Poppenga, R. H., Ziegler, A. F., Habecker, P. L., Singletary, D. L., Walter, M. K., and Miller, P. G. 2005. Zinc phosphide intoxication of wild turkeys (*Meleagris gallopavo*). *J. Wildl. Dis.* 41: 218–223.

Porte, C., Janer, G., Lorusso, L. C., Ortiz-Zarragoitia, M., Cajaraville, M. P., Fossi, M. C., and Canesi, L. 2006. Endocrine disruptors in marine organisms: Approaches and perspectives. *Comp. Biochem. Physiol. C Toxicol. Pharmacol.* 143: 303–315.

Poulin, B., Lefebvre, G., and Paz, L. 2010. Red flag for green spray: Adverse trophic effects of Bti on breeding birds. *J. Appl. Ecol.* 47: 884–889.

Proudfoot, A. T. 2009. Aluminium and zinc phosphide poisoning. *Clin. Toxicol. (Phila).* 47: 89–100.

Qiu, X., Zhu, T., Yao, B., Hu, J., and Hu, S. 2005. Contribution of dicofol to the current DDT pollution in China. *Environ. Sci. Technol.* 39: 4385–4390.

Rainwater, T. R., Sauther, M. L., Rainwater, K. A., Mills, R. E., Cuozzo, F. P., Zhang, B., McDaniel, L. N., Abel, M. T., Marsland, E. J., Weber, M. A., Youssouf Jacky, I. A., Platt, S. G., Cobb, G. P., and Anderson, T. A. 2009. Assessment of organochlorine pesticides and metals in ring-tailed lemurs (*Lemur catta*) at Beza Mahafaly Special Reserve, Madagascar. *Am. J. Primatol.* 71: 998–1010.

Rajaei, F., Bahramifar, N., Esmaili Sari, A., Ghasempouri, S. M., and Savabieasfahani, M. 2010a. PCBs and organochlorine pesticides in ducks of Fereydoon-kenar wildlife refuge in Iran. *Bull. Environ. Contam. Toxicol.* 84: 577–581.

Rajaei, F., Esmaili-Sari, A., Bahramifar, N., Ghasempouri, M., and Savabieasfahani, M. 2010b. Avian liver organochlorine and PCB from South coast of the Caspian Sea, Iran. *Ecotoxicology* 19: 329–337.

Ramírez, F., Jover, L., Sanpera, C., Ruiz, X., Piqué, E., and Guitart R. 2009. Combined measurements of egg fatty acids and stable isotopes as indicators of feeding ecology in lake-dwelling birds. *Freshwater Biol.* 54: 1832–1842.

Ramu, K., Kajiwara, N., Tanabe, S., Lam, P. K., and Jefferson, T. A. 2005. Polybrominated diphenyl ethers (PBDEs) and organochlorines in small cetaceans from Hong Kong waters: Levels, profiles and distribution. *Mar. Pollut. Bull.* 51: 669–676.

Ramu, K., Kajiwara, N., Lam, P. K., Jefferson, T. A., Zhou, K., and Tanabe, S. 2006. Temporal variation and biomagnification of organohalogen compounds in finless porpoises (*Neophocaena phocaenoides*) from the South China Sea. *Environ. Pollut.* 144: 516–523.

Ratcliffe, D. A. 1967. Decrease in eggshell weight in certain birds of prey. *Nature* 215: 208–210.

Rattner, B. A. 2009. History of wildlife toxicology. *Ecotoxicology* 18: 773–783.

Riget, F., Dietz, R., Vorkamp, K., Johansen, P., and Muir, D. 2004. Levels and spatial and temporal trends of contaminants in Greenland biota: An updated review. *Sci. Total Environ.* 331: 29–52.

Rigét, F., Bignert, A., Braune, B., Stow, J., and Wilson, S. 2010. Temporal trends of legacy POPs in Arctic biota, an update. *Sci. Total. Environ.* 408: 2874–2884.

Robben, J. H., Kuijpers, E. A., and Mout, H. C. 1998. Plasma superwarfarin levels and vitamin treatment in dogs with anticoagulant rodenticide poisoning. *Vet. Quart.* 20: 24–27.

Ross, P. S. 2000. Marine mammals as sentinels in ecological risk assessment. *Hum. Ecol. Risk Assess.* 6: 29–46.

Roubies, N., Giadinis, N. D., Polizopoulou, Z., and Argiroudis, S. 2008. A retrospective study of chronic copper poisoning in 79 sheep flocks in Greece (1987–2007). *J. Vet. Pharmacol. Ther.* 31: 181–183.

Routti, H., Arukwe, A., Jenssen, B. M., Letcher, R. J., Nyman, M., Bäckman, C., and Gabrielsen, G. W. 2010. Comparative endocrine disruptive effects of contaminants in ringed seals (*Phoca hispida*) from Svalbard and the Baltic Sea. *Comp. Biochem. Physiol. C Toxicol. Pharmacol.* 152: 306–312.

Sage, M., Fourel, I., Cœurdassier, M., Barrat, J., Berny, P., and Giraudoux, P. 2010. Determination of bromadiolone residues in fox faeces by LC/ESI-MS in relationship with toxicological data and clinical signs after repeated exposure. *Environ. Res.* 110: 664–674.

Sakellarides, T. M., Konstantinou, I. K., Hela, D. G., Lambropoulou, D., Dimou, A., and Albanis, T. A. 2006. Accumulation profiles of persistent organochlorines in liver and fat tissues of various waterbird species from Greece. *Chemosphere* 63: 1392–1409.

Salas, J. H., González, M. M., Noa, M., Pérez, N. A., Díaz, G., Gutiérrez, R., Zazueta, H., and Osuna, I. 2003. Organophosphorus pesticide residues in Mexican commercial pasteurized milk. *J. Agric. Food Chem.* 51: 4468–4471.

Sandala, G. M., Sonne-Hansen, C., Dietz, R., Muir, D. C., Valters, K., Bennett, E. R., Born, E. W., and Letcher, R. J. 2004. Hydroxylated and methyl sulfone PCB metabolites in adipose and whole blood of polar bear (*Ursus maritimus*) from East Greenland. *Sci. Total. Environ.* 331: 125–141.

Sanpera, C., Ruiz, X., Llorente, G. A., Jover, L., and Jabeen, R. 2002. Persistent organochlorine compounds in sediment and biota from the Haleji lake: A wildlife sanctuary in South Pakistan. *Bull. Environ. Contam. Toxicol.* 68: 237–244.

Sanpera, C., Ruiz, X., Jover, L., Llorente, G., Jabeen, R., Muhammad, A., Boncompagni, E., and Fasola, M. 2003. Persistent organic pollutants in little egret eggs from selected wetlands in Pakistan. *Arch. Environ. Contam. Toxicol.* 44: 360–368.

Sarabia, J., Sánchez-Barbudo, I., Siqueira, W., Mateo, R., Rollán, E., and Pizarro, M. 2008. Lesions associated with the plexus venosus subcutaneus collaris of pigeons with chlorophacinone toxicosis. *Avian Dis.* 52: 540–543.

Schiavone, A., Corsolini, S., Borghesi, N., and Focardi, S. 2009. Contamination profiles of selected PCB congeners, chlorinated pesticides, PCDD/Fs in Antarctic fur seal pups and penguin eggs. *Chemosphere* 76: 264–269.

Schmidt, P. L. 2009. Companion animals as sentinels for public health. *Vet. Clin. Small Anim.* 39: 241–250.

Sengupta, D., Aktar, M. W., Alam, S., and Chowdhury, A. 2010. Impact assessment and decontamination of pesticides from meat under different culinary processes. *Environ. Monit. Assess.* 169: 37–43.

Sereda, B., Bouwman, H., and Kylin, H. 2009. Comparing water, bovine milk, and indoor residual spraying as possible sources of DDT and pyrethroid residues in breast milk. *J. Toxicol. Environ. Health A* 72: 842–851.

Sheffield, S. R. 1997. Owls as biomonitors of environmental contamination. In *General Technical Report NC-190*. U.S. Forest Service, North Central Research Station: St. Paul.

Shoham-Frider, E., Kress, N., Wynne, D., Scheinin, A., Roditi-Elsar, M., and Kerem, D. 2009. Persistent organochlorine pollutants and heavy metals in tissues of common bottlenose dolphin (*Tursiops truncatus*) from the Levantine Basin of the Eastern Mediterranean. *Chemosphere* 77: 621–627.

Shore, R. F., Birks, J. D. S., and Freestone, P. 1999. Exposure of non-target vertebrates to second-generation rodenticides in Britain, with particular reference to the polecat *Mustela putorius*. *N.Z. J. Ecol.* 23: 199–206

Shore, R. F., Birks, J. D., Freestone, P., and Kitchener, A. C. 1996. Second-generation rodenticides and polecats (*Mustela putorius*) in Britain. *Environ. Pollut.* 91: 279–282.

Shore, R. F., Birks, J. D., Afsar, A., Wienburg, C. L., and Kitchener, A. C. 2003. Spatial and temporal analysis of second-generation anticoagulant rodenticide residues in polecats (*Mustela putorius*) from throughout their range in Britain, 1992–1999. *Environ. Pollut.* 122: 183–193.

Shore, R. F., Malcolm, H. M., McLennan, D., Turk, A., Walker, L. A., Wienburg, C. L., and Burn, A. 2006a. Did Foot and Mouth Disease control operations affect rodenticide exposure in raptors? *Wildl. Manag.* 70: 588–593.

Shore, R. F., Walker, L. A., Turk, A., Wienburg, C. L., Wright, J., Murk, A., and Wanless, S. 2006b. *Wildlife and Pollution: 2003/04 Annual Report.* JNCC Report, No. 391.

Skarphedinsdottir, H., Gunnarsson, K., Gudmundsson, G. A., and Nfon, E. 2010. Bioaccumulation and biomagnification of organochlorines in a marine food web at a pristine site in Iceland. *Arch. Environ. Contam. Toxicol.* 58: 800–809.

Smith, M. R., Thomas, N. J., and Hulse, C. 1995. Application of brain cholinesterase reactivation to differentiate between organophosphorus and carbamate pesticide exposure in wild birds. *J. Wildl. Dis.* 31: 263–267.

Smith, P. N., Cobb, G. P., Godard-Codding, C., Hoff, D., McMurry, S. T., Rainwater, T. R., and Reynolds, K. D. 2007. Contaminant exposure in terrestrial vertebrates. *Environ. Pollut.* 150: 41–64.

Sogorb, M. A., Ganga, R., Vilanova, E., and Soler, F. 2007. Plasma phenylacetate and 1–naphthyl acetate hydrolyzing activities of wild birds as possible non-invasive biomarkers of exposure to organophosphorus and carbamate insecticides. *Toxicol. Lett.* 168: 278–285.

Sonne, C. 2010. Health effects from long-range transported contaminants in Arctic top predators: An integrated review based on studies of polar bears and relevant model species. *Environ. Int.* 36: 461–491.

Sonne, C., Wolkers, H., Leifsson, P. S., Jenssen, B. M., Fuglei, E., Ahlstrøm, O., Dietz, R., Kirkegaard, M., Muir, D. C., and Jørgensen, E. 2008. Organochlorine-induced histopathology in kidney and liver tissue from Arctic fox (*Vulpes lagopus*). *Chemosphere* 71: 1214–1224.

Sonne, C., Gustavson, K., Rigét, F. F., Dietz, R., Birkved, M., Letcher, R. J., Bossi, R., Vorkamp, K., Born, E. W., and Petersen, G. 2009. Reproductive performance in East Greenland polar bears (*Ursus maritimus*) may be affected by organohalogen contaminants as shown by physiologically-based pharmacokinetic (PBPK) modelling. *Chemosphere* 77: 1558–1168.

Spurr, E. B., Maitland, M. J., Taylor, G. E., Wright, G. R. G., Radford, C. D., and Brown, L. E. 2005. Residues of brodifacoum and other anticoagulant pesticides in target and non-target species, Nelson Lakes National Park, New Zealand. *N. Z. J. Zool.* 32: 237–249.

Stern, G. A., Macdonald, C. R., Armstrong, D., Dunn, B., Fuchs, C., Harwood, L., Muir, D. C., and Rosenberg, B. 2005. Spatial trends and factors affecting variation of organochlorine contaminants levels in Canadian Arctic beluga (*Delphinapterus leucas*). *Sci. Total. Environ.* 351–352: 344–368.

Sterner, R. T. and Mauldin, R. E. 1995. Regressors of whole-carcass zinc phosphide/phosphine residues in voles: Indirect evidence of low hazards to predators/scavengers. *Arch. Environ. Contam. Toxicol.* 28: 519–523.

Stone, W. B., Okoniewski, J. C., and Stedelin, J. R. 1999. Poisoning of wildlife with anticoagulant rodenticides in New York. *J. Wildl. Dis.* 35: 187–193.

Stone, W. B., Okoniewski, J. C., and Stedelin, J. R. 2003. Anticoagulant rodenticides and raptors: Recent findings from New York, 1998–2001. *Bull. Environ. Contam. Toxicol.* 70: 34–40.

Storelli, M. M., Storelli, A., Barone, G., and Franchini, D. 2009. Accumulation of polychlorinated biphenyls and organochlorine pesticide in pet cats and dogs: Assessment of toxicological status. *Sci. Total. Environ.* 408: 64–68.

Strause, K. D., Zwiernik, M. J., Im, S. H., Bradley, P. W., Moseley, P. P., Kay, D. P., Park, C. S., Jones, P. D., Blankenship, A. L., Newsted, J. L., and Giesy, J. P. 2007. Risk assessment of great horned owls (*Bubo virginianus*) exposed to polychlorinated biphenyls and DDT along the Kalamazoo River, Michigan, USA. *Environ. Toxicol. Chem.* 26: 1386–1398.

Strum, K. M., Hooper, M. J., Johnson, K. A., Lanctot, R. B., Zaccagnini, M. E., and Sandercock, B. K. 2010. Exposure of migratory shorebirds to cholinesterase-inhibiting contaminants in the Western Hemisphere. *Condor* 112: 15–28.

Sullivan, T. P. and Sullivan, D. S. 2003. Vegetation management and ecosystem disturbance: Impact of glyphosate herbicide on plant and animal diversity in terrestrial systems. *Environ. Rev.* 11: 37–59.

Sun, F., Lin, F. Y., Wong, S. S., and Li, G. C. 2003. The screening of organophosphorus, organochlorine and synthetic pyrethroid pesticides residues in beef fat by tandem solid-phase extraction technique. *J. Food Drug Anal.* 11: 258–265.

Swiggart, R. C., Whitehead C. J. Jr., Curley, A., and Kellogg, F. E. 1972. Wildlife kill resulting from the misuse of arsenic acid herbicide. *Bull. Environ. Contam. Toxicol.* 8: 122–128.

Tanabe, S. 2002. Contamination and toxic effects of persistent endocrine disrupters in marine mammals and birds. *Mar. Pollut. Bull.* 45: 69–77.

Tanabe, S., Iwata, H., and Tatsukawa, R. 1994. Global contamination by persistent organochlorines and their ecotoxicological impact on marine mammals. *Sci. Total. Environ.* 154: 163–177.

Tanabe, S., Madhusree, B., Öztürk, A. A., Tatsukawa, R., Miyazaki, N., Özdamar, E., Aral, O., Samsun, O., and Öztürk, B. 1997. Persistent organochlorine residues in harbour porpoise (*Phocoena phocoena*) from the Black Sea. *Mar. Pollut. Bull.* 34: 338–347.

Tanabe, S., Niimi, S., Minh, T. B., Miyazaki, N., and Petrov, E. A. 2003. Temporal trends of persistent organochlorine contamination in Russia: A case study of Baikal and Caspian Seal. *Arch. Environ. Contam. Toxicol.* 44: 533–545.

Tao, S., Liu, W. X., Li, X. Q., Zhou, D. X., Li, X., Yang, Y. F., Yue, D. P., and Coveney, R. M. 2009. Organochlorine pesticide residuals in chickens and eggs at a poultry farm in Beijing, China. *Environ. Pollut.* 157: 497–502. *Sci. Total. Environ.*

Taylor, R. H., Kaiser, G. W., and Drever, M. C. 2000. Eradication of Norway rats for recovery of seabird habitat on Langara Island, British Columbia. *Restor. Ecol.* 8: 151–160.

Thorsen, M., Shorten R., Lucking R., and Lucking, V. 2000. Norway rats (*Rattus norvegicus*) on Frégate Island, Seychelles: The invasion; subsequent eradication attempts and implications for the island's fauna. *Biol. Conserv.* 96: 133–138.

Timbrell, J. A. 1998. Biomarkers in toxicology. *Toxicology* 129: 1–12.

Towns, D. R. and Broome, K. G. 2003. From small Maria to massive Campbell: Forty years of rat eradications from New Zealand islands. *N. Z. J. Zool.* 30: 377–398.

Turusov, V., Rakitsky, V., and Tomatis, L. 2002. Dichlorodiphenyltrichloroethane (DDT): Ubiquity, persistence, and risks. *Environ. Health Perspect.* 110: 125–128.

Vandenbroucke, V., Bousquet-Melou, A., De Backer, P., and Croubels, S. 2008. Pharmacokinetics of eight anticoagulant rodenticides in mice after single oral administration. *J. Vet. Pharmacol. Ther.* 31: 437–445.

Van den Steen, E., Dauwe, T., Covaci, A., Jaspers, V. L., Pinxten, R., and Eens, M. 2006. Within- and among-clutch variation of organohalogenated contaminants in eggs of great tits (*Parus major*). *Environ. Pollut.* 144: 355–359.

Van den Steen, E., Eens, M., Jaspers, V. L., Covaci, A., and Pinxten, R. 2009a. Effects of laying order and experimentally increased egg production on organic pollutants in eggs of a terrestrial songbird species, the great tit (*Parus major*). *Sci. Total. Environ.* 407: 4764–4770.

Van den Steen, E., Jaspers, V. L., Covaci, A., Neels, H., Eens, M., and Pinxten, R. 2009b. Maternal transfer of organochlorines and brominated flame retardants in blue tits (*Cyanistes caeruleus*). *Environ. Int.* 35: 69–75.

Van den Steen, E., Pinxten, R., Covaci, A., Carere, C., Eeva, T., Heeb, P., Kempenaers, B., Lifjeld, J. T., Massa, B., Norte, A. C., Orell, M., Sanz, J. J., Senar, J. C., Sorace, A., and Eens, M. 2010. The use of blue tit eggs as a biomonitoring tool for organohalogenated pollutants in the European environment. *Sci. Total. Environ.* 408: 1451–1457.

van Drooge, B., Mateo, R., Vives, I., Cardiel, I., and Guitart, R. 2008. Organochlorine residue levels in livers of birds of prey from Spain: Inter-species comparison in relation with diet and migratory patterns. *Environ. Pollut.* 153: 84–91.

Van Overmeire, I., Pussemier, L., Waegeneers, N., Hanot, V., Windal, I., Boxus, L., Covaci, A., Eppe, G., Scippo, M. L., Sioen, I., Bilau, M., Gellynck, X., De Steur, H., Tangni, E. K., and Goeyens, L. 2009. Assessment of the chemical contamination in home-produced eggs in Belgium: General overview of the CONTEGG study. *Sci. Total. Environ.* 407: 4403–4410.

Vergara, P., Fargallo, J. A., Banda, E., Parejo, D., Lemus, J. A., and García-Montijano, M. 2008. Low frequency of anti-acetylcholinesterase pesticide poisoning in lesser and Eurasian kestrels of Spanish grassland and farmland populations. *Biol. Conserv.* 141: 499–505.

Vermeer, K., Risebrough, R. W., Spaans, A. L., and Reynolds, L. M. 1974. Pesticide effects on fishes and birds in rice fields of Surinam, South America. *Environ. Pollut.* 7: 217–236.

Verreault, J., Muir, D. C., Norstrom, R. J., Stirling, I., Fisk, A. T., Gabrielsen, G. W., Derocher, A. E., Evans, T. J., Dietz, R., Sonne, C., Sandala, G. M., Gebbink, W., Riget, F. F., Born, E. W., Taylor, M. K., Nagy, J., and Letcher, R. J. 2005. Chlorinated hydrocarbon contaminants and metabolites in polar bears (*Ursus maritimus*) from Alaska, Canada, East Greenland, and Svalbard: 1996–2002. *Sci. Total. Environ.* 351–352: 369–390.

Verreault, J., Villa, R. A., Gabrielsen, G. W., Skaare, J. U., and Letcher, R. 2006. Maternal transfer of organohalogen contaminants and metabolites to eggs of Arctic-breeding glaucous gulls. *Environ. Pollut.* 144: 1053–1060.

Volmer, P. A., Merola, V., Osborne, T., Bailey, K. L., and Meerdink, G. 2006. Thallium toxicosis in a Pit Bull Terrier. *Vet. Diagn. Invest.* 18: 134–137.

Vorkamp, K., Christensen, J. H., Glasius, M., and Riget, F. F. 2004. Persistent halogenated compounds in black guillemots (*Cepphus grylle*) from Greenland: Levels, compound patterns and spatial trends. *Mar. Pollut. Bull.* 48: 111–121.

Vorkamp, K., Rigét, F. F., Glasius, M., Muir, D. C., and Dietz, R. 2008. Levels and trends of persistent organic pollutants in ringed seals (*Phoca hispida*) from Central West Greenland, with particular focus on polybrominated diphenyl ethers (PBDEs). *Environ. Int.* 34: 499–508.

Vorkamp, K., Thomsen, M., Møller, S., Falk, K., and Sørensen, P. B. 2009. Persistent organochlorine compounds in peregrine falcon (*Falco peregrinus*) eggs from South Greenland: Levels and temporal changes between 1986 and 2003. *Environ. Int.* 35: 336–341.

Vos, J. G., Dybing, E., Greim, H. A., Ladefoged, O., Lambre, C., Tarazona, J. V., Brandt, I., and Vethaak, A. D. 2000. Health effects of endocrine-disrupting chemicals on wildlife, with special reference to the European situation. *Crit. Rev. Toxicol.* 30: 71–133.

Vyas, N. B., Spann, J. W., Hulse, C. S., Gentry, S., and Borges, S. L. 2007. Dermal insecticide residues from birds inhabiting an orchard. *Environ. Monit. Assess.* 133: 209–214.

Wainwright, S. E., Mora, M. A., Sericano, J. L., and Thomas, P. 2001. Chlorinated hydrocarbons and biomarkers of exposure in wading birds and fish of the lower Rio Grande Valley, Texas. *Arch. Environ. Contam. Toxicol.* 40: 101–111.

Walker, C. H. 1995. Biochemical biomarkers in ecotoxicology; some recent developments. *Sci. Total. Environ.* 171: 189–195.

Walker, C. H. 2003. Neurotoxic pesticides and behavioural effects upon birds. *Ecotoxicology* 12: 307–316.

Walker, C. H., Hopkin, S. P., Sibly, R. M., and Peakall, D. B. 1996. *Principles of Ecotoxicology*. Taylor & Francis: London.

Walker, L. A., Shore, R. F., Turk, A., Pereira, M. G., and Best, J. 2008a. The Predatory Bird Monitoring Scheme: Identifying chemical risks to top predators in Britain. *Ambio* 37: 466–471.

Walker, L. A., Turk, A., Long, S. M., Wienburg, C. L., Best, J., and Shore, R. F. 2008b. Second generation anticoagulant rodenticides in tawny owls (*Strix aluco*) from Great Britain. *Sci. Total. Environ.* 392: 93–98.

Walker, L. A., Lawlor, A. J., Llewellyn, N., Peréira, M. G., Potter, E., Townsend, J., Turk, A., and Shore, R. F. 2010. *The Predatory Bird Monitoring Scheme (PBMS) Report 2006–7*. A contract report from the Centre for Ecology & Hydrology to Natural England.

Wang, Y., Kruzik, P., Helsberg, A., Helsberg, I., and Rausch, W. D. 2007. Pesticide poisoning in domestic animals and livestock in Austria: A 6 years retrospective study. *Forensic Sci. Int.* 169: 157–160.

Watanabe, K. P., Saengtienchai, A., Tanaka, K. D., Ikenaka, Y., and Ishizuka, M. 2010. Comparison of warfarin sensitivity between rat and bird species. *Comp. Biochem. Physiol. C Toxicol. Pharmacol.* 152: 114–119.

Weber, J., Halsall, C. J., Muir, D., Teixeira, C., Small, J., Solomon, K., Hermanson, M., Hung, H., and Bidleman, T. 2010. Endosulfan, a global pesticide: A review of its fate in the environment and occurrence in the Arctic. *Sci. Total. Environ.* 408: 2966–2984.

Weijs, L., Das, K., Neels, H., Blust, R., and Covaci, A. 2010. Occurrence of anthropogenic and naturally-produced organohalogenated compounds in tissues of Black Sea harbour porpoises. *Mar. Pollut. Bull.* 60: 725–731.

Wiemeyer, S. N. 1996. Other organochlorine pesticides in birds. In: Beyer, W. N., Heinz, G. H., and Redmon-Norwood, A. W. (eds),_ *Environmental Contaminants in Wildlife: Interpreting Tissue Concentrations*, pp. 99–115. Lewis Publishers: Boca Raton, FL.

Wiemeyer, S. N., Clark D. R. Jr., Spann, J. W., Belisle, A. A., and Bunck, C. M. 2001. Dicofol residues in eggs and carcasses of captive American kestrels. *Environ. Toxicol. Chem.* 20: 2848–2851.

Wiemeyer S. N., Miesner, J. F., Tuttle, P. L., and Murphy, E. C. 2005. Organochlorine contaminants in the American white pelican breeding at Pyramid Lake, Nevada. *Waterbirds* 28(Special Publication 1): 95–101.

Wiktelius, S. and Edwards, C. A. 1997. Organochlorine insecticide residues in African Fauna: 1971–1995. *Rev. Environ. Contam. Toxicol.* 151: 1–37.

Wilson, L., Martin, P. A., Elliott, J. E., Mineau, P., and Cheng, K. M. 2001. Exposure of California quail to organophosphorus insecticides in apple orchards in the Okanagan Valley, British Columbia. *Ecotoxicology* 10: 79–90.

Winters, A. M., Rumbeiha, W. K., Winterstein, S. R., Fine, A. E., Munkhtsog, B., and Hickling, G. J. 2010. Residues in Brandt's voles (*Microtus brandti*) exposed to bromadiolone-impregnated baits in Mongolia. *Ecotoxicol. Environ. Safe.* 73: 1071–1077.

Wobeser, G., Bollinger, T., Leighton, F. A., Blakley, B., and Mineau, P. 2004. Secondary poisoning of eagles following intentional poisoning of coyotes with anticholinesterase pesticides in western Canada. *J. Wildl. Dis.* 40: 163–172.

Wolkers, H., Lydersen, C., and Kovacs, K. M. 2004. Accumulation and lactational transfer of PCBs and pesticides in harbor seals (*Phoca vitulina*) from Svalbard, Norway. *Sci. Total. Environ.* 319: 137–146.

Wolkers, H., van Bavel, B., Ericson, I., Skoglund, E., Kovacs, K. M., and Lydersen, C. 2006. Congener-specific accumulation and patterns of chlorinated and brominated contaminants in adult male walruses from Svalbard, Norway: Indications for individual-specific prey selection. *Sci. Total. Environ.* 370: 70–79.

Wren, C. D. 1986. A review of metal accumulation and toxicity in wild mammals. I: Mercury. *Environ. Res.* 40: 210–244.

Yates, M. A., Fuller, M. R., Henny, C. J., Seegar, W. S., and Garcia, J. 2010. Wintering area DDE source to migratory white-faced ibis revealed by satellite telemetry and prey sampling. *Ecotoxicology* 19: 153–162.

Yogui, G. T., Santos, M. C., Bertozzi, C. P., and Montone, R. C. 2010. Levels of persistent organic pollutants and residual pattern of DDTs in small cetaceans from the coast of São Paulo, Brazil. *Mar. Pollut. Bull.* 60: 1862–1867.

Zitko, V. 2003a. Chlorinated pesticides: Aldrin, DDT, endrin, dieldrin, mirex. In: Fiedler, H. (ed.), *Persistent Organic Pollutants*, pp. 47–90. Springer-Verlag: Berlin Heidelberg.

Zitko, V. 2003b. Hexachlorobenzene. In: Fiedler, H. (ed.), *Persistent Organic Pollutants*, pp. 91–122. Springer-Verlag: Berlin Heidelberg.

Section VI

Pesticides and Men

15

Pesticide Residues in Man

Sameeh A. Mansour

CONTENTS

15.1 Introduction

Currently, the world population is growing at an annual rate of 1.2%, that is, 77 million people per year. Six countries account for half of this annual increment: India, China, Pakistan, Nigeria, Bangladesh, and Indonesia. The world population grew from 2.5 billion in 1950 to 6.1 billion in the year 2000. Current projections show a continued increase in population (but a steady decline in the population growth rate), with the population expected to reach between 8 and 10.5 billion in the year 2050 (IDB 2010).

The worldwide consumption of pesticides is about 2 million tons per year, of which 24% is consumed in the United States alone, 45% in Europe, and 25% in the rest of the world. India's share, for example, is just 3.75%. The usage of pesticides in India is only 0.5 kg/ha, while in Korea and Japan, it is 6.6 and 12.0 kg/ha, respectively (Gupta 2004).

At present, approximately 1500 active ingredients have been registered as pesticides, and the formulators mix these compounds with one or more of some 900 "inert" ingredients to create approximately 50,000 commercial pesticides registered for use. Roughly, 85% of the pesticides currently used in the world are devoted to the agricultural sector; almost 10% are dedicated to sanitary measures against vectors in public health programs, and the rest are applied in specific sites such as buildings, transport media, and residential areas (WHO 1993).

Data on pesticide consumption in three continents, expressed as intensity (metric tons/hectare, MT/ha) (WRI 1996), reveal that the most intensive pesticide users among Latin American countries are Costa Rica and Belize (18.0 and 17.4 MT/ha, respectively), followed by Trinidad and Tobago (13.3 MT/ha). In 1990, Brazil consumed the highest quantity (67,000 MT) in 61,350 hectares; achieving, however, the lowest intensive pesticide use in the continent (1.1 MT/ha). Pesticide consumption data (mostly for the year 1992) collected from WRI (1996) based on FAO and other sources are presented for some African countries in Figure 15.1. With intensity expressed in terms of metric ton per cropped hectare, Morocco (9.9 MT/ha), Tanzania (4.2 MT/ha), Burundi (3.0 MT/ha), Togo (2.5 MT/ha), and Egypt (2.1 MT/ha) are the most intensive pesticide users in the continent among the 13 countries compared.

At present, the pressing challenge is, therefore, to produce more food and ensure food security regionally in order to alleviate poverty and undernourishment and, at the same time, to improve human health and welfare. Pesticides have contributed to dramatic

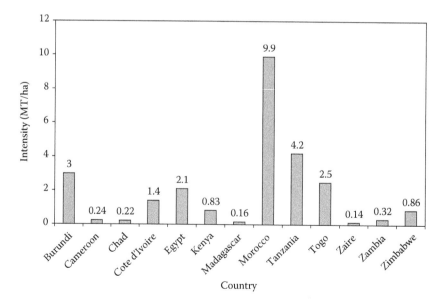

FIGURE 15.1

Pesticide use per cropped hectare, in terms of MT/HA, in some African countries. (Adapted from WRI. 1996. *Pesticides and the Immune System: The Public Health Risks.* World Resources Institute: Washington, DC.)

increases in crop yields and in the quantity and variety of the diet. Also, they have helped to limit the spread of certain diseases. Thus, pesticides cannot be totally abolished from agricultural practice, without causing famine. In a recent report (Whitford et al. 2004), the following has been mentioned in favor of pesticides

> *"For decades, discussions among scientists and the public have focused on the real, predicted, and perceived risks that pesticides pose to people and the environment. Each use of a pesticide poses some level of risk, so it is not surprising that scientists, the regulated community, government officials, and the public need a realistic understanding of the risks associated with pesticide use. We must analyze how risk is assessed, identify the risks, and determine an appropriate level of concern. There are significant risks associated with leaving certain pests uncontrolled; and, in some cases, pesticides are the only viable alternative. Properly used, pesticides provide benefits essential to our way of life. Uncontrolled pests can cause serious consequences:*
>
> - *A person bitten by mosquitoes carrying West Nile virus may die.*
> - *A child stung by bees, wasps, or ants may suffer a severe allergic reaction.*
> - *A dog infested with fleas may become stressed to the point of illness.*
> - *A farmer's diseased tomatoes may be declined by the cannery.*
> - *A load of wheat contaminated with wild garlic may be rejected by the mill.*
> - *A homeowner may have to spend thousands of dollars to repair structural damage caused by termites."*

Despite the benefits of pesticide usages, they can cause injury to human health as well as the environment. The range of these adverse health effects includes acute and persistent injury to the nervous system, lung damage, injury to the reproductive organs, and dysfunction of the immune and endocrine systems, birth defects, and cancer (Mansour 2004).

The pesticides are detrimental to people's health and environment. It is reported that approximately 3 million people are poisoned and 200,000 die each year around the world from pesticide poisoning, with a majority of them belonging to the developing countries (FAO 2000). It is also believed that in developing countries, the incidence of pesticide poisoning may even be greater than that reported due to underreporting, lack of data, and misdiagnosis. The dilemma is because of those pesticides, which developed countries have banned due to their toxic effects, but are still being used in the developing countries (Wilson and Tisdell 2001).

In this chapter, we discuss "pesticide residues in man." In other words, we are going to talk about "source and sink"; we consider that man is the cause for hosting pesticide residues in his body. In this context, we will try to demonstrate the sources and magnitudes of pesticide residues in the environment and human body; pesticide residues with respect to human sociodemographic characteristics; hazards of occupational exposure with reference to epidemiological data of poisoning disasters; toxic hazards (in general) of pesticide exposure and factors influencing persistence, accumulation, and fate of pesticide residues; ways to remove or minimize pesticidal residues; and, if possible, the future outlook and the needs to magnify pesticide benefits and limit their hazards.

15.2 Environmental Dynamics of Pesticides

Pesticides move through air, soil, and water and find their way into the living tissues where they can bioaccumulate through the food chain, eventually to enter the human diet.

Approximately 85%–90% of the applied agricultural pesticides never reach the target organisms, but disperse through the air, soil, and water (Moses et al. 1993). Persistent pesticides can remain for decades; the half-life of toxaphene in soil, for example, is up to 29 years (PAN 1993). Pesticides that are not bound in soils or taken up into plants and animals can run off into rivers and lakes and move into the aquatic food chain, inducing severe damage to aquatic life. Such environmental mobility can cause contamination of several environmental compartments. Pesticides, thereby, augment other sources of environmental pollution, which include manufacturing processes such as the pulp and paper industry, textile and leather dying, and thermal processes in the metallurgical, cement, motor vehicle, and steel industries (Barakat 2003).

Organochlorine pesticides (OCPs) are a class of chemicals that came into widespread use in the late 1940s. Despite being banned in industrialized countries since the 1970s, or subjected to restrictions in use in many others, they persist to this day in the environment. Distribution and accumulation of these chemicals in different environmental components had been long recognized. Mansour (2004) reviewed this issue in detail, focusing on the situation in Egypt and in Africa (Mansour 2008). In the same issue, Gupta (2004) has published an extensive report on the status in India. From these reports, it can clearly be deduced that the OCP residues are present in all environmental components, however, in varying concentration levels, according to the lipophilic character of the analyzed sample.

15.2.1 Principal Sources of Pesticide Residues in Man

The main source of nonoccupational exposure to pesticides is through the diet. With the exception of occupational exposure to pesticides, most exposure to these chemicals occurs via dietary intake (DeVoto et al. 1998), especially food of animal origin, but also through water, ambient and indoor air, dust, and soil (Dua et al. 2001; Manirakiza et al. 2002). In a nonoccupational pesticide exposure study (NOPES) carried out by US EPA, it was concluded that for 14 of the 25 pesticides tested, food appears to be the major contributor to total exposure, whereas air appears to be the dominant one for six of the other eleven compounds (EPA 1990). Food contamination by pesticides originates from normal use of agricultural pesticides before and/or after crop harvesting, misuse of these chemicals, and unintended environmental contamination (Eilrich 1991). OCPs are lipophilic compounds that accumulate and even biomagnify their concentration along the food chain, especially in fatty foods (Manirakiza et al. 2002). However, pesticide residues in food reported from most surveys are sufficiently below MRLs for the respective compounds. But much of the concern over residue levels is whether the current levels are safe rather than whether they are under the legal tolerance level (Gots 1992).

Some published reports (DeVoto et al. 1998; Hanaoka et al. 2002; Manirakiza et al. 2002) suggest that the serum levels of organochlorines (OCs) are related to the consumption of various foods. Thus, biological monitoring of the exposure can be carried out by the determination of intact compounds or their metabolites in the blood, serum, plasma, and urine. Such biological materials can be conveniently obtained for the study of body burden following chronic exposure to pesticides (Waliszewski et al. 1999a).

15.2.2 Outdoor Pesticide Contamination

A great number of pesticide compounds have been found to contaminate water resources, ambient air, fog, rain, and soils, in numerous studies (Glotfelty et al. 1987; Coupe et al. 2000; Sanusi et al. 2000). Professional uses of pesticides include crop, greenhouse, cattle, and pet treatments, but pest control operations in buildings also contribute to outdoor contamination. The latter can be tracked in by shoes, clothes, and air drift (Thompson et al. 2003).

Contamination of the aquatic environment by chlorinated hydrocarbons is of great concern because their residues reside in multiple compartments of the aquatic ecosystem. OCs share common physical and chemical properties, such as high chemical stability, interactivity, appreciable volatility at ambient temperatures, low water solubility, and high lipid–water partition coefficients (Mackay and Leeinonen 1975). Because of these characteristics, the OC compounds are well known for environmental persistence, their presence in surface waters and groundwater, their ubiquitous distribution throughout the world, and accumulation in fat reserves of marine and terrestrial organisms.

Once a pesticide is released into an aquatic ecosystem, it rapidly distributes preferentially between different compartments of this aquatic ecosystem. Studies conducted on different aquatic ecosystems in Egypt, for example, revealed that the accumulation pattern for pesticides (and heavy metals) in different ecosystem compartments has the following hierarchy: sediment > fish > water (Badawy et al. 1984; Mansour et al. 2001a; Mansour and Sidky 2002; Mansour 2006).

The presence of pesticide residues is not confined to inhabited continents. Hoferkamp et al. (2010) reviewed the levels of selected current use pesticides (CUPs) that have been identified and reported in Arctic media (i.e., air, water, sediment, and biota) since the year 2000. Almost all of the 10 CUPs (chlorothalonil, chlorpyrifos, dacthal, diazinon, dicofol, lindane, methoxychlor, pentachloronitrobenzene (PCNB), pentachlorophenol, and trifluralin) examined in the review currently are, or have been, high production volume chemicals globally. Characteristic travel distances for the 10 chemicals range from 55 km (methoxychlor) to 12,100 km (PCNB). Surveys and long-term monitoring studies have demonstrated the presence of 9 of the 10 CUPs in the Arctic environment. Only dicofol has not been reported. The presence of these chemicals has mainly been reported in high-volume air samples and in snow from Arctic ice caps and lake catchments. There are many other CUPs registered for use, which have not been determined in the Arctic environments. The discovery of the CUPs currently measured in the Arctic has been mainly serendipitous, a result of analyzing some samples using the same suite of analysts as used for studies in midlatitude locations. A more systematic approach is needed to assess whether other CUPs might be accumulating in the Arctic and ultimately to assess whether their presence has any significance biologically or results in risks for human consumers.

15.2.3 Indoor Pesticide Contamination

Pesticides also contaminate the indoor environment, as a consequence of indoor as well as outdoor uses, for occupational and residential purposes. Domestic pesticide uses include pet treatments, extermination of household pests, removal of lice, and garden and lawn treatments. In a French pilot study, the pesticide exposure of nonoccupationally exposed subjects compared with some occupational exposure was investigated (Bouvier et al. 2006). Thirty-eight insecticides, herbicides, and fungicides were measured in indoor air with an air sampler for 24 h and on hands by wiping them with isopropanol-wetted swabs. Seventeen different pesticides were detected at least once in indoor air and twenty-one pesticides were detected on the hands. An average of 4.2 ± 1.7 different pesticides were detected per indoor air sample. The OCs, lindane, α-endosulfan, and α-hexachlorocyclohexanes (α-HCHs) were the most frequently detected compounds in 97%, 69%, and 38% of the samples, respectively. The organophosphates dichlorvos and fenthion, the carbamate propoxur, and the herbicides atrazine and alachlor were detected in more than 20% of the air samples. On the other hand, indoor air concentrations were often low, but could reach 200–300 ng/m^3 in residences for atrazine and propoxur. Propoxur levels significantly differed between the air in veterinary places and that in other places, so did the dieldrin levels between residences and workplaces. There were a greater number of pesticides on the hands than in air, and the most frequently detected were malathion, lindane, and trifluralin, in more than 60% of the subjects. Maximum levels (up to 1000–3000 ng/hands) were observed either in the general population or in workers, depending on the pesticides. However, no significant difference was observed between the workers' and general population's hand-wipe pesticide levels. As expected, gardeners were exposed to pesticides sprayed in greenhouses. Florists and veterinary workers, whose pesticide exposure had not been described until now, were also indirectly exposed to pesticides used for former pest control operations. Overall, the general population was exposed to more various pesticides and at levels sometimes higher than in occupational places (Bouvier et al. 2006). According to Bouvier et al. (2006), the most frequently found pesticides in residences were not the same as those in the US studies but the levels were similar.

Tan et al. (2007) measured OCPs and PCBs in house-dust samples collected from 31 homes across the island-state of Singapore. OCPs such as HCHs, chlordanes, and dichlorodiphenyltrichloroethanes (DDTs) were tested, with a range of <LOD to 240 ng/g dust, <LOD to 110 ng/g dust, and <LOD to 770 ng/g dust, respectively. Of 41 target PCB congeners measured, 28 were detected, and the median level of PCBs was 5.6 ng/g dust. The prevalence of these pesticides and industrial chemicals, which have been banned many years ago, suggests limited indoor degradation and ongoing environmental accumulation. Weak negative correlations between an increased elevation of the residence and the concentration of DDTs and some low-chlorinated PCB congeners were noted, most likely due to increased ventilation and distance from ground-based sources at higher floors. The authors concluded that levels detected in several studies conducted in the United States were several orders of magnitude higher than the concentrations observed in Singapore. Values of daily human intake of OCPs and PCBs via house dust were low compared with dietary data from overseas. Ingestion of dust is indeed an exposure pathway for these OCs, but may not be the predominant one (Tan et al. 2007).

15.3 Pesticide Residues in Human Body

Pesticides entering the human body through various routes are preferentially distributed among different body fluids and tissues. Such distribution is influenced by many factors related to the physicochemical and biological characteristics of the pesticides and the physiological conditions of the exposed individuals. In this respect, we will present research data on pesticide residues in the human blood, urine, seminal fluid, and adipose tissues.

15.3.1 Pesticide Residues in Human Blood

Stehr-Green et al. (1988) measured the levels of 11 pesticide residues and metabolites in the serum samples collected from 85 rural-dwelling white Americans in 1986. This group had a mean age of 36.9 ± 17.3 years and a male:female ratio of approximately 3:2. The average number of years spent living in rural areas was 33 ± 16 years, and 75% of the participants

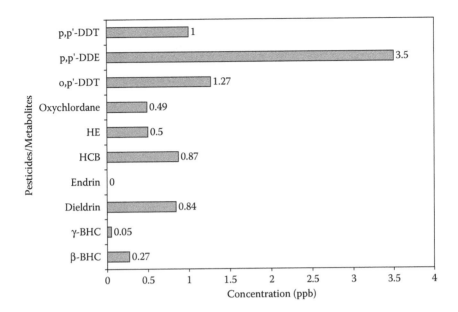

FIGURE 15.2
Distribution of serum pesticide residue levels in a rural population, 1986 (N = 85; white American persons). HE, Heptachlor epoxide. (Adapted from Stehr-Green, P. A., Farrar, J. A., Burse, V. W., Royce, W. G., and Wohlleb, J. C. 1988. A survey of measured levels and dietary sources of selected organochlorine pesticide residues and metabolites in human sera from a rural population. *Am. J. Public Health* 78(7): 828–830.)

had spent more than 90% of their lives living on a farm. In general, the serum levels increased with age, with males having slightly higher levels than females. Consumption of eggs from home-raised hens contributed substantially to increased serum concentrations of trans-nonachlor, heptachlor epoxide, and oxychlordane. Consumption of home-grown root vegetables likewise contributed to increased serum concentrations of trans-nonachlor and oxychlordane. Health risks, if any, that may be attributable to these "background" levels of exposure remain to be fully characterized in this and all other affected populations. The results are represented in Figure 15.2.

According to Becker et al. (2002), blood samples from about 4800 subjects were selected to be representative of the German population with regard to region (East/West Germany), community size, age (18–69 years), and gender. Cadmium, lead, mercury, hexachlorobenzene (HCB), HCH (α-, β-, and γ-HCH), 1,1-dichloro-2,2-bis (p-chlorophenyl) ethylene (p,p'-DDE), and polychlorinated biphenyls (PCBs; PCB-138, PCB-153, PCB-180) were analyzed in the whole blood to document the extent, distribution, and determinants of exposure of the general population to these substances. Focusing on OCs, the authors reported that the mean concentrations of HCB and p,p'-DDE in blood were 0.44 µg/L and 1.6 µg/L, respectively. In East Germany (the former GDR), the DDE concentration in the blood was more than twice as high as that in West Germany. Geometric means for PCB were 0.42, 0.68, and 0.44 µg/L for PCB-138, PCB-153, and PCB-180, respectively. A marked increase of HCB, DDE, and PCB levels with age was observed. α- and γ-HCH were detected in 1.7% and 5.2% of the samples only. β-HCH was quantified in 34% of the samples with a 95th percentile of 0.5 µg/L.

As part of the National Health and Examination Survey (1999–2000), the Centers for Disease Control and Prevention analyzed serum samples from the US population for many environmental chemicals associated with neurodevelopmental effects (Needham et al. 2005).

TABLE 15.1

Concentration (95th Percentile ng/g Serum) of Selected Organochlorine Pollutants in Serum Samples from US Population (1999/2000) Based on Age, Gender, and Race or Ethnicity

Parameter	β-HCH	p,p′-DDE	Oxychlordane	HE[a]	PCBs[b]
Age group					
12–19 years	0.05	2.31	<LOD	<LOD	2.87
20 years and older	0.48	12.33	0.33	0.19	3.04
Gender					
Male	0.29	9.63	0.30	0.15	111.00[c]
Female	0.56	13.20	0.32	0.20	118.00[c]
Race or ethnicity					
Mexican Americans	0.91	31.52	0.28	0.17	0.47[c]
Non-Hispanic Black	0.36	14.56	0.31	0.11	1.27[c]
Non-Hispanic White	0.39	8.04	0.32	0.18	0.70[c]

Source: Adapted from Needham, L. L., Barr, D. B., Caudill, S. P., et al. 2005. Concentrations of environmental chemicals associated with neurodevelopmental effects in U.S. population. *Neurotoxicology* 26: 531–554.

[a] Heptachlor epoxide.
[b] The measured PCB congeners represent 11 compounds of numbers: 74, 99, 105, 118, 138, 146, 153, 156, 170, 180, 187.
[c] The values for PCB 153 only.

Concentration data of selected metals, persistent organic pollutants (POPs), organophosphorus (OP) and carbamate insecticides, and cotinine were presented. For example, the 95th percentile estimates for serum total PCBs (whole weight) in the population aged 20 years and older are about 2.7 ng/g, and for serum dioxin, the total toxic equivalence is between 40 and 50 pg/g lipid basis. The authors stated that the US general population is daily exposed to many environmental chemicals, some of which are associated with neurodevelopmental effects; however, human levels of these chemicals are decreasing over time in the US population. This reflects the effects of legislation, industry efforts, and changes in lifestyle and activity patterns in the US population. Based on the authors' data, Table 15.1 summarizes serum concentrations of β-HCH, p,p′-DDE, oxychlordane, heptachlor epoxide, and PCB congeners. It clarifies that older and female subjects retain higher concentration levels of the measured pollutants than younger and male subjects. Persons of different races or ethnicities differed in their serum contaminant levels based on the measured compounds. For example, β-HCH and p,p′-DDE levels were higher in the Mexican Americans than in the non-Hispanic (black or white) persons; however, the non-Hispanic blacks showed the highest serum concentration of PCB 153 (1.27 ng/g serum).

Serum samples from 83 farmers exposed to pesticides through their work in 14 different fruit and vegetable farm stations, in addition to 5 normal control persons, living in a rural area in Pakistan (Gadap, Karachi) were examined for the presence of pesticide residues and enzyme levels in their blood (Azmi et al. 2006). The analysis indicated that there was a significant difference in the residue levels in exposed and unexposed persons. Moreover, there was a direct relation between the high residue levels and their enzyme levels (e.g., ALT, GOT, GPT, and ALP). The results shown in Table 15.2 indicate the prevalence of cypermethrin and monocrotophos among the subjects in 13 farm stations. DDT and DDE, diazinon, and deltamethrin were less frequently detected. Profenofos was detected at a

TABLE 15.2

Pesticide Residues in Blood Serum Samples from Farmers of 14 Different Fruit and Vegetable Farm Stations in Gadap (Rural Area), Karachi, Pakistan

Station No.	Detected Pesticide	Concentration (µg/mL)
1	Cypermethrin	0.19–9.13
	DDT	0.28–3.70
	DDE	0.67–11.39
	Diazinon	2.64–5.40
	Monocrotophos	2.60–6.00
	Deltamethrin	0.41–0.55
2	Cypermethrin	4.20–4.93
	DDE	11.4
	Monocrotophos	35.40
3	Cypermethrin	7.94
	Monocrotophos	9.6
4	Cypermethrin	16.80
5	Cypermethrin	4.47
	DDE	4.22
6	No data available	
7	Cypermethrin	0.64–17.80
	DDT	3.17–4.98
	DDE	14.20–39.20
	Monocrotophos	11.2
	Profenofos	65.50
8	Cypermethrin	8.58–29.40
	Monocrotophos	29.70
9	Cypermethrin	17.80
	Diazinon	15.80
	Monocrotophos	14.2
10	Cypermethrin	32.40–34.00
	Deltamethrin	40.00
	Monocrotophos	29.60
11	Cypermethrin	0.63–11.32
	DDT	11.70
	DDE	41.30
12	Cypermethrin	2.37–26.40
	Diazinon	28.80
	DDT	1.74
13	Cypermethrin	5.20–15.00
	Deltamethrin	3.18
	Diazinon	6.12
	Monocrotophos	1.32
	DDT	3.98
	DDE	13.70
14	Cypermethrin	2.01–6.03
	Monocrotophos	2.52–13.90
	DDT	7.30

Source: Adapted from Azmi, M. A., Naqvi, S. N. H., Arshad Azmi, M., and Aslam, M. 2006. Effect of pesticide residues on health and different enzyme levels in the blood of farm workers from Gadap (rural area) Karachi, Pakistan. *Chemosphere* 64: 1739–1744.

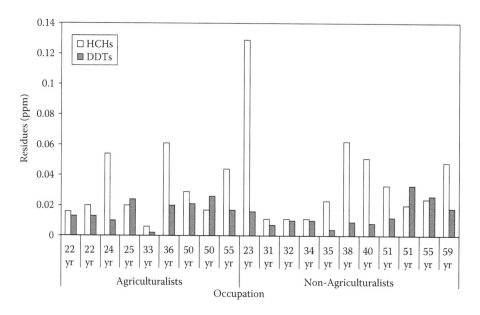

FIGURE 15.3

Concentration levels (ppm) of HCH and DDT residues in serum blood samples taken from two groups of people, from Aandipatti area of Madurai, South India: One group that has direct exposure to pesticides (agriculturists and public health workers), and another group that has indirect exposure to pesticides through the food chain. (Data refer to Subramaniam, K. and Solomon, R. D. J. 2006. Organochlorine pesticides BHC and DDE in human blood in and around Madurai, India. *Indian J. Clin. Biochem.* 21(2): 169–172.)

concentration of 65.5 µg/mL in the farmers from station no. 7 only. The highest concentration of cypermethrin (34.0 µg/mL) was found in the farmers from station no. 10. DDT + DDE ranged 0.95–15.09 µg/mL in the farmers of station no. 1, compared with 17.37–44.18 µg/mL for no. 7 (Table 15.2).

Blood serum from 154 volunteers at 13 UK locations in 2003 was analyzed for a range of PCBs, OCPs, and PBDEs. HCB, p,p'-DDE, p,p'-DDT, and β-HCH were the dominant OC pesticides in most samples. Concentration was age-correlated for the less easily metabolized PCBs, p,p'-DDT and p,p'-DDE, HCBs, and HCHs. With increasing age, females tended to have lower concentrations of the more chlorinated PCBs than males (Thomas et al. 2006).

Subramaniam and Solomon (2006) monitored OCP residues (DDTs and HCHs) in the blood samples taken from two groups of people from Madurai, South India: one that had direct exposure to pesticides (agriculturists and public health workers) compared with the other group that had indirect exposure to pesticides through the food chain. High concentrations of both the contaminants were observed in the serum samples of the people who had direct exposure to the pesticides, with few exceptions. Generally, the pesticide residue concentration in serum ranged from 0.006 to 0.130 ppm for HCHs and 0.002 to 0.033 ppm for DDTs; however, there was variation between professional tasks and ages of the studied limited population (Figure 15.3). Significance of this study reveals the presence of these banned pesticides in human serum.

15.3.2 Pesticide Residues in Human Urine

Hill et al. (1995) measured 12 analytes in the urine of 1000 adults (20–59 years age) living in the United States to establish a reference range of concentrations of pesticide residues. Six

TABLE 15.3

Pesticide Residues[a] in Urine Samples Collected from Adult American Citizens[b]

Pesticide/Metabolite	Frequency (%)	The 95th Percentile Concentration (µg/L)
2,5-Dichlorophenol	98	790
2,4-Dichlorophenol	64	64
1-Naphthol	86	43
2-Naphthol	81	30
3,5,6-Trichloro-2-pyridinol	82	13
Pentachlorophenol	64	8.2
4-Nitrophenol	41	<6
2,4,5-Trichlorophenol	20	<6
2,4,6-Trichlorophenol	9.5	<6
2,4-Dichlorophenoxy acetic acid	12	<6
2-Isopropoxyphenol	6.8	<6
7-Carbofuranphenol	1.5	<6

Source: Adapted from Hill, R. H., Jr., Head, S. L., Baker, S., et al. 1995. Pesticide residues in urine of adults living in the United States: Reference range concentrations. *Environ. Res.* 71: 99–108.

[a] Number of targeted analysts: 12.

[b] Number of subjects: 1000.

analytes were frequently found: 2,5-dichlorophenol (in 98% of adults); 2,4-dichlorophenol (in 64%); 1-naphthol (in 86%); 2-naphthol (in 81%); 3,5,6-trichloro-2-pyredinol (TCPY) (in 82%); and pentachlorophenol (in 64%). The 95th percentile concentrations (95th PCs) for the above metabolites were 790, 64, 43, 30, 13, and 8.2 µg/L, respectively. The 95th PCs for other analytes were found to be less than 6.0 µg/L (Table 15.3). Exposure to p-dichlorobenzene is ubiquitous; naphthalene and chlorpyrifos are also major sources of pesticide exposure. Exposure to chlorpyrifos appears to be increasing. Although pentachlorophenol exposure is frequent, the exposure appears to be decreasing. The obtained reference range of concentrations provides information about pesticide exposure and serves as a basis against which concentrations in the subjects who may have been exposed to pesticides can be compared.

A total of 1146 American inhabitants took part in a voluntary investigation for measuring pesticide residues in their urine (Heudorf and Angerer 2001). All of them stated that they had never used chlorpyrifos in their homes. Spot urine samples of the study participants were analyzed for six metabolites of OP insecticides (dimethylphosphate (DMP), diethylphosphate (DEP), dimethylthiophosphate (DMTP), diethylthiophosphate (DETP), dimethyldithiophosphate (DMDTP), and diethyldithiophosphate (DEDTP)) using a very sensitive gas chromatographic method with mass-selective detection and a limit of detection of 1 µg/L. No evidence of increased internal exposure due to former chlorpyrifos application was found either in children or in adults in these homes (>4 years ago). The median values and 95th percentiles of the urinary metabolite concentrations in 484 adults were (µg/g creatinine) as follows: DMP: 15.5 and 102.5; DMTP: 13.5 and 125.8; DMDTP: <1 and 13.1; DEP: 2.1 and 11.6; DETP: <1 and 6.4; and DEDTP: both <1. The urinary metabolite concentrations in children less than 6 years of age were higher; this was caused mainly by lower creatinine concentrations. To conclude, no increase in the internal exposure due to former indoor application of chlorpyrifos was found, and the reference values published for internal OP exposure in adults in Germany were confirmed. However, as shown in other environmental studies, the urinary excretion of OP metabolites exceeds dietary intake several folds; this has been estimated from the data in various duplicate dietary studies.

Barr et al. (2005) reported population-based concentrations (stratified by age, sex, and composite race/ethnicity variables) of selective metabolites of chlorpyrifos (3,5,6-tri-chloro-2-pyridinol; TCPY), chlorpyrifos-methyl (TCPY), malathion (malathion dicar-boxylic acid, MDA), diazinon (2-isopropyl-4-methyl-6-hydroxypyrimidine, IMPY), methyl parathion (para-nitrophenol, PNP), and parathion (PNP). They measured the concentrations of TCPY, MDA, IMPY, and PNP in 1997 urine samples from participants, aged 6–59 years, of the National Health and Nutrition Examination Survey (1999–2000). The authors detected TCPY in more than 96% of the samples tested. Other OP pesticide metabolites were detected less frequently: MDA, 52%; IMPY, 29%; and PNP, 22%. The geometric means for TCPY were 1.77 mg/L and 1.58 mg/g creatinine. The 95th percentile for TCPY was 9.9 mg/L (8.42 mg/g creatinine). The 95th percentile for MDA was 1.6 mg/L (1.8 mg/g creatinine). The 95th percentiles for IMPY and PNP were 3.7 mg/L (3.4 mg/g creatinine) and 5.0 mg/L (4.2 mg/g creatinine), respectively (Figure 15.4). Multivariate analyses showed that children aged 6–11 years had significantly higher concentrations of TCPY than adults and adolescents. Similarly, adolescents had signifi-cantly higher TCPY concentrations than adults. Although the concentrations between sexes and among composite racial or ethnic groups varied, no significant differences were observed.

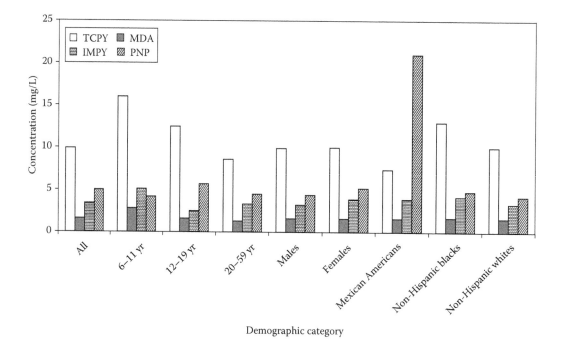

Demographic category

FIGURE 15.4

The 95th percentile concentration (95th PC) of selected metabolites of organophosphorus pesticides in urine samples collected from the US population. *Metabolite abbreviation corresponding to the parental pesticide*: TCPY, Chlorpyrifos; MDA, Malathion; IMPY, Diazinon; PNP, Parathion. (Adapted from Barr, D. B., Allen, R., Olsson, A. O., et al. 2005. Concentrations of selective metabolites of organophosphorus pesticides in the United States population. *Environ. Res.* 99: 314–326.)

Ninety-two Egyptian school children (46 girls and 46 boys), living in urban Giza, were studied for chronic exposure to OCP residues (Sherif et al. 2005). Varying levels of OCP residues were detected in the urine samples of 36 of the 92 studied children; constituting 39% of the total. The most frequently detected residues were p,p'-DDE (36.1%) and p,p'-DDT (22.2%). α-HCH and heptachlor epoxide were not detected in the urine samples.

Panuwet et al. (2009) evaluated the exposure to pesticides among secondary school students aged 12–13 years old in Chiang Mai Province, Thailand. A total of 207 urine samples were analyzed for 18 specific pesticide metabolites. The authors found 14 metabolites in the urine samples tested; 7 of them were detected with a frequency >17%. The most frequently detected metabolites were 2-[(dimethoxyphosphorothioyl)sulfanyl] succinic acid (MDA), PNP, TPCY (metabolite of chlorpyrifos), 2,4-dichlorophenoxyaceticacid (2,4-D), cis- and trans-3-(2,2-dichlorovinyl)-2,2-dimethylcyclopropane-1-carboxylicacids (c-DCCA and t-DCCA; metabolite of permethrin), and 3-phenoxybenzoicacid (3-PBA; metabolite of pyrethroids). The students were classified into four groups according to their parental occupations: farmers (N = 60), merchants and traders (N = 39), government and company employees (N = 52), and laborers (N = 56). Children of the farmers had significantly higher urinary concentrations of pyrethroid insecticide metabolites than did other children (p < 0.05). Similarly, children of the agricultural families had significantly higher pyrethroid metabolite concentrations (Figure 15.5). Males had significantly higher values of PNP; however, no other sex-related differences were observed. Because parental occupation and agricultural activities seemed to have little

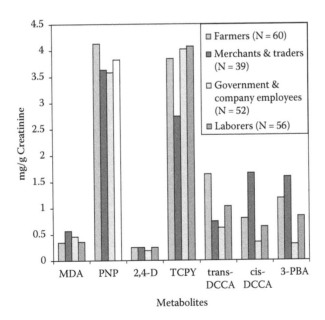

FIGURE 15.5
Major specific pesticide metabolites detected in urine samples of school students from northern Thailand as classified by parental occupation. *Metabolite abbreviation corresponding to the parental pesticide:* MDA, Malathion; PNP, Parathion; 2,4-D, 2,4-Dichlorophenol; TCPY, Chlorpyrifos; cis-DCCA and trans-DCCA, Cyfluthrin, Permethrin, Cypermethrin; 3-PBA, Most of the commercially available pyrethroids. (Adapted from Panuwet, P., Prapamontol, T., Chantara, S., and Barr, D. B. 2009. Urinary pesticide metabolites in school students from northern Thailand. *Int. J. Hyg. Environ. Health* 212: 288–297.)

influence on pesticide levels, dietary sources were the likely contributors to the metabolite levels observed.

15.3.3 Pesticide Residues in Seminal Fluid

There is increasing evidence suggesting that human fertility rates (FRs) are declining not only in Western countries but also in developing countries (Skakkebaek et al. 2006). Recent studies have associated various environmental pollutants including chlorinated pesticides and heavy metals with declining semen quality (Talamanca et al. 2001; Kumar et al. 2000a; Pant et al. 2003). Hauser et al. (2002) suggested that the presence of PCBs p,p′-DDE and phthalate esters in human seminal plasma might be responsible for the deterioration of semen quality without an obvious etiology.

In a recent study, Pant et al. (2007) explored the possibility of a correlation between the chlorinated pesticide level and the sperm count and motility in both fertile and infertile men. A total of 50 infertile men (30.68 ± 0.49 years old) plus another 50 fertile men (29.43 ± 3.80 years old) were included in the study. Based on the authors' findings, Table 15.4 summarizes the difference between the pesticide and semen contents in both the investigated groups. It appears that the infertile group had approximately five times more of the total HCH than the fertile group was, and approximately two times more of total DDT. The authors added that although conflicting results exist in the literature regarding the effect of chlorinated pesticides on semen quality, their positive effect should be considered as many of them have antiandrogenic or antiestrogenic properties and could therefore hypothetically inhibit normal negative feedback of hormones, leading to an increase in hormone production in the testes.

Comparisons of the FRs failed to detect significant differences among populations from three selected areas in Northeast Italy, while regression analysis showed a significant decrease in FR relative to the total amount of pesticides used in the studied areas (Clementi et al. 2008).

TABLE 15.4

Pattern of Pesticide Residues in Human Seminal Plasma of Fertile and Infertile Men from Lucknow, India

Pesticide	Mean Concentration (µg/L) in Semen from Fertile Men	Relative Concentration (× times) in Semen from Infertile Men
α-HCH	4.69	3.7
β-HCH	9.24	5.3
γ-HCH	2.44	8.4
δ-HCH	2.54	1.9
Total HCH	18.94	4.9
Aldrin	0.36	0.70
P,p′DDT	3.07	0.7
O,p′-DDT	0.12	24.3
P,p′-DDE	7.24	2.8
p,p′-DDD	13.14	1.6
Total DDT	23.32	1.9

Source: Adapted from Pant, N., Kumara, R., Mathur, N., Srivastava, S. P., Saxena, D. K., and Gujrati, V. R. 2007. Chlorinated pesticide concentration in semen of fertile and infertile men and correlation with sperm quality. *Environ. Toxicol. Pharmacol.* 23: 135–139.

15.3.4 Pesticide Residues in Adipose Tissues

As a matter of fact, OCP residues persist in adipose tissues over a period of many decades causing immune suppression and inhibition of various enzymes, leading to higher susceptibility to cancer (Amaral Mendes 2002).

Kutz et al. (1976) summarized the results of a large monitoring program conducted to determine the incidence and level of exposure to pesticides experienced by the general population of the United States and to identify changes and trends in these parameters when they occur. Selected OC residues were measured in human adipose tissues during the fiscal years 1970, 1971, and 1972. The program was limited to the collection and chemical analysis of adipose tissue samples that were obtained with the cooperation of pathologists in 32 cities (of >25,000 people) selected in the contiguous 48 states according to a statistical experimental design. The most detected OCP residues throughout the 3 years, in terms of general geometric mean and frequency, were total DDTs (7.6 ppm lipid basis; with 99.90% frequency), BHC (0.31 ppm lipid basis; with 99.23% frequency), dieldrin (0.19 ppm lipid basis; with 97.98% frequency), and heptachlor epoxide (0.09 ppm lipid basis; with 93.76% frequency).

Kang et al. (1997) determined HCHs, DDTs, HCBs, PCBs, polychlorinated dibenzo-p-dioxins, and dibenzofurans (PCDDs/DFs) in human adipose tissue samples collected from western Kyungnam, Korea. The residue levels of OC compounds were in the order of DDTs followed by PCBs > HCHs > HCBs > PCDDs/DFs. The mean concentrations of 2,3,7,8-TeCDD in male and female tissues were 2.8 and 1.7 pg/g on lipid weight bases, respectively. No significant difference was found in the residue levels of PCDDs/DFs between sexes. By contrast, PCBs and DDTs showed a significant difference between sexes. Unlike the age trend observed for HCHs, PCBs, and DDTs, PCDDs/DFs revealed a constant or rather decreasing pattern with increasing age. The authors reported that OC concentrations in human adipose tissues in western Kyungnam were generally much lower than those in other countries.

Concentrations of persistent OCPs such as DDTs, HCHs, chlordane compounds (CHLs), HCBs, and PCBs were determined in a wide variety of foodstuffs and human tissues collected from Shanghai and its vicinity in China in 2000/2001 (Nakata et al. 2002). Table 15.5

TABLE 15.5

Organochlorine Pesticide Residues (ng/g Lipid Weight) in Selected Foodstuffs Collected from Shanghai and Yixing cities, China, in 2001 and 2002

	Residues (ng/g Lipid Weight)			
Foodstuffs	**DDTs**[a]	**HCHs**[b]	**CHLs**[c]	**HCB**
Seafood	4726.0	6.8	7.2	3.0
Meat	2.5	4.7	0.2	1.6
Vegetables and Fruits	26.4	79.4	17.1	31.7
Tea Leaves	368.0	59.0	13.0	199.0
Milk[d]	111.5	26.0	0.5	9.0

Source: Data refer to Nakata, H., Kawazoe, M., Arizono, K., et al. 2002. Organochlorine pesticides and polychlorinated biphenyl residues in foodstuffs and human tissues from China: Status of contamination, historical trend, and human dietary exposure. *Arch. Environ. Contam. Toxicol.* 43: 473–480.

[a] Sum of p,p′-DDE, p,p′-DDD, and p,p′-DDT.
[b] Sum of α-, β-, and γ-HCH.
[c] Sum of trans- and cis-chlordanes and nonachlors and oxychlordane.
[d] Milk, normal and processed

outlines the general picture of these pollutants in different foodstuffs. Seafood contained the highest level of DDTs (4726 ng/g lipid weight), while the highest level of HCB (199 ng/g lipid weight) was found in tea leaves. Based on the authors' report, DDT and its metabolites were prominent compounds in most of the foodstuffs. In particular, mussels contained noticeable residues of DDTs (34,000 ng/g lipid weight), which are one to three orders greater than the reported levels in the bivalves in other Asian countries. Concentrations of HCHs, CHLs, HCB, and PCBs in foodstuffs were generally low, suggesting small amounts of inputs into the environment. Considering that foodstuffs are a main source of human exposure to contaminants, the greater concentrations of DDTs and HCHs in Chinese people might be due to extensive usage of these compounds as agricultural pesticides in the past. Continuous monitoring and epidemiological studies of OCPs in humans are warranted in China.

In Mexico, eighty human abdominal adipose tissue samples from Veracruz and 80 samples from Puebla were analyzed and the obtained results were compared among both populations (Waliszewski et al. 2010). The results from Veracruz showed higher contamination levels (mg/kg on lipid base) compared with Puebla: β-HCH: 0.072 versus 0.029; p,p'-DDE (dichlorodiphenyldichloroethylene): 2.364 versus 0.726; o,p'-DDT: 0.022 versus 0.025; p,p'-DDT: 0.192 versus 0.061; and $\sum$-DDT: 2.589 versus 0.806. The population of Veracruz and of Puebla divided by sex, origin, and cause of death presented no statistical differences. The comparison between sexes (women and men groups) in Veracruz and Puebla indicated significantly higher levels in Veracruz and resulted in statistically significant differences. Calculating possible risks (odds ratios, ORs), p,p'-DDE (OR = 5.04) and o,p'-DDT (OR = 2.93) revealed a significantly higher risk for the Veracruz population. The study indicated prolonged DDT exposure of Mexicans caused by its sanitary use in the past and the persistence of its residues in soil and air.

15.4 Pesticide Residues with Respect to Human Sociodemographic Characteristics

This issue has been partially demonstrated in Section 15.3. More data on age and race effects, ethnic differences, social classes, and geographical variations are given below.

Although most human exposure to OCPs can probably be linked to dietary contamination, other factors, such as age (Zumbado et al. 2005), gender (Lino and da Silveira 2006; Petrik et al. 2006), lifestyle (Deutch et al. 2003), and type of habitat (Zumbado et al. 2005; Lino and da Silveira 2006; Petrik et al. 2006), have been reported to be good predictors of OC levels in human populations.

The public health risks of pesticides depend not only on how toxic various compounds are, but also on how many people are exposed, what their risk-related demographic, socioeconomic, and health profiles are, what kinds of pesticides they are exposed to, and the extent and routes of exposure (WRI 1996). Four large groups are potentially exposed to pesticide hazards: (1) Farmers and farm workers; (2) Workers and laborers in pesticide factories; (3) Populations that live in areas of intensive pesticide use or production; and (4) Populations exposed to persistent pesticides that bioaccumulate in food. Groups within the general population that can be exposed to high levels of bioaccumulated pesticides include (1) Habitual consumers of fish, livestock, and dairy products; (2) Fetuses and nursing infants whose mothers' bodies have accumulated substantial levels of persistent pesticides; and (3) Sick people who metabolize pesticide-bioaccumulated fatty tissues while ill. The general population can be exposed to low levels of pesticides in three general ways: (1) Vector control for public health and other nonagricultural purposes; (2) Environmental residues; and (3) Food residues (WRI 1996).

Karalliedde et al. (2003) have reviewed the literature to identify the variables that need to be considered following exposure to OPs in humans. These include factors related to the OPs (physicochemical properties, solvents, and impurities), duration and routes of exposure, and factors related to the individual(s) exposed. Individual variables include variations in metabolic, sequestration, and excretory processes and health status (age, gender, environmental factors, concurrent medications, and cholinergic status). Exposure to xenobiotics through the environment, occupation, or following therapy is an unavoidable aspect of modern life. The assessment of ill-health following exposure is critical to the development and compliance with guidelines and to the adoption of the best instrumentation.

It was reported that occupational pesticide poisoning among children less than 18 years of age is roughly 10%–20% of all poisonings (Henao et al. 1993). In the Philippines, children accounted for nearly one in eight poisonings (Rola 1989). Similar conditions were found in other developing countries (WRI 1996). In California, children of migrant farm workers living near sprayed fields experienced depressed cholinesterase activity and symptoms of acute pesticide exposure. Of the children who did field work, 40% had abnormally low cholinesterase levels due to exposure to pesticide residues in drift (Richter 1992). Six million children live in poverty in America's inner cities. These children are at a high risk of exposure to pesticides that are used in urban schools, homes, and day-care centers for control of cockroaches, rats, and other vermin. There is growing evidence that indoor pesticide exposure is of considerable magnitude in the United States and that pesticide concentrations may be higher in urban than in rural areas. Of particular concern is exposure of pregnant women and their fetuses to household pesticides (Berkowitz et al. 2003).

Desi et al. (1998) carried out epidemiological examinations on plant protection Hungarian workers to find early symptoms (biomarkers) of moderate contamination by pesticides. Measurement of the changes in cholinesterase and γ-glutamyltransferase enzyme activity did not reveal that the people were affected. Determination of chromosome abnormalities from peripheral lymphocytes disclosed numerical aberrations at an early stage and lasted after cessation of work for several months. There were differences in the results among different groups of plant protection workers, as well as among those working in open field or in glasshouses. Women, who carried out horticultural work in glasshouses after spraying showed some chromosome abnormalities too.

Adipose tissues of 50 Jordanian human patients (5–96 years old) were taken and analyzed by gas chromatography equipped with electron capture detector (^{63}Ni) for the determination of storage levels of HCH isomers, DDTs and their metabolites, and cyclodiene compounds. The data are reported according to age groups and sexes (Figure 15.6). The results showed relatively moderate concentrations of all studied pesticides and that Jordanian men are more exposed to DDTs and HCHs than women (Alawi et al. 1999).

Cruz et al. (2003) measured OCP residues in human serum from an urban and two rural populations in Portugal in an attempt to evaluate the contamination levels of Portuguese rural and urban populations and to establish a relation with sex and age of the individuals. Except total HCH, the other measured OCs were found at higher concentration levels in urban population than in rural population. Total DDT recorded was 93.5 µg/L in urban population compared with 43.9 in rural population. Mean total DDT levels were always

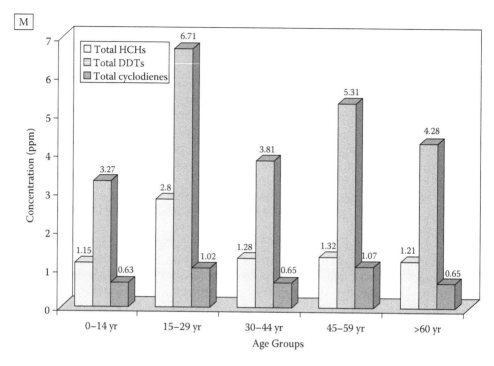

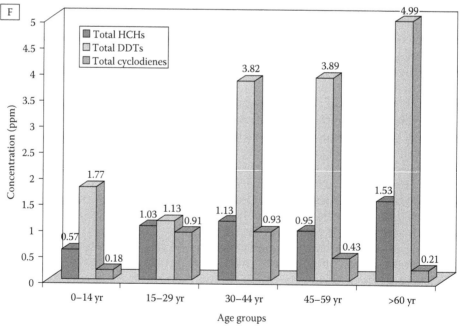

FIGURE 15.6

Concentration of organochlorine pesticide residues in Jordanian adipose tissue samples from male (M) and female (F) subjects. Total HCHs (α-, β-, and γ-HCH); Total DDTs (o,p'-DDT, p,p'-DDT, o,p'-DDE, p,p'-DDE, o,p'-DDD, and p,p'-DDD). Total cyclodienes (heptachlor, aldrin, heptachlor epoxide, dieldrin, and endrin). (Adapted from Alawi, M. A., Tamimi, S., and Jaghabir, M. 1999. Storage of organochlorine pesticides in human adipose tissues of Jordanian males and females. *Chemosphere* 38(12): 2865–2873.)

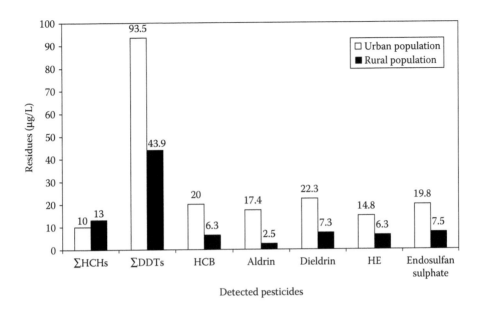

FIGURE 15.7
Mean concentration (μg/L) of organochlorine pesticide residues in human serum samples from urban and rural populations in Portugal. ∑HCH includes α-, β-, and γ-isomers. ∑DDT includes p,p'-DDE, p,p'-DDD, o,p'-DDT, and p,p'-DDT. HE, Heptachlor epoxide. (Adapted from Cruz, S., Lino, C., and Silveira, M. I. 2003. Evaluation of organochlorine pesticide residues in human serum from an urban and two rural populations in Portugal. *Sci. Total Environ.* 317: 23–35.)

higher than mean total HCH levels, and HCH levels were slightly higher in rural population, while all the other estimated pesticide residues were higher in urban than rural population (Figure 15.7). This situation can be due to higher consumption of animal-origin foods and dairy products by urban population (DeVoto et al. 1998). According to Cruz et al. (2003), Portugal imports animal- and vegetable-origin foods from Asia and Latin America where DDTs are still in use. Urban populations preferentially consume these imported products, once they have higher accessibility to this kind of products, while rural populations tend to consume the products they grow.

Focusing on the data illustrated in Figure 15.8, it appears that the levels of total HCH and total DDT tend to be higher in men than in women either in urban or rural Portuguese population. Other studies, in Germany, Japan, and Senegambian region, also reported higher levels of HCH isomers in males (DeVoto et al. 1998; Hanaoka et al. 2002). Studies conducted in the United States, Belgium, and Egypt (Ahlborg et al. 1995; Charlier and Plomteux 2002; Soliman et al. 1979) have also reported mean total DDT levels in men higher than those in women. According to Cruz et al. (2003), such results could suggest that lactation is a route of excretion, as there are studies that indicate the transfer of OC residues to breast milk (Waliszewski et al. 1999b). Even OP-toxic residues, such as chlorpyrifos, have shown to be transferred through lactation to suckling rats (Mansour and Mossa 2010).

Previous reports established a relationship between age and OC residue levels (Stehr-Green et al. 1988; Soliman et al. 1979; Voorspoels et al. 2002). Figures 15.9 and 15.10 illustrate the total concentration levels of HCHs and DDTs in urban and rural populations in Portugal (Cruz et al. 2003). The latter investigators reported that younger groups have higher levels of detection of the analyzed OCP residues, causing suspicion that exposure to pesticide residues continues to happen.

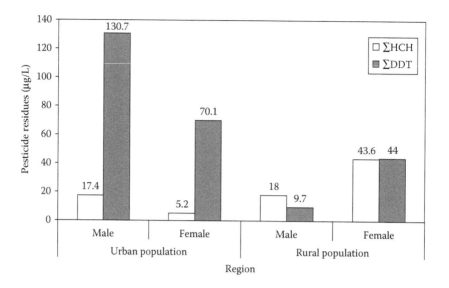

FIGURE 15.8
Mean concentration (μg/L) of organochlorine pesticide residues in human serum samples of both sexes from urban and rural populations in Portugal. $\sum$HCH includes α-, β-, and γ-isomers. $\sum$DDT includes p,p'-DDE, p,p'-DDD, o,p'-DDT, and p,p'-DDT. (Adapted from Cruz, S., Lino, C., and Silveira, M. I. 2003. Evaluation of organochlorine pesticide residues in human serum from an urban and two rural populations in Portugal. *Sci. Total Environ.* 317: 23–35.)

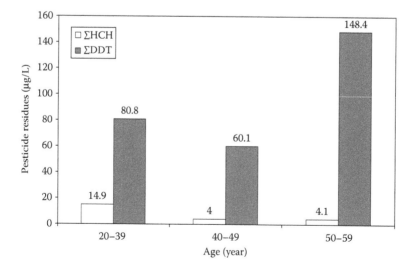

FIGURE 15.9
Mean concentration (μg/L) of organochlorine pesticide residues detected in human serum samples of different age groups from an urban population in Coimbra, Portugal. $\sum$HCH includes α-, β-, and γ-isomers. $\sum$DDT includes p,p'-DDE, p,p'-DDD, o,p'-DDT, and p,p'-DDT. (Adapted from Cruz, S., Lino, C., and Silveira, M. I. 2003. Evaluation of organochlorine pesticide residues in human serum from an urban and two rural populations in Portugal. *Sci. Total Environ.* 317: 23–35.)

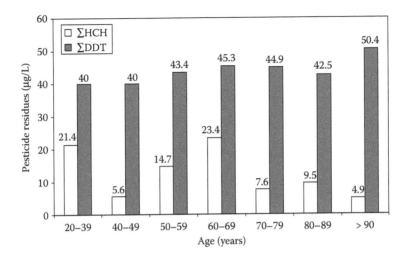

FIGURE 15.10
Mean concentration (μg/L) of organochlorine pesticide residues detected in human serum samples of different age groups from a rural population in Verride, Portugal. $\sum$HCH includes α-, β-, and γ-isomers. $\sum$DDT includes p,p'-DDE, p,p'-DDD, o,p'-DDT, and p,p'-DDT. (Adapted from Cruz, S., Lino, C., and Silveira, M. I. 2003. Evaluation of organochlorine pesticide residues in human serum from an urban and two rural populations in Portugal. *Sci. Total Environ.* 317: 23–35.)

Variation between different countries in the levels of human serum OCP residues is illustrated in Figure 15.11, based on the results of the previous investigators (e.g., Krawinkel et al. 1989; Leoni et al. 1989; Guardino et al. 1996; DeVoto et al. 1998; Stellman et al. 1998; Waliszewski et al. 1999a; Wicklund Glynn et al. 2000; Dua et al. 2001; Charlier and Plomteux 2002; Hanaoka et al. 2002; Cruz et al. 2003). It appears that India and Sweden occupy high ranks regarding contamination levels of OCP residues in human serum.

In a recent study, Lino and da Silveira (2006) monitored OCP residues in serum samples collected from Portuguese students. Concentrations of selected OC pollutants (α-, β-, and γ-HCH, aldrin, dieldrin, HE, HCB, p,p'-DDT, o,p'-DDT, p,p'-DDE, p,p'-DDD) and endosulfan sulfate were measured in the serum samples collected from 160 students. Endosulfan sulfate, p,p'-DDE, o,p'-DDT, and p,p'-DDD were the most frequently identified residues. Mean total DDT levels were higher than mean total HCH levels (Figures 15.12 and 15.13). For $\sum$DDT, the highest levels were found among females (Figure 15.12) and in urban samples (Figure 15.13). An opposite situation was found in $\sum$HCH: males presented higher levels than females. The authors mentioned that the mean concentration of OCP residues, present in the student populations, showed that it is among the highest levels of contamination, when compared with others from Europe, Asia, and America.

In 1998, one of the largest determinations of OCPs in a representative sample of a Spanish population (682 serum samples from the Canary Islands) was made in the context of the Canary Islands Nutrition Survey (ENCA) (Zumbado et al. 2005). The objective was to point out the differences in pesticide contamination between islands, and along with this, to analyze if a connection could be established with gender, age, or habitat of the subjects. Concentrations of selected persistent OC pollutants (p,p'-DDT, o,p'-DDT, p,p'-DDE, o,p'-DDE, p,p'-DDD, and o,p'-DDD) were measured in the collected serum samples. Almost all of the samples (99.3%) presented detectable levels of some DDT derivatives,

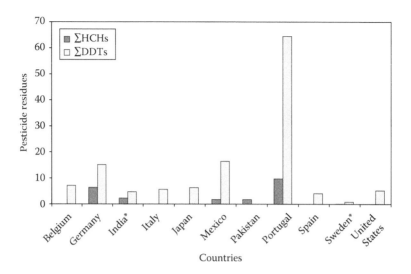

FIGURE 15.11

Comparison of organochlorine pesticide residue mean levels in blood serum samples of different countries. *Note*: *Values of India and Sweden are in mg/kg; multiply by 1000 to change into μg/L (or μg/kg) reported for the other compared countries. (Data with respect to countries and investigators are: Belgium: Charlier and Plomteux (2002); mean value for data on men and women; no available data for HCHs. Germany: DeVoto et al. (1998), estimation in whole blood. India: Dua et al. (2001), mean of results of three different intervals (e.g., March 2.14; July 3.12, and November 1.32 mg/L for ΣHCH, and 2.73, 6.92, and 4.59 mg/kg for ΣDDT, respectively) for the above-mentioned months. Italy: Leoni et al. (1989), estimation conducted on women serum samples; no available data for HCHs. Japan: Hanaoka et al. (2002); no available data for HCHs. Mexico: Waliszewski et al. (1999a). Pakistan: Krawinkel et al. (1989), no available data for DDTs. Portugal: Cruz et al. (2003), mean values for population from one urban and two rural regions. Spain: Guardino et al. (1996), estimations in whole blood, no available data for HCHs. Sweden: Wicklund Glynn et al. (2000). United States: Stellman et al. (1998); excluding p,p′-DDD from measured isomers of DDT; no available data for HCHs.)

with p,p′-DDE being the most frequently detected OC. Based on the authors' data, the median concentration of total DDT body burden present in the Canary Islands (370 ng/g fat) was similar to that found in other European countries, although it was noteworthy that a fourth of the population showed a total DDT body burden higher than 715 ng/g fat. Interestingly, statistically significant differences were found in the serum levels of OC pesticides between islands, these levels being higher in people from Tenerife and Gran Canaria (415 and 612 ng/g fat, respectively; Figure 15.14a), the islands that represent both the highest population and the highest surface devoted to intensive agriculture. Samples from La Palma showed the highest concentration level of p,p′-DDE (ca. 140 ng/g fat, Figure 15.14b). As expected, the serum levels of both total DDT body burden (Figure 15.14a) and p,p′-DDE (Figure 15.14b) increased with age, and urban inhabitants showed the highest levels of total DDT body burden (542 ng/g fat, Figure 15.14a) compared with ca.120 ng/g fat of the metabolite p,p′-DDE (Figure 15.14b). Statistically significant differences were also found in relation to gender, with women showing higher levels of these OCPs than men (Figure 15.14a and 15.14b). Most studies showed an association between p,p′-DDE levels and age (e.g., Voorspoels et al. 2002). DDT and its derivatives are considered as xenoestrogens, and they have been linked to breast cancer. Zumbado et al. (2005) stated that the Canary Islands present high incidence of and mortality from breast cancer, a matter which points to the possibility that these environmental contaminants may be playing a determining role in this respect.

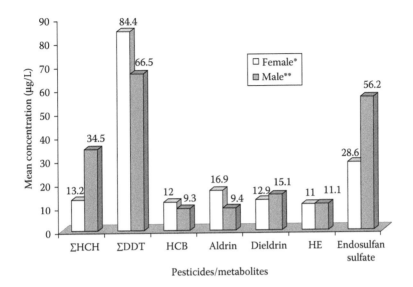

FIGURE 15.12
Mean concentration (μg/L) of OCP residues in human serum samples from Portuguese students by gender.
*Number of females = 80 and **number of males = 80. ∑HCH (α-, β-, and γ-HCH. ∑DDT (p,p'-DDE, p,p'-DDD,
o,p'-DDT, and p,p'-DDT). HE, Heptachlor epoxide. (Adapted from Lino, C. M. and da Silveira, M. I. N. 2006.
Evaluation of organochlorine pesticides in serum from students in Coimbra, Portugal: 1997–2001. *Environ. Res.*
102: 339–351.)

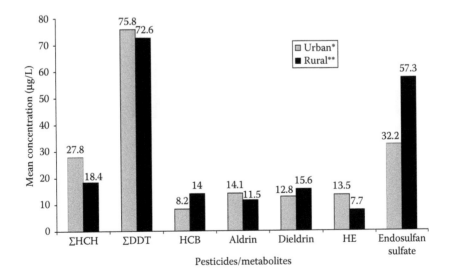

FIGURE 15.13
Mean concentration (μg/L) of OCP residues in human serum samples from Portuguese students by residence
area. *Number of females = 93 and **number of males = 67. ∑HCH (α-, β-, and γ-HCH), ∑DDT (p,p'-DDE, p,p'-
DDD, o,p'-DDT, and p,p'-DDT). HE, Heptachlor epoxide. (Adapted from Lino, C. M. and da Silveira, M. I. N.
2006. Evaluation of organochlorine pesticides in serum from students in Coimbra, Portugal: 1997–2001. *Environ.
Res.* 102: 339–351.)

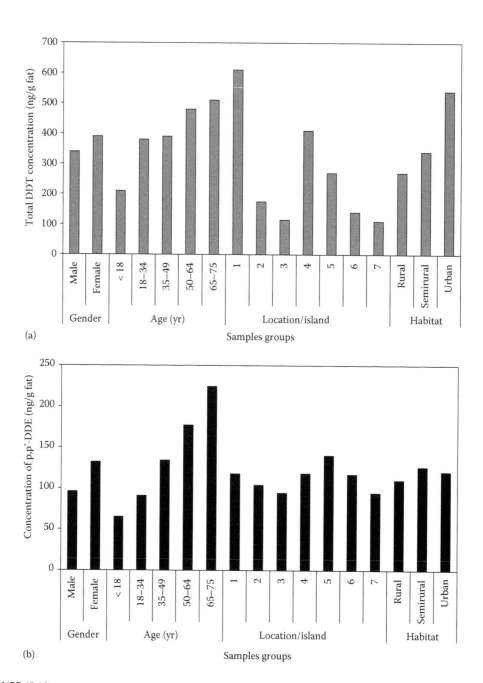

FIGURE 15.14

Total DDT (a) and p,p'-DDE (b) concentrations (ng/g fat) in human serum samples from the Canary Islands (Spain) as affected by sociodemographic and geographical factors (e.g., gender, age, habitat). *Note:* Total DDT includes six derivatives (o,p'-DDE, p,p'-DDE, o,p'-DDD, p,p'-DDD, o,p'-DDT, and p,p'-DDT). Total number of serum samples = 682. Locations are Islands of Gran Canaria (1), Lanzarote (2), Fuerteventura (3), Tenerife (4), La Palma (5), La Gomera (6); and El Hierro (7). (Data adapted from Zumbado, M., Goethals, M., Álvarez-León, E. E., et al. 2005. Inadvertent exposure to organochlorine pesticides DDT and derivatives in people from the Canary Islands (Spain). *Sci. Total Environ.* 339: 49–62.)

Carreño et al. (2007) monitored the levels of 14 OCPs in the blood of 220 young males (mean age: 20.75 years) in Southern Spain. Aldrin, dieldrin, endrin, lindane, methoxychlor, endosulfans, and DDT and its metabolites were identified. Detectable concentrations of p,p'-DDE were found in 96% of serum samples. Among the remaining DDTs, o,p'-DDD was the most prevalent, detected in 65% of serum samples. Detectable concentrations of endosulfan I or II or their metabolites endosulfan diol or endosulfan sulfate were found in all the samples; endosulfan diol was the most frequently detected metabolite (92%), followed by endosulfan sulfate. Results indicate that men of reproductive age in Southern Spain have been and are exposed to OCPs. HCB, dieldrin, endosulfan sulfate, and p,p'-DDT levels were associated with the mother's employment in agriculture during pregnancy. On the other hand, the ratio of p,p'-DDT to p,p'-DDE, which may indicate whether the exposure was in the distant (low ratio) or recent (high ratio) past, was estimated at 0.7 in their study. This ratio suggests the influence of the restriction and prohibition of DDT and the decrease in exposure to these compounds over the past few decades.

Also in Southern Spain, another study (Lopez-Espinosa et al. 2008) was conducted to assess the presence of 16 OCP residues in 52 fat samples collected from boys with a mean age of 7 years (0–15 years). No pesticide was found in more than 50% of the samples, except for p,p'-DDE (79% of the samples; median, 710 ng/g lipid). After this compound, the most frequent pesticides were o,p'-DDT (17%; median, 330 ng/g lipid) and o,p'-DDD (15%; median, 1510 ng/g lipid). No statistically significant association was found between p,p'-DDE or ∑DDTs and the birth year, birth weight, gestational age, infant feeding history or the age, weight, height, or Quetelet Index at the time of sampling. The lack of correlation between the presence of the main metabolite, p,p'-DDE, and that of the parent compounds, o,p'-DDT and p,p'-DDT, suggests that children were exposed mainly to the metabolite rather than to the commercial pesticide, which was banned 30 years ago. By contrast, among the currently used OCs, endosulfan was positively correlated with the presence of its metabolites, suggesting exposure to the commercial products. In this respect, it may be useful to mention that children can be exposed to OCPs in utero via the placenta (Lopez-Espinosa et al. 2007) and after birth via lactation (Solomon and Weiss 2002) and infant formula (Cressey and Vannoort 2003). Diet is an important potential route of pesticide exposure in older children (Patandin et al. 1999). Children are especially vulnerable to pesticide exposure because they consume a larger amount of food and water relative to their body weight as compared with adults. Also, the types of food consumed by children differ from those consumed by adults, that is, children's diet contains more milk products and more fruits and vegetables (Bearer 1995), which are known sources of exposure to fat-soluble or persistent and less persistent pesticides, respectively. Infants may also bear additional exposure risks because of their tendency to explore their environment with their mouths, their closeness to the ground, and the longer time they spend playing outdoors (Garry 2004; Bearer 1995). Finally, the developing organ systems of children are more sensitive and their bodies have a limited ability to detoxify pesticides (Bearer 1995).

Table 15.6 shows data of OCP residues in the blood of 220 young males aged 18–23 years (Carreño et al. 2007) and in the adipose tissues of 52 boys aged 0–15 years. (Lopez-Espinosa et al. 2008).

Porta et al. (2008) reviewed the adverse effects of POPs on human health and the impact of policies aiming to reduce human exposure to POPs, which warrant monitoring the body concentrations of POPs in representative samples of the subjects. While numerous *ad hoc* studies are being conducted to understand the effects of POPs, only a few countries conduct nationwide surveillance programs on human concentrations of POPs, and even lesser

TABLE 15.6

Residues of Organochlorine Pesticides in Serum from Young Males (N = 220) and Adipose Tissues from Children (N = 52) Living in Southern Spain

Pesticide/Metabolite	Blood Serum[a]		Adipose Tissues[b]	
	Mean Concentration (ng/mL)	Frequency (%)	Mean Concentration (ng/g Lipid)	Frequency (%)
∑DDT	12.77	99.1	4180	94.0
∑Endosulfans	25.76	100.0	239.0	34.0
Aldrin	3.75	79.0	310.0	12
Endrin	5.04	60.70	<LOD	<LOD
Dieldrin	1.85	40.7	130	8.0
Lindane	1.84	64.7	11.0	12.0
Methoxychlor	2.84	60.7	16.0	6.0
Hexachlorobenzene	3.88	79.9	—	—
Mirex	—	—	38.0	10.0
Chlordane	—	—	490.0	14.0

Source: [a]Adapted from Carreño, J., Rivas, A., Granada, A., and Lopez-Espinosa, M. J. 2007. Exposure of young men to organochlorine pesticides in Southern Spain. *Environ. Res.* 103: 55–61; and [b]Adapted from Lopez-Espinosa, M. J., Lopez-Navarrete, E., Rivas, A., et al. 2008. Organochlorine pesticide exposure in children living in southern Spain. *Environ. Res.* 106: 1–6. ∑DDT, (o,p'-DDT, p,p'-DDT, o,p'-DDD, p,p'-DDE). ∑Endosulfans, (endosulfan I, endosulfan II, endosulfan diol, endosulfan sulfate). LOD, limit of detection. –, no data.

number of countries do so in representative samples of the general population. In this context, Mansour (2009) reviewed the current status of POPs in Africa with respect to their environmental distribution, stockpiles of some pesticidal POPs, and action plans of their elimination in some African countries.

Recently, Porta et al. (2010) evaluated the distribution of serum concentrations of 19 POPs and their main predictors in a representative sample (N = 919) of the general population (18–74 years) of Catalonia, Spain. Through multivariate statistical models, the authors investigated the influence of sex, age, body mass index (BMI), socioeconomic status, and gender on serum POP concentrations. The following compounds were detected in more than 85% of the subjects: p,p'-DDT, p,p'-DDE, PCB congeners (118, 138, 153 and 180), HCB, and β-HCH. p,p'-DDE, HCB, and β-HCH showed the highest concentrations (median = 399.3, 159.4, and 91.9 ng/g lipid, respectively; Table 15.7). Distributions were highly skewed and

TABLE 15.7

Concentration (ng/g Lipid) of POPs in Serum Samples from Catalan (Spain) Population (N = 919 Participants)

Parameter	p,p'-DDT	p,p'-DDE	PCB 118	PCB 138	PCB 153	PCB 180	HCB	β-HCH
Mean concentration[a]	29.3	399.3	22.8	69.5	100.1	77.3	159.4	91.9
Percentage of detection	88.2	100.0	88.6	98.2	99.2	100.0	97.9	96.9

Source: Adapted from Porta, M., Gasull, M., Puigdomènech, E., et al. 2010. Distribution of blood concentrations of persistent organic pollutants in a representative sample of the population of Catalonia. *Environ. Int.* 36: 655–664.

[a] Concentrations are expressed as mean values for all participants.

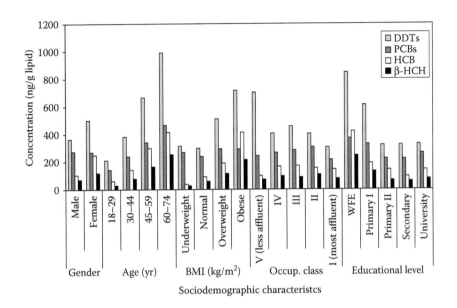

FIGURE 15.15
Serum concentration of POPs in Catalan (Spain) population by sociodemographic characteristics. *Note:* DDTs = (p,p'-DDT and p,p'-DDE). PCBs = (PCB congeners No. 118, 138, 153, and 180). BMI, body mass index. Occupational classes I, II, and III are nonmanual workers, and classes IV and V are manual workers. Educational level: WFE, without formal education; Primary I and II: first and second primary schooling stages. (Data refer to Porta, M., Gasull, M., Puigdomènech, E., et al. 2010. Distribution of blood concentrations of persistent organic pollutants in a representative sample of the population of Catalonia. *Environ. Int.* 36: 655–664.)

interindividual differences were up to 7700-fold. POP levels differed significantly by gender, age, BMI, educational level, and parity (Figure 15.15). The authors concluded that exposure to POPs remains common, a vast majority of the population has much lower blood concentrations than a relative minority, and the population distributions of POP are hence highly skewed to the right. Shifting distributions toward lower concentrations requires more energetic policies and population strategies.

The blood serum of the cacao farmers and their domestic water sources were analyzed for insecticide residues in selected cacao growing communities of Southwestern Nigeria (Sosan et al. 2008). The farmers were grouped into five exposure periods based on their years of involvement in insecticide application: <5 years, 5–9 years, 10–14 years, 15–19 years, and >20 years. The residue analyses revealed that 42 out of the 76 farmers had residues of diazinon, endosulfan, propoxur, and lindane in their blood; and 47.6% out of these farmers belonged to the >20-year exposure period. About 34% of the farmers had diazinon with a mean concentration of 0.067 mg/kg, 29% endosulfan (mean = 0.033 mg/kg), 23% propoxur (mean = 0.095 mg/kg), and 17% lindane (mean = 0.080 mg/kg) in their blood (Figure 15.16). The residues of lindane, endosulfan, and propoxur in all the exposure duration categories were found to be far below the no observable adverse effect level (NOAEL), while diazinon residues detected in the blood serum of the farmers in all the exposure duration categories exceeded the NOAEL of 0.02 mg/kg for the insecticide. The authors concluded that cacao farmers in Southwestern Nigeria may have been occupationally exposed due to insecticide application for mired control in their cacao plantations; and the exposure at times was of such magnitude so as to be hazardous to the farmers and their respective communities.

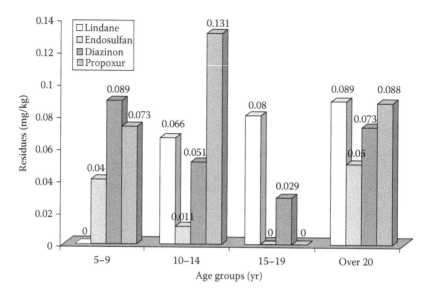

FIGURE 15.16
Mean concentration of insecticide residues in the blood in a number of selected cacao farmers in Southwestern Nigeria, based on their years of exposure. (Adapted from Sosan, M. B., Akingbohungbe, A. E., Ojo, I. A. O., and Durosinmi, M. A. 2008. Insecticide residues in the blood serum and domestic water source of cacao farmers in Southwestern Nigeria. *Chemosphere* 72: 781–784.)

15.5 Pesticide Epidemiology and Health Hazards of Occupational Exposure

Problems associated with pesticide hazards to man and to the environment are not confined to the developing countries. Developed nations have already suffered these problems and are still facing some problems in certain locations. For many reasons, the severity of pesticide hazards is much pronounced in the third world countries. A number of long-persistent OCs and highly toxic organophosphates, which have been banned or severely restricted, are still marketed and used in many developing countries (Mansour 2008).

The general population could be exposed to low levels of pesticides, mainly through food and environmental residues in air and water. In addition to this, workers dealing occupationally with pesticides are much exposed to these substances through agriculture and household practices, as well as formulation, packaging, distribution, and selling (WRI 1996). Pesticides' epidemiological data and some reports of occupational health hazards in selected countries (e.g., Egypt, India, and Pakistan) are briefly described next.

15.5.1 Egypt

For about 25 years, the use of DDT and many other OCPs in Egyptian agriculture has been banned. However, these long-persistent compounds are still detectable in many different types of environmental samples (e.g., water, fish, sediment, vegetables, fruits, milk, foodstuffs, etc.).

Epidemiological investigations increasingly address pesticides and their potential association with human diseases. This increased concern for human toxicity potential addresses various levels of exposure (high, medium, low, absent) through various routes of exposure, either directly or indirectly (e.g., food, air, water, soil). Therefore, the availability of data on pesticide consumption and use patterns, levels of pesticide residues in different environmental components, health risks associated with occupational exposure to pesticides in workplaces, and hospital discharge data of pesticide poisoning incidents is needed (Hodgson and Levi 1996).

Human exposure to pesticides in Egypt may be excessive, especially through ground application of insecticides in cotton fields, where compounds of high toxicity are often used. A very large number of workers, laborers, and spraying supervisors are involved in the spraying of cotton fields three to five times each season. Unfortunately, they are often not equipped with protective clothes or masks. Moreover, over a period of 40 days in hot climate, many thousands of children are engaged in collecting egg masses of the cotton leafworm, *Spodoptera littoralis*, in the cotton fields in Egypt. Furthermore, they help spray pesticides in cotton plantations three to five times in the season. Some of these children are under 15 years of age (Mansour 2004). Similarly, in Costa Rica, one in five individuals who handle pesticides is under 18 years of age (Wesseling et al. 1997).

Furthermore, the large numbers of workers involved in pesticide formulation plants and in greenhouse cultivation have to be also considered. Therefore, this may reflect the great magnitude of human exposure to pesticide intoxication in Egypt. Official reports on pesticide poisoning are lacking, except for some drastic cases released to the public and a few reports published by scientists. The only available data to us are those published by El-Gamal (1983), who reported cases of poisoning and a number of fatalities each year from 1966 to 1982 (Figure 15.17). He stated that after the year 1977, accidents due to acute intoxication had decreased because of the introduction of some preventive measures. But, he added that over 60% of the workers engaged in pesticide applications suffer from chronic toxicity.

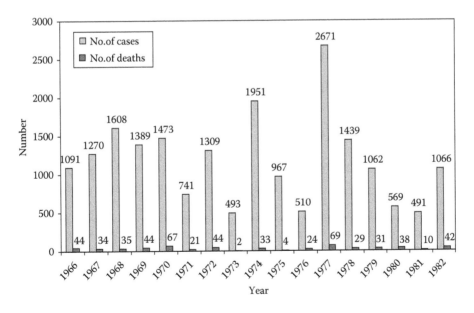

FIGURE 15.17
Incidences of pesticide poisoning in Egypt during the period 1966–1982. (Adapted from El-Gamal, A. 1983. Persistence of some pesticides in semi-arid conditions in Egypt. *Proceedings, International Conference on Environmental Hazards in Agrochemistry, Alexandria, Egypt,* 8–12 November 1983, vol. I, pp. 54–75.)

Recently, Mansour (2010) analyzed the pattern of intoxication based on the data obtained from different poison control centers (PCCs) in Egypt. He estimated the total acute intoxication to be as much as 48,000 cases annually, including about 7200 cases (~15%) of pesticide poisoning. This means that acute poisoning at present accounts for 61 persons/100,000 capita. Contribution of acute pesticide poisoning (APP), mainly by cholinesterase inhibitors, represents approximately 9.2/100,000 in the general population and doubles if related to the country's workforce dependent on the agricultural subsector. However, the author stated that the above-mentioned data does not reflect the overall figure of poisoning in Egypt, since the data of the other poison control services are not included due to unavailability and/or underreporting. The reasons for underreporting are many. In some areas, people suffering from acute poisoning may lack access to medical care and may not even report the illness to the medical system (WHO 2004). Studies in developed countries have demonstrated the annual incidence rate of APP in agricultural workers to be as much as 18.2/100,000 persons (Calvert et al. 2004). For many reasons, the incidences are expected to be higher in developing countries.

A number of studies on workers occupationally exposed to pesticides in Egypt were carried out by many investigators. The hazard of exposure to the OP insecticide phosfolan was estimated in terms of the amount of the insecticide retained on the workmen's body pads during field spraying and the acetylcholinesterase (AChE) inhibition in the red blood cells (Soliman et al. 1979). The authors found that the toxic dose received every spraying day by each worker varied with the type of job. The body of the mixer received the maximum exposure with 10–12-fold that of the assistants. The highly exposed group of workers suffered from 31% to 44% AChE inhibition in RBCs. About half of the inhibited enzyme activity recovered after 48h, and then it took more than 3–4 weeks to reach complete recovery.

Two spray men occupationally working in public health in Alexandria city, Egypt, experienced acute toxicity resulting from exposure to diazinon (60% EC). AChE activity showed a marked reduction up to 18 days after exposure, and then the enzyme activity was recovered 20–28 days after the poisoning incident. The diazinon that was stored in tin-plated sheet steel containers was found to be completely converted into diazinon transformation products, which are much more toxic than diazinon (Soliman et al. 1982).

A study was conducted to evaluate the impact on the health of workers exposed to pesticides in large- and small-scale Egyptian formulation plants (Amr and Halim 1997). Dermatitis and neuropsychiatry manifestations were the most prevalent health effects in this exposed population, compared with the controls, especially in workers with a longer duration of employment.

In a study at Sharkia Governorate, a total of 150 workers occupationally exposed to pesticides and 50 control subjects were subjected to clinical and dermatological examinations, patch tests, tests of liver and renal functions, and complete blood count, blood sugar, and urine analysis. Activities of the antioxidant enzymes superoxide dismutase, glutathione peroxidase, and glutathione reductase were also evaluated (Amer et al. 2002). Dermatological findings were positive in 78%, 76%, and 54% of the workers exposed to OP, pyrethroid, and Carbamate (CM) pesticides, respectively. The patch test was positive in 70% of the workers exposed to pyrethroids and 64% exposed to CM pesticides. Liver enzyme levels were generally increased in the workers, while the antioxidant enzyme activity was significantly decreased in all the workers compared with the controls.

In a study at Kafr El-Sheikh Governorate, Egypt, involving 240 individuals, pesticide applicators (PAs) had greater reduction of semen quality compared with nonfarm workers (NFWs). Also, biochemical markers (e.g., uric acid, urea, creatinine, and aspartate aminotransferase (AST)) in PAs were near the upper limits of the normal range (Attia 2005).

Farm workers occupationally exposed to OP pesticides through their application to cotton fields in Menoufiya Governorate, Egypt, were examined (Farahat et al. 2007). Results indicated that exposed individuals exhibited significantly lower performance than did nonexposed controls on six neurological tests (similarities, digital symbol, trail making part A and B, letter cancellation, digital span, and Benton visual retention). Serum AChE was significantly lower in exposed (87.34 U/mL) than control (108.25 U/mL) participants, and longer duration of work with pesticides was associated with a lower AChE level.

15.5.2 India

According to Gupta (2004), there have been a number of outbreaks of accidental poisoning by pesticides, in India, that deserve special mention. The first report of poisoning due to pesticides was from Kerala in 1958, where over 100 people died after consuming wheat flour contaminated with parathion (Karunakaran 1958). In the same year, poisoning in Kerala caused deaths of 102 people. This was mainly due to careless handling and storage of wheat. Subsequently, several cases of human and animal poisonings, in addition to deaths of birds and fishes, have been reported. Out of the 35 cases of malathion (diazole) poisoning reported during 1967–1968, five died. Electrocardiography (ECG) changes were recorded in all the cases. Autopsy and histopathological studies revealed damage to the myocardium (Sethuraman 1977).

In another report from Madhya Pradesh, 12 humans who consumed wheat contaminated with aldrin dust and Gammexane for 6–12 months developed symptoms of poisoning, which consisted of myoclonic jerks, generalized colonic convulsions, and weakness in the extremities (Gupta 1975). In another outbreak in 1977, eight cases of grand mal seizers were reported from a village of Uttar Pradesh, following accidental ingestion of HCH-contaminated wheat (Nag et al. 1977).

Epidemiological studies conducted on occupational pesticide workers have shown that exposure to dibromochloropropane (DBCP), chlordecone, endosulfan, ethyldibromide, DDT, carbaryl, and ethylmethiniphos causes abortion, still births, defects in offspring, male infertility, neonatal deaths, congenital defects, testicular dysfunction, and abnormalities (Kumar et al. 2000b).

In India, cotton field workers exposed to various pesticides had an increased risk of male infertility and male-mediated adverse reproductive outcomes such as abortions, stillbirths, neonatal deaths, and congenital defects (Rupa et al. 1991). DDT and its metabolites, HCH and its isomers, and PCBs have been detected in the seminal fluid, cervical mucus, and follicular fluid (Talamanca et al. 2001). In another study, an association between the serum PCBs and p,p'-DDE and abnormal sperm motility, sperm count, and sperm morphology was established in populations without specific exposure to these chemicals (Hauser et al. 2002). An *in vitro* study has shown that lindane depolarizes the human sperm membrane and inhibits the sperm cytological responsiveness to progesterone, the physiological agonist that stimulates the onset of acrosome reaction at the site of fertilization (Silvestroni and Palleschi 1999).

Studies on 356 workers in four units manufacturing HCH in India revealed neurological symptoms (21%) with significant increase in liver enzymes, which were related to the intensity of exposure (Nigam et al. 1993). Observations confined to health surveillance in male formulators engaged in the production of dust and liquid formulations of various pesticides (malathion, methyl parathion, DDT, and lindane) revealed several types of adverse effects and reproductive problems (Gupta et al. 1984).

The workers spraying methomyl, a carbamate insecticide, showed significant ECG changes, indicating cardiotoxic effect of methomyl (Sayed et al. 1992). In addition to imbalances in the immune system, various kinds of skin diseases including pigmentation on the exposed parts have been reported in the workers handling pesticides (Sharma et al. 1986).

Data on reproductive toxicity collected from 1106 couples engaged in spraying of OC, OP, and carbamate insecticides in cotton fields and from unexposed workers (1020) gave the following reproductive performances for the former and the latter groups (Rupa et al. 1991): (a) Abortions (26% vs. 15%); (b) Still births (8.7% vs. 2.6%); (c) Neonatal deaths (9.2% vs. 2.2%); and (d) Congenital effects (3% vs. 0.1%).

A cytogenetic study revealed a significant increase in chromatid breaks and gaps in the chromosomes in the peripheral blood in the grape garden workers exposed to pesticides (Rita et al. 1987). The workers spraying HCH, DDT, malathion, and cyfluthrin for malaria control showed increased levels of serum IgG (malathion exposure) and serum IgA (cyfluthrin exposure) (Karnik et al. 1993).

In another study, the thyroid function of the formulators exposed to a combination of pesticides in the organized sector was examined. Total T3 was significantly suppressed in the formulators, while a marginal decrease (7%) was noticed in the T4 level. The TSH levels were also elevated by 28%, but the rise was statistically insignificant (Zaidi et al. 2000). These formulators had significantly low serum cholinesterase activity and high serum BHC content, indicating appreciable exposure to their working environment.

Twenty blood samples randomly selected from four different villages of Punjab were analyzed for 14 OCs and 14 OP pesticides by Mathur et al. (2005). The results revealed that the total number of pesticides detected in the blood samples from Punjab was 15 out of the 28 pesticides analyzed, which indicated that each person is exposed to and carries a body burden of multiple pesticides due to a combination of direct and indirect exposure to these pesticides. The total OCP residues accounted to approximately 0.14 mg/L, of which total DDT metabolites reached approximately 0.065 mg/L. Total OP pesticide residues accounted to approximately 0.23 mg/L, of which monocrotophos recorded the highest concentration level (~0.095 mg/L) among the other three detected OP pesticides (chlorpyrifos, malathion, and phosphamidon) (Table 15.8). The authors concluded from their study that human pesticide residue is a biological index of pesticide exposure and

TABLE 15.8

Organochlorine and Organophosphorus Pesticide Residues (mg/L) in Human Blood Samples from Villages of Punjab, India

Organochlorines		Organophosphorus	
Pesticide/Metabolite	Concentration	Pesticide/Metabolite	Concentration
t-HCH	0.057	Chlorpyrifos	0.0662
Heptachlor	0.0006	Malathion	0.0301
Aldrin	0.0062	Monocrotophos	0.0948
Dieldrin	ND	Phosphamidon	0.0366
Chlordane	0.009	Total OPs	0.2278
t-DDT	0.0652		
t-Endosulfan	0.0046		
Total OCPs	0.1424		

Source: Adapted from Mathur, H. B., Agarwal, H. C., Johnson, S., and Saikia, N. 2005. Analysis of pesticide residues in blood samples from villages of Punjab. CSE/PML/PR-21/2005, 41, Tughlakabad Institutional Area, New Delhi 110062, pp. 21 + Annex III.

studies on the blood can be used for assessing the total body burden data of pesticides in the occupationally exposed and unexposed populations.

15.5.3 Pakistan

In 1976, epidemic organophosphate insecticide poisoning due to malathion occurred among 7500 field workers in the Pakistan malaria control program. In July, the peak month of the epidemic, it was estimated that there were about 2800 cases. In field studies, low red-cell cholinesterase activities were associated with the signs and symptoms of organophosphate insecticide intoxication. Toxicity was seen with three different formulations of the insecticide and was greatest with the products containing increased amounts of isomalathion, a toxic malathion degradation product. Poor work practices, which had developed when DDT was the primary insecticide for malaria control, resulted in excessive skin contact with and percutaneous absorption of the pesticide. Airborne malathion concentrations were very low. Implementation of good work practices and proscription of use of the two pesticide formulations most contaminated with isomalathion halted the epidemic in September. An extensive training program and surveillance system for pesticide toxicity preceded the 1977 spraying operations (Baker et al. 1978).

OCP residues were monitored in the samples of human blood (N = 32) obtained from concerned volunteers suspecting oral, dermal, or inhalation exposure to endosulfan and heptachlor, either environmentally or occupationally, either recently or in the recent past of Multan Division, (Pakistan) during 1995–1996. Residues of endosulfan were found to be higher when compared to heptachlor in all the samples. Concentration of endosulfan residues in the blood in the population of Multan and Mailsi regions ranged between 58.13–90.29 and 60.13–82.14 g/kg, respectively, while heptachlor ranged between 0.37–12.978 and 1.23–9.997 g/kg, respectively (Ansari et al. 1997).

Ahmad et al. 2002 have documented nonoccupational pesticide exposure accidents due to pesticides left on commercial foods, home germicides, toilet cleaners, termiticides, and suicides. The same situation has already been recognized that people living in agricultural areas may have significant public health consequences by nonoccupational exposure (Simcox et al. 1995).

Bunggush and Anwar (2000) reported the results of pesticide poisoning survey from 72 medical centers in Pakistan. A total of 408 cases were recorded. Of these, 165 (40.4%) were accidental and 243 (59.6%) were suicidal. The total number of deaths due to suicidal poisoning was 63 (26%).

Tariq et al. (2007) introduced an integrated picture of pesticide exposure to humans, animals, plants, waters, soils/sediments, and the atmosphere in Pakistan. They reported that in some areas of Punjab and Sindh, groundwater has been found contaminated due to pesticide use. There is considerable evidence that farmers have overused and misused the pesticides especially in cotton-growing areas. It is evident from the biological monitoring studies that farmers are at a higher risk for acute and chronic health effects associated with pesticides due to occupational exposure. Furthermore, the intensive use of pesticides (higher sprays more than the recommended dose) in cotton-growing areas involves a special risk to the field workers and pickers, in addition to unacceptable residue concentration in cottonseed oils and cakes. The authors reported that data limitations are still the major obstacle toward establishing clear environmental trends in Pakistan. They suggest that reliable monitoring, assessment, and reporting procedures should be implemented in accordance with appropriate environmental policies, laws, and regulations, in order to minimize the pesticide exposure.

15.6 Toxic Hazards of Pesticide Exposure in Man (General)

Exposure to pesticides, either directly or indirectly, may induce acute or chronic health impairment to humans.

15.6.1 Acute Intoxication

Almost half of the world's 6.5 billion people live in rural areas, mostly in farm households. Of the roughly 2 billion workers in the world, in 1990, 1.1 billion (over half) were farmers, the largest occupationally exposed group. Of these, 95% lived and worked in developing countries (WRI 1994).

Toxic outbreaks or collective poisonings have resulted from misuse of almost every type of pesticide: OCPs such as DDT, lindane, and chlordane; chlorinated camphenes such as toxaphene; cyclodienes such as aldrin and dieldrin; OP and carbamate cholinesterase inhibitors; organomercury fungicides; inorganics; and others (Ferrer and Cabral 1993). Acute poisoning incidents can occur in any country. In recent years, they have become less common in developed countries than in developing countries. But unfortunately, reliable statistics on individual poisonings by pesticides are difficult to obtain (Hodgson and Levi 1996).

Acute human pesticide poisoning represents a significant cause of morbidity and mortality in underdeveloped and developing countries. The World Health Organization (WHO) estimated that there are approximately 3 million cases of APP annually, with 220,000 deaths. On the basis of a survey of self-reported minor poisoning carried out in the Asian regions, it is estimated that there could be as many as 25 million agricultural workers in the developing world suffering an episode of poisoning each year (Jeyaratnam 1990). About 95% of fatal pesticide poisoning occurs in less developed countries (Ellenhorns et al. 1997). OP and CM pesticides are believed to cause tens of thousands of deaths and many more clinical poisonings every year. The principal mechanism of their action, namely the inhibition of cholinesterase group of enzymes, is also responsible for numerous and differing toxic effects that are mediated by other mechanisms, such as inflammation, immunotoxicity, myopathy, genetic toxicity, oncogenicity, and developmental and reproductive toxicity (Ballantyne and Marrs 1992). Moreover, reported use of certain OP and CM pesticides was associated with increased non-Hodgkin's lymphoma (NHL) incidences (Dreiher and Kordysh 2006).

Many researchers tried to correlate various enzymes with the harmful effects of pesticides, especially in the case of glutamate oxaloacetate transaminase (GOT), glutamate pyruvate transaminase (GPT), and alkaline phosphatase (ALP) (Misra et al. 1985; Kamal et al. 1990; Azmi et al. 2006).

15.6.2 Chronic Effects

Recently, public concern over potential adverse health effects has focused on a number of chronic end points—carcinogenesis, developmental and reproductive effects, immunological effects, neurotoxicity, cytogenetic damage, and endocrine disruption. Next is a brief survey of the most reported chronic effects of pesticides in humans.

  i. *Carcinogenesis*: Through linear risk extrapolation of animal data and maximum exposure levels of 550 million people, it is reported that there are 37,000 cancer cases yearly associated with pesticide use in developing countries (WHO 1990).

In recent years, there has been an increase in public concern that chronic low-level exposure to pesticide residues in food and water might pose a serious cancer risk to the general population (Hodgson and Levi 1996). Recent reports suggested an association between exposure to pesticides and different types of human cancer (Amaral Mendes 2002; Safi 2002). While epidemiological studies have often implicated pesticides as causative agents in human cancer (Blair et al. 1993), it has usually been at a marginal level of significance. It is suspected that DDT and its breakdown product DDE, persisting in the environment long after being banned, may be involved in the causation of breast cancer as a result of estrogenic activity (e.g., Krieger et al. 1994). In his study on the association between chronic exposure to pesticides and recorded cases of human malignancy, Safi (2002) reported that heavy use and misuse of pesticides and other toxic substances in the Gaza (Palestine) environment are suspected to be correlated with the growing incidence of cancer. Higher level of OCPs, especially β-HCH, γ-HCH, and p,p′-DDE, might be associated with prostate cancer risk (Kumar et al. 2010). How pesticides increase these cancer risks is not thoroughly understood. Only about eight pesticides are judged likely to be direct carcinogens by the International Agency for Research on Cancer (IARC) (Vainio et al. 1994). However, some carcinogenic agents activate cancers indirectly by altering the genetic materials of cells, disrupting cell division, and inducing chromosomal damage in the lymphocytes (Cuneo et al. 1992; Garry et al. 1992). In 1993, Pesticides News listed 70 possible carcinogens—now the list has grown to over 160. Many of the pesticides included are obsolete chemicals but may be found in stockpiles. Other pesticides are still in use, especially those listed by the US EPA (PAN 2001).

ii. *Hematopoietic cancers*: Hematopoiesis is the formation of blood cellular components from the hematopoietic stem cells located in the bone marrow. This process has a very high level of continuous replacement of specialized cells. This implies that a tremendous number of mitoses occur and there is an increased risk of mutations. The bone marrow also contains fat tissues that have a high storage capacity and may accumulate lipophilic compounds resulting in continuous exposure of hematopoietic progenitors and dendritic cells to xenobiotics. Hardell and coworkers were among the first to propose a link between exposure to phenoxy herbicides and the emergence of lymphoma (Hardell 1993). Thereafter, many epidemiological studies were conducted in populations exposed to pesticides. Although a clear correlation has been established between the agricultural practices and the occurrence of hematological malignancies (Alavanja et al. 2004), a specific link with the use of pesticides is not always observed. Nevertheless, in a recent meta-analysis, Merhi et al. (2010) showed a significant relationship between the professional exposure to pesticides and the risk of developing hematopoietic malignancies, particularly NHL. In AML (acute myeloid leukemia) patients professionally exposed to pesticides or organic solvents, the clinicobiologic characteristics are similar to the features of AML arising in patients with prior chemotherapy for another tumor, thus suggesting that similar transformation pathways may underlie leukemogenesis induced by cytotoxic drugs and by environmental exposure to certain pesticides or organic solvents (Cuneo et al. 1992).

iii. *Neurotoxicity*: Because of the basic similarities between the mammalian and the insect nervous systems, the insecticides designed to attack the insect nervous system are capable of producing acute and chronic neurotoxic effects in mammals (Ecobichon et al. 1990). Specifically, OP insecticides are potent neurotoxicants.

iv. *Reproductive effects*: A number of pesticides clearly have the potential to cause reproductive toxicity in animals, and several compounds (e.g., DBCP, EDB, chlordecone, carbaryl) are known to affect human reproduction (Mattison et al. 1990). Kelce et al. (1995) demonstrated that the primary DDT metabolite, p,p'-DDE, interferes with the action of male sex hormones, or androgens, suggesting that the feminizing action of DDT resulted from the antiandrogenic action of the DDT metabolite.

v. *Infertility of men*: It is well known that exposure to pesticides at occupational levels can adversely affect the semen quality. In the late 1970s, the nematocide DBCP affected more than 26,000 plantation workers in 12 countries; 64% had low sperm concentrations and 28% were involuntarily childless (Slutsky et al. 1999). Ethylene dibromide was an active component of approximately 100 pesticides. Its use was severely restricted in 1984 because of reduced sperm counts and semen volume in exposed workers (Schrader et al. 1988). More recently, a small study of herbicide sprayers in Argentina showed decreased sperm concentration and morphology related to high urinary levels of 2,4-D metabolites (Lerda and Rizzi 1991). Pesticides may affect male fertility by several mechanisms, including induction of developmental abnormalities of the male reproductive tract and by directly or indirectly affecting the function of normally developed gonads. It can be of significance in this context to note that albeit endocrine disruption is currently regarded as the prominent mechanism, whereby pesticides possibly mediate disruption of male fertility, and also other mechanisms, including excessive radical oxygen species (ROS) generation, may be involved (Meeker et al. 2004). It is noteworthy in this respect that the sperm cells appear to be particularly vulnerable to ROS, being deprived of the cytoplasmic defense enzyme with scavenging properties (Meeker et al. 2004).

vi. *Developmental effects*: Mammalian developmental toxicity does not refer only to the effects during in utero development, but also includes adverse effects on the developing organism that may have resulted from exposure of either parent before conception, of the mother during prenatal development, or postnatal to the time of sexual maturation. The embryo is most vulnerable to the initiation of major birth defects between 3 weeks and 2 months of gestation, the critical period of organogenesis. Exposure to toxic chemicals during the first 2 weeks typically leads to fetal death, while exposure after organogenesis is more likely to cause growth retardation and functional deficits (Hodgson and Levi 1996). In humans and rodents, exposure to hormonally active chemicals, such as methoxychlor (Gray 1992), DDT (Kelce et al. 1995), and vinclozolin (Gray et al. 1994), during the period of sex differentiation can produce a wide range of abnormal sexual phenotypes, including masculinized and defeminized females and feminized and demasculinized males.

vii. *Endocrine disrupting effects*: An endocrine disrupting chemical (EDC) is defined as an exogenous substance that causes adverse health effects in an intact organism, or its progeny, secondary to the changes in the endocrine function. Their effects on humans, wildlife, and the environment have been the subject of much attention by the international scientific community, since concerns about them were first raised by Colborn et al. (1993). Recently, the potential of certain pesticides to act as

EDCs has been confirmed (Amaral Mendes 2002; Charlier and Plomteux 2002). In addition to most OCPs, other pesticides, such as organophosphates, carbamates, triazines, and pyrethroids, that are less persistent and less toxic than the OCs were used to replace them, but many are now confirmed or suspected EDCs (Andersen et al. 2002).

viii. *Cytogenetic effects*: Cytogenetic damage related to pesticide exposure has been reported in various populations. Some investigators (e.g., De Ferrari et al. 1991; Kourakis et al. 1992; Joksic et al. 1997) have reported significant differences in the percentage of chromosomal aberrations (CAs) in exposed individuals (range, 2.66%–10.30%) compared with the control (range, 0.53%–5.52%). In workers of flower plantations located at Quito, Ecuador, South America, exposed to 27 pesticides (e.g., aldicarb, benomyl, captan, carbendazim, carbofuran, cartap, chlorothalonil, cypermethrin, deltamethrin, endosulfan, fenamiphos, fosetyl, iprodione, profenofos, propineb, vinclozolin, etc.), CA showed an increased frequency compared with the control group (20.59% vs. 2.73%; $p < 0.001$). Levels of erythrocyte AChE below the optimal level (>28 U/mL blood) were found in 88% of the exposed individuals (Mino et al. 2002).

ix. *Immunotoxicity*: The scientific evidence suggesting that many pesticides damage the immune system is impressive. Animal studies have found that pesticides alter the immune system's normal structure, disturb the immune responses, and reduce the animal's resistance to antigens and infectious agents. There is convincing direct and indirect evidence that these findings carry over to the human populations exposed to pesticides (WRI 1996). All mammalian (and avian and fish) immune systems are structurally similar, and considerable evidence shows that animal models are valid for testing human immunotoxicity (Turner 1994). Hundreds of studies (including compounds from different classes) have shown that pesticides can induce changes in the immune system structure and function, and these changes correlate closely in experimental animals with altered host resistance to pathogens (Vos et al. 1994). Malathion, which is considered a compound with very low toxicity (oral LD50 = 2100 mg/kg body weight), for example, dysregulates the immune system, especially affecting the nonspecific immune mechanisms. Chronic exposure at low doses over prolonged periods can also depress the humeral immune responses (Barnett and Rogers 1994). In addition to the active ingredients, pesticide solvents, inert materials, impurities, and contaminants cause measurable immunosuppression in several species (Kerkvliet 1994).

x. *Cancer and immunosuppression*: Studies have shown that pesticide exposure significantly reduces resistance to bacterial, viral, and parasitic infections and promotes tumor growth in many animal species (Bradley 1995). People exposed to pesticides are at an increased risk of contracting certain cancers known to be associated with immune suppression (Blair et al. 1992). Studies that specifically focused on pesticide exposure (occupationally and nonoccupationally) have found that phenoxy acid herbicides and other pesticides are associated with NHL and soft tissue sarcoma—two cancers associated with immunosuppression—and also with leukemia, multiple myeloma, and brain cancer (Leiss and Savity 1995). The fact that the farmers and others exposed to pesticides experience higher risks for the same cancers that afflict patients with clear immune deficiencies suggests that pesticides suppress the immune system and its self-regulating capabilities and thus raise cancer risks (WRI 1996).

xi. *Compounding risk factors*: Malnutrition prevalent in many developing country populations could bring about an increased susceptibility to (pesticide) intoxication, especially in women and children. Malnourishment, infectious diseases, and toxic chemicals interact with each other and with the immune system (Jamall and Davis 1991). In addition, exposure to more than one toxic compound (e.g., pesticides) is common in real life. Humans ingesting food preparations contaminated with pesticide residues, workers in pesticide manufacturing and packing units, and agricultural workers who prepare, mix, and apply pesticides in the fields are all potentially exposed to more than one pesticide on the same or on successive days (Krishnan and Brodeur 1994). Such exposure may induce a wide array of health effects, ranging from myelotoxicity to cytogenetic changes and carcinogenic effects (Hrelia et al. 1994). Interaction between different types of pesticides has received considerable attention of several investigators (e.g., Krishnan and Brodeur 1991; Cantalamessa 1993; Dikshith and Raizada 1994; Mansour et al. 1997, 2001b; Mansour and Refaie 2000; Mansour and Heikal 2001). In animal studies, a number of commercial pesticides has been reported to interact, inducing supra-additive effects (e.g., malathion + carbaryl; malathion + endosulfan; DDT + dimethoate) or infra-additive effects (e.g., malathion + parathion; malathion + aldrin; chlordane + carbaryl) (Krishnan and Brodeur 1991). Among 12 binary mixtures of different insecticides, including avermectin, imidacloprid, and buprofezin, with traditional insecticides such as chlorpyrifos, methomyl, deltamethrin, and triazophos, antagonism was the most frequently observed effect (67%), whatever the biochemical criteria used in the assessment process. Additive and potentiation effects occurred in 26% and 7% of the cases, respectively (Mansour and Refaie 2000; Mansour and Heikal 2001). The joint action analysis, based on the growth rate of the experimental rats, revealed that the most effective mixture showing potentiating effect against the male rats was deltamethrin + chlorpyrifos; while the least effective one showing antagonistic effect was avermectin + deltamethrin (Mansour et al. 2001b).

xii. *Other health hazard*: Diabetes mellitus or type 2 diabetes (T2D) is recognized as a global major public health problem. The prevalence of diabetes in all age groups worldwide has increased rapidly, in most parts of the world. It was estimated to be 2.8% in 2000 and 4.4% in 2030. The total number of people with diabetes is projected to rise from 171 million in 2000 to 366 million in 2030 (Wild et al. 2004). In particular, T2D has recently been linked with background exposure to POPs at low dose. Cross-sectional studies in the US general populations observed that the current background exposure to POPs was associated with T2D, insulin resistance, and dyslipidemia (Lee et al. 2006, 2007). Similar cross-sectional findings concerning OC pesticides and prevalence of diabetes were observed among Mexican Americans, Canadian First Nations, adults of Native Americans, and fish consumers (Cox et al. 2007; Philibert et al. 2009; Turyk et al. 2009). However, OC pesticides were not associated with T2D among the Inuit population in Greenland, which is highly exposed to POPs due to a high intake of marine mammals (Jorgensen et al. 2008). Son et al. (2010) investigated the association between OC pesticides and T2D in Koreans (randomly selected 40 diabetic patients and 40 normal controls). Ten OC pesticides (β-HCH, HCB, heptachlor epoxide, p,p'-DDE, p,p'-DDD, p,p'-DDT, o,p'-DDT, oxychlordane, trans-nonachlor, and mirex) detectable in greater than or equal to 70% of the controls were analyzed in relation to diabetes. Most OC pesticides showed strong associations with T2D after adjusting for age, sex, BMI, alcohol consumption, and cigarette smoking. Compared with the subjects in the lowest tertile of each OC pesticide, adjusted ORs in the third tertile ranged from 3.1 for heptachlor epoxide to 26.0 for oxychlordane. In the case of chemicals belonging to the DDT family, adjusted ORs in the third tertile were in the range from 10.6 for p,p'-DDT to 12.7 for p,p'-DDE. In this exploratory study with small sample size, low-dose background exposure to OC pesticides was strongly associated with prevalent T2D in Koreans, even though absolute concentrations of OC pesticides were not higher than that in other populations. The authors concluded that Asians may be more susceptible to the adverse effects of OC pesticides than other races. On the other hand, Rezg et al. (2010), in an extensive review, mentioned that OP pesticides may present an important risk factor to T2D induction. Although the authors do not regard their findings as definitive evidence, but, the potential implications nonetheless deserve serious consideration and may require substantial revision.

15.7 Factors Contributing to Health Risks of Pesticide Exposure

It has been long recognized that the inability of the farmers to understand and follow label instructions, due to illiteracy, in addition to unsafe use or misuse of pesticides, is among the major factors of pesticide poisoning in developing countries (Karisson 2004). The potential for pesticide poisoning in developing countries is, however, not limited to the unsafe use or misuse of pesticides but it is also from accidental contamination. Such type of contamination or poisoning may result from improper storage of pesticides, improper disposal of pesticide containers, the use of empty containers for storing foodstuffs and water, and the repacking of pesticides in smaller sizes and subsequent sale in open air markets next to farm produce or foodstuffs (Al-Saleh 1994). Other sources of accidental poisoning to farmers and community members include drift from sprayed fields, early reentry to sprayed farms, and eating crops that have recently been sprayed or treated with pesticides (Clarke et al. 1997).

In the light of a comparative study of pesticide use and exposure patterns in different countries (e.g., Brazil, Costa Rica, Ecuador, Paraguay, Venezuela, Egypt, South Africa, India, and Malaysia), Dinham (1993) concluded that in the third world countries, pesticides cannot be used safely. Health and safety issues are exacerbated by a general lack of hazard awareness; the lack of protective clothing or difficulty in wearing protective clothing in tropical climates; shortage of facilities for washing after use or in case of accidents; the value of containers for reuse in storing food and drink; illiteracy; labeling difficulties related to language, complexity, or misleading information; lack of regulatory authorities; and lack of enforcement. In many developing countries, families share the farm work. Men, women, and children typically have different agricultural work roles. In the Benguet district of the Philippines, nearly one third of all children and one half of all wives of all farm workers help apply pesticides (Rola 1989).

In this context, it may be useful to report here the results of the questionnaires designed to investigate the attitude and the behavior of the Egyptian farmers regarding pesticide use and handling in two large provinces, for example, Sharkia Governorate (Ibitayo 2006) and Beheira Governorate (Mansour 2008). The main purpose was to get the answers concerning the levels of education, the farmers' knowledge of pesticides, and the sources they used to obtain information about use and risk avoidance, as well as ways in which they dispose of empty pesticide containers. Figure 15.18(a–e) summarizes the results of the two investigations, which may highlight the necessity of paying more attention to this concern.

The participants in the questionnaire of the year 2008 differed greatly in how long they had been employed in farm work (including pesticide applications). Because health hazards are proportionate to duration of exposure to toxicants, Mansour (2008) estimated the ratio (years of employment: age; E/A) for each participant and expressed the products in percentages. The resulting values were used in rating the degree of occupational exposure to pesticides among the studied group in five categories such as (a) Excessive occupational exposure for individuals of E/A = >70%; (b) Extreme occupational exposure for individuals of E/A = 56%–70%; (c) High occupational exposure for individuals of E/A = 36%–55%; (d) Moderate occupational exposure for individuals of

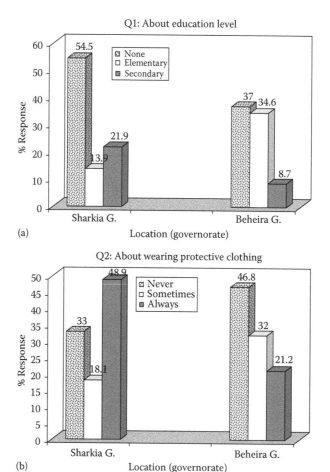

FIGURE 15.18
(a–e) Results of a questionnaire regarding the attitude and behavior of Egyptian farmers in two different governorates, Sharkia and Beheira, with respect to pesticide use and handling. (From Ibitayo, O. O., Risk Anal., 26, 989–995, 2006 and Mansour, S. A., *Rev. Environ. Contam. Toxicol.*, 196, 1–51, 2008.)

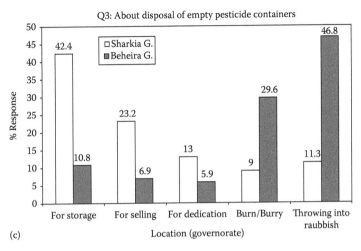

(c)

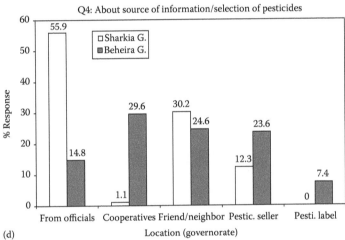

(d)

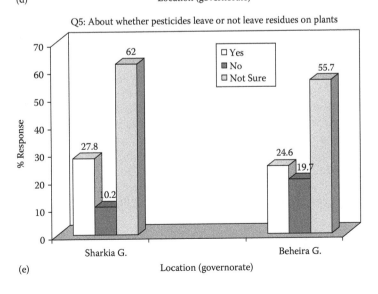

(e)

FIGURE 15.18 (Continued)

E/A = 20%–35%; (e) Slight occupational exposure for individuals of E/A = 20%–35%; and (f) Low occupational exposure for individuals of E/A = < 20%.

Information on the hazard awareness and practice of pesticides among farmers from other African countries were provided by Williamson et al. (2008). Trends in pesticide use were surveyed for 45 farmers per crop from Benin, and 25 or 50 farmers from Senegal per cotton or vegetable field, respectively. Approximately, 56% of the Senegal cotton farmers corresponding to 86% of Senegal vegetable farmers claimed not using protective clothing. In case of Benin farmers, only 10%–12% destroy empty the insecticide containers, the remainder reuse these containers for different purposes, and 2%–8% store the pesticides in the fields, while the others keep them in bedroom, kitchen, separate store, and under granary. Health effects of pesticides are negligible for 19%–28% of the Benin farmers. Comparing results of Williamson et al. (2008) with those of Ibitayo (2006) and Mansour (2008) may indicate similar attitude of pesticide misuse among the African farmers.

Poisoning surveillance systems are usually maintained only at large urban hospitals. Village health centers may be completely excluded from monitoring reports. Most of the severe poisoning cases never reach the hospital and many of those that do are misdiagnosed as stroke or respiratory and cardiovascular diseases (Loevinsohn 1987).

Taking into consideration the above mentioned facts, Mansour (2008) has summarized factors contributing to health risks due to pesticides, especially in developing countries, into three major items: (1) Factors assigned to governmental authorities; (2) Factors referred to pesticide users; and (3) Other factors (e.g., malnutrition, infectious diseases, multiple exposure to toxic chemicals and mixtures, poverty, etc.).

15.8 Factors Affecting Persistence and Fate of Pesticide Residues

When a pesticide is used in the environment, it becomes distributed among four major compartments: water, air, soil, and biota (living organisms). The fraction of the chemical that moves into each compartment is governed by the physicochemical properties of that chemical. The physicochemical properties (e.g., solubility, adsorption, and volatilization), the environmental conditions (e.g., wind, temperature, and location), and the metabolic processes all have powerful effects on the persistence and toxicity of the pesticides.

This means that the chemicals released into the environment rarely remain in the form, or at the location, of release. For example, pesticides used as sprays may drift from the point of application as air contaminants or enter runoff water as water contaminants. Many of these chemicals are susceptible to microbial degradation and are rapidly detoxified, frequently being broken down to products that can enter the carbon, nitrogen, and oxygen cycles. However, as mentioned before, OCPs are recalcitrant to a greater or lesser degree to metabolism by microorganisms and persist in soil and water as contaminants. Moreover, they may enter biologic food chains and move to higher trophic levels or persist in processed crops as food contaminants (Hodgson 2010).

15.8.1 Physicochemical Factors

Pesticides are distributed in the environment by physical processes such as sedimentation, adsorption, and volatilization. They can then be degraded by chemical and/or biological processes. Chemical processes generally occur in water or in the atmosphere and follow one of four reactions: oxidation, reduction, hydrolysis, and photolysis. Biological mechanisms in soil and living organisms utilize oxidation, reduction, hydrolysis, and conjugation to degrade the chemicals. The process of degradation will largely be governed by the compartment (water, soil, atmosphere, biota) in which the pesticide is distributed, and this distribution is governed by the physical processes previously mentioned (Linde 1994).

Pesticides that have high solubility in water will remain in water and tend to not be adsorbed onto soil and living organisms. Among the factors influencing the water solubility and hydrolysis of the chemicals are polarity, molecular size, temperature, and pH. Most pesticides are less polar than water, so they tend to accumulate in soil or living organisms that contain organic matter; based on their octanol–water partition coefficient (K_{ow}) (Linde 1994).

Adsorption of pesticides onto soils or sediments is a major factor in the transportation and eventual degradation of chemicals. Chemicals that are nonpolar tend to be pushed out of water and onto soils that contain nonpolar carbon material. Organic carbon content of the soil, pesticide polarity, and soil water's pH and salinity are among the factors influencing pesticide adsorption onto soil. There are two major ways by which pesticides can reach surface waters and groundwater: runoff and leaching. Runoff will occur if the chemical does not adsorb onto soil. Leaching occurs when the chemical is weakly adsorbed by soil and can easily move through the soil. Weak acid pesticides are bound weakly to soil, so they can easily move downward to groundwater (Linde 1994).

Bioconcentration factor (BCF) is indicative of the accumulation of a chemical in living organisms (biota) compared with the concentration in water. It is an indicator of how much a chemical will accumulate in living organisms, such as fish. Once absorbed into an organism, chemicals can move through the food chain ending up in humans. BCF values are unitless and generally range from one to a million. Bioconcentration is based on the law of equilibrium. For example, if a fish exposed to atrazine has a BCF of 110, the chemical would be absorbed according to the relation

$$BCF = \text{concentration in fish/concentration in water} = 110.$$

According to Linde (1994), BCF values for some pesticides are aldicarb (2), carbaryl (46), simazine (100), atrazine (110), lindane (169), and DDT (29700).

Pesticides with high vapor pressures (P_{vp}) may become environmental problems because they can volatilize and disperse over a large area. A chemical with a low vapor pressure does not move into air, so there is a potential for accumulation in water if it is water-soluble. If it is not water-soluble, the chemical may accumulate in soil or biota. Therefore, volatilization is one of the main transport pathways by which pesticides move from water and soil surfaces into the atmosphere. Volatilization from these surfaces is governed by various physical, chemical, and environmental factors, which mean that the rates will be different for each chemical depending on the environmental conditions (e.g., wind, temperature, and location; Linde 1994).

Large quantities of pesticides are lost by volatilization into the atmosphere. In the atmosphere, there are two major degradation pathways that occur (i.e., photochemical reactions and free radical reactions). The products formed may or may not be more toxic than the parent chemical. Once a pesticide has been degraded, a major removal process for chemicals is to precipitate out of air and return to the Earth's surface. Another removal process is for the products to be dissolved in rain and fall back to Earth (Linde 1994).

15.8.2 Metabolic Processes

Microbial metabolism is a process by which a chemical is absorbed into the cell membrane of the microbe. Enzymes within the microbe break down the chemical into smaller fragments. Pesticides are generally degraded by microbes along with the material excreted by the roots of plants. Bacteria account for 65% of the total biomass in soil, so they generally account for most degradation processes. Bacteria dominate the degradation process in alkaline soils and water (pH > 5.5). Fungi dominate the degradation process in acidic soils. The surrounding conditions will determine whether aerobic or anaerobic metabolism will occur in the degradation of a pesticide. In soils, the first layer degrades the chemicals via aerobic metabolism because air (O_2) flows over the surface. Below this layer, anaerobic metabolism occurs because of the lack of oxygen (Linde 1994).

The toxicity of pesticides, as with that of any toxicants, involves the interaction of the pesticide or one or more of its metabolites with a target macromolecule. Therefore, toxicity cannot be thought of as a single defining molecular event. On the contrary, the expression of a toxic endpoint is the final event in a cascade of events that begins with exposure. This cascade involves absorption, distribution, metabolism to reactive metabolites, further distribution, and interaction with target molecules. At the same time, pesticides may be metabolically detoxified or excreted or the potentially toxic lesion may be repaired. Factors that affect any of these interactions may affect the ultimate expression of toxicity, although, in general, metabolism and interaction with target molecules play the most important roles. Thus, any of the factors that affect pesticide metabolism may affect the ultimate toxicity of the pesticide (Hodgson 2001). On the other hand, the major processes affecting the fate or persistence of a pesticide in the human body are governed by several factors, which play an important role in the magnitude of the endpoint effect. For example, absorption of the pesticide through the skin differs with age (Shah et al. 1987) and thickness of the skin of the individuals (Feldman and Maibach 1974) as well as the chemical properties of the toxicant and environmental factors. These factors may all affect the dermal absorption and the rate of penetration of pesticides (Baynes and Rivier 2001).

In this respect, it may be convenient to mention the chemical group known as "persistent organic pollutants (POPs)." These POPs have key characteristics in common, which cause their ubiquitous distribution and accumulation in the environment and living organisms including humans. Although many different forms of POPs may exist, both natural and anthropogenic, POPs that are noted for their persistence and bioaccumulative characteristics include many of the first-generation OC insecticides such as dieldrin, DDT, toxaphene, and chlordane and several industrial chemical products or by-products, including PCBs, dibenzo-p-dioxins (dioxins), and dibenzo-p-furans (furans). Many of these compounds have been or continue to be used in large quantities and, due to their environmental persistence, have the ability to bioaccumulate and biomagnify (Anonymous 2001). POPs may have the following properties, generally (UNIDO 2003):

i. POPs are environmentally persistent compounds; they resist breakdown by natural processes, and in some cases, remain in the environment for decades.

ii. POPs are characterized by low water solubility and high fat solubility, which make them bioavailable to mammals. They bioaccumulate exponentially up to the top of the food chain. Some of these compounds, such as PCBs, may persist in the environment for long periods of time and may bioaccumulate by factors up to 70,000 fold.

iii. POPs are semivolatile compounds, enabling them to move long distances in the atmosphere before deposition occurs. They are volatilized at warm temperatures and condense at cooler temperatures, reaching their highest concentrations in the cooler regions of the world (northern latitudes and high altitudes).

iv. POPs are toxic compounds, causing adverse health effects, such as birth defects and damage to the immune and respiratory systems and critical organs. Hormone system dysfunction associated with POPs includes damage to the reproductive system, sex-linked disorders, and shortened lactation periods for nursing mothers. Also, endocrine disruption can have developmental and carcinogenic effects. Moreover, exposure to POPs can result in death in humans (including aborted fetuses) and wildlife. Exposure of the fetus to minute concentrations of some POPs can result in neurophysiological effects, such as attention deficits, learning disorders, behavioral problems (e.g., increased aggression), and poor gross and fine motor coordination.

15.8.3 Environmental Factors

It is known that the environmental conditions (e.g., wind, temperature, and location) have profound effects on the persistence and toxicity of pesticides. Nowadays, climatic change has arisen as a subject of major concern to ecologists and toxicologists. There is extensive peer-reviewed literature on the environmental impacts of climate change, particularly ecological impacts; however, there have been few studies investigating the relationships between pesticides and climate change. Based on the source–pathway–receptor analysis, Bloomfield et al. (2006) estimated that the main climate drivers for changing pesticide fate and behavior are changing rainfall patterns (changes in seasonality and intensity) and increased temperatures. The increase in global temperature will lead to greater volatilization and, consequently, to higher concentrations of chlorinated POPs in the atmosphere and in aquatic systems and ultimately in the human body (Grimalt et al. 2010). The overall effect of climate change on pesticide fate and transport is likely to be very variable and difficult to predict because of the uncertainties associated with the climate predictions. Some important source, pathway and receptor responses to climate change have been identified, but many of these responses have conflicting implications for pesticide fate and transport and the system needs to be assessed in a holistic manner if a full picture of the pesticide fate is to be obtained. In the long term, land-use change driven by changes in climate may have a more significant effect on the pesticides in the environment than the direct impacts of climate change on specific pesticide fate and transport processes (Grimalt et al. 2010).

15.9 Minimizing and Removing Pesticide Residues

Taking into consideration the large number of workers involved in pesticide application, especially in developing countries, and their attitudes and behaviors on dealing with these chemicals, solutions to the problem of health risks among those workers must be addressed. The farm workers are often careless about wearing protective clothes or masks, and thus they are acutely exposed to the pesticides applied. Illiteracy among the workers also causes overdosing of the sprayed pesticides and harvesting their crops before suitable waiting periods. Under such conditions, one could expect health hazards to the workers and consumers of their crops. This matter highlights the role of agricultural extension services to guide and educate the farm workers sector, and develop a system of "Spray-man

License." Additionally, agricultural inspectors, for example, "agricultural engineers," should receive frequent training programs on principles of "good agricultural practices, GAP"; thus they can transfer practices to the farmers in a simple way.

Even under conventional or organic farming, vegetables and fruits may contain pesticide residues and heavy metals, often below the maximum residue limits (MRLs). Several investigators (e.g., Bhanti and Taneja 2007; Peris et al. 2007) have reported that residues of OCPs, OPPs, and certain heavy metals in foods at levels currently regarded as safe adversely affect human health. On the other hand, such fruits may contain residues of multichemical contaminants, such as pesticides and heavy metals (Mansour et al. 2009a,b); a matter which may pose higher health hazards to the consumers (Mumtaz 1995; Caldas et al. 2006).

Ever since synthetic organic pesticides were first used in agriculture, investigators have been concerned with the fate of the residues during normal food processing. Washing, peeling, and cooking were soon recognized as important, but not necessarily reliable, means of reducing the residue levels in fruits and vegetables (Crosby 1965). At present, the literature offers more information about the potential role of different food processes (e.g., washing, peeling, cooking, acid or alkaline treatment, etc.) in removing most of the chemical contaminants from vegetables and fruits (e.g., Abou-Arab et al. 1998; Abou-Arab and Abou-Donia 2001; Randhawa et al. 2007; Klinhom et al. 2008).

In a monitoring study in India, as an example in this respect, Kumari (2008) estimated the residue levels of OCs, synthetic pyrethroids (SPs), organophosphates (OPs), and carbamates in three unprocessed and processed vegetables, namely, eggplant, cauliflower, and okra, to determine the residue levels and evaluate the effects of different household processes (washing and boiling or cooking) on the reduction of residues. In all the three vegetables, washing reduced the residues by 20%–77% and boiling reduced them by 32%–100%. Maximum (77%) reduction of OP insecticides was observed in eggplant, followed by 74% in cauliflower and 50% in okra by washing. The same trend was observed by boiling process, where maximum (100%) reduction of OP insecticides was observed in eggplant, followed by 92% in cauliflower and 75% in okra. Boiling was found comparatively more effective than washing in dislodging the residues.

15.10 The Future Outlook and Needs

In this respect, we will briefly touch upon some aspects related to the area of pest control practices that would reflect on better managing of pesticide performance and/or introduce safer alternatives.

Integrated Pest Control Management: It is agreed that without pest control practices, there will be big losses in the yield of agricultural crops. Chemical pesticides still represent the rapidly effective pest controlling weapon. Other controlling tools, such as biological, physical, and mechanical controls, are effective but at slower rate. This critical situation caused the need to use chemical pesticides wisely and in conjunction with other controlling tools within the concept of so-called integrated pest management (IPM). As we are talking here about the future outlook and needs with respect to pesticide residues in human body, IPM may due to minimizing pesticide residues in the agricultural crops. From our perspective, IPM as a strategy should be encouraged and predominated. Other future outlook and needs in the course of achieving more pesticide safety is to improve pesticide application technologies, as demonstrated next.

Pesticide Application Technologies: All users of pesticides are confronted with several potential hazards. Those who mix, load, apply, and handle pesticides have a risk of exposure. Misapplication, spills, and drift are all major sources of human and environmental contamination. Because pesticides are likely to be a part of the pest management system for the foreseeable future, ways to reduce risks in the use of pesticides must be practiced. Subsequently, marked improvements in the application technologies of pesticides have been developed. Variable rate applications, prescription rates of crop protection products, direct injection, closed handling systems, on-board dry and liquid application systems, control systems, spot sprayers, shielded sprayers, air assist systems, new nozzle designs, and tank-rinsing devices are examples of the technological developments that have affected the pesticide application industry. There has also been a major effort to reduce the amount of chemicals used. Chemical companies are developing new products that are effective at very low rates and designed for targeted applications with equipment that can apply precisely the correct amount when and where it is needed (Wolf 2002).

Organic Food: Chemophobia is the most common reason for the public to choose organic food on the assumption that such food is free of synthetic pesticides. Organic farms are distinguished from all other forms of farming by a rejection of soluble minerals as fertilizers and synthetic pesticides in favor of natural ones (Trewavas 2004). Subsequently, this has resulted in increasing demand for organic produce of different varieties of vegetables, fruits, milk products, and cereals in many parts of the world. While organic food accounts for 1%–2% of the total food sales worldwide, the organic food market is growing rapidly, far ahead of the rest of the food industry, in both developed and developing nations. World organic food sales jumped from US $23 billion in 2002 (Organic Monitor 2002) to US $52 billion in 2008 (Datamonitor 2009). The world organic market has been growing by 20% a year since the early 1990s, with future growth estimates ranging from 10% to 50% annually, depending on the country. It is supposed that organically produced vegetables and fruits contain less (or nil) pesticide residues as compared with that in conventionally produced ones. In recent studies, organically farmed cucumber fruits and potato tubers were found to contain total pesticide residues and heavy metals much lower than those found in conventionally farmed products (Mansour et al. 2009a,b).

Genetically Modified Crops (e.g., Transgenic Bt Crops): These are pesticidal substances produced by plants from the genetic material that has been added to the plants. For example, the gene for the *Bacillus thuringiensis* (Bt) pesticidal protein can be isolated and introduced into the plant's own genetic material. Then the plant, instead of the Bt bacterium, manufactures the substance that destroys the pest. So, there will be no need to use chemical pesticides to combat the pests attacking such genetically modified plants. Since 1996, a wide range of crop plants have been genetically engineered to contain the δ-endotoxin gene from Bt. These "Bt crops" are now available commercially in the United States and in many other countries. Some developing countries, such as Egypt, have been dealing with this technology since 1990 (Mansour 2009). EPA regulates both the protein and its genetic material; the plant itself is not regulated. Using this technology, scientists have modified corn, cotton, and potatoes to produce a pesticidal protein that is toxic when ingested by specific insect pests. Such GM crops are known as "plant-incorporated protectants (PIPs)." Since March 1995, EPA has registered 12 PIPs. Also, scientists have applied biotechnology to produce crops that are resistant to certain herbicides. The most common herbicide-tolerant crops (HTC or HRC) are those resistant to glyphosate herbicide (e.g., cotton, corn, soybean, and canola) (EPA and SEG 2003; EPA 2004). Benefits versus risks of Bt transgenic plants, however, still face a big debate between the supporters and the challengers (IUPAC 2004; EPA 2004).

15.11 Conclusion

The use of agrochemicals, including pesticides, remains a common practice especially in tropical regions and countries in the south. Cheap compounds, such as DDT, HCH, and lindane, that are environmentally persistent, are today banned from agricultural use in developed countries, but remain popular in some developing countries, especially for malaria vector control in mosquito-endemic regions. As a consequence, residues of these chemicals contaminate food and disperse in the environment. The presence of such pesticide residues is not confined to inhabited continents. Several pesticidal POPs have the characteristics of "global distillation or the grasshopper effect," which have resulted in accumulation of these compounds even in the Arctic media and in regions where they have never been used. There is strong scientific evidence that pesticides, as a whole, can induce a wide array of human health effects ranging from myelotoxicity to cytogenetic damage and carcinogenicity. Their occurrence in the human body is evident in people either occupationally or nonoccupationally exposed; in population living either in rural or urban areas; and in individuals of different ethnic routes. This is because the food that we eat is a major source of hosting pesticide residues in our bodies. The water that we drink and the air that we breathe are other sources of pesticide pathway in our bodies. The developed countries have already addressed the pesticide problem, but are still facing some problems in certain locations. Therefore, it can be deduced that the problems associated with pesticide hazards are not confined to the developing countries. Generally, in the third world countries, pesticides should be used carefully since toxic outbreaks are often attributed to misuse of these substances. Magnifying the severity of pesticide misuse is the availability of banned compounds in local markets; a matter that the national authorities must address. Since all the Earth inhabitants are sailing in one boat, it is hoped that the spirit of cooperation, mutual interest, and regard for one another's integrity could be adopted before pesticides are regulated out of existence and the entire world suffers. Coordinated efforts are needed to increase the production of food, but with a view to enhancing food quality and safety as well as to controlling residues of pesticides in humans and in the environment. Proper application of IPM programs within the framework of GAP, developing new technologies in pesticide application, moving toward organic food production, and shifting wisely into genetically engineered crops would positively reflect on human health, agriculture, biodiversity, and ecosystem management and thus support the sustainable development themes of the UN World Summits.

References

Abou-Arab, A. A. K. and Abou-Donia, M. A. 2001. Pesticide residues in some Egyptian spices and medicinal plants as affected by processing. *Food Chem.* 72: 439–445.

Abou-Arab, A. A. K., Soliman, K. M., and Naguib, Kh. 1998. Pesticide residue contents in Egyptian vegetables and fruits and removal by washing. *Bull. Nutr. Inst. Cairo Egypt* 18(1): 117–137.

Ahlborg, U. G., Lipworth, L., Titus-Ernstoff, L., et al. 1995. Organochlorine compounds in relation to breast cancer, endometrial cancer, and endometriosis: An assessment for the biological and epidemiological evidence. *Crit. Rev. Toxicol.* 25(6): 463–531.

Ahmad, R., Ahad, K., Iqbal, R., and Muhammad, A. 2002. Acute poisoning due to commercial pesticides in Multan. *Pak. J. Med. Sci.* 18: 227–231.

Alavanja, M. C., Hoppin, J. A., and Kamel, F. 2004. Health effects of chronic pesticide exposure: Cancer and neurotoxicity. *Annu. Rev. Public Health* 25: 155–197.

Alawi, M. A., Tamimi, S., and Jaghabir, M. 1999. Storage of organochlorine pesticides in human adipose tissues of Jordanian males and females. *Chemosphere* 38(12): 2865–2873.

Al-Saleh, I. 1994. Pesticides: A review article. *J. Environ. Pathol. Oncol.* 13(3): 151–161.

Amaral Mendes, J. J. 2002. The endocrine disrupters: A major medical challenge. *Food Chem. Toxicol.* 40: 781–788.

Amer, M., Metwalli, M., and Abu el-Magd, Y. 2002. Skin diseases and enzymatic antioxidant activity among workers exposed to pesticides. *East Mediterr. Health J.* 8(2–3): 363–373.

Amr, M. M. and Halim, Z. S. 1997. Psychiatric disorders among Egyptian pesticide applicators and formulators. *Environ. Res.* 73: 193–199.

Andersen, H. R., Cook, S. J., and Waldbillig, D. 2002. Effects of currently used pesticides in assays for estrogenicity, androgenicity, and aromatase activity in vitro. *Toxicol. Appl. Pharmacol.* 179: 1–12.

Anonymous. 2001. Persistent organic pollutants and the Stockholm Convention: A resource guide. A Report Prepared by Resource Futures International for the World Bank and CIDA, 22 pp.

Ansari, M. T., Iqbal, Z., and Ahmad, B. 1997. Organochlorine pesticide residues in human blood in the population of Multan (Pakistan). *Pak. J. Pharm. Sci.* 10(1): 19–28.

Attia, A. M. 2005. Risk assessment of occupational exposure to pesticides. In: Linkov, I. and Ramadan, A. B. (eds), *Comparative Risk Assessment and Environmental Decision Making*, NATO Science Series, vol. 38(3), pp. 349–362. Springer: Netherlands.

Azmi, M. A., Naqvi, S. N. H., Arshad Azmi, M., and Aslam, M. 2006. Effect of pesticide residues on health and different enzyme levels in the blood of farm workers from Gadap (rural area) Karachi—Pakistan. *Chemosphere* 64: 1739–1744.

Badawy, M. I., El-Dib, M. A., and Aly, O. A. 1984. Spill of methyl parathion in the Mediterranean Sea: A case study at Port-Said, Egypt. *Bull. Environ. Contam. Toxicol.* 32: 469–477.

Baker, E. L. J., Dobbin, R. D., Zack, M., et al. 1978. Epidemic malathion poisoning in Pakistan malaria workers. *Lancet* 311(8054): 31–34.

Ballantyne, B. and Marrs, T. C. 1992. Overview of ecological and clinical aspects of organophosphates and carbamates. In: Ballantyne, B. and Marrs, T. C. (eds), *Clinical and Experimental Toxicology of Organophosphates and Carbamates*, pp. 3–14. Butterworth-Heinemann: Oxford.

Barakat, A. O. 2003. Persistent organic pollutants in smoke particles emitted during open burning of municipal solid wastes. *Bull. Environ. Contam. Toxicol.* 70(1): 174–181.

Barnett, J. and Rogers, K. 1994. Pesticides. In: Dean, J. H., et al. (eds), *Immunotoxicity and Immunopharmacology*, 2nd edn, 191–213. New York: Raven Press, Ltd.

Barr, D. B., Allen, R., Olsson, A. O., et al. 2005. Concentrations of selective metabolites of organophosphorus pesticides in the United States population. *Environ. Res.* 99: 314–326.

Baynes, R. E. and Rivier, J. E. 2001. Pesticide disposition: Dermal absorption. In: Krieger, R. I. (ed.), *Handbook of Pesticide Toxicology, Vol. 1: Principles*, Chapter 22, pp. 515–530. Academic Press: New York.

Bearer, C. F. 1995. How are children different from adults? *Environ. Health Perspect.* 103: 7–12.

Becker, K., Kaus, S., Krause, C., et al. 2002. German Environmental Survey 1998. (GerES III): Environmental pollutants in blood of the German population. *Int. J. Hyg. Environ. Health* 205(4): 297–308.

Berkowitz, G. S., Obel, J., Deych, E., et al. 2003. Exposure to indoor pesticides during pregnancy in a multiethnic, urban cohort. *Environ. Health Perspect.* 111: 79–84.

Bhanti, M. and Taneja, A. 2007. Contamination of vegetables of different seasons with organophosphorous pesticides and related health assessment in northern India. *Chemosphere* 69: 63–68.

Blair, A., Zahm, S. H., Pearce, N. E., Heineman, E. F., and Fraumeni, J. F., Jr. 1992. Clues to cancer etiology from studies of farmers. *Scand. J. Work Environ. Health* 18: 209–215.

Blair, A., Dosemeci, M., and Heineman, E. F. 1993. Cancer and other causes of death among male and female farmers from twenty three states. *Am. J. Ind. Med.* 23: 729–742.

Bloomfield, J. P., Williams, R. J., Gooddy, D. C., Cape, J. N., and Guha, P. 2006. Impacts of climate change on the fate and behaviour of pesticides in surface and groundwater—A UK perspective. *Sci. Total Environ.* 369: 163–177.

Bouvier, G., Blanchard, O., Momas, I., and Seta, N. 2006. Pesticide exposure of non-occupationally exposed subjects compared to some occupational exposure: A French pilot study. *Sci. Total Environ.* 366: 74–91.

Bradley, S. G. 1995. Introduction to animal models in immunotoxicology: Host resistance. In: Burleson, G. R., Dean, J. H., and Munson, A. E. (eds), *Methods Immunotoxicology*, vol. 2, pp. 135–141. Wiley-Liss Inc.: New York.

Bunggush, R. A. and Anwar, T. 2000. Preliminary survey for pesticide poisoning in Pakistan. *Pak. J. Biol. Sci.* 3(11): 1976–1978.

Caldas, E. D., Boon, P. E., and Tressou, J. 2006. Probabilistic assessment of the cumulative acute exposure to organophosphorus and carbamate insecticides in the Brazilian diet. *Toxicology* 222: 132–142.

Calvert, G. M., Plate, D. K., Das, R., Rosales, R., Shafey, O., and Thomsen, C. 2004. Acute occupational pesticide-related illness in the US, 1998–1999: Surveillance findings from the SENSOR-pesticide program. *Am. J. Ind. Med.* 45: 14–23.

Cantalamessa, F. 1993. Acute toxicity of two pyrethroids, permethrin, and cypermethrin in neonatal and adult rats. *Arch. Toxicol.* 67(7): 510–513.

Carreño, J., Rivas, A., Granada, A., and Lopez-Espinosa, M. J. 2007. Exposure of young men to organochlorine pesticides in Southern Spain. *Environ. Res.* 103: 55–61.

Charlier, C. J. and Plomteux, G. J. 2002. Determination of organochlorine pesticide residues in the blood of healthy individuals. *Clin. Chem. Lab. Med.* 40(4): 361–364.

Clarke, E. E. K., Levy, L. S., Spurgeon, A., and Calvert, L. A. 1997. The problems associated with pesticide use by irrigation workers in Ghana. *Occup. Med.* 47(5): 301–308.

Clementi, M., Tiboni, G. M., Causin, R., et al. 2008. Pesticides and fertility: An epidemiological study in Northeast Italy and review of the literature. *Reprod. Toxicol.* 26: 13–18.

Colborn, T., vom Saal, F. S., and Soto, A. M. 1993. Developmental effects of endocrine-disrupting chemicals in wildlife and humans. *Environ. Health Perspect.* 101: 378–84.

Coupe, R. H., Manning, M. A., Foreman Goolsby, D. A., and Majewski, M. S. 2000. Occurrence of pesticides in rain and air in urban and agricultural areas of Mississippi, April–September 1995. *Sci. Total Environ.* 248(2–3): 227–240.

Cox, S., Niskar, A. S., Narayan, K. M., and Marcus, M. 2007. Prevalence of self-reported diabetes and exposure to organochlorine pesticides among Mexican Americans: Hispanic health and nutrition examination survey, 1982–1984. *Environ. Health Perspect.* 115: 1747–1752.

Cressey, P. J. and Vannoort, R. W. 2003. Pesticide content of infant formulae and weaning foods available in New Zealand. *Food Addit. Contam.* 20: 57–64.

Crosby, D. G. 1965. The intentional removal of pesticide residues. In: Chichester, C. O. (ed.), *Research in Pesticides: Proceedings of the Conference on Research Needs and Approaches to the Use of Agricultural Chemicals from a Public Health Viewpoint*, 213 pp. University of California, Davis, CA and Academic Press Inc.: New York.

Cruz, S., Lino, C., and Silveira, M. I. 2003. Evaluation of organochlorine pesticide residues in human serum from an urban and two rural populations in Portugal. *Sci. Total Environ.* 317: 23–35.

Cuneo, A., Fagioli, F., Pazzi, I., et al. 1992. Morphologic, immunologic and cytogenetic studies in acute myeloid leukemia following occupational exposure to pesticides and organic solvents. *Leuk. Res.* 16: 789–796.

Datamonitor. 2009. Food: Global Industry Guide. http://www.researchandmarkets.com/research/18f9c2/food_global_indus (Retrieved 28 August 2008).

De Ferrari, M., Artuso, M., Bonassi, S., et al. 1991. Cytogenetic biomonitoring of an Italian population exposed to pesticides: Chromosome aberration and sister-chromatid exchange analysis in peripheral blood lymphocytes. *Mut. Res.* 260: 105–113.

Desi, I., Nagymajtenyi, L., Schulz, H., and Nehez, M. 1998. Epidemiological investigations and experimental model studies on exposure of pesticides. *Toxicol. Lett.* 96(97): 351–359.

Deutch, B., Pedersen, H. S., Jorgensen, E. C., and Hansen, J. C. 2003. Smoking as a determinant of high organochlorine levels in Greenland. *Arch. Environ. Health* 58: 30–36.

DeVoto, E., Kohlmeier, L. and Heeschen, W. 1998. Some dietary predictors of plasma organochlorine concentrations in an elderly German population. *Arch. Environ. Health* 53(2): 147–155.

Dikshith, T. S. S. and Raizada, R. B. 1994. Response of carbon tetrachloride pretreated rats to endosulfan, carbaryl and phosphamidon. *Ind. Health* 21(4): 263–272.

Dinham, B. 1993. *The Pesticide Hazard: A Global Health and Environmental Audit*. Zed Books for The Pesticides Trust: London, UK.

Dreiher, J. and Kordysh, E. 2006. Non-Hodgkin lymphoma and pesticide exposure: 25 years of research. *Acta Haematol.* 116(3): 153–164.

Dua, V. K., Kumari, R., Sharma, V. P. and Subbarao, S. K. 2001. Organochlorine residues in human blood from Nainital (U.P.), India. *Bull. Environ. Contam. Toxicol.* 67: 42–45.

Ecobichon, D. J., Davis, J. E., Doull, J., et al. 1990. Neurotoxic effects of pesticides. In: Baker, S. R. and Wilkinson, C. F. (eds), *The Effects of Pesticides on Human Health*, pp. 131–199. Princeton Scientific Publishers: New Jersey.

Eilrich, G. L. 1991. Tracking the fate of residues from farm gate to the table. In: Tweedy, B. G., Dishburger, H. J., Ballantine, L. G., and McCarthy, J. (eds), *Pesticide Residues and Good Safety: A Harvest of Viewpoints*, pp. 202–212. American Chemists' Society: Washington, DC.

El-Gamal, A. 1983. Persistence of some pesticides in semi-arid conditions in Egypt. *Proceedings, International Conference on Environmental Hazards in Agrochemistry, Alexandria, Egypt*, 8–12 November 1983, vol. I, pp. 54–75.

Ellenhorns, M. J., Schonwald, S., Ordog, G., and Wasserberger, J. 1997. *Ellenhorn's Medical Toxicology: Diagnosis and Treatment of Human Poisoning*, 2nd edn. Williams & Wilkins: Maryland.

EPA. 1990. Pesticide Exposure Study (NOPES), U.S. Environmental Protection Agency, Research Triangle Park, NC.

EPA. 2004. Biopesticides. URL: http://www.epa.gov/agriculture/tbio.html (30 June 2004).

EPA and SEG (Safe Environment Group). 2003. Fact findings on DDT use in Ethiopia, Addis Ababa. The Federal Democratic Republic of Ethiopia Environmental Protection Authority.

FAO. 2000. Project concept paper. HEALTH: Health in ecological agricultural learning. Prepared by the FAO programme for community IPM in Asia. Food and Agricultural Organization of the United Nations, Rome. http://www.fao.org/nars/partners/2nrm/proposal/9-2-6.doc.

Farahat, T. M., Abdelrasoul, G. M., Amr, M. M., Shebl, M. M., Farahat, F. M., and Anger, W. K. 2007. Neurobehavioral effects among workers occupationally exposed to organophosphorus pesticides. *Occup. Environ. Med.* 60: 279–286.

Feldman, R. J. and Maibach, H. I. 1974. Percutaneous penetration of some pesticides and herbicides in man. *Toxicol. Appl. Pharmacol.* 28: 126–132.

Ferrer, A. and Cabral, R. 1993. Collective poisonings caused by pesticides: mechanisms of production and mechanisms of prevention. *Rev. Environ. Toxicol.* 5: 161–201.

Garry, V. F. 2004. Pesticides and children. *Toxicol. Appl. Pharmacol.* 198: 152–163.

Garry, V. F., Danzl, T. J., Tarone, R., et al. 1992. Chromosome rearrangements in fumigant appliers: possible relationship to non-Hodgkin's lymphoma risk. *Cancer Epidmiol. Biomarkers Prev.* 1: 287–291.

Glotfelty, D. E., Seiber, J. N., and Liljedahl, L. A. 1987. Pesticides in fog. *Nature* 325(6105): 602–605.

Gots, R. E. 1992. *Toxic Risks: Science, Regulation and Perception*. Lewis Publishers: Florida.

Gray, L. E. 1992. Chemical-induced alterations of sexual differentiation: A review of effects in humans and rodents. In: Colborn, T. and Clement, C. (eds), *Chemically-Induced Alterations in Sexual and Functional Development: The Wildlife/Human Connection*, pp. 203–230. Princeton Scientific Publishers: New Jersey.

Gray, L. E., Ostby, J. S., and Kelce, W. R. 1994. Developmental effects of an environmental antiandrogen: The fungicide vinclozolin alters sex differentiation of the male rat. *Toxicol. Appl. Pharmacol.* 129: 46–52.

Grimalt, J. O., Catalan, J., and Fernandez, P. 2010. Distribution of persistent organic pollutants and mercury in freshwater ecosystems under changing climate conditions. In: Kernan, M., Battarbee, R. W., and Moss, B. (eds), *Climate Change Impacts on Freshwater Ecosystems*, pp. 180–202. Wiley-Blackwell and John Wiley & Sons, Ltd.

Guardino, X., Serra, C., Obiolis, J., et al. 1996. Determination of DDT and related compounds in blood samples from agricultural workers. *J. Chromatogr. A* 719: 141–147.

Gupta, P. C. 1975. Neurotoxicity of chronic chlorinated hydrocarbon insecticide poisoning – A clinical and electroencephalographic study in man. *Ind. J. Med. Res.* 63: 601–606.

Gupta, P. K. 2004. Pesticide exposure – Indian Scene. *Toxicology* 198: 83–90.

Gupta, S. K., Jain, J. P., Sayed, H. N., and Kashyap, S. K. 1984. Health hazards in pesticide formulators exposed to a combination of pesticides. *Ind. J. Med. Res.* 79: 666–672.

Hanaoka, T., Takahashi, Y., Kobayashi, M., et al. 2002. Residuals of beta-hexachlorocyclohexane, dichlorodiphenyltrichloroethane, and hexachlorobenzene in serum, and relations with consumption of dietary components in rural residents in Japan. *Sci. Total Environ.* 286: 119–127.

Hardell, L. 1993. Phenoxy herbicides, chlorophenols, soft-tissue sarcoma (STS) and malignant lymphoma. *Br. J. Cancer* 67: 1154–1156.

Hauser, R., Altshul, L., Chen, Z., Ryan, L., Overstreet, J., and Schiff, I. 2002. Environmental organochlorines and semen quality: Results of a pilot study. *Environ. Health Perspect.* 110: 229–233.

Henao, S., Finkelman, J., Albert, L., and de Konig, H. W. 1993. Pesticides and Health in the America. Environ. Ser. No. 12, Plant American Health Organization, Division of Health & Environment: Washington, DC.

Heudorf, U. and Angerer, J. 2001. Metabolites of organophosphorous insecticides in urine specimens from inhabitants of a residential area. *Environ. Res.* 86(1): 80–87.

Hill, R. H., Jr., Head, S. L., Baker, S., et al. 1995. Pesticide residues in urine of adults living in the United States: Reference range concentrations. *Environ. Res.* 71: 99–108.

Hodgson, E. 2001. Factors that affect pesticide metabolism and toxicity. In: Krieger, R. I. (ed.), *Handbook of Pesticide Toxicology, vol. 1: Principles*, Chapter 21, pp. 507–513. Academic Press: New York.

Hodgson, E. 2010. Introduction to toxicology. In: Hodgson, E. (ed.), *A text book of modern toxicology*, 4th edn, pp. 3–14. John Wiley & Sons: New York.

Hodgson, E. and Levi, P. E. 1996. Pesticides: An important but underused model for the environmental health sciences. *Environ. Health Perspect.* 104(Suppl. 1): 97–106.

Hoferkamp, L., Hermanson, M. H., and Muir, D. C. G. 2010. Current use pesticides in Arctic media; 2000–2007. *Sci. Total Environ.* 408: 2985–2994.

Hrelia, P., Maffei, F., Vigagni, F., et al. 1994. Interactive effects between trichloroethylene and pesticides at metabolic and genetic level in mice. *Environ. Health Perspect.* 102(9): 31–34.

Ibitayo, O. O. 2006. Egyptian farmers' attitudes and behaviors regarding agricultural pesticides: Implications for pesticide risk communication. *Risk Anal.* 26(4): 989–995.

IDB. 2010. International Data Base (IDB)—World Population. Census.gov. 28 June 2010. http://www.census.gov/ipc/www/idb/worldpopinfo.php (Retrieved 1 August 2010).

IUPAC. 2004. Impact of transgenic crops on the use of agrochemicals and the environment. URL: http://www.iupac.org/projects/2001/200 1-024-2-600.html.

Jamall, I. S. and Davis, B. 1991. Chemicals and environmentally caused diseases in developing countries. *Infect. Dis. Clin. North Am.* 5(2): 365–375.

Jeyaratnam, J. 1990. Acute pesticide poisonings: A major global problem. *Wld. Health Statist. Quart.* 43(3): 139–144.

Joksic, G., Vidakovi, A., and Spasojevi-Tima, V. 1997. Cytogenetic monitoring of pesticide sprayers. *Environ. Res.* 75(2): 113–118.

Jorgensen, M. E., Borch-Johnsen, K., and Bjerregaard, P. 2008. A cross-sectional study of the association between persistent organic pollutants and glucose intolerance among Greenland Inuit. *Diabetologia* 51:1416–1422.

Kamal, A. A., Elgarhy, M. T., Maklady, F., Mostafa, M. A., and Massoud, A. 1990. Serum cholinesterase and liver function among a group of organophosphorus pesticides sprayers in Egypt. *J. Toxicol. Clin. Exp.* 10: 427–435.

Kang, Y. S., Matsuda, M., Kawano, M., Wakimoto, T., and Min, B. Y. 1997. Organochlorine pesticides, polychlorinated biphenyls, polychlorinated dibenzo-p-dioxins and dibenzofurans in human adipose tissue from Western Kyungnam, Korea. *Chemosphere* 35(10): 2107–2117.

Karalliedde, L. D., Edwards, P., and Marrs, T. C. 2003. Variables influencing the toxic response to organophosphates in humans. *Food Chem. Toxicol.* 41: 1–13.

Karisson, S. I. 2004. Agricultural pesticides in developing countries – A multilevel governance challenge. *Environment* 46(4): 22–41.

Karnik, A. B., Suthar, A. M., Patel, M. M., Sadhu, H. G., Venkaiah, K., and Nigam, S. K. 1993. Immunological profile in workers exposed to pesticides. *Ind. J. Ind. Med.* 39: 110–112.

Karunakaran, C. O. 1958. The Kerala food poisoning. *J. Ind Med Assoc* 31: 204–205.

Kelce, W. R., Stone, C. R., Laws, S. C., Gray, L. E., Kempainen, I. A., and Wilson, E. M. 1995. Persistent DDT metabolite p,p'-DDE is a potent androgen receptor antagonist. *Nature* 375: 581–585.

Kerkvliet, N. I. 1994. Immunotoxicology of dioxins and related chemicals. In: Schecter, A. (ed.), *Dioxin and Health*, pp. 199–221. Plenum Press: New York.

Klinhom, P., Halee, A., and Methawiwat, S. 2008. The effectiveness of household chemicals in residue removal of methomyl and carbaryl pesticides on Chinese-Kale. *Kasetsart. J. (Nat. Sci.)* 42: 136–143.

Kourakis, A., Mouratidou, M., Kokkinos, G., et al. 1992. Frequencies of chromosomal aberrations in pesticide sprayers working in plastic green houses. *Mut. Res.* 279: 145–148.

Krawinkel, M. B., Plehn, G., Kruse, H., and Kasi, A. M. 1989. Organochlorine residues in Baluchistan Pakistan blood and fat concentrations in humans. *Bull. Environ. Contam. Toxicol.* 43: 821–826.

Krieger, N., Wolff, M. S., Hiatt, R. A., Rivera, M., Vogelman, J., and Orentreich, N. 1994. Breast cancer and serum organochlorines: A perspective study among White, Black and Asian women. *J. Natl. Cancer Inst.* 88(8): 589–599.

Krishnan, K. and Brodeur, J. 1991. Toxicological consequences of combined exposure to environmental pollutants. *Arch. Complex Environ. Stud.* 3(3): 1–106.

Krishnan, K. and Brodeur, J. 1994. Toxic interactions among environmental pollutants: Corroborating laboratory observations with human experience. *Environ. Health Prespec.* 102(Suppl. 9): 11–17.

Kumar, R., Pant, N., and Srivastava, S. P. 2000a. Chlorinated pesticides and heavy metals in human semen. *Int. J. Androl.* 23: 145–149.

Kumar, S., Gautam, A. K., and Saiyed, H. W. 2000b. Occupational exposure and male reproductive dysfunction a growing concern. *Indian J. Occup. Environ. Med.* 4: 89–94.

Kumar,V., Yadav, C. S., Singh, S., et al. 2010. CYP 1A1 polymorphism and organochlorine pesticides levels in the etiology of prostate cancer. *Chemosphere* 81: 464–468.

Kumari, B. 2008. Effects of household processing on reduction of pesticide residues in vegetables. *ARPN J. Agric. Biol. Sci.* 3(4): 46–51.

Kutz, F. W., Yobs, A. R., and Strassman, S. C. 1976. Organochlorine pesticide residues in human adipose tissue. *Toxicol. Pathol.* 4(1): 17–19.

Lee, D. H., Jacobs, D. R., Jr., and Porta, M. 2006. Could low-level background exposure to persistent organic pollutants contribute to the social burden of type 2 diabetes? *J. Epidemiol. Community Health* 60: 1006–1008.

Lee, D. H., Lee, I. K., Jin, S. H., Steffes, M., and Jacobs, D. R., Jr. 2007. Association between serum concentrations of persistent organic pollutants and insulin resistance among nondiabetic adults: results from the National Health and Nutrition Examination Survey 1999–2002. *Diab. Care* 30: 622–628.

Leiss, J. K. and Savity, D. A. 1995. Home pesticide use and childhood cancer: A case control study. *Am. J. Public Health* 85: 249–252.

Leoni, V., Fabiani, L., Marinelli, G., et al. 1989. PCB and other organochlorine compounds in blood of women with or without miscarriage: A hypothesis of correlation. *Ecotoxicol. Environ. Safe.* 17: 1–11.

Lerda, D. and Rizzi, R. 1991. Study of reproductive function in persons occupationally exposed to 2,4-dichlorophenoxyacetic acid (2,4-D). *Mutat. Res.* 262: 47–50.

Linde, C. D. 1994. Physico-chemical Properties and Environmental Fate of Pesticides. Environmental Hazards Assessment Program State of California, Environmental Protection Agency, Department of Pesticide Regulation, Environmental Monitoring and Pest Management Branch, Sacramento, CA 95814, EH 94-03: 53p.

Lino, C. M. and da Silveira, M. I. N. 2006. Evaluation of organochlorine pesticides in serum from students in Coimbra, Portugal: 1997–2001. *Environ. Res.* 102: 339–351.

Loevinsohn, M. E. 1987. Insecticide use and increased mortality in rural Central Luzon, Philippines. *Lancet* 13: 1359–1362.

Lopez-Espinosa, M. J., Granada, A., Carreño, J., Salvatierra, M., Olea-Serrano, F., and Olea, N. 2007. Organochlorine pesticides in placentas from Southern Spain and some related factors. *Placenta* 28: 631–638.

Lopez-Espinosa, M. J., Lopez-Navarrete, E., Rivas, A., et al. 2008. Organochlorine pesticide exposure in children living in southern Spain. *Environ. Res.* 106: 1–6.

Mackay, D. and Leeinonen, P. J. 1975. Rate of evaporation of low solubility contaminants from water bodies to atmosphere. *Environ. Sci. Technol.* 9: 1178–1180.

Manirakiza, P., Akimbamijo, O., Covaci, A., et al. 2002. Persistent chlorinated pesticides in fish and cattle fat and their implications for human serum concentrations from the Sene-Gambian region. *J. Environ. Monitor.* 4: 609–617.

Mansour, S. A. 2004. Pesticide exposure – Egyptian Scene. *Toxicology* 198: 91–115.

Mansour, S. A. 2006. Monitoring of pesticides and heavy metals in the western desert lakes of Egypt and bioassaying potential toxicity of their waters. *Proceedings, 4th Environmental Conference on Pesticides and Related Organic Micropollutants in the Environment/10th Symposium on the Chemical Fate of Modern Pesticides, University of Almeria, Almeria, Spain,* 26–29 November 2006, pp. 113–117.

Mansour, S. A. 2008. Environmental impact of pesticides in Egypt. *Rev. Environ. Contam. Toxicol.* 196: 1–51.

Mansour, S. A. 2009. An overview to the present status of biopesticides in Egypt. In: Latos, T. H. (ed.), *Cover Crops and Crop Yields*, pp. 175–203. Nova Science Publishers, Inc.: Hauppauge, NY.

Mansour, S. A. 2010. Epidemiological studies, Egypt. In: Satoh, T. and Gupta, R. C. (eds), *Anticholineseterase Pesticides: Metabolism, Neurotoxicity, and Epidemiology*, Chapter 28, pp. 379–401. John Wiley & Sons: ISBN: 978-0-470-41030-1.

Mansour, S. A. and Heikal, T. M. 2001. Xenobiotics Interaction. 3: A further study on the use of biochemical markers to analyze the joint action of insecticide mixtures in the rat. *J. Egypt Ger. Soc. Zool.* 35(E): 31–49.

Mansour, S. A. and Mossa, A. H. 2010. Adverse effects of lactational exposure to chlorpyrifos in suckling rats. *Hum. Exp. Toxicol.* 29(2): 77–92.

Mansour, S. A. and Refaie, A. A. 2000. Xenobiotics Interaction. 2: An approach to the use of biochemical data measurements for interpreting interaction of insecticide mixtures in rat. *Adv. Pharmacol. Toxicol.* 1(1): 1–20.

Mansour, S. A. and Sidky, M. S. M. 2002. Ecotoxicological studies. 3: Heavy metals contaminating water and fish from Fayoum Governorate, Egypt. *Food Chem.* 78: 15–22.

Mansour, S. A., Samaan, H. A., Abdel-Mageed, F. A., and Nessiem, A. L. 1997. Xenobiotics Interaction. I: Effect of different insecticides, with and without an anti-bilharzial drug (Praziquantel) on albinorats. *J. Egypt Ger. Soc. Zool.* 23(E): 93–106.

Mansour, S. A., Mahran, M. R., and Sidky, M. S. M. 2001a. Ecotoxicological studies. 4: Monitoring of pesticide residues in the major components of Lake Qarun. *Egypt. J. Egypt Acad. Soc. Environ. Dev.* 2(1): 83–116.

Mansour, S. A., Refaie, A. A., and Nada, S. A. 2001b. Xenobiotics Interaction. 4: Effect of some pesticides and their mixtures on the growth rate of albino rats. *Adv. Pharmacol. Toxicol.* 2(2): 9–24.

Mansour, S. A., Belal, M. H., Abou-Arab, A. K., Ashour, H. M., and Gad, M. F. 2009a. Evaluation of some pollutant levels in conventionally and organically farmed potato tubers and their risks to human health. *Food Chem. Toxicol.* 47: 615–624.

Mansour, S. A., Belal, M. H., Abou-Arab, A. K., and Gad, M. F. 2009b. Monitoring of pesticides and heavy metals in cucumber fruits produced from different farming systems. *Chemosphere* 75: 601–609.

Mathur, H. B., Agarwal, H. C., Johnson, S., and Saikia, N. 2005. Analysis of pesticide residues in blood samples from villages of Punjab. CSE/PML/PR-21/2005, 41, Tughlakabad Institutional Area, New Delhi 110062, pp. 21 + Annex III.

Mattison, D. R., Bogumil, R. J., Chapin, R., et al. 1990. Reproductive effects of pesticides. In: Baker, S. R. and Wilkinson, C. F. (eds), *The Effects of Pesticides on Human Health*, pp. 297–389. Princeton Scientific Publishers: New Jersey.

Meeker, J. D., Ryan, L., Barr, D. B., et al. 2004. The relationship of urinary metabolites of carbaryl/naphthalene and chlorpyrifos with human semen quality. *Environ. Health Perspect.* 112: 1665–1670.

Merhi, M., Demurc, C., Racaud-Sultan, C., et al. 2010. Gender-linked haematopoietic and metabolic disturbances induced by a pesticide mixture administered at low dose to mice. *Toxicology* 267: 80–90.

Mino, C. P., Bustamante, G., Sánchez, M. E., and Leone, P. E. 2002. Cytogenetic monitoring in a population occupationally exposed to pesticides in Ecuador. *Environ. Health Perspect.* 110(11): 1077–1080.

Misra, U. K., Nag, D., Bhushan, V., and Ray, P. K. 1985. Clinical and biochemical changes in chronically exposed organophosphate workers. *Toxicol. Lett.* 244: 187–193.

Moses, M., Johnson, E. S., Anger, W. K., et al. 1993. Environmental equity and pesticide exposure. *Toxicol. Ind. Health* 9(5): 913–959.

Mumtaz, M. M. 1995. Risk assessment of chemical mixtures from a public health perspective. *Toxicol. Lett.* 527–532.

Nag, D., Singh, C. C. and Senon, S. 1977. Epilepsy endemic due to benzene hexachloride. *Trop. Geogr. Med.* 29: 229–232.

Nakata, H., Kawazoe, M., Arizono, K., et al. 2002. Organochlorine pesticides and polychlorinated biphenyl residues in foodstuffs and human tissues from China: Status of contamination, historical trend, and human dietary exposure. *Arch. Environ. Contam. Toxicol.* 43: 473–480.

Needham, L. L., Barr, D. B., Caudill, S. P., et al. 2005. Concentrations of environmental chemicals associated with neurodevelopmental effects in U.S. population. *Neurotoxicology* 26: 531–554.

Nigam, S. K., Karnik, A. B., Chattopadhyay, P., Lakkad, B. C., Venkaiah, K., and Kashyap, S. K. 1993. Clinical and biochemical investigations to evolve early diagnosis in workers involved in the manufacture of hexachlorocyclohexane. *Int. Arch. Occup. Environ. Health* 65(1): 193–196.

Organic Monitor. 2002. The Global Market for Organic Food & Drink. http://www.organicmonitor.com/700140.htm (Retrieved 2006-06-20).

PAN. 1993. Toxaphene in North Sea Fish. Global Pesticide Campaigner. *Pestic. Act. Netw.* 3(4): 1–17.

PAN. 2001. The list of the lists: A catalogue of lists of pesticides identifying those associated with particularly harmful health or environmental impacts. *Pestic. Act. Netw.* 3 November, p. 9.

Pant, N., Upadhyay, G., Pandey, S., Mathur, N., Saxena, D. K., and Srivastava, S. P. 2003. Lead and cadmium concentration in the seminal plasma of general population and their correlation with sperm quality. *Reprod. Toxicol.* 17: 447–450.

Pant, N., Kumara, R., Mathur, N., Srivastava, S. P., Saxena, D. K., and Gujrati, V. R. 2007. Chlorinated pesticide concentration in semen of fertile and infertile men and correlation with sperm quality. *Environ. Toxicol. Pharmacol.* 23: 135–139.

Panuwet, P., Prapamontol, T., Chantara, S., and Barr, D. B. 2009. Urinary pesticide metabolites in school students from northern Thailand. *Int. J. Hyg. Environ. Health* 212: 288–297.

Patandin, S., Dagnelie, P. C., Mulder, P. G., et al. 1999. Dietary exposure to polychlorinated biphenyls and dioxins from infancy until adulthood: A comparison between breast-feeding, toddler, and long-term exposure. *Environ. Health Perspect.* 107: 45–51.

Peris, M., Micó, C., Recatalá, L., Sánchez, R., and Sánchez, J. 2007. Heavy metal contents in horticultural crops of a representative area of the European Mediterranean region. *Sci. Total Environ.* 378: 42–48.

Petrik, J., Drobna, B., Pavuk, M., Jursa, S., Wimmerova, S., and Chovancova, J. 2006. Serum PCBs and organochlorine pesticides in Slovakia: Age, gender, and residence as determinants of organochlorine concentrations. *Chemosphere* 65(3): 410–418.

Philibert, A., Schwartz, H., and Mergler, D. 2009. An exploratory study of diabetes in a First Nation community with respect to serum concentrations of p,p'-DDE and PCBs and fish consumption. *Int. J. Environ. Res. Public Health* 6: 3179–3189.

Porta, M., Puigdomènech, F., Ballester, E., et al. 2008. Monitoring concentrations of persistent organic pollutants in the general population: The international experience-Review. *Environ. Int.* 34: 546–561.

Porta, M., Gasull, M., Puigdomènech, E., et al. 2010. Distribution of blood concentrations of persistent organic pollutants in a representative sample of the population of Catalonia. *Environ. Int.* 36: 655–664.

Randhawa, M. A., Anjum, F. M., Asi, M. R., Butt, M. S., Ahmed, A., and Randhawa, M. S. 2007. Removal of endosulfan residues from vegetables by household processing. *J. Sci. Ind. Res.* 66: 849–852.

Rezg, R., Mornagui, B., El-Fazaa, S., and Gharbi, N. 2010. Organophosphorus pesticides as food chain contaminants and type 2 diabetes: A review. *Trends Food Sci. Technol.* 21: 345–357.

Richter, E. D. 1992. Aerial application and spray drift of anticholinesterases: Protective measures. In: Ballatyne, B. and Aldridge, T. (eds), *Clinical and Experimental Toxicology of Organophosphates and Carbamates*, pp. 623–631. Butterworth-Heinemann Ltd.: Oxford.

Rita, P., Reddy, P. P., and Reddy, S. V. 1987. Monitoring of workers occupationally exposed to pesticides in grape gardens of Andhra Pradesh. *Environ. Res.* 44: 1–7.

Rola, A. C. 1989. Pesticides, Health risks and farm productivity: A Philippine experience. Monogr No. 89-01, *Agricultural Policy Research Programme*, University of Philippines, Los Banos: Los Banos.

Rupa, D. S., Reddy, P. P., and Reddy, O. S. 1991. Reproductive performance in population exposed to pesticides in cotton fields in India. *Environ. Res.* 55: 123–128.

Safi, J. M. 2002. Association between chronic exposure to pesticides and recorded cases of human malignancy in Gaza Governorates (1990–1999). *Sci. Total Environ.* 284: 75–84.

Sanusi, A., Millet, M., Mirabe, P., and Wortham, H. 2000. Comparison of atmospheric pesticide concentrations measured at three sampling sites: Local, regional and long-range transport. *Sci. Total Environ.* 263(1–3): 263–277.

Sayed, H. N., Sadhu, H. G., Bhatnagar, V. K., Dewan, A., Venkaiah, K., and Kashyap, S. K. 1992. Cardiac toxicity following short term exposure to methomyl in spray-men and rabbits. *Hum. Exp. Toxicol.* 11: 93–95.

Schrader, S. M., Turner, T. W., and Ratcliffe, J. M. 1988. The effects of ethylene dibromide on semen quality: A comparison of short term and chronic exposure. *Reprod. Toxicol.* 2: 191–198.

Sethuraman, V. A. 1977. Case of BHC poisoning in a heifer calf. *Ind. Vet. J.* 54: 486–487.

Shah, P. V., Fisher, H. L., Sumler, M. R., Monroe, R. J., Chernoff, N., and Hall, L. L. 1987. Comparison of the penetration of 14 pesticides through the skin of young and adult rats. *J. Toxicol. Environ. Health* 21: 353–366.

Sharma, R. P., Bhatnagar, V. K., Saigal, S., Gupta, M. M., Bansal, G., and Malviya, A. N. 1986. Skin diseases in workers handling pesticides. *Environ. Ecol.* 4: 175–177.

Sherif, S. O., Abdulla, E. A., Morsy, A. F., and Ahmad, R. T. 2005. Chronic exposure to organochlorine pesticides in urban Giza school children. *Med. J. Cairo Univ.* 73(2): 241–255.

Silvestroni, L. and Palleschi, S. 1999. Effects of organochlorine xenobiotics on human spermatozoa. *Chemosphere* 39(8): 1249–1252.

Simcox, N. J., Fenske, R. A., Wolz, S. A., Lee, I. C., and Kalman, D. A. 1995. Pesticides in household dust and soil: Exposure pathways for children of agricultural families. *Environ. Health Perspect.* 103: 1126–1134.

Skakkebaek, N. E., Jørgensen, N., Main, K. M., et al. 2006. Is human fecundity declining? *Int. J. Androl.* 29: 2–11.

Slutsky, M., Levin, J. L., and Levy, B. S. 1999. Azoospermia and oligospermia among a large cohort of DBCP applicators in 12 countries. *Int. J. Occup. Environ. Health* 5: 116–122.

Soliman, A. S, Smith, M. A., Cooper, S. P., et al. 1979. Serum organochlorine pesticide levels in patients with colorectal cancer in Egypt. *Arch. Environ. Health* 52(6): 409–415.

Soliman, S. A., Sovocool, G. W., Curley, A., Ahmed, N. S., El-Fiki, S., and El-Sebae, A. H. 1982. Two acute human poisoning cases resulting from exposure to diazinon transformation products in Egypt. *Arch. Environ. Health* 37(4): 207–212.

Solomon, G. M. and Weiss, P. M. 2002. Chemical contaminants in breast milk: Time trends and regional variability. *Environ. Health Perspect.* 110: 339–347.

Son, H. K., Kim, S. A., Kang, J. H., et al. 2010. Strong associations between low-dose organochlorine pesticides and type 2 diabetes in Korea. *Environ. Int.* 36: 410–414.

Sosan, M. B., Akingbohungbe, A. E., Ojo, I. A. O., and Durosinmi, M. A. 2008. Insecticide residues in the blood serum and domestic water source of cacao farmers in Southwestern Nigeria. *Chemosphere* 72: 781–784.

Stehr-Green, P. A., Farrar, J. A., Burse, V. W., Royce, W. G., and Wohlleb, J. C. 1988. A survey of measured levels and dietary sources of selected organochlorine pesticide residues and metabolites in human sera from a rural population. *Am. J. Public Health* 78(7): 828–830.

Stellman, S. D., Djordjevic, M. V., Muscat, J. E., et al. 1998. Relative abundance of organochlorine pesticides and polychlorinated biphenyls in adipose tissue and serum of women in Long Island, New York. *Cancer Epidemiol. Biomark Prev.* 7: 489–496.

Subramaniam, K. and Solomon, R. D. J. 2006. Organochlorine pesticides BHC and DDE in human blood in and around Madurai, India. *Indian J. Clin. Biochem.* 21(2): 169–172.

Talamanca, F., Traina, M., and Urbani, E. 2001. Occupational exposure to metals, solvents and pesticides recent evidence on male reproductive effects and biological markers. *Occup. Med.* 51: 174–178.

Tan, J., Cheng, S. M., Loganath, A., Chong, Y. S., and Obbard, J. P. 2007. Selected organochlorine pesticide and polychlorinated biphenyl residues in house dust in Singapore. *Chemosphere* 68: 1675–1682.

Tariq, M. I., Afzal, S., Hussain, I., and Sultana, N. 2007. Pesticides exposure in Pakistan: A review. *Environ. Int.* 33: 1107–1122.

Thomas, G. O., Wilkinson, M., Hodson, S., and Jones, K. C. 2006. Organohalogen chemicals in human blood from the United Kingdom. *Environ. Poll.* 141(1): 30–41.

Thompson, B., Coronado, G. D., Grossman, J. E., et al. 2003. Pesticide take-home pathway among children of agricultural workers: Study design, methods, and baseline findings. *J. Occup. Environ. Med.* 45(1): 42–53.

Trewavas, A. 2004. A critical assessment of organic farming and food assertions with particular respect to the UK and the potential environmental benefits of no-till agriculture. *Crop. Protect.* 23: 757–781.

Turner, R. J. (ed.). 1994. *Immunology: A Comparative Approach*. John Wiley & Sons Ltd.: Chichester, UK.

Turyk, M., Anderson, H. A., Knobeloch, L., Imm, P., and Persky, V. W. 2009. Organochlorine exposure and incidence of diabetes in a cohort of Great Lakes sport fish consumers. *Environ. Health Perspect.* 117: 1076–1082.

UNIDO. 2003. Enhancing chemical management for improved productivity, market access and environment. Persistent Organic Pollutants (POPs). Vienna, Austria: United Nations Industrial Development Organization, V.03-88748, 7 pp.

Vainio, H., Matos, E., and Kogevinas, M. 1994. Identification of occupational carcinogens. In: Pearce, N., Matos, E., Vainio, H., Boffetta, P., and Kogevinas, M. (eds), *Occupational Cancer in Developing Countries*, No. 129, pp. 41–59. International Agency for Research on Cancer: Lyon.

Voorspoels, S., Covaci, A., Maervoet, J., and Schepens, P. 2002. Relationship between age and levels of organochlorine contaminants in human serum of a Belgian population. *Bull. Environ. Contam. Toxicol.* 69: 22–29.

Vos, J. G., Smialowicz, R., and Van Loveren, H. 1994. Animal models for assessment. In: Dean, J. H., Luster, M. I., Munson, A. E., and Kimber, I. (eds), *Immunotoxicology and Immunopharmacology*, 2nd edn, pp. 19–30. Raven Press Ltd.: New York.

Waliszewski, S. M., Aguirre, A. A., Benitez, A., Infanzon, R. M., Infanzon, R., and Rivera, J. 1999a. Organochlorine pesticide residues in human blood serum of inhabitants of Veracruz, Mexico. *Bull. Environ. Contam. Toxicol.* 62: 397–402.

Waliszewski, S. M., Aguirre, A. A., Infanzon, R. M., Benitez, A., and Rivera, J. 1999b. Comparison of organochlorine pesticide levels in adipose tissue and human milk of mothers living in Veracruz, Mexico. *Bull. Environ. Contam. Toxicol.* 62: 685–690.

Waliszewski, S. M., Quintana, R. V., Corona, C. A., et al. 2010. Comparison of organochlorine pesticide levels in human adipose tissue of inhabitants from Veracruz and Puebla, Mexico. *Arch. Environ. Contam. Toxicol.* 58(1): 230–236.

Wesseling, C., Hogstedt, C., Picado, A., and Johansson, L. 1997. Unintentional fatal paraquat poisonings among agricultural workers in Costa Rica: A report of 15 cases. *Am. J. Ind. Med.* 32(5): 433–441.

Whitford, F., Pike, D., Hanger, G., Burroughs, F., Johnson, B., Blessing, A. 2004. *The Benefits of Pesticides: A Story Worth Telling*. Purdue University, Purdue Extension, PPP-70: pp. 33. http://www.ces.purdue.edu/new.

WHO. 1990. *Cancer: Causes, Occurrence, and Control*. IARC Scientific Publication No. 100: Lyon. WRI. 1994. World Resources Institute in Collaboration with the UN Environmental Programme, World Resources 1994–1995, Washington, DC.

WHO. 1993. Pesticides and health in the Americas. Environment Series No. 12. WHO: Washington, DC.

WHO. 2004. The impact of pesticides on health: Preventing intentional and unintentional deaths from pesticide poisoning. WHO: Geneva. http:www.who.int/mental... health/prevention/suicide/en/PesticidesHealth2.pdf.

Wicklund Glynn, A., Wolk, A., Aune, M., et al. 2000. Serum concentrations of organochlorines in men: A search for markers of exposure. *Sci. Total Environ.* 263: 197–208.

Wild, S., Roglic, G., Green, A., Sicree, R., and King, H. 2004. Global prevalence of diabetes: estimates for the year 2000 and projections for 2030. *Diabetes Care* 27(5): 1047–1053.

Williamson, S., Ball, A., and Pretty, J. 2008. Trends in pesticide use and drivers for safer pest management in four African countries. *Crop Protect.* 27: 1327–1334.

Wilson, C. and Tisdell, C. 2001. Why farmers continue to use pesticides despite environmental, health and sustainability costs. *Ecol. Econ.* 39: 449–462.

Wolf, R. E. 2002. New technologies for the delivery of pesticides in agriculture. In: Wheeler, W. B. (ed.), *Pesticides in Agriculture and the Environment*, 360 pp. Marcel Dekker, Inc.: New York.

WRI. 1994. World Resources Institute (WRI) in collaboration with the UN Environmental Programme, World Resources 1994–1995. Oxford University Press: UK.

WRI. 1996. *Pesticides and the Immune System: The Public Health Risks*. World Resources Institute: Washington, DC.

Zaidi, S. S. A., Bhatnagar, V. K., Gandhi, S. J., Shah, M. P., Kulkarni, P. K., and Sayed, H. N. 2000. Assessment of thyroid functions in pesticide formulators. *Hum. Exp. Toxicol.* 19: 497–500.

Zumbado, M., Goethals, M., Álvarez-León, E. E., et al. 2005. Inadvertent exposure to organochlorine pesticides DDT and derivatives in people from the Canary Islands (Spain). *Sci. Total Environ.* 339: 49–62.

16

Pesticides and Skin Diseases in Man

Radoslaw Spiewak

CONTENTS

16.1 Introduction

The skin is the largest organ of the human body. With a surface area of 1.5–2 m² and thickness of 0.5–5 mm, it amounts to 15% of the total body weight. Thanks to its complexity and unique multilayer structure, this versatile organ forms an extremely effective barrier against a range of external factors—from solar irradiation to microbes and from mechanical trauma to xenobiotics and irritants (Spiewak 1998). Its durability is warranted by its continuous regeneration. The outermost layer of the skin—the epidermis—is fully renewed every 3–4 weeks, a process referred to as "epidermal cycle." The function of an effective barrier between the internal environment and the external influence is not the only purpose of the largest organ—the skin also takes active part in metabolism, homeostasis, thermoregulation, and hormonal regulation. Next to the digestive tract, the skin is populated by the largest number of immune cells and plays a pivotal role in innate and adaptive immune response.

16.2 Problems with Terminology

Considering the above-mentioned complexity and importance of the skin functions, one can wonder why some people consider the skin to be merely a kind of "encasing" for the internal organs. Specifically, it is impossible to defend some toxicologists' views on the skin as a kind of semipermeable membrane through which pesticides enter the body. This idea is reflected in the popular misnomer "dermal toxicity" that has been used through decades for internal or systemic toxicity of pesticides absorbed via the skin (Noakes and Sanderson 1969; Zendzian 2003). The logics would dictate that the term should be used in reference to toxic damage caused by pesticides to the skin itself, in analogy to, for example, "hepatotoxicity," "cardiotoxicity," or "neurotoxicity" (meaning toxicity to the respective organs, rather than toxicity of the agents absorbed through the liver, heart, or nerves). In order to avoid further confusion, the misleading term "dermal toxicity" should be ultimately and definitely abandoned. In accordance with scientific clarity, the toxic effects of dermally absorbed pesticides to the internal organs could be referred to as "percutaneous toxicity" (Whitesell et al. 1947), "toxicity of dermally absorbed pesticides" (Latuszynska et al. 1999), or perhaps "transdermal toxicity." The word "dermatotoxicity" (Kimura et al. 1998; Kataranovski et al. 2005) should be instead promoted as the actual term for describing skin damage caused by chemical substances, also when ingested, inhaled, or administered parenterally (Ali and Oehme 1992; Muhammad et al. 2010).

16.3 Effects of Pesticides in the Skin

Pesticides may act upon the skin as irritants, sensitizing haptens, photohaptens, and carcinogens. They may also possess pharmacological properties that can alter the physiological functions of the skin. The underlying mechanisms are only partially known, and in most cases, we have to retreat to the general understating of the skin effects caused by xenobiotics, rather than specific data from studies on specific pesticides. The reader is directed elsewhere for an overview of the general mechanisms underlying irritant and allergic contact dermatitis (ACD) (English 2004; Gober and Gaspari 2008; Nosbaum et al. 2009), photosensitivity (Lankerani and Baron 2004; Kerr and Ferguson 2010), chemically induced porphyrias (Downey 1999), and carcinogenesis (Bowden et al. 1993; Slaga et al. 1995). Only research specifically devoted to studying the responses to pesticide within the skin will be discussed further in this paragraph. At present, information on this topic is scarce and limited to a few pesticides only. As many other xenobiotics, pesticides may act as sensitizing haptens or autoimmunogenic compounds that trigger pathologic reactions via thymus-dependent mechanisms. One of the pesticides that causes adverse immune effects in both man and rat in the T cell–dependent manner is hexachlorobenzene

(HCB)—a fungicide formerly used for seed treatment (Ezendam et al. 2004). It appears, however, that skin lesions can develop in response to HCB also in the T cell–independent manner. Lesions in the T cell–depleted rats, though developed more slowly, were similar in appearance; histopathological examination indicated the important roles of mononuclear phagocytes and eosinophils in skin reactions to HCB (Michielsen et al. 1999). Exposure of mice to parathion or methoxychlor enhanced the rates of sensitization upon a subsequent exposure to environmental haptens, which may seem somewhat paradoxical as both pesticides alone possess immunosuppressive rather than immunostimulatory effects, probably via apoptosis of thymus-dependent lymphocytes (Fukuyama et al. 2010a,b). Pesticides may also possess pharmacological properties, for example, the organophosphate insecticide malathion causes long-term modulation in the physiological function of the cutaneous vasculature through its influence on both neuronal and nonneuronal acetylcholinesterase (Boutsiouki and Clough 2004). The insecticide bisphenol-methoxychlor has estrogen-like properties (Sakabe et al. 1998). Chlorinated pesticides may interfere with the metabolism of vitamin A (retinoid) in the skin, resulting in disturbances of keratinization within the pilosebaceous duct, which in turn result in occupational acne (Coenraads et al. 1994). The herbicide propanil induces chloracne in humans and hyperkeratosis in rabbits (Kimbrough 1980). Although, typically referred to as "chloracne," the comedogenic effect seems not restricted to organochlorine pesticides as it was also observed in the case of zineb (Kimura et al. 1998).

Although the undisputed major risk factor for skin cancer is UV exposure, animal studies indicate that inflammation and histological changes are more pronounced in the pesticide-exposed skin (Kimura et al. 1998). Arsenic, but possibly also other pesticides, is capable of inducing cancer independently from sunlight exposure. Arsenic causes DNA damage in keratinocytes via generation of hydroxyl radicals (Shi et al. 2004). This carcinogenic effect may occur via upregulation of mitogen-activated protein (MAP) kinase cascades and subsequent upregulation of genes including c-fos and c-jun (Cooper et al. 2004). Furthermore, arsenic pesticides cause increases in granulocyte/macrophage colony-stimulating factor (GM-CSF) in the epidermis and transforming growth factor-alpha (TGF-α)—both factors known to be promoters of skin tumor formation (Luster et al. 1995; Germolec et al. 1998). Upregulation of matrix-degrading proteases such as matrix metalloproteinase 9 (MMP-9) via epidermal growth factor (EGF) may also play an important role in arsenic-induced skin carcinogenesis (Cooper et al. 2004). Also, the nonarsenic herbicide paraquat was demonstrated to cause genotoxic effects on human lymphocytes, manifested by an increased number of chromosomal aberrations (Jovtchev et al. 2010).

16.4 Pesticides and Autoimmunity

Autoimmunity is an unfavorable breakage of the immunologic tolerance toward the body's own antigens. This is a systemic phenomenon; however, numerous autoimmune diseases involve the skin. Pesticide exposure is a suspected risk factor for autoimmune diseases (Holsapple 2002); and a higher prevalence of antinuclear antibodies (ANA) related to connective tissue diseases was observed among rural residents (Spiewak and Stojek 2003). Long-term exposure to insecticides was demonstrated to be a significant risk factor for two autoimmune connective tissue diseases—rheumatoid arthritis and systemic lupus erythematosus (SLE) (Parks et al. 2011). However, this effect seems to depend on the type of pesticide, as the risk for developing autoantibodies shows a positive correlation with occupational exposure to trifluralin, but a negative correlation with exposure to bromoxynil and 2,4-dichlorophenoxyacetic acid (Semchuk et al. 2007). Bisphenol-methoxychlor stimulates the human keratinocytes to express the SS-A/Ro autoantigens that are associated with SLE, Sjögren's syndrome, and other autoimmune connective tissue diseases with prominent skin involvement (Sakabe et al. 1998).

16.5 Pesticide-Related Skin Diseases

Pesticides form a very heterogeneous group of substances that are classified according to utilitarian criteria; therefore, their chemical structures and biological effects are very diverse. An apparently common feature of these substances is their high biological activity, which may affect also the human skin. An overview of pesticide-related skin disorders is given in Table 16.1.

16.5.1 Chemical Burns

Skin contact with corrosive chemicals may cause direct destruction (necrosis) of the exposed tissues. As corrosive properties of the chemicals become apparent at an early stage of the safety assessment, skin burns are mainly the results of accidental exposures to concentrated pesticides, their criminal (homicidal or suicidal) use, or extremely negligent handling.

16.5.2 Irritant Contact Dermatitis

Irritant contact dermatitis (ICD) is among the most frequent occupational diseases worldwide. Its clinical diagnosis is relatively difficult to establish with an undisputed certainty, mainly due to the following factors: (1) Similarity of symptoms and frequent overlapping with ACD; (2) On individual basis, there is little possibility of confirming the causal link between the disease and a particular exposure; and, most importantly, (3) The diagnosis of "irritant contact dermatitis" is made by exclusion: there is no other way of establishing the diagnosis of ICD than by excluding all other types of dermatitis. Altogether, this leads to a considerable underreporting of ICD, both at the individual and population levels (Spiewak 2003). Table 16.2 presents a proposal of working criteria for the diagnosis of work-related ICD, which are a compilation of own experience (Spiewak 1999) and criteria proposed by Penagos (2002).

16.5.3 Phototoxic Contact Dermatitis

The terms "phototoxic reaction" and "phototoxicity" refer to an inflammatory reaction of the skin, resulting from direct cellular damage produced by the photochemical reaction initiated by photoactive chemicals (photosensitizers) and the active spectrum of radiation on the skin, mainly in the UVA range (Spiewak 2009b). The two main pathways of phototoxicity are reactive oxygen species (ROS) pathway and reactive nitrogen species (RNS) pathway. The most common clinical manifestation of phototoxicity is an exaggerated,

TABLE 16.1

An Overview of Skin Diseases Caused by Pesticides, Their Precursors, and Contaminants

Skin Disease	Examples of Causative Pesticides (Source)
Chemical burns	Calcium polysulfide (Horiuchi et al. 2008)
	Dazomet (Horiuchi et al. 2008)
	Diazinon (Horiuchi et al. 2008)
	Dichlorvos (Horiuchi et al. 2008)
	Diquat (Manoguerra 1990)
	Fenitrothion (Horiuchi et al. 2008)
	Glyphosate (Bradberry et al. 2004; Horiuchi et al. 2008)
	Gramoxone (Penagos 2002)
	Methomyl (Horiuchi et al. 2008)
	Methyl bromide (Horiuchi et al. 2008)
	Paraquat (Rahman et al. 2007; Horiuchi et al. 2008; Premaratna et al. 2008)
	Phenothrin (Choudhary 1999)
	Quintozene (Horiuchi et al. 2008)
	Zineb (Horiuchi et al. 2008)
Irritant contact dermatitis	Captafol (Stoke 1979)
	2-Chloro-N,N-diallyl-acetamide (Spencer 1966)
	Chlorothalonil (Lensen et al. 2007)
	D–D mixture[a] (Nater and Gooskens 1976)
	Dichlorvos (Mathias 1983)
	Glyphosate (Heras-Mendaza et al. 2008)
	Lindane (Yu et al. 2005)
	Paraquat (Howard 1979)
	Propargite (CDC 1986)
Phototoxic contact dermatitis	Chlorothalonil (Penagos 2002)
	Glyphosate (Hindson and Diffey 1984)
Allergic contact dermatitis	Acephate (Nakamura and Miyachi 2002)
	Anilazin (Jung 1993b)
	Benomyl (Savitt 1972; Jung et al. 1987)
	Bupirimate (McFadden et al. 1993)
	Captafol (Camarasa 1975; Brown 1984; Mark et al. 1999)
	Captan (Jung et al. 1987; Lisi et al. 1987; Guo et al. 1996; Verma et al. 2007)
	Carbendazim (Jung 1996)
	4-Chloro-2-butynyl N-(3-chlorophenyl) carbamate (Brancaccio et al. 1977)
	Chlorothalonil (Bruynzeel and Ketel 1986)
	Chlorpyriphos-methyl (Bonamonte et al. 2001)
	Cypermethrin (Jung 1996)
	Dazomet (Jung et al. 1987)
	D-D mixture[a] (Nater and Gooskens 1976)
	Dichlorvos (Jung et al. 1987)
	2,4-Dichloro-6-(o-chloroanilino)-s-triazine anilazine (Schuman and Dobson 1985)
	2,4-Dichlorophenoxyacetic acid (Jung et al. 1987)
	1,3-Dichloropropene (Bousema et al. 1991)
	Fenpropimorph (Jung 1993a)
	Fenthion (Bonamonte et al. 2001)
	Fluazinam (Bruynzeel et al. 1995)
	Folpet (Mark et al. 1999)
	Imazalil (Penagos 2002)
	Lindane (Jung 1987)
	Malathion (Guo et al. 1996; Bonamonte et al. 2001)
	Mancozeb (Burry 1976; Lisi and Caraffini 1985; Koch 1996)
	Maneb (Nater et al. 1979; Jung et al. 1987; Manuzzi 1988; Piraccini et al. 1991)
	Metham sodium (Jung et al. 1987)

(continued)

TABLE 16.1 (Continued)

An Overview of Skin Diseases Caused by Pesticides, Their Precursors, and Contaminants

Skin Disease	Examples of Causative Pesticides (Source)
	Methidathion (Bonamonte et al. 2001)
	Methyl parathione (Jung et al. 1987)
	Metiram (Koch 1996)
	Nematin (Helmdach and Schlenzka 1984)
	Norflurazon (Leow and Maibach 1996)
	Parathion (Pevny 1980)
	Parathion-methyl (Bonamonte et al. 2001)
	Phaltan (Fregert 1967)
	Phenmedipham (Jung et al. 1987)
	Propachlor (Schubert 1979; Jung et al. 1987)
	Propagite (Saunders et al. 1987)
	Propineb (Nishioka and Takahata 2000)
	Pyrethrum (McCord and Kilkee 1921)
	Thiabendazole (Penagos 2002)
	Thiuram (Spiewak 2004)
	Trichlorfon methyl (Jung et al. 1987)
	2,4,5-trichlorophenoxyacetic acid (Jung et al. 1987)
	Zineb (Jung et al. 1987)
	Ziram (Manuzzi 1988)
Photoallergic contact dermatitis	Captan (Mark et al. 1999)
	Chlorothalonil (Penagos 2002; Matsushita et al. 1996)
	Fenitrothion (Nakamura et al. 1999)
	Folpet (Mark et al. 1999)
	Mancozeb (Higo et al. 1996)
	Maneb (Lisi et al. 1987; Nakamura et al. 1999)
	Phoxim (Nakamura and Miyachi 2003)
Urticaria	Captan (Croy 1973)
	Chlorothalonil (Dannaker et al. 1993)
	Pentachlorophenol (Kentor 1986)
	Permethrin (Fuortes 1999)
Erythema multiforme[b]	Dimethoate (Schena and Barba 1992)
	Imazalil (Penagos 2002)
	Methyl parathion (Bhargava et al. 1977)
	Pyrethrum (Garcia-Bravo et al. 1995)
	Thiabendazole (Penagos 2002)
Erythema dyschromicum perstans (Ashy dermatosis)	Chlorothalonil (Penagos et al. 1996)
Acne	2,4-Dichlorophenol (Bleiberg et al. 1964; Poland et al. 1971)
	Hexachlorocyclohexane, HCH (Braun 1970)
	Neburone (Barriere et al. 1985)
	Pentachlorophenol, PCP (Cole et al. 1986; Hryhorczuk et al. 1998)
	Propanil (Kimbrough 1980)
	3,4,3′,4′-Tetrachloroazoxybenzene (Taylor et al. 1977)
	2,3,6,7-Tetrachlorodibenzodioxin (Jirasek et al. 1976)
	2,3,7,8-Tetrachlorodibenzop-dioxin, TCDD (Li 1986)
	1,1,1-Trichloro-2,2-di(4-chlorophenyl)ethane, DDT (Braun 1970)
	2,4,5-Trichlorophenol (Bleiberg et al. 1964; Poland et al. 1971)
	2,4,5-Trichlorophenoxyacetic acid (Braun 1970)
Porphyria cutanea tarda	Diazinon (Bopp and and Kosminsky 1975)
	2,4-Dichlorophenol (Bleiberg et al. 1964)
	Hexachlorobenzene (Schmid 1960)
	2,3,6,7-Tetrachlordibenzodioxin (Jirasek et al. 1976)
	2,4,5-Trichlorophenol (Bleiberg et al. 1964)

TABLE 16.1 (Continued)

An Overview of Skin Diseases Caused by Pesticides, Their Precursors, and Contaminants

Skin Disease	Examples of Causative Pesticides (Source)
Pigmentation disorders (leukoderma, melanoderma, leukomelanoderma)	Bipyridyl (Wang et al. 1987) Calcium polysulfide (Horiuchi et al. 2008) 4-Chloro-2-butynyl N-(3-chlorophenyl) carbamate (Brancaccio et al. 1977) Chlorpyrifos (Horiuchi et al. 2008) Methomyl (Horiuchi et al. 2008)
Nail disorders	Dinitro-ortho-cresol (Baran 1974) Diquat (Baran 1974) Paraquat (Hearn and Keir 1971; Baran 1974; Howard 1979; Botella et al. 1985)
Hair disorders	1,1,1-Trichloro-2,2-di(4-chlorophenyl)ethane, DDT (Haustein 1968; Kwiatkowska and Plonka 1971) Lindane (Haustein 1968)
Nonmelanoma skin cancer	Arsenic pesticides (Roth 1956; Braun 1958; Col et al. 1999; Braun 1958; Thiers et al. 1967; Jampel and Jerdan 1987; Wesseling et al. 1999) Paraquat (Wesseling et al. 1999)
Melanoma	Arsenic pesticides (Dennis et al. 2010) Benomyl (Dennis et al. 2010) Carbaryl (Dennis et al. 2010) 1,1,1-Trichloro-2,2-di(4-chlorophenyl)ethane, DDT (Barthel 1985) Maneb (Dennis et al. 2010) Mancozeb (Dennis et al. 2010) Parathion (Dennis et al. 2010)

[a] D–D mixture content: dichloropropanes, dichloropropenes, epichlorohydrin.

[b] Some researchers consider this disease a variant of ACD rather than a "real" erythema multiforme—for more explanation please refer to the paragraph on the disease.

sunburn-like reaction in the exposed areas of the skin, typically followed by a localized hyperpigmentation. In contrast to typical sunburns, phototoxic reactions are provoked by UV doses that normally are well tolerated. No individual- or photosensitizer-specific predisposition is required for phototoxic reactions. Pesticides capable of inducing phototoxic reactions are listed in Table 16.1.

TABLE 16.2

A Proposed Set of Criteria for Diagnosing Irritant Contact Dermatitis in Epidemiological Studies

Criterion	Description
Onset of the disease	Onset of symptoms within minutes or hours of the present exposure and/or within 2 weeks since the first exposure to the chemical in question
Subjective symptoms	Early in the clinical course: pain, burning, stinging sensation, or discomfort of the skin rather than pruritus (there is no urge to scratch the diseased skin)
Clinical picture	Macular erythema (dark-red, livid-red), glazed, parched, or scalded appearance of the epidermis, hyperkeratoses, fissuring, or chemical burns predominating
Patch test with the chemical in question[a]	Patient: negative reaction within nontoxic concentrations, irritant reaction within toxic concentrations Healthy (previously nonexposed) controls: negative reaction within nontoxic concentrations, irritant reactions to toxic concentrations in at least 50% of the controls

[a] Extreme care should be exercised when patch testing with suspected irritants. Material safety data sheets and other available toxicological data must be carefully consulted before undertaking the tests. From an ethical point of view, it is advisable that the researcher performing these tests serves self as a first control subject.

16.5.4 Allergic Contact Dermatitis

When analyzing the data on contact allergy (CA) and allergic contact dermatitis (ACD) to pesticides, one must not forget the difference between these two terms, which are too frequently mixed up—contact allergy (synonym: contact hypersensitivity) and allergic contact dermatitis (synonym: allergic contact eczema). CA is the body's readiness to develop an inflammatory reaction against specific haptens (low-molecular-weight chemicals). This state of hypersensitivity is acquired during previous skin exposures to the hapten. The presence of altered immune reactivity does not necessarily imply the existence of any disease. A group of people with CA will never develop clinical symptoms. Among those symptomatic, a vast majority will indeed develop the inflammatory skin disease—ACD; however, some sensitized people may also develop allergic contact stomatitis, conjunctivitis, urticaria, asthma, allergic rhinitis, or systemic reactions (Spiewak 2008). Typically, numerous exposures are necessary to induce hypersensitivity to haptens (Vandenberg and Epstein 1963). The duration of this induction phase depends on the individual predisposition of the exposed person and the sensitizing potency of the hapten (Schlede et al. 2003; Basketter et al. 2006). Some pesticides are potent sensitizers; parathion was even utilized as a model hapten in animal studies (Lisi et al. 1987). Strong-to-extreme sensitizing potency in animal models was also reported for maneb, mancozeb, zineb, and benomyl (Matsushita et al. 1976, 1977). This property is reflected in the reports on "epidemics" of ACD among farmers caused by nematin (Helmdach and Schlenzka 1984), 2,4-dichloro-6-(o-chloroanilino)-s-triazine anilazine (Schuman and Dobson 1985), and propargite (Saunders et al. 1987). In Japan, out of 815 pesticide-related diseases and poisonings registered from 1968 to 1970, ACD was diagnosed in 34% (Matsushita et al. 1980). Among all the occupational dermatoses compensated by the Polish Farmers' Social Insurance Fund, 18% were caused by pesticides (Spiewak 2003). In a study of Taiwanese fruit farmers spraying pesticides on a regular basis, contact hypersensitivity to pesticides was found in 40% and clinical symptoms of contact dermatitis was found in 30% (Guo et al. 1996). An anecdotal yet striking report suggests that susceptibility to pesticide sensitization might be race-dependent: in a Californian flower plantation, all seven coworkers of Japanese ancestry developed ACD to benomyl, while all ten Mexican coworkers remained unaffected. Moreover, one Japanese volunteer in the control group developed positive reaction to benomyl, while two Caucasians remained negative (Savitt 1972).

The above observations indicate that CA and ACD to pesticides are a relevant problem. However, prophylactic screening and further diagnosis of pesticide-related contact dermatitis is complicated by the dynamic changes in pesticide use in agriculture, industry, and households. New pesticides are continuously allowed while old ones are being withdrawn from the market. However, even if a pesticide is no longer on the market, sensitization of humans may persist for years and decades, eventually to reappear as cross-reactivity with a new pesticide structurally related to the primary sensitizer (Matsushita et al. 1976; Matsushita and Aoyama 1981; Peluso et al. 1991; Koch 1996). Numerous products of daily use (rubber, medications, housekeeping products) contain pesticides or chemically related substances that may provoke relapses of the disease. Allergizing properties may be due to pesticides themselves or due to additives such as emulgators or preservatives. Skin reactions may be provoked by degradation products of the active substances or additives, as in the case of workers sorting potatoes previously treated with metham sodium (sodium N-methyl dithiocarbamate), which underwent hydrolytic decomposition to form sensitizing methyl isothiocyanate (Schubert et al. 1993). The diagnosis should be based on clinical symptoms, history of exposure, and positive patch test reaction with the substance in question (Spiewak 2008). Hints on patch testing with pesticides (concentrations, vehicles) may be found elsewhere (de Groot 2008; Spiewak 2011).

16.5.5 Photoallergic Contact Dermatitis

Photoallergic contact dermatitis (PhotoACD) is an inflammatory disease that develops in the skin of sensitized individuals following exposure to sensitizing hapten ("photohapten" in this case) and the subsequent irradiation of the light. The light supplies the energy necessary for the conversion of precursors (prohaptens or prehaptens) into the actual sensitizing photohaptens or for the initiation of coupling (photobinding) the hapten with an endogenous carrier protein into a full antigen. Further, pathomechanisms and clinical appearance of the disease seem the same as in ACD, with the difference that it affects the body sites that are exposed both to the photohapten and subsequently to the light (Spiewak 2009a). This may sometimes complicate the diagnosis of PhotoACD, due to the irregular clinical pattern. The photosensitizing pesticide could be tolerated on cloudy days, and the Sun could be tolerated when not exposed to the photohapten. Thus, the risk of underdiagnosing this disease seems considerably high. Pesticides with documented photohaptenic properties are listed in Table 16.1.

16.5.6 Urticaria

Urticaria is characterized by the appearance of wheals—the clinical expression of localized vascular edema in the upper dermis caused by dilatation and increased permeability of the capillaries in response to the eliciting agent. The clinical expression of the reaction is wheals developing in the site of contact with the provoking chemical (contact urticaria) or sometimes involving also other body sites (generalized urticaria). The mechanisms of contact urticaria involve specific immune reactions or nonimmunologic mechanisms (Bourrain 2006). Pesticides are among the so-called urticants, that is, xenobiotics capable of provoking urticaria (Table 16.1).

16.5.7 Erythema Multiforme

A characteristic of this entity is a sudden onset of concentric erythematous papules with two or more concentric rings of slightly differing colors resembling the iris or bull's eye. Some authors suggest that erythema multiforme (EM) caused by external chemicals is a variant of ACD; however, taking into account that the etiology of EM remains obscure, it is hard to make any definite statement on its nature (Meneghini and Angelini 1981).

16.5.8 Erythema Dyschromicum Perstans (Ashy Dermatosis)

This skin disease occurs predominantly in dark-skinned individuals, and is characterized by the presence of single or multiple ashen macules of variable size and shape. The cause of the disease remains unknown; however, in a study of 39 farm workers, the disease could be related to chlorothalonil exposure (Penagos et al. 1996).

16.5.9 Acne

The occupationally related variant of acne is frequently referred to as chloracne, because it is mainly caused by certain chlorinated cyclic hydrocarbons (not by gaseous form of chlorine).

534

Pesticides: Evaluation of Environmental Pollution

Skin symptoms may be accompanied by signs of systemic toxicity. Acnegenic properties of chlorine compounds become most apparent under regular exposure; therefore, workers of pesticide factories and pesticide sprayers are at a highest risk. Chloracne was found in over 80% of the workers involved in the production of pentachlorophenol (Cheng et al. 1993; Jirasek et al. 1976). A case of severe therapy-resistant chloracne, coexisting with palmoplantar keratoderma and scleroderma, was reported in a man with a lifelong exposure to chlorinated pesticides (Poskitt et al. 1994). Next to the declared content, the assessment of the acnegenic potential of a pesticide must take into account traces of raw materials, intermediate compounds, and decay products that are not specified on the label. For example, heating of chlorophenols in the process of pesticide production leads to the formation of unwanted chlorodioxins with strong acnegenic properties that may contaminate the final product (Delvaux et al. 1975). TCDD (2,3,4,8-tetrachlorodibenzo-p-dioxin) is a strong acnegenic substance, contaminating, as an impurity, the herbicide 2,4,5-trichlorophenol (Hogan and Tanglertsampan 1992). Acnegenic activity of chloraniline pesticides propanil, 3,4-dichloroaniline, and methazole was attributed to their contaminants 3,4,3′,4′-tetrachloroazobenzene and 3,4,3′,4′-tetrachloroazoxybenzene (Kimbrough 1980). The clinical picture of chloracne includes straw-colored cysts, comedones, pustules, and abscesses located on the predilection sites, that is, the cheeks, the ears, the postauricular regions, the genitalia, the chest, and the back (Li 1986). Chloracne should be differentiated from elastosis with comedones, which is not uncommon among outdoor workers (Hogan and Tanglertsampan 1992).

16.5.10 Porphyria Cutanea Tarda

Porphyria cutanea tarda (PCT) is the most common form of porphyria, characterized by cutaneous photosensitivity that causes scarring bullae, hyperpigmentation, excessive hair growth on the face, sometimes also skin thickening, and hair loss. Porphyrinogens influence heme synthesis and porphyrin metabolism, which, under heavy exposure, may lead to clinical symptoms accompanied by elevated porphyrins in the urine (Sciarra et al. 2010). An outbreak of PCT from 1955 to 1959 in Turkey ("Turkish porphyria") was caused by eating wheat seeds that were preserved with hexachlorobenzene (HCB) (Peters et al. 1987).

16.5.11 Pigmentation Disorders

This group includes skin disorders with an abnormal (diminished or excessive) production of the natural skin dye melanin by melanocytes. Pigmentation disorders (melasma, vitiligo, and leukoderma) were found in 25% of the Ecuadorean pesticide applicators, as compared with 10% of unexposed controls (Cole et al. 1997). The putative mechanisms of diminished melanin production could be either a direct cytotoxic effect on the melanocytes (apoptosis) or pesticide-induced damage to the sympathetic nerves that control the melanocytes via chemical synapses, resulting in functional disturbances of melanin production. Furthermore, hyperpigmentation is a frequent outcome of phototoxic reactions. Postinflammatory hypopigmentation that persisted for over 2 years was reported in a pesticide sprayer, whose patches of depigmented skin emerged at the sites of the previous contact dermatitis to 4-chloro-2-butynyl N-(3-chlorophenyl) carbamate (Brancaccio and Chamales 1977).

16.5.12 Nail Disorders

Permanent nail dystrophy was described in persons exposed to herbicides and insecticides (Table 16.1). Pesticide-related nail dystrophy may start as a deformity with the breakdown of the nail bed, transverse bands, or discolorations, eventually leading to the loss of the nails (Botella et al. 1985).

16.5.13 Hair Disorders

Hair loss is a systemic symptom provoked by a range of external factors, including acute and chronic intoxications, chemotherapy, and infections. Therefore, the negligible number of reports on diseases of the hair induced by pesticides seems somewhat surprising (Table 16.1). It could be due to the actual lack of the influence of pesticides on hair physiology, or perhaps due to the lack of interest by affected workers and occupational physicians.

16.5.14 Nonmelanoma Skin Cancer

This group includes squamous cell carcinoma (SCC), basal cell carcinoma (BCC), and Bowen's disease, which is considered carcinoma in situ. Seven out of eight studies analyzed by Blair and Zahm demonstrated high incidence of nonmelanoma skin cancers in farmers, which, in the first place, was attributed to sunlight exposure but also to other environmental factors (Blair and Zahm 1991). There is a strong evidence for carcinogenic effects of arsenic pesticides. Arsenic is a metal with a clear predilection for the skin (Lansdown 1995). Until 1960s, arsenic salts were widely used as insecticides. At that time, acute poisonings were very common, which suggests high exposure in many pesticide sprayers. Arsenic compounds were first introduced and widely used in vineyards, therefore the first reported cases of arsenic-related cancer were vintners (Roth 1956; Braun 1958; Thiers et al. 1967). A more recent report of skin cancer related to long-term arsenic exposure was published in 1987 (Jampel and Jerdan 1987). In an epidemiological study conducted in Costa Rica, skin cancers (lip, melanoma, nonmelanocytic skin, and penile cancer) occurred increasingly in coffee-growing areas with extensive use of paraquat and lead arsenate (Wesseling et al. 1999).

The effect of arsenic compounds on the skin may become apparent many years after the exposure. First case reports of palmoplantar keratosis appeared 10–15 years after the introduction of arsenic into agriculture (Roth 1957). However, arsenic-related skin cancers may emerge more than five decades after the exposure (Spiewak 2001). Therefore, every farmer presenting with skin cancer to a doctor should be questioned for possible contacts with arsenic compounds and other potentially carcinogenic pesticides. Typical signs of long-term arsenic exposure are palmoplantar keratosis (excessive thickening of the hand palms and foot soles) and chronic skin inflammation of distal body parts (acrodermatitis atrophicans). Later in the course of the disease, Bowen's disease (carcinoma in situ), multiple BCC, and SCC with ulceration may appear (Braun 1958; Jackson and Grainge 1975). Occupational skin cancer should be in first range considered in workers with multiple or recurrent skin cancers, who were in the past involved in the production or spraying of arsenic insecticides. Such patients must be checked for other epithelial tumors, especially lung cancer.

Exposures to nonarsenical insecticides may also pose a risk for skin and lip cancer (IARC 1991). In a study of 20 workers manufacturing paraquat from 4,4′-bipyridyl, six were diagnosed with SCC (Bowra et al. 1982). In a large study of 228 workers exposed to bipyridyl at 21 paraquat-producing factories, two were diagnosed with Bowen's disease. Exposure to sunlight appeared to be an important cofactor for the development of this disease (Wang et al. 1987).

16.5.15 Melanoma

A high incidence of melanoma among farmers had been known for a long time; however, it was not clear whether this was exclusively caused by exposure to the sunlight during outdoor work, or perhaps other environmental factors may also play a role (Blair 1982). First epidemiological data linking the increased risk for melanoma with DDT exposure originated from studies of Eastern German pesticide applicators (Barthel 1985). Two recent studies link melanoma with the exposure to maneb, mancozeb, parathion, and carbaryl; each pesticide causing an increase in the risk for developing melanoma by more than 50 times (Dennis et al. 2010). Pesticide-related risk for melanoma may also be increased in people exposed to pesticides during home use (Fortes et al. 2007).

16.6 Concluding Remarks

This review clearly demonstrates that the skin is not a mere "gate of entry" for the pesticides to cause internal damage, but a living organ that is first affected by their biological effects and moreover, is involved in the metabolism of pesticides. This strengthens the need for abandoning the misnomer "dermal toxicity" that is improperly used by many authors referring, in fact, to systemic toxicity of dermally absorbed pesticides while overlooking the damage caused to the skin itself. After discussing the spectrum of adverse skin reactions to pesticides, one has to remember that some pesticides, such as lindane or pyrethrin, have also been used in the treatment of skin diseases, especially those caused by arthopods (e.g., lice, demodicosis, etc.). Like with any drug, adverse skin reactions may develop to pesticide-containing medical preparations, including burns to phenothrin in antilice lotions (Choudhary 1999) or ulcerative ICD from lindane in antiscabies lotion (Yu et al. 2005).

Acknowledgment

This study was financed in part by the Statutory Grant No. K/ZDS/001906 of the Jagiellonian University Medical College, Krakow, Poland.

References

Ali, N. and Oehme, F. W. 1992. A literature review of dermatotoxicity. *Vet. Hum. Toxicol.* 34: 428–437.
Baran, R. L. 1974. Nail damage caused by weed killers and insecticides. *Arch. Dermatol.* 110: 467.
Barriere, H., Gerault, C., Bureau, B., and Mousset, S. 1985. Acne chlorique par manipulation d'herbicides. *Ann. Dermatol. Venereol.* 112: 369–370.
Barthel, E. 1985. Erhöhte Mortalität an Ösophaguskrebs, Magenkrebs und Hautmelanom bei Pestizid exponierten Schädlingsbekämpfern in der DDR. *Arch. Geschwulstforsch* 55: 481–488.
Basketter, D. A., Jefferies, D., Safford, B. J., et al. 2006. The impact of exposure variables on the induction of skin sensitization. *Contact Dermatitis* 55: 178–185.

Bhargava, R. K., Singh, V., and Soni, V. 1977. Erythema multiforme resulting from insecticide spray. *Arch. Dermatol.* 113: 686–687.

Blair, A. 1982. Cancer risks associated with agriculture: epidemiologic evidence. *Basic Life Sci.* 21: 93–111.

Blair, A. and Zahm, S. H. 1991. Cancer among farmers. *Occup. Med.* 6: 335–354.

Bleiberg, J., Wallen, M., Brodkin, R., and Applebaum, I. L. 1964. Industrially acquired porphyria. *Arch. Dermatol.* 89: 793–797.

Bonamonte, D., Foti, C., Cassano, N., Rigano, L., and Angelini, G. 2001. Contact dermatitis from organophosphorus pesticides. *Contact Dermatitis* 44: 179–180.

Bopp, C. and Kosminsky, B. 1975. Toxic hepato-cutaneous porphyria (occupational disease). *Med. Cutan. Ibero. Lat. Am.* 3: 271–279.

Botella, R., Sastre, A., and Castells, A. 1985. Contact dermatitis to paraquat. *Contact Dermatitis* 13: 123–124.

Bourrain, J. L. 2006. Occupational contact urticaria. *Clin. Rev. Allergy Immunol.* 30: 39–46.

Bousema, M. T., Wiemer, G. R., and van Joost, T. 1991. A classic case of sensitization to DD-95. *Contact Dermatitis* 24: 132.

Boutsiouki, P. and Clough, G. F. 2004. Modulation of microvascular function following low-dose exposure to the organophosphorous compound malathion in human skin in vivo. *J. Appl. Physiol.* 97: 1091–1097.

Bowden, G. T., Nelson, M. A., Levy, J. P., Finch, J., and Krieg, P. 1993. Molecular mechanisms of skin carcinogenesis induced by chemicals and ionizing radiation. *Recent Results Cancer Res.* 128: 215–230.

Bowra, G. T., Duffield, D. P., Osborn, A. J., and Purchase, I. F. 1982. Premalignant and neoplastic skin lesions associated with occupational exposure to "tarry" byproducts during manufacture of 4,4'-bipyridyl. *Br. J. Ind. Med.* 39: 76–81.

Bradberry, S. M., Proudfoot, A. T., and Vale, J. A. 2004. Glyphosate poisoning. *Toxicol. Rev.* 23: 159–167.

Brancaccio, R. and Chamales, M. H. 1977. Contact allergy and depigmentation produced by the herbicide Carbyne. *Contact Dermatitis* 3: 108–109.

Braun, W. 1958. Carcinoma of the skin and the internal organs caused by arsenic. *Germ. Med. Monthly* 3: 321–324.

Braun, W. 1970. Die Chlorakne. *Ther. Umsch.* 27: 541–546.

Brown, R. 1984. Contact sensitivity to Difolatan (Captafol). *Contact Dermatitis* 10: 181–182.

Bruynzeel, D. P. and van Ketel, W. G. 1986. Contact dermatitis due to chlorothalonil in floriculture. *Contact Dermatitis* 14: 67–68.

Bruynzeel, D. P., Tafelkruijer, J., and Wilks, M. F. 1995. Contact dermatitis due to a new fungicide used in the tulip bulb industry. *Contact Dermatitis* 33: 8–11.

Burry, J. N. 1976. Contact dermatitis from agricultural fungicide in South Australia. *Contact Dermatitis* 2: 289.

Camarasa, G. 1975. Difolatan dermatitis. *Contact Dermatitis* 1: 127.

CDC. 1986. Outbreak of Severe Dermatitis among Orange Pickers – California. *MMWR* 35: 465–467.

Cheng, W. N., Coenraads, P. J., Hao, Z. H., and Liu, G. F. 1993. A health survey of workers in the pentachlorophenol section of a chemical manufacturing plant. *Am. J. Ind. Med.* 24: 81–92.

Choudhary, S. 1999. Burns due to anti-lice lotion. *Burns* 25: 184–185.

Coenraads, P. J., Brouwer, A., Olie, K., and Tang, N. 1994. Chloracne. Some recent issues. *Dermatol. Clin.* 12: 569–576.

Col, M., Col, C., Soran, A., Sayli, B. S., and Ozturk, S. 1999. Arsenic-related Bowen's disease, palmar keratosis, and skin cancer. *Environ. Health Perspect.* 107: 687–689.

Cole, G. W., Stone, O., Gates, D., and Culver, D. 1986. Chloracne from pentachlorophenol-preserved wood. *Contact Dermatitis* 15: 164–168.

Cole, D. C., Carpio, F., Math, J. J., and Leon, N. 1997. Dermatitis in Ecuadorean farm workers. *Contact Dermatitis* 37: 1–8.

Cooper, K. L., Myers, T. A., Rosenberg, M., Chavez, M., and Hudson, L. G. 2004. Roles of mitogen activated protein kinases and EGF receptor in arsenite-stimulated matrix metalloproteinase-9 production. *Toxicol. Appl. Pharmacol.* 200: 177–185.

Croy, I. 1973. Ein kasuistischer Beitrag zur Ätiologie der Urtikaria durch Pilzbekämpfungsmittel auf Captan-Basis. *Z. Gesamte. Hyg.* 19: 710–711.

Dannaker, C. J., Maibach, H. I., and O'Malley, M. 1993. Contact urticaria and anaphylaxis to the fungicide chlorothalonil. *Cutis* 52: 312–315.

de Groot, A. C. 2008. *Patch Testing. Test Concentrations and Vehicles for 4350 Chemicals.* Acdegroot Publishing: Wapserveen.

Delvaux, E. L., Verstraete, J., Hautfenne, A., De Sart, F., and Goffin, G. 1975. Les polychloro dibenzo-p-dioxines. *Toxicology* 3: 187–206.

Dennis, L. K., Lynch, C. F., Sandler, D. P., and Alavanja, M. C. 2010. Pesticide use and cutaneous melanoma in pesticide applicators in the agricultural heath study. *Environ. Health Perspect.* 118: 812–817.

Downey, D. C. 1999. Porphyria and chemicals. *Med. Hypotheses* 53: 166–171.

English, J. S. 2004. Current concepts of irritant contact dermatitis. *Occup. Environ. Med.* 61: 722–726.

Ezendam, J., Hassing, I., Bleumink, R., Vos, J. G., and Pieters, R. 2004. Hexachlorobenzene-induced immunopathology in Brown Norway rats is partly mediated by T cells. *Toxicol. Sci.* 78: 88–95.

Fortes, C., Mastroeni, S., Melchi, F., et al. 2007. The association between residential pesticide use and cutaneous melanoma. *Eur. J. Cancer* 43: 1066–1075.

Fregert, S. 1967. Allergic contact dermatitis from the pesticides captan and phaltan. *Contact Dermatitis Newsletter* (2): 11.

Fukuyama, T., Kosaka, T., Tajima, Y., et al. 2010a. Prior exposure to organophosphorus and organochlorine pesticides increases the allergic potential of environmental chemical allergens in a local lymph node assay. *Toxicol. Lett.* 199: 347–356.

Fukuyama, T., Tajima, Y., Ueda, H., et al. 2010b. Apoptosis in immunocytes induced by several types of pesticides. *J. Immunotoxicol.* 7: 39–56.

Fuortes, L. 1999. Urticaria due to airborne permethrin exposure. *Vet. Hum. Toxicol.* 41: 92–93.

Garcia-Bravo, B., Rodriguez-Pichardo, A., de Pierola, S. F., and Camacho, F. 1995. Airborne erythema-multiforme-like eruption due to pyrethrum. *Contact Dermatitis* 33: 433.

Germolec, D. R., Spalding, J., Yu, H. S., et al. 1998. Arsenic enhancement of skin neoplasia by chronic stimulation of growth factors. *Am. J. Pathol.* 153: 1775–1785.

Gober, M. D. and Gaspari, A. A. 2008. Allergic contact dermatitis. *Curr. Dir. Autoimmun.* 10: 1–26.

Guo, Y. L., Wang, B. J., Lee, C. C., Wang, J. D. 1996. Prevalence of dermatoses and skin sensitisation associated with use of pesticides in fruit farmers of southern Taiwan. *Occup. Environ. Med.* 53: 427–431.

Haustein, U. F. 1968. Chlorkohlenwasserstoffhaltiges Pestizid als Ursache einer toxischen Alopezie. *Z. Haut. Geschlechtskr.* 43: 105–108.

Hearn, C. E. and Keir, W. 1971. Nail damage in spray operators exposed to paraquat. *Br. J. Ind. Med.* 28: 399–403.

Helmdach, F. and Schlenzka, K. 1984. Die Entwicklung der Berufsekzeme in der Landwirtschaft des Bezirkes Magdeburg seit 1966. 1. Mitteilung. *Dermatol. Monatsschr.* 170: 625–631.

Heras-Mendaza, F., Casado-Farinas, I., Paredes-Gascon, M., and Conde-Salazar, L. 2008. Erythema multiforme-like eruption due to an irritant contact dermatitis from a glyphosate pesticide. *Contact Dermatitis* 59: 54–56.

Higo, A., Ohtake, N., Saruwatari, K., and Kanzaki, T. 1996. Photoallergic contact dermatitis from mancozeb, an agricultural fungicide. *Contact Dermatitis* 35: 183.

Hindson, C. and Diffey, B. 1984. Phototoxicity of glyphosate in a weedkiller. *Contact Dermatitis* 10: 51–52.

Hogan, D. J. and Tanglertsampan, C. 1992. The less common occupational dermatoses. *Occup. Med.* 7: 385–401.

Holsapple, M. P. 2002. Autoimmunity by pesticides: A critical review of the state of the science. *Toxicol. Lett.* 127: 101–109.

Horiuchi, N., Oguchi, S., Nagami, H., and Nishigaki, Y. 2008. Pesticide-related dermatitis in Saku district, Japan, 1975–2000. *Int. J. Occup. Environ. Health* 14: 25–34.

Howard, J. K. 1979. A clinical survey of paraquat formulation workers. *Br. J. Ind. Med.* 36: 220–223.

Hryhorczuk, D. O., Wallace, W. H., Persky, V., et al. 1998. A morbidity study of former pentachloro-phenol-production workers. *Environ. Health Perspect.* 106: 401–408.

IARC Working Group. 1991. Occupational exposures in spraying and application of insecticides. *IARC Monogr. Eval. Carcinog. Risk Chem. Hum.* 53: 45–92.

Jackson, R. and Grainge, J. W. 1975. Arsenic and cancer. *Can. Med. Assoc. J.* 113: 396–401.

Jampel, R. M. and Jerdan, M. S. 1987. Palmar lesions and a nonhealing ulcer of the ear in a former agricultural worker. Arsenical keratoses; invasive squamous cell carcinoma of the right ear; and bowenoid keratoses. *Arch. Dermatol.* 123: 253–256.

Jirasek, L., Kalensky, J., Kubec, K., Pazderova, J., and Lukas, E. 1976. Chlorakne, Porphyria cutanea tarda und andere Intoxikationen durch Herbizide. *Hautarzt* 27: 328–333.

Jovtchev, G., Gateva, S., Stergios, M., and Kulekova, S. 2010. Cytotoxic and genotoxic effects of para-quat in Hordeum vulgare and human lymphocytes in vitro. *Environ. Toxicol.* 25: 294–303.

Jung, H.-D. 1987. Kontaktekzeme durch Pflanzenschutzmittel. *Dtsch. Derm.* 35: 761.

Jung, H.-D. 1993a. Kontaktdermatitis durch das Pestizid Fenpropimorph in Corbel. *Dtsch. Derm.* 41: 627–628.

Jung, H.-D. 1993b. Kontaktdermatitis durch das Pestizid Anilazin im Handelspräparat Dyrene. *Dtsch. Derm.* 41: 818–21.

Jung, H.-D. 1996. Epikutantestung von Pestiziden. *Dtsch. Derm.* 44: 129–131.

Jung, H.-D., Rothe, A., and Heise, H. 1987. Zur Epikutantestung mit Pflanzenschutz- und Schädlingsbekämpfungsmitteln (Pestiziden). *Dermatosen* 35: 43–51.

Kataranovski, M., Prokic, V., Kataranovski, D., Zolotarevski, L., and Majstorovic, I. 2005. Dermatotoxicity of epicutaneously applied anticoagulant warfarin. *Toxicology* 212: 206–218.

Kentor, P. M. 1986. Urticaria from contact with pentachlorophenate. *J. Am. Med. Assoc.* 256: 3350.

Kerr, A. and Ferguson, J. 2010. Photoallergic contact dermatitis. *Photodermatol. Photoimmunol. Photomed.* 26: 56–65.

Kimbrough, R. D. 1980. Human health effects of selected pesticides, chloroaniline derivatives. *J. Environ. Sci. Health B* 15: 977–992.

Kimura, T., Kuroki, K., and Doi, K. 1998. Dermatotoxicity of agricultural chemicals in the dorsal skin of hairless dogs. *Toxicol. Pathol.* 26: 442–447.

Koch, P. 1996. Occupational allergic contact dermatitis and airborne contact dermatitis from 5 fungi-cides in a wineyard worker. Cross-reactions between fungicides of the dithiocarbamate group? *Contact Dermatitis* 34: 324–329.

Kwiatkowska, E. and Plonka, T. 1971. Przypadek wylysienia wywolanego przez DDT. *Przegl. Dermatol.* 58: 185–190.

Lankerani, L. and Baron, E. D. 2004. Photosensitivity to exogenous agents. *J. Cutan. Med. Surg.* 8: 424–431.

Lansdown, A. B. 1995. Physiological and toxicological changes in the skin resulting from the action and interaction of metal ions. *Crit. Rev. Toxicol.* 25: 397–462.

Latuszynska, J., Luty, S., Halliop, J., et al. 1999. Studies of toxicity of dermally-absorbed nurelle D 550 EC preparations. *Ann. Agric. Environ. Med.* 6: 151–159.

Lensen, G., Jungbauer, F., Goncalo, M., and Coenraads, P. J. 2007. Airborne irritant contact dermatitis and conjunctivitis after occupational exposure to chlorothalonil in textiles. *Contact Dermatitis* 57: 181–186.

Leow, Y. H. and Maibach, H. I. 1996. Allergic contact dermatitis from norflurazon (Predict). *Contact Dermatitis* 35: 369–370.

Li, W.-M. 1986. The role of pesticides in skin disease. *Int. J. Dermatol.* 25: 295–297.

Lisi, P. and Caraffini, S. 1985. Pellagroid dermatitis from mancozeb with vitiligo. *Contact Dermatitis* 13(2): 124–125.

Lisi, P., Caraffini, S., and Assalve, D. 1987. Irritation and sensitization potential of pesticides. *Contact Dermatitis* 17, 212–218.

Luster, M. I., Wilmer, J. L., Germolec, D. R., et al. 1995. Role of keratinocyte-derived cytokines in chemical toxicity. *Toxicol. Lett.* 82–83: 471–476.

Manoguerra, A. S. 1990. Full thickness skin burns secondary to an unusual exposure to diquat dibro-mide. *J. Toxicol. Clin. Toxicol.* 28: 107–110.

Manuzzi, P. 1988. Contact dermatitis due to Ziram and Maneb. *Contact Dermatitis* 19: 148.

Mark, K. A., Brancaccio, R. R., Soter, N. A., and Cohen, D. E. 1999. Allergic contact and photoallergic contact dermatitis to plant and pesticide allergens. *Arch. Dermatol.* 135: 67–70.

Mathias, C. G. 1983. Persistent contact dermatitis from the insecticide dichlorvos. *Contact Dermatitis* 9: 217–218.

Matsushita, S., Kanekura, T., Saruwatari, K., and Kanzaki, T. 1996. Photoallergic contact dermatitis due to Daconil. *Contact Dermatitis* 35: 115–116.

Matsushita, T. and Aoyama, K. 1981. Cross reactions between some pesticides and the fungicide benomyl in contact allergy. *Ind. Health* 19(2): 77–83.

Matsushita, T., Arimatsu, Y., and Nomura, S. 1976. Experimental study on contact dermatitis caused by dithiocarbamates maneb, mancozeb, zineb, and their related compounds. *Int. Arch. Occup. Environ. Health* 37(3): 169–178.

Matsushita, T., Yoshioka, M., Aoyama, K., Arimatsu, Y., and Nomura S. 1977. Experimental study on contact dermatitis caused by fungicides benomyl and thiophanate methyl. *Ind. Health* 15: 141–147.

Matsushita, T., Nomura, S., and Wakatsuki, T. 1980. Epidemiology of contact dermatitis from pesticides in Japan. *Contact Dermatitis* 6: 255–259.

McCord, C. P. and Kilkee, C. H. 1921. Pyrethrum dermatitis. *J. Am. Med. Assoc.* 77: 448–449.

McFadden, J. P., Kinoulty, M, and Rycroft, R. J. 1993. Allergic contact dermatitis from the fungicide bupirimate. *Contact Dermatitis* 28: 47.

Meneghini, C. L. and Angelini, G. 1981. Secondary polymorphic eruptions in allergic contact dermatitis. *Dermatologica* 163: 63–70.

Michielsen, C. C., Bloksma, N., Klatter, F. A., Rozing, J., Vos, J. G., and van Dijk, J. E. 1999. The role of thymus-dependent T cells in hexachlorobenzene-induced inflammatory skin and lung lesions. *Toxicol. Appl. Pharmacol.* 161: 180–191.

Muhammad, F., Haider, H., Rahman, Z. U., et al. 2010. Dermatotoxic effects of orally administered ciprofloxacin in sweating and nonsweating animal models. *Cutan. Ocul. Toxicol.* 29: 254–260.

Nakamura, M., Arima, Y., Nobuhara, S., and Miyachi, Y. 1999. Airborne photocontact dermatitis due to the pesticides maneb and fenitrothion. *Contact Dermatitis* 40: 222–223.

Nakamura, M. and Miyachi, Y. 2002. Airborne contact dermatitis caused by the pesticide acephate. *Contact Dermatitis* 47: 121–122.

Nakamura, M. and Miyachi, Y. 2003. Airborne photocontact dermatitis due to the insecticide phoxim. *Contact Dermatitis* 49: 105–106.

Nater, J. P. and Gooskens, V. H. 1976. Occupational dermatosis due to a soil fumigant. *Contact Dermatitis* 2: 227–229.

Nater, J. P., Terpstra, H., and Bleumink, E. 1979. Allergic contact sensitization to the fungicide Maneb. *Contact Dermatitis* 5: 24–26.

Nishioka, K. and Takahata, H. 2000. Contact allergy due to propineb. *Contact Dermatitis* 43: 310.

Noakes, D. N. and Sanderson, D. M. 1969. A method for determining the dermal toxicity of pesticides. *Br. J. Ind. Med.* 26: 59–64.

Nosbaum, A., Vocanson, M., Rozieres, A., Hennino, A., and Nicolas, J. F. 2009. Allergic and irritant contact dermatitis. *Eur. J. Dermatol.* 19: 325–332.

Parks, C. G., Walitt, B. T., Pettinger, M., et al. 2011. Insecticide use and risk of rheumatoid arthritis and systemic lupus erythematosus in the Women's Health Initiative Observational Study. *Arthritis Care Res. (Hoboken)* 63: 184–94.

Peluso, A. M., Tardio, M., Adamo, F., and Venturo, N. 1991. Multiple sensitization due to bis-dithiocarbamate and thiophthalimide pesticides. *Contact Dermatitis* 25(5): 327.

Penagos, H. G. 2002. Contact dermatitis caused by pesticides among banana plantation workers in Panama. *Int. J. Occup. Environ. Health* 8: 14–18.

Penagos, H., Jimenez, V., Fallas, V., O'Malley, M., and Maibach, H. I. 1996. Chlorothalonil, a possible cause of erythema dyschromicum perstans (ashy dermatitis). *Contact Dermatitis* 35: 214–218.

Peters, H., Cripps, D., Göcmen, A., Bryan, G., Ertürk, E., and Morris, C. 1987. Turkish epidemic hexachlorobenzene porphyria. A 30-year study. *Ann. N. Y. Acad. Sci.* 514: 183–190.

Pevny, I. 1980. Pestizid-Allergie. Allergisches Kontaktekzem bei einer Winzerin. *Derm. Beruf. Umwelt.* 28: 186–189.

Piraccini, B. M., Cameli, N., Peluso, A. M., and Tardio, M. 1991. A case of allergic contact dermatitis due to the pesticide maneb. *Contact Dermatitis* 24: 381–382.

Poland, A. P., Smith, D., Metter, G., and Possick, P. 1971. A health survey of workers in a 2,4-D and 2,4,5-T plant with special attention to chloracne, porphyria cutanea tarda, and psychologic parameters. *Arch. Environ. Health* 22: 316–327.

Poskitt, L. B., Duffill, M. B., and Rademaker, M. 1994. Chloracne, palmoplantar keratoderma and localized scleroderma in a weed sprayer. *Clin. Exp. Dermatol.* 19: 264–267.

Premaratna, R., Rathnasena, B. G., and de Silva, H. J. 2008. Accidental scrotal burns from paraquat while handling a patient. *Ceylon Med. J.* 53: 102–103.

Rahman, M., Lewis, D. M., and Allison, K. 2007. A case of paraquat burns following an industrial accident. *Emerg. Med. J.* 24(11): 777.

Roth, F. 1956. Über die chronische Arsenvergiftung der Moselwinzer unter besonderer Berücksichtigung des Arsenkrebses. *Z. Krebsforsch.* 61: 287–319.

Roth, F. 1957. The sequelae of chronic arsenic poisoning in Moselle vintners. *Germ. Med. Monthly* 2: 172–175.

Sakabe, K., Yoshida, T., Furuya, H., Kayama, F., and Chan, E. K. 1998. Estrogenic xenobiotics increase expression of SS-A/Ro autoantigens in cultured human epidermal cells. *Acta Derm. Venereol.* 78: 420–423.

Saunders, L. D., Ames, R. G., Knaak, J. B., and Jackson, R. J. 1987. Outbreak of Omite CR-induced dermatitis among orange pickers in Tulare County, California. *J. Occup. Med.* 29, 409–413.

Savitt, L. E. 1972. Contact dermatitis due to benomyl insecticide. *Arch. Dermatol.* 105: 926–967.

Schena, D. and Barba, A. 1992. Erythema-multiforme-like contact dermatitis from dimethoate. *Contact Dermatitis* 27: 116–117.

Schlede, E., Aberer, W., Fuchs, T., et al. 2003. Chemical substances and contact allergy – 244 substances ranked according to allergenic potency. *Toxicology* 193: 219–259.

Schmid, R. 1960. Cutaneous porphyria in Turkey. *N. Eng. J. Med.* 263: 397–398.

Schubert, H. 1979. Allergisches Kontaktekzem durch Propachlor. *Dermatol. Monatsschr.* 165: 495–498.

Schubert, H., Würbach, G., Prater, E., Jung, H.-D., and Tarnick, M. 1993. Kontaktdermatitis auf Metham-Natrium. *Dermatosen* 41: 28–33.

Schuman, S. H. and Dobson, R. L. 1985. An outbreak of contact dermatitis in farm workers. *J. Am. Acad. Dermatol.* 13: 220–223.

Sciarra, G., Cenni, A., Amati, R., Giomarelli, A., and Maurello, M. T. 2010. Alteration of urinary porphyrin pattern as indicator of early effects of occupational and environmental exposure to dioxins and PCBs. *Ital. J. Occup. Environ. Hyg.* 1: 10–17.

Semchuk, K. M., Rosenberg, A. M., McDuffie, H. H., Cessna, A. J., Pahwa, P., and Irvine, D. G. 2007. Antinuclear antibodies and bromoxynil exposure in a rural sample. *J. Toxicol. Environ. Health A* 70: 638–657.

Shi, H., Hudson, L. G., Ding, W., et al. 2004. Arsenite causes DNA damage in keratinocytes via generation of hydroxyl radicals. *Chem. Res. Toxicol.* 17: 871–878.

Slaga, T. J., DiGiovanni, J., Winberg, L. D., and Budunova, I. V. 1995. Skin carcinogenesis: characteristics, mechanisms, and prevention. *Prog. Clin. Biol. Res.* 391: 1–20.

Spencer, M. C. 1966. Herbicide dermatitis. *J. Am. Med. Assoc.* 198: 1307–1308.

Spiewak, R. 1998. Occupational dermatoses in agriculture. *J. Agric. Safety Health* 4: 77–79.

Spiewak, R. 1999. Occupational dermatoses in farmers – A proposal for diagnostic procedure. *Ann. Agric. Environ. Med.* 6: 63–72.

Spiewak, R. 2001. Pesticides as a cause of occupational skin disease in farmers. *Ann. Agric. Environ. Med.* 8: 1–5.

Spiewak, R. 2003. Occupational dermatoses among Polish private farmers, 1991–1999. *Am. J. Ind. Med.* 43: 647–655.

Spiewak, R. 2004. Köbnerizing occupational contact allergy to thiuram in a farmer with psoriasis. *Contact Dermatitis* 51: 214–215.

Spiewak, R. 2008. Patch testing for contact allergy and allergic contact dermatitis. *Open Allergy J.* 1: 42–51.

Spiewak, R. 2009a. Pathomechanisms of photoallergic dermatitis. In: Spiewak, R. (ed.), *Photoallergy and Photopatch Testing*, pp. 23–26. Institute of Dermatology: Krakow.

Spiewak, R. 2009b. Pathomechanisms of phototoxic dermatitis. Pathomechanisms of photoallergic dermatitis. In: Spiewak, R. (ed.), *Photoallergy and Photopatch Testing*, pp. 20–22. Institute of Dermatology: Krakow.

Spiewak, R. 2012. Farmers and Farm Workers. In: Rustemeyer, T., Elsner, P., John, S. M., and Maibach, H. I. (eds), *Kanerva's Occupational Skin Diseases*, 2nd edn. Springer Verlag: Heidelberg (in print).

Spiewak, R. and Stojek, N. M. 2003. Antinuclear antibodies among eastern-Polish rural inhabitants. *Ann. Agric. Environ. Med.* 10: 207–209.

Stoke, J. C. 1979. Captafol dermatitis in the timber industry. *Contact Dermatitis* 5: 284–292.

Taylor, J. S., Wuthrich, R. C., Lloyd, K. M., and Poland, A. 1977. Chloracne from manufacture of a new herbicide. *Arch. Dermatol.* 113: 616–619.

Thiers, H., Colomb, D., Moulin, G., and Colin, L. 1967. Le cancer cutane arsenical des viticulteurs du Beaujolais. *Ann. Dermatol. Syphiligr.* 94: 133–158.

Vandenberg, J. J. and Epstein, W. L. 1963. Experimental nickel contact sensitization in man. *J. Invest. Dermatol.* 41: 413–418.

Verma, G., Sharma, N. L., Shanker, V., Mahajan, V. K., and Tegta, G. R. 2007. Pesticide contact dermatitis in fruit and vegetable farmers of Himachal Pradesh (India). *Contact Dermatitis* 57: 316–320.

Wang, J. D., Li, W. E., Hu, F. C., and Hu, K. H. 1987. Occupational risk and the development of premalignant skin lesions among paraquat manufacturers. *Br. J. Ind. Med.* 44: 196–200.

Wesseling, C., Antich, D., Hogstedt, C., Rodriguez, A. C., and Ahlbom, A. 1999. Geographical differences of cancer incidence in Costa Rica in relation to environmental and occupational pesticide exposure. *Int. J. Epidemiol.* 28: 365–374.

Whitesell, M. F., Alvarez, E., and Draize, J. H. 1947. The percutaneous toxicity of thioglycolates. *Fed. Proc.* 6: 383.

Yu, K. J., Chen, H. H., Chang, Y. C., Hong, H. S., and Ho, H. C. 2005. Ulcerative irritant contact dermatitis from lindane. *Contact Dermatitis* 52: 118–119.

Zendzian, R. P. 2003. Pesticide residue on/in the washed skin and its potential contribution to dermal toxicity. *J. Appl. Toxicol.* 23: 121–136.

17

Pesticide Residues in Mother's Milk

M. Paramasivam, J. Rajeswaran, S. Chandrasekaran, and S. Kuttalam

CONTENTS

17.1 Introduction

Many chemical contaminants, particularly those that are lipophilic and of relatively low molecular weight, can accumulate in breast milk (Zhou et al. 2011). These contaminants arise from two broad categories: persistent organic pollutants (POPs) and organochlorine pesticides (OCPs). Both of them are generally stable toxicants and bioaccumulative and thus often ubiquitous in the environment. OCPs and POPs were extensively used in the past for agricultural, vector control, and industrial purposes for several decades. The potential health effects of these contaminants on both mother and child are of great concern, making it important to carefully monitor the contaminant levels and trends. More studies have been conducted in this area, but most of them are confined to small locations; therefore, it may not necessarily represent the actual population of the country where sampling was done. In addition, majority of these studies are restricted to the same panel of POPs, and therefore, the new or rising trends in contaminants that may impede effective public health responses are ignored.

Levels of contaminants in mother's milk are associated with age, duration of lactation, food habits, and local use pattern of the chemical, and hence mixing of breast milk samples from women makes it difficult to identify differences related to actual exposure conditions. The wide variation in the methods of chemical analysis adopted in different studies across the world hinders the comparison of data. So, it is perilous to extrapolate from the observed levels of contaminants in breast milk around the world to predict potential health effects or to declare specific levels as safe. Despite the difficulties in generalizing across studies, heavier local use of contaminants is the major predictor of the levels of contaminants in breast milk, but contamination is even detectable in the areas with no local use of chemicals (Saeed et al. 2000). In general, ban on the production or use of POPs has been associated with decreasing residues of these chemicals in breast milk samples over the subsequent decades. The following sections review some of the data on specific chemicals to illustrate time trends and regional differences in breast milk levels of OCP contaminants.

17.2 Characteristics of Organochlorine Pesticides

Organochlorine pesticides are a kind of synthetic chemical pesticides, which can be divided into two groups: one takes benzene as the raw material and the other takes cyclopentadiene. These pesticides are very stable. Both persistence and bioaccumulation are strong. They are difficult to degrade naturally in the environment, and they can be a threat to ecosystems and human health through evaporation, migration, food chain transfer, and other pathways (Tan et al. 2008; Katsoyiannis and Salnara 2005).

The Stockholm Convention listed the OCPs among the POPs (Table 17.1; Wong et al. 2005). The Convention is a global treaty to protect human health and the environment from POPs. These chemicals are difficult to degrade into less hazardous substances in the environment. They are lipophilic compounds that tend to bioaccumulate in fatty tissues through the food chain.

Dichlorodiphenyltrichloroethane: Dichlorodiphenyltrichloroethane (DDT) is a commercial organochlorine insecticide that has been widely used on agricultural crops as well as for vector control (Hui et al. 2008). Technical-grade DDT is a mixture of up to 14 compounds. The active ingredient is p,p'-DDT (65%–80%). The other compounds include 15%–21% of o,p'-DDT, up to 4% of p,p'-DDD, and other substances (Wong et al. 2005). DDT can be degraded to DDD compounds under anaerobic conditions, while it can be degraded to dichlorodiphenyldichloroethylene (DDE) under aerobic conditions (Chen et al. 2007). These degradation products are more persistent than the parent compound and are bioaccumulative, are transported over long ranges, and have adverse effects on humans, animals, and the environment. DDT has been banned for all uses in 49 countries.

TABLE 17.1

Persistent Organic Pollutants Listed in the Stockholm Convention

Category	Chemical
Organochlorine pesticides	Aldrin, chlordane, dieldrin, DDT, endrin, heptachlor, hexachlorobenzene (HCB), mirex, and toxaphene
Industrial chemicals	Hexachlorobenzene and polychlorinated biphenyls (PCBs)
Unintended by-products	Dioxins and furans

DDT elimination from the body can take some time; its half-life in humans is approximately 4 years. DDT's major metabolite, DDE, has a half-life of approximately 6 years. The relative proportions of DDT and DDE detected in human tissues can be an indication of the length of time since exposure (Salem and Ahmed 2002). In areas where DDT exposure is recent, the DDE/DDT ratio is low, whereas in areas where substantial time has passed since its use, the DDE/DDT value is higher. Levels of these contaminants in breast milk are often six to seven times higher than in the blood.

Hexachlorocyclohexane: Hexachlorocyclohexane (HCH) is an insecticide made up of a mixture of eight isomers. The components are 58%–80% of α-HCH, 5%–14% of β-HCH, 12%–14% of γ-HCH, 2%–10% of δ-HCH, and 3%–5% of other isomers (Chen et al. 2007; Devanathan et al. 2009). Different isomer forms have different levels of persistence and bioaccumulate in breast milk differently. The γ-isomer of HCH, also known as lindane, is widely used as an insecticide directly applied to the body and scalp to treat head and body lice. The β-isomer of HCH is the most persistent and bioaccumulative form (Salem and Ahmed 2002; Ennaceur et al. 2007). The α- and γ-isomers of HCH are converted into the β-isomer in organisms. The dechlorination rate sequence in HCHs is α-HCH>γ-HCH>δ -HCH> β-HCH, in which the α-HCH is the most unstable isomer and has the fastest degradation rate; γ-HCH in agricultural soils can be easily transformed into other HCHs by decomposition or biotransformation; β-HCH is stable, with the lowest solubility, and does not evaporate easily (Chen et al. 2007). As a result of this conversion, as much as 90% of HCH detected in human tissues and breast milk is β-HCH (Sonawane 1995).

Chlordane: Chlordane (CHL), a mixture of more than 26 compounds, is an organochlorine cyclodiene pesticide and typically consists of 15% cis-chlordane, 15% trans-chlordane, 9.7% trans-nonachlor, 3.9% heptachlor, 3.8% cis-nonachlor, and other chlorinated hydrocarbons and by-products (Solomon and Weiss 2002). Chlordane has been used as an agricultural pesticide, on home lawns and gardens, and against termite control in buildings. Like most POPs, the breakdown of chlordane once it has attached to soil particles or sediment is very slow; in some cases, it has been found in the soil up to 20 years after the initial treatment. Chlordane is rapidly metabolized in organisms into oxychlordane and γ-chlordane or into impurities such as trans-nonachlor or cis-nonachlor. These breakdown products that persist in the tissue of fish, birds, and mammals are found in breast milk (Kunisue et al. 2004b; Tanabe and Kunisue 2007; Subramanian et al. 2007; Devanathan et al. 2009).

Heptachlor: Heptachlor is an organochlorine cyclodiene pesticide that has been used to control termites and as an insecticide on seed grains and food crops. Heptachlor rapidly oxidizes by both photochemical and biological processes to heptachlor epoxide, which is extremely persistent in the soil. In some cases, trace amounts of heptachlor epoxide have been found in the soil 14–16 years after application. Plants can draw heptachlor epoxide directly from the soil, and the chemical bioaccumulates in animals (Solomon and Weiss 2002).

Dieldrin and aldrin: Dieldrin and aldrin are closely related organochlorine insecticides that are extremely persistent in the environment. Both pesticides have been used in agriculture, and dieldrin was also used for vector control, veterinary purposes, and termite control (Solomon and Weiss 2002; Mustafa et al. 2010). In both plants and animals, aldrin, once present in soil or water, is transformed to dieldrin. Dieldrin breaks down very slowly, does not easily evaporate into the air, and binds to soil particles. Plants take up aldrin and dieldrin residues directly from the soil. In animals, including humans, dieldrin is stored in the fat and leaves the body very slowly. Because of low water solubility and a tendency to bind strongly to soil, both aldrin and dieldrin migrate downward very slowly through the soil or into the surface water or groundwater.

Endosulfan: Endosulfan, an organochlorine insecticide of the cyclodiene subgroup, acts as a poison to a wide variety of insects and mites on contact and as a stomach poison upon ingestion. Technical-grade endosulfan contains 94% α- and β-endosulfan. The α- and β-isomers are present in the ratio of 7:3, respectively (Cerrillo et al. 2005). The α-isomer has been shown to be three times more toxic than the β-isomer. Endosulfan sulfate, a reaction product found in technical-grade endosulfan as a result of oxidation, is considered to be equally toxic and more persistent than the parent compound (Shetty et al. 2000; Pathak et al. 2008).

Hexachlorobenzene: Hexachlorobenzene (HCB) is a fungicide mainly recommended for seed treatment. Therefore, the presence of HCB in human milk samples might be due to the contamination from other pesticide formulations present as an impurity and/or as a by-product of various chlorination processes and combustion of municipal solid waste incinerators and industrial processes (Sonawane 1995; Devanathan et al. 2009). HCB binds strongly to soil particles as well as to sediment and builds up in plants when it is present in the soil.

17.3 Sources of Pesticide Residues in Mother's Milk

Human exposure to OCPs can be through various pathways (Figure 17.1) such as dietary intake, occupational exposure via dust, air, consumer products, dermal absorption, and inhalation. It could also be transferred from mother to infant via breast milk and umbilical cord blood (Suzuki et al. 2005).

Sources of pesticide residues in mother's milk are mainly from the pesticides being used in agricultural tracts that are released into the environment and come in human contact directly or indirectly. Organohalogen compounds may enter the human body via

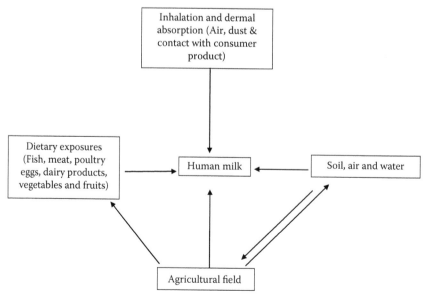

FIGURE 17.1
Sources of pesticide residues in mother's milk.

the external environment, food, water, and air (Malarvannan et al. 2008). Different routes of exposure are dietary intake, particularly from the consumption of fish (Campoy et al. 2001a; Harris et al. 2001; Yu et al. 2007; Raab et al. 2008), meat (Campoy et al. 2001a; Yu et al. 2007; Raab et al. 2008; Tutu et al. 2011), milk (Schinas et al. 2000; Raab et al. 2008), and vegetables (Campoy et al. 2001a); contaminated diet (Kumar et al. 2006a); food commodities (Campoy et al. 2001a; Kumar et al. 2006a; Alle et al. 2009); agricultural field (Kumar et al. 2006a); soil and air pollution (Waliszewski et al. 2009); sediments, water, biota, crops, and human fluid (Tutu et al. 2011); and food, ambient and indoor air, dust, water, and soil (Devanathan et al. 2009). Contamination of mother's milk by OCPs has been reported throughout the world. As a general rule, milk and milk products followed by edible oil were found to contain the highest levels of pesticide residues in India (Devanathan et al. 2009).

17.4 Hazards of Pesticide Residues in Mother's Milk

17.4.1 Types of Exposure and Hazards

Different groups and segments of a population are exposed to pesticides in different ways and in different degrees. Humans can be exposed to these pollutants through dietary intake, inhalation, dermal adsorption, occupational exposure, and prenatal transfer from mother to fetus (Suzuki et al. 2005).

17.4.2 Assessment of Human Exposure

In human beings, the pesticide residue level is an index of exposure, which may be acute, occupational, or incidental. In the general population, the residue level is a measure of the incidental exposure and/or average levels of the persistent pesticide, which is mainly through the diet. The main dietary source is fatty food, such as meat, fish, poultry, and dairy products (Sonawane 1995; Campoy et al. 2001a; Harris et al. 2001; Yu et al. 2007; Raab et al. 2008; Tutu et al. 2011). Continuous exposure primarily occurs through the consumption of food irrespective of the levels of pesticide residues detected in that area. Food imported from other countries that still use OCPs may be a source of human exposure (Saeed et al. 2000). Exposure via inhalation of the ambient air is thought to be insignificant when compared with dietary intake. Prenatal exposure to OCPs is possible via the placenta. Breast-feeding is the major source of infant exposure.

A list of the 12 most dangerous POPs and their uses and known health effects is presented in Table 17.2 (Adeola 2004).

17.4.3 Bioaccumulation and Biomagnification of Pesticides

The continued use of huge amounts of different kinds of poisonous agricultural pesticides increases their concentration in an organism and multiplies it through the food chain (biomagnification). Environmental chemicals find their way into breast milk through the dietary route through the food chain. These chemicals begin their journey in the environment by being absorbed into water, air, and soil. As smaller animals ingest these chemicals, some of them are stored in the animals' fat tissues for a long

TABLE 17.2

POPs: Uses and Their Adverse Health Effects

Class	Chemical	Uses	Adverse Health Effects
Organochlorine pesticide	Aldrin	Insecticide	Cancer, malaise, dizziness, and nausea
	Chlordane	Insecticide, termiticide	Cancer
	DDT	Insecticide	Cancer of the liver and immune system suppression
	Dieldrin	Insecticide	Liver and biliary cancer
	Endrin	Insecticide, rodenticide	Cancers
	Heptachlor	Insecticide, termiticide	Cancers, mutations, stillbirths, birth defects, and liver disease
	HCB	Fungicide	Cancers, mutations, birth defects, fetal and embryo toxicity, nervous disorder, and liver disease
	Mirex	Insecticide, termiticide	Acute toxicity and possibly cancers
	Toxaphene	Insecticide	Cancer, chromosomal aberrations, and liver and kidney problems
Industrial chemicals	PCB	Industry manufacture, coplanar	Cancers, mutations, births defects, fetal and embryo toxicity, neurological disorder, and liver damage
	Dioxins	By-product	Peripheral neuropathies, fatigue, depression, liver disease, and embryo toxicity
	Furans	By-product	Peripheral neuropathies, embryo toxicity, and liver problems

time. This is due to the fact that these are fat-soluble or lipophilic (fat-loving) chemicals and are not excreted in urine. Animal bodies clear these lipophilic compounds slowly since their half-life is many years. When larger animals feed on the smaller ones, these stored chemicals are passed along. Eventually, higher concentrations of chemicals are stored in human fat tissue. For the breast-feeding mother, this is of particular concern, as chemicals stored in the fat tissue are released into the milk and are passed on to her infant (Figure 17.2).

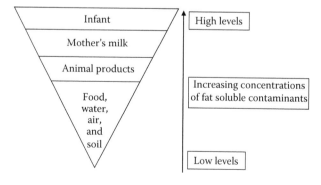

FIGURE 17.2
Biomagnification through the food chain.

17.5 Methods for Analysis of Organochlorine Pesticides in Mother's Milk

The literature describes several preparation methods to measure OCPs in mother's milk (Table 17.3). The standard method consists of three steps: extraction, cleanup, and estimation. Gas chromatography (GC) with electron capture detection (ECD) has been frequently used for the determination of OCPs, due to its high sensitivity, excellent separation efficiency, and speed of analysis. Extraction and enrichment steps are always necessary prior to instrumental determination because of the large-scale dilution of contaminants in biological fluid samples. In addition, a cleanup step is necessary in order to obtain clean solutions and to eliminate interfering compounds that limit the performance of capillary columns and analysis. A crucial aspect, in current biological studies at trace levels, is that the identity of analytical results must be categorically established in order to avoid false-positive results. However, this technique cannot differentiate coeluted compounds.

Gas chromatography with mass spectrometry can resolve the problem of coelution and avoid the misidentification of analytes if operated in the selected ion monitoring mode. However, the concentrations of some OCPs in the general population are too low for quantification by gas chromatography with an electron capture detector or gas chromatography with mass spectrometry. Mass spectrometry allows quantification by isotope dilution. Gas chromatography–high-resolution mass spectrometric analysis with an isotope dilution method enables better identification and quantification than the other methods mentioned here. To increase the sensitivity, the selected ion monitoring mode is also used during gas chromatography–high-resolution mass spectrometric analysis.

17.6 Residues in Human Milk

Human breast milk could be used as a sort of indicator for understanding the biological specificity in the accumulation of organochlorines in the tropical environment as well as assessing the extent of environmental pollution. Monitoring of human milk is important from two standpoints. First, pesticides tend to accumulate in the fat tissue and are relatively easy to isolate and measure; second, their potential risks to infants, who rely solely on mother's milk for a substantial period, need to be evaluated. Residues of these compounds in human milk have been reported from different parts of the world and from India.

17.6.1 Pesticide Residues in the General Population

Everybody has OCP residues in the body. Irrespective of the age, gender, socioeconomic status, and country, these pesticides and their metabolites are detectable in blood or tissue, although the exposure levels differ according to various factors. The concentrations of these pesticides in human milk show regional differences because of the regional variations in their use and the different times of discontinuation. The concentration of organochlorine molecules in mother's milk mainly depends on their accumulation in the maternal fatty tissue and their subsequent mobilization (Campoy et al. 2001b).

TABLE 17.3

The Standard Method Consists of Three Steps: Extraction, Cleanup, and Estimation of Organochlorine Pesticides in Mother's Milk

Extraction Solvents	References	Cleanup Methods	References	Determination Methods	References
Acetone: Hexane	Burke et al. (2003), Mueller et al. (2008), and Raab et al. (2008)	Florisil	Saeed et al. (2000), Bates et al. (2002), Kunisue et al. (2004b), Kumar et al. (2006a), Ennaceur et al. (2007), Sharaf et al. (2008) and Devanathan et al. (2009)	Electron capture detector	Burke et al. (2003), Kumar et al. (2006a), Johnson-Restrepo et al. (2007), Subramanian et al. (2007), Malarvannan et al. (2008), Okonkwo et al. (2008), Alle et al. (2009), Behrooz, et al. (2009), Devanathan et al. (2009), Waliszewski et al. (2009), Cok et al. (2011), and Tutu et al. (2011)
Hexane: Acetonitrile: Ethanol	Ennaceur et al. (2007)	Florisil-Silica Gel column	Yu et al. (2007)	Mass spectrometry	Wong et al. (2002), Burke et al. (2003), Minh et al. (2004), Poon et al. (2005), Hui et al. (2008), Mueller et al. (2008), Haraguchi et al. (2009), Zhou et al. (2011), Srivastava et al. (2011) and Fujii et al. (2011)
Hexane	Rodas-Ortíz et al. (2008)	Silica	Tutu et al. (2011)	High-resolution mass spectrometry	Saleh et al. (2002), Gonzalez-Rodriguez et al. (2005), and Mueller et al. (2008)
Ethanol	Yu et al. (2007)	GPC	Sudaryanto et al. (2005), Kunisue et al. (2006), Subramanian et al. (2007), Raab et al. (2008), Hui et al. (2008), Devanathan et al. (2009), Zhou et al. (2011), and Fujii et al. (2011)	Bench-topped double focusing mass selective detector	Sudaryanto et al. (2005)
Ethanol/toluene	Hui et al. (2008)	—	—	Selective ion monitoring mode	Bates et al. (2002)
Dichloromethane	Kumar et al. (2006a)	—	—	Fourier transform infrared (FTIR)	Srivastava et al. (2011)
Ethyl acetate: methanol: acetone	Schinas et al. (2000)	—	—	—	—
Formic acid and sulfuric acid	Covaci et al. (2001), Rodas-Ortíz et al. (2008), Behrooz et al. (2009) and Waliszewski et al. (2009)	—	—	—	—

As stated earlier, many of these lipid-soluble compounds bioaccumulate and are not cleared rapidly. Human milk is one route of elimination for the mother's body burden, but unfortunately that route also increases the exposure of infants.

17.6.2 Pesticide Residues in the Indian Population

The mother's milk was collected from general public in different regions of India such as Chennai, Mumbai, New Delhi, Kolkata, and Agra (Subramanian et al. 2007; Devanathan et al. 2009; Someya et al. 2010; Kumar et al. 2006b). The occurrence of these chemicals was noticed in the human breast milk collected from mothers living in and near various locations such as agricultural and fishing villages, small towns, metropolitan cities, municipal dumpsites, etc. In general, relatively high concentrations of DDT were detected in mother's milk from different parts of India, followed by (in order) HCHs, CHLs, and HCB. In the samples collected during 2002–2003 (Subramanian et al. 2007), it was found that all samples from Chennai (metropolitan city), Perungudi (municipal dumpsite), Parangipettai (fishing village), and Chidambaram (agricultural town) contained measurable quantities of organochlorines (HCHs, DDTs, CHLs, and HCB; Table 17.4). The concentrations of DDT in mother's milk from Delhi, Kolkata, and Chennai were relatively higher than those from other regions (Table 17.4). DDTs in Chennai mothers were few times higher than those from Perungudi, Parangipettai, and Chidambaram, which can be explained by the possible use of DDT for vector control within city limits (Subramanian et al. 2007). Ironically, Chennai, which does not have any agricultural activity, had one order higher HCHs, the chemical that is used predominantly for agricultural purposes, than places like Chidambaram and Parangipettai, areas with agricultural activities. Whether HCHs are still in use by the local government inside the city limits for insect control or by the private entrepreneurs and public inside their premises is yet to be known.

TABLE 17.4

Residues of OCPs (ng/g Lipid Weight) in Mother's Milk Samples in the General Population in India

Location	Survey Year	DDTs	HCHs	CHLs	HCB	References
Chennai (city)	2002–03	1200[a]	4500[b]	7.3[c]	4.2	Subramanian et al. (2007)
Perungudi	2002–03	450[a]	650[b]	8.2[c]	1.7	Subramanian et al. (2007)
Parangipettai	2002–03	330[a]	560[b]	1.6[c]	1.5	Subramanian et al. (2007)
Chidambaram	2002–03	240[a]	410[b]	0.99[c]	1.4	Subramanian et al. (2007)
Mumbai	2005–06	450[a]	220[b]	3.4[c]	1.7	Devanathan et al. (2009)
Delhi	2005–06	1500[a]	340[b]	2.6[c]	3.2	Devanathan et al. (2009)
Kolkata	2004–05	1100[a]	670[b]	7.3[c]	4.2	Subramanian et al. (2007)
	2004–05	665[a]	265[b]	5.7[c]	2.85	Someya et al. (2010)
Agra-Bakhoti	2005	179[d]	123[b]	—	—	Kumar et al. (2006b)
Chiraigaon	2005	170[d]	129[b]	—	—	Kumar et al. (2006b)
Ghodhakhas	2005	174[d]	131[b]	—	—	Kumar et al. (2006b)
Minahas	2005	179[d]	127[b]	—	—	Kumar et al. (2006b)

[a] p, p'-DDE+p,p'-DDD+p,p'-DDT.
[b] α-HCH+β -HCH+γ -HCH.
[c] Oxychlordane+trans-nonachlor+cis-nonachlor.
[d] o,p'-DDT+p,p'DDE+p,p'-DDD.

Among the locations, no significant difference in the concentrations of DDTs was found between New Delhi (1500 ng/g lipid weight) and Kolkata (1100 ng/g lipid weight), but the levels of DDTs in Mumbai (450 ng/g lipid weight) were higher than in the Agra region. Levels of DDTs in the major cities (urban) were higher than those reported in rural areas near Agra (180 ng/g lipid weight; Kumar et al. 2006b) and Chidambaram (240 ng/g lipid weight; Subramanian et al. 2007), indicating higher exposures to DDTs in urban areas than rural areas. This may probably be due to the ongoing usage of DDT for the eradication of vector-borne diseases, applied largely in urbanized areas (Subramanian et al. 2007). In addition, differences in the dietary habits between urban and rural people may also be responsible for the higher concentrations of DDTs in urban areas because the urban residents are relatively richer and consume more animal products and protein-rich food that have more of these lipophilic contaminants (Devanathan et al. 2009).

DDT levels were lower than those in the same locations observed previously; indicating that the concentration of DDTs has been declining in the Indian environment. For example, DDTs in human breast milk from New Delhi collected in 1989 were 3700 ng/g lipid wt. and have declined to 1500 ng/g lipid weight; the DDT levels in Mumbai have declined from 8000 ng/g lipid weight in 1984 to 450 ng/g lipid weight; and, in Kolkata, there was a decline from 4800 ng/g lipid weight in 1984 to 1100 ng/g lipid weight (Devanathan et al. 2009). However, there is an exception: DDT concentrations in Chennai show an increase from 760 ng/g lipid weight in 1988 to 1200 ng/g lipid weight in 2003 (Subramanian et al. 2007). Past and ongoing usage of DDT for controlling vector-borne diseases and/or the continuing intake of contaminated foods may be a plausible reason for such an increasing trend (Subramanian et al. 2007).

Breast milk samples from Chennai had higher HCH concentrations (4500 mg/g fat weight; Subramanian et al. 2007) than the samples from Kolkata (670), Delhi (340), Mumbai (220) (Devanathan et al. 2009), and Agra (127.25) (Kumar et al. 2006b). These values indicate a continued usage of technical HCH mixture in Chennai and possibly in some other places in India (Subramanian et al. 2007; Devanathan et al. 2009). It seems that this reflects a huge amount of usage in the past and/or recent illegal use of technical HCH (Tanabe and Kuinisue 2007; Kumar et al. 2006b).

Further, it could be seen in Table 17.4 that there is a clear temporal increase in the levels of HCH and DDT compounds in the human milk samples collected from Chennai during 2002–2003 (Subramanian et al. 2007) when compared with the results published by Tanabe et al. (1990) from the samples collected during 1988. These results reveal that Chennai mothers have been exposed to higher amounts of HCHs and DDTs all through these years than the people living in other parts of the state. The levels of both the chemicals drastically decreased in the breast milk samples from Parangipettai and Chidambaram, indicating the effective implication of the bans imposed by the Government of India, whereas the exemptions permitted for the use of these compounds for vector control in Chennai boosted their levels in mother's milk. Of the four metropolis evaluated, Mumbai seems to be less contaminated (DDTs: 450 ng/g; HCHs: 220 ng/g) with respect to these two contaminants than the other three cities.

Chlordane compounds and HCB were the least prevalent pollutants detected in all the regions of India (Table 17.4). CHL levels in human milk (ng/g lipid weight) are relatively similar among the regions, New Delhi (2.6), Mumbai (3.4), and Kolkata (7.3). The levels were also relatively similar to those in the southern part of India such as Chennai (7.3 ng/g lipid weight; Subramanian et al. 2007). Similar to CHLs, contamination with HCB was also usually low and relatively similar among the regions (Chennai, New Delhi, Mumbai, and Kolkata were 4.2, 3.2, 1.7, and 4.4 ng/g lipid weight, respectively), probably due to its minimal usage in India (Devanathan et al. 2009).

Dieldrin, heptachlor, and endosulfan are closely related to cyclodiene pesticides. Surveys conducted in Rajasthan (India) have demonstrated that the human milk samples analyzed had quantifiable concentrations of dieldrin, heptachlor, and endosulfan (Kumar et al. 2006a).

The comparison of Indian data with those of the DDTs, CHILs and HCB in mother's milk from some other Asian countries (Tanabe and Kunisue 2007) showed that Indian levels of were lower than some Asian countries and comparable to some others, whereas HCH levels were higher than almost all the countries except a location in China (Table 17.5), indicating the prominence of HCH contamination in India.

HCH isomers, endosulfan, malathion, chlorpyrifos, and methyl-parathion were monitored in human milk samples from Bhopal, Madhya Pradesh, by Sanghi et al. (2003). The endosulfan concentrations were the highest and exceeded the total HCH, chlorpyrifos, and malathion concentrations by 3.5-, 1.5-, and 8.4-fold, respectively. Through breast milk, infants consumed 8.6 times more endosulfan and 4.1 times more malathion than the average daily intake levels recommended by the World Health Organization. A correlation analysis (r values) between mothers' age and the content of the chemicals accumulated in breast milk indicated a substantial degree of correlation for malathion. The other chemicals showed low to negligible correlation with donor age.

17.6.3 Levels of OCPs in the Asian Population

Levels of OCPs in mother's milk collected from the general public in Asian countries such as Cambodia, Vietnam, Philippines, Malaysia, Indonesia, China, and Japan (Kunisue et al. 2002, 2004a,b; Minh et al. 2004; Tue et al. 2010; Malarvannan et al. 2008; Sudaryanto et al. 2005, 2006; Zhou et al. 2011; Haraguchi et al. 2009; Hui et al. 2008; Wong et al. 2002; Kunisue et al. 2006) are illustrated in Table 17.5. In general, in Asian developing countries, DDT levels are relatively high, followed by HCHs, HCB, and CHLs. Among Asian developing countries, the concentrations of DDTs in human breast milk from China (Guangzhou), Vietnam, Cambodia, and Malaysia are relatively higher than those from other countries (Table 17.5).

Mean concentrations of DDTs in human breast milk were 2100 ng/g and 2300 ng/g in Hanoi and Ho Chi Minh City, respectively, and the levels in the primiparas group were higher than those in the multiparas group. This difference indicates that excretion via milk during lactation is an important factor that reduces DDT burden in nursing mothers. Further examination of the composition of DDTs revealed that p,p'-DDE is the predominant compound accounting for 85%–90% of the total DDT concentrations. Interestingly, DDT concentrations in Vietnamese human breast milk were among the highest values reported for the countries surveyed. This observation again suggests the recent use of DDTs in Vietnam and that Vietnam may be a potential source of DDTs in the south Asian region (Minh et al. 2004). In Vietnam, higher levels of DDTs in sediments from populated locations as compared to those from paddy field were reported, indicating recent application of DDTs for public health purposes such as malaria control rather than for agriculture. In addition, the elevated residue levels of DDT in wild avian species and foodstuffs from Vietnam were probably related to more recent exposure to technical mixtures of these pesticides used in agricultural and suburban/rural areas (Tue et al. 2010). This similarity in the contamination pattern of OCPs in the human milk of Cambodian population and fish population of Cambodia indicates that DDT is the dominant contaminant in the Cambodian ecosystem (Kunisue et al. 2004b).

TABLE 17.5

Concentrations of OCPs (ng/g Lipid Weight) in Mother's Milk Collected from the General Population in Asian Countries Except India

Country/Region	Survey Year	DDTs	HCHs	CHLs	HCB	References
China (Dalian)	2002	2100[a]	1400[b]	16[c]	81	Kunisue et al. (2004b)
	2007	527.2[d]	231.8[b]	3.0[e]	32.8	Zhou et al. (2011)
Beijing	2000	641.6[a]	169[b]	—	—	Yao et al. (2005)
	2007	1300[a]	570[d]	3.8	86	Haraguchi et al. (2009)
Hong Kong	2002–2003	1500[d]	—	—	—	Hui et al. (2008)
Shenyang	2002	870[f]	550[d]	6.7	56	Kunisue et al. (2004b)
Guangzhou	2002	3500[f]	1110	6.7	56	Wong et al. (2002)
Japan	2001	390a	110[g]	67[c]	16	Kunisue et al. (2006)
	2002	380[a]	140[g]	63[c]	16	Kunisue et al. (2006)
	2003	230[a]	83[g]	77[c]	10	Kunisue et al. (2006)
	2004	400[a]	100[g]	95[c]	13	Kunisue et al. (2006)
Japan (Sendai)	2007	260[a]	190[d]	41	18	Haraguchi et al. (2009)
Japan (Kyoto)	2007	160[a]	77[d]	31	13	Haraguchi et al. (2009)
Japan (Takayama)	2007	97[a]	49[d]	17	8.1	Haraguchi et al. (2009)
Korea (Seoul)	2007	180[a]	110[d]	14	13	Haraguchi et al. (2009)
Malaysia	2003	1600[a]	230[b]	23[c]	11	Sudaryanto et al. (2005)
Philippines	2000	190[a]	4.7[b]	15[c]	<0.56	Kunisue et al. (2002)
	2004	314[a]	7.7[b]	17.14e	2.84	Malarvannan et al. (2008)
Indonesia	2001	640[a]	14[eg]	2.0[c]	2.2	Sudaryanto et al. (2006)
	2000–2003	1000[a]	16[b]	3.8[c]	2.0	Sudaryanto et al. (2006)
Vietnam (South)	2000–2001	2100[a]	58[b]	2.0[c]	3.9	Minh et al. (2004)
Vietnam (North)	2000–2001	2300[a]	14[b]	6.9[c]	2.5	Minh et al. (2004)
Vietnam	2007	495[a]	11.9[b]	0.68[c]	2.4	Tue et al. (2010)
Vietnam	2007	1200[a]	140	0.75	7.4	Haraguchi et al. (2009)
Vietnam	2010	86.9	78[b]	—	—	Si and Gia (2010)
Cambodia (urban)	2000	1800[a]	5.6b	1.6[h]	1.6	Kunisue et al. (2004b(Suburban))
	2000	1200[a]	4.8	1.7	1.8	Kunisue et al. (2004b)

[a] p,p′-DDE + p,p′-DDD + p,p′-DDT.
[b] α-HCH + β-HCH + γ-HCH.
[c] Oxychlordane+tans-nonachlor+cis-nonachlor.
[d] o,p′-DDE + p,p′-DDE + o,p′-DDT + p,p′-DDT + o,p′-DDD + p,p′-DDD.
[e] Oxychlordane + tans-nonachlor + cis-nonachlor + cis-chlordane.
[f] p,p′-DDE + p,p′-DDT.
[g] α-HCH + β-HCH.

Another survey showed that DDT compounds were the predominant organochlorine contaminants in human milk samples from Vietnam (Tue et al. 2010). The total levels of p,p′-DDT and its metabolites were an order of magnitude higher than the total of HCHs > HCB > CHLs. The levels of CHLs were low: oxychlordane, trans-nonachlor, and cis-nonachlor were not detected in 27%, 39%, and 21% of the samples, respectively. These results may reflect the usage pattern of OCPs in Vietnam, as the usage of HCB and CHLs was very limited compared with other countries (Minh et al. 2004). The higher accumulation of OCs in urban donors may be explained by their richer diets and shorter total lactation times, as levels of the contaminants were found to increase with the consumption of food from animal origin and decrease with the duration of lactation. These two factors represented respectively the main intake and depuration pathways of OCs. Although DDT levels in human breast milk from Vietnam are still in the highest range in our worldwide comparison, it should be noted that their concentrations have declined over the past 10 years (Minh et al. 2004).

All breast milk samples contained an excess of HCH with its isomers, α-HCH, β-HCH, and γ-HCH. Among them, the average content of γ-HCH (32.3 ng/g lipid) was higher than that of α-HCH (23.3 ng/g lipid) and β-HCH (22.4 ng/g lipid). The total content of HCHs in breast milk in their study was higher than that reported by Minh et al. (2004) in Hanoi (with 58 ng/g lipid β-HCH) and in Ho Chi Minh city (13.5 ng/g lipid β-HCH). Table 17.5 shows that the total content of HCH in breast milk in this study was lower than that in studies in countries such as China, Japan, and Malaysia. However, the total content of HCHs in the study was higher than that in studies in Indonesia, Cambodia, and Philippines. The average concentrations of p,p'-DDE, o,p'-DDT, p,p'-DDT, and DDTs in breast milk were 40.3, 11.4, 35.2, and 86.9 ng/g lipid, respectively. Among them, the content of p,p'-DDE was higher than that of o,p'-DDT and p,p'-DDT. This result is much lower than that of Minh et al. (2004) in Hanoi and Ho Chi Minh city (the content of DDTs was 2100 and 2300 ng/g lipid, respectively). However, they were lower than those reported in other studies in Cambodia (1500 ng/g lipid), The Philippines (190 ng/g lipid), Malaysia (1600 ng/g lipid), Indonesia (630 ng/g lipid), and Japan (290 ng/g lipid; Si and Gia 2010).

A survey showed that in Malaysia, the residue levels of DDTs, HCHs, CHLs, and HCB were detected with concentrations of 1600, 230, 23, and 11 ng/g lipid weight, respectively, due to the widespread use of DDT in both agriculture and malaria control (Sudaryanto et al. 2005). However, the composition of DDTs were dominated by p,p'-DDE (>95%). In addition, β-HCH was mostly the only HCH isomer detected whereas α- and γ-HCH were found in trace quantities. This might be due to the restriction in the usage of technical HCH or lindane not having been used recently in Malaysia.

The national survey on the levels of persistent OCPs in the breast milk of mothers in China (Zhou et al. 2011) showed that the levels of DDTs, HCHs, and HCB varied from 527.2, 231.8, and 32.8 ng/g lipid respectively. The mean concentrations of DDTs, HCHs and HCB in breast milk from rural and urban (539.3 ng/g lipid and 629.2 ng/g lipid; 197.9 ng/g lipid and 265.6 ng/g lipid; and 33.8 ng/g lipid and 32.3 ng/g lipid, respectively. Chlordane and mirex were present in samples at relatively lower concentrations. In Chinese and Indian human breast milk samples, notably higher concentrations of HCHs were observed compared with other Asian countries (Tables 17.4 and 17.5). It seems that this reflects a huge amount of usage in the past and/or recent illegal use of technical HCH in these two countries. These results indicate that relatively high sources of HCHs are still present in the environment of these countries and that technical HCH may be illegally used for public health purposes in these two countries exposing residents to HCHs (Tanabe and Kunisue 2007).

Milk samples (115 from Hong Kong and 54 from Guangzhou), in the lactation period from 3–5 weeks, were analyzed (Wong et al. 2002). The results demonstrated that the mean levels μg/g of fat of p,p'-DDT (Hong Kong: 0.39; Guangzhou: 0.70), p,p'-DDE (2.48; 2.85), and β-HCH (0.95; 1.11) were 2–15-fold higher when compared with studies conducted elsewhere (i.e., United Kingdom, Germany, Sweden, Spain, and Canada). When compared to a similar study conducted 10 years ago in Hong Kong (p,p'-DDT 2.17 μg/g of fat, p,p'-DDE 11.67 ng/g of fat, and β-HCH 15.96 ng/g of fat), a considerable reduction in the levels of their contaminations was observed.

The mean concentrations of DDTs were 1300 ng/g lipid (n = 25) in China and 1200 ng/g lipid (n = 20) in Vietnam, both of which were 5–7 times higher than values in Korea (180 ng/g lipid) and Japan (170 ng/g lipid; Haraguchi et al. 2009). DDT composition was dominated by p,p'-DDE (94–96%), followed by p,p'-DDT (3–5%) and p,p'-DDD (1–2%) in most cases. The levels of p,p'-DDD were the highest in mothers from Hanoi, whereas the levels of o,p'-DDT were the highest in mothers from Beijing. Thus, the ratios of p,p'-DDE/p,p'-DDT and o,p'-DDT/p,p'-DDT were significantly higher in Beijing than in the other regions (Table 17.5).

Burke et al. (2003) reported that all samples contained detectable residues of p,p'-DDT (urban mean 0.11 ± 0.18 mg/kg (n = 5), rural mean 0.07 ± 0.03 mg/kg (n = 5)) and p,p'-DDE (urban mean 0.05 ± 0.04 mg/kg (n = 5), rural mean 0.76 ± 1.46 mg/kg (n = 5)). Residues of HCB, β-HCH, α-endosulfan, and dieldrin were also detected in some samples. There was no significant difference (at the 95% confidence level) in the levels of pesticides between urban and rural areas. The levels of OCPs in the milk from Indonesian women were lower than those reported from Indonesian women exposed to DDT through malaria eradication schemes and were similar to levels reported from the United Kingdom and Japan.

In human breast milk from China, HCB levels were also higher than those in other Asian countries. HCB has been produced in a large chemical factory along Ya–Er Lake, which is located in the eastern part of Wuhan, Hubei province, and in an investigation, high levels of HCB were detected in soils and sediments from this lake. In addition, high concentrations of HCB were detected in tea leaves, indicating the recent use of this chemical as a fungicide (Tanabe and Kunisue 2007). These results imply that the use of HCB is still continuing in China.

The results showed that the median levels (ranges) of total HCHs (four HCH isomers) in human milk were 54.7 (5.7–159.3) ng/g lipid; and the median concentration of total DDTs (DDT and its metabolites) were 2114.6 (329.1–6164.6) ng/g lipid. The predominant pollutant in the HCH family was β-HCH. p,p'-DDE was a predominant pollutant in all DDEs, DDDs, and DDTs for all the samples tested and accounted for more than 80% of total HCHs and DDTs (Qu et al. 2010).

Human milk specimens from 55 women in Shijiazhuang urban and Tangshan rural areas in Hebei Province in northern China were collected and analyzed for persistent organic pollutants, such as p,p'-DDE, HCB, β-HCH, and dioxins. The concentrations of p,p'-DDE and β-HCH in human milk from the primiparous mothers in Shijiazhuang (3330 and 108.8 ng/g fat, respectively) were higher than those in samples from mothers from Tangshan (1916 and 21.2 ng/g fat, respectively). In addition, the level of p,p'-DDE correlated positively with a dietary history of meat consumption or consumption of internal organs of pigs and sheep (r = 0.38 and r = 0.52, respectively) but not with the intake frequencies of vegetables and fruits in primiparous mothers. The higher level for p,p'-DDE may be due to a greater intake of contaminated fatty foods in the industrialized areas (Sun et al. 2005).

Dicofol residues have been measured in human breast milk from China, Korea, and Japan (Fujii et al. 2011). The geometric mean concentration (ng/g lipid) of dicofol in the Japanese breast milk samples was 0.3 and significantly lower than that in Chinese (9.6) or Korean breast milk samples (1.9). Although dicofol has not been used in Japan since 2004, residual dicofol may be present in imported food products from China, Korea, and other countries where dicofol is still in usage.

In Japan, contamination levels of OCPs such as DDTs, HCHs, CHLs, and HCB were detected in human breast milk, and DDTs were dominant followed by HCH > CHLs > HCB. The mean concentration of OCPs in primiparas was higher than multiparas, implying elimination of OCPs via lactation (Kunisue et al. 2006; Haraguchi et al. 2009).

Human breast milk samples were analyzed in Philippines during 2004 (Malarvannan et al. 2008, 2009). The p,p'-DDE contributed to contamination, followed by p,p'-DDT and p,p'-DDD. The HCH concentrations ranged between 3.6 and 12 ng/g, and CHL showed higher levels than those reported in other Asian developing countries like China, Indonesia, and Cambodia but lower levels than those in Japan.

Hedley et al. (2010) studied the concentrations of 24 POPs in 10 pools of human milk samples, collected at 2–6 weeks postpartum from 238 primiparous women living in Hong Kong and south China, who participated in the 2002–2003 WHO exposure study. The mean levels of α-HCH (mean 0.6 ng/g fat), β-HCH (940 ng/g fat), γ-HCH (1.8 ng/g fat), dieldrin (1.0 ng/g fat), and HCB (21.8 ng/g fat) were much lower than the 1985 estimates. Mean levels of α-HCH, γ-HCH, dieldrin, cis-heptachlor epoxide (0.7 ng/g fat), sum-chlordane (6.1 ng/g fat), and trans-nonachlor (12.0 ng/g fat) were comparable to the international median levels of the 15 other countries participating in the 2002–2003 WHO exposure study. Continuous monitoring of POPs in human milk, especially β-HCH, is needed for surveillance and interpretation of time trends and for linkage to strict enforcement of agricultural regulations.

Residues of OCPs in human milk were monitored in central Taiwan between December 2000 and November 2001 by Chao et al. (2006). The OCPs were analyzed by GC/MS for 36 human milk samples from healthy women ranging between 20 and 36 years of age. The predominant OCPs were p,p'-DDE, p,p'-DDT, α-CHL, heptachlor epoxide, heptachlor, β-HCH, and γ-HCH, with median levels of 228, 19, 7.4, 4.0, 2.3, 1.2, and 0.8 ng/g lipid, respectively. The residues of OCPs in human milk samples from central Taiwan were comparable to those described in the results obtained from Japan and were significantly lower than those from investigations in Asian countries, including China, Thailand, Indonesia, and Vietnam. Low DDE/DDT ratio (mean = 13.6, SD = 6.54) indicated that residual OCPs in human milk mainly originated from past exposure. A notable decrease in DDT levels (ΣDDT = 333 ng/g lipid) in human milk was found in this study compared to results from the previous two decades (ΣDDT = 3595 ng/g lipid).

A survey was carried out on the levels of OCPs in human milk from Hmong hill tribes in northern Thailand (Stuetz et al. 2001). DDT was detected in all samples with a median and maximum level of 209 and 2012 ng of total DDT isomers per milliliter of milk, respectively. The median and highest percentages of p,p'-DDT were 23.2 and 44.7%, respectively. In 15 samples, heptachlor was detected in the metabolized form of heptachlor epoxide with a median value of 4.4 ng/mL. The estimated daily intakes of DDT, heptachlor, and heptachlor epoxide by the infants exceeded up to 20 times the acceptable daily intakes as recommended by the FAO and WHO. In nine samples, HCB was detected with a median value of 5.4 ng/mL; seven of the nine samples also had the highest values of DDT residues. The γ-isomer of HCH was found only in one sample with 3.6 ng/mL. The mean sum-DDT residues of 14.96 mg/kg milk fat as well as the estimated daily intake by the infants are some of the highest reported values in the 1990s. The fact that the mother breast-feeds her first child and the fact that she originally comes from a region where DDT is still in use as a vector-control agent, as well as the former use of OCPs in agriculture, seem to be the main factors for high DDT and other OCP residues in the mother's milk.

Among human breast milk samples from Asian developing countries, levels of DDTs in Vietnam, China, Cambodia, and Malaysia; HCHs in China and India; and HCB in China were relatively higher compared with other countries, suggesting again the recent use of these organochlorine insecticides and fungicides and that these Asian developing countries may be potential sources of these contaminants to the global environment. In Asian developing countries, further investigations on human exposure and temporal trends of POPs, especially organochlorine insecticides, are needed to elucidate future pollution trends.

It has been previously reported that DDTs were predominant in environmental media, biota, and foodstuff from these Asian developing countries (Tanabe and Kunisue 2007).

In China, it has been recently reported that not only DDTs but also HCHs were predominant in various environmental samples along the coastal areas and foodstuffs. The investigation on the pollution of DDTs and HCHs in sediment cores indicated the recent input of these contaminants into coastal areas from inland. In addition, the investigation on the pollution of DDTs in fish and mussels collected from aquaculture cages in coastal waters showed high proportions of p,p'-DDT and also indicated the recent releases of this chemical to the environment.

17.6.4 Levels of OCPs in the World Population

The organochlorine pesticides for which available data on the levels in human milk in the United States exist include chlordane, DDTs, HCHs, and HCB (Johnson-Restrepo et al. 2007). DDT showed the highest concentration among the OCPs with an overall mean concentration of 64.5 ng/g lipid followed by CHLs (32.4 ng/g). The levels of DDT compounds in human milk have declined substantially where the use of DDT compounds has been restricted or banned for many years (LaKind et al. 2004). Oxychlordane, which is the most stable metabolite arising from the biotransformation of technical chlordane, was the most abundant compound in breast milk samples, accounting for 59% of the total CHL concentration (Johnson-Restrepo et al. 2007). The mean concentration of HCHs in breast milk samples in the United States in 2004 was 18.9 ng/g lipid weight (Johnson-Restrepo et al. 2007). In general, the mean concentration of OCPs in breast milk from the United States was lower than those reported for Japan, China, and Russia (Johnson-Restrepo et al. 2007). Other countries, including Germany and Japan, have reported a steady decline in HCHs in human milk (Solomon and Wiess 2002).

The mean concentrations of HCB, DDTs, dieldrin, and oxychlordane in breast milk were 27, 180, 4, and 5 ng/g, respectively, when compared to levels measured in previous years in Bavaria, Germany (Raab et al. 2008). In another study, over a time span of 8 years, a free analysis of their milk was offered to lactating mothers residing in the state of Lower Saxony, Germany (Zietz et al. 2008). The human milk was analyzed for a number of organic chemicals including DDT, HCB, and β-HCH. In total, 4314 samples were collected in the years 1999–2006 and analyzed for their content of these POPs. A clear downward trend of median total, DDT, β-HCH, and HCB values in all participants and also in different selected subgroups could be observed. The median value of calculated total DDT in the year 2006 including all participants was 0.0815 mg/kg lipid, that of β-HCH was 0.0116 mg/kg lipid, and that of HCB was 0.0229 mg/kg lipid. There were reductions between 40.9% and 47.1% compared to the year 1999. Among other influencing factors, median concentrations of DDT, β-HCH, and HCB showed a clear rise with increasing age of mothers, whereas an increasing number of breast-fed infants per mother led to a decrease. The proportions of other measured substances exceeding the limits of quantification were as follows: dieldrin 68.6%, α-HCH 1.3%, γ-HCH 60.1%, heptachlor epoxide 41.5%, musk xylene 15.6%, and musk ambrette 0.4%. It has been concluded that the known declining trend of important xenobiotic substances in the human milk of German mothers has continued (Zietz et al. 2008).

The mean concentrations were 343.4 ng/g lipid for p,p'-DDE, 72 ng/g for p,p'-DDT, and 42.1 ng/g for p,p'-DDD as reported in surveys of human breast milk in Brazil (Azeredo et al. 2008). 19 of the 20 localities presented total DDT contamination ranges from 118 to 771.4 ng of total DDT/g of lipid. Just one locality (Cachoeirinha) presented a geometric mean for total DDT higher than 1000 ng of total DDT/g of lipid (Azeredo et al. 2008). There is sufficient evidence that the presence of DDT and its metabolites in the analyzed milk samples is due to the diet rich in fish, considering the statement of Harris et al. (2001) that the consumption of contaminated food represents an important source of organochlorine pesticides to humans. The population living in the Madeira River region is characterized by a high consumption of fish in their diet (Azeredo et al. 2008).

A total of 173 milk samples were collected from 12 regions of Australia during 2002–2003 (Mueller et al. 2008). The OCPs ranged from 200 to 1600 ng/g lipid. The median concentration of DDE, β-HCH, dieldrin, HCB, trans-nonachlor, and p,p'-DDT were 311, 80, 19, 18, 11, and 9 ng/g lipid, respectively. The concentration of p,p'-DDE showed a relatively wide range from 150 to 870 ng/g lipid. Overall, DDE was the dominating OCP detected, with a contribution to the sum of OCPs between 53% and 88% followed by β-HCH ranging between 3% and 42%. This study indicated a substantial decline of OCP concentration from the early 1980s to the 1990s; little decline could be observed since then (Mueller et al. 2008). Furthermore, the use of multivariate statistics indicated some regional trends with slightly higher levels of the broadly used insecticides, DDT and HCH, in both historic and recent samples from Melbourne, whereas, sample pools collected from mothers who lived in rural Queensland and New South Wales as well as Adelaide and Sydney showed comparatively higher levels of heptachlor and dieldrin, both of which have been used for termite treatment. These results indicate that even 20 years after the discontinuation of usage, historical use of OCPs rather than exposure via global transport of OCPs is responsible for continuous low exposure in Australia.

Data on the exposure of breast-fed infants to polychlorinated biphenyls (PCBs) and selected OCPs in the Wielkopolska province were collected (Poland) (Szyrwinska and Lulek 2007). The levels of indicator PCBs, and selected OCPs, including DDT metabolites (HCB, p,p'-DDT, p,p'-DDD, p,p'-DDE, and α-, β-, and γ-HCH) were determined in 27 human milk samples, collected in 2000–2001, according to the WHO protocol. The estimated daily intakes (EDIs) of all analytes were calculated. Authors compared results with those obtained by an analysis of human milk samples from other European and non-European countries, collected in the same period of time. They stated that median exposure of Wielkopolska first breast-fed infants to OCPs is comparable (EDI_{HCB} = 0.086 μg/kg bw/day; $EDI_{\beta\text{-HCH}}$ = 0.063 μg/kg bw/day) or higher ($EDI_{p,p'\text{-DDE}}$ = 3.495 μg/kg bw/day) than in other European countries, while exposure to PCBs ($EDI\Sigma7PCB$ = 0.364 μg/kg bw/day) is situated at the lower end of the intake of these xenobiotics by breast-fed infants from different regions of Europe.

Human breast milk samples from primiparous women from Northern (Tromsø) (N = 10) and Southern Norway (Oslo) (N = 19) collected in 2000–2001 were analyzed with respect to hexachlorobenzene (HCB), hexachlorocyclohexane (HCH), chlordane (CHL), DDT, mirex, and toxaphene (CHBs). The sum of DDTs was the major organochlorine (OC) (170 and 110 ng/g lipid weight, respectively). Other OCs were found in lower levels. The concentration of the sum of HCHs was significantly higher in the breast milk samples from Oslo compared to Tromsø (p < 0.05). The present study showed that concentrations of OCs in the breast milk of primiparous women decreased 50–60% since 1991 and that this trend is continuing (Polder et al. 2008).

The OCPs were determined in 423 breast milk samples from women living in Norway (Polder et al. 2009). The samples were collected in six countries, representing South, Central, and North Norway in 2002–2006. Initial results showed significantly lower levels of OCPs in breast milk from ethnic Norwegians (N = 377) compared to ethnic non-Norwegians (N = 46). Median concentrations (range) of p,p'-DDE, HCB, β-HCH, and oxychlordane in the breast milk samples of the Norwegian women, all parities included, were 41 (5.4–492), 11 (3.6–24), 4.7 (0.9–37), and 2.8 (0.5–16) ng/g lipid weight, respectively. Results indicated that the sum of p,p'-DDE and β-HCH are good predictors for monitoring DDT and HCH levels in the breast milk of Norwegian women. Multivariable linear regression analyses showed that age was strongly associated with increasing OC levels (P < 0.001), whereas parity was associated with decreasing OC levels (p < 0.001). Smoking was associated with higher levels of p,p'-DDE and β-HCH. Median levels of OCs in the present Norwegian primiparas seemed to be 29%–62% lower than the corresponding results found in a Norwegian study from 2000 to 2002.

Tadevosyan et al. (2007) conducted a study to evaluate organochlorine pesticide (OCP) exposure and the possible relationship with adverse birth outcomes through analysis of breast milk samples from a rural Armenian population in 1993–2000, and 266 samples were randomly collected during the first 2–3 days after delivery. Residues of OCP (lindane, DDT, DDD/DDE) in breast milk were measured using gas chromatography. DDE consistently ranged from undetectable to 0.14 mg/L. DDT was detected twice (7%) in 2000, providing evidence of the illegal use of this banned pesticide. Total frequency of DDT/DDE detection was 77% and lindane 51%. Almost 8% of the breast milk samples contained lindane, exceeding the acceptable daily intake (ADI) estimated by the WHO, and 20% exceeded the ADI for DDT/DDE. A doubling in the frequency of pregnancy or delivery complications for those with OCP in their breast milk, though not statistically significant, is worrying. Further research is necessary, involving a larger sample size, possibly from rural and urban regions.

The levels of persistent organochlorine contaminants in 43 breast milk samples from Czech Republic were analyzed (Cajka and Hajslova 2003). The mean concentrations of β-HCH, p,p'-DDT, p,p'-DDE, and HCB were 56, 81, 1017, and 318 and 64, 86, 921, and 425 ng/g in primiparous and secundiparous donors, respectively. The concentrations of both groups of measured analytes were higher in secundiparous mothers compared to primiparous donors.

Breast milk samples, randomly collected from 32 Kuwaiti donors, were analyzed for chlorinated pesticides. DDE residues ranged from 127 to 3333 µg/kg, averaging 833 µg/kg, expressed on a fat-weight basis. DDT levels ranged from 0.6 to 67 µg/kg fat and averaged at 12.4 µg/kg fat. High DDE/DDT ratios were found, which indicated that the exposure to DDT, in most cases, happened quite some time ago. In addition, many of the samples also contained isomers of HCH, heptachlor epoxide, aldrin, dieldrin, and endrin. Residue levels of non-DDT pesticides were generally low. Overall levels of DDT pesticides in Kuwaiti milk samples were lower than the levels reported from other Middle Eastern countries, although methodologies may not be directly comparable (Saeed et al. 2000).

A total of 54 human breast milk samples from lactating mothers were collected from the Pediatric Department of the Assiut University Hospital (Egypt) during January–August 2001 (Salem and Ahmed 2002). The samples were analyzed for OCP residues by using gas chromatography–electron capture detector (GC-ECD). The results revealed that all detected organochlorine residue levels were markedly lower in comparison with the previously documented levels, although they still had the same frequency percentages as recorded previously. p,p'-DDE and β-HCH were found to be the main contaminants and were detected in all analyzed samples (100%), followed by HCB (88.9%), p,p'-DDT (83.3%), and oxychlordane (33.3%). The mean total DDT levels in breast milk samples was 16.166 ± 14.82 µg/L, formed mainly of p,p'-DDE (95%). The levels ranged from 3.353 to 67.159 µg/L, with levels exceeding the Extraneous Residue Limits (ERLs) issued by FAO/WHO, in 1994 and 1996 (50 µg/L), in only three samples (5.6%). The average of total HCH isomers was 1.599 ± 1.595 µg/L milk, formed mainly of β-HCH. HCB and oxychlordane were detected at lower average levels of 0.389 ± 0.156 and 0.285 ± 0.175 µg/L milk, respectively.

Within the detection limits, no sample was found to contain aldrin, dieldrin, heptachlor, heptachlor epoxide, endrin, or γ- and α-chlordane residues. The estimated daily intakes (EDIs) for total DDTs, γ-HCH, and total HCH residues never exceeded the recommended FAO/WHO acceptable daily intake (ADIs). The possible health impact of these pollutants was reviewed.

Cerrillo et al. (2005) investigated the presence of endosulfan I, endosulfan II, and endosulfan metabolites in fatty and nonfatty tissues and fluids from women of reproductive age and children in Southern Spain. All of the human milk samples analyzed contained endosulfan ether. The second most frequently found compound was endosulfan lactone, which was present in 91.3% of the human milk samples. The most abundant was endosulfan II, with a mean concentration of 10.70 ng/mL of human milk. The highest concentration of commercial endosulfan I and endosulfan II was found in human milk, with a mean value (I+II) of 11.38 ng/mL milk. These findings support the lipophilicity of these chemicals and their elimination by milk secretion. Therefore, women of reproductive age in Southern Spain appear to be currently exposed to endosulfan compounds. Because these chemicals can be mobilized during pregnancy and lactation, further research is necessary to investigate the health consequence in children resulting from exposure to chemicals suspected of immunotoxic, neurotoxic, or endocrine-disrupting effects.

Human breast milk samples, collected during April 2007 in Tabriz, Iran, were analyzed for organochlorine pesticides (Behrooz et al. 2009). The OCs present at the highest concentration were HCB (1020 ng/g lipid weight), DDT (1930 ng/g lipid weight), and HCH (1660 ng/g lipid weight; Table 17.6). OCPs were the major contaminants in the samples from Tabriz, Iran, with p,p'-DDE being measured in all samples, and β-HCH having a detection frequency of 90%. β-HCH was the most prevalent HCH isomer with a mean of 735 ng/g lipid weight. The mean concentration of β-HCH was 580 ng/g lipid weight, while the mean concentration of β-HCH was 350 ng/g lipid weight. HCB was found in 14 human milk samples (46%), with a mean concentration of 1020 ng/g lipid weight.

A study was conducted to determine the types and levels of OCP residues in the breast milk of 21 primiparous mothers in La, a suburb of Accra urban community in the Greater Accra region of Ghana by Tutu et al. (2011). The concentrations of OCP residues in the human breast milk samples ranged from 1.839 to 99.05 μg/kg fat. With the exception of endosulfan sulfate, whose mean concentration (99.052 μg/kg) was above the Australian Maximum Residue Limit (MRL) of 20 μg/kg for milk, the mean concentrations for all the other organochlorines detected were below their respective limits.

The levels of 13 OCPs were measured in the breast milk from 87 Tunisian mothers throughout their lactation periods (Ennaceur et al. 2007). All samples contained detectable residues of p,p'-DDE (2.421 mg/kg milk fat) and HCB (0.260 mg/kg milk fat). The other OCP residues were detected in the following order of frequency: dieldrin 38%, β-HCH 22.9%, and γ-HCH 6.8%.

17.6.5 Variation Associated with Parity and Age

In lactating women, several factors such as the number of previous children, age of mother, food intake preferences, and breast-feeding parameters can affect the concentrations of OCs in human milk (Harris et al. 2001). The levels of excreted OCPs are influenced by several factors such as the donor's age, number of children, and duration of breast-feeding (Czaja et al. 2001). Despite this, concentrations of OCs decrease with the increase in the number of previous children nursed (Minh et al. 2004). There was no association (p > 0.05) between the concentrations of OCs and the number of children, which might be due to

TABLE 17.6

Concentrations of OCPs (ng/g Lipid Weight) in Mother's Milk Collected from the General Population in the World Other than Asia

Country/Region	Survey Year	DDTs	HCHs	CHLs	HCB	Reference
Egypt	2001	517.31[a]	47.97[b]	7.13[c]	9.26	Salem and Ahmed (2002)
	2009	1315	225	—	—	Rahman (2010)
United Kingdom (Lancaster)	2001–2003	156[d]	16b	0.3[e]	17	Kalantzi et al. (2004)
London	2001–2003	330	59.6	0.46	24.6	Kalantzi et al. (2004)
Taiwan	2000–2001	333[a]	4[d]	10	—	Chao et al. (2006)
Russia	2003–2004	660[a]	810[d]	19	100	Tsydenova et al. (2007)
United States	2004	65[d]	19	32	3.2	Johnson-Restrepo et al. (2007)
Mexico	2005–2006	3064.63[f]	753.67	974.62	—	Rodas-Ortíz et al. (2008)
Turkey (Kahramanmaras)	2003	2571	175	31	24	Erdogrul et al. (2004)
Antalya	2007–2008	1400[f]	160[d]	—	39	Cok et al. (2011)
Ghana	2008–2009	18.05[f]	26.44	—	—	Tutu et al. (2011)
Iran	2007	1930[a]	1660[d]	—	1020	Behrooz et al. (2009)
Germany	2005	180[a]	17[d]	5	27	Raab et al. (2008)
Australia	2002	319.62[a]	80.29[d]	20	18	Mueller et al. (2008)
Tunisia	2002–2002	3863[a]	67[g]	—	260	Ennaceur et al. (2007)
Norway	2000–2002	110[a]	14[g]	14[c]	—	Polder et al. (2008)
Belgium	2006	156[a]	12[g]	7.8[c]	—	Colles et al. (2008)
Poland (Wielkopolska)	2004	868	14.1	3.3	32.3	Jaraczewska et al. (2006)

[a] p,p'-DDE + p,p'-DDD + p,p'-DDT.
[b] α-HCH + β-HCH + γ-HCH.
[c] Oxychlordane + trans-nonachlor + cis-nonachlor.
[d] o,p'-DDE + p,p'-DDE + o,p'-DDT + p,p'-DDT + o,p'-DDD + p,p'-DDD.
[e] Oxychlordane + trans-nonachlor + cis-nonachlor + cis-chlordane.
[f] p,p'-DDE + p,p'-DDT.
[g] α-HCH + β-HCH.

ongoing exposures (Devanathan et al. 2009). No significant correlation was observed between concentrations of OCs in human breast milk and the age of mothers, and there was no difference in the concentrations between primipara and multipara. The estimated infant daily intake of OCs showed that the intake of HCHs through lactation exceeded the TDI, which is of primary concern on infant health (Devanathan et al. 2009). Furthermore, an examination of the relationship between concentrations of OCs in human breast milk and the age of mothers also showed no significant correlation ($p > 0.05$; Sudaryanto et al. 2005; Devanathan et al. 2009). A similar finding of no such correlation between OC levels and the age of mothers was observed in the milk samples collected from Dalian and Shenyang, Northeastern China (Kunisue et al. 2004b). Similarly, the study carried out on milk samples from Russia (Tsydenova et al. 2007) also found no statistically significant difference in OC levels between primiparas and multiparas, although the milk of primiparas should have higher OC concentrations than that of multiparas, as OCs accumulate throughout the years before pregnancy.

A positive correlation was observed between the concentration of OCPs in human breast milk and mothers' age with primiparas. Primiparous women having higher OCP concentrations than multiparous women suggest that these parameters influence the OCP burden in lactating women (Behrooz et al. 2009). Zhou et al. (2011) found that no statistical difference was found in the residual level of DDTs in breast milk between rural and urban areas in China. In China, with its rapid urbanization and industrialization in rural areas, the difference in the level of environmental contamination between rural areas and urban areas has narrowed gradually. The highest level of DDTs in breast milk was from the Hubei rural area, followed by Shanghai urban area and Hubei urban area. The levels of DDTs from Ningxia rural and urban areas and Heilongjiang rural and urban areas were notably lower than those from other provinces. Concentrations of DDTs were significantly higher in southern areas (761.7 ng/g lipid) than those in northern areas (434.1 ng/g lipid), if the twelve provinces were divided into two groups from the Yangtze River (Changjiang), indicating considerable exposure to DDTs in southern areas. No significant correlations were found to total DDT in the function of age and parity in Brazilian human breast milk (Azeredo et al. 2008). A recent study confirmed that point sources such as dicofol and direct technical DDT usage from antifouling paint may have contributed DDTs to soil, freshwater sediment, and freshwater fish species.

The mean level of HCHs in breast milk samples from urban areas was higher than those from rural areas in China. Different patterns of food composition and food consumption between urban and rural areas may be responsible for the variation of HCHs in breast milk (Sudaryanto et al. 2006). A significant difference in the levels of HCHs in breast milk between the southern part of the country and the northern part of the country was not observed.

Residual levels of HCHs in breast milk samples from the Hubei urban area were the highest, followed by the Liaoning rural area and Shanghai urban area. The levels of HCHs in breast milk samples from the Shanxi rural and urban areas were the lowest. The major isomer of HCHs found in samples was β-HCH, with a mean value of 220.4 ng/g lipid that accounted for 95.08% of the total HCH. The highest level of HCB in breast milk was from the Liaoning rural area, followed by Hubei rural area and Hebei urban area. The lowest level of HCB was from the Jiangxi urban area. A significant difference in the concentrations of HCB and other OCPs between urban areas and rural areas as well as between the southern part of the country and the northern part of the country were not observed.

It is known that adult females excrete lipophilic contaminants such as OCs during lactation, thus reducing the body burden of such contaminants. Minh et al. observed higher concentrations of OCs in human breast milk of primiparous mothers as compared to multiparas mothers. In both groups, strong correlations between contaminant levels and age were observed. However, only DDT levels of the multiparous group in Ho Chi Minh City significantly correlated with age (Minh et al. 2004). Nevertheless, slopes of the regression lines for DDTs and mother's age are quite similar regardless of the differences in groups and areas. This result would probably be interpreted as an evidence of similar intake rates of DDTs in Vietnamese mothers. In addition, it should be noted that other factors such as the length of lactation, time of sampling during breast-feeding, etc. could also influence the OC levels in women. Nevertheless, the available data from this study probably indicate that parity and age play important roles in controlling the OC burden in humans.

Age and exposure to persistent contaminants usually generate a positive correlation. Therefore, the lower the average age, the less the individual is likely to be exposed to contaminants. No significant correlation was found in this study between age and POP concentrations, neither were the differences between age classes statistically significant, even though the 35–40 age class showed the highest concentrations of OC compounds on average.

No significant relationship between the number of childbirths and concentrations of HCB, HCH, dieldrin, and p,p'-DDT was found (Ennaceur et al. 2007). These results support the theory that the total body load of organochlorines is shed via the breast milk with each successive pregnancy and lactation. Rahman (2010) found in a study conducted in Egypt that there was no significant difference between the levels of OCPs and the mother's age, while there was a significant difference and correlation between the levels of OCPs and the number of times the mother had breast-fed (primiparous and multiparous).

Although it cannot be clearly explained why no significant correlation was observed between the levels of OCs in human breast milk and the age or parity, the narrow range of mothers' age and the low number of samples analyzed in the present study might be the plausible explanations.

17.7 Trends in Mother's Milk Residues

In general, the levels of organochlorine pesticides in countries where these chemicals have been banned have dropped well below all current benchmark levels (Solomon and Weiss 2002). The levels of these chemicals in milk of the general population have been decreasing over time, after the periods of their peak production and use. Relatively higher concentrations of DDTs were detected in mother's milk from Asian developing countries than in samples from other countries (Sudaryanto et al. 2005; Tanabe and Kunisue 2007).

The more recent surveys of pesticides in human milk have demonstrated that the concentrations are lower than those observed in previous surveys. Despite this decrease in concentrations, more effort is needed to characterize the potential adverse effects of the low concentrations of chlorinated pesticides found in human milk. Differences in the accumulation patterns of OCs indicate region-specific usage of chemicals in India. DDT levels were lower than those in the same locations observed previously, indicating that the concentration of DDT has been declining in the Indian environment. For example, the concentration of DDT in human breast milk samples from New Delhi collected in 1989 was 3700 ng/g lipid weight and has declined to 1500 ng/g lipid weight; the DDT levels in Mumbai have declined from 8000 ng/g lipid weight in 1984 to 450 ng/g lipid weight; in Kolkata, there was a decline from 4800 ng/g lipid weight in 1984 to 1100 ng/g lipid weight (Devanathan et al. 2009). Except in Chennai, levels of DDT and HCH were lower when compared to previous observations, indicating the effects of bans and restrictions imposed. The general declining trends found in the present study confirm the positive effects of governmental and voluntary restrictions and prohibitions on the usage of DDT and other measures taken to minimize the pollution. Comprehensive studies on the OC contamination in India are therefore necessary to understand the source and evaluation of possible long-term impacts of OCs (Devanathan et al. 2009).

The present comprehensive study on OCP residues in breast milk showed that DDT was the most abundant pesticide, followed by HCH and HCB. These concentrations of DDT and HCH dramatically declined with time compared with data from regional breast milk surveys in China in the past 20 years (Zhou et al. 2011).

The DDT levels in breast milk indicate a marked decline in DDT human body load in Hong Kong in the past 30 years, assuming reasonable comparability of assay results in different previous surveys (Hui et al. 2008). The sum of DDT levels was also slightly lower than those in more recent breast milk samples (Wong et al. 2002). Both p,p'-DDT and p,p'-DDE concentrations in breast milk in Hong Kong decreased over time (Hui et al. 2008). This observation is consistent with a worldwide downward trend in DDT body load (Smith

1999) and it was even suggested that the decline in average levels of DDT in breast milk in most countries was strongly correlated with the length of time since DDT restriction.

The study from Australia indicated that following a substantial decline of OCP concentrations from the early 1980s to the 1990s, little decline could be observed since then (Mueller et al. 2008). The distribution and time trend of organochlorine pesticide concentrations in human milk samples from Croatia collected in 1981–2003 are available. Between 1981/1982 and 1987/1989, the concentrations of HCB, β-HCH, DDE, and total PCBs decreased about 50%, while for the last decade, the concentrations have been decreasing very slowly (Krauthacker et al. 2009).

Devi et al. (2011) reviewed the available data/information on OCP contamination of breast milk in the female population of Hong Kong, China. Concentrations of DDT and HCH in human breast milk samples from Dalian were similar to those from Hong Kong and Guangzhou, while those from Shenyang were somewhat lower. In 1976, the concentration of β-HCH was much higher than γ-HCH, dieldrin, and HCB. Further monitoring in 1985 and 2002 demonstrated that organochlorine pesticide concentrations in Hong Kong breast milk was found to be much lower than before. It demonstrated that the use of pesticides in Hong Kong were lower and people were following the permissible limit according to world health organization.

In a study in Mexico, the total OCPs showed a decreasing tendency with the number of births (primipara and multipara and age ranks), but these differences were not significant (Rodas-Ortíz et al. 2008). This conclusion was the same in a study of Tadevosyan et al., where no differences in pesticide content were detected between the milk of primiparous and multiparous Armenian women (Tadevosyan et al. 2007). No correlation was found between the levels of pesticide body burden and frequency of pregnancy or delivery complications, infant gender ratio, birth defects, or infant weight and height. There was some decrease of body mass and a statistically significant change of chest circumference among the infants of mothers with detectable levels of OCPs.

17.8 Perspective and Recommendation

Human milk is the best source of nutrition for infants. Breast milk contains the optimal balance of fats, carbohydrates, and proteins for developing babies, and it provides a range of benefits for growth, immunity, and development. In the last few decades, the continued efforts of scientists to measure environmental pollutants in human milk have been important for defining the true toxic contribution of these chemicals to public health, especially to the infant population's health. However, it also shows that there are several gaps in current knowledge including the following: (a) insufficient information on the nature and levels of contaminants in breast milk, (b) lack of consistent protocols for collecting and analyzing breast milk samples, (c) lack of toxicokinetic data, and (d) lack of data on health outcomes that may be produced in infants by exposure to chemicals in breast milk.

Additional data are needed on the chemical contaminants in mother's milk and other foods and food consumption patterns of infants and children. To be nationally representative, such an effort would need to include women of various socioeconomic backgrounds and geographic locations. Comprehensive monitoring of the levels of contaminants in breast milk with standardized protocols for specimen collection and analysis is needed worldwide. Only with more reliable and better standardized approaches to the selection of subjects, milk sampling and collection, and analytical methods, conclusions can be drawn about global patterns of contamination, trends over time, and emerging hazards.

Good data on time trends and geographic patterns would aid in generating hypotheses and would lead to more definitive studies. Such information would also provide a sound basis for evidence-based public health policies. Without such data, it is difficult to provide advice to health-care professionals and to new mothers on the potential risks and benefits of breast-feeding. Another need is to study lactating women prospectively to determine the rates of decrease in concentrations of chemicals over the course of lactation. It will also be necessary to develop data that will permit comparison of breast milk contamination levels with contaminant levels associated with other infant food sources, such as infant formula and cow's milk. Such data will permit us to compare the risks associated with each source of infant nutrition. A centralized database for chemical residue data and standardized analytical procedures may be needed to improve the ability to characterize chemical exposures. Frequent and continuous monitoring of these pesticides will help to take precautions in consuming food of animal origin, fish, vegetables, fruits, and grains from contaminated areas.

References

Adeola, F. O. 2004. Boon or bane? The environmental and health impacts of persistent organic pollutants (POPs). *Hum. Ecol. Rev.* 11(1): 27–35.

Alle, A., Dembelle, A., Yao, B., and Ado, G. 2009. Distribution of organochlorine pesticides in human breast milk and adipose tissue from two locations in Cote d'Ivoire. *Asian J. Appl. Sci.* 2(5): 456–463.

Azeredo, A., Torres, J. P. M., Fonseca, M. F., Britto, J. L., Bastos, W. R., Azevedoe Silva, C. E., Cavalcanti, G., Meire, R. O., Sarcinelli, P. N., Claudio, L., Markowitz, S., and Malm, O. 2008. DDT and its metabolites in breast milk from the Madeira River basin in the Amazon, Brazil. *Chemosphere* 73(1): S246–S251.

Bates, M. N., Thomsonb, B., and Garrett, N. 2002. Reduction in organochlorine levels in the milk of New Zealand women. *Arch. Environ. Health: An Int. J.* 57(6): 591–597.

Behrooz, R. D., Sari, A. E., Bahramifar, N., Naghdi, F., and Shahriyari, A. R. 2009. Organochlorine pesticide and polychlorinated biphenyl residues in human milk from Tabriz, Iran. *Toxicol. Environ. Chem.* 91(8): 1455–1468.

Burke, E. R., Holden, A. J., and Shaw, I. C. 2003. A method to determine residue levels of persistent organochlorine pesticides in human milk from Indonesian women. *Chemosphere* 50(4): 529–535.

Cajka, T. and Hajslova, J. 2003. Polychlorinated biphenyls and organochlorine pesticides in human milk from the locality Prague, Czech Republic: A comparative study. *Bull. Environ. Contam. Toxicol.* 70: 913–919.

Campoy, C., Olea-Serranob, F., Jiménez, M., Bayés, R., Cañabate, F., Rosales, M. J., Blanca, E., and Olea, N. 2001a. Diet and organochlorine contaminants in women of reproductive age under 40 years old. *Early Hum. Develop.* 65: S173–S182.

Campoy, C., Jiménez, M., Olea-Serrano, M. F., Frias, M. M., Cañabate, F., Olea, N., Bayés, R., and Molina-Font, J. A. 2001b. Analysis of organochlorine pesticides in human milk: Preliminary results. *Early Hum. Develop.* 65: S183–S190.

Cerrillo, I., Granada, A., Lopez-Espinosa, M. J., Olmos, B., Jimenez, M., Cano, A., Oleaa, N., and Olea-Serrano, M. F. 2005. Endosulfan and its metabolites in fertile women, placenta, cord blood, and human milk. *Environ. Res.* 98: 233–239.

Chao, H. R., Wang, S. L., Lin, T. C., and Chung, X. H. 2006. Levels of organochlorine pesticides in human milk from central Taiwan. *Chemosphere* 62(11): 1774–1785.

Chen, S., Shi, L., Shan, Z., and Hu, Q. 2007. Analytical, nutritional and clinical methods determination of organochlorine pesticide residues in rice and human and fish fat by simplified two-dimensional gas chromatography. *Food Chem.* 104: 1315–1319.

Cok, I., Yelken, C., Durmaz, E., Uner, M., Sever, B., and Satır, F. 2011. Polychlorinated biphenyl and organochlorine pesticide levels in human breast milk from the Mediterranean city Antalya, Turkey. *Bull. Environ. Contam. Toxicol.* 86: 423–427.

Colles, A., Koppen, G., Hanot, V., Nelen, V., Dewolf, M. C., Noel, E., Malisch, R., Kotz, A., Kypke, K., Biot, P., Vinkx, C., and Schoeters, G. 2008. Fourth WHO-coordinated survey of human milk for persistent organic pollutants (POPs): Belgian results. *Chemosphere* 73(6): 907–914.

Covaci, A., Hura, C., and Schepens, R. 2001. Determination of selected persistent organochlorine pollutants in human milk using solid phase disk extraction and narrow bore capillary GC-MS. *Chromatographia* 54: 247–252.

Czaja, K., Ludwicki, J. K., Góralczyk, K., and Struciński, P. 2001. Relationship between two consecutive lactations and fat level in persistent organochlorine compound concentrations in human breast milk. *Chemosphere* 43: 889–893.

Devanathan, G., Subramanian, A., Someya, M., Sudaryanto, A., Isobe, T., Takahashi, S., Chakraborty, P., and Tanabe, S. 2009. Persistent organochlorines in human breast milk from major metropolitan cities in India. *Environ. Pollut.* 157(1): 148–154.

Devi, N. L., Shihua, Q., Yadav, I. C., and Wei, C. 2011. Organochlorine pesticides (OCPs) in breast milk in Hong Kong-review. *J. Am. Sci.* 7(1): 334–340.

Ennaceur, S., Gandoura, N., and Driss, M. R. 2007. Organochlorine pesticide residues in human milk of mothers living in Northern Tunisia. *Bull. Environ. Contamin. Toxicol.* 78: 325–329.

Erdogrul, O., Covaci, A., Kurtul, N., and Schepens, P. 2004. Levels of organohalogenated persistent pollutant in human milk from Kaharamanmaras region, Turkey. *Environ. Int.* 30: 659–666.

Fujii, Y., Haraguchi, K., Harada, K. H., Hitomi, T., Inoue, K., Itoh, Y., Watanabe, T., Takenaka, K., Uehara, S., Yang, H. R., Kim, M. Y., Moon, C. S., Kim, H. S., Wang, P., Liu, A., Hung, N. N., and Koizumi, A. 2011. Detection of dicofol and related pesticides in human breast milk from China, Korea and Japan. *Chemosphere* 82: 25–31.

Gonzalez-Rodrıguez, M. J., Arrebola Liebanas, F. J., Garrido Frenich, A., Martınez Vidal, J. L., and Sanchez Lopez, F. J. 2005. Determination of pesticides and some metabolites in different kinds of milk by solid-phase microextraction and low-pressure gas chromatography-tandem mass spectrometry. *Anal. Bioanal. Chem.* 382: 164–172.

Haraguchi, K., Koizumi, A., Inoue, K., Harada, K. H., Hitomi, T., Minata, M., Tanabe, M., Kato, Y., Nishimura, E., Yamamoto, Y., Watanabe, T., Takenaka, K., Uehara, S., Yang, H. R., Kim, M. Y., Moon, C. S., Kim, H. S., Wang, P., Liu, A., and Hung, N. N. 2009. Levels and regional trends of persistent organochlorines and polybrominated diphenyl ethers in Asian breast milk demonstrate POPs signatures unique to individual countries. *Environ. Int.* 35: 1072–1079.

Harris, C. A., Woolridge, M. W., and Hay, A. W. M. 2001. Factors affecting the transfer of organochlorine pesticide residues in breast milk. *Chemosphere* 43: 243–256.

Hedley, A. J., Hui, L. L., Kypke, K., Malisch, R., Van Leeuwen, F. X. R., Moy, G., Wong, T. W., and. Nelson, E. A. S. 2010. Residues of persistent organic pollutants (POPs) in human milk in Hong Kong. *Chemosphere* 79(3): 259–265.

Hui, L. L., Hedley, A. J., Kypke, K., Cowling, B. J., Nelson, E. A. S., Wong, T. W., Van Leeuwen, F. X. R., and Malisch, R. 2008. DDT levels in human milk in Hong Kong, 2001–02. *Chemosphere* 73: 50–55.

Jaraczewska, K., Lulek, J., Covaci, A., Voorspoels, S., Kaluba-Skotarczak, A., Drews, K., and Schepens, P. 2006. Distribution of polychlorinated biphenyls, organochlorine pesticides and polybrominated diphenyl ethers in human umbilical cord serum, maternal serum and milk from Wielkopolska region, Poland. *Sci. Total Environ.* 372(1): 20–31.

Johnson-Restrepo, B., Addink, R., Wong, C., Arcarob, K., and Kannan, K. 2007. Polybrominated diphenyl ethers and organochlorine pesticides in human breast milk from Massachusetts, USA. *J. Environ. Monit.* 9: 1205–1212.

Kalantzi, O. I., Martin, F. L., Thomas, G. O., Alcock, R. E., Tang, H. R., Drury, S. C., Carmichael, P. L., Nicholson, J. K., and Jones, K. C. 2004. Different levels of polybrominated diphenyl ethers (PBDEs) and chlorinated compounds in breast milk from two U.K. regions. *Environ. Health Perspect.* 112: 1085–1091.

Katsoyiannis, A. and Salnara, C. J. 2005. Persistent organic Pollutants (POPs) in the conventional activated sludge treatment process: Fate and mass balance. *Environ. Res.* 97(3): 245–257.

Krauthacker, B., Votava-Raić, A., Herceg Romanić, S., Tješić-Drinković, D., Tješić-Drinković, D., and Reiner, E. 2009. Persistent organochlorine compounds in human milk collected in Croatia over two decades. *Arch. Environ. Contamin. Toxicol.* 57: 616–622.

Kumar, A., Baroth, A., Soni, I., Bhatnagar, P., and John, P. J. 2006a. Organochlorine pesticide residues in milk and blood of women from Anupgarh, Rajasthan, India. *Environ. Monit. Assess.* 116: 1–7.

Kumar, A., Dayal, P., Shukla, G., Singh, G., and Joseph, P. E. 2006b. DDT and HCH residue load in mother's breast milk: A survey of lactating mother's from remote villages in Agra region. *Environ. Int.* 32: 248–251.

Kunisue, T., Watanabe, M., Someya, M., Monirith, I., Minh, T. B., Subramanian, A., Tana, T. S., Viet, P. H., Prudente, M., and Tanabe, S. 2002. PCDDs, PCDFs, PCBs and organochlorine insecticides in human breast milk collected from Asian developing countries: Risk assessment for infants. *Organohalogen Comp.* 58: 285–287.

Kunisue, T., Someya, M., Monirith, I., Watanabe, M., Tana, T. S., and Tanabe, S. 2004a. Occurrence of PCBs, organochlorine insecticides, tris (4-Chlorophenyl) methane, and tris (4-Chlorophenyl) methanol in human breast milk collected from Cambodia. *Arch. Environ. Contam. Toxicol.* 46: 405–412.

Kunisue, T., Someya, M., Kayama, F., Jin, Y., and Tanabe, S. 2004b. Persistent organochlorines in human breast milk collected from primiparae in Dalian and Shenyang, China. *Environ. Pollut.* 131: 381–392.

Kunisue, T., Muraoka, M., Ohtake, M., Sudaryanto, A., Minh, N. H., Ueno, D., Higaki, Y., Ochi, M., Tsydenova, O., Kamikawa, S., Tonegi, T., Nakamura, Y., Shimomura, H., Nagayama, Y., and Tanabe, J. 2006. Contamination status of persistent organochlorines in human breast milk from Japan: Recent levels and temporal trend. *Chemosphere* 64: 1601–1608.

LaKind, J. S., Wilkins, A. A., and Berlin, C. M. Jr. 2004. Review: Environmental chemicals in human milk: A review of levels, infant exposures and health, and guidance for future research. *Toxicol. Appl. Pharmacol.* 198: 184–208.

Malarvannan, G., Kunisue, T., Isobe, T., Sudaryanto, A., Takahashi, S., Prudente, M., and Tanabe, S. 2008. Spatial distribution and accumulation of organohalogen compounds in human breast milk from the Philippines, interdisciplinary studies on environmental chemistry-biological responses to chemical pollutants, Murakami, Y., Nakayama, K., Kitamura, S.-I., Iwata, H., and Tanabe, S. (eds), pp. 339–347.

Malarvannan, G., Kunisue, T., Isobe, T., Sudaryanto, A., Takahashi, S., Prudente, M., Subramanian, A., and Tanabe, S. 2009. Organohalogen compounds in human breast milk from mothers living in Payatas and Malate, the Philippines: Levels, accumulation kinetics and infant health risk. *Environ. Pollut.* 157(6): 1924–1932.

Minh, N. H., Someya, M., Minh, T. B., Kunisue, T., Iwata, H., Watanabe, M., Tanabe, S., Viet, P. H., and Tuyen, B. C. 2004. Persistent organochlorine residues in human breast milk from Hanoi and Hochiminh city, Vietnam: Contamination, accumulation kinetics and risk assessment for infants. *Environ. Pollut.* 129: 431–441.

Mueller, J. F., Harden, F., Toms, L. M., Symons, R., and Furst, P. 2008. Persistent organochlorine pesticides in human milk samples from Australia. *Chemosphere* 70: 712–720.

Mustafa, M. D., Pathak, R., Tripathi, A. K., Ahmed, R. S., Guleria, K., and Banerjee, B. D. 2010. Maternal and cord blood levels of Aldrin and Dieldrin in Delhi population. *Environ. Monit. Assess.* 171: 633–638.

Okonkwo, J. O., Mutshatshi, T. N., Botha, B., and Agyei, N. 2008. DDT, DDE and DDD in Human Milk from South Africa. *Bull. Environ. Contam. Toxicol.* 81: 348–354.

Pathak, P., Suke, S. G., Ahmed, R. S., Tripathi, A. K., Guleria, K., Sharma, C. S., Makhijani, S. D., Mishra, M., and Banerjee, B. D. 2008. Endosulfan and Other Organochlorine Pesticide Residues in Maternal and Cord Blood in North Indian Population. *Bull. Environ. Contam. Toxicol.* 81: 216–219.

Polder, A., Thomsen, C., Lindström, G., Løken, K. B., and Skaare, J. U. 2008. Levels and temporal trends of chlorinated pesticides, polychlorinated biphenyls and brominated flame retardants in individual human breast milk samples from Northern and Southern Norway. *Chemosphere* 73(1): 14–23.

Polder, A., Skaare, J. U., Skjerve, E., Løken, K. B., and Eggesbø, M. 2009. Levels of chlorinated pesticides and polychlorinated biphenyls in Norwegian breast milk (2002–2006), and factors that may predict the level of contamination. *Sci. Total Environ.* 407(16): 4584–4590.

Poon, B. H. T., Leung, C. K. M., Wong, C. K. C., and Wong, M. H. 2005. Polychlorinated Biphenyls and Organochlorine Pesticides in Human AdiposeTissue and Breast Milk Collected in Hong Kong. *Arch. Environ. Contam. Toxicol.* 49: 274–282.

Qu, W., Suri, R. P. S., Bi, X., Sheng, G., and Fu, J. 2010. Exposure of young mothers and newborns to organochlorine pesticides (OCPs) in Guangzhou, China. *Sci. Total Environ.* 408: 3133–3138.

Raab, U., Preiss, U., Albrecht, M., Shahin, N., Parlar, H., and Fromme, H. 2008. Concentrations of polybrominated diphenyl ethers, organochlorine compounds and nitro musks in mother's milk from Germany (Bavaria). *Chemosphere* 72: 87–94.

Rahman, S. H. A. Al. 2010. Persistent organochlorine in human breast milk from Al-Sharkia Governorate, Egypt. *Egypt. Acad. J. Biol. Sci.* 2(1): 21–30.

Rodas-Ortíz, J. P., Ceja-Moreno, V., González-Navarrete, R. L., Alvarado-Mejía, J., Rodríguez-Hernández, M. E., and Gold-Bouchot, G. 2008. Organochlorine Pesticides and Polychlorinated Biphenyls Levels in Human Milk from Chelem, Yucatán, México. *Bull. Environ. Contam. Toxicol.* 80: 255–259.

Saeed, T., Sawaya, W. N., Ahmad, N., Rajagopal, S., Dashti, B., and Awadhi, S. Al. 2000. Assessment of the levels of chlorinated pesticides in breast milk in Kuwait. *Food Addit Contam.* 17(12): 1013–1018.

Saleh, I. A. I., Zahrani, M. A. I., Shanshoury, M. A. I., Mohammed, G., Shinwari, N., AI-Doush, I., and Basile, R. 2002. Comparative Quantitative Study of DDT and its Metabolites in Human Milk by Gas Chromatography with Electron Capture and Mass Spectrometric Detection. *Chromatographia* 55: 457–461.

Salem, D. A. and Ahmed, M. M. 2002. Evaluation of Some Organochlorine Pesticides in Human Breast Milk and Infants' Dietary Intake in Middle and Upper Egypt. *Alexandria J. Pediatr.* 16(2): 259–265.

Sanghi, R., Pillai, M. K., Jayalekshmi, T. R., and Nair, A. 2003. Organochlorine and organophosphorus pesticide residues in breast milk from Bhopal, Madhya Pradesh, India. *Hum. Exp. Toxicol.* 22(2): 73–76.

Schade, G. and Heinzow, B. 1998. Organochlorine pesticides and polychlorinated biphenyls in human milk of mothers living in northern Germany: Current extent of contamination, time trend from 1986 to 1997 and factors that influence the levels of contamination. *Sci. Total Environ.* 215: 31–39.

Schinas, V., Leotsinidis, M., Alexopoulos, A., Tsapanos, V., and Kondakis, X. G. 2000. Organochlorine Pesticide Residues in Human Breast Milk from Southwest Greece: Associations with Weekly Food Consumption Patterns of Mothers. *Arch. Environ. Health* 55(6): 411–417.

Sharaf, N. E., Elserougy, S. M., and Hussein, A. E. D. A. 2008. Organochlorine Pesticdes in Breast Milk and other related Tissues of Some Egyptian Mothers. *Am. Eur. J. Agric. Environ. Sci.* 4: 434–442.

Shetty, P. K., Mitra, J., Murthy, N. B. K., Namitha, K. K., Savitha, K. N., and Raghu, K. 2000. Biodegradation of cyclodiene insecticide endosulfan by Mucor thermo-hyalospora MTCC 1384. *Curr. Sci.* 79(9): 1381–1383.

Si, H. T. and Gia, N. H. 2010. Organochlorine pesticides and polychlorinated biphenyls in human breast milk in the suburbs of Hue city, Vietnam: Preliminary results. *J. Sci.* 61: 393–401.

Smith, D. 1999. Worldwide trends in DDT levels in human breast milk. *Int. J. Epidemiol.* 28: 179–188.

Solomon, G. M. and Weiss, P. M. 2002. Chemical contaminants in breast milk: time trends and regional variability. *Environ. Health Perspect.* 110: A339–A347.

Someya, M., Ohtake, M., Kunisue, T., Subramanian, A., Takahashi, S., Chakraborty, P., Ramachandran, R., and Tanabe, S. 2010. Persistent organic pollutants in breast milk of mothers residing around an open dumping site in Kolkata, India: Specific dioxin-like PCB levels and fish as a potential source. *Environ. Int.* 36: 27–35.

Sonawane, B. R. 1995. Chemical Contaminants in Human Milk: An Overview. *Environ. Health Perspect.* 103(6): 197–205.

Srivastava, S., Narvi, S. S., and Prasad, S. C. 2011. Levels of select organophosphates in human colostrum and mature milk samples in rural region of Faizabad district, Uttar Pradesh, India. *Hum. Exp. Toxicol.* 30(10): 1458–1463.

Stuetz, W., Prapamontol, T., Erhardt, J. G., and Classen, H. G. 2001. Organochlorin pesticide residues in human milk of a Hmong hill tribe in northern Thailand. *Sci. Total Environ.* 273: 53–60.

Subramanian, A., Ohtake, M., Kunisue, T., and Tanabe, S. 2007. High levels of organochlorines in mothers' milk from Chennai (Madras) city, India. *Chemosphere* 68: 928–939.

Sudaryanto, A., Kunisue, T., Kajiwara, N., Iwata, H., Adibroto, T. A., Hartono, P., and Tanabe, S. 2006. Specific accumulation of organochlorines in human breast milk from Indonesia: Levels, distribution, accumulation kinetics and infant health risk. *Environ. Pollut.* 139(1): 107–117.

Sudaryanto, A., Kunisue, T., Tanabe, S., Niida, M., and Hashim, H. 2005. Persistent Organochlorine Compounds in Human Breast Milk from Mothers Living in Penang and Kedah, Malaysia. *Arch. Environ. Contam. Toxicol.* 49: 429–437.

Sun, S. J., Zhao, J. H., Koga, M., Ma, Y. X., Liu, D. W., Nakamura, M., Liu, H. J., Horiguchi, H., Clark, G. C., and Kayama, F. 2005. Persistent organic pollutants in human milk in women from urban and rural areas in northern China. *Environ. Res.* 99(3): 285–293.

Suzuki, G., Nakano, M., and Nakano, S. 2005. Distribution of PCDDs/PCDFs and Co-PCBs in human maternal blood, cord blood, placenta, milk, and adipose tissue: Dioxins showing high toxic equivalency factor accumulate in the placenta. *Biosci. Biotech. Biochem.* 69: 1836–1847.

Szyrwinska, K. and Lulek, J. 2007. Exposure to specific polychlorinated biphenyls and some chlorinated pesticides via breast milk in Poland. *Chemosphere* 66: 1895–1903.

Tadevosyan, A., Reynolds, S. J., Kelly, K. M., Fuortes, L., Mairapetyan, A., Tadevosyan, N., Petrosyan, M., and Beglaryan, S. 2007. Organochlorine pesticide residues in breast milk in Armenia. *J. PreClin. Clin. Res.* 1(1): 84–88.

Tan, Z., Li, C. H., Yu, L. W., Xie, Y. F., and Lei, Z. X. 2008. Vertical distribution of organochlorine pesticide residue along soil profile in agricultural soils of Huizhou. *Chin. J. Eco Agric.* 16(5): 1134–1137.

Tanabe, S. and Kunisue, T. 2007. Review: Persistent organic pollutants in human breast milk from Asian countries. *Environ. Pollut.* 146: 400–413.

Tanabe, S., Gondaira, F., Subramanian, A., Ramesh, A., Mohan, D., Kumaran, P. L., Tasukawa, R., and Venugopalan, V. K. 1990. Specific pattern of persistent organochlorine residues in human breast milk from South India. *J. Agric. Food Chem.* 38: 899–903.

Tsydenova, O. V., Sudaryanto, A., Kajiwara, N., Kunisue, T., Batoev, V. B., and Tanabe, S. 2007. Organohalogen compounds in human breast milk from Republic of Buryatia, Russia. *Environ. Pollut.* 146(1): 225–232.

Tue, N. M., Sudaryanto, A., Minh, T. B., Nhat, B. H., Isobe, T., Takahashi, S., Viet, P. H., and Tanabe, S. 2010. Kinetic differences of legacy organochlorine pesticides and polychlorinated biphenyls in Vietnamese human breast milk. *Chemosphere* 81: 1006–1011.

Tutu, A. O., Yeboah, P. O., Golow, A. A., Denutsui, D., and Blankson-Arthur, S. 2011. Organochlorine Pesticides Residues in the Breast Milk of Some Primiparae Mothers in La Community, Accra, Ghana. *Res. J. Environ. Earth Sci.* 3(2): 153–159.

Waliszewski, S. M., Santiesteban, M. G., Pietrini, R. V., Arroyo, S. G., Munoz, O. A., Herrero-Mercado, M., and Carvajal, O. 2009. Breast Milk Excretion Kinetic of b-HCH, pp' DDE and pp' DDT. *Bull. Environ. Contam. Toxicol.* 83: 869–873.

Wong, C. K. C., Leung, K. M., Poon, B. H. T., Lan, C. Y., and Wong, M. H. 2002. Organochlorine Hydrocarbons in Human Breast Milk Collected in Hong Kong and Guangzhou. *Arch. Environ. Contam. Toxicol.*43: 364–372.

Wong, M. H., Leung, A. O. W., Chan, J. K. Y., and Choi, M. P. K. 2005. A review on the usage of POP pesticides in China, with emphasis on DDT loadings in human milk. *Chemosphere* 60: 740–752.

Yao, Z., Zhang, Y., and Jiang, G. 2005. Residues of Organochlorine Compounds in Human Breast Milk Collected from Beijing, People's Republic of China. *Bull. Environ. Contam. Toxicol.* 74: 155–161.

Yu, Z., Palkovicova, L., Drobna, B., Petrik, J., Kocan, A., Trnovec, T., and Hertz-Picciotto, I. 2007. Comparison of organochlorine compound concentrations in colostrum and mature milk. *Chemosphere* 66: 1012–1018.

Zhou, P., Wu, Y., Yin, S., Li, J., Zhao, Y., Zhang, L., Chen, H., Liu, Y., Yang, X., and Li, X. 2011. National survey of the levels of persistent organochlorine pesticides in the breast milk of mothers in China. *Environ. Pollut.* 159: 524–531.

Zietz, B. P., Hoopmann, M., Funcke, M., Huppmann, R., Suchenwirth, R., and Gierden, E. 2008. Long-term biomonitoring of polychlorinated biphenyls and organochlorine pesticides in human milk from mothers living in northern Germany. *Int. J. Hyg. Environ. Health* 211: 624–638.

Section VII

Pesticides and Food

18

Pesticide Residues in Milk and Milk Products

N. C. Basantia and S. K. Saxena

CONTENTS

18.1 Introduction

Milk is an essential food for infants and aged persons. It is a good source of protein, fat, and major minerals. The cow is a highly efficient food processor and is one of the few species that convert grass into healthy nutrition for humans: milk. The dairy farming has a problem with greenhouse gases. On the one hand, milk production has an effect on the environment, and on the other hand, the environment can have an impact on milk production through environmental contaminants such as pesticides, dioxins, etc. that may lead to residues in milk.

A dairy cow can be exposed to pesticides through the air it breathes, the water it drinks, and the forage it consumes. Depending on the type and properties of the pesticide, the residue is excreted through urine, adsorbed in the adipose tissue, or excreted into the milk. Pesticides may be applied directly to livestock or to structures (e.g., barns and housing units) or to control pests (e.g., parasites and vectors) using dipping vats, sprayers, and foggers (Bertrand 2010). The potential pollutants from pesticides include the active and inert ingredients, diluents, and persistent degradation products.

Pesticides and their degradation products may enter groundwater and surface water in solution, in emulsion, or bound to soil particles. Pesticides may, in some instances, impair the use of surface waters and groundwater. Some pesticides are suspected or known to cause chronic or acute health hazards to humans as well as adverse ecological impacts. Pesticides should be managed to avoid their migration into off-site land or water environment by implementing an integrated pest management (IPM) strategy and as documented in a pesticide management plan (PMP). When designing and implementing an IPM strategy, preference should be given to alternative pest management strategies, and the use of synthetic chemical pesticides should be the last option.

18.2 Entry of Pesticide Residues into Milk

Milk can be contaminated by the residues of pesticides through a variety of sources. The major source of organochlorine pesticide (OCP) residues is fodder and soil, while organophosphorus residues are mainly associated with ingestion through licking of insecticides used to control parasites on animals.

After ingestion, the active ingredient is mostly metabolized in the liver or stored in the fat. Excretion through milk can lead to residues in milk and milk products. The active ingredients that are soluble in water will be mostly eliminated through urine and feces. However, the substances that are soluble in fat will be eliminated mainly through milk. Fat solubility can be estimated by log P, which is the logarithm of the octanol–water partition coefficient. log P is the prime indicator of fat solubility and, by consequence, of the potential transfer to milk. When the log P exceeds 3, the compound is designated as fat-soluble (Bertrand 2010).

In Table 18.1, physicochemical properties of some pesticide residues are summed up (Tomlin 2000).

18.3 Hazardous Nature of Pesticide Residues in Milk Compared to Other Sources

Different studies have reported the presence of toxic residues in milk as well as in vegetables and meat products, which may cause serious, acute or chronic health hazards, depending on the length of exposure.

A higher level of contamination (63.38% of 325 samples) by OCPs in bovine milk has been reported in India (Nag and Raikwar 2008). Similar findings for dairy products (milk, cheese, labneh, yogurt, and butter) have been reported in Jordan (Salem et al. 2009). OCPs used in agriculture may reach aquatic bodies and finally humans through the food web. OCPs, for instance, are suspected to have a high affinity for the testicular androgen receptors, causing the dysfunction of the reproductive system (Singh et al. 2008), and to be associated with Parkinson disease, type 2 diabetes, and malarial infection (through the depletion of liver reserves of vitamin A) (Stuetz et al. 2006). In addition, DDT, mirex, toxaphene, HCH, and HCB are considered possible carcinogens (IARC 2006). Fat-soluble pesticides are accumulated in animal tissues and are excreted into milk but not into urine. These pesticides are more resistant to degradation compared to polar pesticides. Therefore, the pesticides present in milk are more hazardous compared to those in other matrices.

TABLE 18.1

Physicochemical Properties of Some Pesticide Residues

Type of Compound	Water Solubility mg/L at 25°C	log P	Excretion in Mammals
Organochlorine Residue			
Lindane	8.5	3.5	It is fairly readily metabolized by animals to pentachlorocyclohexane, 1,2,4-trichlorobenzene, and isomeric trichlorophenols and excreted as glucuronic acid derivatives.
Dieldrin	0.186	4.32–5.4	Some accumulation of dieldrin occurs in animal tissue, particularly fat; dieldrin is very slowly metabolized to water-soluble products that are excreted from the body.
DDT	0.005	6.19	It is metabolized (very slowly) to a range of saturated and unsaturated products by progressive dechlorination.
Heptachlor	0.056	5.44	Residues accumulate in fatty tissues and are excreted in milk.
Endosulfan	0.32–0.33	4.74[a]	In animals, heptachlor metabolizes to the epoxide, which can be found in most body organs, but it particularly accumulates in body fat. Rapidly metabolized and excreted by mammals.
Organophosphorus residues			
Fenitrothion	14[b]	3.43	Rapid mammalian metabolism and excretion. The most important metabolites are dimethylfenitrooxon and 3-methyl-4-nitrophenol.
Fenthion	4.2	4.84	Elimination of residues in mammals by excretion of hydrolysis products. Major metabolites are fenthion sulfoxide and sulfone and their oxygen analogs. Further degradation of these metabolites to the corresponding phenols can occur.
2,4-D	620	2.62	Rapidly eliminated from mammals (as parent compound), often within 24 h. Maximum organ concentration is reached in approximately 12 h.
Atrazine	33	2.3	In mammals, rapid and complete metabolism of ingested residues is primarily by oxidative dealkylation of the amino groups.
Carbamate residues			
Aldicarb	4930[b]	1.36	Aldicarb is toxic to mammals, but sublethal doses are metabolized rapidly with over 90% excreted in 3–4 days.
Carbaryl	120	1.59	In mammals, carbaryl does not accumulate and is rapidly metabolized to nontoxic substances, particularly 1-napthol, and excreted.
Propoxur	0.14	1.9	In mammals, metabolism, principally to 2-hydroxyphenyl-N-methylcarbamate and 2-isopropoxyphenol, and excretion in the urine are rapid.
Pyrethroids			
Cypermethrin	0.004	6.6	In rats, it is virtually eliminated from the body within 8 days with extensive metabolism occurring.
Deltamethrin	0.006	6.1[b]	In insects and mammals, elimination is by hydrolysis, hydroxylation, and elimination as glucoside conjugate.
Permethrin	<0.2 μg	4.6	In rats, it is virtually eliminated from the body within 8 days with extensive metabolism occurring.

Source: From Tomlin, C. 2000. *The Pesticide Manual: A World Compendium*, 15th edn. British Crop Protection Council.

[a] 22°C.

[b] 20°C.

18.4 Current Levels of Pesticide Residues in Milk and Milk Products in Different Regions

Milk and milk products are mostly contaminated with organochlorine and organophosphorus pesticide residues, as reported in a number of studies carried out in various parts of the world. Among OCP residues, DDT, HCH, heptachlor, chlordane, aldrin, and dieldrin are detected in most of the cases. Around 355 samples of cow's milk and 448 samples of national butter brands in Mexico were analyzed for OCP residues. HCH was found in concentrations of 0.094 mg/kg in cow's milk and 0.093 mg/kg in butter, whereas DDT was found in concentrations of 0.159 mg/kg and 0.049 mg/kg in milk and butter, respectively (Waliszewski et al. 1997). In another study from the tropical part of Mexico, the milk sample was found to be contaminated with β-HCH at a concentration of 0.106 mg/kg followed by pp'-DDT at a concentration of 0.078 mg/kg on fat basis and pp'-DDE at a concentration of 0.051 mg/kg on fat basis. Some of the high-distribution brands of liquid milk in Mexico were also contaminated with detectable levels of organophosphorus pesticide residues including dichlorvos, phorate, chlorpyrifos, and chlorfenvinphos (Salas et al. 2003). Trace levels of DDT and HCH were found in milk and milk product samples from Maharashtra in India (Pandit et al. 2002). Commercially available butter samples from Punjab region showed detectable levels of HCH (0.092–0.645 mg/kg) and total DDT residues (0.469 mg/kg). The HCH residue includes α-, β-, γ-, and δ-HCH, and the DDT residue includes pp'-DDD, pp'-DDE, and pp'-DDT (Gill et al. 2009). Buffalo's and cow's milk samples from different animal farms in Egypt, during the years 2008 and 2009, were also contaminated with OCP residues: HCB, aldrin, heptachlor, heptachlor epoxide, chlordane, and DDT and its derivatives. The concentration of HCB residue in buffalo's milk was 0.162 mg/kg fat and that in cow's milk was 0.150 mg/kg fat. The concentrations of aldrin were 0.066 mg/kg fat and 0.05 mg/kg fat in cow's and buffalo's milk, respectively (Abou Donia et al. 2010). In Argentina, buffalo's milk samples were contaminated with heptachlor and heptachlor epoxides, γ-HCH, and γ-chlordane (Lenardon et al. 2004).

18.5 Effects of Processing Treatments on Pesticide Residues in Milk and Milk Products

Raw milk undergoes various processing treatments before consumption and is consumed in different forms of milk products. So, it is important to know the effect of processing treatments on residues and the distribution of pesticide residues in different milk products.

Raw milk undergoes a number of processing treatments like pasteurization, sterilization, concentration, separation of fat, fermentation, coagulation, and drying. Depending on the type and nature of the pesticide compound, they are transferred from one phase to another or degraded totally or partially to other compounds that may be more or less toxic than the parent compound. Pasteurization has little effect on pesticide residues.

Pasteurization of raw milk at 93°C–100°C for a few seconds shows a small effect on β-BHC and pp'-DDT. Sterilization at 121°C for 15 min showed 83.25%, 91.67%, and 68.7% loss of β-BHC, lindane, and p,p'-DDT, respectively. However, p,p'-DDD was found in significantly higher concentrations in UHT milk (Heck et al. 2007). The amounts of endosulfan sulfate and chlordane decreased by 17.1% and 11.2% on fat basis, respectively, in condensed milk compared to the levels in raw milk (Liaska 1968). Endosulfan sulfate found in spray-dried, evaporated, and roller-dried milk showed around 42%, 42%, and 70% decrease compared to raw milk, whereas concentration decrease of chlordane for these products was approximately 25.0%, 45%, and 50%, respectively (Liaska 1968). The skimming of cow's and buffalo's milk led to a slight reduction in the pesticide (β-BHC, lindane, DDT) residue concentration (Liaska 1968).

Preparation of butter from churning buffalo's or cow's cream reduced the concentration of HCB, lindane, Aldrin, heptachlor epoxide, chlordane, and total DDT. This might be due to the transfer of most of the fat globules to the butter milk carrying considerable pesticide residues (Langlois et al. 1974; Waliszewski et al. 1997).

In cheese-making, heating and salting caused a great amount of degradation of leptophos residues. The reduction in total DDT in Ras cheese, made from contaminated raw milk spiked at different levels (0.1, 1, and 10 ppm), was 40.6, 33.9, and 25.5%, respectively, at the end of a 6-month storage. This reduction was effected by Streptococci, Lactobacilli, and yeasts. The reduction of lindane was higher in Domiati cheese made by acid–enzyme coagulation than Ras cheese made by enzyme coagulation. The reduction of pesticides in cheese may be due to the microorganisms in ripening cheese as well as interference of pesticide residues with cellular metabolism of the microorganism (Chacko and Lackwood 1967; Hantke and Bradley 1972; Kim et al. 1970).

The manufacture of yoghurt from buffalo's or cow's milk resulted in the decrease of levels of pesticide residues. This reduction might be due to heat treatment as well as the activity of the starter culture.

18.6 Maximum Residual Limits

Maximum Residual limits (MRLs) for various pesticide residues have been fixed by different International and National Regulatory bodies (Table 18.2).

18.7 Quantification of Pesticide Residues in Milk and Milk Products

Pesticide analysis plays an important role in estimating human and environmental exposure to various compounds in diverse agricultural practices. Since the analyte concentrations are generally very low and the sample matrix is complex, the interference of matrix should be considered when performing a measurement. So, the analysis of the sample requires studious sample preparation steps, extraction methods, and cleanup steps to minimize the interfering of the matrix. The analysis of pesticide residues in milk and milk products involves the following steps:

1. Sample preparation
2. Extraction
3. Cleanup or isolation
4. Separation and detection.

TABLE 18.2

Maximum Residual Limits (MRLs) for Pesticide Residues in Milk and Milk Products

Compound	As per FAO/WHO,[a] (mg/kg)	As per EU,[b] (mg/kg)	As per USFDA,[c] (mg/kg)	As per PFA,[d] India (mg/kg)
Alachlor	0.01	—	0.02	—
Aldrin	0.006	0.006	—	0.15[e]
Aldicarb	0.01	0.01	—	—
Azinphos-ethyl	0.01	0.05	—	—
Azinphos-methyl	0.01	—	—	—
Captafol	0.01	0.01	—	—
Carbaryl	0.05	—	1.0	—
Carbofuran	0.1	0.1	0.1	0.05
Carbosulfan	0.05	0.05	—	—
Chlordane	0.002	0.002	0.05	0.05[e]
Chlorfenvinphos	0.01	—	—	0.2
Chlortoluron	0.05	—	—	—
Chlorpyrifos	0.01	0.01	—	0.01[e]
Chlorpyrifos methyl	0.01	0.01	0.05	—
Cypermethrin	0.05	0.02	—	0.01[e]
DDT(sum of pp'-DDT, op'-DDD, and pp'--DDE)	0.04	0.04	—	1.25[e]
Deltamethrin	0.05	—	0.1[e]	—
Diazinon	0.01	0.01	—	—
Diclofop	0.01	—	—	—
Dicofol	0.02	0.02	—	—
Diquat	0.05	0.05	—	—
Disulfoton	0.02	0.02	—	—
Dithiocarbamates	0.05	—	—	—
Endrin	—	0.0008	—	—
Endosulfan	—	0.004	—	—
Ethion	—	—	0.5	0.5[e]
Glyphosate	0.05	0.1	0.1	—
Heptachlor (sum of heptachlor and heptachlor epoxide)	0.004	0.004	—	0.15[e]
Hexachlor Cyclohexane	0.004 0.003[1]	0.007	—	0.05 0.02[1]
Hexachlorobenzene	0.01	0.01	—	—
Imazalil	0.05	0.02	—	—
Imidacloprid	0.1	—	—	—
Lindane	0.001	0.001	0.05	0.01
Parathion	0.05	—	—	—
Parathion methyl	0.02	—	—	—
Permethrin	0.05	0.05	—	—

TABLE 18.2 (Continued)

Maximum Residual Limits (Mrls) for Pesticide Residues in Milk and Milk Products

Compound	As per FAO/WHO,[a] (mg/kg)	As per EU,[b] (mg/kg)	As per USFDA,[c] (mg/kg)	As per PFA,[d] India (mg/kg)
Phorate	0.02	0.02	—	0.05[e]
Phosalone	0.01	—	—	—
Propoxur	0.05	0.05	—	—
Quintozene	0.01	0.01	—	—
Simazine	0.05	—	—	—

Source: [a]http://www.codexalimentarius.net/pestres/data/commodities/details.html; [b]EEC Council Directives 86/363/EEC & 90/642/EEC. 1986. Maximum levels for Pesticide residues in and on Food stuffs of animal origin. *Off. J. Eur.* L221: 43, 7 August 1986; [c]http//www.epa.gov/pesticides/food/viewtools.htm; [d]PFA. 2011. The Prevention of Food Adulteration Act, 1954, R65, India, Commercial Law Publishers (India) Pvt. Ltd., Delhi (Edition 2011).
Weights on fat basis per β-isomers.

18.7.1 Sample Preparation

This is the first step to provide a representative and composite sample from which sub-samples can be taken for analysis. Milk is an emulsion of the oil-in-water type. Fat has a tendency to come to the surface, so attention has to be paid to get homogenized and representative samples. The sample should be handled carefully so that possible contamination is prevented and the loss of volatile pesticides is avoided.

18.7.2 Extraction

Pesticides are extracted from the sample employing solvents such as acetonitrile, petroleum ether, hexane, dichloromethane, or acetone. The solvent is blended with the sample and homogenized. In the case of milk, pesticides are either present in the fat phase in free form or bound to the lipoprotein or protein. So, acetonitrile is used to deproteinize the sample to release any pesticide bound to the protein and, at the same time, to precipitate the fat and protein and bring pesticides to a dissolved form. Then, the pesticides are partitioned to the petroleum ether phase. The extraction timing depends on the type of pesticide, matrix, and physicochemical properties of the solvent. The most common problem faced is the incomplete recovery and formation of emulsion during the partitioning process. This can be avoided by using a suitable solvent with sodium chloride added or by one or more solvent combinations.

Different approaches and extraction methods may be used, such as solid-phase extraction (SPE), solid-phase microextraction (SPME), accelerated solvent extraction (ASE), matrix solid-phase dispersion (MSPD), and supercritical fluid extraction (SFE) (Table 18.3).

18.7.3 Cleanup

Two types of preparative column chromatography are used for cleanup and purification: adsorption and size exclusion. Materials used in adsorption chromatography are Florisil, alumina, and silica gel.

TABLE 18.3

Residual Analysis of Pesticides In Milk and Milk Products (Selected Methods)

Analyte	Extraction	Separation	Analytical System	Recovery (%)	LOD (µg/kg)	References
Organochlorine pesticides	Extraction with petroleum ether	Capillary DB-5 column	GC-ECD	91.0–99.1	0.01–0.03	Waliszewski et al. (1997)
Organochlorine pesticides	Cleanup of hexane extract with official method	DC-200 Packed on chromosorb	GC-ECD	84.5–98.2	1	Losada et al. (1996)
Herbicides and fungicides	SPE extraction	C-18	LC-MS	82–120	0.008–1.4	Bogialli et al. (2006)
Organochlorine pesticides	Acetonitrile petroleum ether followed by Florisil cleanup	HP-1 Capillary column 30 m length	GC-ECD	90–94	1	Abou Donia et al. (2010)
Cypermethrin, fipronil, chlorfenvinphos	QuEChERS (MSPD)	GC-MS	DB-5 Capillary column	70–120	20	Brondi et al. (2010)
Organochlorine pesticides in butter	Extraction with hexane partitioned with acetonitrile and dichloromethane	GC-ECD	Chromosorb OV-17	85–92	100	Gill et al. (2009)
48 compounds in milk	Acetonitrile acetone	UPLC/ Q-TOF and GC-MS/MS	C-18, 2.1 × 100 mm and ZB-50	70–85 70–100	—	Hakala et al. (2009)

New technologies such as SPE, SPME, and MSPD are suitable to perform extraction and cleanup in a single step.

18.7.4 Separation and Detection

The quantification of the target analytes is usually performed with gas chromatography or liquid chromatography separation, followed by detection using various detectors.

Gas chromatography appears to be the most useful technique for the quantitative determination of pesticide residues in milk and milk products.

Possible detectors in combination with GC are electron capture detector (ECD), thermionic detectors such as nitrogen and phosphorus detector (NPD), flame photometric detector (FPD) or flame ionization detector, and mass selective detector (MSD).

High-performance liquid chromatography is the second most frequently used technique to determine very polar and low volatile pesticides. The separation mostly occurs on reverse-phase packed columns. Different types of detectors for HPLC to determine pesticide residues are UV absorption, fluorescence, conductivity, electrochemical, and mass spectrometer detectors.

Liquid chromatography, along with a mass selective detector (MSD), is a very powerful technique for the quantification and confirmation of pesticide compounds (see Bogialli and Di Corcia (2010) for further details).

References

Abou Donia, M. A., Abou-Arab, A. A. K., Enb, A., El-Senaily, M. H., and Abd-Robou, N. S. 2010. Chemical composition of raw milk and accumulation of pesticide residues in milk products. *Global Vet.* 4(1): 6–14.

Bertrand, S. 2010. Pesticide consumption at farm level and residue in environment and in milk. *Bull. Int. Dairy Fed.* 44: 33–40.

Bogialli, S., Curini, R., Di Corcia, A., Lagana, A., Stabile, A., and Sturchio, E. 2006. Development of a multi residue method for analyzing herbicide and fungicide residue in bovine milk based on solid phase extraction and Liquid chromatography tandem mass spectrometry. *J. Chromatogr. A* 1102: 1–10.

Bogialli, S. and Di Corcia, A. 2010. Environmental Contaminants. In: Nollet, L. M. L. and Toldrá, F. (eds), *Handbook of Dairy Foods Analysis*. CRC Press: Boca Raton, FL.

Brondi, S. H. G., Macedo, A. N., Sartarelli, N. C., and Noggueria, A. R. A. 2010. Application of QuEChERS method and gas chromatography – mass spectrometry for the analysis of pesticide residues in cattle's milk, meat and fat. In: 8th European Pesticide Residue Workshop – EPRW 2010, Strasbourg.

Chacko, C. I. and Lackwood, J. L. 1967. Accumulation of DDT and dieldrin by microorganisms. *Can. J. Microbiol.* 13: 1123–1126.

EEC Council Directives 86/363/EEC & 90/642/EEC. 1986. Maximum levels for Pesticide residues in and on Food stuffs of animal origin. *Off. J. Eur.* L221: 43, 7 August 1986.

Gill, J. P. S., Sharma, J. K., and Aulakh, R. S. 2009. Studies on Organochlorine Pesticide residues in Butter in Punjab. *Toxicol. Int.* 6(2): 133–136.

Hakala, K. and Hyvönen, E. 2009. Monitoring of pesticides in animal origin products with UPLCQ-TOF and GCMS/MS. Finnish Food Safety Authority Evira. http://www.eprw2010.com/download/Poster%20Monitoring/PM%20032%20Hakala/PM%20032%20Hakala.pdf.

Hantke, W. E. and Bradley, R. 1972. Effect of diedrin on bacteria producing lactic acid. *J. Milk Food Technol.* 35: 655–659.

Heck, M. C., Santos, J. S., Bogusz Tunior, S., Costabebe, I., and Ernanuelli, T. 2007. Estimation of children exposure to Organochlorine compounds through milk in Rio Grando Do Sul, Brazil. *Food Chem.* 102: 288–294.

http//www.epa.gov/pesticides/food/viewtools.htm.

http://www.codexalimentarius.net/pestres/data/commodities/details.html.

International Agency for Research on Cancer (IARC). 2006. Overall evaluation of Carcinogenicity to humans. http://www.monographs.iarc.fr/ENG/Classification/index.php>. Update 13/04/2011.

Kim, S. C. and Harmon, L. G. 1970. Relationship between some chlorinated hydrocarbon insecticides and lactic culture organism in milk. *J. Dairy Sci.* 53: 155–160.

Langlois, B. E., Liska, B. J., and Hill, D. L. 1974. The effects of processing and storage of dairy products on chlorinated insecticide residues. I: DDT and lindane. *J. Milk Food Technol.* 27: 264–269.

Lenardon, A., Lorenzatti, E., and Althaus, R. L. 2004. Residues of Organochlorine Pesticide Residues in Buffalo Milk in Argentina. *Buffalo Bull.* 23(1): 10–15.

Liaska, B. J. 1968. Effect of processing on pesticide residues in milk. *J. Anim. Sci.* 27: 827–830.

Losada, A., Fernandez, N., Diez, M. J., Terán, M. T., Garcia, J. J., Sierra, M. 1996. Organochlorine Pesticides residue in bovine milk from León, Spain. *Sci. Total Environ.* 181(2): 133–135.

Nag, S. K. and Raikwar, M. K. 2008. Organochlorine pesticide residues in bovine milk. *Bull. Environ. Contam. Toxicol.* 80: 5–9.

Pandit, C. G., Sharma, S., Srivastava, P. K., and Sahu, S. K. 2002. Persistent organochlorine pesticide residues in milk and dairy products from India. *Food Addit. Contam.* 19: 153–157.

PFA. 2011. The Prevention of Food Adulteration Act, 1954, R65, India, Commercial Law Publishers (India) Pvt. Ltd., Delhi (Edition 2011).

Salas, J. H., González, M. M., Noa, M., Pérez, N. A., Díaz, G., Gutiérrez, R., Zazueta, H., and Osuna, I. 2003. Organophosphorus pesticide residues in Mexican commercial pasteurized milk. *J. Agric. Food Chem.* 51(15); 4468–4471.

Salem, N. M., Ahmad, R., and Estaitieh H. 2009. Organochlorine pesticide residues in dairy products in Jordan. *Chemosphere* 77: 673–678.

Singh, P. B., Singh, V., and Nayak, P. K. 2008. Pesticide residues and reproductive dysfunction in different vertebrates from north India. *Food Chem. Toxicol.* 46: 2533–2539.

Stuetz, W., McGready, R., Cho, T., Prapamostal, T., Biesalski, H. K., Stepniewska, K., and Nosten, I. 2006. Relation of DDT residues to plasma retinol, α-tocopherol and β-carotene during pregnancy and malaria infection: A case-control study in Karen women in northern Thailand. *Sci. Total Environ.* 363: 78–86.

Tomlin, C. 2000. *The Pesticide Manual: A World Compendium*, 15th edn. British Crop Protection Council.

Waliszewski, S. M., Parodio, V. T., Waliszewski, K. N., Chantiri, J. N., Aguirre, A. A., Infanzon, R. M., and Rivera, J. 1997. Organochlorine pesticide residues in cow's milk and butter in Mexico. *Sci. Total Environ.* 208: 127–132.

Section VIII

Biopesticides

19

Biopesticides

Rajendra Singh Tanwar, Prem Dureja, and Hamir Singh Rathore

CONTENTS

The introduction of DDT in 1945, followed by the use of various other synthetic chemical pesticides, has played a key role in the increase in agricultural productivity, in the protection of crops and forests, and in the control of insect vectors of human diseases. However, the unscrupulous use of chemical insecticides has led to serious problems such as development of resistance in insects, insect resurgence, and occurrence of higher levels of pesticide residues in food and other agricultural commodities, environment contamination, destruction of beneficial insects such as honeybees, pollinators, parasitoids, and predators (Table 19.1).

Increased public concern about the potential adverse environmental effects associated with the heavy use of chemical insecticides has prompted the examination of alternative methods for insect pest control. One such alternative is the use of biopesticides.

TABLE 19.1

Problems Associated with Chemical Pesticides

- Persistence
- Bioaccumulation
- Biomagnifications
- Chronic and acute toxicity
- Resistance
- Secondary pest outbreak
- Environmental contamination
- Effect on nontarget and beneficial organisms

TABLE 19.2

Different Types of Biopesticides

- Microbial organisms
- Plant-derived pesticides
- Secondary metabolites from microbes
- Insect growth and behavioral regulator
- Genetic modification

TABLE 19.3

Pros and Cons of Biopesticides in Comparison with Conventional Pesticides

Advantages	Disadvantages
Less toxic	Short shelf-life
Biodegrade quickly	Limited field persistence
More targeted to specific pest	Narrow target range
Specific mode of action	Narrower target range
Mange rather than irradiate	Slower acting

Biopesticides: According to US EPA, biopesticides are pesticides derived from natural materials such as animals, plants, bacteria, and minerals. They include microbial pesticides, entomopathogenic nematodes, baculoviruses, plant-derived pesticides, and insect pheromones, the latter when used as mating disruption agents receive increased exposure in scientific annals (Menn 1996, 1997; Table 19.2).

Some common benefits and disadvantages of biopesticides in comparison with conventional pesticides are shown in Table 19.3.

19.1 Global Biopesticide Market

Annual sales of microbial pesticides are reported to be US $750 million globally, amounting only to 2.5% of the chemical market (Evans 2008). The global market for biopesticides

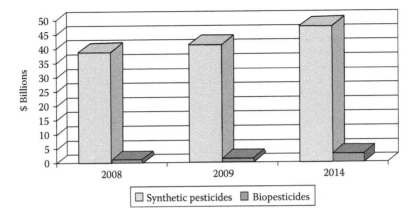

FIGURE 19.1
Projected global pesticide market by segment, 2008–2014.

is forecast to reach US $2.8 billion by the year 2015. This segment is expected to grow at a 15.6% compound annual growth rate (CAGR) from $1.6 billion in 2009 to $3.3 billion in 2014. Key factors driving market growth include increasing environmental concerns and consumer inclination toward chemical-free crops. Other factors propelling market growth include introduction of new products, acceptance of substitutes to conventional pesticides, and declining market for harmful organophosphate insecticides.

The United States represents the largest region for biopesticides worldwide, as stated by the new market research report on biopesticides (Figure 19.1). On the other hand, with a CAGR of more than 15% over the analysis period, Europe represents the fastest growing regional market for biopesticides. Europe lags behind the United States in terms of registered biopesticides with 77 active substances as of 2008 against 279 in the United States. The wide difference between the two leading biopesticides consumers could be attributed to highly fragmented nature of European market and biopesticides being relatively newer concept for the consumers in the region. Further, the government support for capacity development of biopesticides places the US market ahead of any other region in the world. Asia-Pacific constitutes the other leading market, with biopesticides sales projected to reach US $362 million in 2012 (Figure 19.2).

19.2 Microbial Pesticides

Microbial biopesticides are products derived from various microscopic organisms. Microbial products may consist of the organisms themselves and/or the metabolites they produce. They suppress pests either by producing a toxin specific to the pests, causing disease, preventing establishment of other microorganisms through competition, or various other modes of action (Clemson HGIC 2007). For all crop types, bacterial biopesticides claim about 74% of the market; fungal biopesticides, about 10%; viral biopesticides, 5%; predator biopesticides, 8%; and "other" biopesticides, 3% (Thakore 2006). At present there are approximately 73 microbial active ingredients that have been registered by the US EPA.

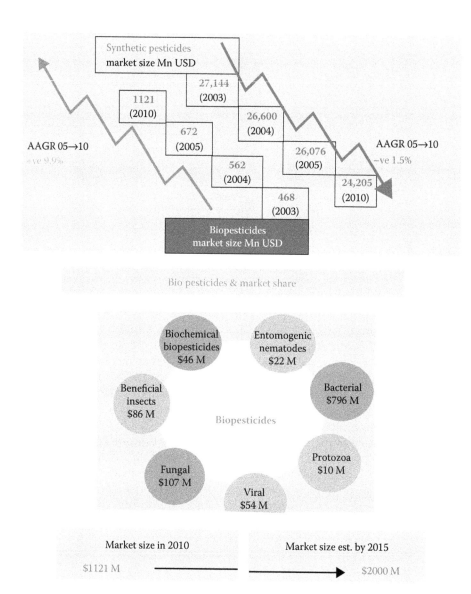

FIGURE 19.2
Biopesticides and market share.

The registered microbial biopesticides include 35 bacterial products, 15 fungi, 6 nonviable (genetically engineered) microbial pesticides, 8 plant incorporated protectants, 1 protozoa, 1 yeast, and 6 viruses (Steinwand 2008). They are generally divided into six different sub-categories of products:

- Bacteria
- Fungi
- Protozoa
- Viruses
- Yeast.

19.3 Bacteria Biopesticides

Biopesticides based on bacteria have been used to control plant diseases, nematodes, insects, and weeds. Bacteria are present in all soils and are the most abundant microorganisms in soil samples. Many spore forming and nonspore forming bacteria are known to be effective against a wide spectrum of insects and diseases (Garczynski and Siegel 2007).

The most well-known and widely used of all biopesticides are insecticides based on *Bacillus thuringiensis*, commonly referred to as "Bt." During spore formation, Bt produces insecticidal proteins (known as δ-endotoxins) that kill caterpillar pests, fly and mosquito larvae, or beetles (depending on the subspecies and strain of Bt) that ingest them through feeding in Bt-treated areas. The highly specific δ-endotoxins bind to and destroy the cellular lining of the insect digestive tract, causing the insect to stop feeding and die. Bt has been in continuous commercial use for over 50 years, a record not exceeded by any other insecticide active ingredient. *Bacillus sphaericus* is another insecticidal bacterium that has been used successfully to control certain mosquito species (Copping 2009).

Bt is registered for use against 90 different insects as Bt formulations listed in Table 19.4. Bt can be used at any time, upto harvest, and is also used as a cleanup spray before taking the produce to market. *Bacillus thuringiensis israelensis* (BtI) is another strain of BtI isolated from Israel and extremely lethal to insects belonging to Diptera. This BtI strain will find increasing use in future.

Other bacteria are used for the control of plant pathogens. Certain strains of *Bacillus subtilis*, *Bacillus pumilus*, *Pseudomonas* spp., and *Streptomyces* spp. increase yield and prevent plant diseases by outcompeting plant pathogens in the rhizosphere, producing antifungal compounds, and by promoting plant and root growth.

19.4 Fungal Biopesticides

Different fungal biopesticides can be used to control plant diseases (caused by other fungi, bacteria, or nematodes), as well as some insect pests and weeds. Fungi are a diverse group of organisms and can be found in almost every environment on the Earth. Most have complex lifecycles, and some are parasitic to various eukaryotes, including plants and insects. Some species have proven useful as microbial biopesticides. However, because they are living organisms, they do require specific environmental conditions to proliferate (Hibbett et al. 2007; Humber 2007, 2008).

TABLE 19.4

Some Commercially Available Bt Formulations

Product	Producer	Trade name
Bacillus thuringiensis HD-1	1. Abbott	Dipe** HG & 4L
	2. Biochem products	Bactospeine, Novabac
	3. Sandoz	Thuricide + HP, HPCL 32B & 16B
	4. Upjohn	SOK-BT
	5. Agriculture & Nutrition Co.	Bactur TH
	6. Pennwalt, Holland N.V.	Tribactur

**Copping 2009.

Because fungal biopesticides are so diverse in nature, their means of affecting the target pests can be equally diverse. The most common modes of action are through competitive exclusion, mycoparasitism, and production of metabolites. Some fungi can exhibit all of these modes of action. Two of the most common commercial fungal biopesticides are *Trichoderma* spp. and *Beauveria bassiana*. Each are frequently used in the nursery, ornamental, vegetable, field crop, and forestry industries to control a variety of pests.

Trichoderma spp. are some of the most common fungi in nature. Many beneficial *Trichoderma* have the ability to readily colonize plant roots, without harming the plants. It is this close relationship with the plants that makes these species excellent biocontrol agents. These microbial biofungicides can outcompete pathogenic fungi for food and space and, in the process, can stimulate plant host defenses and affect root growth. In addition, they have the ability to attack and parasitize plant pathogens under certain environmental conditions.

Beauveria bassiana is a fungus that acts as a parasite on many insect species. *B. bassiana* has a broad host range, although individual strains may be restricted in the number of insects it can attack (Inglis et al. 2001). *B. bassiana* spores adhere directly to the host cuticle, where they will germinate, produce enzymes that attack and dissolve the cuticle, penetrate, and grow into the insect's body, feeding on internal tissues and releasing an insect toxin (Feng et al. 1994). As the insect dies, it changes color to pink or brown and eventually the entire body cavity is filled with fungal mass. *B. bassiana* has proved effective in controlling troublesome crop pests such as aphids, thrips and whitefly—even chemical pesticide-resistant strains such as Q-Biotype Whitefly (Pell et al. 2001).

19.5 Protozoa Biopesticides

Protozoa are single-celled eukaryotic organisms that exist in both water and soil. While most protozoa feed on bacteria and decaying organic matter, a wide range of protozoan species are insect parasites (Jehle et al. 2006).

For example, *Nosema locustae* are known to be natural biocontrol agents of many grasshopper species. *Nosema* infects at least 90 species of grasshoppers. It is nontoxic to humans and other mammals, as well as to over 250 natural predators of grasshoppers. *Nosema* is particularly important in areas where environmental issues preclude the use of traditional insecticides.

Grasshoppers cause millions of dollars of damage to forage crops each year. The protozoan *Nosema locustae* has been developed into a commercially available, grasshopper control agent that infects and weakens young grasshoppers and adversely affects the female grasshoppers' ability to reproduce.

After consuming *Nosema locustae*, grasshopper feeding is typically curtailed within a week. Within 2 weeks or longer, as many as 50% of the infected insects die and approximately half of the surviving population remain weak, consuming 75% less forage than a healthy insect.

Grasshoppers are notorious scavengers. An important function in the transmission of Nosema spores to healthy grasshoppers occurs as the insects scavenge and cannibalize infected cadavers. Since infected grasshoppers develop a large number of the spores within them, the cannibalizing grasshoppers get a much greater dose of the disease-causing organisms than through the initial *Nosema* application.

19.6 Viruses Biopesticides

Microbial biopesticides known as baculoviruses are a family of naturally occurring viruses known to infect only insects and some related arthropods. Most are so specific in their action that they infect and kill only one or a few species of *Lepidoptera* larvae (caterpillars), making them good candidates for management of crop pests with minimal off-target effects. Baculoviruses used as microbial biopesticides consist of DNA surrounded by a protein coat (nucleocapsid), which is itself embedded in a protein "microcapsule" or occlusion body (OB) that provides some protection from degradation in the environment. Depending on the virus, OBs may contain a single nucleocapsid (a granulovirus, GV) or multiple nucleocapsids (nucleopolyhedrovirus, NPV).

Upon ingestion by a susceptible caterpillar, OBs are dissolved within the alkaline midgut, releasing nucleocapsids, which infect the cells lining the midgut. The viral DNA replicates in the nuclei of the host cells and then spreads throughout the body of the larvae, essentially turning it into a "virus factory." The infected insect stops feeding within a few days, dies and disintegrates, releasing billions of new OBs, which can be ingested and can cause new infection of the neighboring larvae.

The granulovirus of the codling moth *Cydia pomonella* (CpGV) is a good example of a commercially successful viral insecticide. First discovered in the 1960s, it is now the active ingredient of about half a dozen products sold worldwide. Often used in conjunction pheromone-based mating disruption, CpGV limits codling moth populations and damage in pome fruits while preserving the beneficial insects and minimizing the chemical residues. Although accepted for use in organic farming, most CpGV applications occur in conventional orchards where its unique mode of action can minimize the risk of resistance to chemical insecticides.

19.7 Yeast Biopesticides

A variety of yeasts have been investigated for their usefulness in controlling plant diseases. Nonpathogenic *Cryptococcus* and *Candida* species naturally occur on plant tissues and in water. Isolates from a variety of crops have been investigated for their biocontrol capacities. For example, *C. olephila* Strain O, first isolated from golden delicious apples, has been developed into an effective biopesticide for the control of postharvest fruit rots. It is applied to apples and pears after harvest—but before storage—to control particular fungal pathogens. The yeast serves as an antagonist to fungal pathogens such as gray mold (*Botrytis cinerea*) and blue mold (*Penicillium expansum*) that cause postharvest decay.

Candida oleophila strain works primarily through competition for nutrients and precolonization of plant wound sites. However, there is evidence that it produces enzymes that can degrade the fungal cell walls and stimulate plant host defense pathways in freshly harvested fruits, both of which may also make a restricted contribution to the strain's antagonistic activity.

19.8 Botanical Biopesticides

The use of plant derivatives or botanical pesticides in agriculture dates back at least 2 millennia in ancient China, Egypt, Greece, and India (Thacker 2002; Ware 1983). Even in Europe and North America, the documented use of botanicals extends back more than 150 years. At present, there are four major types of botanical products used for insect control (pyrethrum, rotenone, neem, and essential oils), along with three others in limited use (ryania, nicotine, and sabadilla) (Figure 19.3).

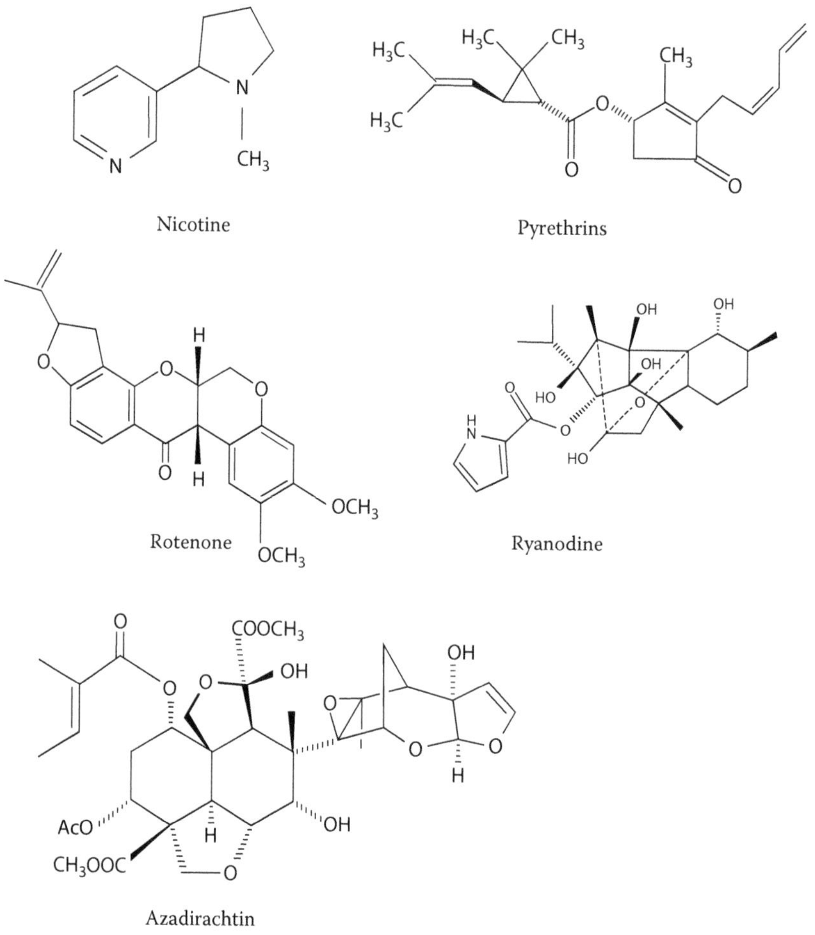

FIGURE 19.3
Botanical products used for insect control.

19.8.1 Pyrethrum

Pyrethrum refers to the oleoresin extracted from the dried flowers of the pyrethrum daisy, *Tanacetum cinerariifolium* (Asteraceae). The flowers of the plant are harvested shortly after blooming and are either dried or powdered or the oils within the flowers are extracted with the solvents. There are three esters of chrysanthemic acid and three esters of pyrethric acid. The extracts usually have an active ingredient content of about 30%. These active insecticidal components are collectively known as pyrethrins. Two pyrethrins are most prominent, pyrethrin-I and pyrethrin-II. The pyrethrins have another four different active ingredients, Cinerin I and II and Jasmolin I and II. Pyrethrin compounds have been used primarily to control human lice, mosquitoes, cockroaches, beetles, and flies. Some "pyrethrin dusts," used to control insects in horticultural crops, are only 0.3%–0.5% pyrethrins and are used at rates of up to 50 lb/A. Other pyrethrin compounds may be used in grain storage and in poultry pens and on dogs and cats to control lice and fleas.

The insecticidal action of the pyrethrins is characterized by a rapid knockdown effect, particularly in flying insects, and hyperactivity and convulsions in most insects. These symptoms are a result of the neurotoxic action of the pyrethrins, which block voltage-gated sodium channels in nerve axons. As such, the mechanism of action of pyrethrins is qualitatively similar to that of DDT and many synthetic organochlorine insecticides. In purity, pyrethrins are moderately toxic to mammals (rat oral acute LD_{50} values range from 350 to 500 mg/kg), but technical grade pyrethrum is considerably less toxic (~1500 mg/kg) (Casida and Quistad 1995).

Pyrethrins are especially labile in the presence of the UV component of sunlight, a fact that has greatly limited their use outdoors. A recent study indicated that the half-lives of pyrethrins on field-grown tomato and bell pepper fruits were 2 h or less (Antonious 2004).

For many years, world production of pyrethrum was led by Kenya, with lesser quantities produced in Tanzania and Ecuador. In the past 5 years, Botanical Resources Australia, with plantings in Tasmania, has become the second largest producer in the world (30% of world production at present). Pyrethrum produced in Tasmania is qualitatively similar to that produced in East Africa and elsewhere.

19.8.2 Neem Pesticides

The key insecticidal ingredient found in the neem tree is azadirachtin, a naturally occurring substance that belongs to an organic molecule class called tetranortriterpenoids (Grace-Sierra 1990). Azadirachtin is just one of more than 70 limonoids produced by the neem tree. It occurs in all parts of the neem tree, but is concentrated in the kernel. The concentration of azadirachtin in neem kernels varies with environmental factors and genetic makeup, but may be as high as 10 g/kg of seed kernels. A single tree yields about 2 kg of kernels each year. It is a powerful insect antifeedant and growth regulator. It is structurally similar to insect hormones called "ecdysones," which control the process of metamorphosis as the insects grow from larvae to pupae to adults. Metamorphosis requires the careful synchrony of many hormones and other physiological changes to be successful, and azadirachtin seems to be an "ecdysone blocker." It blocks the insect's production and release of these vital hormones. Insects then will not molt, thus breaking their life cycle (National Research Council 1992; AgriDyne Technologies 1994).

Azadirachtin may also serve as a feeding deterrent for some insects. Depending on the stage of life cycle, insect death may not occur for several days. However, upon ingestion of minute quantities, insects become quiescent and stop feeding. Residual insecticidal activity is evident for 7–10 days or longer, depending on the insect and the application rate (*Farm Chemicals Handbook* 1995; Thomson 1992).

Azadirachtin is used to control whiteflies, aphids, thrips, fungus gnats, caterpillars, beetles, mushroom flies, mealybugs, leaf miners, gypsy moths, and others on food, greenhouse crops, ornamentals, and turf (Thomson 1992).

19.8.3 Rotenone

Rotenone is a naturally occurring chemical with insecticidal, acaricidal (mite and spider-killing), and piscicidal (fish-killing) properties, obtained from the roots of several tropical and subtropical plant species belonging to the genus *Lonchocarpus* or *Derris*. It is a selective, nonspecific insecticide, used in home gardens for insect control, for lice and tick control on pets, and for fish eradications as part of water body management. Both a contact and stomach poison to insects, it kills them slowly, but causes them to stop their feeding almost immediately. It exerts its toxic action by acting as a general inhibitor of cellular respiration.

Rotenoids, the rotenone-related materials, have been used as crop insecticides since 1848, when they were applied to plants to control leaf-eating caterpillars. However, they have been used for centuries (at least since 1649) in South America to paralyze fish, causing them to surface (5). *Derris* root has long been used as a fish poison and its insecticidal properties were known to the Chinese well before it was isolated by E. Geoffrey in 1895. The use of the ground root of certain species of *Derris* was patented in 1912, since when it has been established that the active compounds are rotenoids, of which the main insecticide is rotenone.

Rotenone is a mitochondrial poison, which blocks the electron transport chain and prevents energy production (Hollingworth et al. 1994). As an insecticide, it is considered a stomach poison because it must be ingested to be effective. Pure rotenone is comparable to DDT and other synthetic insecticides in terms of its acute toxicity to mammals (rat oral LD_{50} is 132 mg/kg), although it is much less toxic at the levels seen in formulated products.

19.8.4 Sabadilla

Sapodilla is a botanical insecticide obtained from the seeds of the South American lily *Schoenocaulon officinale*. In purity, the active principals, cevadine-type alkaloids, are extremely toxic to mammals (rat oral LD_{50} is 13 mg/kg), but commercial preparations typically contain less than 1% active ingredient, providing a margin of safety. The mode of action of these alkaloids is remarkably similar to that of the pyrethrins, despite their lack of structural similarity.

19.8.5 Ryanodine

Ryanodine, the toxic principle of the roots and stems of the shrub *R. speciosa*, has a wide spectrum of activities against many insect pests. It is relatively nontoxic to mammals and is capable of controlling important larval lepidopteran pests at 3–16 g alkaloid equivalents per acre, making it one of the most potent natural insecticides. At least eleven ryanoids have been isolated and characterized from various Ryania extracts (Ruest et al. 1985; Jefferies et al. 1991, 1992a,b). Structurally related ryanodines active against *Macaronesia fortunata* and *Helicoverpa armigera* have also been reported from Persia indica (Gonzalez-Coloma et al. 1990, 1993, 1996). Ryanodines reportedly affect insect muscles by binding to the calcium channels in the sarcoplasmic reticulum. This causes calcium ion flow into the cells and death follows rapidly. Ryania extract–based wettable powders exhibit low mammalian toxicity with an acute oral LD_{50} of 1200 mg/kg channels in the sarcoplasmic reticulum in rats. This causes calcium ion flow into the cells and death follows rapidly. Ryania extract–based wettable powders exhibit low mammalian toxicity with an acute oral LD_{50} of 1200 mg/kg in rats.

19.8.6 Nicotine

Like pyrethrum and rotenone, nicotine, an alkaloid obtained from the foliage of tobacco plants (*Nicotiana tabacum*) and related species, has a long history as an insecticide. Nicotine and two closely related alkaloids, nornicotine and anabasine, are synaptic poisons that mimic the neurotransmitter acetylcholine. As such, they cause symptoms of poisoning similar to those seen with organophosphate and carbamate insecticides (Hayes 1982). Owing to the extreme toxicity of pure nicotine to mammals (rat oral LD_{50} is 50 mg/kg) and its rapid dermal absorption in humans, nicotine has seen declining use, primarily as a fumigant in greenhouses against soft-bodied pests. However, there remains some interest in preparing stable nicotine fatty acid soaps, presumably with reduced bioavailability and toxicity to humans (Casanova et al. 2002).

19.9 Microbial Metabolites

Living organisms generally produce a mixture of structurally related compounds, of which one or more are found in abundance. Advances in the chemical and biochemical sciences from 1950s made possible the isolation and biological characterization of the main active ingredients of all currently used pest control agents. Screening of fermentation broths for antibacterial activity has resulted in the discovery of thousands of antibiotics, a few of which have commercial utility.

19.9.1 Avermectins and Milbemycins

Avermectin and milbemycin groups of biopesticides are derived from a family of macrocyclic lactones produced by the novel soil actinomycetes *Streptomyces avermitilis* (Campbell et al. 1983; Campbell 1989) and *Streptomyces hygroscopicus* (Takiguchi et al. 1983; Ono et al. 1983). They are among the most potent anthelmintic, insecticidal and acaricidal compounds known. These compounds exhibited a broad spectrum of activities against agricultural pests such as aphids, mites, caterpillars, intestinal worms, and other parasites that prey on crops and ornamental plants. These are also used as veterinarian medicine for the treatment of internal and external parasites and mites of pets and livestock. It has a wide range of activities against lepidopteran pest species with LC_{50} values ranging from 0.01 to 6 ppm against tobacco hornworm *Manduca sexta*, corn earworm *Heliothis zea*, and southern armyworm *Spodoptera eridania*. In the commercial fermented culture of abamectin (avermectin B1), avermectin B1a is the major component constituting 80%–85% of avermectins and 15–20% avermectin B1b. While avermectin B1 is widely used as an agricultural miticide, its 22,23-dihydro derivative ivermectin is used worldwide as a broad-spectrum parasiticide in animals and humans. This semisynthetic derivative was the first macrocyclic lactone developed and introduced commercially in 1981 for the treatment of helminthic parasitic infections in farm animals as well as for the treatment of river blindness in humans. It is available for use in various formulations as liquid, paste, tablet, chewable, feed premix, pour-on liquid, and as subcutaneous injection. It is a mixture of two avermectins, namely, 22,23-dihydroxy avermectin B1a and 22,23-dihydroavermectin B1b. The first macrocyclic lactone was developed and introduced commercially in 1981 for the treatment of helminthic parasitic infections in farm animals as well as for the treatment of river blindness in humans. It is available for use in various formulations as liquid, paste, tablet, chewable, feed premix, pour-on liquid, and as subcutaneous injection. It is a mixture of two avermectins, namely, 22,23-dihydroxy avermectin B1a and 22,23-dihydroavermectin B1b.

19.9.2 Spinosyns

Spinosyns are a relatively recent example of new broad-spectrum natural secondary metabolites derived from the aerobic fermentation of the soil actinomycete *Saccharopolyspora spinosa*. These are effective against a broad range of lepidopteran insect pests (Thompson and Hutchins 1999). Successful introduction of spinosad represents an important milestone in the use of natural products for commercial pest control. These were discovered through screening of culture broth of *Saccharopolyspora spinosa* obtained from a Caribbean island in the 1980s (Hoehn et al. 1991). Subsequent isolation and structural elucidation work led to the identification of macrocyclic lactones comprising a unique tetracyclic ring system with two attached monosaccharide units. The major constituents spinosyn-A and spinosyn-D present in 85:15 ratio together constitute over 90% of the active material. A large number of synthetic analogs with improved physiochemical properties and field performance and broader activity spectrum against pests have been developed modifying different portions of the spinosyn molecule (Sparks et al. 2001). Several of the spinosyn-based products such as Tracer, Spino-Tor, and Success exhibited excellent insect control properties against a variety of insect pests including thrips, fleas, and pest species of the lepidoptera, diptera, and hymenoptera. In general, spinosyns exhibit activity at par with some of the synthetic pyrethroid insecticides.

Spinosyns have been found particularly effective against economic pests on crops such as cotton and vegetables. They are also effective against several stored grain pests of wheat (Fang et al. 2002). Extensive field evaluation of the spinosyns has demonstrated their adequate efficacy, stability, and environmental suitability to enable their use without chemical modification. They exert their effects on insects through hyperexcitation and disruption of the insect nervous system (Salgado et al. 1997, 1998). Spinosyns are slow to penetrate into the larvae of *H. virescens*; however, once inside the insects, these do not readily metabolize and therefore, account for high level of activity.

19.9.3 Strobilurins

One of the most promising new groups of fungicides discovered through random screening is the strobilurins, also referred to as β-methoxyacrylates produced by the fungus *Strobilurus tenacellus*, which feeds on decaying plant materials in the soil. Musilek (1969) was the first to report the isolation of strobilurin-A, originally named as mucidin from *Oudemansiella mucida*. Since its first use as an agricultural fungicide, several additional strobilurins and their related products, namely, oudemansins and myxothiazoles have been reported. Other strobilurins such as strobilurin-B, -C, -D, -E, -F1, -F2, -G, -H, -X, and hydroxystrobilurin-D are characterized by the presence of various substituents in the basic β-methoxyacrylate nucleus. These reduced-risk fungicides are active at low application rates and have low mammalian and avian toxicity. The strobilurin fungicides inhibit cell growth at low concentrations by preventing the cells from generating energy efficiently in their mitochondria. Strobilurins are active against a broad range of fungi but are nontoxic to mammals and birds. Since they act on a single target site, these are susceptible to the development of resistance. Oudemansins, first reported from mycelial cultures of *Oudemansiella mucida* were subsequently isolated from submerged cultures of *Xerula melanotricha* and from another basidiomycetes fungus *Oudemansiella radicata*. They coexist along with strobilurin-A in *Oudemansiella mucida* and their biological activity closely resembles that of the strobilurins. At least three oudemansins (A, B and X) known so far exhibit considerable fungicidal activity (Clough 1993).

19.10 Biochemical Pesticides

Biochemical pesticides are naturally occurring substances that control pests by nontoxic mechanisms. Biochemical pesticides include substances, such as insect sex pheromones, which interfere with mating, as well as various scented plant extracts that attract insect pests to traps. They include

1. Insect pheromones
2. Plant extracts and oils
3. Plant growth regulators
4. Insect growth regulators.

Biochemical pesticides typically fall into distinct biologically functional classes, including semiochemicals, plant extracts, natural plant growth regulators, and natural insect growth regulators (Figure 19.4). These classes of biochemical pesticides are described below with examples provided for each category. There are almost 122 biochemical

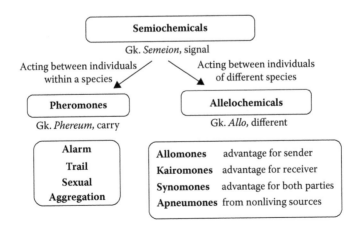

FIGURE 19.4
Semiochemicals.

pesticide active ingredients registered with the EPA, which include 18 floral attractants, 20 plant growth regulators, 6 insect growth regulators, 19 repellents, and 36 pheromones (Steinwand 2008).

19.10.1 Semiochemicals

Semiochemicals (Gk. *semeion*, a signal) are chemicals that mediate interactions between organisms. Semiochemicals are subdivided into allelochemicals or pheromones depending on whether the interactions are interspecific or intraspecific, respectively (Anonymous 1981). Allelochemicals are chemicals that are significant to individuals of a species different from the source species. They are subdivided into several groups depending on whether the response of the receiver is adaptively favorable to the emitter but to not the receiver (allomones), is favorable to the receiver but not to the emitter (kairomones), or is favorable to both the emitter and the receiver (synomones). Within both allelochemicals and pheromones, it is sometimes useful to refer to chemicals as arrestants, attractants, repellents, deterrents, stimulants, or other descriptive terms. These terms can indicate what behavior is involved in the response such as a feeding stimulant or flight arrestant.

19.10.2 Insect Pheromones

Insect pheromones are chemicals used by an insect to communicate with other members of the same species. Structurally these chemicals are often very similar to substances used in flavors and fragrances. The insect pheromones themselves do not kill a target pest. When used for pest management, two common uses are to attract an insect to a trap containing a lethal pesticide or to disrupt mating. With mating disruption, proportionately large concentrations of the sex pheromones are present in the air, thus confusing the males and decreasing their success rate at locating a female with which to mate. Pheromones can also be used to monitor pest populations as part of larger integrated pest management (IPM) systems, particularly to determine appropriate timing and application of pesticides. Insect pheromones account for a large percentage of the biochemical pesticides on the market. In mid-2002, EPA had registered 36 pheromones, which comprised over 200 individual products (Ware and Whitacre 2004). Insect sex pheromones can be used alone to manage pest populations when pest pressure is moderate to low, such as after several years of consecutive use. Other practical uses include "in survey traps to provide information about population levels, to delineate infestations, to monitor control or eradication programs, and to warn of new pest introductions" (Ware and Whitacre 2004).

Advantages to the use of insect pheromones include their high species specificity and relatively low toxicity. Sex pheromones tend to be specific to a particular species or even strain of insect, making them one of the most targeted pest management strategies. This specificity thus maintains an ecological balance by leaving undisturbed populations of other insect species and nontarget organisms.

A disadvantage of insect pheromones is that they often must be used in combination with other pest management strategies to achieve the efficacy desired. This is particularly true when pest pressure is high. With high pest pressure, the male is more likely to locate a mate by simply bumping into the female rather than by using pheromones to communicate over long distances. However, the combination of pest management strategies typically lowers the pest pressure in subsequent years, creating the opportunity for the insect pheromones to be used alone.

Pheromones (Gk. *phereum*, to carry; *horman*, to excite or stimulate) are released by one member of a species to cause a specific interaction with another member of the same species. Pheromones may be further classified on the basis of the interaction mediated, such as alarm, aggregation, or sex pheromone. It is the sex pheromones of insects that are of particular interest to agricultural integrated pest management (IPM) practitioners. The first insect sex pheromones were identified in the late 1950s and initially were scientific curiosities.

Kairomones are also members of the group of semiochemicals where the signal detected is beneficial to the perceiving species. For example, a mountain pine beetle attacks pine trees and not alder trees because it recognizes both the kairomones of appropriate hosts and nonhost species.

19.11 Conclusion

Effective biopesticides are similar to chemical pesticides and work with good efficacy against the pest for which they are labeled. There has been considerable consumer and governmental pressure to minimize the use of chemical pesticides. This can be seen by the desire of the consumer to purchase organic produce and by supermarkets demanding that growers utilize environmental sensitive crop production strategies. The unique value for biopesticides is based on the three "R's": resistance management, restricted entry intervals, and residues. Since biopesticides have multimodes of action, there is less chance of developing resistance in a particular insect. They can be an excellent component of an integrated pest management (IPM) program. Restricted entry intervals (REIs) determine the timing of application relative to a number of factors. The majority of biopesticides have low REIs, mostly around 0–4 h with no preharvest interval. After a biopesticide is applied, a farmer can go into the field and harvest immediately. Residues are a regulatory issue with pesticides. Since biopesticides tend to pose fewer risks compared to conventional chemical pesticides, the EPA generally requires less data to achieve new product registrations (<1 year). Conventional chemical pesticides can take up to 3 years for regulatory approval. The cost of registration of a biopesticide product is usually an order of magnitude less than for a chemical pesticide product. In the United States, the cost of registration of a biopesticide can usually be measured in hundreds of thousands of dollars. By contrast, registration of a chemical pesticide costs millions of dollars. This reflects the regulator's preference for biopesticides, especially Bt products. Because of high chemical registration costs, many high-value crop and niche markets are now being left without effective chemical crop protection solutions, leaving Bt as the economical choice for many producers.

References

AgriDyne Technologies, Inc. 1994. *Greenhouse Grower*. Floritech report: Tough on pests, easy on crops—And the environment. AgriDyne Technologies, Inc.: Salt Lake City, UT.

Anonymous. 1981. Semiochemicals: Their role in pest control. In: Nordlund, D. A., Jones, R. L., and Lewis, W. J. (eds), 306 p. John Wiley & Sons: New York.

Antonious, G. F. 2004. Residues and halflives of pyrethrins on field-grown pepper and tomato. *J. Environ. Sci. Health* B39: 491–503.

Campbell, W. C. 1989. *Ivermectin and Abamectin*, 363 p. Springer Verlag: New York.

Campbell, W. C., Fischer, M. H., Stapley, E. O., Albers-Schonberg, G., and Jacob, T. A. 1983. Ivermectin: A potent new antiparasitic agent. *Science* 221: 823–828.

Casanova, H., Ortiz, C., Peláez, C., Vallejo, A., Moreno, M. E., and Acevedo, M. 2002. Insecticide formulations based on nicotine oleate stabilized by sodium caseinate. *J. Agric. Food Chem.* 50: 6389–6394.

Casida, J. E. and Quistad, G. B. 1995. *Pyrethrum Flowers: Production, Chemistry, Toxicology and Uses*, 356 pp. Oxford University Press: Oxford, UK.

Clemson Extension [Clemson HGIC], home and garden information center, 2007. Organic Pesticides and Biopesticides. URL:http://hgic.clemson.edu/factsheets/HGIC2756.htm. Clemson University.

Clough, J. M. 1993. The Strobilurins, oudemansins, and myxothiazols: Fungicidal derivatives of 2-methoxyacrylic acid. *Nat. Prod. Rep.* 565–574.

Copping, L. G. 2009. *The Manual of Biocontrol Agents*, 4th edn. BCPC Publications: Alton.

Evans, J. 2008. Biopesticides: From cult to mainstream. *Agrow Mag.* 27: 11–14.

Fang, L. A., Subramanyam, B., and Arthur, F. H. 2002. Effectiveness of spinosad on four classes of wheat against five stored-product insects. *J. Econ. Entomol.* 95: 640–650.

Farm Chemicals Handbook. 1995. Meister Publishing Co.: Willoughby, OH.

Feng, M. G., Poprawski, T. J., and Khachatourians, G. G. 1994. Production, formulation and application of the entomopathogenic fungus *Beauveria bassiana* for insect control: Current status. *Biocontrol. Sci. Technol.* 4(1): 3–34.

Garczynski, S. F. and Siegel, J. P. 2007. Bacteria. In: Lacey, L. A. and Kaya, H. K. (eds), *Field Manual of Techniques in Invertebrate Pathology*, 2nd edn, pp. 175–197. Springer: Dordrecht.

Gonzalez-Coloma, A., Hernandez, M. G., Perales, A., and Fraga, B. M. 1990. Chemical ecology of Canariam lauret forest. *Chem. Ecol.* 16: 2723–2733.

Gonzalez-Coloma, A., Cabrera, R., Soccorro Monzon, A. R., and Fraga, B. M. 1993. Persiea indica as a natural source of the insecticidal ryanodol. *Phytochemistry* 34: 397–400.

Gonzalez-Coloma, A., Terrero, D., Perales, A., Escoubas, P., and Fraga, B. N. 1996. Insect antifeedant ryanodine diterpenes from Persiea indica. *J. Agric. Food Chem.* 44: 296–300.

Grace-Sierra Crop Protection Co. 1990. Margosan-O technical bulletin. Grace-Sierra Crop Protection Co.: Milpitas, CA.

Hayes, W. J. Jr. 1982. *Pesticides Studied in Man*, 672 pp. Baltimore: Williams & Wilkins.

Hibbett, D. S., et al. 2007. A higher-level phylogenetic classification of the fungi. *Mycol. Res.* 111: 509–547.

Hoehn, M. M., Michel, K. H., and Yao, R. C. 1991. European Patent Appl. 398, 588.

Hollingsworth, R., Ahmmadsahib, K., Gedelhak, G., and McLaughlin, J. 1994. New inhibitors of complex I of the mitochondrial electron transport chain with activity as pesticides. *Biochem. Soc. Trans.* 22: 230–233.

Humber, R. A. 2007. Recent phylogenetically based reclassifications of fungal pathogens of invertebrates. Insect Mycologist and Curator, ARSEF. November 2007. http://arsef.fpsnl.cornell.edu/mycology/phyloreclass.pdf. Cited 20 December 2009.

Humber, R. A. 2008. Evolution of entomopathogenicity in fungi. *J. Invertebr. Pathol.* 98: 262–288.

Inglis, G. D., Goettel, M. S., Butt, T. M., and Strasser, H. 2001. Use of hyphomycetous fungi for managing insect pests. In: Butt, T. M., Jackson, C., Magan N. (eds), *Fungi as Biocontrol Agents: Progress, Problems and Potential*, pp. 23–69. CAB International: Wallingford.

Jefferies, P. R., Tora, R. F., and Casida, J. E. 1991. Ryanodol 3-pyridine-3- carboxylate: A novel ryanoid from Ryania insecticide. *J. Nat. Prod.* 54: 1147–1149.

Jefferies, P. R., Tora, R. F., Brannigan, E., Pessah, I., and Casida, J. E. 1992a. Ryania insecticide: Analysis and biological activity of 10 natural ryanoids. *J. Agric. Food Chem.* 40: 142–146.

Jefferies, P. R., Lam, W., Tora, R. F., and Casida, J. E. 1992b. Ryania insecticide: Structural assignment of four natural 8ax-hydroxy-10-eparyanoids. *J. Agric. Food Chem.* 40: 509–512.

Jehle, J. A., Blissard, G. W., Bonning, B. C., Cory, J. S., Herniou, E. A., Rohrmann, G. F., Theilmann, D. A., Thiem, S. M., and Vlak, J. M. 2006. On the classification and nomenclature of baculoviruses: A proposal for revision. *Arch. Virol.* 151: 1257–1266.

Menn, J. J. 1996. Biopesticides: Has their time come? *J. Environ. Health B* 31(3): 383–389.

Menn, J. J. 1997. Biopesticides—are they relevant? In Focus on Biopesticides, pp. 1, 2. The Royal Society of Chemistry: Cambridge, UK.

Musilek, V. 1969. Czech. Pat. CS 136492 (British Pat. GB 1163910) 1965. Chem Abstr. 70, 18900y and 1971, 74, 123689s.

National Research Council. 1992. Neem: A tree for solving global problems. National Academy Press: Washington, DC.

Ono, M., Mishima, H., Takiguchi, Y., and Terao, M. 1983. Milbemycins, a new family of macrolide antibiotics: fermentation, isolation, physico-chemical properties and bioconversion of milbemycin-J and milbemycin-K. *J. Antibiot.* 36: 509.

Pell, J. K., Eilenberg, J., Hajek, A. E., and Steinkraus, D. C. 2001. Biology, ecology and pest management potential of Entomophthorales. In: Butt, T. M., Jackson, C., and Magan, N. (eds), *Fungi as Biocontrol Agents: Progress, Problems and Potential*, pp. 71–153. CAB International: Wallingford.

Ruest, L., Taylor, D. R., and Deslongchamps, P. 1985. Investigation of the constituents of Ryania speciosa. *Can. J. Chem.* 63: 2840–2843.

Salgado, V. L., Watson G. B., and Sheets J. J. 1997. Studies on the mode of action of spinosad, the active ingredient in Tracer. *Proceedings of 1997 Beltwide Cotton Production Conference*, pp. 1082–1086. National Cotton Council Memphis, TN, USA.

Salgado, V. L., Sheets, J. J., Watson G. B., and Schmidt A. L. 1998. Studies on the mode of action of spinosad: The internal effective concentration dependence on neural excitation. *Pestic. Biochem. Physiol.* 60: 103–110.

Sparks, T. C., Crouse, G. D., and Durst, G. 2001. Natural products as insecticides: The biology, biochemistry and quantitative structure-activity relationships of spinosyns and spinosads. *Pest Manag. Sci.* 57: 896–905.

Steinwand, B. 2008. Personal Communication. Washington, DC: US Environmental Protection Agency.

Takiguchi, Y., Ono, M., Muramatsu, S., Ide, J., Mishima, H., and Terao, M. 1983. Milbemycins, a new family of macrolide antibiotics: Fermentation, isolation and physicochemical properties of milbemycin-D, E, F, G and H. *J. Antibiot.* 36: 502.

Thacker, J. M. R. 2002. *An Introduction to Arthropod Pest Control*, 343 pp. Cambridge University Press: Cambridge, UK.

Thakore, Y. 2006. The biopesticide market for global agricultural use. *Ind. Biotechnol.* 2(3): 192–208.

Thompson, G. D. and Hutchins, S. 1999. Spinosad – A new class of fermentation-derived insect control agents. *Pestic. Outlook* 10: 78–82.

Thomson, W. T. 1992. *Agricultural Chemicals. Book I: Insecticides.* Thomson Publications: Fresno, CA.

Ware, G. W. 1983. *Pesticides. Theory and Application*, 308 pp. Freeman: San Francisco.

Ware, G. W. and Whitacre, D. M. 2004. *The Pesticide Book*, 6th ed, 488 pp. Meister Media Worldwide: Willoughby, Ohio.

Section IX

Endocrine-Disrupting Pesticides

20

Endocrine-Disrupting Pesticides

Shafiullah, H. S. Rathore, and Leo M. L. Nollet

CONTENTS

20.1 Introduction

Compounds that interfere with natural hormone functions have many names: environmental estrogens, xenoestrogens, endocrine disruptors, endocrine modulators, or environmental hormones. Although all scientists do not accept one specific term, endocrine disruptors (EDs) and endocrine-disrupting chemicals (EDCs) are most likely used most widely.

A number of pesticides are known to interfere with the endocrine system and thereby impair fertility and the development of animals and possibly also of humans.

The following definition is mentioned in the *Special Report on Environmental Endocrine Disruption: An Effects Assessment and Analysis* (EPA 1997): "An environmental endocrine or hormone disruptor may be defined as an exogenous agent that interferes with the synthesis, secretion, transport, binding, action, or elimination of natural hormones in the body that are responsible for the maintenance of homeostasis, reproduction, development, and/or behavior."

During the last decade, considerable amounts of scientific and financial resources have been employed to clarify the potential risk to human health of EDCs. Pesticides and some of their metabolites are an important group of EDCs.

Agriculture in all countries has become strongly dependent on chemicals for crop protection. Farmers are exposed to pesticides during planting and cultivation of crops, pesticide applications, mixing, loading, and cleaning of application equipment. Farmers may be at greater risk if they incorrectly handle, store, or dispose of pesticides or if they do not wear protective clothes. Developing countries consume many times more (80% of all) pesticides produced worldwide. Agricultural workers from developing countries are likely to be exposed to higher levels of even more dangerous classes of pesticides. It seems due to the lack of specific legislation, poor living conditions, shortage of foodstuffs, limited subsidies for farmers, and insufficient pesticide market regulation.

A number of pesticides are persistent, lipid soluble, nonbiodegradable, accumulative, and also endocrine disruptors. Exposure to pesticides displaying a hormone-like molecular structure has been attracting scientific interest as a consequence of the observed association with several biological hazards in humans and animals. In some cases, pesticides may bind to sex hormone receptors, activate them, and thus lead to responses similar to endogenous estrogens and androgens. They may also bind to hormone receptors without activating them. Thus, they block the binding of endogenous hormones, which therefore become inactive. On the other hand, a number of indirect (anti)estrogenic and (anti)androgenic reactions are possible, for example, changes in the concentration of hormones at the receptors in the target organs, interference with the biosynthesis of hormones in the endocrine organs, or effects on the biotransformation in the liver. As a consequence, the binding of hormones to proteins in the blood plasma as well as the endocrine activity of pituitary and hypothalamus may be influenced. This may be due to the complex nature of the combined effects of contaminants present in the environment.

Therefore, numerous papers, several reviews, and many books have been published on the subject.

Endocrine systems and hormones are briefly discussed before discussing pesticides and their endocrine-disrupting effects. In the last part, endocrine-disrupting effects of the well-known pesticide DDT and its metabolites are detailed.

Figure 20.1 shows the position of the endocrine glands and health hazards associated with the extensive use of pesticides.

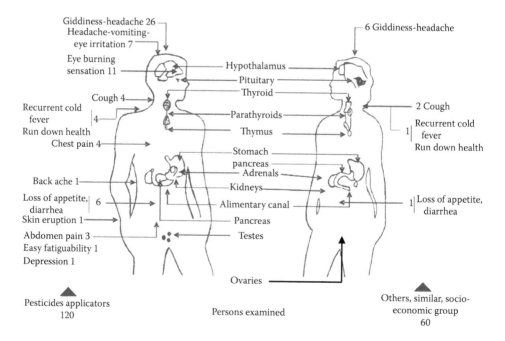

FIGURE 20.1
Position of the endocrine glands and health hazards associated with extensive use of pesticides. (*Note:* Figures indicate number of incidents.)

20.2 Endocrine Systems of Animals and Men

Glands are organs of an animal body that manufacture some liquid products that are secreted from the cells (both internal and external secretions).

There are two types of glands:

1. *Ducted glands*: Ducted glands secrete their product through well-defined ducts, for example, sweat glands. These are numerous small, tubular glands that are found nearly everywhere in the skin.

2. *Ductless glands*: These glands are also called endocrine glands or internally secreting glands. They secrete hormones directly into the bloodstream in response to instructions from the brain. The blood then carries them around the body where they control the internal chemical processes. Hence, they are linked with the nervous system and also with the circulatory system.

The various endocrine glands are interrelated. Some even regulate the secretion of hormones from other glands. An abundance of a particular hormone ensures that the production of the hormone that stimulates its production is inhibited. Together with the nervous system, the endocrine system controls the activities of all organs and systems in the human body.

Secretions from endocrine glands are known as hormones. The ductless glands are briefly described below (the list is not exhaustive):

- *Thymus:* The thymus is situated in front of the trachea, behind the top of the breastbone. It secretes thymic hormones. In early childhood, it plays a role in building up resistance (immunity) to diseases and physical development.

- *Thyroid:* This two-lobed gland in all vertebrates is located in front of and on either side of the trachea. It secretes thyroxine, which contains 65% iodine. The rate of production is regulated by the pituitary gland. Hormones of the thyroid regulate metabolism. A deficiency of these hormones causes dwarfness in children called cretinism or myxedema in adults. Other malfunctions of the thyroid are goiter or Grave's disease, that is, enlargement of the thyroid gland.

- *Parathyroid:* The four, small kidney-shaped glands lie in pairs near or within the lateral lobes of the thyroid gland and secrete parathyroid hormones, also called parathormone. The function of parathormone is to control the level of calcium in blood.

- *Gonads:* The gonads are the reproductive glands, that is, ovaries in females and testes in males. These glands secrete sex hormones and are related to the reproductive system.

- *Adrenal:* These are small dissimilar glands, one above each kidney, consisting of a cortex and medulla. The cortex secretes corticosteroids and sex hormones; the medulla secretes adrenaline. Adrenaline causes acceleration of the breath, contraction of small blood vessels, and dilation of the pupil and increases the metabolic rate. It also causes heightened emotion and a sudden increase in physical strength, as during fear or anger.

- *Pancreas:* The pancreas is a soft, irregular gland lying near the stomach, 15 cm long. It secretes pancreatic juice into the duodenum and contains the Islets of

Langerhans, producing insulin. Insulin is an aid in the metabolism of carbohydrates. Its failure results in diabetes.

- *Pituitary*: The pituitary is a small oval-shaped gland at the base of the brain
  secreting different hormones. It is the master gland as it controls the other endocrine glands and influences growth and metabolism. The hormones secreted
  by it control mental, sexual, and physical growth of the body. Deficiency may
  cause dwarfism and an excess result in much increased body growth called
  acromegaly.

- *Hypothalamus*: The hypothalamus is a part of the brain and is involved in all
  aspects of the emotional, reproductive, vegetative (including circulatory, digestive, and respiratory), endocrine, hormonal, visceral (intestine), and autonomic
  functions, and has a predominantly regulatory function.

20.2.1 Axes Within the Endocrine System

The hypothalamic–pituitary–thyroid axis (HPT axis) is a part of the endocrine system and
is responsible for the regulation of metabolism. The hypothalamus measures low levels
of thyroid hormone and reacts with the release of thyrotropin-releasing hormone (TRH).
TRH stimulates the pituitary to produce thyroid stimulating hormone (TSH). TSH in turn
stimulates the thyroid to make thyroid hormones until normal levels in the blood are
found. At high levels of thyroid hormone, the negative feedback starts.

The hypothalamic–pituitary–adrenal axis (HPA or HPTA axis) is a complex set of control and feedback between the hypothalamus, pituitary, and adrenal glands. They are an
important part of the neuroendocrine system that regulates not only stress responses but
also digestion, immune system, mood, sexuality, and energy consumption. When needed,
the production of corticotropin is stimulated in the pituitary via corticotropin-releasing
hormone in the hypothalamus. Negative feedback mechanisms come into effect at high
concentrations of glucocorticoids, such as cortisol and corticosterone, in a reduced production of corticotropin by the pituitary.

The hypothalamic–pituitary–gonadal axis (HPG axis) consists of hypothalamus, pituitary, and gonads. This system plays an important role in the development and regulation
of certain body functions such as reproduction and immune system.

The hypothalamus secretes gonadotropin-releasing hormone (GnRH). This stimulates
the pituitary to manufacture luteinizing hormone (LH) and follicle-stimulating hormone
(FSH). Both LH and FSH act on the gonads to release sex hormones. (Too) high levels of
androgens and estrogens are put into effect as a negative feedback to slow down the pituitary and hypothalamus.

20.3 Secretion from Glands: Hormones

All hormones have regulating functions and are part of sometimes simple, but sometimes
complex, regulating systems. The body controls the quantities of hormones to release by
negative feedback. One part of the brain involved in that system is the hypothalamus. The
pituitary is regulated by the hypothalamus. Endocrine glands are controlled by the releasing factors of the pituitary. Between endocrine glands and pituitary, feedback systems
exist. The hypothalamus registers quantities of different hormones in the blood and reacts

with hormonal or neural messages to the endocrine glands, especially to the pituitary. The messages indicate if the production of hormones has to be stimulated or inhibited.

Hormones are chemical substances formed in an endocrine gland and conveyed by the bloodstream to a specific organ or tissue elsewhere (sites remote from that of its production) whose function they modify by means of chemical activity. Hormones are produced in extremely small quantities, but their action is very rapid. They control growth and various life processes of the body. For example, in the ovaries and testes, female and male hormones are produced, respectively, which are responsible for female and male characteristics such as voice, facial and body hair, stature, appearance, etc.

Homeostasis is a regulation within a body to achieve or maintain a balance. The body comes into action when this balance is compromised, think of body temperature and blood sugar level. Homeostasis is set within certain limits. Setting and recording of those boundaries occur during fetal and neonatal periods.

To learn more about endocrine systems and hormones, see Henry and Norman (2003), Norman et al. (2011), and Neal and Sompayrac (2001).

20.4 Endocrine System and Hormones of Insects

In insects, the endocrine system comprises four main organs, namely neurosecretory cells, corpora cardiaca, corpora allata, and prothoracic glands, which are described in the following paragraphs.

Neurosecretory cells: The neurosecretory cells (NSC) are neurons in the central nervous system that are specialized to secret hormones. Typically, two groups of such cells are located in the protocerebral lobes of the brain: a pair of median groups and a pair of lateral groups present respectively in the *pars medialis* and *pars lateralis*. Whereas the axons of the median groups make a cross before emerging from the brain as the *Nervi corporis cardiaci* I (NCC 1), those of the lateral groups pass out without decussating as the *Nervi corporis cardiaci* II (NCC II). In most insects, the two NCCs enter the corpus cardiacum of their side to unload the hormone brought from NSC for storage and release into circulation (hemolymph) as and when needed.

Corpora cardiaca: The corpora cardiaca (CC) are small paired bodies situated immediately behind the brain and are believed to be modified ganglia. They are made up of their own intrinsic secretary cells and nerve fibers received from the brain through the NCCs. They function as storage and release organs for the hormone synthesized in the NSC and since the hormone is released in the hemolymph, they are called neurohemal organs. While the storage and release of hormone constitute their main function, they are also believed to produce their own (indigenous) hormone/hormones.

Corpora allata: The corpora allata (CA) are also paired organs though they may fuse into one in some insects. They lie immediately behind the CC with which they are connected by *Nervi corporis allati* (NCA) not conspicuous in many cases. They are ectodermal in origin and clearly glandular.

Prothoracic glands: The prothoracic glands (PTG) are again paired and flattened organs lying close to the prothoracic spiracles. They are ectodermal, glandular, and larval organs being lost in the adult. They are innervated in some insects and noninnervated in others.

In the larvae of *Cyclorrhaphous diptera*, the CC, CA, and PTG are all fused into a single structure called the ring gland or Weismann's ring after the name of the discoverer.

Several hormones are produced by insects. A few of them, namely the brain hormone, juvenile hormone, and ecdysone produced by endocrine organs, are described in the following lines.

Brain hormone: The NSC of the brain synthesizes it, so it is called a neurohormone. It is believed to be peptidergic (proteinaceous) in nature, but its chemical structure is still to be determined. It is also known as activation hormone (AH) due to the fact that it possesses two adenotropic (gland-activating) factors—the allatotropic hormone or allatotropin and the prothoracicotropic hormone or prothoracicotropin—which activate the CA and PTG, respectively, to produce their own hormones.

Juvenile hormone: The CA produces a hormone that is known as the juvenile hormone (JH), which was initially known as neotinin, due to its ability to keep the larva in permanent juvenile condition. Chemically, the JH is an isoprenoid.

Ecdysone: It is produced by the PTG. It is a steroid, and it brings about molting in insects, so it is also known as the molting hormone (MH). The MH contains two hormones: a-ecdysone and b-ecdysone. The a-ecdysone is a prohormone produced by the PTG that is converted into b-ecdysone in the peripheral tissues of the body. The latter is also called 20-hydroxyecdysone, ecdysterone, and crustecdysone. In fact, it has been found to be responsible for bringing about molting in insects and is the true MH.

For further information, see Nijhout (1994).

20.5 Chemical Communication Between Cells

The regulation of growth and metabolism of complex multicellular organisms depends heavily upon chemical messages sent between cells. This includes secretion of hormones into the circulatory system, chemical transfer of information through communicating cell junctions, and passage of signals between neurons in the brain.

20.5.1 Chemical Regulators

Hormones belong to the group of chemical regulators. In addition to hormones, there are two categories of regulators: neurocrines and cytocrines. Neurocrine regulators are chemicals secreted by the nervous system. Neurocrines are divided in two groups. The first group, consisting of neurohormones, is secreted into the blood. The second group consists of neurotransmitters and neuromodulators. These are secreted in the extracellular fluid of the synapses between neurons and their target cells such as other neurons, muscle cells, and gland cells. A neurotransmitter triggers or inhibits an action in the target cells of the synapse. A neuromodulator increases or decreases the response of target cells to the neurotransmitter.

Cytocrines are local chemical regulators. They are issued in the extracellular tissue fluid and can diffuse a short distance to their target cells. Cytocrines are, in turn, divided into autocrine and paracrine regulators. Paracrine regulators affect other cell types, and autocrine regulators affect the secretory cell type.

A chemical regulator goes through several steps: synthesis, release, transport (free or bound to a protein), binding to a specific receptor, feedback, degradation, and excretion (by binding to other substances).

A neurotransmitter is a molecule that is used for signal transmission between nerve cells (neurons) in the nervous system. The synapse is the place where this signal is transmitted.

20.5.2 Receptor

A receptor is usually a protein that selectively binds a chemical signal (ligand). The binding of an endogenous ligand leads to a specific cell response. The binding of a hormone-disrupting substance with a receptor can activate the receptor in the same way as an endogenous ligand, or it may interfere with the binding and action of endogenous ligands without activating the receptor itself.

The result of a biological ligand–receptor interaction depends on many factors besides the concentration of the ligand and the binding affinity of the chemical. Cell responses will be different depending on the receptor types and transduction (transmission of signals or genetic material) mechanisms in the cell and also on whether the ligand acts as a receptor agonist, antagonist, inverse agonist, partial agonist, or gives a mixed response. An agonist may activate a receptor resulting in a maximum response. Most natural ligands are agonists. Antagonists bind to the receptor but do not activate it. The receptor is blocked from binding of agonists. Inverse agonists are antagonists that further reduce the activation of the receptor. Partial agonists activate the receptor not fully, and the biological response is only partial.

Hormone disruption is a consequence of the binding of a hormone to its receptor on the cell surface, cytoplasm, or nucleus. This is followed by a series of events that lead to changes in gene expression. There are very few data available on the link between the whole range of events and the adverse health effects including cancer and reproductive disorders.

20.6 Endocrine Disruption

The mechanisms behind endocrine disruptors are very diverse.

To start, we must not forget that endogenous hormones act themselves through different mechanisms. The classic mechanism of action of hormones, such as estrogens, androgens, thyroid hormones, and progesterone, is a binding of the hormone to its receptor, the interaction of this hormone–receptor complex with other cofactors in the cell, and the activation or inactivation of transcription of a gene.

Hormone receptors can also bind to chemicals (xenobiotics) other than the normal body chemicals. An important note is in order here: a single hormone, such as estradiol, triggers a whole variety of receptors, signaling mechanisms, and may interact with completely different cofactors depending on the phenotype of the target cell. Furthermore, specific enzymes are involved through synthesis, degradation, and inactivation of hormones. A single, several, or all of these enzymes may be targets of endocrine disruptors.

Endogenous hormones, especially estrogen, androgen, and thyroid hormone, bind to proteins in the blood which reduces their bioavailability. EDCs do not bind to the same protein which increases their bioavailability with respect to endogenous hormones. Enzymes that break down endogenous hormones do not break down EDCs not always even quickly. So they remain available for a longer duration and can be stored in reserve tissues, usually fat, because most EDCs are fat-soluble.

There are other mechanisms of disruption of the system and functioning of steroid hormones. Since very low doses of EDCs may have an adverse effect, traditional dose–response curves are not applicable. The question raises here if thresholds for EDCs are rightly in place. Exposure is usually for several endocrine disruptors, not for one. One cannot speak of a conventional exposure. The specific mechanisms underlying the disruption by chemicals of the endocrine systems are quite complex and certainly not for all chemicals elucidated. Substances can in many ways disrupt normal functions:

- They can act as natural hormones and bind to a receptor. This provokes a response from the cell, similar to that following a stimulus of a natural hormone (agonist response).
- They can bind to a receptor and prevent a normal response (antagonistic response).
- The substance may bind to the hormone receptor without initiating a response and thus the operation of the natural hormone is blocked. This is called an antiestrogenic or antiandrogenic activity.
- The endocrine-disrupting substance affects hormone production. The substance has an inhibitory or enhancing effect on the hormone production.
- The endocrine-disrupting substance affects the degradation of hormones. Examples of such pesticides are DDT, DDE, lindane, endosulfan, kepone, or triazine.
- EDCs can accelerate the degradation of endogenous hormones. The accelerated degradation of hormones can cause a serious disruption of hormonal balance. The synthetic estrogen DES, for example, stimulates the cytochrome P450 leading to an enhanced degradation of hormones.
- EDCs can affect aromatase function. Many fungicides inhibit aromatase activity.
- The substance may cause changes in the feedback mechanism.
- Chemicals may provoke changes in the enzymes involved in the biosynthesis and metabolism of circulating hormones.
- Structural changes may occur in the endocrine glands.

The impact of endocrine disruptors on homeostasis depends on several factors:

- Exposure during the adult period may be offset by normal homeostatic mechanisms. Ultimately, there is no observable effect.
- Exposure during the period of programming of homeostasis can result in a permanent change of reaction or sensitivity to signals.
- Exposure to EDCs at different stages of life, or during different seasons, results in different effects.
- An interaction exists between the various endocrine glands, so it is unpredictable in which gland the effects will occur (not necessarily in the expected system).

A lot of caution is needed to extrapolate in vitro results in vivo situations.

Xenoestrogens are chemicals that mimic the effects of natural estrogens. They are a part of the very heterogeneous group of endocrine disruptors. Xenoestrogens are exogenous estrogens with hormonal side effects. They can work synergistically to enhance the effect of estrogen, which can lead to an estrogen dominance. Examples are atrazine, dieldrin, DDT, endosulfan, lindane, and methoxychlor.

The cytochrome P450 enzyme system (CYP, P450, or CYP450) is a collection of enzymes some of which are important in the degradation of various xenobiotics, such as medications, toxins, and endocrine disruptors.

Aromatase is a complex of enzymes from the group of cytochrome P450. Aromatase converts anabolic steroids such as androstenedione to testosterone or a steroid with an aromatic ring such as estrone and estradiol. Aromatase converts androgens into estrogens.

For further information, the reader is directed to Gore (2007), Metzler (2010), Norris and Carr (2006), Hester and Harrison (1999), and Damstra et al. (2002).

A literature survey shows that there are vociferous discussions about endocrine-disrupting pesticides and about endocrine-disrupting compounds in general.

20.7 Endocrine-Disrupting Pesticides

There is a genuine concern about endocrine-disrupting pesticides that are lipophilic, resistant to metabolism, and able to bioconcentrate up the food chain. Therefore, they accumulate in body fats and can be transferred to the developing offspring via the placenta or via the egg. Predator animals (and humans) feeding at the top of the food chain are at increased risk, particularly mammals, because during breast feeding contaminants are again mobilized and transferred to the newly born infant. Marine mammals may be most vulnerable, because not only do they carry large amounts of body fat, but also the oceans are the final sink for many persistent pollutants.

Following paragraphs are taken from Damstra et al. (2002). These authors are very well pointing out the impact of endocrine-disrupting chemicals.

The last two decades have witnessed growing scientific concerns and public debate over the potential adverse effects that may result from exposure to a group of chemicals that have the potential to alter the normal functioning of the endocrine system in wildlife and humans. Concerns regarding exposure to these EDCs are due primarily to (1) adverse effects observed in certain wildlife, fish, and ecosystems; (2) the increased incidence of certain endocrine-related human diseases; and (3) endocrine disruption resulting from exposure to certain environmental chemicals observed in laboratory experimental animals. These concerns have stimulated many national governments, international organizations, scientific societies, the chemical industry, and public interest groups to establish research programs, organize conferences and workshops, and form expert groups and committees to address and evaluate EDC-related issues.

Since the publication of Rachel Carson's Silent Spring (Carson 1962), there has been increasing awareness that chemicals in the environment can exert profound and deleterious effects on wildlife populations and that human health is inextricably linked to the health of the environment. The last two decades, in particular, have witnessed a growing scientific concern, public debate, and media attention over the possible deleterious effects in humans and wildlife that may result from exposure to chemicals that have the potential to interfere with the endocrine system. The intensity of the concerns and lack of consensus among scientists can best be ameliorated by an objective evaluation of the available scientific data on the potential adverse effects of these chemicals from a global perspective.

Some suspected endocrine-disrupting pesticides that are reported in the literature are detailed in Table 20.1.

TABLE 20.1

Some Suspected Endocrine-Disrupting Pesticides

Compound	Affected Hormone System	Mechanism	References
DDT metabolite, p,p′-DDE	Androgen	Inhibits androgen binding to the androgen receptor, androgen-induced transcriptional activity, and androgen action in developing, pubertal, and adult male rats	Kelce (1995)
Linuron	Androgen	Androgen-receptor antagonist	Waller et al. (1996), Soto et al. (1994), McLachlan (1980)
Procymidone	Androgen	Androgen-receptor antagonist	Ostby et al. (1999) and Gray at al. (1999)
Sumithrin	Androgen		Go et al. (1999)
Vinclozolin	Androgen	Antiandrogenic (competes with androgens for the androgen receptor (AR), inhibits AR-DNA binding, and alters androgen-dependent gene expression)	Soto et al. (1994), Soto et al. (1995), Kelce et al. (1994), and Gray et al. (1999)
Fenitrothion	Antiandrogen	Competitive androgen–receptor antagonist	Tamura et al. (2001)
Mirex	Antiandrogenic activity inhibits production of LH Potentially thyroid		Porter et al. (1993), Chen et al. (1986), and Chernoff et al. (1976)
Cypermethrin	Disruption of reproductive function		Willingham et al. (2000) and Moore and Waring (2001)
Ketoconazole	Effect on reproductive systems		Marty et al. (1999) and Marty et al. (2001)
Aldrin	Estrogen	Binds to estrogen receptors, competes with estradiol	Jorgenson (2001)
Allethrin, d-trans	Estrogen		Go et al. (1999)
DDT	Estrogen	DDT and related compounds act in a number of ways to disrupt endocrine functions by binding with the estrogen receptor, including estrogen mimicry and antagonism, altering the pattern of synthesis or metabolism of hormones, and modifying hormone receptor levels	Soto et al. (1994), Lascombe et al. (2000), Kupfer et al. (1980), and Rajapakse et al. (2001)
Dicofol (Kelthane)	Estrogen		Vinggaard et al. (1999)
Dieldrin	Estrogen	Binds to the estrogen receptor, competes with estradiol	Soto et al. (1994) and Jorgenson (2001)
Endosulfan	Estrogen		Soto et al. (1994, 1995)
Fenarimol	Estrogen	Estrogen-receptor agonist	Vinggaard et al. (1999)
Fenvalerate	Estrogen		Go et al. (1999)

TABLE 20.1 (Continued)

Some Suspected Endocrine-Disrupting Pesticides

Compound	Affected Hormone System	Mechanism	References
Kepone (chlordecone)	Estrogen	Displays androgen- and estrogen-receptor-binding affinities	Waller et al. (1996), Lambright et al. (2000), and Gray et al. (1999)
Lindane (hexachlorocyclohexane)	Estrogen Androgen	Inhibits ligand binding to androgen and estrogen receptors	Danzo (1997)
Methoxychlor	Estrogen	Through mechanisms other than receptor antagonism Precise mechanism still unclear	Pickford and Morris (1999)
Nonachlor, trans-	Estrogen	Estrogen-receptor agonist	Willingham et al. (2000)
Toxaphene	Estrogen Thyroid		Soto et al. (1994)
Triadimefon	Estrogen	Estrogen-receptor agonist	Vinggaard et al. (1999)
Triadimenol	Estrogen	Estrogen-receptor agonist	Vinggaard et al. (1999)
Carbaryl	Estrogen and progesterone		Klotz et al. (1997b)
Permethrin	Estrogenic		Go et al. (1999)
Iprodione	Inhibition of testosterone synthesis		Benhamed (1996)
Atrazine	Neuroendocrine pituitary (depression of LH surge), testosterone metabolism	Inhibits ligand binding to androgen and estrogen receptors	Danzo (1997)
Oxychlordane	Reproductive		Guillette et al. (1999)
Tributyltin	Reproductive		Horiguchi et al. (2000)
Trifluralin	Reproductive metabolic		Rawlings et al. (1998)
Chlordane	Testosterone and progesterone		Willingham et al. (2000)
Amitrole	Thyroid	Thyroid peroxidase inhibitor inhibits thyroid hormone synthesis	Hurley et al. (1998)
Clofentezine	Thyroid	Enhances the secretion of thyroid hormone	Hurley et al. (1998)
Ethylene thiourea	Thyroid	Thyroid peroxidase inhibitor	Hurley et al. (1998)
Fenbuconazole	Thyroid	Enhances the secretion of thyroid hormone	Hurley et al. (1998)
Fipronil	Thyroid	Enhances the secretion of thyroid hormone	Hurley et al. (1998)
Heptachlor	Thyroid		Akhtar et al. (1996) and Reuber (1987)
Heptachlor epoxide	Thyroid Reproductive		Reuber (1987)
Karate	Thyroid	Decrease of thyroid hormone in serum; direct effect on the thyroid gland?	Akhtar et al. (1996)

(continued)

TABLE 20.1 (Continued)

Some Suspected Endocrine-Disrupting Pesticides

Compound	Affected Hormone System	Mechanism	References
Malathion	Thyroid	Decrease of thyroid hormone in serum, with perhaps a direct effect on the thyroid gland	Akhtar et al. (1996)
Mancozeb	Thyroid	Thyroid peroxidase inhibitor	Hurley et al. (1998)
Maneb	Thyroid	The metabolite ethylenethiourea inhibits thyroid hormone synthesis	Toppari et al. (1995)
Methomyl	Thyroid		Porter et al. (1993) and Klotz et al. (1997b)
Metribuzin	Thyroid		Porter et al. (1993)
Nitrofen	Thyroid	Structural similarities to the thyroid hormones; nitrofen or its metabolite may have thyroid hormone activities	Stevens and Summer (1991)
Pendimethalin	Thyroid	Enhances secretion of thyroid hormone	Hurley et al. (1998)
Pentachloronitrobenzene	Thyroid	Enhances secretion of thyroid hormone	Hurley et al. (1998)
Prodiamine	Thyroid	Enhances secretion of thyroid hormone	Hurley et al. (1998)
Pyrimethanil	Thyroid	Enhances secretion of thyroid hormone	Hurley et al. (1998)
Talstar	Thyroid	A decrease of thyroid hormone in serum; direct effect on the thyroid gland?	Akhtar et al. (1996)
Thiazopyr	Thyroid	Enhances secretion of thyroid hormone	Hurley et al. (1998)
Zineb	Thyroid	The metabolite ethylenethiourea inhibits thyroid hormone synthesis	Toppari et al. (1995)
Ziram	Thyroid	Inhibits the iodide peroxidase Structural similarities between ziram and thiram; ziram can be metabolized to thiram in the environment	Marinovich et al. (1997)
Acetochlor	Thyroid (decrease of thyroid hormone levels, increase in TSH)		Hurley et al. (1998)
Alachlor	Thyroid (decrease of thyroid hormone levels, increase in TSH)		Wilson et al. (1996)

Source: Taken from the website *Our Stolen Future.*

20.8 Current Literature

A nonexhaustive selection of recent literature on endocrine disruption and pesticides is given in the following paragraphs.

Propper (2005) has systematically reviewed literature on the mixes of endocrine compounds. The literature may be distributed in three groups, that is, the hormone investigated as being targets for disruption, the biological mechanisms of endocrine disruption, and physiological systems that are affected by endocrine-disrupting compounds. This format may be useful in testing the hypothesis that the disruption literature focuses on first, estrogens and androgens as the targets of disruption; secondly, receptor binding as the mechanism of disruption and thirdly, reproduction and development as a physiological aspect.

Piersma et al. (1998) have tested six known and alleged disruptors namely ethinylestradiol (contraceptive), coumestrol (phytoestrogen), 4-tert-octylphenol (surfactant), bisphenol (plastic monomer), vinclozolin (fungicide), and butylbenzyl phthalate (plasticizer) by using an existing reproductive toxicology screening test. All the compounds excluding ethinylestradiol were clearly scored as reproductive toxicants, affecting one or more parameters such as fertility, luteinization, spermatogenesis, and fetal development. They have concluded that endocrine disruptors are likely to be effective in existing OECD reproductive toxicity test system used for human risk evaluation at dosages below the maximal dosages used in former test.

Martinez-Valenzuela et al. (2009) have evaluated the genotoxic effects of pesticide mixtures on agricultural workers using as biomarkers the sister chromatid exchangers (SCE), mitotic index, cell proliferated kinetics, and replication index in lymphocytes of the peripheral blood. They have also analyzed the micronuclei frequency and other nuclear anomalies in exfoliated buccal cells of the individuals who had been exposed to pesticides in Las Grullas, Ahome, Sinaloa, Mexico. This study afforded valuable data to estimate the possible risk to health associated with pesticide exposure.

Quassinti et al. (2009) have studied in vitro the influence of paraquat and glyphosate on ovarian and testicular steroidogenesis in amphibian. They tracked the production of 17β-estradiol and testosterone when ovarian tissue and testis of the water frog *Rana esculenta* were incubated in presence of the two herbicides. These results indicate that paraquat acts on gonadal steroidogenesis through a mechanism involving reactive oxygen species. Glyphosate showed no effect on gonadal steroidogenesis, and it may interfere in amphibian reproductive processes.

Salaberria et al. (2009) investigated the endocrine-disrupting effect of atrazine (ATZ) herbicide. They have also studied the effects of ATZ on hepatic metabolism at the onset of puberty in rainbow trout. The results showed that ATZ can alter hepatic metabolism, induce estrogenic effects, and cause oxidative stress in vivo and that these effects are linked.

Orton et al. (2009) have investigated the effects of common herbicides (isoproturon, diuron, linuron, 4-chloro-2-methylphenoxy acetic acid (MCPP), mecoprop, atrazine, simazine, trifluralin, chlorpropham, bentazone, 2,4-dichlorophenoxy acetic acid (2,4-D), and pentachlorophenol (PCP)) on endocrine end points at environmentally relevant concentrations. The most common effects were antiestrogenic/antiandrogenic activity in the yeast screen, and inhibition of ovulation in vitro, accompanied by decreased testosterone production. They concluded that novel effects of herbicides and PCP at environmentally relevant concentrations were found, while the effects of these compounds on humans and/or wildlife warrant further investigations.

Darras (2008) has discussed the interference of organohalogens with different processes of brain development. He has concluded that it is essential to accurately assess the long term risks of early exposure to organohalogens. Several of the mechanisms involved are similar to those of polychlorinated biphenyls (PCBs), each group also works via own specific pathways. The fact that persistent organohalogens can amplify the neurotoxic effects of other environmental pollutants such as heavy metals further increases their risk for human and animal neurodevelopment.

Benotti et al. (2009) have described results of a comprehensive survey of 20 pharmaceuticals, 25 known or potential EDCs and 6 other water contaminants including atrazine in source water, finished drinking water, and distribution system (tap) water from 19 US drinking water treatment plants sampled during 2006–2007. The results provide an assessment of the actual concentrations to which people are exposed from drinking water. Occurrence data were used to propose a set of indication compounds that can predict the presence of pharmaceuticals and EDCs as well as monitor the efficacy of treatment processes.

Ejaz et al. (2004) have developed a possible correlation between pesticides exposure and growing incidences of cancer disrupting in Pakistan. They have claimed that majority of endocrine-disrupting pesticides are carcinogenic. Farmers may therefore be at higher risk for acute and chronic health effects associated with pesticides. Cancer of breast, ovary, prostate, testis, and thyroid are hormone-dependent, which fostered research on the potential risk associated with occupational and environmental exposure to the so-called endocrine-disrupting pesticides. Professional as well as public exposure to pesticides raises cancer risk. Interaction with adjuvant and with other toxicants increases the actual risk. Furthermore, organochlorine pesticides and triazine herbicides require further investigation for possible etiologic role in some hormone-dependent cancers.

Prins (2008) has published commentary on increasing evidence both from epidemiology studies and animal models that specific endocrine-disrupting compounds may influence the development or progression of prostate cancer. In large part, these effects appear to be linked to interference with estrogen signaling, either through interacting with estrogen receptors (ERs) or by influencing steroid metabolism and altering estrogen levels within the body. In humans, epidemiologic evidence links specific pesticides, PCBs, and inorganic arsenic exposures to elevated prostate cancer risk. In animal models, there was also augmentation of prostate carcinogenesis with several other environmental estrogenic compounds including cadmium, UV filters, and bisphenol A. He has concluded that infants and children may be considered a highly susceptible population for endocrine-disruptor exposure and increased possible risk of prostate cancer with aging.

McKinlay et al. (2008) have reviewed the known routes of human pesticide exposure with particular reference to endocrine-disrupting pesticides (EDPs). Many EDPs are harmful at very low doses, especially if exposure occurs during sensitive stages of developments via epigenetic changes. They also reviewed the available deterministic and probabilistic models commonly used to calculate human exposure and the creation of more holistic models of human pesticide exposure including a requirement for new quantitative data sets.

Eskenazi et al. (2007) have reported an adverse association of prenatal organophosphate pesticide exposure as measured by dialkylphosphates (DAPs) with mental development problems at 24 months of age.

20.9 Endocrine Action of DDT

As already discussed in paragraph 6, endocrine disruptors affect in many ways (Hester and Harrison 1999).

TABLE 20.2

Representative Endocrine Actions of DDT and Its Metabolites in Vertebrates

Compound	Label Effect	Effect	Study Type	References
Not Specified DDT				
DDT	Antiandrogenic	Altered testosterone metabolism	Birds	Peakall (1967)
DDT		Altered thyroxine metabolism	Birds	Bastomsky (1974)
DDT		Altered progesterone metabolism	Birds	Peakall (1967)
Technical DDT				
DDT	Estrogenic	Estrogenic response cells	E-screen	Sonnenschein and Soto (1998)
DDT	Estrogenic	Increase uterine weight	Rat	Welch et al. (1969)
DDT	Antiestrogenic	Antagonized estrogenic treatment	Tiger salamander	Clark et al. (1998)
DDT		Increases steroid metabolism	Rat (liver)	Welch et al. (1969)
DDT (2 isomers)				
o,p′-DDT	Estrogenic?	ER interaction	Alligator ER	Vonier et al. (1996)
o,p′-DDT	Estrogenic?	Weakly binds ER	Human ER	Gaido et al. (1997)
o,p′-DDT	Estrogenic	Induction of zona radiata proteins	Atlantic salmon	Arukwe et al. (1997)
o,p′-DDT	Estrogenic	Estrogenic response cells	E-screen	Sonnenschein and Soto (1998)
o,p′-DDT	Estrogenic	ER binding + response	Mice ER	Shelby et al. (1996)
o,p′-DDT	Estrogenic	Increase in uterine weight	Rat	Welch et al. (1969)
o,p′-DDT	Estrogenic	Vtg synthesis in males	Red eared turtle	Palmer and Palmer (1995)
o,p′-DDT	Estrogenic	Induction of Vtg/hepatic EBS	Trout	Donohoe and Curtis (1996)
o,p′-DDT	Estrogenic	ER binding + response	YES	Arnold et al. (1996)
o,p′-DDT	Estrogenic	ER binding + response	YES/CB/MCF-7	Klotz et al. (1996)
o,p′-DDT	Antiestrogenic	Strong antagonist	Rabbit ER (CB)	Danzo (1997)
o,p′-DDT	Antiandrogenic	Antagonist	Human AR	Maness et al. (1998)
o,p′-DDT	Antiandrogenic	Strong antagonist	Rat AR	Danzo (1997)
o,p′-DDT	Antiandrogenic	Antagonist	YES AR	Sohoni and Sumpter (1998)
o,p′-DDT	Hyperthyroidism	Increased thyroid weight	Bengal finch	Jefferies (1969)
o,p′-DDT	Estrogenic/ antiandrogenic	Skewed sex ratio toward female	Seagull	Fry and Toone (1981)
o,p′-DDT	Estrogenic/ antiandrogenic	Development of right oviduct	Seagull	Fry and Toone (1981)
o,p′-DDT	Estrogenic/ antiandrogenic	Feminized testis	Seagull	Fry and Toone (1981)
p,p′-DDT	Estrogenic?	Weak ER interaction	Alligator ER	Vonier et al. (1996)
p,p′-DDT	Estrogenic	Estrogenic response cells	E-screen	Sonnenschein and Soto (1998)

(continued)

TABLE 20.2 (Continued)

Representative Endocrine Actions of DDT and Its Metabolites in Vertebrates

Compound	Label Effect	Effect	Study Type	References
p,p′-DDT	Antiprogesterone	Antiprogesterone (antiprogestin)	Human PR (CB)	Klotz et al. (1997b)
p,p′-DDT	Antiestrogenic	Weak antagonist	Rabbit ER (CB)	Danzo (1997)
p,p′-DDT	Antiandrogenic	Decrease testicular growth	Cockerel	Burlington and Lindeman (1950)
p,p′-DDT	Antiandrogenic	Antagonist	Human AR	Maness et al. (1998)
p,p′-DDT	Antiandrogenic	Strong antagonist	Rat AR	Danzo (1997)
p,p′-DDT	Corticosterone	Induces stress-related response	Senegal walking frog	Hayes et al. (1997)
DDE (2 isomers)				
o,p′-DDE	Estrogenic?	ER interaction	Alligator ER	Vonier et al. (1996)
o,p′-DDE	Estrogenic?	Weakly binds ER	Human ER	Gaido et al. (1997)
o,p′-DDE	Estrogenic	Estrogenic response cells	E-screen	Sonnenschein and Soto (1998)
o,p′-DDE	Estrogenic	Induction of Vtg/hepatic EBS	Trout	Donohoe and Curtis (1996)
o,p′-DDE	Antiprogesterone	Antiprogesterone	Fowl PR (CB)	Lundhom (1997)
o,p′-DDE	Antiandrogenic	Antagonist	Fowl AR	Lundhom (1997)
o,p′-DDE	Antiandrogenic	Antagonist	Human AR	Maness et al. (1998)
p,p′-DDE	Estrogenic?	Weak ER interaction	Alligator ER	Vonier et al. (1996)
p,p′-DDE	Estrogenic	Development of gonaducts	Tiger salamander	Clark et al. (1998)
p,p′-DDE	Antiestrogenic?	Weak antagonist	Rabbit ER (CB)	Danzo (1997)
p,p′-DDE	Antiandrogenic	Inhibits transcription	Human AR/CB	Kelce et al. (1995)
p,p′-DDE	Antiandrogenic	Reduction anogenital distance	Male rats (offspring)	Kelce et al. (1995)
p,p′-DDE	Antiandrogenic	Retention of thoracic nipples	Male rats (offspring)	Kelce et al. (1995)
p,p′-DDE	Antiandrogenic	Strong antagonist	Rat AR	Danzo (1997)
p,p′-DDE	Antiandrogenic	Antagonist	YES AR	Sohoni and Sumpter (1998)
p,p′-DDE	Androgenic?	Antagonist, agonist	Human AR	Maness et al. (1998)
p,p′-DDE	Androgenic?	AR interaction	Human AR	Gaido et al. (1997)
DDD (2 isomers)				
o,p′-DDD	Estrogenic?	ER interaction	Alligator ER	Vonier et al. (1996)
o,p′-DDD	Estrogenic?	Weakly binds ER	Human ER	Gaido et al. (1997)
o,p′-DDD	Estrogenic	ER binding + response (weak)	YES/CB/MCF-7	Klotz et al. (1996)
o,p′-DDD	Antiandrogenic	Antagonist	Human AR	Gaido et al. (1997)
p,p′-DDD	Estrogenic?	Weak ER-interaction	Alligator AR	Vonier et al. (1996)
p,p′-DDD	Estrogenic	Sex reversal	American alligator	Crain (1997)
p,p′-DDD	Estrogenic	ER binding + response	YES CB/MCF-7	Klotz et al. (1996)

Source: From Guillette, L. J., Kools, S. A. E., Gunderson, M. P., Bermudez, D. S. 2006. DDT and its analogues: New insights into their endocrine-disrupting effects on wildlife. In: Norris, D. O. and Carr, J. A. (eds), *Endocrine Disruption.* Oxford University Press: Oxford, UK.

ER, estrogen receptor; Vtg, vitellogenin; YES, yeast estrogen screen; AR, androgen receptor; PR, progesterone receptor; EBS, estrogen-binding sites; CB, cell bioassay; MCF-7, Michigan Cancer Foundation – 7; ?, possible effect.

They can either bind to the hormone's receptor or mimic the hormone, or they can block the action of the hormone. Secondly, they can stimulate or inhibit the enzymes responsible for the synthesis or clearance of a hormone and thereby give rise to an increased or decreased action of the hormone.

Numerous studies have documented the endocrine actions of DDT and its metabolites in vertebrates (Guillette et al. 2006; see Table 20.2). DDT has been taken as an example for the wide variety of effects of a pesticide.

References

Akhtar, N., Kayani, S. A., Ahmad, M. M., and Shahab, M. 1996. Insecticide-induced changes in secretory activity of the thyroid gland in rats. *J. Appl. Toxicol.* 16(5): 397–400.

Arnold, S., Robinson, M. K., Notides, A. C., Guillette, L. J., and McLachlan, J. A. 1996. A yeast estrogen screen for examining the relative exposure of cells to natural and xenoestrogens. *Environ. Health Perspect.* 104: 544–548.

Arukwe, A., Knudsen, F. R., and Goksoyr, A. 1997. Fish zona radiata (eggshell) protein: A sensitive biomarker for environmental estrogens. *Environ. Health Perspect.* 105: 418–422.

Bastomsky, C. H. 1974. Effects of a polychlorinated biphenyl mixture (Arochlor 1254) and DDT on biliary thyroxine excretion in rats. *Endocrinology* 95: 1150–1155.

Benhamed, M. 1996. Report INSERM Unité 407/96001 Lyon, of October 21, 1996. Effects of iprodione and its metabolites RP36112 and RP36115 on testosterone secretion in cultured Leydig cells: sites of action. Unpublished. Ownership: Rhône-Poulenc Agro.

Benotti, M. J., Trenholm, R. A., Vanderford, B. J., Holady, J. C., Stanford, B. D., and Snyder, S. A. 2009. Pharmaceuticals and endocrine compounds in U.S. drinking water. *Environ. Sci. Technol.* 43(3): 597–603.

Burlington, H. and Lindeman, V. 1950. Effect of DDT on testes and secondary sex characters of white leghorn cockerels. *Proc. Soc. Exp. Biol. Med.* 74: 48–51.

Carson, R. 1962. *Silent Spring.* Mariner Books: New York.

Chen, T. T., Reid, P. C., Van Beneden, R., and Sonstegard, R. A. 1986. Effect of Arochlor 1254 and Mirex on estradiol-induced vittellogenin production in juvenile rainbow trout. *Can. J. Fish. Aquat. Sci.* 43: 169–73.

Chernoff, N., Scotti, T. M., and Linder, R. E. 1976. Cataractogenic proterties of mirex in rats and mice with notes on kepone. *Toxicol. Appl. Pharmacol.* 37: 1883.

Clark, E. J., Norris, D. O., and Jones, R. E. 1998. Interactions of gonadal steroids and pesticides (DDT, DDE) on gonaduct growth in larval tiger salamanders, *Ambystoma trigrinum. Gen. Comp. Endocr.* 109: 94–105.

Crain, D. A. 1997. Effects of endocrine-disrupting contaminants on reproduction in the American alligator, *Alligator mississippiensis.* Ph. D. Dissertation. University of Florida, Gainesville.

Damstra, T., Barlow, S., Bergman, A., Kavlock, R., and Van Der Kraak, G. 2002. Global Assessment of the State-of-the-Science of Endocrine Disruptors. International Programme of Chemical safety, WHO/PCS/EDC/02.2.

Danzo, B. J. 1997. Environmental xenobiotics may disrupt normal endocrine function by interfering with the binding of physiological ligands to steroid receptors and binding proteins. *Environ. Health Perspect.* 105: 294–301.

Darras V. M. 2008. Endocrine disrupting polyhalogenated organic pollutants interfere with thyroid hormone signaling in the developing brain. *The Cerebellum* 7(1): 26–37.

Donohoe, R. M. and Curtis, L. R. 1996. Estrogenic activity of chlordecone, o,p'-DDT and o,p'-DDE in juvenile rainbow trout: Induction of vitellogenesis and interaction with hepatic estrogen binding sites. *Aquat. Toxicol.* 36: 31–52.

Ejaz, S., Akram, W., Lim, C. W., Lee, J. J., and Husain, I. 2004. Endocrine disrupting pesticides: A leading cause of cancer among people in Pakistan. *Exp. Oncol.* 26(2): 98–105.

EPA Special Report on Environmental Endocrine Disruption: An Effects Assessment and Analysis by EPA, US, 1997 (http://www.epa.gov/raf/publications/pdfs/ENDOCRINE.PDF).

Eskenazi, B., Marks, A. R., Bradman, A., Harley, K., Barr, D. B., Johnson, C., Morga, N., and Jewell, N. P. 2007. Organophosphate pesticide exposure and neurodevelopment in young Mexican-American children. *Environ. Health Perspect.* 15(5): 792–798.

Fry, D. M. and Toone, C. K. 1981. DDT-induced feminization of gull embryos. *Science*. 213: 922–924.

Gaido, K. W., Leonard, L. S., Lovell, S., Gould, J. C., Babai, D., Portier, C. J., and McDonnell, D. P. 1997. Evaluation of chemicals with endocrine modulating activity in a yeast-based steroid hormone receptor gene transcription assay. *Toxicol. Appl. Pharmacol.* 143: 205.

Go, V., Garey, J., Wolff, M. S., and Pogo, B. G. T. 1999. Estrogenic potential of certain pyrethroid compounds in the MCF-7 human breast carcinoma cell line. *Environ. Health Perspect.* 107: 173–177.

Gore, A. C. 2007. *Endocrine-Disrupting Chemicals: From Basic Research to Clinical Practice (Contemporary Endocrinology)*, 1st edn. Humana Press: Totowa, NJ.

Gray, L. E., Wolf, C., Lambright, C., Mann, P., Price, M., Cooper, R. L., and Ostby, J. 1999. Administration of potentially antiandrogenic pesticides (procymidone, linuron, iprodione, chlozolinate, p,p'-DDE, and ketoconazole) and toxic substances (dibutyl- and diethylhexyl phthalate, PCB 169, and ethane dimethane sulphonate) during sexual differentiation produces diverse profiles of reproductive malformations in the male rat. *Toxicol. Ind. Health.* 15: 94–118.

Guillette, L. J. Jr., Brock, J. W., Rooney, A. A., and Woodward, A. R. 1999. Serum concentrations of various environmental contaminants and their relationship to sex steroid concentrations and phallus size in juvenile American alligators. *Arch. Environ. Contam. Toxicol.* 36: 447–455; Springer-Verlag: New York.

Guillette, L. J., Kools, S. A. E., Gunderson, M. P., and Bermudez, D. S. 2006. DDT and its analogues: New insights into their endocrine-disrupting effects on wildlife. In: Norris, D. O. and Carr, J. A. (eds), *Endocrine Disruption*. Oxford University Press: Oxford, UK.

Hayes, T. B., Wu, T. H., and Gill, T. N. 1997. DDT-like effects as a result of corticosterone treatment in an anuran amphibian: Is DDT a corticoid mimic or a stressor? *Environ. Toxicol. Chem.* 16: 1948–1953.

Henry, H. and Norman, A. W. 2003. *Encyclopedia of Hormones, Three-Volume Set*, 1st edn. Academic Press: San Diego, CA.

Hester, R. E. and Harrison, R. M. 1999. *Endocrine Disrupting Chemicals*. The Royal Society of Chemistry: Cambridge.

Horiguchi, T., Takiguchi, N., Cho, H. S., Kojima, M., Kaya, M., Shiraishi, H., Morita, M., Hirose, H., and Shimizu, M. 2000. Ovo-testis and disturbed reproductive cycle in the giant abalone, Haliotis madaka: Possible linkage with organotin contamination in a site of population decline. *Mar. Environ. Res.* 50(1–5): 223–229.

Hurley, P. M., Hill, R. N., and Whiting, R. J. 1998. Mode of carcinogenic action of pesticides inducing thyroid follicular cell tumors in rodents. *Environ. Health Perspect.* 106: 437–445.

Jefferies, D. J. 1969. Induction of apparent hyperthyroidism in birds fed DDT. *Nature* 222: 578–579.

Jorgenson, J. L. 2001. Aldrin and dieldrin: A review of research on their production, environmental deposition and fate, bioaccumulation, toxicology, and epidemiology in the United States. *Environ. Health Perspect.* 109(suppl 1): 113–139.

Kelce, W. R. 1995. Persistent DDT metabolite p,p'-DDE is a potent androgen receptor antagonist. *Nature* 375(6532): 581–585.

Kelce, W. R., Monosson, E., Gamcsik, M. P., Laws, S. C., and Gray, L. E. 1994. Environmental hormone disruptors: Evidence that vinclozolin developmental toxicity is mediated by antiandrogenic metabolites. *Toxicol. Appl. Pharmacol.* 126: 276–285.

Kelce, W. R., Stone, C. R., Laws, S. C., Gray, L. E., Kemppainen, J. A., and Wilson, E. M. 1995. Persistent DDT metabolite p,p'-DDE is a potent androgen receptor antagonist. *Nature* 375: 581–585.

Klotz, D. M., Beckman, B. S., Hill, S. M., McLachlan, J. A., Walters, M. R., and Arnold, S. A. 1996. Identification of environmental chemicals with estrogenic activity using a combination of in vitro assays. *Environ. Health Perspect.* 104: 1084–1089.

Klotz, D. M., Ladlie, B. L., Vonier, P. M., McLachlan, J. A., and Arnold, S. F. 1997a. o,p'-DDT and its metabolites inhibit progesterone-dependent responses in yeast and human cells. *Mol. Cell. Endocrinol.* 129: 63–71.

Klotz, D. M., Arnold, S. F., and McLachlan, J. A. 1997b. Inhibition of 17 beta-estradiol and progesterone activity in human breast and endometrial cancer cells by carbamate insecticides. *Life Sci.* 60(17): 1467–1475.

Kupfer, D., et al. 1980. Estrogenic properties of DDT and its analogs. *Toxicol. Appl. Pharmacol.* 14: 358–367.

Lambright, C., Ostby, J., Bobseine, K., Wilson, V., Hotchkiss, A. K., Mann, P. C., and Gray, L. E. Jr. 2000. Cellular and molecular mechanisms of action of Linuron: An antiandrogenic herbicide that produces reproductive malformations in male rats. *Toxicol. Sci.* 56: 389–399.

Lascombe, I., Beffa, D., Rüegg, U., Tarradellas, J., and Wahli, W. 2000. Estrogenic activity assessment of environmental chemicals using in vitro assays: Identification of two new estrogenic compounds. *Environ. Health Perspect.* 108: 621–629.

Lundhom, C. E. 1997. DDE-Induced eggshell thinning in birds: Effects of p,p'-DDE on calcium and prostaglandin metabolism of the eggshell gland. *Comp. Biochem. Physiol.* 118C: 113–128.

Maness, S. C., McDonnell, D. P., and Gaido, K. W. 1998. Inhibition of androgen receptor-dependent transcriptional activity by DDT isomers and methoxychlor in HepG2 human hepatoma cells. *Toxicol. Appl. Pharmacol.* 151: 135–142.

Marinovich, M. M., Guizzetti, F., Ghilardi, F., Viviani, B., Corsini, E., and Galli, C. L. 1997. Thyroid peroxidase as toxicity target for dithiocarbamates. *Arch. Toxicol.* 71: 508–512.

Martinez-Valenzuela, C., Gómez-Arroyo, S., Villalobos-Pietrini, R., Waliszewski, S., Maria Elena Calderón-Segura, M. E., Félix-Gastélum, R., and Álvarez-Torres, A. 2009. Genotoxic biomonitoring of agricultural workers exposed to pesticides in north of Sinaloa State, Mexico. *Environ. Int.* 35(8): 1155–1159.

Marty, M. S., Crissman, J. W., and Carney, E. W. 1999. Evaluation of the EDSTAC female pubertal assay in CD rats using 17beta-estradiol, steroid biosynthesis inhibitors, and a thyroid inhibitor. *Toxicol. Sci.* 52: 269–277.

Marty, M. S., Crissman, J. W., and Carney, E. W. 2001. Evaluation of the male pubertal onset assay to detect testosterone and steroid biosynthesis inhibitors in CD rats. *Toxicol. Sci.* 60: 285–295.

McKinlay, R., Plant, J. A., Bell, J. N. B., and Voulvoulis, N. 2008. Calculating human exposure to endocrine disrupting pesticides via agricultural and non-agricultural exposure routes. *Sci. Total Environ.* 398(1–3): 1–12.

McLachlan, J. A. (ed.). 1980. Estrogenicity of kepone in birds and mammals. In: *Estrogens in the Environment*, pp. 239–264. Elsevier/North Holland: New York.

Metzler, M. 2010. *Endocrine Disruptors Part II (The Handbook of Environmental Chemistry/Anthropogenic Compounds)*, 1st edn. Springer: New York.

Moore, A. and Waring, C. P. 2001. The effects of a synthetic pyrethroid pesticide on some aspects of reproduction in Atlantic salmon (Salmo salar L.). *Aquat. Toxicol.* 52(1): 1–12.

Neal, J. M. and Sompayrac, L. L. 2001. *How the Endocrine System Works*, 1st edn. Wiley-Blackwell: New York.

Nijhout, H. F. 1994. *Insect Hormones*. Princeton University Press: Princeton, NJ.

Norman, A. W., Henry, H., and Litwack, G. 2011. Hormones, 3rd edn. Academic Press: New York.

Norris, D. O. and Carr, J. A. 2006. *Endocrine Disruption*. Oxford University Press: Oxford, UK.

Orton, F., Lutz, I., Kloas, W., and Routledge, E. J. 2009. Endocrine disrupting effects of herbicides and pentachlorophenol: In vitro and in vivo evidence. *Environ. Sci. Technol.* 43(6): 2144–2150.

Ostby, J., Kelce, W. R., Lambright, C., Wolf, C. J., Mann, P., and Gray, L. E. Jr. 1999. The fungicide procymidone alters sexual differentiation in the male rat by acting as an androgen-receptor antagonist in vivo and in vitro. *Toxicol. Ind. Health* 15(1–2): 80–93.

Our Stolen Future. http://www.ourstolenfuture.org/basics/chemlist.htm (November 2010).

Palmer, B. and Palmer, S. 1995. Vitellogenin induction by xenobiotic estrogens in the red-eared turtle and African clawed frog. *Envron. Health Perspect.* 103(suppl 4): 19–25.

Peakall, D. B. 1967. Pesticide-induced enzyme breakdown of steroids in birds. *Nature* 216: 505–506.

Pickford, D. B. and Morris, I. D. 1999. Effects of endocrine-disrupting contaminants on amphibian oogenesis: Methoxychlor inhibits progesterone-induced maturation of Xenopus laevis oocytes in vitro. *Environ. Health Perspect.* 107: 285–292.

Piersma, A. H., Verhoef, A., Elvers, L. H., and Wester, P. W. 1998. Toxicity of compounds with endocrine activity in the OECD 421 reproductive toxicity screening test. Rijksinstituut voor Volksgezondheid en Milieu RIVM Rapport nr. 650030002.

Porter, W. P., Green, S. M., Debbink, N. L., and Carlson, I. 1993. Groundwater pesticides: interactive effects of low concentrations of carbamates aldicarb and methomyl and the triazine metribuzin on thyroxine and somatotropin levels in white rats. *J. Toxicol. Environ. Health* 40(1): 15–34.

Prins, G. S. 2008. Endocrine disruptors and prostate cancer risk. *Endocr. Relat. Cancer* 15(3): 649–656.

Propper, R. C. 2005. The study of endocrine-disrupting compounds: Past approaches and new directions. *Integr. Comp. Biol.* 45(1): 194–200.

Quassinti, L., Maccari, E., Murri, O., and Bromucci, M. 2009. Effects of paraquat and glyphosate on steroidogenesis in gonads of frog rana esculenta in vitro. *Pestic. Biochem. Physiol.* 93(2): 91–95.

Rajapakse, N., Ong, D., and Kortenkamp, A. 2001. Defining the impact of weakly estrogenic chemicals on the action of steroidal estrogens. *Toxicol. Sci.* 60: 296–304.

Rawlings, N. C., Cook, S. J., and Waldbillig, D. 1998. Effects of the pesticides carbofuran, chlorpyrifos, dimethoate, lindane, triallate, trifluralin, 2,4-D, and pentachlorophenol on the metabolic endocrine and reproductive endocrine system in ewes. *J. Toxicol. Environ. Health Part A* 54(1): 21–36.

Reuber, M. D. 1987. Carcinogenicity of heptachlor and heptachlor epoxide. *J. Environ. Pathol. Toxicol. Oncol.* 7(3): 85–114.

Salaberria, I., Hansen, B. H., Asensio, V., Olsvik, P. A., Rolf, A., Asensio, V., Andersen, R. A., and Jenssen, B. M. 2009. Effects of atrazin on hepatic metabolism and endocrine homeostasis in rainbow trout (Oncorhynchus mykiss). *Toxicol. Pharmacol.* 234(1): 98–106.

Shelby, M. D., Newbold, R. R., Tully, D. B., Chae, K., and Davis, V. L. 1996. Assessing environmental chemicals for estrogenicity using a combination of in vitro and in vivo assays. *Environ. Health Perspect.* 104: 1296–1300.

Sohoni, P. and Sumpter, J. P. 1998. Several environmental oestrogens are also anti-androgens. *J. Endocrinol.* 158: 327–339.

Sonnenschein, C. and Soto, A. M. 1998. An updated review of environmental estrogen and androgen mimics and antagonists. *J. Steroid Biochem. Mol. Biol.* 65: 143–153.

Soto, A. M., Chung, K. L., and Sonnenschein, C. 1994. The pesticides endosulfan, toxaphene, and dieldrin have estrogenic effects on human estrogen-sensitive cells. *Environ. Health Perspect.* 102(4): 380–383.

Soto, A. M., Sonnenschein, C., Chung, K. L., Fernandez, M. F., Olea, N., and Serrano, F. O. 1995. The E-SCREEN assay as a tool to identify estrogens: An update on estrogenic environmental pollutants. *Environ. Health Perspect.* 103(Suppl 7): 113–122.

Stevens, J. T. and Summer, D. D. 1991. Herbicides. In: Hayes, W. J. Jr. and Laws, E. R. Jr. (eds), *Handbook of Pesticide Toxicology*, pp. 1991–1408. Academic Press: San Diego, CA.

Tamura, H., Maness, S. C., Reischmann, K., Dorman, D. C., Gray, L. E., and Gaido, K. W. 2001. Androgen receptor antagonism by the organophosphate insecticide fenitrothion. *Toxicol. Sci.* 60: 56–62.

Toppari, J., Larsen, J. C., Christiansen, P., Giwercman, A., and Grandjean, P. 1995. Male reproductive health and environmental chemicals with estrogenic effects. Miljoproject no. 290. Report of the Ministry of Environment and Energy, Danish Protection Agency.

Vinggaard, A. M., Breinholt, V., and Larsen, J. C. 1999. Screening of selected pesticides for oestrogen receptor activation in vitro. *Food Addit. Contam.* 16(12): 533–542.

Vonier, P. M., Crain, D. A., McLachlan, J. A., Guilette, L. J., and Arnold, S. F. 1996. Interaction of environmental chemicals with the estrogen and progesterone receptors from the oviduct of the American alligator. *Environ. Health Perspect.* 104: 1318–1322.

Waller, C. L., Juma, B. W., Gray, L. E. Jr., and Kelce, W. R. 1996. Three-dimensional quantitative structure-activity relationships for androgen receptor ligands. *Toxicol. Appl. Pharmacol.* 137(2): 219–227.

Welch, R. M., Levin, W., and Conney, A. H. 1969. Estrogen action of DDT and its analogs. *Toxicol. Appl. Pharmacol.* 14: 358–367.

Willingham, E., Rhen, T., Sakata, J. T., and Crews, D. 2000. Embryonic treatment with xenobiotics disrupts steroid hormone profiles in hatchling red-eared slider turtles (Trachemys scripta elegans). *Environ. Health Perspect.* 108: 329–332.

Wilson, A. G. E., Thake, D. C., Heydens, W. E., Brewster, D. W., and Hotz, K. J. 1996. Mode of action of thyroid tumor formation in the male long-evans rat administered high doses of alachlor. *Fundam. Appl. Toxicol.* 33(1): 16–23.

Index

Milton Keynes UK
Ingram Content Group UK Ltd.
UKHW051902071024
449327UK00025B/2064